The Chemistry of Gold Extraction

Second Edition

John O. Marsden and C. Iain House

Published by
Society for Mining, Metallurgy, and Exploration

Society for Mining, Metallurgy, and Exploration, Inc. (SME)
12999 E. Adam Aircraft Circle
Englewood, Colorado, USA 80112
(303) 948-4200 / (800) 763-3132
www.smenet.org

SME advances the worldwide mining and minerals community through information exchange and professional development. With members in more than 70 countries, SME is the world's largest association of mining and minerals professionals.

Copyright © 2006 Society for Mining, Metallurgy, and Exploration, Inc.
Electronic edition published 2009.

All Rights Reserved. Printed in the United States of America.

Information contained in this work has been obtained by SME from sources believed to be reliable. However, neither SME nor its authors and editors guarantee the accuracy or completeness of any information published herein, and neither SME nor its authors and editors shall be responsible for any errors, omissions, or damages arising out of use of this information. This work is published with the understanding that SME and its authors and editors are supplying information but are not attempting to render engineering or other professional services. Any statement or views presented herein are those of individual authors and editors and are nor necessarily those of SME. The mention of trade names for commercial products does not imply the approval or endorsement of SME.

No part of this publication may be reproduced, stored in a retrieval system, or transmitted in any form or by any means, electronic, mechanical, photocopying, recording, or otherwise, without the prior written permission of the publisher.

ISBN 978-0-87335-240-6
Ebook 978-0-87335-274-1

Library of Congress Cataloging-in-Publication Data

Marsden, John, 1960-
  The chemistry of gold extraction / John Marsden, Iain House.-- 2nd ed.
      p. cm.
  Includes bibliographical references and index.
   ISBN-13: 978-0-87335-240-6 (print) -- ISBN 978-0-87335-274-1 (ebook)
   1. Gold--Metallurgy. 2. Gold ores. 3. Hydrometallurgy. 4. Extraction (Chemistry). I. House, Iain, 1959- II. Title.
TN760.M37 2006
669'.22--dc22
                             2005057609

Contents

FOREWORD TO THE SECOND EDITION xi

FOREWORD TO THE FIRST EDITION xiii

PREFACE TO SECOND EDITION xv

PREFACE TO FIRST EDITION xvii

CHAPTER 1   **HISTORICAL DEVELOPMENTS**   **1**
1.1 Precyanidation: Pre-1888   **1**
    1.1.1 Early History   **1**
    1.1.2 European Developments to 1848   **2**
    1.1.3 Gold Rush Era   **3**
    1.1.4 Early Hydrometallurgy   **4**
    1.1.5 Early Pyrometallurgy   **6**
1.2 Cyanidation: 1889–1971   **8**
    1.2.1 Invention of Cyanidation   **8**
    1.2.2 Flowsheet Development   **8**
    1.2.3 Activated Carbon   **9**
    1.2.4 Changing Economic Climate   **10**
1.3 Era of Major Technological Development: 1972–2000   **10**
    1.3.1 CIP Revolution   **12**
    1.3.2 Heap Leaching   **13**
    1.3.3 Refractory Ores   **13**
1.4 Into the 21st Century   **16**
References   **17**

CHAPTER 2   **ORE DEPOSITS AND PROCESS MINERALOGY**   **19**
2.1 Gold Minerals   **19**
    2.1.1 Native Gold   **20**
    2.1.2 Electrum   **21**
    2.1.3 Gold Tellurides   **21**
    2.1.4 Other Gold Minerals   **24**
    2.1.5 Gold with Sulfides   **25**
2.2 Classification of Gold-Bearing Materials   **26**
2.3 Placers   **27**
    2.3.1 Formation of Placers   **27**
    2.3.2 Commercial Significance   **28**
    2.3.3 Gold Mineralogy   **28**
2.4 Free-Milling Ores   **30**
    2.4.1 Palaeoplacers and Quartz Vein Gold Ores   **30**
    2.4.2 Other Hard Rock Ores   **33**
2.5 Oxidized Ores   **37**
2.6 Silver-Rich Ores   **39**

    2.7 Iron Sulfides   **41**
        2.7.1 Pyrite   **42**
        2.7.2 Marcasite   **43**
        2.7.3 Pyrrhotite   **43**
    2.8 Arsenic Sulfides   **44**
        2.8.1 Arsenopyrite   **44**
        2.8.2 Orpiment   **45**
        2.8.3 Realgar   **45**
    2.9 Copper Sulfides   **45**
        2.9.1 Chalcopyrite   **45**
        2.9.2 Other Copper Sulfides   **45**
    2.10 Antimony Sulfides   **46**
        2.10.1 Aurostibnite   **46**
        2.10.2 Stibnite   **46**
    2.11 Tellurides   **46**
    2.12 Carbonaceous Ores   **47**
    2.13 Gravity Concentrates   **48**
    2.14 Flotation Concentrates   **48**
    2.15 Tailings   **49**
        2.15.1 Gravity Concentration Tailings   **50**
        2.15.2 Cyanidation Tailings   **50**
        2.15.3 Flotation Tailings   **51**
    2.16 Refinery Materials   **51**
        2.16.1 Calcine   **52**
        2.16.2 Roaster Dust   **53**
        2.16.3 Anode Slime   **53**
        2.16.4 Slag   **53**
    2.17 Recycled Gold   **53**
    2.18 Determinative Methods   **54**
        2.18.1 Ore Composition   **55**
        2.18.2 Textural Characteristics   **58**
        2.18.3 Special Factors in Placer Ore Evaluation   **64**
        2.18.4 Process Stream Mineralogy   **64**
    References   **65**

CHAPTER 3   **PROCESS SELECTION**   **69**
    3.1 Factors Affecting Process Selection   **69**
        3.1.1 Geological   **70**
        3.1.2 Mineralogical   **71**
        3.1.3 Metallurgical   **71**
        3.1.4 Environmental   **72**
        3.1.5 Geographical   **77**
        3.1.6 Economic and Political   **77**
    3.2 Unit Process Options   **77**
        3.2.1 Comminution   **78**
        3.2.2 Classification   **80**
        3.2.3 Solid–Liquid Separation   **80**
        3.2.4 Ore Concentration   **82**
        3.2.5 Oxidative Pretreatment   **84**
        3.2.6 Leaching   **85**
        3.2.7 Solution Purification and Concentration   **86**

3.2.8 Recovery 86
  3.2.9 Refining 87
  3.2.10 Effluent Treatment 88
 3.3 Flowsheet Options 89
  3.3.1 Placers 89
  3.3.2 Free-Milling and Oxidized Ores 90
  3.3.3 Nonrefractory Sulfidic Gold Ores 94
  3.3.4 Refractory Sulfidic Gold Ores 96
  3.3.5 Silver-Rich Ores 98
  3.3.6 Carbonaceous Ores 99
  3.3.7 Gold-Telluride Ores 100
  3.3.8 Copper–Gold Ores 101
  3.3.9 Gravity Concentrates 102
  3.3.10 Flotation Concentrates 102
  3.3.11 Gold Recovery from Leach Solutions and Slurries 103
  3.3.12 Refining 105
 3.4 Cost Considerations 105
 References 107

CHAPTER 4 **PRINCIPLES OF GOLD HYDROMETALLURGY** 111
 4.1 Reaction Chemistry of Gold 111
  4.1.1 Gold–Water Reactions 112
  4.1.2 Gold Complexes 113
 4.2 Chemical Equilibria 113
  4.2.1 Definition of Equilibrium 113
  4.2.2 Electrochemical Considerations 116
  4.2.3 Activities and Concentrations 118
  4.2.4 pH Scale and pH Modification 121
  4.2.5 Complexation 122
  4.2.6 Solubility of Solids 123
  4.2.7 Solubility of Gases 124
  4.2.8 Deposition of Gold from Solution 125
  4.2.9 Graphical Representation of Equilibria 126
 4.3 Reaction Kinetics 131
  4.3.1 Modeling of Kinetics 132
  4.3.2 Mass Transport 134
  4.3.3 Absorption of Gases in Liquids 136
  4.3.4 Electrochemical Reactions 137
  4.3.5 Particulate Factors in Solid–Liquid Systems 138
 4.4 Experimental Methods 141
  4.4.1 Measurement of Solution Potentials 142
  4.4.2 Rotating Disc Electrodes 142
  4.4.3 Potential Sweep Methods 143
 References 144

CHAPTER 5 **OXIDATIVE PRETREATMENT** 147
 5.1 Hydrometallurgical Sulfide Oxidation 147
  5.1.1 Iron Sulfides 149
  5.1.2 Arsenic Sulfides 152
  5.1.3 Copper Sulfides 152
  5.1.4 Other Sulfides 153

    5.1.5 Sulfur **154**
    5.1.6 Precipitation Reactions **158**
5.2 Oxygen: Low-Pressure Oxidation **161**
    5.2.1 Reaction Chemistry **161**
    5.2.2 Reaction Kinetics **162**
    5.2.3 Process Considerations **162**
5.3 Oxygen: High-Pressure Acidic Oxidation **163**
    5.3.1 Reaction Chemistry **164**
    5.3.2 Reaction Kinetics **165**
    5.3.3 Behavior of Other Species **169**
    5.3.4 Process Considerations **170**
5.4 Oxygen: High-Pressure Nonacidic Oxidation **175**
    5.4.1 Reaction Chemistry and Conditions **175**
    5.4.2 Reaction Kinetics **177**
    5.4.3 Process Considerations **177**
5.5 Nitric Acid Oxidation **178**
    5.5.1 Reaction Chemistry **178**
    5.5.2 Reaction Kinetics **180**
    5.5.3 Process Considerations **182**
5.6 Chlorine Oxidation **185**
    5.6.1 Chlorine Chemistry **186**
    5.6.2 Deactivation of Carbonaceous Material **187**
    5.6.3 Sulfide Oxidation **188**
    5.6.4 Effect of Other Ore Constituents **189**
    5.6.5 Process Considerations **190**
5.7 Biological Oxidation **190**
    5.7.1 Reaction Chemistry and Mechanism **192**
    5.7.2 Reaction Kinetics and Operating Conditions **194**
    5.7.3 Process Considerations **200**
5.8 Pyrometallurgical Oxidation **205**
    5.8.1 Roasting Reaction Chemistry **206**
    5.8.2 Roasting Kinetics and Efficiency **212**
    5.8.3 Roasting Process Considerations **217**
    5.8.4 Microwave Energy **224**
References **224**

CHAPTER 6 **LEACHING 233**
6.1 Cyanidation **233**
    6.1.1 Chemistry of Cyanide Solutions **233**
    6.1.2 Gold Dissolution **236**
    6.1.3 Reaction Kinetics **241**
    6.1.4 Behavior of Other Minerals in Alkaline Cyanide Solutions **251**
    6.1.5 Process Considerations **263**
6.2 Chlorination **271**
    6.2.1 Mechanism of Gold Dissolution **272**
    6.2.2 Reaction Kinetics **272**
    6.2.3 Behavior of Other Minerals in Chloride Solution **273**
    6.2.4 Process Considerations **274**
6.3 Thiosulfate **276**
    6.3.1 Reaction Chemistry and Kinetics **276**
    6.3.2 Process Considerations **280**

	6.4 Thiourea **281**
	6.4.1 Reaction Chemistry and Kinetics **282**
	6.4.2 Process Considerations **284**
	6.5 Thiocyanate **284**
	6.6 Ammonia **286**
	6.7 Other Lixiviants **287**
	References **288**
CHAPTER 7	**SOLUTION PURIFICATION AND CONCENTRATION 297**
	7.1 Carbon Adsorption **297**
	7.1.1 Properties of Activated Carbon **297**
	7.1.2 Adsorption from Cyanide Solutions **303**
	7.1.3 Elution **312**
	7.1.4 Carbon Fouling and Reactivation **318**
	7.1.5 Process Considerations **324**
	7.1.6 Adsorption from Noncyanide Solutions **334**
	7.2 Ion Exchange Resins **335**
	7.2.1 Properties of Resins **336**
	7.2.2 Adsorption from Cyanide Solutions **336**
	7.2.3 Elution and Regeneration **343**
	7.2.4 Process Considerations **348**
	7.2.5 Adsorption from Noncyanide Solutions **351**
	7.3 Solvent Extraction **353**
	7.3.1 General Principles **354**
	7.3.2 Extraction Systems **354**
	7.3.3 Process Considerations **358**
	References **358**
CHAPTER 8	**RECOVERY 365**
	8.1 Zinc Precipitation **365**
	8.1.1 Reaction Chemistry **366**
	8.1.2 Reaction Kinetics and Factors Affecting Efficiency **371**
	8.1.3 Process Considerations **383**
	8.2 Aluminum Precipitation **386**
	8.3 Electrowinning **387**
	8.3.1 Electrowinning Fundamentals **387**
	8.3.2 Reaction Chemistry **389**
	8.3.3 Reaction Kinetics and Factors Affecting Efficiency **392**
	8.3.4 Process Considerations **396**
	8.4 Recovery from Noncyanide Solutions **400**
	References **403**
CHAPTER 9	**SURFACE CHEMICAL METHODS 409**
	9.1 Principles of Surface Chemistry **409**
	9.1.1 Mineral–Water Interface **409**
	9.1.2 Hydrophobicity **412**
	9.1.3 Surface Chemistry of Gold **413**
	9.1.4 Reagents **413**
	9.2 Flotation **419**
	9.2.1 Application of Flotation **419**
	9.2.2 Gold **421**
	9.2.3 Gold Tellurides **428**

	9.2.4 Sulfide Minerals **428**
	9.2.5 Carbonaceous Matter **433**
	9.2.6 Silicates **435**
	9.2.7 Process Considerations **435**
	9.3 Amalgamation **438**
	9.3.1 Properties of Mercury **438**
	9.3.2 Factors Affecting Amalgamation **439**
	9.3.3 Process Considerations **441**
	9.4 Coal–Gold Agglomeration **441**
	References **443**
CHAPTER 10	**REFINING 449**
	10.1 Pyrometallurgy of Gold **449**
	10.2 Acid Leaching **451**
	10.2.1 Zinc Precipitates and High-Grade Sludges **451**
	10.2.2 Loaded Cathodes **452**
	10.2.3 Other Materials **453**
	10.3 Pyrometallurgical Methods for Crude Bullion Production **453**
	10.3.1 Mercury Removal by Retorting **453**
	10.3.2 Roasting (Calcining) **454**
	10.3.3 Smelting **455**
	10.4 Bullion Refining **459**
	10.4.1 Pyrometallurgical Refining **459**
	10.4.2 Electrolytic Refining **461**
	10.4.3 Hydrometallurgical Refining **462**
	10.4.4 Refining Operations **464**
	References **465**
CHAPTER 11	**EFFLUENT TREATMENT 467**
	11.1 Types of Waste and Effluent Control Parameters **467**
	11.1.1 Gases **468**
	11.1.2 Solids **468**
	11.1.3 Liquids **470**
	11.2 Reagent and Metals Recovery **473**
	11.2.1 Direct Solution Recycle **473**
	11.2.2 Acidification, Volatilization, and Reneutralization **475**
	11.2.3 Ion Exchange **477**
	11.2.4 Activated Carbon **480**
	11.2.5 Electrolytic Treatment **481**
	11.2.6 Sulfide Precipitation **481**
	11.2.7 Ion Precipitate Flotation **482**
	11.3 Detoxification **482**
	11.3.1 Dilution **482**
	11.3.2 Cyanide **484**
	11.3.3 Metals **494**
	References **498**
CHAPTER 12	**INDUSTRIAL APPLICATIONS 503**
	12.1 Distribution of Process Technology **503**
	12.2 Industrial Process Flowsheets **512**
	12.2.1 Placers **512**
	12.2.2 Free-Milling Ores **514**

Preface to Second Edition

It is a great pleasure to present the revised, updated 2006 edition of *The Chemistry of Gold Extraction*. The first edition, published in 1992, was well received, and the modest printing of slightly fewer than 1,000 copies sold out by 1996. After a hiatus from direct involvement in the gold extraction industry, in 2004 the authors decided to produce a second edition.

The preface to the first edition contained a request for comment and criticism from readers and, over the past 13 years, feedback was received from many people—most of it was positive and, occasionally, some was negative (usually justifiably). In all cases, metallurgists and other industry specialists provided constructive and healthy input, and the authors have tried to incorporate this wherever possible and practical in the second edition. Preparing this text also gave the authors the chance to become reacquainted with the latest developments in technology. We found ourselves reabsorbed and fascinated by the chemistry of this unique metal that continues to simultaneously excite, challenge, frustrate, and entice all of us involved with its metallurgy and extraction.

Additionally, based on the extent of developments in the gold extraction industry in the years between editions, updates to some key sections of the book were warranted. Several chapters have more modest updates and revisions, while others have been significantly revised. The "Oxidative Pretreatment" chapter (5), for example, was updated to reflect the latest industrial developments in pressure oxidation, nitric acid oxidation, biological oxidation, and roasting. The "Leaching" chapter (6) has been revised and reorganized to include recent research and developments in cyanidation and chlorination processes, and to give more comprehensive coverage of thiosulfate leaching chemistry. Thiosulfate leaching has been the subject of intense research and development over the past 15 years or so, and the chemistry of thiosulfate systems is now better understood. Chapters on "Solution Purification and Concentration" (7) and "Recovery" (8) now reflect advances in recovery from noncyanide systems. The "Surface Chemical Methods" chapter (9) was expanded to include excellent recent research and development on the flotation of gold-bearing sulfide minerals and free gold.

"Effluent Treatment" (11) has become increasingly important in gold extraction with the publication of the International Cyanide Management Code (2002), as well as the efforts of the Global Mining Initiative (1999–2002) and the International Council for Mining and Metallurgy (established in 2002). The chapter for this topic therefore was updated and expanded to include several emerging options for effluent treatment, including improved ion exchange resin-based technology. The chapter on "Industrial Applications" (12) now includes examples of process plant flowsheets developed during the past 15 years, and provides a comprehensive database of flowsheets: Process descriptions and flowsheets have been provided for 44 major gold operations worldwide. Additionally, the layout of this chapter has been improved to provide readers with quicker access to relevant industrial plant examples.

In addition to many professionals in the industry who have assisted in the creation of this second edition, we are particularly grateful to Corby Anderson, Jim Arnold, David Baughman, Wolfgang Baum, Richard Beck, Bob Brewer (posthumously), Dario Clement, Steve Dixon, Rob Dunne, Chris Fleming, Maurie Fuerstenau, Rick Gilbert, Ralph Hackl,

Doug Halbe, Marilyn Hames, Nick Hazen, Jinxing Ji, Rob La Nauze, Marc Le Vier, John Mansanti, Brad Marchant, Basie Maree, Peter Mason, Terry McNulty, David Menne (posthumously), Jan Miller, Terry Mudder, Joe Pease, Allen Phillips, Robert Shoemaker, Gary Simmons, Patrick Taylor, Phil Thompson, Larry Todd, and Hans Von Michaelis for their input, contributions (sometimes informally), stimulation, and friendship. We would also like to thank Kay Ramsey for her careful assistance with the manuscript, Kathy Kaiser and Diane Serafin for their editorial work, Rick Frye for page design and composition, Andrew Peterson for technical illustration, Mike Coney for tracking down obscure reference citations, and Jane Olivier for project management at SME.

Finally, we would like to thank our wives, Dana and Julia, and our families for their tolerance and endurance throughout this project, as this undertaking would not have been possible without their support and encouragement.

The authors have made every effort to ensure the accuracy and completeness of the material presented in this book, however, errors and omissions may have occurred. It also should be remembered that no two gold ores are alike. Thorough sampling, ore characterization, analysis, metallurgical test work, evaluation, process selection, and process development should be completed to ensure that any new mineral extraction project, or any change to an existing operation, is successful. Fundamental to an understanding of these steps in practical application is the chemistry of the metal itself. We hope this second edition of *The Chemistry of Gold Extraction* helps all industry professionals in meeting this challenge.

John O. Marsden
Phoenix
Arizona

Iain House
Aberdeen
Scotland

October 2005

  12.2.3 Silver-Rich Ores **538**
  12.2.4 Refractory Iron Sulfide Ores **542**
  12.2.5 Refractory Arsenopyritic Ores **564**
  12.2.6 Copper-Rich Ores **578**
  12.2.7 Refractory Antimony Sulfide Ores **596**
  12.2.8 Telluride Ores **598**
  12.2.9 Carbonaceous Ores **600**
  12.2.10 Tailings **608**
  12.2.11 Refining **610**
 References **614**

APPENDIX A **SYMBOLS AND ABBREVIATIONS** **619**

APPENDIX B **UNITS AND CONVERSION FACTORS** **625**

SELECTED BIBLIOGRAPHY **627**

INDEX **629**

ABOUT THE AUTHORS **651**

Foreword to the Second Edition

The cyanide process for recovering gold from its ores was patented in 1887 by three men from the British Isles: John MacArthur and the Forrest brothers, Robert and William. It seems fitting now, 118 years later, that the nonpareil and second edition of cyanide chemistry texts has been published by two more Britons: John Marsden and Iain House.

The technical literature of the precious metals over the years has been somewhat unusual in that, although a very large number of technical papers have been published (particularly in the last few years), only a relatively few books have been written. On the other hand, these books were quite remarkable because of their uniformly high quality. Authors such as Rose, Clennell, Hamilton, Dorr, Bosqui, King, Adamson, and others who were published over the years are familiar to virtually every gold metallurgist, and their volumes have become prized possessions. This volume will take its place alongside those classics.

This new edition, and its format of emphasizing the chemistry of every type of gold ore treatment, is superb, and the result is literally a gold mine of information. For the gold metallurgist, reading this volume will be like reading a novel that can't be put down, and this reviewer is certain that it will be the standard reference for gold and silver metallurgists and millmen for many years to come. My thanks and congratulations are offered to these two very dedicated men.

Robert S. Shoemaker
Grass Valley, California
November 10, 2005

Foreword to the First Edition

Some twelve years ago, F.W. McQuiston Jr. and I published our second volume of *Gold and Silver Cyanidation Plant Practice*. The foreword to that volume was contributed by Donald H. McLaughlin and commenced as follows: "The cyanide process of extracting gold and silver from their ores is now in the final decade of the first century since its discovery and still shows no sign of senility or damage from aggressive new competitors." Today, the cyanide process is well into the first decade of its second century and again the basic process has remained unchanged. Certainly such longevity must be unique in the rapidly changing worlds of chemistry and metallurgy.

My first exposure to gold processing was in 1964 when as Chief Metallurgical Engineer for Bechtel I had the privilege of working with Frank McQuiston on the design of Newmont's Carlin Gold plant in Nevada. He was one of the world's great metallurgists and our association continued through many years to include other plant designs, co-authoring three technical volumes and a partnership in a heap leach operation. In a large part it was due to that association that I have been involved with over 40 gold/silver mills and 80-odd heap leaching operations. Consequently I was most pleased to accept the invitation to prepare a Foreword to this volume when it was kindly offered me by Iain House and John Marsden as it gave me an opportunity to reflect and consider the challenges that have been encountered by metallurgists during these last twenty-seven years and which are to be encountered in the immediate future.

The Carlin plant was the first to be built in the United States after World War II. It was only unique by its use of mechanically agitated cyanidation tanks and "Autojet" filters for solution clarification, features which were borrowed from the uranium industry, but Carlin was probably the first gold plant to use an anti-scale reagent and later it was there that chlorine was first used to overcome the preg-robbing effects of refractory, carbonaceous ores. "Refractory ores" are currently the buzz words in the precious metals industry but all gold/silver ores have been refractory at some point in time. For instance, the slime fraction of the Homestake ore, which had for many years been aerated and leached in Merrill plate and frame filters, became refractory due to its high treatment cost but was then successfully treated by carbon-in-pulp. Subsequently both CIP and CIL have been so successful that probably no one today would build a CCD plant unless the ore had a high silver content.

The most remarkable refractory ore treatment that has been developed, however, is that of heap and dump leaching which has resulted in recovering millions of ounces from ores that were refractory only because of their low gold contents. It is very doubtful if there will ever again be such an advance in applied metallurgy in the field of precious metals.

The refractory nature of sulphide ores and concentrates was solved long ago by the development of roasting to destroy the sulphide lattice and liberate the gold. The treatment, however, was expensive and only a few of the roasting plants have survived. While recently there has been a renewed interest in roasting, at the same time there have been other roasters replaced by pressure and bio-oxidation plants as pressures for more environmentally acceptable operations increase. Metallurgists must learn that environmental regulations can be promulgated much more rapidly than processes and equipment can be developed and ores that are treatable today can become completely refractory tomorrow.

Because precious metal ores are becoming more refractory from several causes, this volume will be invaluable to the engineers who will have to cope with these ores. The technical literature of the precious metals has been somewhat unusual in that although there have been a very large number of technical papers published (particularly in the last few years) there have been only a relatively few books written. These technical books have been quite remarkable because of their uniformly high quality and such authors as Rose, Clennell, Hamilton, Dorr, Bosqui and others who published from the 1890s to the 1930s are familiar to virtually every gold metallurgist and their volumes have become prized possessions. Most of these books have, however, been heavily weighted in descriptive material about extraction plants and processes and to obtain much detailed chemical and mineralogic data which is so basic to gold metallurgy one has had to conduct extensive literature searches. This volume, as its title indicates, is uniquely different in that it first approaches the subject of gold through the mineralogy and the chemistry of the minerals with which it is associated and then addresses process selection from the mineralogic viewpoint. Subsequent sections on gold hydrometallurgy, oxidation treatments, leaching, solution purification, recovery, surface chemistry, refining, effluent treatment and industrial applications cover their subjects completely and comprehensively. A large number of graphs, illustrations and tables bring the text into clear focus and each section contains a copious bibliography. Altogether it is literally a gold mine of information and a remarkable piece of work.

The two young men who have written this book have both had rapid advancement in their professions and I am sure they will continue to do so in future years. All involved in the metallurgical profession should be grateful to these two men who have taken the time to prepare this most valuable volume for the minerals industry and I am particularly grateful to have had this chance to express to them both my admiration and appreciation.

Robert S. Shoemaker
Grass Valley, California
May 1, 1991

Preface to First Edition

The extraction of gold from the earth has been pursued with vigor since it developed intrinsic value early in the history of mankind as a result of its unique physical and chemical properties, its decorative appeal, and its scarcity. Despite man's "accursed thirst for gold," it is estimated that all the gold ever produced in a refined form would fill a cube with a side length of 20 m—the volume of a large two-story family home. The amount of ore processed to recover this amount of gold would occupy a volume approximately 100 km × 100 km × 1 m thick, assuming an average ore grade of 6 g/t. It is truly remarkable that so much effort has been devoted to recovering such a small amount of the metal; an effort that has become increasingly reliant on the ability to extract and recover gold from its ores by chemical means.

The purpose of this book is to provide a text on the chemistry of gold extraction, which is of practical use to the many individuals and groups involved in the fascinating subject of finding and producing gold. This work has been motivated by major developments in gold extraction technology that have occurred during the 1980s, including the widespread use of activated carbon for gold recovery, heap leaching, and the development of processes to treat refractory ores. These events were largely encouraged by a sustained favorable gold price following the readoption of the gold standard by the United States in 1972, coupled with the need to process lower-grade ores with more diverse and complex mineralogy.

World gold production (excluding countries with centrally planned economies) increased by 50% between 1980 and 1990. Gold remains a principal target of mineral exploration programs, and continued growth has been forecast for the foreseeable future. Production in almost all regions of the world has increased, with the notable exception of South Africa, previously the world's predominant producer. This geographical variety is creating a greater diversity of technical challenges for the gold metallurgist because of a wider range of ore types, new environmental regulations, and a requirement for the most appropriate technology to suit local needs.

A primary objective of the book has been to bridge the gap between research and the gold mining industry, thereby increasing the transfer of new technology. This book aims to provide the wider range of knowledge which is now required by those working in gold extraction and the gold milling industry in general.

Hydrometallurgical methods form the basis for most gold extraction processes, and we have attempted to relate the chemical theory and principles to industrial practice. The chemistry of gold hydrometallurgy has been divided in four main sections: principles of gold hydrometallurgy, leaching, solution purification and concentration, and recovery. Sections on oxidative pretreatment and surface chemical methods reflect the need to treat increasingly complex sulfidic and carbonaceous gold ores, reviewing both new technology as well as methods that have been applied for many decades. The characteristics of the mined ore dictate the tactics of the gold metallurgist, and therefore gold ore mineralogy is considered from the process engineer's viewpoint. The historical development of the processes used in gold extraction are reviewed to put the technology into perspective and speculate on trends for the future.

Effluent treatment is afforded a separate chapter since this field of technology is rapidly gaining importance in the gold extraction industry in parallel with increasing environmental pressures. The final stage of extraction—refining—is considered in the context of processes used for on-site bullion production at mining locations and at dedicated precious metal refineries.

Throughout the book, emphasis has been placed on the practical application of chemical principles and techniques. Accordingly, a chapter has been devoted to process selection, which discusses the criteria for selection of available processes and the conditions under which they can be most effective. Finally, examples of industrial process flowsheets are used to demonstrate the development of a number of complete process routes and the interaction between the unit processes employed. This chapter also considers the actual distribution of extraction technology around the world, which helps the reader to gain an accurate perspective on the relative importance of the different technology.

Physical and engineering aspects of gold extraction, such as comminution, classification, and solid–liquid separation, are only considered in as far as they affect the chemistry, as these subjects are broadly similar for most metals and have been covered extensively in the literature.

This book is intended for all professionals involved in the precious metals industries. It will be of particular interest and use to scientists and engineers (extraction metallurgists, minerals/metallurgical processing engineers, electro-chemists, chemical engineers, mineral technologists, mining engineers, and material scientists) working in gold extraction in either production, research, or consulting capacities.

ACKNOWLEDGMENTS

The completion of a book of this type is to a large extent a review of previous work, although the authors naturally feel that they have given their personal view and contributed to an increased understanding of gold extraction. We are therefore indebted to all whose work has been read and, in some cases, incorporated in this book. Considerable effort has been made to ensure accuracy and balance, and we hope that previous authors are tolerant of any imperfections. We would welcome corrections and comments from people in research and industry for inclusion if, and when, future editions are warranted.

A special acknowledgment is warranted to Bob Shoemaker (R.S. Shoemaker, Ltd., United States), who not only encouraged this project but also kindly contributed the Foreword. The authors are indebted to Gold Fields Mining Corporation (United States), without whose support this book would not have been completed. We are particularly grateful to Wolfgang Baum (vice president, Pittsburg Mineral & Environmental Technology, Inc., United States), Doug Halbe (manager of metallurgy, Kalgoorlie Consolidated Mines, Ltd., Australia), and Peter Mason (senior process engineer, Fluor Daniel-Wright Engineers, Canada) for reviewing, and providing substantial information for Chapters 2, 8, and 5, respectively. Many other sections of the book were also sent to willing reviewers, whom we wish to thank for their contributions, namely, Jim Arnold (Gold Fields Operating Co., Chimney Creek, United States), Simon Buckley (Leeds Mineral Services Group, UK), Miguel Diaz (Royal School of Mines, UK), Steve Dixon (Bond Gold-Richmond Hill, United States), John Mansanti (Gold Fields Mining Corporation, United States), Brad Marchant and John Chapman (both at Coastech Research, Canada), and Ian Townsend (Larox, UK, formerly with BP Minerals International). Gary Halverson (Giant Yellowknife, Canada), Yvon Sylvestre (Agnico-Eagle, Joutel Division, Canada), and Jerry Gill (Johnson Matthey Inc., Salt Lake City, United States) provided useful contributions for Chapter 12. Jim Arnold, John Mansanti, Mike Gleason (all in the United States), Richard Atkinson and Noel Peverett (both in South Africa) deserve special mention for much stimulating and informative debate on many aspects of gold extraction technology and chemistry. We are grateful to Phelps Dodge Corporation and BP Exploration for their cooperation and consideration during this project.

Finally, we acknowledge the efforts and progress made by the many extractive metallurgy departments at universities and colleges around the world, and particularly at the applied mineral research centers of Mintek (South Africa), U.S. Bureau of Mines (United States), and CSIRO (Australia), among others, who have formed a valuable link between fundamental research and industry. However, the success of these efforts is due in large part to devoted and innovative plant operators at locations where such technology is applied. It is their dedication, persistence, and ingenuity that brings new processes into fruition and often rejuvenates ailing technology. These people know who they are and they have our respect and admiration.

J.O. Marsden
Phoenix, Arizona
United States

C.I. House
London
England

May 31, 1991

CHAPTER 1

Historical Developments

Many of the methods now used for gold extraction are based on techniques that have been known or established for centuries. Gravity concentration, amalgamation, cyanide leaching, chlorination, zinc precipitation, and carbon/charcoal adsorption are all processes that have been used for at least a hundred years, and combinations of these remain as the basis for most gold recovery flowsheets.

The ingenuity of the early pioneers in developing technology with such longevity is admirable. On numerous occasions over the years, rediscovery or reevaluation of this technology has led to its commercial redevelopment, while incorporating the advantages of contemporary improvements. Consequently, it is important to know the historical background of gold extraction processes and technology, as it may well help to shape the future.

The predominance of cyanidation as the principal gold extraction technique since the late 19th century and the commercial acceptance of other important hydrometallurgical processes, such as heap leaching and carbon adsorption in the 1970s and 1980s, naturally divides the history of gold extraction into four main eras:

- Precyanidation: pre-1888
- Cyanidation: 1889–1971
- Era of major technological development: 1972–2000
- Into the 21st century

The key dates in this categorization are the initial application of cyanidation and zinc precipitation in 1889 and the boom in new technology, which started in 1972, when the price of gold was disengaged from its official selling price and allowed to move with market forces.

1.1 PRECYANIDATION: PRE-1888

1.1.1 Early History

Gold and copper were the first metals used by humans because of their occurrence in the native state and their malleable and ductile properties, which meant they could be easily worked with primitive tools. The earliest uses of gold were in the Middle East, during the Neolithic age, where gold was collected from streambeds either manually or by crude gravity concentration methods.

In Egypt, during the reign of Menes in 3050 BC, gold was used as a means of monetary payment—in the form of grains and small bars. However, since this time, the major application for gold has been in decoration and jewelry, as is the case today (Figure 1.1). Finely worked gold ornaments have been found in graves in Mesopotamia originating from about 2700 BC. Similarly, gold mining in Egypt started with alluvial workings and was followed by shallow underground vein mining in Nubia in about 1300 BC. Activity

2 | THE CHEMISTRY OF GOLD EXTRACTION

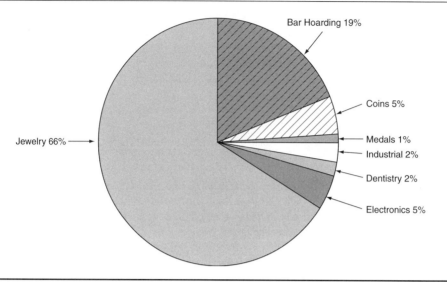

FIGURE 1.1 Approximate gold distribution by end use in 1989 [1]

was centered around Hammamet, with mines up to 90 m deep and sometimes extending 500 m along veins. Early gold recovery methods were varieties of gravity concentration and manual sorting.

The legend of Jason and the Argonauts probably described the voyage of prospectors to gold placers near the Black Sea in about 1300 BC. The miners used sheep's fleeces in sluices to trap gold. Oils were rubbed into the fleece to help collect gold, adding to the efficiency of this technique—one of the earliest applications of surface chemistry in mineral processing. It is likely that gold wetting by mercury (amalgamation) was known in 1000 BC, although it was not commonly used as a commercial gold recovery process until much later.

In Turkey, around 700 BC, the first gold coins were produced, but it was only after a process for fire refining of gold was developed in 560 BC that pure gold coins were minted [2]. This process used salt to remove silver, as silver chloride, from gold metal. Alloying of gold and silver was known to the Egyptians as early as 500 BC.

1.1.2 European Developments to 1848

Roman exploitation of their empire for its mineral wealth was widespread, with gold used as the principal form of payment for imports, notably from China. In Spain, mining was particularly well developed, with new technology such as hydraulic mining, water wheels, and the Archimedian screw in use. The Romans used a sluicing technique whereby broken rock was washed through channels containing prickly shrubs, which caught the gold. At this time there were also thriving mining schools throughout Europe that provided the expertise for the applications of this extraction technology. During the Roman and Greek eras, the vast majority of mine workers were slaves who were subjected to atrocious working conditions and incurred a high mortality rate.

The decline of the Roman Empire resulted in diminished mining activity until a revival in the 11th century, based in Central Europe. Gold mining developments were centered around Harz, which is now part of eastern Germany, and the eastern Alps. By 1400 AD amalgamation and retorting processes were used widely in gold extraction [3]. The classic book *De Re Metallica* by Georg Bauer, who was usually known by his Latin name of Georgius Agricola, provided a good description of mining and mineral dressing practice in 16th century Germany [4].

The prosperity of Central European gold and silver mines came to an abrupt end in the 1550s, when Mexico and the parts of South America now known as Colombia, Peru, and Bolivia were conquered by Spain and European metallurgical practices spread to South America. One of the most significant early finds was the Choco alluvial deposits in Colombia. Production costs were lower than those of European mines, even with the greater transportation distances to Europe, largely because South American operations used primitive, cheap, and labor-intensive panning of streambed material. In 1693 the Minas Gerais deposits in Brazil were discovered and have been in production to the present day.

One curiosity of this Conquistador era is that Spanish metallurgists in South America encountered impure Colombian gold, which occurred with platinum, an element unknown in Europe at the time. This material was thought to be "unripe" gold, and the Spanish treasury ordered it to be discarded into the sea as waste.

The South American competition depressed the European mining industry until the industrial revolution in Britain in the 19th century. In Europe during this period a common gold recovery technique was amalgamation using copper plates. Gold was also produced from other sources of European exploration, such as West Africa, which produced 1 million oz for export to Europe between 1400 and 1600. (All ounces cited in this book are troy ounces.) Production was also significant in China, Japan, and India during this time.

1.1.3 Gold Rush Era

In the first half of the 19th century, Russia was the main source of gold, supplying 60% of the world's production. Underground gold mining started in 1744 near Ekaterinburg, and output increased following the discovery and development of numerous alluvial deposits nearby. A discovery in Siberia in 1838 on the Ulderey River resulted in a gold rush. From 1846, significant mining activity also occurred in the Lena Basin (see Lena Goldfield in Figure 1.2). The main gold-recovery method was primitive gravity concentration by panning, although steam and water power-driven trommels and strakes were also introduced.

Russian gold output was eclipsed by a series of gold rushes in California (United States), South America, Victoria (Australia), and New Zealand in the mid-1800s. The 1848 gold rush in California (Figure 1.3), following the discovery of gold in the previous year, was perhaps the most important of these, opening up the western United States to settlement and contributing significantly to the establishment of the nation. Early Californian miners found gold in dry streambeds where the grade was so high that processing simply consisted of tossing sand and gravels in a blanket—crude dry gravity concentration. Panning techniques were common but were superseded by improved wet gravity concentration equipment, such as cradles and long toms, which consisted of screens and sluices. Similar alluvial deposits were exploited in Alaska, Colorado, Idaho, Montana, Nevada, and South Dakota in the United States, and British Columbia in Canada. However, the most economic and easily minable reserves were soon exhausted, and the importance of hard rock quartz vein mining increased rapidly.

The Australian gold rush started in 1851 with an initial find at Bathurst, New South Wales, and subsequent larger discoveries near Ballarat and Bendigo in Victoria (Figure 1.4). Water shortages restricted the use of the long tom, and consequently gravity concentration equipment requiring less water, such as the rocker, was used and supplemented by puddling tubs to break up clay in the ore. Again, as surface alluvial gold was exhausted, underground mining developed rapidly.

Further gold finds were made in New South Wales, Queensland, Western Australia, and New Zealand in subsequent years. In 1882, the use of dredging to recover submerged

4 | THE CHEMISTRY OF GOLD EXTRACTION

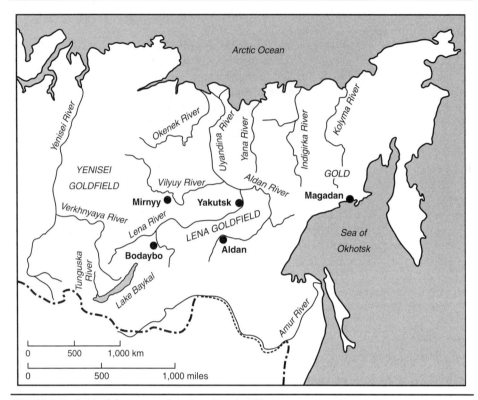

FIGURE 1.2 Gold mining areas of eastern Russia [5]

alluvial material was first reported in New Zealand [3]. The first U.S. dredge was employed at Bannack, Montana, in 1897 (Figure 1.3).

During the gold rush era, gravity concentration equipment was developed to treat a wider range of ore types on progressively larger scales. Amalgamation circuits were modified to include the use of potassium cyanide to clean the mercury and gold surfaces (e.g., the Patio process applied at Comstock, Nevada) [6]. Gravity concentration and amalgamation were used in crushing circuits to recover gold at the earliest possible stage in the flowsheet—a principle of flowsheet design that is still valid.

With the discovery of the Mother Lode (California) and Comstock Lode (Nevada) in the 1850s, attention turned to underground deposits. The gold rushes of 1897 in the Klondike (Yukon, Canada) and of 1898 in Nome (Alaska, United States) marked the end of the prospector era. By this time, the Witwatersrand in South Africa was in production, following the discovery of gold in 1886, heralding the next great gold mining era (see Section 1.2.1). The Witwatersrand basin, which hosts the greatest gold resource in the world, is centered in Johannesburg (Transvaal) and stretches from Welkom (the Orange Free State gold district) in the west, to the Evander district, about 120 km east of Johannesburg (see Figure 1.5).

1.1.4 Early Hydrometallurgy

Despite the advances in gravity concentration and amalgamation, these processes were unsuitable for the recovery of fine gold and gold associated with sulfide minerals. These drawbacks prompted the search for an effective hydrometallurgical process.

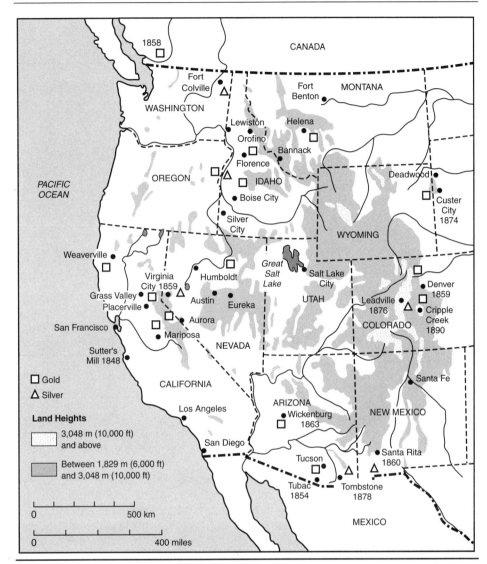

FIGURE 1.3 Major gold and silver discoveries in the western United States, 1848–1890 [5]

Chlorine gas was discovered in 1774 and soon became a commercially available commodity. In 1848 Plattner proposed a process for the treatment of gold ores, which consisted of passing chlorine gas over crushed ore to produce a soluble gold chloride that could be dissolved in water. The gold was then precipitated from solution by ferrous sulfate, hydrogen sulfide, or charcoal. Chlorination was first used commercially to treat Deetken ore (California) in 1858 [6]. By the mid-1860s various chlorination processes were used in the United States, South Africa, and Australia, often to supplement existing gravity concentration circuits or to treat sulfide-rich concentrates. These included entirely hydrometallurgical routes in which chlorine was added in the solution phase.

Chlorination was rarely applied to whole ores directly, mainly because of the high treatment cost, which resulted in high gold cutoff grades (about 50 g/t). In addition, ores containing arsenides, antimonides, and large amounts of sulfides had to be oxidized prior to chlorination, as was also to be the case for cyanidation in the next century [6].

6 THE CHEMISTRY OF GOLD EXTRACTION

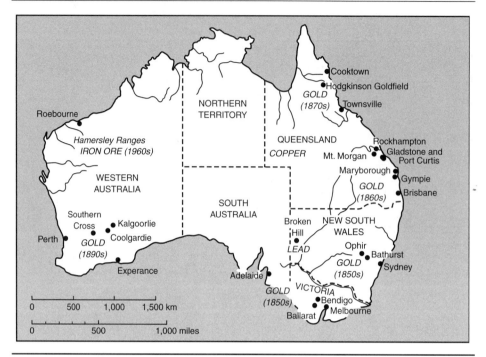

FIGURE 1.4 Early gold discoveries in Australia [5]

In a process that would have been effective though probably expensive and potentially dangerous, Molesworth in 1891 proposed that gold could be extracted from pyrite by roasting at relatively low temperatures with oxygen injection. The calcine could then be amalgamated, although a 15- to 30-min aqua regia leach was envisioned, with gold recovered from solution using charcoal [7].

The properties of several other chemicals for dissolving gold and silver were known in the late 1800s and early 1900s, including bromine/bromide, cyanide, thiosulfate, and thiourea solutions. It has also been reported that between 1900 and 1920 some ores containing arsenopyrite and/or gold tellurides were treated with bromide–cyanide solutions [3].

1.1.5 Early Pyrometallurgy

From the time of the discovery of amalgamation, gold-rich mercury was retorted for mercury removal, and the resulting sponge gold smelted with fluxes to produce gold bullion. Likewise, gravity concentrates, which often contained magnetite, ilmenite, chromite, and other heavy minerals, were smelted with potash, borax, and nitre to remove the contaminants.

Throughout the 19th century various high-grade auriferous lead, silver, and copper ores and concentrates were treated directly by pyrometallurgical methods. These included the following:

- Direct fusion in a bath of lead
- Direct smelting with lead-rich fluxes
- Smelting with fluxes to produce a matte followed by resmelting with lead-rich fluxes.

These processes generated lead–gold–silver alloys, in a manner similar in principle to the well-known fire assay analytical technique in use in the early 21st century. The products

FIGURE 1.5 Gold mining in South Africa

were cupelled to remove lead, leaving a precious metal alloy. The precious metals were separated using one of four methods available at the time [6]:

- Nitric acid dissolution of silver to leave a gold residue.
- Sulfurizing with either sulfur, pyrite, or an antimony sulfide, sulfur, and litharge mixture to produce silver sulfide and a gold–silver alloy. The alloy could be purified by nitric acid treatment.
- Chlorination (Miller process) to remove silver as insoluble silver chloride, leaving gold as soluble gold chloride.
- Electrolytic refining in a potassium cyanide bath.

These processes were practiced in Hungary, Germany, and other areas of eastern Europe.

In 1868 a combined roasting–smelting operation was started at Black Hawk (Colorado) to treat pyritic gold ore. The matte produced was shipped to Wales, United Kingdom, for refining. By 1876 a copper-refining technique was developed in the United States to produce refined gold as a by-product of copper. This was the precursor to anode casting and electrorefining in the copper industry [3].

In parallel with the development of the Plattner chlorination process (Section 1.1.4), roasting pretreatment was developed to oxidize concentrates, and occasionally ores, prior to leaching. Various types of roasters were employed at a number of operations, including Treadwell (Alaska); Amador, Bunker Hill, and Eureka (California); Gibbonsville (Montana); Deloro (Mexico); and Mount Morgan (Queensland, Australia).

1.2 CYANIDATION: 1889–1971

1.2.1 Invention of Cyanidation

The solubility of gold in cyanide solutions was recognized as early as 1783 by Scheele (Sweden) and was studied in the 1840s and 1850s by Elkington and Bagration (Russia), Elsner (Germany), and Faraday (England). Elkington also held a patent for the use of potassium cyanide solutions for electroplating of gold and silver.

The dissolution of gold in aerated cyanide solutions and the role of oxygen in the mechanism were investigated by Elsner in 1846 [8] and the reaction reported as follows:

$$2Au + 4KCN + O + H_2O = 2AuK(CN)_2 + 2KOH \qquad (EQ\ 1.1)$$

Elsner's equation, which is now thought to apply directly to only a minor portion of gold during dissolution, is still quoted in present-day publications on gold leaching. The mechanism is now better understood and is considered in Section 6.1.

In 1867, Rae (United States) patented a process for cyanide leaching of gold and silver ores, although this was never used [6, 10]. The cyanidation process, as it is now known, was patented between 1887 and 1888 by MacArthur and the Forrest brothers and was rapidly developed into a commercial process, first at Crown Mine (New Zealand) in 1889 [3, 9]. The technology spread rapidly and was used at Robinson Deep (South Africa) in 1890; Mercur (Utah) and Calumet (California) in 1891; El Oro (Mexico) in 1900; and La Belliere (France) in 1904.

The development of cyanidation was timely to the exploitation of the deep Witwatersrand ores, which had lower grades than those previously worked, and much of the gold occurred as fine grains in hard rock. The cyanidation process, incorporating cementation with zinc, replaced gravity concentration techniques and generally increased gold recoveries from about 70% to 95%, rescuing a declining industry (see Section 1.2.2). This is indicated by the increase in South African gold production from less than 300,000 oz in 1888 to more than 3 million oz in 1898 [11].

The early Witwatersrand mines were a hotbed for technical developments and revolutionized the gold mining industry. The widespread use of cyanidation led to a decline in the use of gravity concentration and the establishment of hydrometallurgy as a distinct subject within mineral and metal processing. Several excellent texts on South African gold metallurgical practice are available, published in 1949 [12], 1972 [13], and 1987 [11].

1.2.2 Flowsheet Development

The flowsheets of the earliest gold plants on the Central Rand (South Africa) typically included screening, crushing, manual sorting of waste rock, stamp milling, amalgamation, cyanide leaching, solid–liquid separation, and recovery of gold by precipitation with zinc. Closed-circuit grinding and classification were achieved using dewatering cones and rake classifiers. From 1904, comminution circuits further improved with the use of tube mills. Between 1904 and 1908, cyanidation plant equipment was revolutionized by Dorr (United States) through his inventions of continuous, large-scale classification, filtration, and thickening equipment. This technology, developed specifically for the gold industry, was applied subsequently throughout the mining industry in general. At about the same time, Oliver (United States) developed the continuous vacuum filter while air-agitated Browns (New Zealand) and Pachuca (Mexico) tanks were introduced for the agitation of slurries. Air agitation in tall, narrow tanks is still used in some plants, although large, cost-effective, mechanical agitators are now preferred.

From 1890 to about 1918, sand and slime fractions from the milling circuits were often treated separately. Subsequent improvements in filtration and fines-handling equipment allowed so-called "all slimes" plants to be introduced. By 1946, 53 large plants were operating on the Witwatersrand, of which 29 practiced all-sliming and 24 operated sand and slime plants, with only 13% of the tonnage treated as sand. Slimes from the milling circuit were cyanide leached in air-agitated tanks. The leached slurry was then filtered to produce a gold-bearing solution. This slurry leaching practice still forms the basis for many presently operating gold plants and has only significantly changed following the introduction of the carbon-in-pulp (CIP) process in about 1980. By 1922 the practice of direct amalgamation of stamp mill product was replaced by the use of corduroy strakes, which preconcentrated the amalgamation feed and significantly reduced the amount of mercury that was used. Encouraged largely for health and security reasons, this change led to many other advances in gravity concentration techniques within the grinding circuit, such as the use of jigs, Johnson drums, and shaking tables for the recovery of coarse gold. During the late 1980s, about 20% of South African gold was produced from gravity concentrates.

Throughout the development of cyanidation, the recovery of precious metal values from cyanide solutions received much attention. Initial recovery methods included cementation using zinc shavings (MacArthur process) and the electrolytic cell (Siemens–Halske and Tainton processes). The efficiency of the zinc-precipitation process was increased by the following:

- Use of zinc dust rather than zinc shavings in zinc boxes
- Introduction of deaeration
- Addition of small quantities of soluble lead salts to solutions prior to precipitation

Zinc precipitation, sometimes referred to as the Merrill–Crowe process, after the instigators of these improvements, is still widely used today and is described in Section 8.1.

Between 1894 and 1899 the Siemens–Halske electrolytic process was developed and implemented at several plants, particularly for the treatment of dilute solutions produced from slimes decantation. An iron anode and lead cathode were used in an open tank system. In 1915 Bosqui introduced an alternative electrolytic cell, invented by Tainton, which used a closed tank with an iron anode and a carbon sheet cathode. Although these cells were a technical success, they were not economically viable. Even now, a commercial process for electrowinning of gold from dilute and impure run-of-mine leach solutions has not been achieved.

Between 1910 and 1930 flotation was introduced for the treatment of base metal sulfide ores. This quickly led to flotation also being used for recovery of gold-bearing sulfides and free gold concentrates [14, 15]. Early examples include the Empire mine (California), Mount Morgan (Queensland), and Le Roi (British Columbia). Flotation was used in South Africa in 1935, where the Brakpan and Government Mining Areas plants were modified to include flotation of a sand fraction, to yield a sulfide-rich concentrate. This was then reground and leached by cyanidation. The flotation tailings were also leached, with only a short residence time required. This route produced marginally higher gold recoveries at significantly lower cost, which enabled lower-grade sulfide ores to be treated economically.

1.2.3 Activated Carbon

The adsorption of precious metal ions or complexes from aqueous solutions onto activated carbon was first noted in the early 19th century. In the 1890s activated carbon was considered as a possible alternative to zinc cementation for the Witwatersrand gold plants. However, at that time, the only known way of recovering the gold was by combustion of the carbon and smelting of the resulting ash. The inability to reuse the carbon, coupled with advances in zinc cementation technology, made the process uncompetitive.

The first CIP plant to use granular carbon was probably the 250-tpd San Andreas de Copan plant (Honduras) in 1949. The loaded carbon was sold to a smelter. In 1950 the Getchell Mine in northern Nevada operated a 500-tpd CIP plant, and a patent was granted to McQuiston and Chapman in 1951 for a CIP technique [9].

At this time much important work was done at the U.S. Bureau of Mines (USBM), notably by Zadra, Salisbury, and Ross to devise a process for gold recovery from carbon, which would then allow the carbon to be reused. The gold removal process, called elution, desorbed gold values to produce a concentrated gold solution from which gold could be recovered by electrowinning onto steel wool cathodes. Initially, the USBM used a caustic sodium sulfide solution for elution, but this method did not remove silver from the carbon. However, this led to the successful development of the Zadra atmospheric pressure stripping process, using sodium cyanide and sodium hydroxide, which was immediately installed at the San Andreas and Getchell mines. These circuits operated until the plants shut down in the late 1950s and early 1960s, respectively.

Activated carbon was used commercially in 1952 at the Carlton mine in Cripple Creek, Colorado, which operated until 1961 when the plant closed largely due to the low fixed gold price of $35/oz. An important feature of the Carlton flowsheet was the use of a reactivation kiln for carbon regeneration. The combination of efficient carbon regeneration and elution made carbon adsorption methods more competitive with the Merrill–Crowe process and was an important precursor to the widespread development of CIP in the 1980s (see Section 1.3.1).

1.2.4 Changing Economic Climate

The U.S. government set an official gold selling price of $35/oz in January 1934. From 1950 to 1972 the gold mining industry suffered severely under this limitation as production costs increased worldwide with inflation, and consequently profitability decreased. This depressed the gold industry and resulted in limited exploration, little research and development, and the start-up of very few new mines. During this time the major proportion of gold and silver was produced as a by-product of copper and lead processing.

In the late 1960s the Homestake mine (South Dakota) produced 30% of U.S. annual production. In 1965, when the Carlin mine in Nevada opened, it was the first large gold mine to be commissioned in the United States in 50 years. In December 1971 and again in September 1973 the U.S. dollar was devalued, effectively increasing the gold price to $38/oz and then $47/oz, respectively. These devaluations, the end of the official dollar–gold conversion, and turbulent political events (e.g., the 1973 oil price shock) led to dramatically increased prices during the mid- to late-1970s (Figure 1.6), reaching a peak of $850/oz at one point in 1980. These economic conditions greatly improved profitability and increased exploration and production efforts worldwide.

1.3 ERA OF MAJOR TECHNOLOGICAL DEVELOPMENT: 1972–2000

The favorable economic climate for gold, particularly during the late 1970s, led to an almost universal interest of mining companies in precious metals, resulting in a boom in exploration and the rapid development of gold mines on all continents, ranging from small-scale prospecting to multimillion-ounce-per-year producers. For example, Australian gold production reached an all-time low of 502,000 oz of gold in 1976 [18] but then increased tenfold to more than 5 million oz in 1989. The distribution of production by country is shown in Figure 1.7.

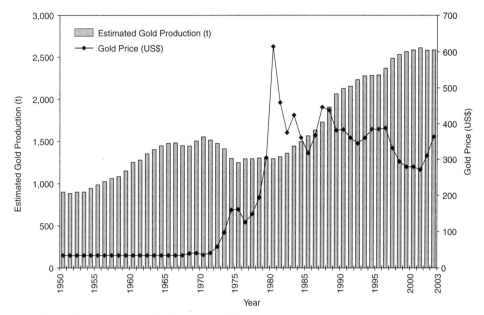

NOTE: 1984–2003 data adapted from [16]. Data prior to 1984 include an estimate for production from former Commonwealth of Independent States countries based on [16] and [17].

FIGURE 1.6 Gold price and estimated world gold mine production [adapted from 16, 17]

Although much good research, development, and industrial application of new technology (notably heap leaching) occurred during the 1970s, the decade of the 1980s produced more technical developments in gold extraction than any other period since the development of cyanidation. The major chemical process technology accepted by the industry included CIP (and carbon-in-leach [CIL]) processing, heap and dump leaching of low-grade ores, electrowinning and replating, pressure oxidation of sulfides, biological oxidation of concentrates, and intensive cyanidation.

Other processes such as biological heap oxidation, whole-ore biological oxidation processes, and resin-in-pulp (RIP) have been demonstrated at pilot plant scale or by small industrial operations. A large number of techniques have shown promise in the laboratory including pressure leaching with various oxidants, alternative gold leaching systems, nitric acid oxidation, and coal–gold agglomeration.

In addition there have been many advances in important ancillary processes; for example:

- Carbon regeneration
- CIP/CIL interstage screening, including the introduction of "pump-cell" technology
- Electrowinning cell design
- Autoclave materials and reactor design
- Roasting equipment
- Process control

This development in industrial chemistry and technology has played an important role in the boom in gold extraction during the 1980s and will provide the competition for emerging technology in the future.

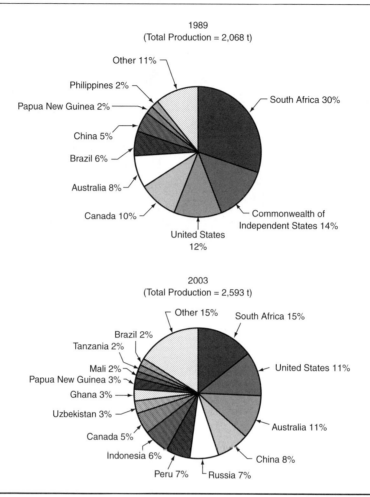

FIGURE 1.7 1989 and 2003 estimated gold mine production by country [16]

1.3.1 CIP Revolution

Although initial work on the development of the CIP process took place in the 1970s, this technology didn't come of age until the 1980s. The precursor to the CIP revolution in the 1980s was the successful operation of Homestake's Lead gold mine, which replaced its conventional cyanidation slime treatment plant with CIP in 1973. In the mid-1970s, Mintek and the Anglo American Research Laboratories (AARL) in South Africa performed significant research and development to improve the CIP process and to adapt it to South African ores. As part of this work, Davidson patented the AARL carbon elution process in 1973, which has become a popular elution method. A small (90,000 tpy) plant was installed at Modderfontein in 1978, and in 1980 three larger plants were installed at President Brand (to treat calcine), Randfontein Estates, and Western Areas (all in South Africa). Between 1981 and 1984, 11 major (>1 Mtpy) CIP and CIL plants were commissioned in South Africa alone, with many others in the United States and Australia establishing activated carbon gold recovery systems as the first choice process route. The first CIP plant in Australia was commissioned at Kambalda in 1981. Other early Australian uses of CIP were at Norseman and Haveluck.

As the application of the CIP process gained momentum, a number of associated technological developments took place. For example, carbon manufacturers developed improved carbons with suitable activity and abrasion resistance for use in gold ore slurries, and various interstage screens were developed (e.g., EPAC, Kambalda, Derrick, NKM). Methods and equipment for on-site reactivation of carbon were improved throughout the 1980s, with continued controversy over the choice of reactivation kilns and stripping methods.

1.3.2 Heap Leaching

Heap leaching was developed by Heinen, Lindstrom, and others at the USBM during the late 1960s and early 1970s as a low-cost treatment method for low-grade ores in Nevada. The first large-scale heap leach operation was installed at the Carlin mine in 1970 and treated ores with grades below the conventional mill cutoff. Heap leaching was subsequently installed at Cortez and Smoky Valley (both in Nevada) in the late 1970s, and thereafter heap leaching boomed in Nevada and the western United States. At this time an agglomeration process was developed, also at the USBM, to allow the treatment of high-clay-content ores, which would otherwise adversely affect leach pad permeability. Agglomeration was achieved by mixing the ore with lime and/or cement and water in an agglomeration drum or by conveyor handling in the heap stacking system.

Heap leaching took advantage of the concurrent development of carbon adsorption processes by using carbon-in-columns for gold recovery from solution, followed by electrowinning. Some plants still prefer to use zinc precipitation in the case of high silver concentrations or when the cost of a carbon stripping and reactivation circuit cannot be justified.

The success of heap leaching was related to the suitability of the Nevada climate, the hydrothermal breccia-type ores (with gold mineralization along cracks and fissures, accessible to solution), the development of agglomeration, and innovative operators. Gold recoveries of 50% to 80% were achieved at much lower capital and operating costs than those of milling/CIP plants. In addition, plant start-up times were extremely short, which led to the application of heap leaching to generate revenue early in new projects to help fund construction of the main processing and/or mining operations.

In 1986, 30% of U.S. gold production resulted from heap leaching, with operations ranging in size from 500 to 10,000 tpd. The use of very large scale heap leaching is maturing with improvements in earth-moving equipment, liner designs, stacker designs, and agglomeration procedures (Table 1.1). The largest operation in Nevada in 1990 was at Round Mountain, which treated approximately 40,000 tpd.

1.3.3 Refractory Ores

During the 1970s and through the 1990s, major efforts were directed at the treatment of so-called refractory ores, that is, ores that could not be effectively treated by simple cyanidation (see Chapters 2 and 5). Historically, severely refractory ores and concentrates, whether sulfidic, carbonaceous, telluride, or a combination of these, have been roasted to completely oxidize the refractory portion of the ore and render the contained gold leachable. Examples of some of the established roasting operations are Fairview (South Africa), La Belliere (France), Getchell (Nevada, United States), Mount Morgan (Australia), and Campbell Red Lake and Giant Yellowknife (both in Canada).

In 1971, Carlin successfully treated carbonaceous ore by chlorination, followed by cyanidation and zinc precipitation (later replaced by CIP). In 1977 the process was altered to a double-oxidation circuit by incorporating a preaeration stage ahead of chlorination. This process was later used at Jerritt Canyon (Nevada), but, due to rising costs, was subsequently discontinued at both locations in favor of the original simple chlorination

TABLE 1.1 Chronology of major events in gold extraction chemistry

Date BC	Event
1000	Amalgamation of gold with mercury discovered.
750	Jabir bin Hayyan establishes aqua regia dissolution of gold.
500	Alloying of gold and silver discovered.

Date AD	Event
1300	Magnus develops nitric acid parting of gold and silver.
1704	Dippel and Diesbach discover Prussian Blue.
1783	Scheele establishes dissolution of gold by aqueous cyanide.
1790	Loweitz discovers gold adsorption by charcoal.
1802	d'Arcet develops sulfuric acid parting of gold and silver.
1840	Elkington develops electroplating from cyanide solutions.
1844	Bagration studies gold cyanidation.
1846	Elsner studies role of oxygen in cyanidation.
1848	Plattner develops chlorination and ferrous sulfate recovery process for gold.
1856	Faraday develops on Elsner's work.
1858	First commercial use of Plattner process, on Deetken ore (Grass Valley, California).
1863	Plattner process widely used in United States, South Africa, and Australia.
1867	Rae patents cyanide leaching process but not used.
1869	Percy proposes charcoal for gold precipitation.
1880	Davis patents gold recovery from chlorinated solutions using charcoal, commercially used in South Carolina (United States).
1886	Harrison discovers gold on Witwatersrand (South Africa).
1887	MacArthur and Forrest brothers patent cyanidation for gold and silver dissolution.
1888	MacArthur and Forrest brothers patent precipitation with zinc shavings.
1889	First cyanidation plant: Crown Mine (New Zealand).
1890	First cyanidation plant in South Africa: Robinson Deep.
1891	First cyanidation plant in United States; Mercur (Utah).
1891	Davis uses charcoal adsorption process at Mount Morgan (Australia) on chlorinated solutions.
1894	Johnson patents charcoal adsorption from cyanide solutions.
1894	Siemens–Halske process for electrolytic recovery of gold applied at Worcester works (Transvaal, South Africa).
1896	Bodlander proposes mechanism of cyanidation leaching involving hydrogen peroxide.
1898	Caldecott uses air-agitated leach tanks.
1904–08	Dorr invents and implements new classifier, thickener, and agitator in United States.
1904–08	Oliver develops continuous vacuum filter.
1904	Merrill introduces zinc dust for gold precipitation.
1906	Crowe applies vacuum deaeration to zinc precipitation.
1908	Aluminum precipitation of silver at Deloro (Mexico).
1916	Charcoal replaces zinc for precipitation at Youanmi (Western Australia).
1934	Chapman and Endquist propose flotation of loaded activated carbon, patented by Chapman (1939).
1947	Thompson explains gold dissolution as corrosion process.
1949	First CIP plant at San Andreas (Honduras).
1951	USBM develops carbon (alkaline sulfide) stripping process.
1951	McQuiston and Chapman patent CIP process.
1952	At USBM, Zadra improves carbon elution process and electrowinning.
1954	Carlton Mill (Colorado, United States) uses CIP with carbon reactivation.
1968	Heinen and Lindstrom perform early work on heap leaching.
1970	First heap leach operation (Carlin, Nevada).
1971	Chlorination of carbonaceous ore at Carlin.

TABLE 1.1 Chronology of major events in gold extraction chemistry (continued)

Date AD	Event
1971	Reverse leach used at West Driefontein (South Africa) for uranium and gold recovery.
1973	Large CIP plant installed at Homestake Lead mine (United States).
1973	Davidson develops AARL elution process.
1975	Mintek starts work on CIP for South African ores.
1976	USBM develops alcohol carbon stripping.
1977	Carlin adopts double-oxidation process.
1978	First South African CIP plant (Modderfontein).
1979	Heinen, McClelland, and Lindstrom develop agglomeration process for heap leaching of clay ores.
1979	Guay patents preaeration and chlorination.
1980–84	Rand Mines Mining and Milling, Simmergo, Ergo, and JMS-OFS build large-scale tailings retreatment plants using CIP in South Africa.
1985	First (acidic) pressure oxidation plant commissioned at Homestake McLaughlin (United States).
1986	Sao Bento (Brazil) starts pressure oxidation plant.
1986	Start-up of concentrate biological oxidation plant in Fairview (South Africa).
1988	First (nonacidic) pressure oxidation plant for gold ore, in Mercur (United States).
1990	Biological oxidation plant for whole-ore treatment started up in Tonkin Springs (United States).
1990	Roasters to treat whole ore commissioned at Big Springs, Jerritt Canyon, and Cortez (United States).
1991	Biological oxidation that precedes pressure oxidation commissioned at Sao Bento (Brazil).
1994	Newmont commissions whole-ore roaster at Carlin.
1994	Biological oxidation plant commissioned successfully at Ashanti Sansu (Ghana) to treat 720 tpd sulfide concentrate.

process. Pressure oxidation of sulfide minerals has been practiced for many years in the nickel and zinc industries, and was investigated for the treatment of gold-bearing sulfides in the 1970s and 1980s. In 1985 a nonacidic pressure oxidation was commissioned at Homestake McLaughlin (California, United States) to treat a pyritic ore. This was followed in 1986 by a similar scheme to treat arsenopyritic flotation concentrates at Sao Bento (Brazil). In 1988, a nonacidic oxidation circuit was started up at Mercur. In the latter case, nonacidic conditions were required because of the high carbonate content of the ore. These plants established pressure oxidation as a viable, albeit higher-cost, method for treating a range of refractory ores. A number of large pressure oxidation plants and expansions were commissioned between 1988 and 2000, including Goldstrike, Getchell, Lone Tree, and Twin Creeks (all in Nevada), Campbell and Con (Canada), and Lihir and Porgera (Papua New Guinea).

Throughout the 1970s and 1980s, researchers at Cardiff University (Wales), University of British Columbia (Canada), University of New Mexico (United States), and Gencor (South Africa), among others, worked on the development of bacterial oxidation for sulfide refractory gold ores. In 1986, a 10-tpd plant was commissioned at Fairview (South Africa) to treat flotation concentrates and has since been operating successfully. In 1990 a 1,500-tpd whole-ore biological oxidation plant was started up at Tonkin Springs (Nevada), though operation ceased due to financial problems; hence, the whole-ore process has not yet been conclusively proven at this scale. To increase plant throughput, a biological oxidation process to partially oxidize flotation concentrates prior to pressure oxidation was commissioned at Sao Bento in 1991. A large biological oxidation facility designed to process about 720 tpd of refractory gold–bearing sulfide concentrates was successfully commissioned at Ashanti in Ghana in 1994–1995 and later expanded to 960 tpd, firmly establishing the technology as a viable commercial process. Subsequently, biological oxidation plants to treat sulfide flotation concentrates were installed at Harbour Lights,

Wiluna, Youanmi, and Beaconsfield (all in Australia), at Tamboraque (Peru), and Laizhou (Shandong, China). The motivation for the continued development of bio-oxidation processes for sulfide concentrate treatment is the potential cost savings over pressure oxidation and the considerable environmental advantages over roasting. Newmont commissioned a 10,000-tpd biological heap oxidation facility at Carlin as a pretreatment step ahead of conventional cyanidation and CIP recovery (see Section 5.7).

Significant improvements in gas scrubbing and cleaning technology were made in the United States, which brought low-cost roasting back into favor in the late 1980s. Roasters were commissioned at Big Springs, Cortez, and Jerritt Canyon (all in Nevada) between 1988 and 1990 to treat whole ores rather than concentrates. A large whole-ore roaster utilizing circulating fluidized bed technology was installed at Newmont's Carlin operations in 1994, paving the way for a resurgence in roaster technology applied to refractory sulfide and carbonaceous ores. Circulating fluidized bed roasters were subsequently installed at Fimiston (Australia) in 1990, Syama (Mali) in 1994, Minahasa (Indonesia) in 1996, and Barrick Goldstrike (Nevada) in 2000.

The continued development of refractory ore projects has encouraged renewed interest in lixiviants for gold that could be used in acidic media to avoid the high neutralization costs required by alkaline cyanidation of oxidized products. In the 1990s a number of lixiviants were investigated, including thiosulfate, thiourea, thiocyanate, chloride/chlorine, and other halides. Of these, thiosulfate has received the most attention and holds the most promise as a technically and economically viable alternative to cyanide. The major challenges for the application of alternative leaching systems are the reagent consumption and cost (in all cases, higher than cyanide), the lack of selectivity for gold (and silver) over other metal constituents of the ore, and the difficulties in recovering gold from solution/slurry after dissolution.

1.4 INTO THE 21ST CENTURY

Metallurgical research and development work was prolific in the 1980s and 1990s, with many process alternatives and improvements either proposed or implemented. The main driving forces for this to continue in the future are the need to treat lower-grade ores with more complex mineralogy, coupled with the increasing environmental requirements placed on mining operations. Some specific areas likely to receive widespread attention are the following:

- Development of alternative gold-leaching reagents, particularly thiosulfate-based systems
- Commercialization of biological oxidation processes for sulfidic gold ores, especially heap systems
- Improved hydrometallurgical treatment of sulfidic ores
- Replacement of carbon adsorption (CIP, CIL, and carbon-in-solution [CIS]) by resin adsorption systems (RIP, resin-in-leach [RIL], and resin-in-solution [RIS])
- Gold recovery from secondary sources, particularly electronic scrap
- Improved ore characterization and mineralogical diagnostics for gold
- Improved control of effluents
- Reagent recovery and recycling from effluent streams (particularly cyanide and thiosulfate)
- Metals recovery and removal from effluent streams
- Direct, selective electrowinning from dilute solutions

- Effective processing techniques for arid environments
- Innovative use of flotation in flowsheets
- Advanced physical and surface chemical techniques for gold recovery
- Alternative methods for upgrading gravity concentrates
- Improved process analysis and control

These technical areas, and many more, are considered in more detail in later chapters and will provide new challenges to scientists and engineers in the future.

REFERENCES

[1] Gold Fields Mineral Services Ltd. 1990. *Gold.* London: Gold Fields Mineral Services Ltd.

[2] Collender, F.D. 1988. The historical importance of gold in the world's monetary systems. Pages 383–386 in *Proceedings Randol Perth International Gold Conference 1988.* Golden, CO: Randol International Ltd.

[3] McNulty, T. 1989. *A Metallurgical History of Gold.* Paper presented at American Mining Congress. San Francisco, CA, September 20.

[4] Agricola, G. 1912. De Re Metallica. Trans. H.C. Hoover and L.H. Hoover. *The Mining Magazine* 637. (Orig. pub. 1556.)

[5] Temple, J. 1972. *Mining and International History.* London: Ernest Benn.

[6] Schnabel, C. 1921. Pages 936–1134 in *Handbook of Metallurgy.* 3rd edition, Volume 1. London: Macmillan.

[7] Anon. 1891. A new process for gold extraction. *Mining Journal* (February).

[8] Elsner, L. 1846. Beobachtungen über das Verhalten regulinischer Metalle in einer wässrigen Lösung von Cyankalium (Observations on the behavior of pure metals in an aqueous solution of cyanide). *Journal für Praktische Chemie* (Germany) 37(1):441–446.

[9] Shoemaker, R.S. 1984. Gold: Quid non mort alia pectora cogis, auri sacra fames. Pages 4–10 in *Precious Metals: Mining, Extraction and Processing.* Edited by V. Kudryk, D.A. Corrigan, and W.W. Liang. Littleton, CO: SME-AIME.

[10] Habashi, F. 1987. One hundred years of cyanidation. *CIM Bulletin* 80(905):108–114.

[11] Stanley, G.G. 1987. *The Extractive Metallurgy of Gold in South Africa.* Monograph M7. Johannesburg: South African Institute of Mining and Metallurgy.

[12] King, A. 1949. *Gold Metallurgy on the Witwatersrand.* Johannesburg: Transvaal Chamber of Mines.

[13] Adamson, R.J., editor. 1972. *Gold Metallurgy in South Africa.* Johannesburg: Chamber of Mines of South Africa.

[14] Motherwell, W. 1914. Flotation test at Mount Morgan. *Mining and Scientific Press* 53140:1044–1046.

[15] Fahrenwald, A.W. 1933. Flotation of gold from river sand and black sand. *Mining Journal* 16(23):3–4.

[16] Gold Fields Mineral Services Ltd. *Gold Survey Reports, 1981–2004.* London: Gold Fields Mineral Services Ltd.

[17] Green, T. 1984. Pages 63–76 in *The New World of Gold.* New York: Walker & Co.

[18] Thomas, P.R., and E.H. Boyle. 1986. *Gold Availability (World): A Minerals Availability Appraisal.* IC 9070. Washington, DC: U.S. Bureau of Mines.

CHAPTER 2

Ore Deposits and Process Mineralogy

The characteristics of an ore deposit and its mineral assemblages determine the mining method(s), extraction process requirements, and, in particular, the performance of all chemical processes involved in gold extraction. Consequently, a good understanding of the mineralogy of an ore is required to design or operate a gold extraction process for optimum efficiency.

The gold mineralogy in each ore deposit is unique, due to the variations in the following:
- Mineralogical mode of occurrence of gold
- Gold grain size distribution
- Host and gangue mineral type
- Host and gangue mineral grain size distribution
- Mineral associations
- Mineral alterations
- Variations of the above within a deposit or with time

It is therefore important to consider the types of ore deposits and mineralogical factors, which, together with the economic, geopolitical, and engineering factors, discussed in Chapter 3, affect the technology used in the ore processing strategy.

This chapter concentrates on the aspects of process mineralogy that are relevant to gold extraction processes, rather than describing the geology of ore deposits and their formation in any detail.

2.1 GOLD MINERALS

Because gold is inert at ambient temperatures and pressures, there are very few naturally occurring compounds of the metal. The average concentration of gold in the earth's crust is 0.005 g/t, which is much lower than most other metals, for example, silver (0.07 g/t) and copper (50 g/t). The low concentration of gold in primary rocks means that upgrading by a factor of 3,000 to 4,000 is usually required during ore formation processes to achieve commercial concentrations. This may be possible by natural gravity concentration processes or by the leaching of gold with natural fluids from the host rock; for example, by highly oxidizing, acidic and complexing (e.g., chloride) solutions, followed by redeposition in a more concentrated form. Owing to its siderophile properties (i.e., weak affinity for oxygen and sulfur; high affinity for metals), gold tends to concentrate in residual hydrothermal fluids and subsequent metallic or sulfidic phases, rather than silicates, which form at an earlier stage of magma cooling. Rocks that are high in clays and low in carbonates are the best sources for gold, and re-precipitation occurs when the hydrothermal solutions encounter a reducing environment (e.g., a region of high carbonate, carbon, or reducing sulfide content). Examples of gold precipitation reactions* in ore formation are as follows [1]:

Gold precipitation by pyrite:

$$4FeS_2 + 6H^+ + 4H_2O \rightleftharpoons 4Fe^{2+} + 7H_2S + SO_4^{2-} \quad \text{(EQ 2.1)}$$

$$AuCl_2^- + 0.5H_2 \rightleftharpoons [Au] + 2Cl^- + H^+ \quad \text{(EQ 2.2)}$$

$$AuCl_2^- + Fe^{2+} \rightleftharpoons [Au] + Fe^{3+} + 2Cl^- \quad \text{(EQ 2.3)}$$

Gold and quartz precipitation by carbon:

$$[C] + 2H_2O \rightleftharpoons CO_2 + 2H_2 \quad \text{(EQ 2.4)}$$

$$(Au^+)complex + 0.5H_2 \rightleftharpoons [Au] + H^+ \quad \text{(EQ 2.5)}$$

Lowering of activity of water to precipitate quartz:

$$[2C] + 2H_2O \rightleftharpoons CO_2 + CH_4 \text{ (in solution)} \quad \text{(EQ 2.6)}$$

$$H_4SiO_4 \rightleftharpoons [SiO_2] + 2H_2O \quad \text{(EQ 2.7)}$$

A summary of the conditions required for gold precipitation are given in Table 2.1.

The predominant occurrence of gold is as native metal, often alloyed with up to 15% Ag. Other gold minerals include alloys with tellurium, selenium, bismuth, mercury, copper, iron, rhodium, and platinum (Table 2.2). There are no common naturally occurring gold oxides, silicates, carbonates, sulfates, or sulfides. Therefore, gold generally occurs in a mineral form different to most other elements which, *inter alia*, often allows selective gold extraction from other mineral mixtures. The most common gold minerals are considered in Sections 2.1.1 through 2.1.5.

2.1.1 Native Gold

Native gold grains have been known to contain up to 99.8% Au, but most vary between 85% and 95% Au content, with silver as the main impurity. Pure gold has a density of 19,300 kg/m^3, though native gold typically has a density of 15,000 kg/m^3. Hence, if liberated from gangue minerals, it can be readily recovered at particle sizes above 10 μm by gravity concentration—the major method of recovery of gold employed throughout history (see Chapter 1). Gravity concentration can be very selective, as the most common gangue minerals (e.g., quartz and other silicates) have densities in the range of 2,700 to 3,500 kg/m^3.

Gold is very soft, ductile, and malleable (1 oz of gold can be beaten into an area of 30 m^2) with Vickers and Mohs hardness numbers of 40 to 95 kg/mm^2 and 2.5 to 3.0, respectively (all ounces cited are troy ounces). These unusual physical properties are a result of the face-centered cubic crystal structure. Native gold rarely occurs in its cubic crystal form, and the famous rounded masses, known as nuggets, are now only found occasionally. Gold does not display cleavage on breakage. There are many terms to describe the various, and often distinctive, forms of native gold—sponge gold, flakey gold, grain gold, foil gold, moss gold, and tree gold.

---

* All equations representing electrochemical reactions in this book have been expressed as reversible reactions (⇌) with the species of interest shown in its oxidized state on the left and the reduced state on the right.

TABLE 2.1 Mechanisms and complexes that affect the precipitation of gold during ore-formation processes [1]

Mechanism	Complexes		
	AuHS⁻	Au(HS)$_2^-$	AuCl$_2^-$
Pressure↓	Unknown	No at temp. >250°C Unknown at temp. <250°C	Yes
Temperature↓	No	No, except when pH↓	Yes
pH↑	No	No	Yes
Boiling	No	No	Mostly yes
Reduction by carbon	Yes	Yes	Yes
Reaction with pyrite, aresenopyrite	No	No	Yes

The freezing and boiling points of gold are 1,064°C and 2,808°C, respectively. Other properties of gold are listed in Table 2.3. Gold has a metallic luster, and the color is a distinctive deep yellow (golden) but may be light yellow or orange-yellow with high silver and copper contents, respectively. Its distinctive high reflectivity and low hardness can be used in its identification by microscopic examination of polished sections (Figure 2.1). Pure gold is an excellent electrical and thermal conductor (Table 2.3).

2.1.2 Electrum

Gold usually occurs alloyed with some silver; however, when the silver content is between 25% and 55%, the mineral is called electrum. Electrum has a pale yellow color, due to the silver content, as indicated in Figure 2.2, and a lower density (i.e., 13,000 to 16,000 kg/m^3) than gold.

A term commonly used to express the purity of gold or concentration of silver is fineness, defined as:

$$\text{fineness} = \frac{(\text{wt \% Au}) \times 1,000}{(\text{wt \% Au} + \text{wt \% Ag})} \quad \text{(EQ 2.8)}$$

This formula provides a measure of the relative concentration of gold and silver rather than an absolute gold concentration and becomes less meaningful with increasing concentrations of other metals such as iron and copper. Fineness is most usefully applied to an analysis of bullion. As silver tends to report with gold in most process flowsheets, an indication of ultimate bullion fineness can be roughly estimated from the relative gold and silver concentrations in an ore or concentrate sample, or from microprobe analyses of gold grains (see Section 2.18.2); however, recoveries of the two metals can vary significantly, especially by cyanidation.

2.1.3 Gold Tellurides

The chemistry of gold tellurides is relatively complex with a series of identifiable minerals (Table 2.2). The more common gold-bearing tellurides are sylvanite ((Au,Ag)$_2$Te$_4$), calaverite (AuTe$_2$), and petzite (Ag$_3$AuTe$_2$), with krennerite (AuTe$_2$), montbrayite (Au$_2$Te$_3$), and kostovite (CuAuTe$_4$) less common. Gold-telluride occurrence is widespread and is often associated with some free gold and sulfide minerals. The densities of gold tellurides (8,000 to 10,000 kg/m^3) are lower than native gold, and the colors are less distinctive shades of white, gray, and black. The silver mineral, hessite (Ag$_2$Te), is commonly encountered in gold-telluride ores.

TABLE 2.2 Properties of naturally occurring gold minerals [2]

	Native Gold	Electrum	Calaverite	Krennerite	Sylvanite	Montbrayite
Formula	Au	(Au, Ag)	$AuTe_2$	$AuTe_2$	$(Au,Ag)_2Te_4$	Au_2Te_3*
Au content (%)	>75	45–75	39.2–42.8	30.7–43.9	24.2–29.9	38.6–44.3
Crystal system	Cubic	Cubic	Monoclinic	Orthorhombic	Monoclinic	Triclinic
Specific gravity	16.0–19.3	13.0–16.0	9.2	8.6	8.2	9.9
Mohs hardness	2.5–3	2–2.5	2.5–3	2.5	1.5–2	2.5
VHN[†]	41–94	34–44	199–209	117–130	102–221	198–228
Color[‡]	Deep yellow	Pale yellow	White or creamy yellow	Creamy yellow	Creamy white	Creamy white
Bireflectance	—	—	Weak-distinct	Weak	Distinct	Weak
Anisotropy	—	—	Weak-distinct	Strong	Strong	Weak-moderate
Internal reflections	—	—	—	—	—	—
Reflectance:						
White light[§]	74	See below[††]	64.0	72.0	49.0–59.0	64.0
at 480 nm**	35		61.0	64.9	49.1–55.7	55.8
at 540 nm	66		64.4	71.9	49.7–58.2	63.5
at 580 nm	71		65.8	74.6	49.3–59.4	66.1
at 640 nm	82		67.3	76.0	48.5–60.7	67.3
Cleavage	—	—	—	Two—one good, one poor	Two—one good, one poor	Three—all good
Twinning	Growth twins common	—	Lamellar; rare	Lamellar; rare	Lamellar; common	—
Etch tests:						
KCN	Some etched black	Etched	—	—	—	—
HNO_3 concentrate	—	—	Stains purple brown	Stains light brown	Stains dark gray-brown	Stains light gray-brown
HNO_3/H_2O 1:1	—	—	Stains light brown	Stains light brown	Stains brown	Stains light yellow-brown
HCl	—	—	—	—	—	—
$FeCl_3$	—	—	Slowly stains light brown	Stains light yellow	Stains light yellow-brown	—
KOH	—	—	—	Tarnishes light gray-brown	—	—
$HgCl_2$	—	—	—	—	—	—

TABLE 2.2 Properties of naturally occurring gold minerals [2], continued

	Petzite	Hessite	Nagyagite	Kostovite	Aurostibnite	Maldonite
Formula	Ag_3AuTe_2	Ag_2Te	$Au(Pb,Sb,Fe)_8(S,Te)_{11}$ [##]	$CuAuTe_4$	$AuSb_2$	Au_2Bi
Au content (%)	19.0–25.2	4.7	7.4–10.2	25.2	43.5–50.9	64.5–65.1
Crystal system	Cubic	Cubic (probable)	Tetrahedral/Orthorhombic (probable)	Monoclinic (probable)	Cubic	Cubic
Specific gravity	9.1	8.4	7.5		9.9	15.5
Mohs hardness	2.5	2.5–3	1.5	2–2.5	3?	1.5–2
VHN [†]	43–74	24–41	39–110	35–43	248–262	>110
Color [‡]	Grayish-white with violet tint	Grayish-white	Grayish-white	Creamy white	White but tarnishes pink	Gray-white
Bireflectance		Very weak	Weak	Distinct		
Anisotropy	Very weak–weak	Strong	Weak-distinct	Strong		
Internal reflections						
Reflectance:						
White light [§]	~37.0	39.41	~39.0	54.0–60.0	~61.0	50.0–60.0 (approx.)
at 480 nm [**]	42.1	39.9–40.7	42.2	52.1–55.2	61.0	
at 540 nm	37.1	38.7–40.9	38.7	54.2–60.1		
at 580 nm	34.9	37.8–41.3	37.0	53.0–57.9		
at 640 nm	33.3	37.3–43.0	35.1	49.3–55.2		
Cleavage	Three—good		One—perfect	One		
Twinning		Lamellar; common	Cross-hatch; not uncommon	Lamellar; ubiquitous	Unknown	
Etch tests:						
KCN	Some stains light brown	Slowly stains black			Rapidly stains iridescent	
HNO_3 concentrated	Stains dark brown				Stains dark brown	
HNO_3/H_2O 1:1	Stains dark brown-violet	Stains iridescent to black	Slowly stains iridescent		Stains iridescent	
HCl	Some stains iridescent	Slowly stains black			Slowly stains light brown	
$FeCl_3$	Stains iridescent	Stains iridescent				
KOH						
$HgCl_2$	Slowly stains brown	Stains brown to iridescent				

NOTES: A blank space means no data found. A dash (—) indicates nil, not present, or no reaction.
* Also reported as $(Au,Sb)_2(Te,Bi)_3$.
† VHN: Vickers hardness number refers to 100-g load except for krennerite (25 g), petzite, and aurostibnite (15 g).
‡ Color in incident reflected white light.
§ White light reflectances are approximate only.
** Reflectance values are given at wavelengths of 480, 540, 580, and 640 nm except for gold (470, 550, 590, 650 nm), kostovite (482, 559, 589, 668 nm), and aurostibnite (546 nm).
†† Reflectances of native gold of various fineness values are as follows:

Fineness	1,000	900	800	700	600	500
R (470 nm)	36.4	43.5	56.0	66.8	75.1	81.5
R (541 nm)	71.6	77.9	83.1	86.2	88.0	89.4

Also reported as $Pb_5Au(TeSb)_4S_{5-8}$.

TABLE 2.3 Properties of gold

Property	Value
Atomic weight	196.9665
Melting point (K)	1,337 (1,064°C)
Boiling point (K)	3,081 (2,808°C)
Atomic radius, Au lattice (nm)	0.1422
Crystal structure	Face-centered cubic; 4 atoms/unit cell
Lattice constant at ambient temperature (nm)	0.407
Interatomic distance at ambient temperature (nm)	0.2878
Density at 273 K (g/cm^3)	19.32
Brinell hardness (10/500/90) (annealed at 1,013 K) (MPa)	25
Modulus of elasticity at 293 K (annealed at 1,173 K) (MPa)	7.747×10^4
Poisson's ratio, as drawn	0.42
Tensile strength (annealed at 573 K) (MPa)	123.6–137.3
Elongation (annealed at 573 K) (%)	39–45
Compressibility at 300 K (Pa^{-1})	6.01×10^{-12}
Heat of fusion (J/mol)	1.268×10^4
Heat of evaporation at 298 K (J/mol)	3.653×10^5
Vapor pressure (Pa)	
at 1,000 K	5.5×10^{-8}
at 1,500 K	8.5×10^{-2}
at 2,000 K	82.0
at 2,500 K	4.9×10^3
at 3,000 K	7.1×10^5
Specific heat at 298 K (J/(g·K))	1.288×10^{-1}
Thermal conductivity at 273 K (W/(m·K))	311.4
Thermal expansion at 273–373 K (K^{-1})	1.416×10^{-7}
Electrical resistivity at 273 K (Ω·cm)	2.05×10^{-5}
Temperature coefficient of resistivity at 273–373 K (K^{-1})	4.06×10^{-3}
Work function (J)	
Thermionic	$7.69–7.85 \times 10^{-19}$
Photoelectric	$8.17–8.76 \times 10^{-19}$
Thermal; emf (mV)	
at 273 K	0.92
at 373 K	6.40
at 1,073 K	12.35
Total emissivity at 493–893 K	0.018–0.035
Susceptibility (magnetic) at 291 K (cm^3/g) (=emu/g)	1.43×10^{-7}
Entropy at 298 K (J/K)	47.33

2.1.4 Other Gold Minerals

Gold occasionally occurs with bismuth in the mineral maldonite (Au_2Bi), named after the occurrence at Maldon (Victoria, Australia). Maldonite has a density of 15,500 kg/m^3 and a lower Mohs hardness (1.5 to 2.0) than gold. It has a very low solubility in cyanide solutions.

Gold and copper (Cu) form the extremely rare intermetallic compounds auricupride ($AuCu_3$) and tetra-auricupride (AuCu). Natural $AuCu_3$ actually contains only 40% Au, instead of the stoichiometric amount of 50.8%. The crystal system is face-centered cubic with copper at the center and gold at the corners. The copper content imparts a deep red color (Figure 2.2).

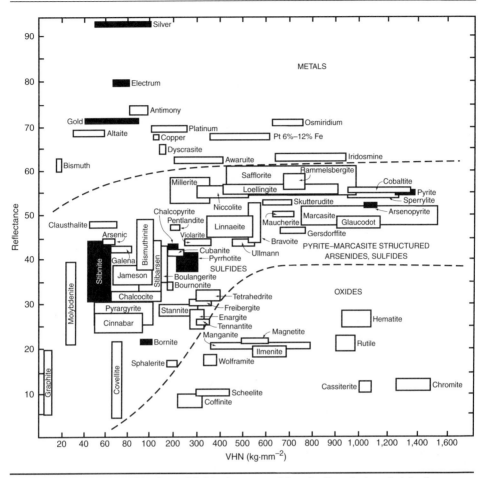

FIGURE 2.1 Diagram to aid reflected light microscopy determination of minerals (shading indicates minerals of particular importance in gold extraction) [3]

2.1.5 Gold with Sulfides

Gold can occur as ultrafine ("invisible") solid solution inclusions within sulfide mineral grain structures [5, 6]. For example, gold content within sulfide mineral structures has been measured as follows:

- Arsenopyrite: <0.2 to 15,200 g/t
- Pyrite: <0.2 to 132 g/t
- Tetrahedrite: <0.2 to 72 g/t
- Chalcopyrite: <0.2 to 7.7 g/t

Such occurrences are important because, for example, in an ore containing 1% by weight of arsenopyrite and a gold ore grade of 10 g/t, all of the gold could be present in solid solution (i.e., invisible) within the arsenopyrite. In this case, the arsenopyrite would only need to have an average gold content of 1,000 g/t, much less than the upper end of the range indicated, and concentrations in this range (i.e., 250 to 1,500 g/t) are commonly encountered in practice. Hence the occurrence of gold in solid solution in sulfide minerals

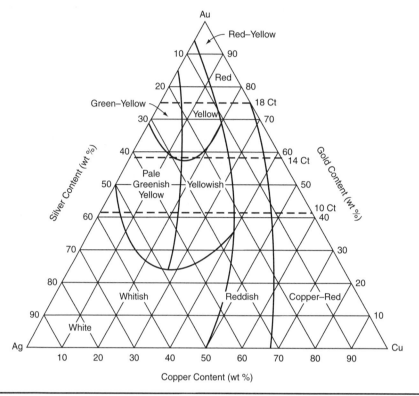

FIGURE 2.2 Gradation of color of Au–Cu–Ag alloys. Courtesy of Academic Press [1]

(especially arsenopyrite, pyrite, and pyrrhotite) is of major importance in many refractory gold systems.

2.2 CLASSIFICATION OF GOLD-BEARING MATERIALS

Primary and secondary gold-bearing materials can be classified into 15 mineral processing–based categories, which are related to their mineralogical and historical characteristics (adapted from [7]):

Primary ores
- Placers
- Free-milling ores
- Oxidized ores
- Silver-rich ores
- Iron sulfides
- Arsenic sulfides
- Copper sulfides
- Antimony sulfides
- Tellurides
- Carbonaceous ores

Secondary materials
- Gravity concentrates
- Flotation concentrates
- Tailings
- Refinery materials
- Recycled gold

Each of these classes of gold-bearing materials has special mineralogical characteristics, which affect their processing and are considered in more detail in Sections 2.3 through 2.17.

2.3 PLACERS

Placer gold ores contain alluvial, eluvial, or colluvial material in active ore-deposit-forming systems and have been classified here as deposits where diagenetic processes have occurred to only a limited extent. Ore crushing and grinding are unlikely to be required in the treatment of such ores. Hard rock palaeoplacers (e.g., Witwatersrand ores of South Africa) have been classed as free-milling ores (Section 2.4), in keeping with convention throughout the industry.

2.3.1 Formation of Placers

Placer deposits are formed as a result of gold liberation by weathering and hydraulic transport of gold particles away from a primary gold deposit. This is possible because gold is chemically inert and dense, resulting in accumulations of gold relatively close to exposed primary deposits. The prerequisites for the formation of placers include the following:

- A primary source of gold (e.g., gold–quartz veins, auriferous sulfide deposits, or former placers)
- A long period of chemical and physical weathering to release gold grains from the host rock
- Concentration of gold particles by gravity, almost certainly involving moving water as the transportation medium
- Stable bedrock and surface conditions over a long period (e.g., no glaciation or folding) to allow significant concentrations of gold to accumulate

There are several classes of placer ores [8, 9], which relate to the means of gold concentration and the distance from the primary gold deposit (Figure 2.3). These are listed in the following sections.

Eluvial (or residual) placers. Usually such deposits are overlying, or located at or very near, the parent deposit and consist of weathered rock from which some of the finer and lighter minerals have been washed away, leaving gold at a higher concentration. Because prolonged mechanical erosion by water has not occurred, the gold grade is typically lower than other placer types. Eluvial gold deposits in tropical regions are commonly lateritized (i.e., with the host rock weathered to form hydrated iron and aluminum oxides with silica).

Colluvial (or deluvial) placers. Gold has been transported some distance from the parent deposit and is not located in an established stream system (e.g., on the slopes surrounding outcropping source rocks.)

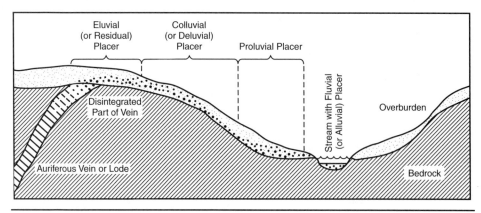

FIGURE 2.3 Outcrop of a gold–quartz vein supplying material to form eluvial and alluvial placers [9]

Fluvial (or alluvial) placers. Occurring in stream or river systems, the gold tends to concentrate upstream of obstructions and in areas of lower fluid velocity. Examples of gold concentration processes in fluvial placers are given in Figure 2.4.

Marine placers. Formed by the natural sorting action of a beach environment, marine placers concentrate the valuable minerals. Gold in such deposits is often associated with other dense minerals such as iron (magnetite), titanium (ilmenite and rutile), or tin (cassiterite) minerals. Gold accumulations can occur in beach (sand) terraces as a result of a drop in sea level relative to the land mass (e.g., South Island, New Zealand) or are present as submerged alluvial formations resulting from a relative increase in sea level.

2.3.2 Commercial Significance

The amount of gold present in placer ores is usually low compared with the associated primary hard rock deposit from which the ores were formed (Figure 2.3). However, due to the ease of operation and low costs, placers are often commercially significant and may be the forerunner to further underground mining. The capital and operating costs of placer operations can be very low, allowing economic mining of ores containing as little as 0.2 g/t Au. However, the contribution of placer gold (excluding palaeoplacers) to annual world production is now small, between 2% and 5%.

In several notable cases, the proportion of gold present in placers exceeds that in the parent deposit (Table 2.4). These deposits are called giant placers [10] and have yielded many millions of ounces of gold. These deposits were particularly significant in the 19th century gold rushes (see Section 1.1.3).

2.3.3 Gold Mineralogy

The gold mineralization in placers differs from all other ore classes because the ore is in a particulate or loosely consolidated form and the gold has been liberated to a large extent by natural processes. Consequently, the savings in crushing and grinding costs compared to other ore types allows very low-grade ores to be treated economically.

Gold grains of several centimeters in diameter occasionally occur, although sizes of 50 to 100 μm, and smaller, are more normal. The relationship between the gold particle size and distance from the parent deposit is usually inverse. In the case of Snake River (Montana, United States), very fine gold has been mined up to 400 km from the source, following periods of flooding [11].

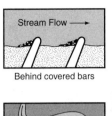
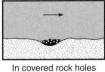

Behind covered bars | In covered rock holes | In potholes below waterfalls

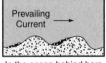

On the inside of meander loops | Downstream from the mouth of a tributary | In the ocean behind bars against the prevailing current

Obstructing or deflecting barriers allow faster-moving waters to carry away the suspended load of light and fine-grained material while trapping the denser and coarser particles, which are moving along the bottom by rolling or by partial suspension. Placers may form wherever moving water occurs, though they are most commonly associated with streams.

FIGURE 2.4 Typical sites for fluvial placer accumulations [9]

TABLE 2.4 Size, discovery, and gold occurrence in giant placer deposits

Location of Gold Placer	Date of Discovery	Approximate Yield of Placer Gold (oz × 10^6)	Ratio of Placer–Lode Gold*	Age of Host Rocks to Parent Mineralization
Otago, New Zealand	1861	8.0	27	Mesozoic
Westland, New Zealand	1864	5.1	2.5	Palaeozoic
California, United States	1849	42.0†	2	Palaeozoic
British Columbia, Canada	1857	6.0	(11)	
Klondike, Yukon Territories	1896	9.0	(9)	Palaeozoic
Fairbanks, Alaska, United States	1903	8.0	38.5	Palaeozoic
Lena-Amur region, Russia		40.0	—	—
South America:				
Colombia	1493	32.0	2	Tertiary (probably)
Peru		4.0		Tertiary (probably)
Bolivia		9.9	Predominantly placers	Tertiary (probably)
Chile		11.0		Tertiary (probably)
Witwatersrand, South Africa	1886	843.0	(840)	Archaean

NOTE: Blank cells indicate information was not available.
* The ratio of placer–lode gold was estimated from available data; parentheses indicate relatively trivial recorded lode production from region.
† Yield in ounces estimated on basis of US$20/oz.

The evaluation of ore grade in placers is difficult because of the low grade and unusually coarse gold grain size. Extremely large samples of several hundreds of tons must be treated in a mineral concentration sampling plant for the evaluation of a placer ore grade and for flowsheet design. Commonly, this is a gravity concentration plant composed of spirals, jigs, and centrifugal concentrators. Estimates of gold grade by this method may differ greatly from the eventual average gold concentration recovered from operations, with ratios of recovered-to-expected gold content varying from 32% to 149% in one study [12]. A guide to sampling and analysis of placer gold ores has been provided by MacDonald [11].

The fineness of gold in young placers depends on the original source and varies between 600 and 900 (i.e., 60% to 90% gold content). Placer gold grains have been found to have an outer rim, which has a higher fineness [13]. This has been attributed to silver dissolution, as silver sulfate (Ag_2SO_4) or silver carbonate (Ag_2CO_3), and is further supported by evidence that fineness increases with distance downstream [14]. The lower surface silver content gives placer gold a deeper gold–yellow appearance than gold in hard rock ores (Figure 2.2).

The degree of gold liberation and the surface chemical properties of placer gold are important for the effectiveness of gravity concentration and amalgamation. As most gangue minerals are lighter than gold, unliberated gold grains are recovered less efficiently by gravity concentration. Detailed surface chemical data on gold grains in placer deposits are not abundant; however, it is known that sulfur and hydrocarbon adsorption can occur, and the presence of impurities in the gold significantly affects amalgamation (see Section 9.3). Plate 1 shows gold grains from placer deposits displaying coatings of mercury, silica, and iron oxides, which are detrimental to recovery by amalgamation.

2.4 FREE-MILLING ORES

Free-milling ores are defined as those from which cyanidation can extract approximately 95% of the gold when the ore is ground to a size of 80% <75 μm, as commonly applied in industrial practice, without incurring prohibitively high reagent consumptions. Frequently, some of the gold is recovered by gravity concentration and/or amalgamation, and gangue mineral composition does not significantly affect the processing requirements.

The two main classes of free-milling ores are palaeoplacers and quartz vein gold ores. Both contain gold mineralization within a hard rock matrix. Some epithermal deposits may be free milling but more commonly contain significant concentrations of sulfide minerals and are therefore considered in subsequent classes separately.

2.4.1 Palaeoplacers and Quartz Vein Gold Ores

Palaeoplacers are literally fossilized placers, the most famous being the Witwatersrand lakebed reefs in South Africa. Others include Jacobina (Brazil), Blind River–Elliot Lake (Canada), and Tarkwa (Ghana). The major examples are of Precambrian age (>570 million years old).

Palaeoplacers consist of lithified (the formation of massive rock from loose sediment) conglomerates which contain small rounded pebbles of quartz in a matrix of pyrite, fine quartz, micaceous materials, and small quantities of heavy, resistant minerals such as magnetite (Fe_3O_4), uraninite (UO_2/U_3O_8), platinum group metals (PGMs), titanium minerals, and gold.

From a mineral processing point of view, palaeoplacers differ from young alluvial placers, as the gold is unliberated and the ore is consolidated. Crushing and grinding is therefore required to liberate the gold to an extent that allows efficient gold extraction. Because palaeoplacer gold deposits have been mined at depths of up to 3 km, mining costs are generally more than an order of magnitude greater than those for young placer deposits. Also, many of these quartz-rich ore types are very hard, resulting in high processing costs in some cases.

2.4.1.1 Witwatersrand Ores

The Witwatersrand region has been the world's predominant gold-producing area for the past 120 years. The gold ores are relatively easy to treat and contain coarse gold recoverable by gravity concentration. The native gold in Witwatersrand ore typically contains between 7.5% and 14.3% silver with an average of about 10%. Values ranging from 0.3% to 30% have been measured, which reflect the presence of gold from different

TABLE 2.5 Mineralogical composition (by mass) of three auriferous Witwatersrand (South Africa) banket reefs [15]

Component	Vaal*	Ventersdorp†	Dominion‡
Gold (ppm)	50	44	202
Silver (ppm)	8	5	—
Uranium oxide (ppm)	870	290	—
Uraninite, thorite (%)	—	—	11.5
Quartz (%)	88.3	88.9	30.6
Chlorite (%)	0.8	4.9	26.2
Muscovite (sericite) (%)	4.4	3.0	0.2
Pyrophyllite (%)	0.1	0.2	—
Zircon (%)	0.1	0.2	1.1
Monazite (%)	—	—	6.0
Chromite (%)	0.2	0.1	0.5
Cassiterite (%)	—	—	0.3
Titanium minerals (%)	0.1	0.1	12.3
Sulfide minerals (%)	6.0	2.6	11.1

NOTE: Dashes indicate that information was not applicable.
* Vaal = Vaal reef, Hartebeestfontein mine, Klerksdorp fluvial fan.
† Ventersdorp = Ventersdorp contact reef, Venters mine, West Rand fluvial fan.
‡ Dominion = Upper reef, Bramley section, Dominion Reefs mine.

source reefs. The mineralogical compositions for three reef samples are given in Table 2.5. A notable feature is the low concentrations of sulfides, particularly those of copper, lead, and zinc.

The Witwatersrand ores are made up of three main types of material: coarse quartz pebble material (called "banket" reef), carbon seams, and pyritic quartzite. The gold occurrence has been classed into the following five groups based on the formation processes:

- Detrital gold particles
- Gold that has been biochemically redistributed into carbonaceous matter
- Gold that has been recrystallized by metamorphic and diagenetic processes
- Primary gold located in detrital allogenic sulfide minerals
- Gold located in secondary quartz veins

The characteristics of these component ore types are given in Table 2.6. Additional information on the mineralogy of Witwatersrand ores is available in the literature [16 to 19].

The gold grain size and morphology in Witwatersrand ores has been studied by dissolving gangue minerals (quartz and pyrite) in hydrofluoric acid [19]. The gold was classified into three categories based on its association, as follows:

- Gold with quartz; flat, crystalline, or porous gold grains
- Gold with thucholite; minute particles enclosed in carbon and coarse carbon particles embedded in carbon seams
- Gold associated with pyrite or as small, altered particles and thin coatings on pyrite (small quantities may be associated with pyrrhotite)

The size of gold grains in Witwatersrand ores varies (Figure 2.5) but tends to average about 80% finer than 75 to 100 μm. This size also provides a general guideline for the size required to achieve an acceptable gold extraction by cyanidation of free-milling Witwatersrand ores.

TABLE 2.6 Major types of gold in Witwatersrand reefs [16,17]

Type of Gold	Estimated Proportion of Total Gold	Texture	Size Range	Intergrowths with Other Minerals	Possible Origin or Genesis	Important Constituents and Trace Elements
Detrital gold	Up to 90% in some reefs	All transitions from solid, angular, and rounded particles to repeatedly distorted flakes	From about 10 to rarely 500 μm; average 50 to 100 μm	Inclusions of, or intergrown with, sphalerite, linneite, gersdorffite, cobaltite, chalcopyrite	Primary deposits or reworked older sediments; no primary deposits have yet been found	5% to 30% Ag; average 10%; 1% to 4% Hg; usually traces of Cu, Sb, Co, Ni, Bi, and others
Biochemically redistributed gold	Not present in all reefs; can locally be the most important source of gold	Filaments or aggregates of filaments, with remnants of cell structures; irregularly formed, often perforated particles	Filaments range in diameter from 0.5 to 2 μm; aggregates up to 1 mm across	Intimately intergrown with carbonaceous matter	Possibly detrital gold redistributed by primitive organisms	Not much different from detrital gold
Gold recrystallized or redistributed by metamorphic processes within the reef	5% to 40% depending on local conditions	Often as overgrowths on secondary pyrite, platy crystal aggregates, polycrystalline aggregates	From about 5 μm on pyrite up to 2 mm as aggregates	Epitaxial growth on pyrite; as crack filling in various minerals	Mobilization by the heat of intruding dykes, sills, or overlying lava; regional metamorphism	Not analyzed for trace elements; silver content not different from detrital gold
Primary gold in detrital allogenic sulfides	Less than 2%, seldom more	Irregular grains	Seldom larger than 20 μm	Mostly in pyrite, rare in arsenopyrite	Probably relict from primary deposits	Usually higher silver content than detrital gold
Gold in secondary quartz veins	Not important, although high concentrations found locally	Irregularly formed crystals and crystal aggregates	From about 10 μm to several mm for large aggregates	Not reported	Hydrothermal and pseudo-hydrothermal mobilization of reef gold	Not reported

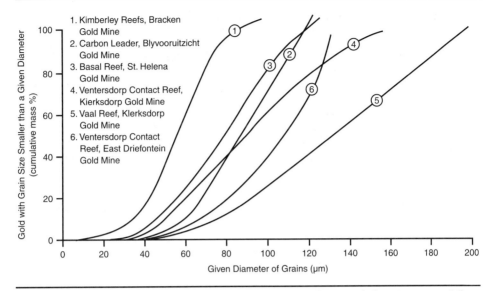

FIGURE 2.5 Grain size of gold in Witwatersrand ores expressed as cumulative mass percent finer than given diameter [15]

The gold grade of mixed Witwatersrand ores is relatively high (5 to 15 g/t), due to the high cutoff grades required as a result of high costs of deep mining, which contribute to overall production costs that are often more than \$250/oz. By comparison, similar per-ounce production costs can also be readily achieved by U.S. heap leach operations, which can compete by treating ores containing only 2 to 5 g/t.

In some Witwatersrand ores, a relationship between gold and uraninite, related to the particular reef, has been exploited by using radiometric sorting to selectively preconcentrate coarse gold- (and uranium-) bearing rock from waste rock.

2.4.1.2 Other Palaeoplacers

The second largest palaeoplacer, Jacobina (Brazil), has gold grades of 5 to 15 g/t, associated with uranium in the Lower Proterozoic conglomerates.

At Tarkwa (Ghana), there are three mineralized conglomerates with most of the gold occurring in the basal levels, which are discontinuous lenses 600 to 1,000 m long in the direction of current flow and 100 to 150 m wide [9]. Pyrite concentrations are low, with hematite, ilmenite, and magnetite typically more abundant than in Witwatersrand ores. Average gold grades are about 5 g/t, and extractions >95% can usually be achieved by gravity concentration and cyanidation.

2.4.2 Other Hard Rock Ores

Various nonplacer gold ore types, usually formed as a result of deposition from hydrothermal solutions, can be classified as free milling. Epithermal deposits (Table 2.7 and Figures 2.6 and 2.7) may fall into this category but quite often have some refractory component(s) and may be considered to be refractory, depending on the mineralogical component with the greatest impact on the gold extraction methods (see Sections 2.5 to 2.12). A schematic classification for South African ores with free-milling characteristics, but containing carbonaceous, submicroscopic gold content and base metal sulfide end series, is shown in Figure 2.8.

TABLE 2.7 Types of hydrothermal gold deposits (according to alteration assemblages) [9]

Deep-Seated Deposits

Chlorite–carbonate alteration type (e.g., Yellowknife, Canada; Mother Lode, California)
  Quartz–carbonate–sericite–chlorite alteration
  Au/Ag ≥1; occurs mainly in metamorphosed mafics
  High in arsenopyrite; Sb, Mo, W common
Silicate alteration type (e.g., Juneau, Alaska; Rossland, British Columbia)
  Quartz–albite–amphibole–pyroxene–biotite alteration
  Au/Ag >1; occurs mainly in metasediments
  High in arsenopyrite; Mo, W, Bi, pyrrhotite, tourmaline common; seems to be an end member of series with chlorite–carbonate type alteration
Iron formation type (e.g., Homestake, South Dakota)

Epithermal Deposits

Au-telluride vein type (e.g., Cripple Creek, Colorado; Fiji; Romania)
  Au/Ag ≈1; sericitic–carbonate alteration
  K-feldspar common in veins
Alunitic vein type (e.g., Goldfield, Nevada; Chinquashih, Taiwan; Kasuga, Japan; El Indio, Chile)
  Au/Ag ≈1; gold with As, Hg, (Te)
  Quartz–alunite–kaolinite alteration grading out to argillic alteration
Normal gold vein type (e.g., Oatman and Balel, Russia; Bagulo, Philippines)
  Au/Ag ≈1; gold with As, Hg; adularia common
  Silicification; argillic alteration
Carlin disseminated type (e.g., Carlin, Jerritt Canyon, Cortez, and Getchell, Nevada)
  Au disseminated in calcareous, carbonaceous sediment
  Au/Ag ≥1; high in As, Hg, Sb, Tl, F
  Silicification; carbonate dissolution
Disseminated gold in volcanics (e.g., Round Mountain and Borealis, Nevada)
  Au/Ag ≈1; high in As, Hg, Sb, Tl, F
  Sinter, lakebeds nearby; silicification; extreme leaching; sericite
Silver-base metal vein type
  Sericitic–argillic alteration ± k-feldspar
  Au/Ag ≈0.01; often Sb, Se rich with some Pb, Zn, Cu
  Occurs mainly in intermediate volcanics
  Most veins contain adularia, calcite, quartz
  Mn rich (e.g., San Juan, Tonopah)
  Mn poor (e.g., Pachuca and Guanajuato, Mexico; Comstock, Nevada)
Disseminated silver type
  Stockworks in intermediate volcanics with argillic to sericitic alteration; silicification
  Mn poor (Delamar and Rochester, Nevada); Mn rich (Candelaria, Nevada)
  Disseminated gold in clastic sediments (lakebeds)
  Silica, barite, Mn oxide, Pb, Nz (e.g., Waterloo, Ontario; Creede, Colorado; Hardshell, Arizona)

Quartz–gold veins or lodes comprise a variety of deposits that are essentially hydrothermal veins of quartz and gold that either replace wall rock or fill open spaces along fractured zones (Figure 2.6). Most are Precambrian or Tertiary in age and can occur to depths in excess of 1 km. The main categories are described in the following sections [22].

Auriferous veins, lodes, sheeted zones, and saddle reefs in faulted or folded sedimentary rocks. These deposits are widespread and have produced large amounts of gold in the past. Typically they consist of folded and metamorphosed sequences of shale, sandstone, and graywacke, often of marine origin. The principal gangue minerals are quartz, feldspar, mica, and chlorite. Arsenopyrite and pyrite are the most common metallic

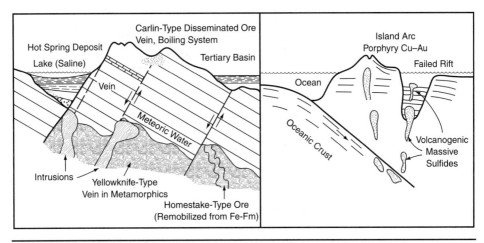

FIGURE 2.6 Schematic description of various types of gold deposits [9]

minerals, although sphalerite, galena, chalcopyrite, and pyrrhotite also occur. The gold present often contains little silver, and the concentrations of gold tellurides, auriferous sulfides, and aurostibnite are low. Examples include Salsigne (France), Pilgrims Rest (South Africa), Bendigo (Australia), and Muruntau (Uzbekistan).

Gold–silver veins, lodes or stockworks, and irregular silicified bodies in fractures, faults, and shear, breccia, or sheeted zones in volcanic rocks.   The most favorable host rocks are basalts, andesites, and rhyolites of Precambrian or Tertiary age. The Precambrian structures are commonly called greenstones. Older structures are generally regionally metamorphosed and gold occurs in lodes and stockworks (Figure 2.7), and irregular masses near fracture and shear zone systems. Younger occurrences are usually confined to fault zones. Mineralization in these deposits is characterized by quartz, carbonate minerals, pyrite, arsenopyrite, base metal sulfides, and a variety of sulfosalt minerals (e.g., tetrahedrite and tennantite). Gold mineralization generally consists of native gold, some tellurides, and occasionally aurostibnite. Example locations are Kolar (India) and Yellowknife (Northwest Territories, Canada).

Gold–silver occurrence in a complex geological environment comprising sedimentary, volcanic, and various igneous intrusive and granitized rocks.   The gold mineralization is commonly as free gold, but some tellurides and disseminated gold–sulfides are present. Examples include Kirkland Lake (Canada) and the Juneau (Alaska) ores.

Lode deposits have contributed between 20% and 25% of the world's total gold production. Examples of the most historically significant deposits are those of Ballarat and Bendigo (Victoria, Australia) and certain goldfields in New Zealand, Mexico, and the western United States. Some of the ores have an extremely high grade and are suitable for direct shipment to a gold refinery. An example of such an ore, from Chile, is given in Plate 2. Plate 3 shows a quartz vein sample containing some coarse gold, amenable to gravity concentration, and finer gold, which would require grinding to adequately liberate the gold.

The fineness of native gold in these deposits is generally >800 and decreases with depth, due to less silver dissolution occurring during ore formation. The gold mineralogy from a mineral processing point of view is usually more straightforward than for low-grade epithermal ores, as the gold grade tends to be higher and the gold grain size is typically coarser. Although gold grades may be high in such materials, a disadvantage is that the

36 THE CHEMISTRY OF GOLD EXTRACTION

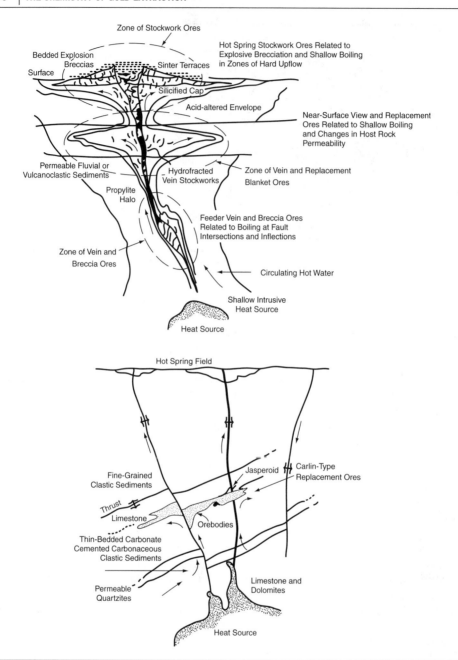

FIGURE 2.7 Schematic description of epithermal gold deposits [20]

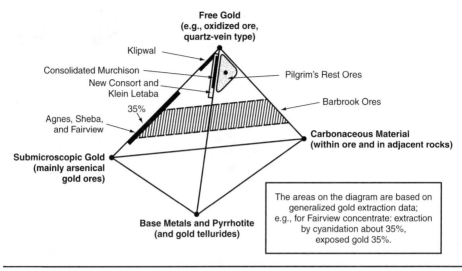

FIGURE 2.8 Diagrammatic representation of the causes of refractoriness of gold ores in South Africa [21]

extent of such gold mineralization is usually limited, resulting in a smaller size of deposits in general.

A list of minerals that are commonly associated with gold is given in Table 2.8.

2.5 OXIDIZED ORES

In an oxidized ore, the ore material has been oxidized or weathered, possibly in a zone that is atypical of the primary sulfide deposit, and for which some special processing may be required.

Oxidation and other hydrothermal alteration processes lead to the breakdown of rock structure, resulting in increased permeability. This often allows high leaching extractions to be achieved by heap leaching of run-of-mine ore, even though the ore particle size may be very coarse.

A detrimental feature of rock oxidation and alteration is the formation of considerable amounts of hydrated, amorphous, and/or poorly crystalline silica, clay minerals, sulfate salts, and oxide and hydroxide gangue phases. Some of these phases have relatively high solubilities in comminution and cyanide leaching, and may act as strong cyanicides (cyanide consumers), due to the generation of extremely large and fresh surface areas with high sorption potential [24]. Other phases, such as clay minerals and amorphous silica, may seriously interfere with processing.

Gold usually occurs either liberated or associated with the alteration products of pyrite and other sulfidic minerals. The most common of these are the iron oxides, such as hematite (Fe_2O_3), magnetite (Fe_3O_4), goethite (FeOOH), and limonite ($FeOOH \cdot nH_2O$), although gold may also be associated with manganese oxides/hydroxides. Generally the degree of gold liberation is increased by oxidation; however, in some cases, protective coatings of secondary and hydrated oxides on gold may be encountered. Plates 4 to 6 show examples of gold coated by iron oxides or hydrated oxides. These gold grains were undissolved in cyanide solutions but would be sufficiently coarse to be recoverable by gravity concentration.

TABLE 2.8 Common and important minerals associated with precious metal ores [6, 23]

	Elemental	Sulfides	Arsenides	Antimonides	Selenides	Tellurides
Antimony		Sb_2S_3, stibnite				
Arsenic		AsS, realgar As_2S_3, orpiment				
Bismuth	Bi, native bismuth	Bi_2S_3, bismuthinite				Bi_2Te_2S, tetradymite
Carbon	C, graphite/ amorphous C					
Cobalt			CoAsS, cobaltite			
Copper	Cu, native copper	Cu_2S, chalcocite; CuS, covellite; Cu_5FeS_4, bornite; $CuFeS_2$, chalcopyrite	Cu_3AsS_4, enargite; $(Cu,Fe)_{12}As_4S_{13}$, tennantite; $Cu_3(As,Sb)S_4$, famatinite	$(Cu,Fe)_{12}Sb_4S_{13}$, tetrahedrite		
Gold	Au, native gold; Au, Ag, electrum			$AuSb_2$, aurostibnite		$AuTe_2$, krennerite, calaverite
Iron		FeS, pyrrhotite; FeS_2, pyrite, marcasite	FeAsS, arsenopyrite			
Lead		PbS, galena				
Mercury		HgS, cinnabar				
Nickel		$(Fe,Ni)_9S_8$, pentlandite				
Silver	Ag, native silver; Ag, Au, electrum	Ag_2S, argentite; (Pb, Ag) S, argentiferous galena	Ag_3AsS_2, proustite; $(Cu,Fe,Ag)_{12}As_4S_{13}$, argentiferous tennantite	Ag_3SbS_3, pyrargyrite; $(Cu,Fe,Ag)_{12}Sb_4S_{13}$, argentiferous tetrahedrite	Ag_2Se, naumannite	Ag_2Te, hessite
Zinc		ZnS, sphalerite				

Oxidized ores differ from primary ores as a large proportion of fines are often generated by crushing and grinding processes or during heap leaching, largely due to the fact that clays and clay-forming minerals are more abundant. The presence of clays, such as pyrophyllite ($Al_2Si_4O_{10}(OH)_2$), talc ($Mg_3Si_4O_{10}(OH)_2$), kaolinite ($Al_4Si_4O_{10}(OH)_8$), and montmorillonite ($Al_4Si_8O_{20}(OH)_4 \cdot nH_2O$) can have important process implications, for example, as follows:

- Decreased heap or dump leach material permeability
- Increased slurry viscosities in processing (e.g., cyanidation or carbon-in-pulp [CIP] process), resulting in increased energy requirements for slurry mixing and/or less efficient chemical reaction
- Blinding of activated carbon, in CIP or carbon-in-leach (CIL)

Carbonate minerals such as calcite ($CaCO_3$), dolomite ($CaMg(CO_3)_2$), and siderite ($FeCO_3$) are also more common in oxidized ores, and these can affect pH control, especially during oxidative pretreatment processes.

In the oxidized zone of vein deposits, native gold generally has a lower silver content, due to the greater solubility of silver. The distribution of gold in supergene-enriched and lateritized deposits is shown schematically in Figure 2.9. Gold is typically present as unaltered, liberated, or partially liberated grains and as redistributed secondary gold at depth, with some lateral mobilization. A zone of supergene enrichment may exist at a level above the primary orebody and below a depleted region [25].

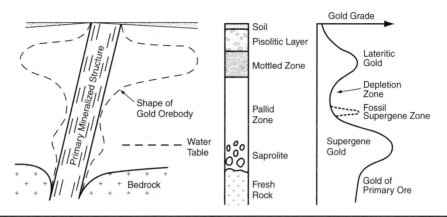

FIGURE 2.9 Distribution of gold occurring in a weathered zone, indicating supergene enrichment [25]

The deposits of the Yilgarn region (Western Australia) have been subjected to prolonged and extensive oxidation to depths of 50 to 100 m. These ores respond well to heap leaching and agitated cyanide leaching and overlie primary sulfidic-telluride deposits. Gold occurs as residual primary gold and redeposited supergene mineralization in iron oxide–silicate gangue. This secondary gold occurs as fine grains (sometimes as octahedral crystals) or with a spongy texture, has low silver content, and is closely associated with iron oxides. The proposed deposition mechanism is by dissolution of gold in saline groundwater and redeposition under reducing (e.g., ferrous ion, Fe^{2+}) conditions. Reviews on the geology and geochemistry of epithermal gold deposits have been provided by Hedenquist and Reid [26] and Berger and Bethke [27].

Oxidized ores may also contain various oxide copper minerals, many of which dissolve in alkaline cyanide solution and may, depending on the concentration, interfere with gold leaching and recovery processes [28].

Figures 2.10 and 2.11 show many of the critical process mineralogical factors in bulk minable, epithermal gold–silver deposits in the western United States [24], indicating that refractoriness, mineralogical ore variation, and clay content are the major factors affecting operations. Considerable process mineralogical work on such gold ores has been performed by Hausen [29, 30].

2.6 SILVER-RICH ORES

Although gold is almost always associated with silver, when the silver grade is high (>10 g/t) and/or the gold is present as electrum, the processing may need to be modified. The greater reactivity of silver particularly influences the behavior of gold in flotation, leaching, and/or recovery processes.

Silver has a value of about one hundredth that of gold but tends to occur at higher grades and therefore may be a significant source of revenue to a gold operation. The drawback is that the larger volume or mass of recovered product may reduce gold recovery (e.g., in CIP circuits and recovery sections) if the gold circuit design has not adequately taken into account the silver mineralization.

Electrum readily tarnishes in the presence of sulfide ions to form a silver sulfide layer of 1 to 2 μm thickness, which can limit the access of cyanide solution, thereby decreasing dissolution kinetics and potentially reducing the gold and silver extraction.

40 | THE CHEMISTRY OF GOLD EXTRACTION

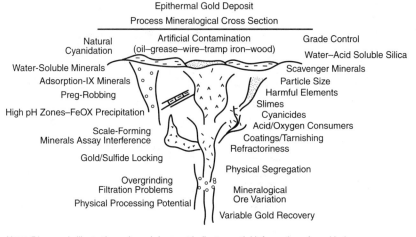

FIGURE 2.10 Factors in epithermal gold deposit that may affect chemical extraction performance [24]

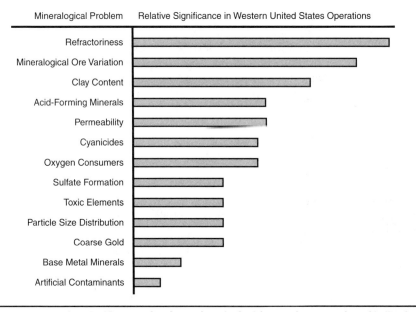

FIGURE 2.11 Relative significance of various mineralogical factors in processing of bulk-minable gold–silver ores in the western United States [24]

Such coatings are less likely to be hydrophobic, particularly if further oxidation to silver sulfate or silver oxide occurs, and therefore recoveries by flotation may also be reduced.

The E_h–pH diagram (see Section 4.2.9.3) for silver in sulfide systems (Figure 2.12) indicates the instability of silver under mildly oxidizing conditions (>0.7 V) at neutral pH by the formation of Ag_2SO_4 and in acidic mildly reducing solutions through the formation of argentite (Ag_2S). In alkaline conditions, Ag_2O and AgO are thermodynamically stable.

Native silver can occur at >95% purity, though this is rare; more commonly, silver is associated with gold, copper, and lead and, to a lesser extent, other metals. It has similar

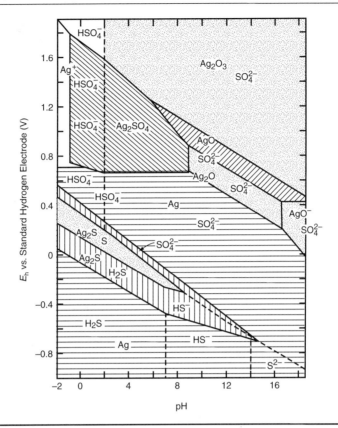

FIGURE 2.12 Potential versus pH diagram for Ag–S–H$_2$O at 25°C and 1 atm pressure

electrical conductivity, malleability, and ductility compared to gold, though it has a higher reflectivity (Figure 2.1). The density of silver is 10,000 to 11,000 kg/m^3, and the melting point is 960.5°C.

Some of the world's richest silver deposits are epithermal, containing hydrothermal veins of quartz, carbonates, and fluorite, often in altered tertiary rocks. The fineness of gold in epithermal deposits is typically low, at about 500 to 800.

The recoveries of silver achieved in processing plants are generally lower than those for gold, primarily due to more complex mineralogy. For example, there are about 75 silver minerals, the most important of which are included in Table 2.8, and about 200 silver-bearing minerals. In addition, few plants, particularly gold plants, have tried to optimize silver recovery based on a thorough knowledge of the mineralogy. In the early 1990s, at the Rochester Coeur d'Alene deposit in Nevada (one of the world's largest silver mines), extractions of 80% Au and 50% Ag were achieved by cyanide heap leaching of crushed ore containing about 2 to 3 g/t Au and 40 to 50 g/t Ag. This is an excellent example of an efficient silver–gold heap leaching operation.

2.7 IRON SULFIDES

In this class of ores, gold is principally unliberated in an iron sulfide matrix, or the behavior of the iron sulfide(s) affects process selection or operating conditions.

The most important iron sulfide minerals are the following:
- Pyrite (FeS_2)
- Marcasite (FeS_2)
- Pyrrhotite ($Fe_{1-x}S$), where x = 0.0 to 0.2

Arsenopyrite (FeAsS) also has many similar characteristics, though it is sufficiently distinctive to warrant separate consideration (see Section 2.8.1).

2.7.1 Pyrite

The sulfide mineral most commonly associated with gold, pyrite is very common throughout the world and is ubiquitous in sulfide orebodies. Although not usually an accessory mineral in primary igneous rocks, it is common in ore veins and metamorphic ores.

Pyrite has a commonly displayed cubic cleavage, a brassy yellow color, and a metallic luster, which is sufficiently close to that of gold to warrant the phrase "fool's gold." The density of pyrite is 4,800 to 5,000 kg/m^3, and it is relatively hard (Figure 2.1), with a value of 6 to 6.5 on the Mohs scale. Pyrite is a semiconductor with either n-type or p-type properties (Table 2.9). It can occur in cubic or framboidal habits, each of which has different reactivity in aqueous solution (see Section 5.1).

Pyrite is a very stable mineral in aqueous solutions, and its high standard reduction potential results in unreactivity under the mildly oxidizing conditions typical of cyanide leaching. Consequently, fine gold inclusions in pyrite require more extreme grinding and/or strongly oxidizing conditions to liberate the gold. Gold can occur in solid solution (i.e., invisible) within pyrite grains at concentrations from <0.2 to 132 ppm [5]. In contrast, when gold is relatively coarse and accessible to cyanide leach solutions (e.g., Witwatersrand ores), this unreactivity is an advantage as reagent consumptions are not increased by a side reaction with pyrite. Consequently, pyrite is usually only a problem in processing if it affects gold liberation; it is rarely a significant cyanicide (cyanide consumer). Leaching of fine gold grains contained within pyrite is a major difficulty in gold ore treatment (see Chapters 3 and 6) and this is an important source of refractory gold.

Gold can occur in many textural associations with pyrite (and arsenopyrite), as shown schematically in Figure 2.13. For gold–sulfide association types 1 to 3 in the figure, gold may be readily liberated. However, for types 5, 6, and possibly 4, gold may remain unliberated even at fine sizes. For example, Plate 7 shows a gold grain within a coarser pyrite grain (type 3). Increasingly, ores containing type 6 mineralization are being treated (e.g., Carlin-type ores) in which both coarse cubic pyrite (10 to 100 μm) and more abundant fine spheroidal pyrite (1 to 10 μm) occur. Gold grains are typically <1 μm in diameter and occur within pyrite grains, as coatings on pyrite, and dispersed in grains of amorphous carbon. These ores may exhibit preg-robbing characteristics due to the presence of both carbon and ultrafine spheroidal pyrite. An oxidative pretreatment step is usually required to increase gold extraction for these ores (see Section 5.6).

Pyrite can also be recovered by flotation as a by-product to gold (see Chapter 9), and the concentrate is sometimes roasted to produce sulfuric acid (H_2SO_4) and to liberate contained gold. In the past, elemental sulfur has also been produced commercially from pyrite.

In the gold ores of the Barberton Mountainland (Transvaal, South Africa), gold is present as free gold, and gold associated with pyrite and arsenopyrite (e.g., at Fairview 50% with pyrite, 20% with arsenopyrite, and 30% free). Although some coarse gold is present, the majority of gold occurs as fine (5 to 30 μm) inclusions in the sulfides, leading to poor cyanidation performance. The remainder occurs interstitially between sulfide

TABLE 2.9 Electronic and structural properties of selected sulfide and oxide minerals [31]

Formula	Name	Resistivity (Ωm)	Usual Conductor Type	Structure	Ionic Structure
Cu_5FeS_4	Bornite	10^{-3} to 10^{-6}	p	Tetragonal	$(Cu^+)_5Fe^{3+}(S^{2-})_4$
Cu_2S	Chalcocite	4×10^{-2} to 8×10^{-5}	p	Orthorhombic	$(Cu^+)_2S^{2-}$
$CuFeS_2$	Chalcopyrite	2×10^{-4} to 9×10^{-3}	n	Tetragonal	$Cu^+Fe^{3+}(S^{2-})_2$
CuS	Covellite	8×10^{-5} to 7×10^{-7}	Metallic	Hexagonal	$(Cu^+)_2(S^{2-})_2$
PbS	Galena	1×10^{-5} to 7×10^{-6}	n & p	Cubic	$Pb^{2+}S^{2-}$
MoS_2	Molybdenite	7.5 to 8×10^{-3}	n & p	Hexagonal	$Mo^{4+}(S^{2-})$
FeS_2	Pyrite	3×10^{-2} to 1×10^{-3}	n & p	Cubic	$Fe^{2+}(S^{2-})_2$
ZnS	Sphalerite	3×10^{-3} to 1×10^{-4}	—	Cubic	$Zn^{2+}S^{2-}$
SnO_2	Cassiterite	10^2 to 10^{-2}	n	Tetragonal	$Sn^{4+}(O^{2-})_2$
Cu_2O	Cuprite	10^{11} to 10	p	Cubic	$(Cu^+)_2O^{2-}$
Fe_2O_3	Hematite	2.5×10^{-1} to 4×10^{-2}	n & p	Trigonal	$(Fe^{3+})_2(O^{2-})_3$
Fe_3O_4	Magnetite	2×10^{-4} to 4×10^{-5}	n & p	Cubic	$Fe^{3+}[Fe^{3+}Fe^{2+}](O^{2-})_4$
MnO_2	Pyrolusite	10^{-1} to 10^{-3}	n	Tetragonal	$Mn^{4+}(O^{2-})_2$
TiO_2	Rutile	10^4 to 10	n & p	Tetragonal	$Ti^{4+}(O^{2-})_2$
UO_2	Uraninite	20 to 4×10^{-1}	—	Cubic	$(U^{4+})_{1-x}(O^{2-})_{2+x}$

NOTE: Dashes indicate information is not applicable.

grains or in the nonsulfide gangue. The gold usually contains about 10% silver, although some pink-colored gold grains containing nickel and antimony have been detected.

2.7.2 Marcasite

Marcasite has the same composition as pyrite (FeS_2), but it has an orthorhombic rather than cubic crystal system. It is formed at lower temperatures than pyrite and usually occurs in secondary rocks. Although marcasite is less common than pyrite, it frequently occurs intergrown with pyrite in sulfide ores and may constitute up to 30% of the total iron sulfides in a typical pyritic gold ore. Plate 8 shows fine gold grains in a coarser marcasite grain. In certain cases, where marcasite is a major sulfide mineral, it can be a significant cyanide and oxygen consumer in cyanidation, a problem that can be remedied by oxidative pretreatment (see Chapter 5).

2.7.3 Pyrrhotite

Pyrrhotite is the name for iron sulfides given the formula $Fe_{1-x}S$, where x can vary between 0 and 0.2. There are two principal types: hexagonal pyrrhotite (Fe_9S_{10}) and a monoclinic (Fe_7S_8) variety. Pyrrhotite is stable under more reducing conditions than pyrite, and hence it tends to oxidize more readily. Monoclinic pyrrhotite has a relatively high magnetic susceptibility and can be recovered readily by industrial magnetic separation equipment.

Gold–pyrrhotite composite particles occur predominantly in greenstone belt gold ores, for example, in several West Australian ores [32] and in a number of Canadian deposits. The main impact on gold recovery is that pyrrhotite consumes cyanide and oxygen in cyanidation, though gold inclusions in pyrrhotite have also been found.

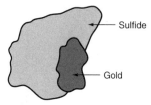

1. Readily Liberated (free-milling) Gold

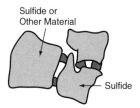

2. Gold Along Crystal Grain Boundaries

3. Gold Grain Enclosed in Pyrite/Sulfide (random position)

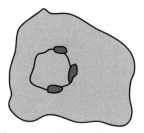

4. Gold Occurrence at the Boundary Between Sulfide Grains

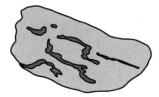

5. Gold in Concretionary Pyrite (or other sulfide) Along Fractures and/or Crystal Defects

6. Gold as Colloidal Particles or in Solid Solution in Sulfide

FIGURE 2.13 Schematic representation of types of gold associations with sulfide minerals (illustrative only)

2.8 ARSENIC SULFIDES

An ore should be considered in this class if the gold is associated with an arsenic sulfide or arsenide matrix, or if the arsenic minerals affect process selection or operating conditions. The most significant arsenic minerals are given in Table 2.8.

2.8.1 Arsenopyrite

Arsenopyrite is a common sulfidic host for gold, second only to pyrite. Gold concentrations as high as 15,200 g/t have been found in an arsenopyrite sample from Villerange in France, and a strong relationship between gold and arsenic content of pyrite has also been established [5]. Gold associations are similar to those for pyrite–gold assemblages (Figure 2.13). Arsenopyrite is marginally less hard (Figure 2.1) though more brittle than pyrite, which results in preferential grinding and a finer product size than pyrite. This can lead to higher flotation recoveries of pyrite over arsenopyrite [33].

In arsenical ores formed at high temperature, gold may be incorporated into the lattice of arsenopyrite either in solid solution or on the growing crystal faces. On cooling, the gold content distorts the structure. The concentration of solid solution gold can be much higher in arsenopyrite than in pyrite, due to its better matched atomic spacing, crystal chemistry, and similar formation temperature to gold. When sulfides are subjected to a thermal event (geological or pyrometallurgical processing), gold can migrate to grain surfaces and fractures, thereby increasing liberation. This is associated with a change from n-type to p-type semiconductors for both pyrite and arsenopyrite [21].

2.8.2 Orpiment

Orpiment (As_2S_3) contains 61% As and is a very minor component in several commercial gold ore deposits. It has a yellow color and a specific gravity of 3,500 kg/m^3. Orpiment tends to form under oxidizing conditions (e.g., in the oxidized zone of mineral veins associated with igneous intrusions). Orpiment is readily soluble in alkaline solutions, which leads to some dissolution during gold cyanidation. This can interfere significantly with cyanidation by consuming cyanide and introducing deleterious arsenic species into the solution.

2.8.3 Realgar

Realgar (As_2S_2 or AsS) is associated with orpiment to which it alters on weathering. It has a red or orange color and a density of 3,600 kg/m^3. Less soluble than orpiment in alkaline cyanide solution, realgar generally has a less significant effect on cyanidation (see Section 6.1).

2.9 COPPER SULFIDES

Ores containing gold associated with copper sulfide minerals that affect process selection or operating conditions are considered in this class. It is relatively uncommon for gold to be associated solely with copper minerals, and there is almost always some pyrite present. In base metal sulfide processing, gold generally reports with copper minerals (e.g., chalcopyrite or bornite) in process streams. This can be attractive from a metallurgical point of view, as there is generally some selectivity against pyrite and the possibility of selling a gold-rich copper concentrate for which a credit is received (typically 90% to 97% of gold value). Although gold grades in copper ores are typically low (usually <1 g/t), gold production as a by-product of copper is relatively large due to the high tonnages of material processed. About 80% of by-product gold comes from copper ores (e.g., Freeport Indonesia's Grasberg operation, Rio Tinto's Bingham Canyon in the United States, and Candelaria in Chile). The majority of the remaining by-product portion is produced from mixed copper–lead–zinc ores.

2.9.1 Chalcopyrite

Chalcopyrite ($CuFeS_2$) is the most abundant copper mineral and contains 34.5% Cu. Commonly associated with pyrite and other copper sulfide minerals, chalcopyrite has the unusual characteristic of containing lower valency copper and higher valency iron (i.e., $Cu^+Fe^{3+}(S^{2-})_2$). A semiconductor (Table 2.9), it has a density of 4,100 to 4,300 kg/m^3. In powdered form, chalcopyrite has a greenish-gray color. Chalcopyrite may be oxidized to yield covellite (CuS) and hematite or may be reduced to chalcocite (Cu_2S), with Fe^{2+} in solution with gaseous hydrogen sulfide (H_2S) evolved [34].

2.9.2 Other Copper Sulfides

Chalcocite and covellite are important copper ore minerals, containing 79.8% and 66.4% Cu, respectively. They are formed by the alteration of primary copper sulfide ores and occur in zones of secondary enrichment in many parts of the world (e.g., in porphyry copper deposits). The color of chalcocite is often dark gray with a blue tinge, and the density is 5,500 to 5,800 kg/m^3. Covellite has a distinctive blue color, which is particularly noticeable in reflected light microscopy.

Bornite has the general formula Cu_5FeS_4, although the exact proportions of copper and iron vary, and it has a distinctive blue-purple color, leading to the name "peacock ore." It occurs in primary copper ores—hence it is often associated with chalcopyrite—and areas of secondary enrichment. Gold–bornite associations are rare, although important examples are the Bougainville (Papua New Guinea) and Olympic Dam (South Australia) deposits.

Copper also occurs with arsenic and sulfur in the tetrahedrite series of minerals, the most important of which are included in Table 2.8.

2.10 ANTIMONY SULFIDES

This class of ores contains either gold associated with antimony minerals, or the antimony minerals present significantly affect process selection or operating conditions.

2.10.1 Aurostibnite

Aurostibnite ($AuSb_3$) can present problems in gold ore treatment due to its low solubility in cyanide solutions and poor amalgamation properties. The mineralogical characteristics are given in Table 2.2. Commonly, aurostibnite is included in stibnite concentrates produced by flotation (see Section 9.2).

2.10.2 Stibnite

Stibnite (Sb_2S_3) is the main source of antimony metal and occurs in quartz veins and with lead–zinc sulfide assemblages. Gold–stibnite associations are rare, though stibnite can occur at sufficiently high concentrations in gold ores to cause problems as a cyanide consumer (see Chapter 6). Gold is closely associated with antimony sulfides at the Blue Spec mine (Pilbara, Western Australia) and at Consolidated Murchison (South Africa) [32].

2.11 TELLURIDES

The processing requirements and behavior of this class of gold ores is affected by the presence of gold tellurides. Tellurides are the only gold minerals other than metallic gold and gold–silver alloys that are of economic significance. There is a wide range of gold–tellurium alloys and mixed metal tellurides, the properties of which are given in Table 2.2. The most important tellurides encountered in gold ores are calaverite, petzite, hessite, krennerite, and maldonite. These minerals are dense (7,500 to 9,500 kg/m^3) due to their gold and silver content, which ranges from 12% to 44%.

Gold-telluride ores usually contain some native gold, together with other metal tellurides, often with complex intergrowths. Excellent examples of telluride gold deposits are Cripple Creek (Colorado, United States), and Emperor and Tavatu (Fiji). Gold-telluride minerals are also found at Kalgoorlie (Western Australia) and Golden Sunlight (Montana, United States) [35].

There are thought to be two main geological environments in which gold–silver tellurides occur:

1. Veins, fissures, and breccia pipes in tertiary rocks (e.g., Cripple Creek, Emperor, and the Carpathian mountain district in east-central Europe). In this type of occurrence, veins are vuggy and composed of quartz and carbonate minerals. Adjacent to the veins is intense wall-rock alteration due to the introduction of large amounts of water, carbon dioxide, and sulfur. Native tellurium is abundant and native gold is rare. Tellurides of metals other than gold are present in lesser amounts. Krennerite is dominant over calaverite. The telluride minerals are fine grained but show good crystalline form.

2. Precambrian rocks or metamorphosed volcanic lava (e.g., Kalgoorlie). In this ore type, wall rocks show some retrograde metamorphism, resulting in rocks of lower-grade metamorphism, coupled with intense structural deformation. Native tellurium is rare, native gold is abundant, and tellurides of mercury, copper, and bismuth are generally present. The stable form of $AuTe_2$ is clearly calaverite, not krennerite, and the telluride minerals lack crystalline form.

Telluride ores may be oxidized in the upper parts of vein deposits to yield, after dissolution of tellurium, fine gold grains, known as mustard gold. The dissolution of tellurium from $AuTeS_2$ is possible at conditions under which gold remains unreacted. This is summarized by the E_h–pH diagram (Figure 2.14), which shows the various tellurium solution species that can be formed under mildly oxidizing conditions.

Ores containing gold-telluride minerals have unique, and generally poorly optimized, processing requirements. Most gold tellurides, with or without silver present, dissolve very slowly in cyanide solutions and usually require an oxidation stage to obtain commercially viable extractions, as discussed in Chapters 5 and 6.

2.12 CARBONACEOUS ORES

Carbonaceous ores contain carbonaceous components that adsorb dissolved gold during leaching, thereby reducing gold extractions by cyanidation. These gold ores sometimes require oxidative pretreatment prior to cyanide leaching (see Chapter 5). Important examples are found at Ashanti and Prestea (both in Ghana), Carlin and Jerritt Canyon, (both in Nevada, United States), and various Western Australian deposits.

Carbonaceous matter is also present in Witwatersrand ores (and is sometimes called thucholite) though it contributes only a small extent to reduced gold extraction. The term *thucholite* is derived from the chemical elements present, that is, Th, U, C, H, O (-lite), and generally describes radioactive hydrocarbon material. Interestingly, this term was originally applied to similar material found in Canadian pegmatite dykes [37].

Carbonaceous matter can adsorb gold from solution in a manner similar to the processes discussed in Section 7.1. Although this material generally does not have as high a specific surface area as activated carbon, in some ores the carbon content is sufficiently high (>5%) and has sufficient adsorption properties to cause a significant reduction in gold extraction. In some cases, as little as 0.1% carbon may produce preg-robbing (removing gold from solution irreversibly) or preg-borrowing (removing gold from solution reversibly) properties. Fractionation and characterization of carbonaceous ores from Carlin, Prestea, and Natalinsk and Bakyrchik (Russia) suggest that such carbonaceous matter consists of three components [38]:

- Hydrocarbon
- Humic acid
- Activated elemental carbon

However, the exact nature of the components is not well established, and variations in carbon activity are great, as may be expected from a surface, rather than a bulk, phenomenon. Reduced gold extractions have also been attributed to gold-adsorbing pyrophyllite and shale in Witwatersrand ores [39], although the adsorption appears to be readily reversible.

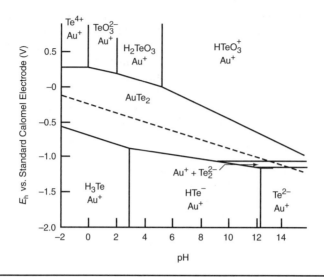

FIGURE 2.14 Partial E_h versus pH diagram for Au-Te-H$_2$O system at 25°C. Activity of Te species taken to be 10^{-4} M [36].

2.13 GRAVITY CONCENTRATES

Gravity concentrates are a special class of material due to their unusual mineral composition, principally coarse gold (i.e., >50 μm to <2 cm) and high concentrations of dense oxide and sulfide minerals. They are recovered using a variety of equipment, for example, sluices, jigs, spiral concentrators, shaking tables, and centrifugal concentrators (e.g., Knelson, Falcon). The concentrates produced typically contain primarily heavy mineral oxides such as magnetite, ilmenite (FeTiO$_3$), and zircon (ZrSiO$_4$), possibly with lesser amounts of sulfides, usually pyrite. The majority of gold present in gravity concentrates is liberated, though some composite grains with pyrite or quartz may be present and some gold may be coated with calcium, iron, and magnesium oxides and carbonates. Fine unliberated gold grains can also occur in ilmenite and rutile, which may not be readily recoverable by amalgamation. Plates 1, 3, and 4 show gold grains that would normally be recovered into a gravity concentrate.

2.14 FLOTATION CONCENTRATES

Flotation concentrates contain gold and other hydrophobic minerals produced by flotation of a primary ore or preconcentrated material. Commonly, gold flotation concentrates have high sulfide content and present a unique class of material. Gold is present either as free gold, which generally floats readily (see Section 9.2.2) or is locked in sulfides. Slurries of sulfide flotation concentrates have the unusual characteristics of being strongly hydrophobic and have rapid settling properties. This often leads to difficulties in handling and subsequent treatment by gravity concentration or cyanidation. The treatment methods for these materials are discussed in Section 3.3. The grain shape and size distributions of gold in Barberton flotation concentrates are described in Table 2.10. Another example is the occurrence of gold in flotation concentrates from the El Indio gold–silver–copper mine (Chile), which is summarized in Table 2.11 [40]. This shows a high proportion of liberated gold and a mixture of associations with sulfide and silicate gangue minerals.

TABLE 2.10 Grain size distribution of the gold in the concentrates treated in the New Consort roaster at Barberton, South Africa [21]

Sample	Size of Gold Particles
Sheba flotation concentrate	5 to 30 µm, rarely >100 µm in equivalent-circle diameter. The larger gold grains, which are commonly elongated*, were possibly more equidimensional grains that were flattened, or grains that originated from veinlets.
Sheba gravity concentrate	10 to 30 µm, maximum of 60 µm.† Larger particles, which are commonly elongated, constitute a significant volume of the gold. Large equidimensional gold grains may require prolonged cyanidation times.
Agnes flotation concentrate‡	5 to 10 µm, maximum of 15 µm. Small grains.
Agnes gravity concentrate‡	5 to 10 µm, maximum of 15 µm. Grain size fine compared to that of Sheba and New Consort concentrates.
New Consort gravity concentrate§	5 to 65 µm, rarely >150 µm.

* High levels of coarse, free, platy gold particles; hence extraction by cyanidation is relatively high.

† Textural relations indicate that the gold is commonly associated with the late-phase base-metal sulfides. These phases commonly occur in structural weaknesses within the sulfide grains, that is, along grain boundaries, cleavages, fractures, etc.

‡ Coarse gold is known to exist in this deposit but was not identified during mineralogical examinations.

§ The New Consort ore is a free-milling ore for the following reasons: (1) no submicroscopic gold is present, (2) coarse gold particles are present, and (3) the gold commonly occurs along fractures and cracks, that is, these are inherent weaknesses in the ore along which the fracture will occur.

TABLE 2.11 Mode of occurrence of gold in El Indio (Chile) flotation concentrates [40]

Mode of Occurrence	Primary Flotation Concentrate	Regrind Flotation Concentrate
	Approximate Percentage Distribution	
Liberated native gold	60	44
Native gold		
Attached to or encapsulated by gangue	15	28
Attached to or encapsulated by enargite	16	10
Attached to or encapsulated by pyrite	5	18
Intergrown with secondary copper minerals	4	<1
Gold-bearing tellurides	<1	<1
Total	**100**	**100**

2.15 TAILINGS

Tailings are materials that have been discharged from either currently operating or disused gold plants as nonvalue-adding products. This class of material can result from any of several extraction processes (e.g., cyanidation, flotation, gravity concentration, amalgamation, etc.), and the history of the material dictates the mineralogy and nature of gold occurrence.

The mineralogical characteristics of plant tailings are extremely diverse and are dependent upon the following factors:

- Type of ore originally treated
- Type of extraction process originally used
- Efficiency of extraction processes
- Age of tailings deposition

In most cases the gold recoveries achievable from tailings by standard cyanidation or other cyanidation methods are low, typically 40% to 70%, and are limited by at least one mineralogical factor. For such tailings, which may contain only 0.5 to 2.0 g/t gold, it is extremely important to understand the gold occurrence in order to be able to propose and optimize an appropriate process flowsheet. Innovative mineralogical techniques have been applied to the characterization of tailings materials [41], many of which are described in Section 2.18.

2.15.1 Gravity Concentration Tailings

Gravity concentration plants are very effective at recovering fully liberated gold of sizes greater than 50 μm, and some equipment is effective at sizes down to about 10 μm. However, gold can be lost in the following forms:

- **Fine gold**, which is too small to be recovered by the installed equipment:

    <500 μm: sluices

    <200 μm: jigs

    <50 to 100 μm: spirals

    <50 μm: shaking tables

    <20 μm: centrifugal concentrators (and down to about 10 μm in some cases)

- **Flakey gold**, which presents a larger surface area in one plane, allowing it to be carried away with lighter particles in flowing film separators, such as shaking tables
- **Hydrophobic gold**, which adheres to the water–air interface, giving the particle an artificially low apparent density and reports to the tailings (e.g., gold coated with fine hydrocarbon or sulfurous material)
- **Unliberated gold**, which is associated with oxide, silicate, or sulfide gangue particle and for which the average particle density is substantially lower than that of gold (i.e., <19,300 kg/m^3)

Gravity concentration tailings usually contain a significant proportion of gold that is recoverable using either a secondary process that employs a separation principle other than gravity concentration (e.g., flotation or cyanidation) or by using a more efficient gravity concentration process, for example the use of spirals and centrifugal separators to recover finer gold particles than the equipment originally used, such as jigs.

In the past, gravity concentration plants often discharged amalgamation tailings. This has resulted in the presence of gold that is partially coated with mercury, which may exist in oxidized forms. Plate 9 shows residual mercury from previous treatment in a sample composed of a mixture of placer and tailings materials.

At the Forrest Hill tailings deposit (Canada), the gravity plant tailings contained 50% of the gold in the <75 μm fraction. Most gold occurred as free grains of electrum or as small inclusions in a wide range of silicates, with some liberated gold (containing 5% to 45% Hg, probably from the previous amalgamation process). Some gold also occurred with pyrite and arsenopyrite, which had been partly replaced by iron hydroxides and arsenates. Subsequent gold extraction from this ore by flotation and cyanidation was >90% [33].

2.15.2 Cyanidation Tailings

Cyanidation is the most efficient and widely applied process for extracting gold from ores. Gold extractions >90% are usually expected and are commonly achieved. The gold remaining in the tailings is generally due to a particular refractory component, whose characteristics dominate any subsequent attempts at additional gold recovery. Common occurrences of gold in cyanidation tailings are the following:

- **Free or exposed gold originating from partial leaching of coarse gold particles.** This occurs when some or all of the gold in the ore is too coarse in size to dissolve during the original leaching period, or insufficient leaching time was provided. Other factors that might contribute to inefficient leaching of coarse gold include the presence of cyanide or oxygen-consuming minerals and operation of the plant at suboptimal conditions.
- **Gold locked within silicate or oxide gangue.** Some gold may be encapsulated in oxide or silicate gangue, in which the gold remains unreacted in cyanide solutions.
- **Gold locked within sulfide minerals.** Fine gold may be present in tailings if it is encapsulated within larger unreactive and nonporous sulfide grains, most commonly pyrite.
- **Coated free or exposed gold.** The existence of coatings on gold in tailings, streams, or dumps has a detrimental effect on subsequent cyanidation. These coatings have been identified as iron oxides or hydroxides (limonite or goethite), which are formed by dissolution and precipitation reactions. In addition, sulfide ions may react with silver in gold grains, which form progressively silver–sulfide, iron–sulfide, and (hydrated) iron–oxide phases. These reactions and products could also occur if ore is stockpiled prior to treatment, thereby reducing extraction efficiency. Coatings that affect the metallurgical response of the ore can be caused by ferruginous clay slimes that retard or prevent dissolution of gold particles.

  Gold grain surfaces, which are soft and malleable, may become coated by embedded gangue grains during crushing and grinding (see Plate 10) These marginally reduce the exposed surface area and may decrease hydrophobicity (and flotability) or leaching rates, which can result in some loss to tailings.

Examination of three Witwatersrand cyanidation plant residues by diagnostic leaching (see also Section 2.18) showed that about half the gold was locked in silicates and the remainder was associated with sulfides. This confirmed that, for this material, flotation was the most appropriate retreatment method [42]. As another illustrative example, the types of gold remaining in South African roaster calcine cyanide leach residues are summarized in Figure 2.15.

2.15.3 Flotation Tailings

Gold may be left in a sulfide flotation plant tailings if it is unliberated, if it is coated with nonfloatable material (e.g., iron oxides, silicates), or if it occurs in composite particles of nonhydrophobic minerals, such as oxides or silicates. This class of material may include particles that contain exposed gold (i.e., types 1 and 2 in Figure 2.13) and hence further gold extraction by cyanidation may be possible. However, the tailings may contain oxidized or partially oxidized sulfide minerals, which can interfere significantly with cyanide leaching.

2.16 REFINERY MATERIALS

Refinery materials are the gold-bearing products of hydrometallurgical or pyrometallurgical methods that must be processed further for additional gold recovery (e.g., calcine, electrowinning by-products, anode slimes, flue dust, and slag materials). There are several

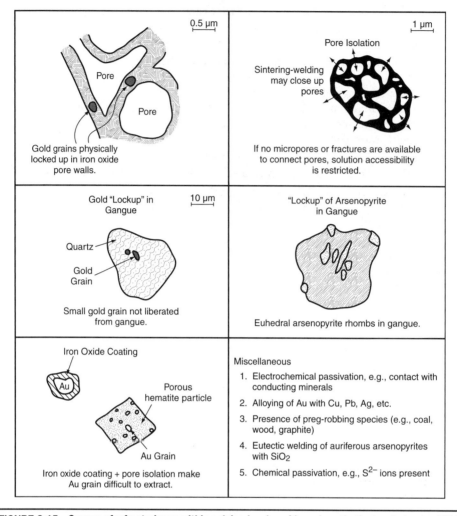

FIGURE 2.15 Causes of refractoriness within calcine leach residues: hypothetical and observed [21]

other materials in this class, for example, zinc precipitates (Plate 11), loaded cathodes, and gold-bearing amalgam (Plates 12 and 13), but since these are important products of unit chemical processes, these are considered in more detail in Chapter 10.

2.16.1 Calcine

Calcine is the name given to the oxidized product of sulfide mineral roasting. Calcine material produced by roasting of pyrite and arsenopyrite is principally hematitic and porous. Small concentrations of magnetite are also often present due to imperfect oxidation (see Section 5.8). The changes in mineral texture that occur during roasting of a pyrite ore are illustrated in Plates 14 and 15 (see also Figure 2.15).

During roasting, gold migration and coalescence occurs to some extent. Gold has also been found to form a film around hematite grains. Thus the gold occurrence changes in a manner that may affect the subsequent processing. Hematite grains can also be formed by a vigorous expansion from the original pyrite, leaving a highly porous (and desirable) popcorn-type structure. Ideally, the only gold left unexposed to leaching is what was originally totally enclosed in silicates [43].

Examination of roaster products from Barberton ores has shown that gold can become alloyed with lead and antimony (from stibnite) to form aurostibnite. Some originally free gold was found to be coated by iron oxide, iron hydroxide, and vitreous layers indicative of higher-temperature areas in the roaster [44].

2.16.2 Roaster Dust

Particulate material in roaster discharge gas is usually collected by filters or precipitators. This material often contains significant amounts of gold or silver and may need to be retreated. The precious metals are present as extremely fine grains and are generally easily leached, although the material may be difficult to wet efficiently. The most important factor affecting the processing is the composition and reactivity of the fine gangue particles, which have a high specific surface area.

2.16.3 Anode Slime

Anode slime is the cell sludge that is generated during electrowinning and electrorefining of base metals that contain precious metal by-products (e.g., copper recovery and refining operations). The insoluble material drops off the anode as the electrode is dissolved and forms a slime, which collects at the bottom of the electrolytic cell. For example, the quantity of anode slime produced at a typical copper electrorefinery is in the range of 2.5 to 25 kg/t cathode processed, and this contains variable quantities of gold and silver depending on the source material [45].

The slimes from anodes at primary copper smelters contain compounds that are not substantially attacked during anodic dissolution, such as Cu_2Te, Cu_2Se, Ag_2Se, $CuAgSe$, $AuAgTe_x$, Cu_2S, NiO, Ag, Au, and PGMs. Other compounds, such as Cu_2O, $AgCl$, $PbSO_4$, and complex oxides of lead, arsenic, antimony, and tin may be formed anodically. The ranges of composition of copper refinery slimes are given in Table 2.12 [45].

At the Kidd Creek copper–silver operation in Canada, the anode slimes contain about 1,200 g/t Au [46]. The majority occurs as tiny (<0.5 μm) particles loosely held together in a poorly defined copper–silver–arsenate–selenate cement, represented by the formula $Ag_3Cu_8(SeO_4)(AsO_4)$ [46].

2.16.4 Slag

Gold in smelter and refinery slag material is present within a low-density matrix which has very low porosity and therefore is not easily amenable to cyanidation. Where further processing is justified, the slag may be crushed, ground, and a concentrate produced by gravity concentration (e.g., by batch processing using a shaking table). This concentrate contains gold, which is primarily free with some gold-bearing sulfide phases and alloys with silver, copper, and antimony. The slag is brittle and shatters, liberating some gold-bearing grains; however, most remain as a gold–silicate composite in which gold is disseminated as fine grains.

2.17 RECYCLED GOLD

Recycled gold is produced from gold-bearing materials resulting from a fabrication process or an end-use application (e.g., jewelry scrap or electrical components). The nature of gold occurrence may range from shavings of high-purity gold to thin coatings of gold

TABLE 2.12 Typical range of composition of copper-refining slimes [45]

Component	Weight (%)
Antimony	0.1–16
Arsenic	0.3–10
Bismuth	Trace–1.0
Copper	3–40
Gold	0–2
Insoluble minerals	0.3–16
Iron	0.1–2
Lead	0.3–35
Nickel	0.1–45
Platinum group metals	0–1
Selenium	0.5–58
Silver	6–30
Tellurium	0.5–10

on electrical components. A preconcentrated (electrostatic separation) metallic fraction from electronic components has been found to contain 1.2 kg/t Au, 13.7 kg/t Ag, together with 38% Cu, 27% Al, and 9% Fe. This fraction contained 50% to 60% of the precious metals in about 15% of the initial mass [47].

2.18 DETERMINATIVE METHODS

To understand gold ore-processing requirements, the important mineralogical properties of the ore must be described, usually in quantitative terms. This requires the following parameters to be determined:

1. Gold ore grade
2. Ore composition (elemental and mineralogical)
3. Concentrations of any other valuable minerals
4. Nature and concentrations of minerals detrimental to processing (e.g., cyanide-consuming minerals, clays, etc.)
5. Gold grain size distribution
6. Gold mineral type
7. Liberation characteristics of all valuable minerals

Determination of the compositional characteristics of parameters 1 to 4 requires samples of a statistically reliable size and from representative locations in the source material (i.e., orebody or other feedstock) followed by the application of standard, well-established analytical and mineralogical techniques. Characteristics 5 through 7 require more sophisticated instrumental methods that can analyze textural aspects of selected samples.

Having determined this information, usually from samples taken from many areas of the deposit under evaluation, test work can begin in order to develop an optimum process flowsheet. The importance of completing a systematic process mineralogical evaluation cannot be overstated. The following list provides some of the benefits from applying such an approach:

- Avoid unnecessary test work
- Avoid biased sample collection

- Avoid overlooking mineralogical factors
- Eliminate duplication of nonoptimum processes from other operations
- Ensure optimal or near-optimal flowsheet development
- Ensure appropriate equipment selection
- Avoid oversizing or undersizing of equipment
- Reduce uncertainties in process design criteria

A summary of determinative techniques used for gold process mineralogy is provided in Table 2.13. These procedures are described in more detail in the following sections.

2.18.1 Ore Composition

2.18.1.1 Gold and Other Precious Metals

The detection and quantification of gold is usually performed by one or more of the following methods:

- Fire assay
- Acid digestion and atomic adsorption spectroscopy (AAS) or inductively coupled plasma (ICP)
- Cyanide leaching and AAS with fire assay of residue
- Physical methods (panning and amalgamation)

Fire assay.   This traditional method of gold analysis can measure gold concentrations down to about 0.1 g/t. Samples containing as much as 50 g/t can be analyzed by this method, in which lead (as litharge, PbO) and glass-forming fluxes are mixed with a finely ground sample (80% to 90% <75 μm). The crucible containing the flux charge is then placed in a muffle furnace at 850°C, and the temperature is raised to 1,000°C over a period of 30 to 40 min or until complete fusion has occurred. Sample weights for the analysis vary depending on precious metal grades but typically between 10 and 30 g are used. To improve precision, larger sample weights (up to 150 g) may be used where the expected gold (and silver) content is low. The lead circulates through the molten charge and collects the precious metals, forming a gold–silver–lead alloy. The alloy is recovered as a button, separated from the glass slag (containing base metal and other impurities), and the lead is removed by cupellation.

For lead removal, the cupel is preheated at 1,000°C. The button is then placed on the hot cupel. Following the initial melting, the temperature is quickly reduced, and the cupelling operation is finished at about 830°C. This produces a precious metal prill, which is either weighed, parted (silver dissolved in nitric acid), and reweighed for gold content, or dissolved entirely in aqua regia (a mixture of nitric (HNO_3) and hydrochloric (HCl) acids), and analyzed for gold and silver by AAS or ICP techniques. The latter method, while slightly more time consuming, improves the overall precision at low gold (and silver) concentrations. The detection limit for gold by ICP with atomic emission spectroscopy (ICP–AES) is 10 ppb (μg/L) with very little interference from other metals. This is an advantage over AAS, which has a similar detection limit but is prone to interference and has a smaller range where the response is linear and most reliable. Inductively coupled plasma with mass spectrometry (ICP–MS) is a developing technique in which detectable ions, rather than plasma, are used as the emission source, with a detection limit of 0.07 μg/L.

Fire assaying is generally regarded as the most accurate, economic, and consistent method for gold analysis and, if done properly, quantifies gold in all forms in the original sample matrix. However, the method is sensitive to fusion firing and cupellation

TABLE 2.13 Summary of techniques commonly used in gold process mineralogy (adapted from [48])

Technique	Abbreviation	Detection Limit	Application
Fire assay	FA	0.1 to 50 g/t Au	Determination of gold in all forms
Quadruple 1 t FA-AA assay	FA-AA	0.001 g/t Au	Determination of gold in all forms
Cyanide leaching	CN	—	Determination of the amount of recoverable gold by cyanidation
Gravity concentration	GC	—	Concentration of gold and gold-bearing minerals
Acid diagnostic leaching	ADL	—	Determination of gold associated with carbonates, sulfides, silicates, and other minerals
Optical microscopy	OMS	~0.5 µm	Systematic scan for gold particles, mineral identification, alteration, and textural characteristics study
Automated digital imaging system	ADIS	—	Gold scan and measurement
Scanning electron microscopy	SEM	Semiquantitative	Gold scan, mineral identification, and surface morphological study
Quantitative evaluation of material by scanning electron microscope	IA	1.5 to 3 µm	Gold scan, mineral identification, liberation, surface morphological study, and modal analysis
Electron microprobe analysis	EMPA	0.1% EDX* 0.02% WDX†	Compositional analysis of gold and associated minerals
Dynamic secondary ion mass spectrometry	D-SIMS	Sub-ppm	Quantification and mapping of gold in sulfides and FeO_x minerals
Microparticle-induced X-ray emission spectroscopy	µ-PIXE	ppm	Quantification and mapping of gold in sulfides and silicates
Laser ablation microprobe inductively coupled plasma mass spectrometry	LAM–ICP–MS	ppb	Quantification of gold in sulfides, silicates, and oxides
Time-of-flight laser ion mass spectrometry	TOF–LIMS	ppm	Quantification of surface gold and analysis of surface chemistry

NOTE: Dashes indicate information is not available.

* EDX = energy dispersive X-rays (for use in SEM analysis).

† WDX = wave dispersive X-rays (for use in microprobe analysis).

temperature and, if not controlled correctly, can result in incomplete removal of gold from mineral components in the sample or loss of precious metals by volatilization. Excellent references are available in the literature [49, 50]. One final comment on fire assaying: There is no such thing as "un-assayable" gold by the fire assay method, provided that the procedure is performed correctly by a trained specialist who is knowledgeable in the art of fire assaying.

Acid digestion. This technique requires the dissolution of precious metal values from a finely divided sample (i.e., ground to 80% to 90% <75 µm) in boiling aqua regia. The slurry produced is filtered and the solution phase either analyzed directly by AAS or ICP, or concentrated by solvent extraction and then analyzed by AAS or ICP. This method is slightly cheaper than the fire assay procedure but has the disadvantage that the gold is not always efficiently dissolved from all mineral components of the original sample (e.g., gold encapsulated in silicates). In this case, hydrofluoric acid can be used to dissolve the silica, but this procedure is more costly and hazardous.

Cyanide leaching. Direct cyanide leaching has been applied for quantification of precious metal values, particularly for larger samples where sample subdivision is undesirable due to potential biasing (e.g., where coarse gold is present). This method may also be used for routine analyses once its efficacy has been well proven against fire assay methods. The sample is leached with cyanide solution at the desired particle size by several methods including:

- Bottle roll leaching (10 to 1,000 g)
- Agitation leaching (500 to 5,000 g)
- Column leaching (100 to 1,000 kg)
- Pilot plant leaching (several tons)

The leach solutions obtained from these procedures are either analyzed directly by AAS or ICP, or may be treated by solvent extraction and then analyzed, as described previously. The solid residues are sometimes also analyzed by fire assay or acid digestion. The cyanide leaching method provides a measure of the concentration of gold that is extractable by cyanidation but does not give an absolute assay value and should be considered as a metallurgical test rather than a true assay technique. A hot cyanide leaching technique has also been developed as an extension of this procedure with the objective of dissolving precious metal values more quickly and effectively than the cold technique described. This method is sometimes popular for the treatment of exploration samples but provides only an indication of recoverable precious metal values and is neither accurate assay nor reliable metallurgical information.

For alluvial ores containing coarse gold, the only statistically meaningful grade analysis is obtained by treating many tons in a pilot plant. This can give a practical measure of the extractable grade, depending on the plant equipment used.

Gold fingerprinting. Innovative gold fingerprinting technology developed by Anglo American Research Laboratories has provided an effective methodology for profiling gold according to its source or provenance. This technique uses a laser ablation procedure combined with ICP–MS to qualitatively determine the minor and trace element impurities present in gold and thereby provide a characteristic "fingerprint" for gold from a particular source. The procedure scans 131 isotopes from ^{45}Sc to ^{238}U as well as providing isotope ratios (e.g., ^{206}Pb, ^{207}Pb, and ^{208}Pb) to generate the trace element fingerprint and distinguish gold from different sources. The fact that the analysis is qualitative is important because trace element concentrations may vary depending on the processing and refining methods used to extract gold from a particular source material. As such, fingerprint signatures must be compared against a database of gold fingerprints. This technique is particularly useful for crime investigations (i.e., gold theft to determine the source of recovered gold), fingerprinting of alluvial-placer gold grains for use in gold exploration, archaeological investigations, and evaluations to determine the source of gold (and other metals) [51].

Physical methods. Laboratory-scale gravity concentration (panning, tabling, spiral classification, etc.) and amalgamation techniques are sometimes used to preconcentrate gold and other heavy metal values prior to analysis by the techniques previously described.

2.18.1.2 Other Metals and Gangue Minerals

The chemical analysis of gangue minerals is generally easier than gold due to their higher concentrations and can be achieved using the following techniques:

- Dissolution and AAS
- Dissolution and ICP
- X-ray fluorescence (XRF)
- X-ray diffraction (XRD)

The dissolution methods used prior to AAS and ICP are standard analytical chemistry techniques [49, 52]. AAS and ICP techniques are based on the use of standard and readily available instrumentation, using well-established measurement procedures. XRF is a well-established method of chemical analysis, which detects the secondary X-rays emitted from a sample subjected to an incident X-ray beam. Detection levels range from about 100 ppm (mg/L) for lighter elements to about 1 ppm for heavy elements, such as gold. When preparation methods are thorough, XRF can be used quantitatively.

Qualitative or semiquantitative mineralogical information, rather than purely elemental composition, can be produced using XRD on crystal or powder samples. XRD analysis is based on the Bragg equation:

$$n\lambda = 2d\sin\theta \qquad (EQ\ 2.9)$$

where
 λ = the wavelength of the incident X-ray beam
 d = the atomic spacing distance
 θ = the angle to the normal under the condition of peak intensity

The "2θ" information obtained is used to calculate the atomic spacing, which is characteristic of a lattice structure and can be compared to standard data to identify the mineralogical components of a sample. The advantage of XRD is that it is the only direct method of determining chemical or mineralogical composition, sample preparation is simple, and the sample does not need to be inserted into a vacuum. Disadvantages are the difficulties in identification of large crystals and very fine powders, and the low limit of detection (typically about 1% to 5%). XRD analysis is of great importance in the determination of alteration mineral phases, especially clays.

Other instrumental techniques described in Section 2.18.2 can also be used to determine gold concentration, although these methods are best suited to describing the textural characteristics of gold occurrence. This is generally most useful for determining the gold mineralogical balance rather than determining a statistically significant gold grade [5].

2.18.2 Textural Characteristics

The most important methods for the determination of information concerning the common associations between elements and, in most cases, specific minerals, are as follows:

- Optical microscopy
- Diagnostic leaching
- Scanning electron microscopy (SEM), including QEMSCAN and Mineral Liberation Analyzer (MLA) technology
- Electron microprobe analysis (EMPA)
- Mossbauer spectroscopy
- X-ray photoelectron spectroscopy (XPS or ESCA)
- Auger electron spectroscopy (AES)
- Ion microprobe and secondary ion mass spectrometry (SIMS and time-of-flight [TOF]-SIMS)
- Laser ion mass spectrometry (LIMS/TOF-LIMS)
- Proton microprobe

Increasingly, microprobe methods are being used to determine gold distributions at sizes finer than those "visible" by electron microscope methods. The characteristics of the available techniques (Table 2.14 and Figure 2.16) allow solid solution and colloidal gold to be detected with limits of less than 1 ppm, compared with electron microprobes which have detection limits of 100 to 200 ppm. Although the determinative mineralogical methods of analysis are described briefly next, more detailed information is available in the literature [6, 35, 53].

Optical microscopy. Reflected light microscopy of gold ores, mineral separation products, leach residues, and other precious metal-bearing materials is the most fundamental and important method of mineralogical analysis. Optical microscopy in reflected light allows the sulfide minerals to be identified by reflectivity, hardness (Figure 2.1), cleavage, and other mineralogical characteristics, and is useful in identifying relatively coarse textural associations (e.g., common sulfide mineral relationships and liberation sizes for sulfides). It is rare to see gold grains, except in flotation or gravity concentrates. An initial optical microscopic examination often saves time in subsequent and more sophisticated analyses. In addition, transmitted light microscopy is a significant analytical tool to determine and quantify gangue mineralogy.

Diagnostic leaching. Diagnostic leaching procedures have been developed for gold ores to determine the association of gold with gangue minerals. These methods involve sequentially leaching gold and ore components with progressively stronger leaching reagents to liberate gold in different mineralogical associations [42, 47, 54, 55, 56]. Such techniques have been applied effectively and successfully for the investigation of many different ore types around the world, including Carlin Trend ores, Brazilian palaeoplacers, and Witwatersrand ores. One example of such a sequential procedure is as follows:

Step 1: Cyanidation to extract free gold.

Step 2: Simple hydrochloric acid leach followed by cyanidation to extract gold within leachable uranium minerals and labile sulfide minerals (e.g., pyrrhotite).

Step 3: Strongly oxidizing nitric acid leach followed by cyanidation to quantify gold in less-soluble minerals (e.g., stable sulfide minerals such as pyrite and arsenopyrite).

Step 4: Acetonitrile treatment followed by cyanidation to determine gold in activated carbon and/or carbonaceous matter (i.e., preg-robbing material).

Step 5: Analysis of residue from steps 1 to 4 indicates gold locked in silicates.

An alternative diagnostic procedure replaces steps 3, 4, and 5 with the following [6]:

Step 3a: Sulfuric acid leach to release gold associated with labile sulfide minerals not attacked by hydrochloric acid (i.e., sphalerite, reactive pyrite, etc.).

Step 4a: Nitric acid leach to release gold associated with refractory pyrite, arsenopyrite, and marcasite.

Step 5a: Hydrofluoric acid leach to release gold encapsulated in silicate minerals.

Another example is a diagnostic leaching technique developed for material containing native gold and gold–silver tellurides (hessite, petzite, and calaverite) at Kalgoorlie:

Step A: Leaching with dilute cyanide (0.1% NaCN [sodium cyanide]) at pH 9.5 for 24 hr to dissolve native gold but not gold–silver telluride minerals.

Step B: Leaching the residue from step A with strong cyanide (2% NaCN) at pH 12.5 for 96 hr to dissolve gold–silver telluride minerals.

The stepwise procedure indicates that in step A native gold dissolved along with about half the hessite and petzite, while none of the refractory calaverite was leached. The second step leached the remainder of the hessite and petzite but very little of the calaverite [56].

TABLE 2.14 Common analytical methods in process mineralogy [45]

Primary Measurement Source	Incident Beam/Resultant Radiation						
	Secondary Electrons	Back-Scattered Electrons	Transmitted Electrons	Photoelectrons	Auger Electrons	X-rays	Ions
Electrons	SEM—topography	SEM—enhanced atomic weight sensitivity	Scanning transmiss on electron microscopy (TEM), electron energy loss spectroscopy-EELS	—	Scanning Auger electron spectroscopy	Energy dispersive spectroscopy (EDS), wavelength dispersive microprobe	—
X-rays	—	—	—	XPS (also known as ESCA)	Auger	XRD-residual stress	—
Ions	SEM	—	—	—	—	PIXE	Mass spectroscopy (SIMS or ion microprobe mass analysis [IMMA]), Rutherford backscatter (RBS)

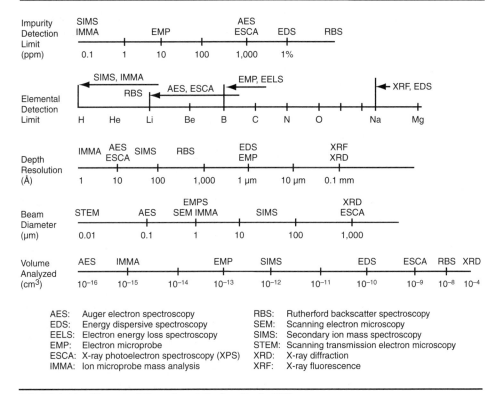

FIGURE 2.16 Characteristics of analytical methods [52]

Many other diagnostic procedures have been proposed and utilized in the gold industry. Although diagnostic leaching procedures can be useful, they can also be quite misleading because different forms of the same mineral can react differently to specific steps of a particular diagnostic leach procedure. Pyrite behaves quite differently in its cubic and framboidal habits (the framboidal form is much more reactive; see Chapter 5), and pyrite containing trace amounts of impurities is known to be more reactive than pure pyrite. Also, careful review of the diagnostic procedures described in the literature (and here), indicates some contradictions in the understanding of the solubility of the various minerals of importance in some steps of these procedures. Care must always be taken when applying diagnostic leach procedures to ensure that effective mineralogical characterization is performed and to correlate the results with more detailed metallurgical evaluation.

Scanning electron microscopy (SEM). Providing very clear imagery of particle textures, such as surface morphology, pore structure, permeability, and coatings, SEM is of most use when coupled with EDS. Analysis by SEM–EDS is fast and requires only simple sample preparation; however, the equipment is relatively expensive and requires specialist operators, as do the other instrumental methods listed here. There have been significant advances in SEM equipment (both hardware and software) in the 1990s and 2000s, summarized as follows:

- QEMSCAN (QEM*SEM) technology, developed by CSIRO (Melbourne, Australia) and marketed by Intellection Pty. Ltd. (Australia)
- MLA, developed and marketed by JKTech (Brisbane, Australia)

The application of these advanced SEM techniques provides the user with accurate and quantitative mineralogical information on a timely basis. The information provided can include the following [57]:

- Type and quantity of minerals present
- Distribution of mineral grains by size fraction and the size distribution of each mineral
- Type and extent of mineral associations, such as locking, degree of encapsulation, and galvanic connection

Analyses can be conducted on samples of feed and tailings from unit process operations, and from intermediate and by-product process streams, to provide important diagnostic information for process design and optimization.

Electron microprobe analysis (EMPA). Used since about 1968 to investigate gold ores [58], EMPA's initial attraction was to exploit the possibility of automated searching for gold grains at low concentrations. EMPA also uses the Bragg equation with θ and d set by the instrument geometry and measurement of λ. Once gold grains have been located, high-quality compositional and textural information (spatial resolution: 1 µm) can be obtained (e.g., gold/gangue associations, coating compositions, and alteration zoning).

Mossbauer spectroscopy. Relying on the absorption and emission of alpha rays, this technique has been used to determine the chemical state of gold present, for example, to distinguish between metallic gold and nonmetallic gold chemically bound within a gangue mineral lattice [43].

X-ray photoelectron spectrometry (XPS). A surface X-ray probe that analyzes emitted electron energy, XPS is used to determine surface compositions and can yield data on the chemical state of the phases studied. Each element is identified by measuring the photoelectron peak energy, which indicates shifts from expected values from which the oxidation state can be determined. All of the elements in the periodic table can be detected. The X-ray beam cannot be focused as closely as an electron beam, and therefore resolution is lower than AES; however, there are no charging effects with insulators, and XPS is nondestructive. A variation of this technique is microparticle-induced X-ray emission spectrometry (µ-PIXE), which is capable of trace element analysis down to 5 to 20 ppm.

Auger emission spectrometry (AES). This surface probe detects elements (atomic numbers 23 and above) in the first few atomic layers of a sample (Figure 2.17).

The technique excites surface atoms with an incident electron beam (1 to 10 keV) and analyzes the energy of secondary and backscattered electrons. The auger electrons can be resolved instrumentally from the backscattered electron spectrum. The primary beam has a penetration depth of about 0.2 to 2 µm and can be detected by EDS. Consequently, this method complements AES and can show differences between bulk and surface phases, for example, coatings on gold particles in tailings streams.

Secondary ion mass spectrometry (SIMS). This technique detects ionized atoms and molecules, produced by ion beam sputtering of a surface, and provides information that cannot be obtained by other microanalytical techniques. The secondary ions are produced from a depth of 0.5 nm into the sample. SIMS can detect all elements in the periodic table, can resolve isotopes for many elements, and has a detection sensitivity of 10^{-4} atomic fraction, that is, in the ppm (g/t) range for many elements and ppb (mg/t) range for some. SIMS has high sensitivity (though quantification is sometimes difficult) up to 4 orders of magnitude higher than XPS, EMPA, and AES, and therefore can be used to detect very fine (<0.1 µm diameter) solid solution gold in refractory ores, for example, gold associated with sulfide, arsenide, and iron oxide minerals [58]. It has been reported that ion

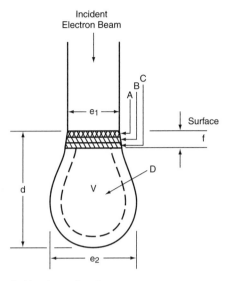

FIGURE 2.17 Schematic of interaction of electron beam with solid and typical escape depths for electron microscopy and microprobe analysis [52]

mass spectrometry (IMS) techniques can quantify the gold content of individual particles with a detection limit of 0.2 to 0.4 g/t for a 60-μm analysis diameter (or 1 g/t at 25 μm). SIMS complements AES and XPS procedures and can be incorporated into a single vacuum system [59]. However, IMS–SIMS is not suitable for insulating minerals due to sample charging. Silicate and carbonate gangue minerals should rather be analyzed by EMPA or proton microprobe analysis.

TOF (Time-of-Flight) SIMS. An advanced technique that allows analysis of surface layers of particles, using brief sputtering of the surface, this method is useful for analyzing species present on the surface of minerals, for example, flotation collectors on particles [6].

Laser ion mass spectrometry (LIMS). This technique is capable of analyzing elements and adsorbed organic compounds on the surfaces of mineral particles, as well as the elemental composition of the particles themselves. Because of the small beam size, the technique can be applied to small mineral grains down to about 5 μm. When applied in conjunction with a TOF-LIMS, the method can compare the surface and subsurface composition of minerals (including gold), representing an advanced and sophisticated option for highly detailed mineralogical investigations [6, 60].

Proton microprobe. Used for multi-element quantification of trace elements in polished or thin sections, this has a detection limit of 2 to 4 g/t depending on the matrix composition [61]. It has been used to examine the gold-bearing microcrystalline quartz–chert and carbonate particles in Carlin ores [5]. It is also useful for analyzing minor elements of interest in gold extraction, for example, arsenic, antimony, selenium, mercury, and tellurium.

An example of the application of combinations of the described methods to determine the mineralogical balance in a typical gold ore sample follows [5]:

- Analysis for gold, silver, arsenic, tellurium, bismuth, antimony, sulfur, and total and organic carbon
- Determination of size distribution, composition, and association of the gold minerals
- Determination of the visible gold fraction by diagnostic leaching
- Identification of the submicroscopic or invisible gold-carrier minerals and determination of their gold content by ion or proton microprobe microanalysis
- Determination of the invisible gold fraction (i.e., ultrafine size or in solid solution)
- Calculation of the gold mineralogical balance
- Characterization of the ore constituents that may be deleterious to processing (e.g., active carbon, pyrrhotite, clays)

This information may be used as a basis for a metallurgical test work program for process design or optimization. Process metallurgists should be aware of all the methods available and have the facilities to employ several complementary methods in conjunction with chemical analysis and metallurgical test work to fully describe the gold mineralogy of a particular ore.

2.18.3 Special Factors in Placer Ore Evaluation

Due to the coarse size of gold in placers and the treatment methods used, a mineralogical examination of placer gold samples should specifically address the following factors:

- The mode of occurrence of the gold and its physical characteristics, including particle size, shape, surface features, and locking and mineralogical aspects that could affect sampling and gravity concentration
- The amount and composition of heavy minerals recovered together with gold in a large gravity concentrate, with particular attention to the presence of additional, potentially valuable minerals
- The presence and quantities of cemented aggregates or clay agglomerates, which may have trapped or accumulated gold values. The extent and type of breakdown and disintegration of these materials by sizing, washing, and attrition scrubbing should be evaluated
- The occurrence of talc, serpentine, graphite, and clay, which can contribute to sickening or flouring of mercury
- Process water and the desliming tailings should be analyzed for fine gold content.

2.18.4 Process Stream Mineralogy

Periodic mineralogical analyses of ore feed, intermediate process streams, and final products of extraction processes should include the following:

- Mineralogical assessment of the reasons for variable gold and silver extraction (e.g., coatings, locking in gangue, coarse gold occurrence, etc.)
- Determination of the type, quantity, and distribution of cyanide, oxygen, and alkali-consuming minerals
- Assessment of the preg-robbing or preg-borrowing potential of gangue minerals
- Quantification of middlings mineralogy (where applicable)

- Determination of the mineralogy, quantity, and size of clays and/or slime-forming minerals
- Determination of type and amount of water-soluble minerals in the plant feed and composition of scale formations
- Assessment of coatings on gold particles

The optimization of gold extraction flowsheets relies on the ability to obtain adequate, accurate, and timely mineralogical information for the gold metallurgist to use along with metallurgical test data and plant operating data, where available.

REFERENCES

[1] Lewis, A. 1982. Gold geochemistry—New ideas about the paragenesis of hydrothermal deposits have implications for finding undiscovered gold ore. *Engineering and Mining Journal* (July):56–60.

[2] Henley, K.J. 1975. Gold ore mineralogy and its relation to metallurgical treatment. *Minerals Science & Engineering* 7(4):289–312.

[3] Tarkian, M. 1974. *Minerals Science & Engineering* 6(2):101–105.

[4] Cohn, J.G., and E.W. Stern. 1979. Gold and gold compounds. In *Encyclopedia of Chemical Technology*. Volume 11. New York: Academic Press.

[5] Chryssoulis, S.L., and L.J. Cabri. 1990. Significance of gold mineralogical balances in mineral processing. *Transactions of the Institution of Mining and Metallurgy* 99:C1–C10.

[6] Petruk, W. 2000. *Applied Mineralogy in the Mining Industry*. Amsterdam: Elsevier Science B.V.

[7] McQuiston, F.W., and R.S. Shoemaker. 1975. *Gold and Silver Cyanidation Plant Practice*. Volume 1. SME-AIME Monograph. Salt Lake City, UT: SME-AIME.

[8] Edwards, R., and K. Atkinson. 1986. Placers and palaeo-placers. Pages 175–214 in *Ore Deposit Geology and Its Influence on Mineral Exploration*. London: Chapman & Hall.

[9] Lewis, A. 1982. Gold geology basics. *Engineering and Mining Journal* (February): 66–72.

[10] Henley, R.W., and J. Adams. 1979. On the evolution of giant gold placers. *Transactions of the Institution of Mining and Metallurgy* 88:B41–B49.

[11] MacDonald, E.H. 1984. Aspects of alluvial gold exploration. Pages 459–472 in *Conference of Gold Mining, Metallurgy and Geology*. Melbourne, Australia: Australasian Institute of Mining and Metallurgy.

[12] Fricker, A.G. 1976. Placer gold-measurement and recovery, sampling. Pages 115–127 in *Proceedings Symposium on Sampling Practices in the Mineral Industry*. Parkville, Australia: Australasian Institute of Mining and Metallurgy.

[13] Desborough, G.A. 1970. Silver depletion indicated by microanalysis of gold from placer occurrences, Western United States. *Economic Geology* 65:304–311.

[14] Hester, B.W. 1973. Geology and evaluation of placer gold deposits in the Klondike area, Yukon Territory. *Transactions of the Institution of Mining and Metallurgy* 79:B60–B67.

[15] De Waal, S.A. 1982. *A Literature Survey of the Metallurgical Aspects of Mineral in Witwatersrand Ores*. Mintek Report M37. Randburg, South Africa: Mintek.

[16] Hallbauer, D.K. 1986. The mineralogy and geochemistry of Witwatersrand pyrite, gold, uranium and carbonaceous matter. Pages 731–752 in *Mineral Deposits of South Africa*. Volume 1. Edited by C.R. Anahaeusser and S. Maske. Johannesburg: Geological Society of South Africa.

[17] James, H.E., R.C. Dunne, S.A. De Waal, and P.A. Laxen. 1982. Established procedures and new advances in the extraction of gold, uranium, and pyrite from Witwatersrand ores. Pages 515–525 in *Proceedings of the 12th Congress of Mining and Metallurgical Institutions Congress*. Edited by H.W. Glen. Johannesburg: South African Institute of Mining and Metallurgy.

[18] Anahaeusser, C.R., C.A. Feather, W.R. Liebenberg, G. Smits, and J.A. Snegg. 1987. Geology and mineralogy of the principal goldfields of South Africa. Pages 1–67 in *The Extractive Metallurgy of Gold*. Edited by G.G. Stanley. Johannesburg: South African Institute of Mining and Metallurgy.

[19] Feather, C.A., and G.M. Koen. 1973. The significance of the mineralogical and surface characteristics of gold grains in the recovery process. *Journal of South African Institute of Mining and Metallurgy* 73:223–234.

[20] Anon. 1988. Epithermal gold. *International Mining* (February):7–12.

[21] Swash, P.M. 1988. A mineralogical investigation of refractory gold ores and their beneficiation, with special reference to arsenical ores. *Journal of South African Institute of Mining and Metallurgy* 88(5):178–180.

[22] Boyle, R.W. 1979. The geochemistry of gold and its deposits. Bulletin 280. Ottawa: Geological Survey of Canada.

[23] Addison, R. 1980. Gold and silver extraction from sulfide ores. *Mining Congress Journal* 66(10):44–54.

[24] Baum, W. 1988. Mineralogy related processing problems of epithermal gold ores. Pages 3–20 in *Process Mineralogy VIII*. Edited by D.J.T. Carson and A.H. Vassiliou. Warrendale, PA: TMS.

[25] Mann, A.W. 1984. Redistribution of the oxidised zone of some W.A. deposits. In *Gold Mining, Metallurgy and Geology*. Melbourne: Australasian Institute of Mining and Metallurgy.

[26] Hedenquist, J., and F. Reid. 1985. *Epithermal Gold*. Sydney, Australia: Earth Science Foundation, University of Sydney.

[27] Berger, B.R., and P.M. Bethke. 1985. Geology and geochemistry of epithermal systems. *Review of Economic Geology* 2:238.

[28] Baum, W. 1998. Copper mineralogy and extraction problems: How to reduce your losses. Pages 41–54 in *Randol Copper Hydrometallurgy Roundtable '98*. Golden, CO: Randol International Ltd.

[29] Hausen, D.M. 1985. Process mineralogy of selected refractory Carlin type gold ores. *Canadian Institute of Mining and Metallurgy Bulletin* 78:83–94.

[30] Hausen, D.M. 1989. Processing of gold quarry refractory gold ores. *Journal of Metals* 41(4):43–45.

[31] Hiskey, J.B., and M.E. Wadsworth. 1981. Electrochemical processes in the leaching of metal sulfides and oxides. Pages 304–325 in *Proceedings and Fundamental Considerations of Selected Hydrometallurgical Systems*. Edited by M.C. Kuhn. New York: SME-AIME.

[32] Vaughan, J.P., and R.C. Dunne. 1987. Mineralogy and processing characteristics of Archean gold ores from Western Australia. Pages 241–256 in *Gold Mining '87*. Littleton, CO: SME.

[33] Cristovici, M.A. 1986. Recovery of gold from old tailings ponds. *Canadian Institute of Mining and Metallurgy Bulletin* 79:27–33.

[34] House, C.I., and G.H. Kelsall. 1985. Hydrometallurgical reduction of SnO_2, $CuFeS_2$ and PbS by electrogenerated Cr(II) and V(II) solutions. Pages 659–682 in *Extraction Metallurgy '85*. London: Institute of Mining and Metallurgy.

[35] Spry, P.G., S. Chryssoulis, and C.G. Ryan. 2004. Process mineralogy of gold: Gold from telluride-bearing ores. *Journal of Metals* (August):60–62.

[36] Jayasekera, S., I.M. Ritchie, and J. Avraamides. 1988. Electrochemical aspects of the leaching of gold telluride. Pages 187–189 in *Perth Gold '88 Randol Conference*. Golden, CO: Randol International Ltd.

[37] Ellsworth, H.V. 1928. Thucholite and uraninite from the Wallingford Mine, near Buckingham, Quebec. *American Mineralogist* 13:442–448.

[38] Osseo-Asare, K., P.M. Afenya, and G.M.K. Abotsi. 1984. Carbonaceous matter in gold ores: Isolation, characterisation and adsorption behavior in aurocyanide solutions. Pages 125–144 in *Precious Metals: Mining, Extraction and Processing*. Edited by V. Kudryk, D.A. Corrigan, and W.W. Liang. Warrendale, PA: TMS.

[39] Corrans, J., and R.C. Dunne. 1985. Optimisation of the recovery of gold and uranium from Witwatersrand residues. *Mintek Review* 2:18–24.

[40] Baum, W., O. Sanhueza, E.H. Smith, and W. Tufar. 1989. The use of mineralogy for plant optimisation at the El Indio gold–silver–copper operation (Chile). *Erzmetall* 42(9):373–378.

[41] Venter, D., S.L. Chryssoulis, and T. Mulpeter. 2004. Using mineralogy to optimize gold recovery by direct cyanidation. *Journal of Metals* (August):53–56.

[42] Tumilty, J.A., A.G. Sweeney, and L. Lorenzen. 1987. Diagnostic leaching in the development of flowsheet for new ore deposits. Pages 157–167 in *International Symposium on Gold Metallurgy*. Edited by R.S. Salter, D.M. Wysouzil, and G.W. MacDonald. New York: Pergamon.

[43] Adam, K., J.M. Prevosteow, A. Kontopoulos, M. Stefakis, and M. Errington. 1990. Application of process mineralogy to the treatment of Olympias pyrite concentrates. Pages 341–350 in *Proceedings of the Gold 1990 Symposium*. Edited by D.M. Hausen, D.N. Halbe, E.U. Petersen, and W.J. Tafuri. Littleton, CO: SME.

[44] Swash, P.M., and P. Ellis. 1986. The roasting of arsenical gold ores: A mineralogical perspective. Pages 235–258 in *Proceedings Gold 100*. Volume 2. Edited by C.E. Fivaz. Johannesburg: South African Institute of Mining and Metallurgy.

[45] Morrison, B.H. 1985. Recovery of silver and gold from refinery slimes at Canadian copper refiners. Pages 259–269 in *Extraction Metallurgy '85*. Melbourne: Australasian Institute of Mining and Metallurgy.

[46] Scott, J.D. 1988. Process mineralogy of silver and gold at Kidd Creek from ore to anode slime. Pages 125–132 in *Proceedings of the International Symposium on Gold Metallurgy*. Volume 1. Edited by R.S. Salter, D.M. Wyslouzil, and G.W. McDonald. New York: Pergamon Press.

[47] Hilliard, H.E., B.W. Dunning, D.A. Kramer, and D.M. Soboroff. 1985. Metallurgical treatment of electronic scrap to recover gold and silver. RI 8940. Washington, DC: U.S. Bureau of Mines.

[48] Zhou, J.Y., and L.J. Cabri. 2004. Gold process mineralogy: Objectives, techniques and applications. *Journal of Metals* (July):49–52.

[49] Lenahan, W.C., and R. de L. Murray-Smith. 1986. *Assay and Analytical Practice in the South African Mining Industry*. Johannesburg: South African Institute of Mining and Metallurgy.

[50] Bugbee, E.E. 1940. *A Textbook of Fire Assaying*. New York: John Wiley & Sons.

[51] Grigorova, B., S. Anderson, J. de Bruyn, W. Smith, K. Stulpner, and A Barzev. 1998. The AARL gold fingerprinting technology. *Gold Bulletin* 31(1):26–29.

[52] Kossowsky, R. 1983. Designing an analytical microscopy laboratory. *Journal of Metals* 35(3):47–54.

[53] Chryssoulis, S., R. Dunne, and A. Coetzee. 2004. Diagnostic microbeam technology in gold ore processing. *Journal of Metals* (July):53–57.

[54] Coetzee, M., M.J. Wilkinson, and J.A. Tumilty. 1988. The degree of comminution required to liberate gold from Witwatersrand quartzites. Pages 43–48 in *Perth Gold '88 Randol Conference*. Golden, CO: Randol International Ltd.

[55] Malhotra, D., and S. Armstrong. 1993. Characterization of refractory gold ores through diagnostic leaching procedures. SME Preprint. Littleton, CO: SME.

[56] Henley, K.J., N.C. Clarke, and P. Sauter. 2001. Evaluation of a diagnostic leaching technique for gold in native gold and gold +/- silver tellurides. *Minerals Engineering* 14(1):1–12.

[57] Butcher, A.R., T.A. Helms, P. Gottlieb, R. Bateman, S. Ellis, and N.W. Johnson. 2000. Advances in the quantification of gold deportment by QemSCAN. Pages 267–271 in *Proceedings of the 7th Mill Operators Conference*. Melbourne: Australasian Institute of Mining and Metallurgy.

[58] Jones, M.P., and J. Gavrilovic. 1968. Automatic searching unit for the quantitative location of rare phases by electron microprobe X-ray microanalysis. *Transactions of the Institution of Mining and Metallurgy* 77:B137–B143.

[59] Chryssoulis, S.L., L.J. Cabri, and R.S. Salter. 1987. Direct determination of invisible gold in refractory sulfide ores. Pages 235–244 in *Proceedings of the International Symposium on Gold Metallurgy*. Edited by R.S. Salter, D.M. Wysouzil, and G.W. MacDonald. New York: Pergamon Press.

[60] Chryssoulis, S.L., and A.H. Winckers. 1996. Effect of lead nitrate on the cyanidation of David Bell Ore. Pages. 127–149 in *Proceedings of the 28th Annual Meeting, Canadian Mineral Processors*. Edited by A. Mular. Ottawa: Canadian Institute of Metallurgy.

[61] Chryssoulis, S.L. 1990. Detection and quantification of "invisible" gold by microprobe techniques. Pages 323–332 in *Proceedings of the Gold 1990 Symposium*. Edited by D.M. Hausen, D.N. Halbe, E.U. Petersen, and W.J. Tafuri. Littleton, CO: SME.

[62] Halverson, G.B. 1990. Fluosolids roasting practice at Giant Yellowknife Mines Ltd. Paper presented at 96th Annual North West Mining Association, Spokane, WA.

PLATE 1 Placer gold particle showing typical mineralogical reasons for recovery problems including old mercury coatings (silver gray); silica coatings (light gray); and iron oxide coatings (brown). Location: California, United States. (Photo courtesy of Wolfgang Baum, Pittsburgh Mineral and Environmental Technology, Pittsburgh, PA.)

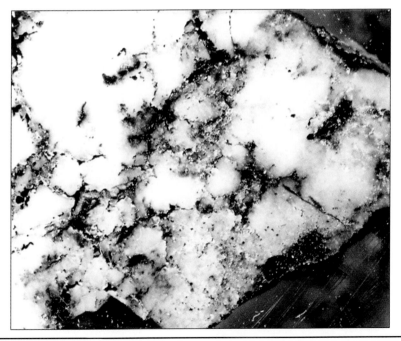

PLATE 2 Coarse-grained native gold in massive quartz matrix (white). This high-grade sample represents a "direct shipping ore." Location: Chile. (Photo courtesy of Wolfgang Baum, Pittsburgh Mineral and Environmental Technology, Pittsburgh, PA.)

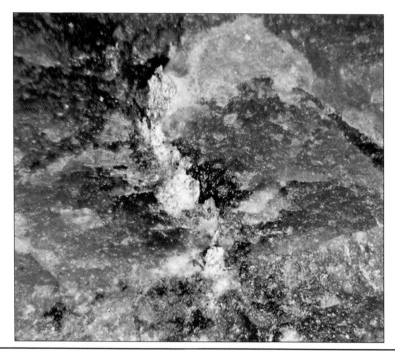

PLATE 3 Coarse- and fine-grained occurrence of native gold in quartz vein material. The coarse gold is amenable to gravity concentration whereas the finely disseminated gold may require fine grinding and subsequent cyanidation. Location: Ontario, Canada. (Photo courtesy of Wolfgang Baum, Pittsburgh Mineral and Environmental Technology, Pittsburgh, PA.)

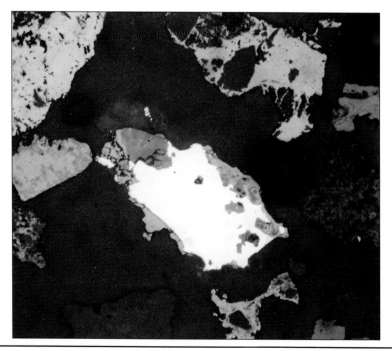

PLATE 4 Coarse-grained native gold (white-yellow) with coatings of hydrous iron oxides (gray). This gold will not respond well to cyanide leaching but is recoverable by gravity methods. Location: California, United States. (Photo courtesy of Wolfgang Baum, Pittsburgh Mineral and Environmental Technology, Pittsburgh, PA.)

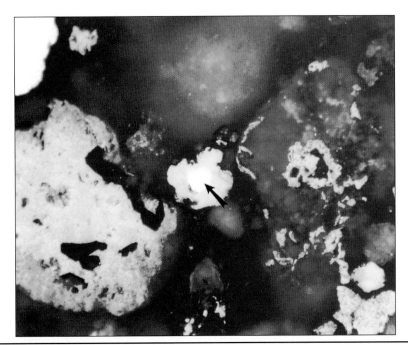

PLATE 5 Iron oxide refractory gold. The native gold (see arrow) is refractory due to complete encapsulation by impervious hydrous iron oxides. The gold–iron oxide particle was recovered by gravity separation from the cyanide leach tailings. Location: Rodalquilar, Spain. (Photo courtesy of Wolfgang Baum, Pittsburgh Mineral and Environmental Technology, Pittsburgh, PA.)

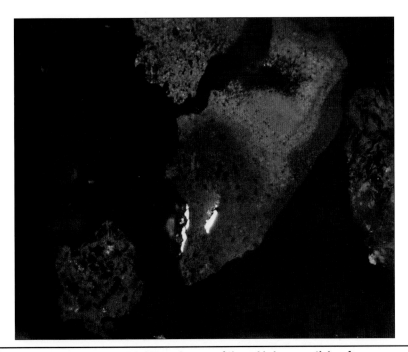

PLATE 6 Iron oxide refractory gold. Although some of the gold shows partial surface exposure and the iron oxide particle exhibits considerable porosity, this gold was not recovered heap leaching due to slimes coatings. Location: Nevada, United States. (Photo courtesy of Wolfgang Baum, Pittsburgh Mineral and Environmental Technology, Pittsburgh, PA.)

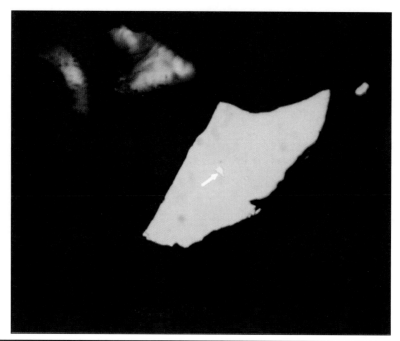

PLATE 7 Sulfide-refractory gold. Native gold (see arrow) is encapsulated in larger pyrite particle. Location: South Carolina, United States. (Photo courtesy of Wolfgang Baum, Pittsburgh Mineral and Environmental Technology, Pittsburgh, PA.)

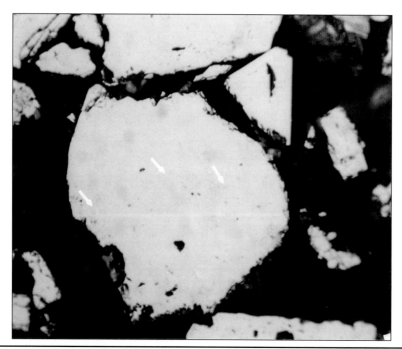

PLATE 8 Fine native gold (see arrows) locked in a marcasite particle. Location: Washington, United States. (Photo courtesy of Wolfgang Baum, Pittsburgh Mineral and Environmental Technology, Pittsburgh, PA.)

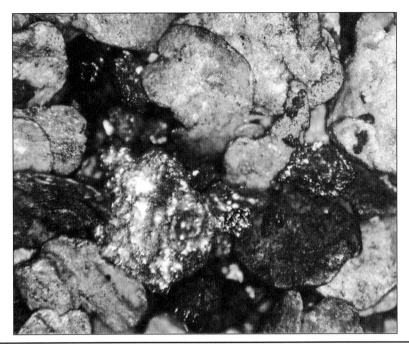

PLATE 9 Gold recovery problems resulting from previous processing techniques employed, in this case, amalgamation. Residual mercury (silver gray) in placer gold ore. Location: California, United States. (Photo courtesy of Wolfgang Baum, Pittsburgh Mineral and Environmental Technology, Pittsburgh, PA.)

PLATE 10 Coarse gold after extended circulation in a mill circuit. Many of the coarser gold particles have been flattened and their surfaces are contaminated with gangue particles. Location: Chile. (Photo courtesy of Wolfgang Baum, Pittsburgh Mineral and Environmental Technology, Pittsburgh, PA.)

PLATE 11 Gold-loaded zinc precipitate (ex solution) (SEM photo courtesy of FMC Gold Co.; Merrill–Crowe precipitate provided by Coeur d'Alene Mines).

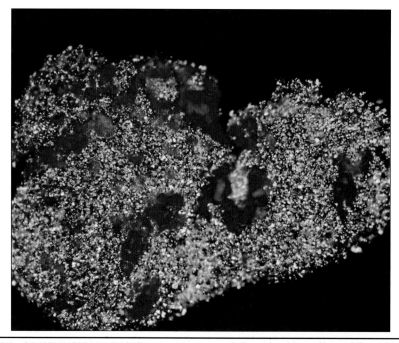

PLATE 12 Amalgamated raw gold after mercury removal. Considerable portions of the gold flakes did not respond to amalgamation due to iron oxide coatings (brown). Location: California, United States. (Photo courtesy of Wolfgang Baum, Pittsburgh Mineral and Environmental Technology, Pittsburgh, PA.)

PLATE 13 Gold crystals after mercury removal from gold amalgam. Location: California, United States. (Photo courtesy of Wolfgang Baum, Pittsburgh Mineral and Environmental Technology, Pittsburgh, PA.)

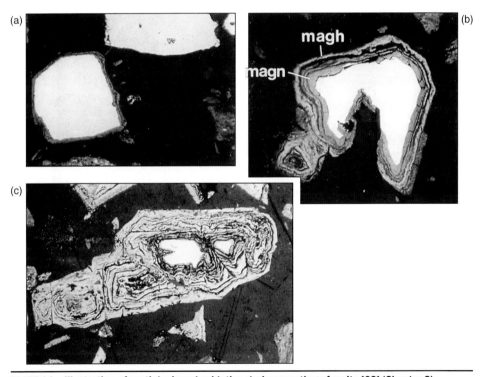

PLATE 14 Illustration of partial mineral oxidation during roasting of pyrite [62] (Chapter 2).
(a) Unoxidized pyrite particle and incipient oxidation: formation of thin magnetite layer;
(b) Intermediate oxidation: double oxidation layer (pyrite–magnetite–hematite); (c) Advanced oxidation: isolated sulfide in the center of the calcine particle.

(a)

(b)

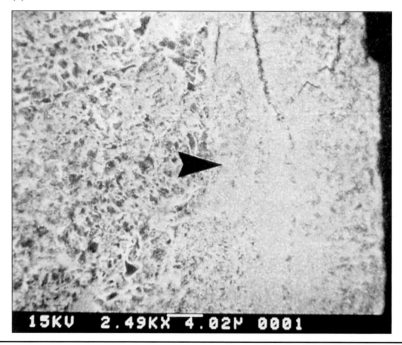

PLATE 15 SEM images showing porosity in a roaster calcine [62] (Chapter 2): (a) fully permeable particle; (b) particle with impermeable coating.

CHAPTER 3

Process Selection

Process selection is the systematic development of the optimum metal extraction route for a particular feed material using the most appropriate technology. In the case of gold this procedure has two main objectives:
- To optimize project economics, principally a function of gold recovery, throughput rate, and processing costs (capital and operating)
- To develop a process that satisfies all of the project requirements, including, for example, political and environmental considerations

The chemical response of a particular gold ore to the various process options plays a key role in achieving these objectives. At present, more than 85% of the world's gold production involves chemical processing (see Chapter 12). Process selection is playing an increasingly important role as the complexity of chemical processing techniques increases with the exploitation of lower-grade and more complex gold ores.

Process selection is an iterative procedure, which usually starts as soon as exploration has established the presence of gold mineralization in sufficient grade and tonnage for the orebody to be considered a potentially economic reserve. The amount of effort devoted to process selection is related to the degree of certainty of the grade and reserve estimations and their absolute values, that is, the overall attractiveness of the deposit.

The risk associated with the development of gold projects can be minimized through well-managed and well-planned metallurgical test work programs, coupled with careful consideration of all the specific project requirements, including available capital, target profitability, acceptable levels of risk, and specific environmental factors.

This chapter provides criteria and methodology for the selection of individual unit chemical processes described in this book and shows how they can be combined in a gold recovery flowsheet.

3.1 FACTORS AFFECTING PROCESS SELECTION

The factors affecting process selection, and the achievement of the listed objectives, can be grouped into six main areas, as follows:
- Geological
- Mineralogical
- Metallurgical
- Environmental
- Geographical
- Economic and political

The role of each of these in a development project is illustrated in Figure 3.1. Two of these factors, mineralogical and metallurgical, have direct impacts on gold extraction

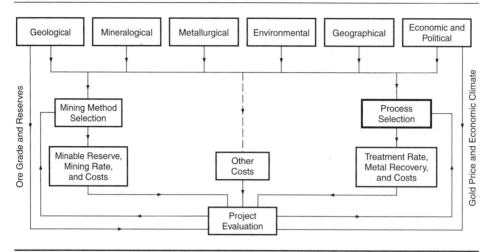

FIGURE 3.1 Factors affecting process selection

chemistry and process selection since they determine the response of the ore to chemical treatment. The other factors have an indirect effect, which depends on specific project conditions and requirements, and on the overall project feasibility.

3.1.1 Geological

3.1.1.1 Ore Grade and Reserves

The grade and tonnage reserves of economic minerals in an orebody determine the type and scale of process technology that can be applied. Low-grade ores and tailings materials (typically <0.5 to 1.5 g/t Au) usually require low-cost treatment, such as heap and dump leaching. Higher-grade ores, typically >1.5 g/t Au, may be treated by higher-cost processes, such as grinding, leaching, and carbon-in-pulp (CIP), for which the additional costs are more than offset by higher gold sales revenue. Complex sulfidic and carbonaceous refractory ores require yet higher grades to justify the additional expense of oxidative pretreatment.

Economies of scale may permit lower-grade ores to be treated at high throughput rates. For example, low-grade tailings materials (<1 g/t Au) may be retreated economically in large-scale agitated leaching circuits, as has been the case at Ergo, Simmergo, and Crown Sands (all in South Africa), among others [1]. Similarly, low-grade ores may be processed on a large scale by grinding, leaching, and CIP, for example at Ridgeway (United States), where 1 g/t Au ore has been successfully treated at a throughput rate of approximately 14,000 tpd [2].

The cutoff grades applied to different extraction processes depend on the metallurgical response of each individual ore, as indicated by gold recovery, processing costs, and throughput rates. The grades and reserves of other minerals of potential economic interest (e.g., silver, uranium, and copper) may also affect this economic evaluation and process selection.

3.1.1.2 Orebody Geometry and Variability

The geometry of an orebody not only affects the mining method but may also dictate the sequence of mining different regions, and possibly different ore types, within the orebody, which can have important processing consequences. Variations in ore properties, such as ore hardness (i.e., work Index or grindability), mineral composition (e.g., sulfide content, nature of gold occurrence, and mineral texture), alteration, degree of fracturing (particle size), and clay content, invariably reduce process efficiency and can significantly affect process selection. For example, orebodies containing pockets of mercury or copper may need special processing techniques, and ores containing highly altered or fractured regions may have particular requirements for materials handling. Such variations can be smoothed out to a large extent by blending, where the variability of the ore determines the amount of blending required. However, the selected process must be able to cope with ore-type variations that are inevitable, even after blending.

3.1.2 Mineralogical

The mineralogical properties of an ore determine its response to the various process options and indicate the potential environmental impact of its treatment. The mineralogical characteristics are determined from the ore composition and textural properties, described in Section 2.18. Such data are used in conjunction with metallurgical test work results and information from other similar orebodies for process selection and flowsheet development (Figure 3.2).

The quality of mineralogical information required for effective process selection depends on the type and variability of the orebody. Ores with "simple" mineralogy, or with similar geological and mineralogical properties to those of other well-understood deposits, require less rigorous analysis than those with complex or unknown mineralogy, although even subtle changes in mineralogy can greatly affect process selection and overall process economics. For example, the mineralogy of ores from the Witwatersrand region in South Africa is well established and relatively consistent, while more complex refractory ores of the Pacific Rim and parts of North America are more variable and generally require more detailed investigation.

3.1.3 Metallurgical

The metallurgical response of an ore to a proposed treatment scheme directly determines the economics of the process, or combination of processes, that may be used. The major factors to be considered in this evaluation are listed as follows:

1. Recovery of gold and other valuable minerals
2. Quality of product and the need for further processing
3. Treatment rate
4. Capital costs
5. Operating costs
6. Environmental impact and permitting requirements
7. Technical risk

Items 1 through 3 affect the revenues generated by the project; items 2 through 6 affect process costs; and item 7 is the level of uncertainty associated with a process. This last factor depends on the track record and complexity of the technology applied, and the ability of the project to absorb unexpected costs associated with the application of higher-risk technology. The optimum flowsheet selection yields the greatest economic benefit, while meeting the other critical project requirements such as compliance with environmental policy and achieving acceptable levels of risk.

The metallurgical response of an ore (or concentrate) to a process, or combination of processes, is determined by a program of metallurgical testing and evaluation. A scheme for such a program appears in Figure 3.3, and a list of commonly applied metallurgical test work procedures is given in Table 3.1. This work often extends beyond the requirements

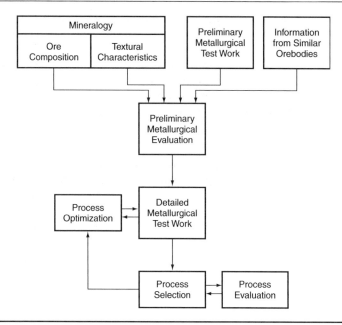

FIGURE 3.2 Schematic process development flowsheet

of process selection and design, and may last well into, or even for the whole duration of, the operating life of the project.

During metallurgical evaluation (Figure 3.2) it is important that all sources of available information are utilized. The results of test work performed on representative ore samples from the project under development usually provides the most accurate data; however, mineralogical information, process design data, and operating experience from other similar operations and orebodies should also be considered.

3.1.4 Environmental

Over the last 25 years, environmental considerations have played an increasingly important role in the development and exploitation of all mineral resources. Legislation has been passed to restrict the use of environmentally unacceptable processes and to control others. This legislation, which has developed at different rates and to varying degrees around the world, can have a major effect on process selection and operation.

In particular, process selection must consider the environmental impact that each unit process has on the following:

- Water quality
- Air quality
- Land degradation
- Visual impact
- Noise
- Flora and fauna
- Rare and endangered species
- Cultural resources
- Sustainable and social development

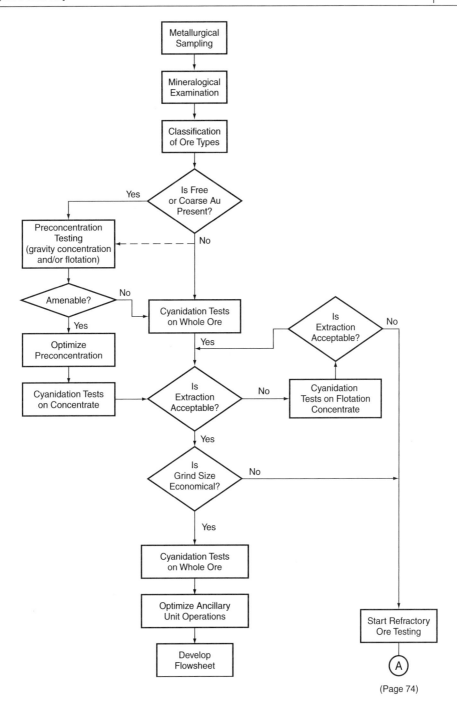

FIGURE 3.3 Schematic flowchart for metallurgical testing of gold ores (continued on pages 74 and 75) (adapted from [3])

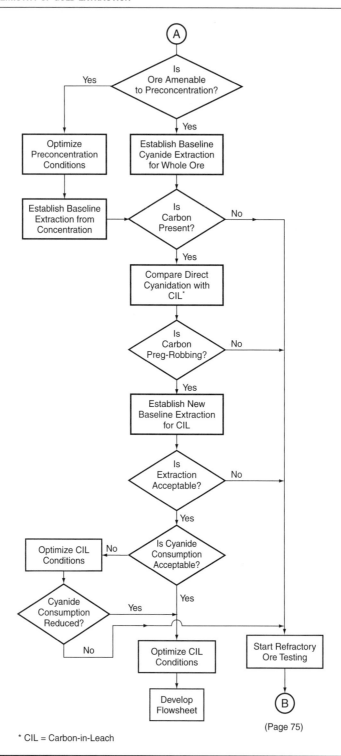

FIGURE 3.3 Schematic flowchart for metallurgical testing of gold ores (continued) (adapted from [3])

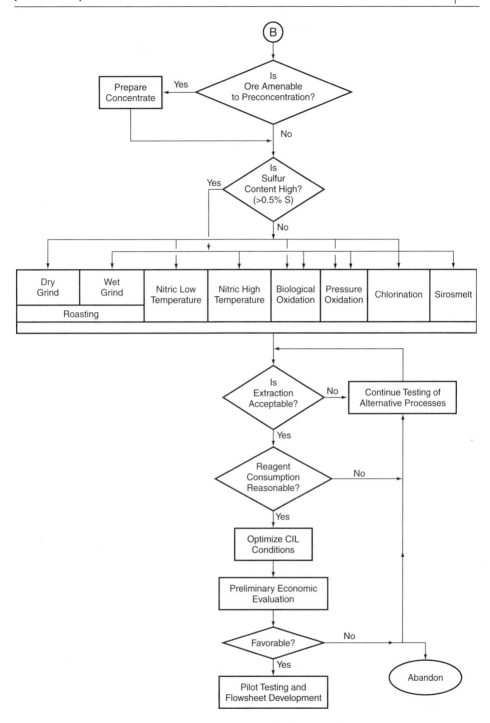

FIGURE 3.3 Schematic flowchart for metallurgical testing of gold ores (continued) (adapted from [3])

TABLE 3.1 Typical metallurgical testing procedures for gold extraction process development

Test	Information Generated
Free-Milling Ores and Tailings	
Screening and analysis	• Gold distribution and potential for different treatment of size fractions • Concentrations of other species that might affect gold extraction
Crushing and grinding	• Crushing, Bond and rod work index, abrasion index • Mill design parameters • Optimum grind size
Gravity concentration	• Grade of concentrate • Recovery to concentrate • Concentrate composition
Flotation	• Grade of concentrate • Recovery to concentrate • Concentrate composition • By-product recovery • Reagent scheme
Leaching	• Gold dissolution • By-products dissolution • Dissolution rate • Optimum leach conditions • Reagent consumptions • Solution composition
Concentration and purification	• Adsorption rate • Adsorption capacity • Other species adsorbed • Fouling of carbon • Attrition losses • Reduction of preg-robbing/borrowing characteristics by CIL
Refractory Ores and Concentrates (in addition to the above)	
Acid generation	• Amount of acid generated
Acid consumption	• Amount of acid consumed
Preg-robbing/borrowing	• Gold adsorption onto ore constituents from standard solution
Oxidative pretreatment: Pressure oxidation, roasting, and biological oxidation	• Percent sulfur oxidation versus percent gold recovery • Oxidation rate • Reagent consumptions • Optimum oxidation conditions

These are affected by the following aspects of chemical extraction processes:

- Type and amount of wastes produced, that is, solids, liquids, or gases
- Short- and long-term stability of waste products
- Degree of alteration of minerals and metals by the process
- Process water balance and the need for discharge, if required
- Method of waste disposal and treatment

Any proposed flowsheet must be capable of conforming to regulatory requirements, and any significant environmental impact(s), whether regulated or not, should be minimized by good process design, effective waste management, the use of appropriate reclamation procedures and, to the extent necessary, by detoxification or treatment of waste streams.

3.1.5 Geographical

The location of the orebody and the proposed treatment facility may have an important effect on process selection. The main factors include the following:

- Climate (rainfall, temperature ranges)
- Water supply
- Topography
- Altitude
- Infrastructure (power supply, site access, etc.)
- Availability of equipment, reagents, and supplies
- Communications
- Availability of skilled and unskilled labor
- Sites of archaeological or religious importance

Of these, climate and water supply generally have the biggest direct impact on process selection. The amount of fresh make-up water required depends on the nature of the process used and climatic conditions that affect the overall water balance, such as rainfall, temperature, humidity, and wind. In some cases process selection may be dominated by climatic conditions and water supply. For example, because of its relatively high water requirement, conventional gravity concentration cannot always be used in particularly arid environments. Conversely, heap leaching may be unsuitable in areas of extremely high rainfall where the water balance results in excessive volumes of solution that must be treated and released.

Extremes of temperature can severely affect rates of chemical reactions. For example, biological oxidation requires close temperature control for optimum bacterial activity, and gold production from heap, dump, or stockpile leaching operations can be severely retarded by extremely cold conditions.

The topography can have a major effect on process capital costs and consequently may have a significant impact on process selection. For example, the costs of heap leaching increase in rugged terrain (although techniques such as valley-fill leaching have helped to overcome this to some extent). The topography is sometimes the determining factor for the location of processing facilities, which has an impact on ore transportation costs, as well as on the costs of other facilities and infrastructure.

All of the factors listed here are site- and project-specific and must be considered independently for each process application.

3.1.6 Economic and Political

Economic and political factors, which may affect process selection, are many and varied, and any detailed discussion is beyond the scope of this book. The most important of these are the price of gold (and other metals of value, e.g., silver, uranium, and platinum group metals), tax rates and structures, and the prevailing economic and political climate, both locally and worldwide. Excellent references on the financial aspects of gold, and to a more limited extent on gold extraction, are available in the literature [4, 5, 6, 7].

3.2 UNIT PROCESS OPTIONS

Ten main unit processes are used in gold extraction circuits, listed in Table 3.2. The options within each of these categories, and indications of how these can be combined in

TABLE 3.2 Unit process operations in gold extraction

Unit Process	Process Type
Size reduction, comminution	Physical
Classification	Physical
Solid–liquid separation	Physical, surface chemical
Concentration	Physical, surface chemical
Oxidative pretreatment	Hydrometallurgical, pyrometallurgical
Leaching	Hydrometallurgical
Purification and concentration	Hydrometallurgical
Recovery	Hydrometallurgical
Refining	Hydrometallurgical, pyrometallurgical
Waste disposal, treatment	Hydrometallurgical

flowsheets, appear in Figure 3.4. The chemical process options, considered in detail in separate chapters (Chapters 5 to 11), and discussed briefly in sections that follow, are:

- Ore concentration (surface chemical methods)
- Oxidative pretreatment
- Leaching
- Solution purification and concentration
- Recovery
- Refining
- Effluent treatment

The other unit processes, namely comminution, classification, solid–liquid separation, and various concentration techniques, are primarily physical processes and are considered briefly in the following sections.

3.2.1 Comminution

Comminution of gold ores and concentrates is primarily required to liberate gold, gold-bearing minerals, and other metals of economic value to make them amenable to subsequent gold extraction steps. However, this preparation may also be necessary to facilitate materials handling between stages.

The degree of comminution required depends on many factors, including the liberation size of gold, the size and nature of the host minerals, and the method(s) to be applied for gold recovery. The optimum particle size is dictated by the economics: a balance between gold recovery, processing costs (i.e., reaction kinetics and reagent consumptions), and comminution costs for a given processing method. Other factors, such as particle fluidization requirements (agitated leaching, flotation, CIP), permeability (e.g., the effect of fines on heap leaching), and solid–liquid separation efficiency, may also play an important role in particle size optimization. The effects of particle size on each of the major extraction processes in use—flotation, cyanide leaching, and oxidative pretreatment followed by cyanide leaching—are considered in the specific chapters (5, 6, and 9) that follow. The type of comminution equipment selected can have an important impact on subsequent processing steps, for example, impact crushing, semiautogenous grinding, and high-pressure roll crushing or high-pressure grinding rolls (HPGRs).

The major uses of comminution in gold extraction flowsheets are for the following:

- Gold liberation before leaching, that is, by crushing prior to heap leaching and crushing and grinding before agitated leaching

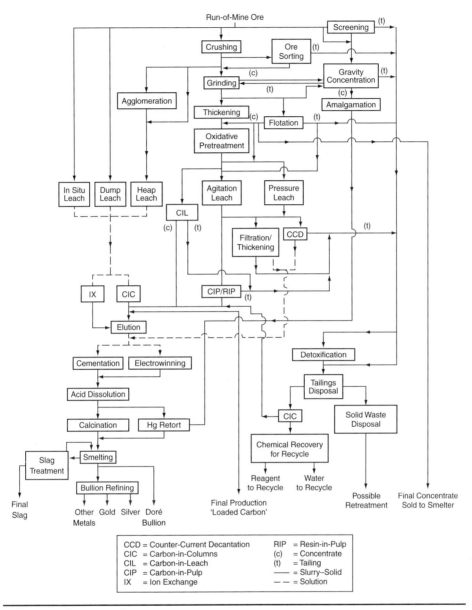

FIGURE 3.4 Process options summary

- Sulfide mineral liberation before flotation
- Gold liberation before flotation and/or gravity concentration
- Optimization of sulfide mineral particle size prior to oxidative pretreatment
- Regrinding of flotation and gravity concentrates or tailings for gold liberation and surface preparation
- Regrinding of roaster calcine for gold liberation and surface preparation
- Ultrafine grinding of gold-bearing sulfide concentrate prior to leaching

Physical aspects of comminution (i.e., procedures and equipment) for treatment of gold ores have been covered comprehensively in the literature [8, 9, 10]. However, comminution processes, and grinding in particular, have several important implications for chemical gold extraction processes and are considered further in the following sections.

3.2.1.1 Mineral Surface Preparation

Due to their high density, gold particles tend to stay in grinding circuits longer than gangue minerals. As a result, and because gold is soft and malleable, particles become flattened, and hard mineral particles (e.g., quartz) may become embedded in gold surfaces. This reduces particle density and hydrophobicity, hindering the response to gravity concentration and flotation, respectively. The problem is usually limited to ores with coarse (>50 µm) gold and may be overcome by the early removal of gold by gravity concentration and/or flotation, within the grinding circuit, if necessary.

Grinding circuits typically consume between 0.5 and 2.5 kg/t iron as grinding media. A portion of this corrodes in the slurry, depending on solution conditions, for example, pH, the presence of cyanide, and other solution species. As a result, gold particles exposed to grinding solutions may become coated with iron oxides, hindering leaching and surface chemical extraction processes. Clay minerals are also known to preferentially coat gold surfaces [11]. In other cases the mineral surfaces may actually be cleaned and fresh surfaces exposed by the grinding action, improving their response to subsequent extraction processes, for example, by regrinding of roaster calcine to expose gold and remove soluble salts [12].

3.2.1.2 Grinding-in-Leach

Grinding mills, slurry pumps, pipelines, and classification equipment can provide good mixing and valuable residence time for leaching reactions, and the addition of cyanide to grinding circuits can achieve >80% gold dissolution for some ores prior to the dedicated leaching stage. This also reduces lockup of coarse gold in grinding equipment.

The major disadvantages of grinding in cyanide solution (the grinding-in-leach process) are that it cannot be used for ores requiring oxidative pretreatment; it increases the amount of iron and iron cyanide complexes in solution; it increases the complexity of metallurgical accounting procedures; and the potential for gold-bearing solution loss as a result of slurry spillage may be increased.

3.2.2 Classification

The most important application of classification in gold extraction flowsheets is the use of cyclones and screens within grinding circuits to optimize grinding efficiency and to obtain the desired particle size for subsequent processing. However, classification may also perform several other important functions:

- Material may be separated based on size for separate treatment in subsequent processes, for example, separate leaching of sand and slime portions of ground slurry.
- Gold adsorbents (i.e., carbon and resins) are separated from slurries and solutions by screening.
- Underground mine backfill preparation.
- Separation of coarse tailings for dam construction.

3.2.3 Solid–Liquid Separation

Solid–liquid separation processes are important in gold extraction flowsheets for the following reasons:

- Gold-rich and barren phases can be separated after leaching, allowing subsequent gold recovery and residue disposal, as applicable.
- Different phases may be treated by various methods for highest process efficiency.
- Chemical equilibria can be shifted to optimize reaction kinetics and thermodynamics.
- Process fluids and reagents may be recycled at various points in the process to optimize water and reagent usage.

The equipment used for solid–liquid separation, particularly thickeners, also provides valuable retention time for chemical reactions. This has been used to good effect in circuits employing counter-current thickening and counter-current filtration in leaching-recovery circuits [13]. The efficiency of solid–liquid separation can also determine the efficiency of subsequent chemical processes, for example:

- Thickener underflow densities determine the residence time of solids in downstream processes (e.g., flotation and leaching).
- Filtration efficiency determines recovery of dissolved gold to the filtrate and affects the gold grade of solid tailings.
- Clarification efficiency determines the effectiveness of Merrill–Crowe zinc precipitation.

The physical separation of solid and liquid phases of a slurry or turbid solution may involve the use of a chemical, or a combination of chemicals, to modify the solution or the surfaces of the solid phase. These include pH modifiers, flocculants, coagulants, and viscosity modifiers. The chemicals used can have a pronounced effect on downstream processes, as discussed in the following sections.

3.2.3.1 pH Modifiers

The modification of pH for solid–liquid separation must take into account the pH requirements of subsequent processes and should try to match these as closely as possible. The most commonly used pH modifiers in gold extraction are calcium hydroxide, sodium hydroxide, and sulfuric acid. The type and concentration of modifier used not only determines the effectiveness of solid–liquid separation but also affects the scale-forming tendency of slurries and solutions downstream. Calcium hydroxide has a greater tendency to form scale than sodium hydroxide, for example, but it is cheaper and has a beneficial coagulating effect on particle settlement, making it more effective for use in solid–liquid separation processes. Sodium hydroxide acts as a dispersant in many solid–liquid systems and can form gelatinous precipitates with silica, which are hard to filter and separate by sedimentation. Also, the use of sulfuric acid for oxidation of calcareous ($CaCO_3$-containing) ores forms gypsum, which can increase slurry viscosities significantly and may form passivating coatings on gold surfaces.

3.2.3.2 Flocculants, Coagulants, Viscosity Modifiers

A variety of organic and inorganic chemicals are commercially available which have very diverse chemical properties. Many of these can adversely affect gold recovery processes by:

- Fouling activated carbon and ion exchange resins
- Fouling precipitation solutions with subsequent loss of precipitation efficiency
- Causing foaming in oxidation processes
- Reducing bacterial activity during biological oxidation

These possible effects should be determined by testing during process development.

3.2.4 Ore Concentration

Ore concentration, or preconcentration as it is often called because of its position ahead of cyanidation in many gold extraction flowsheets, can be used to upgrade ores for one or more of the following reasons:

- To produce a high-grade (gold) concentrate in a small weight fraction of the feed for more economical subsequent treatment
- To reject a portion of the ore that contains no gold in order to reduce the bulk of feed to subsequent processes
- To reject a portion of the ore that is barren but which would otherwise adversely affect subsequent gold extraction, for example, cyanide-consuming sulfide minerals, gold-adsorbing carbonaceous matter, and acid-consuming carbonate constituents

The economic incentive for ore concentration is that the cost savings achieved by treating a smaller amount of material, or by removing deleterious material, is greater than the loss of valuable mineral in the rejected portion. The upgraded fraction is then treated further by various processes, depending on the grade, quantity, mineralogy, and metallurgical properties, as discussed in Section 3.3.

3.2.4.1 Ore Sorting

Ore sorting is the rejection of a barren portion of ore or the acceptance of a gold-rich portion for further treatment. This can either be achieved by manual sorting, based on the visual appearance of the material, or with mechanized ore-sorting equipment, which relies on bulk ore properties such as optical appearance (i.e., color or photometric properties) or radioactivity. Ore sorting has been applied with considerable success to the Witwatersrand pebble–quartz conglomerate ores (South Africa) [1, 14].

3.2.4.2 Gravity Concentration

Gravity concentration is used widely for the recovery of free gold and gold associated with heavier minerals, for example, many sulfide and titanium minerals. A variety of equipment is available for this, and recent developments have enabled the recovery of free gold down to about 10 µm in size. The resulting concentrates may be treated by direct cyanidation, smelting, amalgamation, flotation, or intensive cyanide leaching, depending on their mineralogy.

Gravity concentration techniques evolved significantly during the 1980s and 1990s, largely as a result of the introduction of highly efficient and cost-effective centrifugal concentrating equipment, such as the Knelson and Falcon concentrators [15, 16]. These centrifugal concentrators can be installed in a number of possible configurations in gold extraction flowsheets, to treat the following:

- All or a portion of cyclone underflow slurry in the primary grinding circuit
- All or a portion of a grinding mill discharge
- All or a portion of the cyclone underflow stream in a regrinding circuit (i.e., after rougher flotation)
- Fine portion of an ore feed to a mill, for example, following coarse and fines separation using screening and/or cycloning
- All of the coarse cyclone underflow of a reclaimed tailings stream

In parallel, there have been significant advances in ore characterization and evaluation techniques to predict the response of ores and concentrates to gravity concentration for gold recovery, essential for effective design and operation of gravity concentration equipment in gold process flowsheets. The gravity recoverable gold (GRG) technique, developed at McGill University, is an effective and well-proven technique that measures the natural size distribution of GRG present, addresses the liberation of GRG in a given material, and predicts the maximum amount of gold in such feed material that can potentially be recovered by gravity concentration in a process flowsheet. The method uses a series of tests at progressively finer grind sizes (e.g., 100% <850 μm, 50% <75 μm, and 80% <75 μm) to recover essentially all of the GRG at each size. Most importantly, the test only recovers GRG and not gold present in other forms [17, 18, 19].

The use of centrifugal gravity concentrators in conjunction with flotation where the ore contains a significant proportion of the gold in GRG form (i.e., approximately >40% to 50% GRG at the target grind size) often increases the overall gold recovery by 2% to 5%, depending on the ore mineralogy and the gold particle size. Gravity concentration is particularly useful when there is significant coarse gold present (>250 μm) that is more difficult to recover effectively by flotation. The combination of gravity concentration and flotation is particularly effective for ores containing a wide gold size distribution.

The application of gravity concentration prior to a chemical treatment process (e.g., cyanide leaching) can often be beneficial to overall gold recovery, because the coarse gold particles are recovered prior to leaching and can be treated separately (i.e., by shaking tables, intensive cyanidation, etc.) for maximum recovery. The coarse gold particles take the longest to leach during atmospheric cyanide leaching, and their removal can reduce the leaching retention time and/or increase the overall gold recovery.

3.2.4.3 Flotation

Flotation provides a number of process alternatives for gold ores containing readily floatable minerals, summarized as follows:

- Flotation of free gold and gold-bearing sulfide minerals to produce a gold-rich concentrate. The concentrate can be treated by cyanidation, regrinding and cyanidation, intensive cyanidation, oxidative pretreatment and cyanidation, or by direct smelting.
- Flotation of gold-free sulfide minerals to produce a sulfide-free "tailings" for subsequent cyanidation
- Flotation of carbonaceous material, carbonates, or other material that would otherwise interfere with processing
- Differential flotation, for example, separation of gold, gold-bearing pyrite, arsenopyrite, and pyrite

Many different configurations of flotation circuits utilizing single- and two-stage roughing, regrinding, cleaning, and scavenging have been employed for the recovery of free gold and gold-bearing sulfide minerals. Flash flotation is often an effective way to recover both free gold in the primary grinding circuit before the gold has an opportunity to become overground and flattened, and/or coated with slimes and other products of grinding operations. Flotation is discussed in detail in Chapter 9.

3.2.4.4 Amalgamation

Concerns over the health hazard associated with the use of mercury have greatly reduced the application of amalgamation in the industry. However, it is still used occasionally for the treatment of gravity concentrates because there are few suitable alternatives in some cases. Such applications are frequently found in emerging and/or lesser-developed countries such as Brazil, Colombia, and Indonesia. The amalgam that is produced is retorted and smelted for gold and mercury recovery [1, 8, 9]. Amalgamation is discussed in detail in Chapter 9.

3.2.4.5 Coal–Gold Agglomeration

Coal–gold agglomeration has been developed for the treatment of ores or tailings containing fine free gold which cannot efficiently be recovered by flotation or gravity concentration, and for which cyanidation may be unsuitable for environmental reasons (see Section 9.4).

3.2.4.6 Electrostatic Separation

Electrostatic separation may be used for materials containing gold in a free, or virtually free, state with a nonconductive gangue, for example, crushed slags and other similar refinery by-products. The concentrate produced is either directly smelted, upgraded further by gravity concentration, leached by intensive cyanidation, or a combination of these.

3.2.4.7 Magnetic Separation

Magnetic separation for the concentration of gold–uranium ores and residues of the Witwatersrand has been demonstrated but is not used commercially [20]. High-grade concentrates can be produced from deslimed feed by wet high-intensity magnetic separation, which may then be treated by fine grinding and cyanidation.

3.2.5 Oxidative Pretreatment

Oxidative pretreatment may be required for ores that give poor gold recoveries by conventional leaching or for which reagent consumptions are prohibitively high. This class of ore is frequently termed *refractory* with the extent of refractoriness varying from ore to ore (Chapter 5). Oxidative pretreatment processes either completely or partially oxidize the refractory minerals in the ore, rendering the gold amenable to cyanide leaching. The methods available for oxidation are summarized in Table 5.1.

The degree of oxidation required depends on the ore mineralogy (i.e., the nature of the refractory minerals and the nature of gold mineralization) and the type of oxidation process used. Partial oxidation may be sufficient to passivate the surfaces of refractory minerals, liberate gold associated with a specific mineral, or liberate gold associated at preferential oxidation sites in sulfide minerals. Complete oxidation is usually required when gold is finely dispersed within, or intimately associated with, sulfide minerals.

A slurry preaeration step (see Section 5.2) can be used to oxidize or passivate sulfide minerals that would otherwise react readily in alkaline cyanide solutions, consuming cyanide and oxygen. Without preaeration, such side reactions can significantly reduce the efficiency of gold leaching and increase costs. Ores containing small quantities of pyrrhotite and marcasite are treated successfully by this method.

Roasting has been used to oxidize refractory sulfide, arsenical, carbonaceous, and telluride ores and concentrates for more than 100 years. It can be applied successfully to a wide range of materials, varying greatly in sulfur content and other mineralogical properties. Unfortunately, roasting produces relatively large quantities of gaseous effluents containing a variety of pollutants, for example, sulfur dioxide and arsenic trioxide, which may need to be removed prior to discharging the gas. Increasingly stringent environmental regulations have considerably increased the costs of roaster gas treatment processes in order to ensure compliance with such regulations. This trend is likely to continue in the future, and the application of roasting for the treatment of gold ores and concentrates will probably decline as the associated costs of compliance continue to increase.

Pressure oxidation can also be used to treat various refractory sulfide and arsenical ores and concentrates, but it is generally unsuitable for treatment of carbonaceous material without additional means of reducing the gold-adsorbing properties of these constituents (i.e., by CIL or chlorination). Although the process has relatively high capital and operating costs, it is capable of rapid oxidation of the majority (typically >90%) of sulfide and arsenic minerals in the feed. From an environmental point of view, the process is attractive, because very little noxious gases are produced and any arsenic in the feed can be precipitated as a relatively stable Fe(III) arsenate solid species.

Biological oxidation optimizes the action of naturally occurring bacteria to accelerate sulfide oxidation and has been applied for the treatment of arsenical flotation concentrates (Section 12.2.5.6). The rate of oxidation is relatively slow, compared to pressure oxidation and roasting, and typically 48 to 72 hr of retention time are required to achieve high levels of sulfide oxidation in a well-optimized process. However, the relatively slow kinetics allows partial oxidation of sulfide minerals to occur, which may be desirable for materials in which the gold occurs along fractures or at points of weakness within or between sulfide mineral grains.

Chlorination can be used for passivation of gold-adsorbing, carbonaceous ore constituents prior to cyanide leaching. Such constituents are sometimes called preg-robbing. It has also been used for oxidative leaching of telluride ores, but the method is generally unsuitable for use on high-sulfur materials (i.e., containing greater than approximately 0.5% S to 1.0% S) because of the resulting high chlorine consumptions.

Several nitric acid-based processes have been proposed for the oxidation of sulfide minerals. The oxidation kinetics are fast, but the processes are relatively complex and nitrate ions are introduced into the slurry stream, resulting in some nitrate entrainment in the tailings. This is significant because nitrate species are subject to strict environmental control in many regions of the world.

3.2.6 Leaching

All hydrometallurgical gold extraction routes use a leaching step to produce a gold-bearing solution as an intermediate product. Currently, dilute alkaline cyanide solutions are used exclusively for gold dissolution, although chlorine/chloride media have been used in the past. Other lixiviants, such as thiosulfate, thiocyanate, thiourea, bromide, and iodide solutions are also potential alternatives to cyanide leaching, but none yet has been used commercially. Cyanide leaching can be applied in several forms, summarized as follows:

- Agitated leaching
- Heap or dump (run-of-mine stockpile) leaching
- Vat leaching
- Intensive leaching

Agitated leaching systems are used for the treatment of ground slurries or reclaimed tailings. The product from agitation leaching must either be subjected to one or more stages of solid–liquid separation to allow gold recovery from the solution, or alternatively may be treated "in-pulp" with carbon (CIP) or resin (RIP) for gold recovery. These in-pulp processes can also be incorporated into the leaching circuit for treatment of mildly carbonaceous ores and these configurations are referred to as CIL and resin-in-leach (RIL), respectively.

Heap or dump (stockpile) leaching can be applied to ores where gold occurs in a form that can be at least partially liberated without grinding. The process is performed on crushed or run-of-mine ore and is most suitable for treatment of permeable ore types, although agglomeration processes have been developed to improve the performance of less permeable ores.

Vat leaching is essentially a flooded heap leach with the solution and ore contained within a vessel or other suitable impermeable impoundment. Its application is limited to the leaching of unusual materials that do not respond well to heap or dump leaching but do not require grinding for gold liberation—for example, low-grade oxide and free-milling ores with most of the gold present as liberated, coarse particles. This process is rarely used because of the generally superior economics of heap and agitated leaching systems. A notable exception to this is the Homestake Lead operation (South Dakota, United States) which operated vat leaching throughout the 20th century (see Chapter 12).

Intensive cyanide leaching has been used commercially for the treatment of gravity concentrates containing coarse gold. The leaching kinetics are improved by increasing cyanide and oxygen concentrations and, where necessary, by elevating temperature and pressure.

In situ cyanide leaching of gold ores has been proposed but has not been applied commercially and is not considered to be a viable process option.

Leaching options, including alternative lixiviants, are discussed in detail in Chapter 6.

3.2.7 Solution Purification and Concentration

Solutions produced by leaching typically contain low concentrations of gold, due to the relatively low grade of gold ores. These solutions may be treated directly for gold recovery by a suitable reduction process (see Section 3.2.8), but often the most economic extraction route involves an intermediate concentration step. The choice between these options is discussed in Section 3.3.11. Gold and silver values are adsorbed from the leach solution onto a carrier, such as activated carbon, or, less commonly, ion exchange resin, and then stripped into a smaller volume of "clean" solution. This not only concentrates the gold but also provides an important purification step since it allows:

- Recovery from slurries and unclarified solutions without the need for solid–liquid separation (e.g., in-pulp processing)
- Selective recovery of valuable metals, depending on the carrier used

Activated carbon has been used extensively for concentration and purification of gold leach solutions since about 1980. The major applications are the CIP and CIL processes, which eliminate the need for thickening and/or filtration of leach slurries. Carbon-in-solution (i.e., carbon-in-columns, often abbreviated as CIC or CIS in the industry) has also found wide application for the treatment of leach solutions.

Synthetic ion exchange resins have been used for gold recovery in Uzbekistan for many years and have been applied at Golden Jubilee (South Africa) and Penjom (Malaysia) to replace CIP. Resins will continue to be developed further and will become increasingly important as alternatives to carbon, applied as resin-in-solution (RIS), resin-in-pulp (RIP), or RIL processes.

3.2.8 Recovery

The recovery of gold metal from leach solution, with or without an intermediate concentration and purification stage (see Section 3.3.11), is achieved by reduction processes, either chemically (by zinc precipitation or cementation) or electrolytically (by electrowinning).

3.2.8.1 Dilute Gold Solutions

Leaching processes typically produce low-grade (<3 g/t Au) solutions containing a variety of impurities. Exceptions to this are intensive cyanide leaching and conventional leaching of ores containing approximately 30 g/t Au, or higher, which can produce solution grades >10 g/t Au. Direct recovery of gold from dilute solutions (i.e., no intermediate concentration step) is most effectively achieved by zinc precipitation. Electrowinning is unsuitable because of the poor current-efficiencies obtained from dilute, impure solutions, and the very large size and number of electrowinning cells required to adequately treat large volumes of leach solution.

Direct recovery by zinc precipitation is preferred over carbon adsorption for ores with high silver content (>10:1 Ag–Au ratio) and may have advantages for the treatment of ores with high soluble copper content, that is, above 100 to 200 ppm Cu in solution (see Section 3.3.11).

3.2.8.2 Concentrated Gold Solutions

Carbon- and resin-stripping procedures generate high-grade gold solutions, typically >30 g/t. The purity of these solutions depends on the original leach solution quality and conditions, the selectivity of the carrier (carbon or resin), and the elution procedure. Both electrowinning and zinc precipitation are used for gold recovery from concentrated gold solutions, and there appears to be no clear economic advantage between the two. The choice between these processes is primarily based on the following characteristics:

Electrowinning

- Yields a high-purity product, which requires little refining
- Is easy to operate
- Introduces no zinc impurity into the process
- Typically achieves low single-pass efficiency for equivalent capital expenditure
- Is a relatively clean process

Zinc precipitation

- Achieves high single-pass efficiency for equivalent capital expenditure (compared to electrowinning)
- Precipitates mercury, which can be contained in filtration equipment and recovered as a by-product by retorting
- Generates a lower-purity product than electrowinning (i.e., greater refining requirements)
- Has potentially lower operating labor requirements than electrowinning for large operations
- Introduces zinc species into the system

3.2.9 Refining

The choice of refining method applied at the process site varies greatly according to specific requirements and conditions, such as the following:

- Type of material to be refined
- Size of operation
- Quality of product required for sale
- Availability, proximity, and competitiveness of commercial refineries
- Transportation costs
- Security requirements

In the majority of cases a doré bullion (typically 90% to 99% precious metals) is produced at the mine site, because this can be accurately sampled and weighed for accounting purposes and has a small volume, well suited to secure transportation. Further refining is usually more economically performed by a commercial refiner treating a larger volume of bullion, normally obtained from many suppliers.

The two recovery processes in use yield quite different products which have different fining requirements, summarized in Sections 3.2.9.1 through 3.2.9.3.

3.2.9.1 Zinc Precipitates

The treatment requirements of zinc precipitates depend on their composition, which varies according to the leach solution composition and the method of precipitation. Precipitates obtained directly from dilute leach solutions (Section 3.2.8.1) are typically high in silica and zinc, whereas material produced from concentrated solutions (Section 3.2.8.2) have a lower impurity content.

Generally, precipitates are smelted to produce a bullion product. Often smelting is carried out on-site, but this is not always the case, and high-grade precipitates may be shipped directly to a smelter. Prior to smelting, high zinc materials may need to be treated to avoid excessive fluxing requirements and crucible consumption, although this depends on the smelting method used. This is achieved by leaching with acid (i.e., HCl or H_2SO_4) and/or by calcining to oxidize zinc and other base metals, which are then readily collected in the smelter slag.

Precipitates may require retorting for the removal (and recovery) of mercury prior to any other treatment of the precipitate, since it is a health hazard during refining.

3.2.9.2 Electrowinning Products

Electrowinning products are of three types:

- Foil
- Loaded steel wool (or other type of cathode)
- Cell sludge

Gold foil products are usually of high purity and, depending on the nonprecious metal content, may be shipped directly or melted into bullion buttons or bars at the mine site. Foil-containing contaminants, such as copper, nickel, zinc, and cadmium, may require smelting with fluxes to assist in their removal. Loaded cathodes are either smelted and the iron removed in the slag, or the cathodes are first acid-treated to dissolve excess iron before smelting. Cell sludge is typically less pure than gold foil but is usually amenable to direct smelting without any pretreatment.

3.2.9.3 Carbon

Small operations may not be able to justify the high capital cost of carbon elution, carbon regeneration, and refining processes to produce bullion. In such cases, loaded carbon can be incinerated or ashed, taking appropriate steps to recover any mercury vapor, and the resulting ash smelted with fluxes. This is not an economically attractive option for operations treating more than approximately 0.25 tpd of carbon.

3.2.10 Effluent Treatment

Chemical processes for gold extraction generate a variety of waste products, which must be disposed of, after treatment if necessary, in an economic and environmentally acceptable manner. In some cases, depending on the process characteristics and regulatory requirements, an alternative process that produces acceptable waste products must be developed.

Waste products may be treated to remove or detoxify a particular reagent, and in some cases to recover valuable constituents of the waste stream such as metal values or process reagents. There are two main options for recovering valuable constituents of waste streams, depending on the process method and the nature of the effluent:

1. Recycling all or a portion of the waste stream back to the process following some kind of separation or concentration step
2. Treating all or a portion of the effluent in a dedicated process

The first of these options is common practice in gold extraction flowsheets; for example, by recycling tailing decant solutions to grinding and agitated leaching systems, and by returning barren solution to heap leaching processes. Option 2 is less commonly applied, because of the generally unfavorable economics for metal and reagent recovery processes, and the great majority of effluent treatment schemes are applied primarily for environmental compliance. Specific options for this are considered in Chapter 11 and are not discussed further here.

3.3 FLOWSHEET OPTIONS

A vast number of gold recovery flowsheets are in use around the world for the treatment of a wide range of ores, most of which can be derived from Figure 3.4. In this section, specific flowsheet options are considered for the different ore classifications given in Chapter 2. In a few cases these mineralogical classifications have been combined (e.g., oxidized and free-milling ores, nonrefractory sulfides, etc.), because the approach to process selection for these ores is largely similar. A few general principles apply to the flowsheet alternatives presented in this section:

- Although not shown in each flowsheet, it is generally preferable to recover gold as early as possible in the recovery circuit. This makes downstream portions of the circuit less sensitive to operational fluctuations, reduces the amount of gold held up in the circuit, and often improves overall gold recovery. However, in some cases, it may be undesirable to use concentration techniques (e.g., gravity concentration) because of the security risk of handling high-grade concentrates.
- Gravity concentration should be considered for any ore that contains free gold or gold associated with minerals that are amenable to gravity concentration.
- Alternative (noncyanide) lixiviants are not considered in any of the proposed flowsheet schemes, because these are generally not economically competitive with cyanide leaching.
- The flowsheets described for particular ore types can be combined and modified for ores with more complex mineralogy.
- Selected flowsheets, which are considered to be of particular interest, are discussed in more detail in Chapter 12.

3.3.1 Placers

Placer ores are typically processed by nonchemical, gravity concentration methods, as summarized in Figure 3.5. Often little or no comminution is required except for breaking up of aggregated material. This is then followed by gravity concentration for which a variety of equipment is available. In the past, these methods have only been suitable for the recovery of coarse gold, and the presence of fine gold (<50 μm) reduced recoveries significantly. However, recent equipment developments have improved the recovery of fine gold from sands and gravels, and liberated gold particles of about 10 μm in diameter

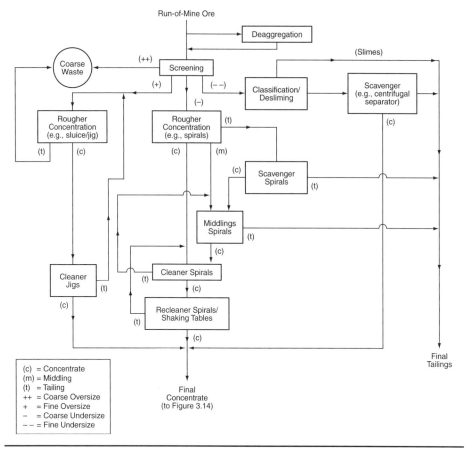

FIGURE 3.5 Flowsheet options for treatment of placer ores

have a reasonable probability of recovery (Section 3.2.4.2). Current trends are to treat large tonnages of low-grade materials, and ores containing as little as 0.1 g/t of recoverable gold have been treated successfully.

Gravity concentration is an environmentally acceptable and effective process that employs relatively simple equipment with few moving parts. The process is therefore often appropriate technology for use in remote areas and undeveloped countries.

Recently there has been considerable investigation into the treatment of placers, which contain fine gold by flotation and/or coal–gold agglomeration processes. Both of these process options have the potential for cost-effective improvement of gold recoveries from placer ores compared to gravity concentration.

3.3.2 Free-Milling and Oxidized Ores

3.3.2.1 Grinding and Agitated Leaching

Grinding to the optimum gold liberation size followed by agitated cyanide leaching of free-milling or oxidized ores is the most common extraction circuit in use (Figure 3.6). Gold recovery from the slurry is usually accomplished either by solid–liquid separation followed by zinc precipitation; CIP followed by electrowinning or zinc precipitation; or a combination of the two, as discussed in Section 3.3.11.

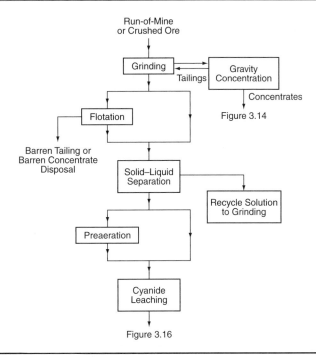

FIGURE 3.6 Flowsheet options for grinding and agitated leaching of free-milling and oxidized ores

Ores containing small quantities of cyanide and oxygen-consuming minerals, such as pyrrhotite and marcasite, may be most economically treated with a preaeration step ahead of leaching. Concentration steps such as gravity concentration and flotation can be included within or immediately after the grinding circuit for ores containing free gold or gold associated with nonrefractory sulfide minerals.

3.3.2.2 Heap Leaching

Low-grade oxidized ores, which could otherwise not be treated by higher-cost grinding and agitation leaching processes, may be effectively treated by heap leaching with cyanide. Some ores require crushing, or crushing and agglomeration, prior to placement on heaps, while others can be treated directly (Figure 3.7). Gold is recovered from heap leaching pregnant solutions either by carbon adsorption and electrowinning/zinc precipitation or by the Merrill–Crowe zinc precipitation process (Section 3.3.11).

The terms *dump* or *stockpile* leaching refer to heap leaching techniques applied to ore at run-of-mine size, as received from the mine.

Heap leaching is less suitable for hard rock, nonoxidized, free-milling ores, for example, Witwatersrand-type ores, as these generally have a finer gold liberation size and are less porous than epithermal and hydrothermal oxidized ores typically encountered in the western United States and Australia. The process is generally unsuitable for ores with more complex mineralogy, although notable exceptions are a number of silver-rich ores (e.g., Coeur-Rochester and Paradise Peak both in Nevada, United States), some mildly refractory sulfidic and carbonaceous ores (Carlin and Jerritt Canyon, both in Nevada, United States), and some copper-bearing ores (Refugio and Marte, Chile). Heap leaching can also be considered for retreatment of old or current nonrefractory tailings.

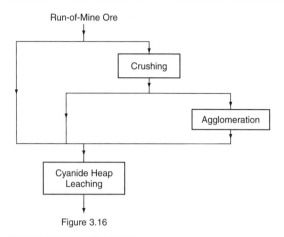

Figure 3.16

FIGURE 3.7 Flowsheet options for heap and dump leaching of free-milling and oxidized ores

3.3.2.3 Combined Heap Leaching, Agitated Leaching, and CIP/CIL

Combined leaching techniques can be applied to the treatment of different fractions of the same ore, either based on the material grade or size. The former is common practice at many plants in the western United States and to a lesser extent in Australia, where high-grade ore is processed by grinding, agitated cyanide leaching, and CIP/CIL, whereas low-grade material is heap or dump leached. Loaded carbon from the two processes can be combined and treated in a common carbon elution-reactivation circuit. An example of a cutoff grade analysis for treatment of a typical Nevada ore by these two processes appears in Figure 3.8. In this example, the cutoff grade between the two processes (obtained from the lower of the two graphs) is approximately 1.4 g/t, with a heap leach/waste cutoff of 0.4 g/t.

Differential treatment based on size may be applicable to ores that are amenable to heap leaching and have a relatively high gold grade, but not sufficiently high to justify grinding, or for which grinding provides little incremental gold recovery. Alternatively, this method may apply where gold is concentrated into the fine size fractions. A size separation is made in the crushing circuit, and the oversize is treated by heap leaching with the leach solution treated by carbon adsorption, as shown in Figure 3.9. The undersize material (e.g., <1 mm) is treated in an agitated leaching-CIP circuit, preceded by grinding if necessary, with subsequent gold recovery from the carbon. This process has been applied at Haveluck (Australia) [21].

The use of the CIP tailings for agglomeration of the coarse, heap leach material, is an interesting variation on this flowsheet [22], and has been applied at Barneys Canyon in Utah (see Section 12.2.4.8).

3.3.2.4 Combined Gold and Uranium Recovery

Uranium-bearing gold ores can be treated by cyanidation for gold extraction followed by sulfuric acid leaching of uranium from the cyanidation residue. Alternatively, cyanide leaching can follow sulfuric acid leaching of uranium in a so-called reverse leaching process. Reverse leaching can yield as much as 2% higher gold recoveries by virtue of the cleaning effect of sulfuric acid on free gold particles and dissolution of carbonates in the ore, which may contain a portion of the gold values [1, 23]. This flowsheet has been used successfully at Hartebeestfontein, Joint Metallurgical Scheme, and West Driefontein (all in South Africa).

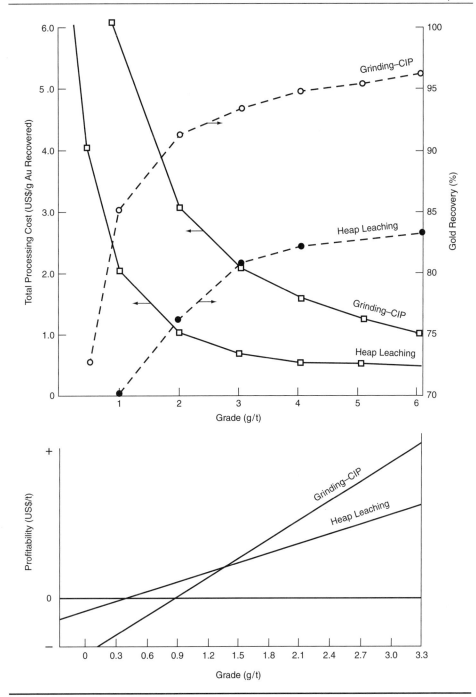

FIGURE 3.8 Example of cutoff grade analysis for processing an ore at identical throughput rates (2,000 tpd) by heap leaching and by grinding followed by CIP

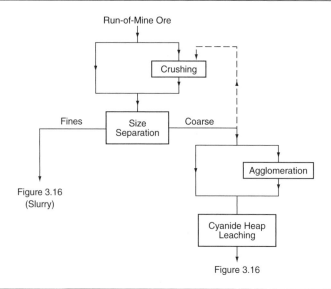

FIGURE 3.9 Flowsheet options for combined heap leaching and agitated leaching of free-milling and oxidized ores

Uranium is recovered from acid leach solution by solvent extraction and precipitation with ammonia to produce ammonium diuranate (yellowcake). The sulfuric acid requirement for uranium leaching can sometimes be produced as a by-product of the gold recovery circuit by roasting of a pyrite flotation concentrate and acid production from the SO_2-rich off-gas, as is the case at several South African gold mines (see Section 5.8.3.8). The roaster calcine may then either be cyanide leached or discarded, depending on the gold grade of the material.

3.3.3 Nonrefractory Sulfidic Gold Ores

Ores that contain sulfide minerals may still yield acceptable gold recoveries (i.e., >90%) by direct cyanidation and are therefore considered to be nonrefractory. In such cases, the gold is not locked in the sulfides and is available for leaching. However, several flowsheet options may be considered (Figure 3.10) that have potential benefits over direction cyanidation.

3.3.3.1 Cyanidation of Whole Ore

The viability of grinding and cyanide leaching of this material will depend on the amount and type of sulfide minerals present and the resulting cyanide, lime, and oxygen consumptions. Preaeration, with or without lead nitrate, may reduce the cyanide and oxygen consumptions. Methods of increasing dissolved oxygen levels in leach slurries, for example, by using pure oxygen or hydrogen peroxide as an oxidant, may also need to be considered to maintain adequate gold dissolution rates. Pure oxygen is used at Rand Mines Milling and Metallurgical Crown Sands' plant in South Africa [1], and hydrogen peroxide is used at Pine Creek (Australia).

Heap leaching may also be applied to this type of material, depending on the nature and amount of sulfides present and their effect on processing costs. In this case, the flowsheet options are identical to those specified in Sections 3.3.2.2 and 3.3.2.3.

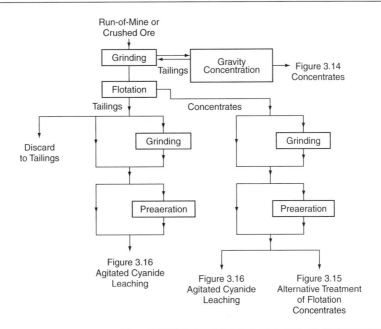

FIGURE 3.10 Flowsheet options for flotation of nonrefractory sulfidic ores

3.3.3.2 Flotation with Cyanidation of the Tailings

This process is applied to ores in which gold is not associated with sulfides, and flotation can be used to produce a barren sulfide concentrate that can either be discarded or sold as a by-product. This reduces the detrimental effect of sulfides on cyanidation. This ore type is rare, unfortunately, as gold has a strong affinity for sulfide minerals, particularly pyrite and arsenopyrite.

3.3.3.3 Flotation with Cyanidation of the Concentrate

Ores that contain free gold and/or gold associated with sulfides and give high gold recovery to a flotation concentrate (the converse of Section 3.3.3.2) may be amenable to this treatment. The flotation concentrate can be reground if necessary, to increase gold liberation, and then cyanide leached. Preaeration may be required, as discussed in Section 3.3.3.1. The flotation tailings are discarded.

A variation that combines this process with that described in Section 3.3.3.2 has been used at Itogon–Suyoc Palidan (Philippines) to recover a chalcopyrite concentrate for sale to a smelter and a separate pyrite concentrate that is directly cyanide leached for gold recovery [13].

In some cases ultrafine grinding of the concentrate may be required to effectively liberate the gold associated with sulfide (and possibly other) minerals. An example of this is at Kalgoorlie Consolidated Gold Mines (Western Australia) where flotation concentrates are subjected to ultrafine grinding to generate a product containing 80% minus 11 to 12 μm, followed by cyanide leaching with CIL recovery to extract >90% of the contained gold values [24].

3.3.3.4 Flotation with Cyanidation of the Concentrate and Tailings

Ores containing gold values distributed between floatable and nonfloatable portions of the ore may be best treated by this method when relatively poor recoveries are obtained to the flotation concentrate. This scheme can offer significant benefits over direct cyanide leaching of the ore because it allows the following:

- Selective grinding of flotation concentrates and direct cyanidation of tailings, which may be required when the gold liberation size is different in the two fractions
- Separate preaeration treatment of two streams
- Different conditions and retention times for leaching of the two fractions
- Different gold (and silver) recovery methods
- Different residue disposal methods
- Differential flotation, which can be used to selectively recover or reject a particular sulfide mineral, for example, flotation of gold-bearing arsenopyrite and depression of barren pyrite

Compared with whole-ore cyanidation, this flowsheet can offer reduced reagent consumptions, better recovery, and lower grinding costs. This has been applied at Paddington (Australia), where the flotation concentrate was reground and cyanide leached with the flotation tailings [25].

3.3.3.5 Flotation of Cyanidation Tailings

Flotation of cyanidation plant tailings may be required when the following occurs:

- Floatable sulfide portion of the tailings contains valuable quantities of gold
- Floatable sulfides are of economic value for reasons other than the gold content, for example, pyrite for sulfuric acid production in South Africa
- Tailings contain floatable carbon, which has valuable gold content

The first two of these are particularly unusual because sulfides are usually detrimental to cyanidation, and it is generally preferable to recover them prior to treatment.

3.3.4 Refractory Sulfidic Gold Ores

The refractory sulfide components of these ores must be oxidized prior to cyanide leaching to achieve acceptable gold recovery. This can be achieved with or without prior concentration by flotation. The flowsheet options are summarized in Figure 3.11.

3.3.4.1 Whole-Ore Sulfide Oxidation and Cyanidation

This is suitable for refractory sulfide ores that are not amenable to concentration processes because of one of the following:

- Gold recovery to the concentrate is unacceptably low and the gold in the tailings is refractory.
- The concentrate produced is less suitable for oxidation than the whole ore, for example, the sulfide sulfur content is too high.

Pretreatment with acid may be required prior to oxidation to neutralize acid consumers. Options available for sulfide oxidation include pressure oxidation, roasting, and biological oxidation (Section 3.2.5). The oxidation products are then neutralized (with or without prior solid–liquid separation) and cyanide leached.

Flotation can be included in this flowsheet as a method of smoothing and optimizing the sulfide content of the feed to the oxidation circuit. Concentrates may be stored and reclaimed as needed to control the feed to the oxidation circuit.

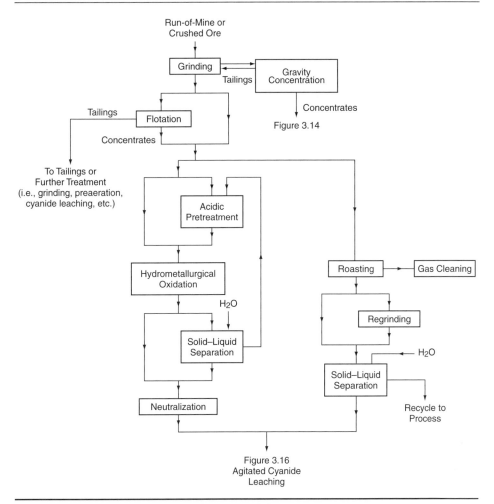

FIGURE 3.11 Flowsheet options for refractory sulfidic ores

3.3.4.2 Flotation with Cyanidation of the Tailings

Certain ores, which are classified as refractory solely as a result of high reagent consumptions associated with their treatment, may be amenable to cyanide leaching following removal of the offending sulfide constituent. The sulfides are removed by flotation and either discarded (if barren or low in gold content) or sold as a by-product.

3.3.4.3 Flotation with Sulfide Oxidation and Cyanidation of Concentrate

Ores that yield a high gold recovery to a flotation concentrate and barren tailings are the most favorable for oxidative pretreatment circuits because the concentrate can be oxidized in a smaller plant than would otherwise be required for whole-ore treatment. This process has the added benefit of potentially allowing the sulfide grade of the concentrate to be controlled within close limits. Flotation parameters can be selected to produce the best possible feed to the oxidation circuit in terms of sulfide content, gold recovery, and volume of material. This flowsheet is shown in Figure 3.11, and the options available for acidic pretreatment, sulfide oxidation, and leaching are identical to those listed in Section 3.3.4.1.

Examples of applications of this type of flowsheet are Sao Bento (Brazil), Fairview (South Africa), and Giant Yellowknife and Campbell Red Lake (both in Canada).

Differential flotation may also be considered to remove barren sulfides from an ore prior to oxidation, in order to reduce the amount of sulfide sulfur that must be oxidized. This can significantly reduce both oxidation and neutralization costs. An example of this is the removal of barren pyrrhotite by flotation followed by secondary flotation to produce a gold-rich arsenopyrite–pyrite concentrate.

Differential flotation has been applied at Harbour Lights (Australia) to separate pyrite and arsenopyrite. The pyrite concentrate was treated by cyanidation, and the arsenopyrite concentrate was directly smelted [25].

Fine regrinding of the concentrate may be beneficial prior to oxidative pretreatment, depending on the ore characteristics and mode of gold occurrence, and the oxidation method used.

3.3.4.4 Flotation with Sulfide Oxidation and Cyanidation of Concentrate and Cyanidation of Tailings

This flowsheet (Figure 3.11) is suitable for ores that yield high sulfide recoveries to a flotation concentrate but relatively poor gold recovery, that is, a significant portion of the gold remains in the flotation tailings. The criteria for production of the flotation concentrate are the same as that specified in Section 3.3.4.3. The oxidation product and the flotation tailings may be cyanide leached separately or together, depending on the mineralogy of the two streams. This flowsheet has been applied at Campbell Red Lake (Canada) historically, where the flotation concentrate was roasted, reground, and cyanide leached, and the flotation tailings were also cyanide leached separately (see Section 12.2.5.2).

Fine regrinding of the concentrate may be beneficial prior to oxidative pretreatment, depending on the ore characteristics and mode of gold occurrence, and the oxidation method used.

3.3.5 Silver-Rich Ores

Ores with high silver contents (>10 to 20:1 silver–gold ratio) generally may be treated similarly to the ore types described in Sections 3.3.2 through 3.3.4, depending on their specific mineralogy. However, some special considerations are required:

- High cyanide concentrations (>0.5 g/L NaCN [sodium cyanide]) are required for leaching, and overall cyanide consumptions are higher than those obtained for leaching of gold ores with low silver content, resulting in a direct cost increase. In addition, the cyanide concentrations in process effluents will also be higher, with a greater likelihood that effluent treatment will be required.
- Pregnant leach solutions produced by silver–gold ores are best treated by solid–liquid separation and Merrill–Crowe zinc precipitation, rather than by CIP. This is because large carbon inventories are required for precious metals recovery, and silver recoveries are relatively poor due to preferential loading of gold on carbon. Carbon stripping and reactivation requirements are also increased due to the greater quantity of carbon that must be transferred.
- A higher silver content requires bigger refining facilities because of the larger bulk of precious metals recovered.

An example of a plant treating a silver-rich ore (Paradise Peak) is given in Section 12.2.3.1. Other examples of high-silver ores are Masbate (Philippines), Rochester, Candelaria, and Tombstone (all in the United States) and La Coipa (Chile).

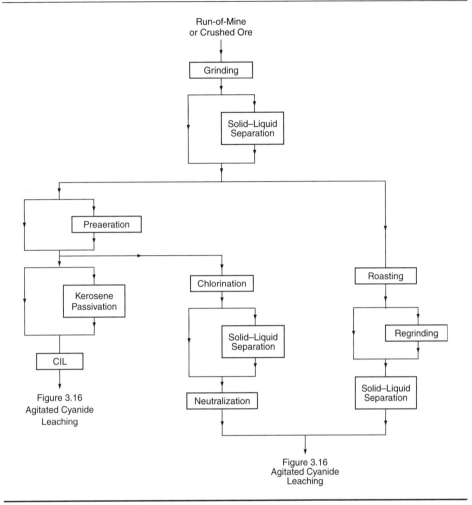

FIGURE 3.12 Flowsheet options for carbonaceous ores

3.3.6 Carbonaceous Ores

These can be divided into two main categories: mildly and highly carbonaceous ores. The two types have different processing requirements (Figure 3.12).

3.3.6.1 Mildly Carbonaceous Ores

Mildly carbonaceous ores contain small quantities of organic or graphitic carbon, typically <1% total organic and graphitic carbon, which adsorb some of the contained gold from solution during leaching. However, this adsorption effect can be reduced by using CIL for simultaneous gold recovery during leaching, or by the addition of a suitable hydrocarbon (e.g., kerosene), which passivates the adsorption surfaces on the carbon. Such ores are sometimes referred to as preg-borrowing, because the adsorption effect can be easily reversed or prevented.

The CIL option was applied specifically for this reason at Mercur (following pressure oxidation of sulfides) and Carlin (both in the United States). In both cases, carbon was added to the slurry as soon as cyanide was added to the circuit. Consequently, this option precludes the use of cyanide in the grinding circuit. Passivation of carbonaceous ore constituents with kerosene has been used at Kerr–Addison (Canada) [13].

3.3.6.2 Highly Carbonaceous Ores

Highly carbonaceous ores contain carbon (typically >1%) in a form that has a strong gold-adsorbing tendency and severely reduces gold extractions to <80% during leaching. The carbonaceous constituents in these ores must generally either be passivated by chlorine treatment (see Section 5.6) or destroyed by roasting (Section 5.8) to enable gold extraction by cyanide leaching. These ores are often called preg-robbing.

Chlorine pretreatment is performed at the optimum grind size for gold liberation. Chlorine and hypochlorite species are either allowed to decay naturally or must be destroyed prior to cyanidation. Roasting is often performed at a coarser size and is typically followed by regrinding of the calcine. Products from both treatments can be cyanide leached (Figure 3.12).

The treatment of carbonaceous ores by chlorination is complicated by the presence of chlorine-consuming sulfides, which increase the cost of the process. Possible flowsheets in these cases are as follows:

- Double-oxidation consisting of sodium carbonate (soda ash) leaching of sulfide-bearing material followed by chlorination and then cyanidation
- Roasting and cyanidation

The double-oxidation process was employed at Carlin and Jerritt Canyon in the 1970s and 1980s, while the second flowsheet has been applied at Jerritt Canyon and Big Springs (both in the United States) since the late 1980s.

A notable exception to the treatment approach described is the process developed to treat Twin Creeks, (Nevada, United States) refractory carbonaceous ore. In this case, ore containing >0.4% organic carbon was found to be highly preg-robbing and could not be treated by conventional means. Acidic pressure oxidation of the material at high temperatures (>210°C), using a feed grind size of 80% −75 µm, gave unacceptably low gold recovery in subsequent cyanidation, due to preg-robbing by organic constituents. However, further testing and development indicated that the oxidation and/or passivation of the organic species in the ore was greatly increased at a feed grind size of 80% <20 to 22 µm using a temperature of 225°C. These conditions oxidized 30% of the organic carbon and 96% to 98% of the sulfide sulfur in 45 to 90 min, resulting in gold recovery of 88% to 90% by subsequent cyanidation [26].

3.3.7 Gold-Telluride Ores

Flowsheet options for telluride ores are illustrated in Figure 3.13. The response of different gold-telluride minerals to cyanide leaching is highly variable and generally not well understood (see Chapter 6); however, some telluride ores can be treated directly by cyanidation.

Many telluride ores respond well to flotation to produce a tellurium-rich concentrate. The concentrate must be oxidized to liberate the contained gold before leaching, and both chlorination and roasting have been used for this purpose commercially at Emperor (Fiji) and Kalgoorlie (Australia). Chlorination is most suitable for ores with low sulfide content (<1% sulfide sulfur) which can be passivated by preaeration, if necessary, to avoid high chlorine consumption.

Tellurium can also be recovered from the oxidized material by sodium sulfide leaching and sodium sulfite precipitation, as practiced at Emperor [13].

Telluride ores and flotation concentrates can be roasted and cyanide leached. This is preferred for high-sulfide materials. Regrinding of flotation concentrates, if required, is

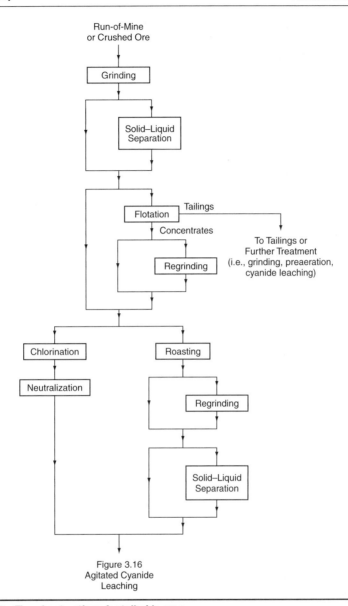

Figure 3.16
Agitated Cyanide
Leaching

FIGURE 3.13 Flowsheet options for telluride ores

usually left until after roasting to better prepare mineral surfaces for cyanide leaching. If economically attractive, the flotation tailings can also be cyanide leached.

3.3.8 Copper–Gold Ores

Gold is produced as a by-product from primary copper ores, especially those in Chile, Indonesia, Canada, and the United States. The major copper minerals in primary deposits are chalcopyrite ($CuFeS_2$) and bornite (Cu_5FeS_4), with minor covellite (CuS), chalcocite (Cu_2S), and tetrahedrite ($Cu_{12}As_4S_{13}$). Some primary ores contain enargite (Cu_3AsS_4) and arsenopyrite (FeAsS). Gold occurs either as free gold, gold associated with copper sulfides (usually in solid solution within chalcopyrite), or as gold associated with iron sulfide minerals (most commonly pyrite, but also with pyrrhotite and other sulfides occasionally). The concentrations of gold in these copper and iron sulfide minerals have been measured in the following ranges [27]:

- Chalcopyrite: 0.01 to 20 g/t
- Tetrahedrite: <0.25 to 59 g/t
- Arsenopyrite: <0.3 g/t to 1.7% by weight
- Pyrite: <0.25 to 800 g/t
- Pyrrhotite: 0.006 to 1.8 g/t

Processing of primary copper–gold ores employs flotation as the primary means of copper and gold recovery. Gravity concentration is often utilized within the grinding circuit if the gold is sufficiently coarse and gives a good response to gravity concentration techniques. Flash flotation is commonly used within the grinding circuit also to recover gold before it has an opportunity to become overground, coated with slimes, and flattened excessively. Rougher and flash flotation concentrates are usually reground and cleaned by flotation, with the objective of maximizing concentrate grade to minimize downstream costs (concentrate transport and smelting) while not significantly compromising metal recovery. Often gravity concentrates are combined with final flotation concentrates; however, sometimes it may be beneficial to leach gravity concentrates on-site by intensive cyanidation (see Section 3.3.9).

Some important copper operations that produce significant gold as a coproduct or by-product include: Freeport-Grasberg and Batu Hijau (both in Indonesia), Escondida, Los Pelambres, and Candelaria (all in Chile), Alumbrera (Argentina), Bingham Canyon (United States), and Olympic Dam and Cadia (both in Australia).

The presence of copper in oxidized ores (Section 3.3.2) presents some specific challenges for processing, considered further in Chapters 6 and 7.

3.3.9 Gravity Concentrates

The options for treating gravity concentrates are summarized in Figure 3.14. Traditionally this material has been treated by amalgamation followed by retorting of the amalgam and smelting of the product. Although this is often still the most cost-effective option, several alternative options should be considered in view of the health and environmental aspects associated with the use of mercury. High-grade gravity concentrates, typically >300 to 500 g/t, can be directly smelted, provided that the material does not contain minerals that adversely affect smelting, for example, sulfide minerals.

Intensive cyanide leaching can be used to dissolve coarse gold in gravity concentrates at fast rates. The chemistry of this process is considered in more detail in Chapter 6. The gold-bearing leach solution can be either treated directly by zinc precipitation or electrowinning, or blended with other process solutions for subsequent gold recovery. This method is being used increasingly in many parts of the world, including South Africa, Australia, and the United States.

Flotation can occasionally be used to upgrade gravity concentrates for direct smelting, but this is rarely applied. The effectiveness of this process depends on the nature of the material and the degree of gold liberation.

3.3.10 Flotation Concentrates

Smelting is an option for treating flotation concentrates; however, the lowest grade that can be considered for smelting is generally between about 300 and 500 g/t, depending on the content of other metal values (e.g., silver, copper, and platinum group metals),

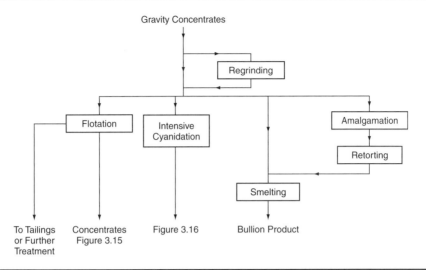

FIGURE 3.14 Flowsheet options for gravity concentrates

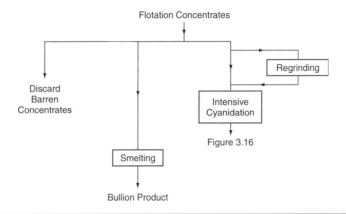

FIGURE 3.15 Flowsheet options for flotation concentrates

the concentrations of any penalty elements, and shipping and transportation distances. Gravity concentration and/or intensive cyanide leaching may be used for concentrates that contain coarse gold (>50 μm). The major options are shown in Figure 3.15. The recovery of gold from copper–gold ores is discussed in Section 3.3.8.

3.3.11 Gold Recovery from Leach Solutions and Slurries

There are three main types of flowsheets for recovering gold from dilute cyanide leach slurries (Figure 3.16):

1. Recovery using activated carbon or resin as an intermediate concentration-purification step
2. Solid–liquid separation and direct recovery from solution by zinc precipitation or electrowinning
3. Combinations of 1 and 2.

104 | THE CHEMISTRY OF GOLD EXTRACTION

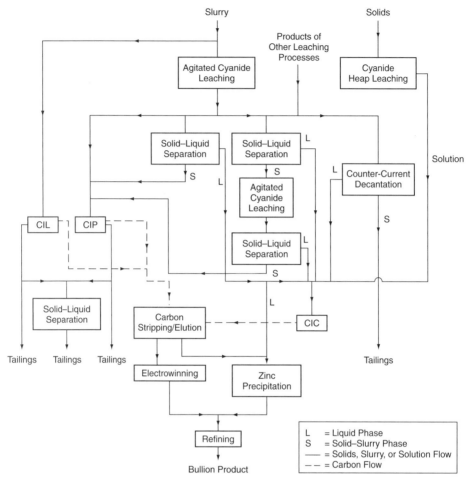

NOTE: RIL, RIP, and RIS can be substituted for CIL, CIP, and CIC, respectively.

FIGURE 3.16 **Flowsheet options for gold recovery from leach solutions**

The type 1 flowsheet is applied widely as an in-pulp or in-leach process. Before the development of CIP, the type 2 circuit was the most widely used option and was applied using either filters or thickeners (i.e., counter-current decantation) for solid–liquid separation followed by zinc precipitation. This method relies on the ability to efficiently wash the slurry in filters or thickeners. Option 3 considers hybrid circuits, which use both solid–liquid separation and concentration techniques; for example, thickening and activated carbon (in-solution and in-pulp). The washing efficiency in the solid–liquid separation stages of these circuits is much less critical than the type 2 flowsheet, and the best features of both flowsheet types can be used to the greatest advantage. These circuits usually have higher capital costs but may achieve better gold recovery, have lower operating costs, and can be less sensitive to process variations [28].

The choice between these options is determined by the economics of each for the entire project life. The cutoff grade for direct recovery versus concentration and purification of solutions is between 6 and 12 mg/L of gold in solution. This value may be lower for smaller operations because of the higher relative cost of carbon stripping and reactivation.

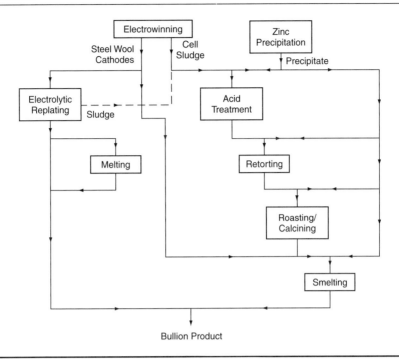

FIGURE 3.17   Flowsheet options for refining

3.3.12 Refining

Flowsheet options for refining are summarized in Figure 3.17 and are discussed in detail in Section 3.2.9 and Chapter 10; they are not considered further here.

3.4 COST CONSIDERATIONS

Relative capital and operating cost factors have been derived for selected process flowsheets, given in Table 3.3. This can be used to calculate approximate, scoping-level capital and operating costs for processes of interest in gold extraction at various throughput rates. In each case an average value and a range have been given, but variations in ore mineralogy and specific project conditions and requirements mean that some industrial processes may fall outside these cost categories. However, the table provides a useful general guide for the comparison of flowsheet options.

The information is based on a combination of operating and engineering data [29 to 33]. Dump leaching of run-of-mine ore was taken as the base case; that is, capital and operating costs = 1.0, assuming a daily throughput rate of 1,000 t. In 2005, 1.0 capital cost unit was valued at approximately US$3.50 million, and 1.0 operating cost unit was valued at US$3.00/t, constant dollar basis (2005).

NOTE: At the time of the preparation of the first edition of this book, in 1990, 1.0 capital cost unit was valued at approximately US$2.6 million, and 1.0 operating cost unit was US$2.60/t money-of-the-day. Capital costs have increased by an estimated 35%, and operating costs have increased by an estimated 15% between 1990 and 2005. The reason for the differential in these cost increases is that inflationary pressures on operating costs have been partially offset by operating efficiency improvements.

TABLE 3.3 Relative capital and operating costs (scoping level) for selected process flowsheets

Flowsheet Option	Relative Cost Factors	
	Capital	Operating
DL, CIC, EW	1.0	1.0
HL (CC), CIC, EW	2.2 (2.0–2.5)	1.9 (1.6–2.1)
HL (MC), CIC, EW	2.8 (2.6–3.1)	1.9 (1.7–2.2)
HL (MC), CIC, Zn	2.9 (2.5–3.2)	1.9 (1.7–2.2)
HL (FC), AGG, CIC, EW	3.9 (3.5–4.5)	2.4 (2.2–2.6)
GR, AL, CIP, EW	6.4 (5.6–8.5)	3.0 (2.8–3.6)
GR, AL, CCD, Zn	8.3 (7.0–9.8)	3.5 (3.0–4.0)
GR, FL, AL, CIP(t), SC	6.7 (5.9–8.5)	3.2 (2.9–3.8)
GR, PO, AL, CIP, EW	13.1 (11–15)	8.5 (8.0–10.0)
GR, RO, AL, CIP, EW	10.8 (9.5–12)	7.5 (7.0–8.5)
GR, BIO, AL, CIP, EW	11.5 (11–12)	9.0 (8.0–10.0)
GR, FL, AL, CIP(c)	4.3 (4.0–5.0)	2.7 (2.5–3.5)
GR, FL, PO, AL, CIP(c), EW	6.7 (6.0–7.5)	5.3 (4.8–6.0)
GR, FL, RO, AL, CIP(c), EW	6.3 (5.6–7.0)	4.1 (3.8–4.7)
GR, FL, BIO, AL, CIP(c), EW	6.0 (5.5–6.5)	6.0 (5.0–6.5)

AGG	Agglomeration		FL	Flotation
AL	Agitation leach		GR	Grinding
BIO	Bio-oxidation		HL	Heap leach
CC	Coarse crush		MC	Medium crush
CCD	Counter-current decantation		PO	Pressure oxidation
CIC	Carbon-in-columns		RO	Roasting
CIP	Carbon-in-pulp		SC	Sale of concentrate
DL	Dump leach		Zn	Zinc precipitation
EW	Electrowinning		(c)	Concentrates
FC	Fine crush		(t)	Tailings

NOTES: 1. Based on 1,000-tpd plants.
2. Numbers in parentheses indicate range of values, while number outside the parentheses is the most likely value.

Other assumptions for these cost estimations are summarized as follows:

- The plant is located in central Nevada, with electricity and water available at the site.
- Costs include plant and ancillary facilities but exclude tailings disposal and effluent treatment.
- Grinding, where applicable, generates a product with a P_{80} = 75 µm. The ore work index is 14 kWh/t. Power cost is US$0.05/kWh.
- Heap and agitated leaching are given, respectively, 45 days and 24 hr leaching time.
- Cyanide consumption is 0.5 kg/t for agitated leaching and 0.25 kg/t for heap leaching. Cyanide cost is US$1.00/kg.
- Flotation, where applied, achieves a concentration ratio of 10:1.
- Ores treated by oxidation contain 2% sulfide sulfur. Biological oxidation achieves 80% sulfur oxidation in 48 hr, and both pressure oxidation and roasting achieve 97% sulfur oxidation within practical time scales.

Table 3.4 contains cost adjustment factors that can be used to scale up the factors in Table 3.3 for higher plant throughputs. These provide relative and approximate indications of treatment costs of the various processes considered in this chapter in order to assist with process selection.

TABLE 3.4   Cost adjustment factors to adjust data in Table 3.3 for various treatment rates

Throughput (tpd)	Capital Cost Adjustment Factor	Operating Cost Adjustment Factor
1,000	1.00	1.00
2,000	1.54	0.98
4,000	2.34	0.92
8,000	3.58	0.81
16,000	5.38	0.67
32,000	8.10	0.50

REFERENCES

[1]   Bosch, D.W. 1987 Retreatment of residues and waste rock. Pages 707–743 in *The Extractive Metallurgy of Gold in South Africa*. Volume 2. Edited by G.G. Stanley. Johannesburg: South African Institute of Mining and Metallurgy.

[2]   Zaburunov, S.A. 1989. Ridgeway: Gold deposit producing big returns through good design. *Engineering and Mining Journal* (August):52–55.

[3]   Foo, K.A., and M.D. Bath. 1989. New gold processing techniques: An engineer's perspective. Pages 233–250 in *Gold Forum on Technology and Practices—World Gold '89*. Edited by R.B. Bhappu and R.J. Harden. Littleton, CO: SME-AIME.

[4]   Murray, S., K. Crisp, P. Klapwijk, and T. Sutton-Pratt. 1990. *Gold 1990*. London: Gold Fields Mineral Services Ltd.

[5]   Green, T. 1984. *The New World of Gold.* New York: Walker & Co.

[6]   Green, T. 1987. *The Prospect for Gold: The View to the Year 2000*. New York: Walker & Co.

[7]   Newcomb, R., and K. Tsuji. 1990. The price of gold. Pages 1–42 in *Gold: Advances in Precious Metals Recovery*. Edited by N. Arbiter and K.N. Han. New York: Gordon & Breach Science Publishers.

[8]   Mular, A.L., and R.B. Bhappu, editors. 1980. *Mineral Processing Plant Design*. 2nd edition. New York: SME-AIME.

[9]   Adamson, R. J. 1972. *Gold Metallurgy in South Africa*. Johannesburg: Chamber of Mines of South Africa.

[10] Weiss, N.L., editor. 1985. *SME Mineral Processing Handbook*. Volumes l and 2. Littleton, CO: SME.

[11] Baum, W. 1988. Mineralogy-related processing problems of epithermal gold–silver ores. Pages 3–20 in *Process Mineralogy VIII*. Edited by D.J.T. Carson and A.H. Vassilon. Warrendale, PA: TMS.

[12] Baum, W., J.O. Sanhueza, E.H. Smith, and W. Tufar. 1989. The use of process mineralogy for plant optimisation at the El Indio gold–silver–copper operation (Chile). *Erzmetall* 42(9):373–378.

[13] McQuiston, F.W., and R.S. Shoemaker. 1975. *Gold and Silver Cyanidation Plant Practice*. Volume 1. SME-AIME Monograph. Salt Lake City, UT: SME-AIME.

[14] Schaffler, M.J. 1988. Combined radiometric-optical sorter and sorting at Buffelsfontein gold mine. *Mine Metallurgical Managers' Association of South Africa* 2(88): 81–106.

[15] Van Kleek, D.M. 2000. Knelson concentrators extreme gravity. Pages 277–281 in *Proceedings of the 7th Mill Operators' Conference*. Melbourne: Australasian Institute of Mining and Metallurgy.

[16] Goulsbra, A., R. Dunne, and S. McAllister. 1998. The application of the continuous Falcon centrifugal gravity concentrator for gold and pyrite recovery. Pages 111–114 in *Proceedings of the Randol Gold & Silver Forum*. Golden, CO: Randol International Ltd.

[17] Laplante, A.R. 2000. Testing requirements and insight for gravity gold circuit design. Pages 73–84 in *Proceedings of the Randol Gold & Silver Forum*. Golden, CO: Randol International Ltd.

[18] Laplante, A.R. 2000 Ten do's and don'ts of gold gravity recovery. Pages 107–118 in *Proceedings of the Randol Gold & Silver Forum*. Golden, CO: Randol International Ltd.

[19] Laplante, A.R. 1995. Predicting gold recovery by gravity. Pages 19–26 in *Proceedings of the XIX International Mineral Processing* Congress. Littleton, CO: SME.

[20] Corrans, I.J., and J. Levin. 1979. Wet high intensity magnetic separation for the concentration of Witwatersrand gold–uranium ores and residues. *Journal of South African Institute of Mining and Metallurgy* 79(8):210–228.

[21] Geldard, D., and L.D. Mann. 1982. The introduction and application of carbon-in-pulp in a heap leach project. Pages 279–288 in *Carbon-in-Pulp Technology for the Extraction of Gold*. Melbourne: Australasian Institute of Mining and Metallurgy.

[22] McGregor, J.P., and G.E. McClelland. 1989. Agglomeration with pulp—a concept to improve the economics of heap leaching. Pages 247–250 in *Proceedings Randol Gold Conference*. Golden, CO: Randol International Ltd.

[23] Wendel, G. 1986. Uranium leach modelling at Hartebeestfontein gold mine. *Mine Metallurgical Managers' Association of South Africa* 2(86):61–68.

[24] Ellis, S. 2003. Ultra fine grinding—a practical alternative to oxidative treatment of refractory gold ores. Pages 11–17 in *Proceedings Eighth Mill Operators' Conference*. Melbourne: Australasian Institute of Mining and Metallurgy.

[25] Avraamides, J.A. 1989. Changing trends in gold ore treatment in Western Australia: The problem of refractory ores. Pages 333–336 in *Proceedings of World Gold '89*. Littleton, CO: SME-AIME.

[26] Simmons, G. 1996. Pressure oxidation process development for treating carbonaceous ores at Twin Creeks. Pages 199–208 in *Proceedings Randol Gold Forum '96*. Golden, CO: Randol International Ltd.

[27] Cabri, L.J. 1992. The distribution of trace precious metals in minerals and mineral products. *Mineralogical Magazine* 56:289–308.

[28] Mansanti, J.G., J.R. Arnold, J.H. Gourdie, and J.O. Marsden. 1989. Double thickener circuit at Gold Fields' Chimney Creek. *Mineral & Metallurgical Processing Journal* (November):179–186.

[29] Carter, R.W., and J.E. Litz. 1988. Comparative economics of refractory gold ore treatment processes. *Proceedings Randol Perth International Gold Conference*. Golden, CO: Randol International Ltd.

[30] Wells, J.A. 1988. A generic study of the capital and operating costs for the recovery of precious metals from refractory ore. Paper presented at Canadian Mineral Processors Conference, Ottawa, Canada, January.

[31] Bhappu, R.B. 1990. Hydrometallurgical processing of precious metal ores. Pages 67–80 in *Gold: Advances in Precious Metals Recovery*. Edited by N. Arbiter and K.N. Han. Littleton, CO: SME.

[32] DeMent, E.R., and N.D. King. 1982. Merrill–Crowe/carbon-in-pulp: An economic evaluation. Pages 89–106 in *Carbon-in-Pulp Technology for the Extraction of Gold*. Melbourne: Australasian Institute of Mining and Metallurgy.

[33] Haines, A.K., and P.C. Van Aswegan. 1990. Process and engineering challenges in the treatment of refractory gold ores. Pages 103–110 *in International Deep Mining Conference–Innovations in Metallurgy Plant Design*. Johannesburg: South African Institute of Mining and Metallurgy.

CHAPTER 4

Principles of Gold Hydrometallurgy

Hydrometallurgical processes are the reactions used to extract, purify, and recover minerals and metals in aqueous systems. The majority of gold extraction flowsheets use hydrometallurgical techniques, the most important of which are leaching, solution purification and concentration, and recovery of gold. Depending on the ore type, hydrometallurgical pretreatment methods such as chlorination, pressure oxidation, and biological oxidation may also be used to increase gold extraction in the subsequent leaching stage by liberating contained gold and by converting interfering constituents into less reactive forms (see Chapter 5).

The two major considerations in hydrometallurgical processes are the extent to which a reaction will proceed and the reaction rate. The former depends on the thermodynamic properties of the chemical system, which determine the overall reaction driving force. The latter, the reaction kinetics, depends on a combination of physical, chemical, and mass transport factors, which can be controlled to some extent by appropriate process selection and plant design.

The most important chemical reactions in gold extraction are those involving gold itself, other metals of value (e.g., silver or platinum group metals), and side reactions involving gangue minerals. These side reactions may result in increased reagent consumption, the dissolution of species which may adversely affect subsequent processes (e.g., copper, mercury), and precipitation of species from solution, all of which can affect the overall gold extraction efficiency. Various chemical reaction types, summarized in Figure 4.1, may be used in the overall gold extraction flowsheet.

Hydrometallurgical extraction processes are complex because of the diversity of species present in ores and in the resulting process solutions. Consequently, general principles, such as theoretical thermodynamics and kinetics, usually provide only an approximation of actual reaction conditions. Nevertheless, these topics are essential to the understanding of gold extraction chemistry. The principles of hydrometallurgical processes relevant to gold extraction are introduced in this general chapter, with subsequent chapters covering each major hydrometallurgical topic.

4.1 REACTION CHEMISTRY OF GOLD

Gold is very stable, as indicated by its lack of reactivity in air and in the majority of aqueous solutions, including strong acids. Gold only dissolves in oxidizing solutions containing certain complexing ligands, for example, cyanide, halides, thiosulfate, thiourea, and thiocyanate. This unique behavior allows gold to be extracted very selectively from ores—a necessity since gold ores are generally low grade.

Gold is classified, with copper and silver, in group IB of the periodic table. Despite the similarity of these metals in electronic structure and ionization potential—the work that must be done to remove an electron from an atom—there are many important differences in their redox chemistry. Much of the chemistry of gold and its compounds, especially its

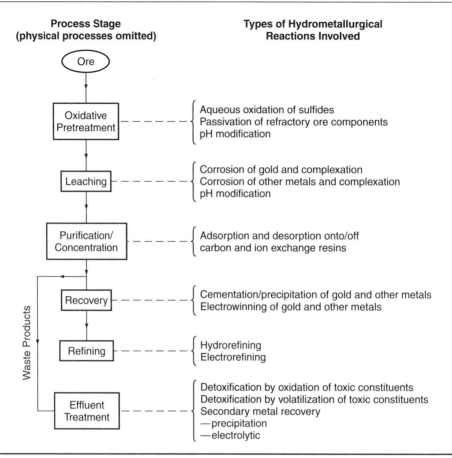

FIGURE 4.1 Reaction chemistry of gold

behavior in aqueous solution, can be related to its relatively high electronegativity, that is, the tendency to attract bonding electrons (see Section 4.1.2).

Gold compounds and solution species exist almost exclusively in the Au(I) and the Au(III) oxidation states. Several Au(II) species have been identified in solution [1], but these are usually transient and are of academic interest only.

4.1.1 Gold–Water Reactions

Gold is unreactive in pure water and over a very wide pH range. At pH <1 and in very oxidizing conditions (solution potentials >1.4 V), Au(III) ions may be formed in solution, as represented by the half-reaction:*

$$Au^{3+} + 3e \rightleftharpoons Au^0; \quad E = 1.52 + 0.0197 \log [Au^{3+}] \text{ (V)} \tag{EQ 4.1}$$

where
  E = the reduction potential defined by the Nernst equation (4.15)
  1.52 V = the standard reduction potential (E^0)

---

* All equations representing electrochemical reactions in this book have been expressed as reversible reactions ($\rightleftharpoons$) with the species of interest shown in its oxidized state on the left and the reduced state on the right.

Au(I) is unstable in water under all potential–pH conditions, as indicated by the half-reaction:

$$Au^+ + e \rightleftharpoons Au^0; \quad E = 1.83 + 0.0591 \log [Au^+] \text{ (V)} \qquad \text{(EQ 4.2)}$$

This reaction has a higher E^0 value than Au(III) reduction, and consequently gold is rapidly oxidized to Au(III) without Au(I) forming to any significant extent. Between approximately pH 1 and 13, and at very high potentials, insoluble Au(III) hydroxide may be formed:

$$Au(OH)_3 + 3H^+ + 3e \rightleftharpoons Au^0 + 3H_2O; \quad E = 1.457 - 0.0591 \text{ pH (V)} \qquad \text{(EQ 4.3)}$$

At pH >13 and above approximately 0.6 V the soluble hydroaurate species, $HAuO_3^{2-}$ (or $Au(OH)_5^{2-}$), is formed. Reactions (4.1) to (4.3) all occur at potentials greater than that required for the reduction of oxygen:

$$O_2 + 4H^+ + 4e \rightleftharpoons 2H_2O; \quad E^0 = 1.229 \text{ (V)} \qquad \text{(EQ 4.4)}$$

This accounts for the stability of gold in aqueous solution. Other oxidizing agents such as nitric, sulfuric, and perchloric acids are also ineffective at dissolving gold in the absence of complexing ligands.

4.1.2 Gold Complexes

The stability of gold is reduced in the presence of certain complexing ligands, such as cyanide, chloride, thiourea, thiocyanate, and thiosulfate ions, by the formation of stable complexes. As a result, gold can be dissolved in relatively mild oxidizing solutions, for example, aerated, aqueous cyanide solutions. In this case, gold oxidation is accompanied by the reduction of dissolved oxygen, for example, Equation (4.4), which provides an essential driving force for the dissolution, as discussed under leaching (see Chapter 6).

Although gold forms stable complexes with chloride ions, hydrochloric acid (HCl) alone is not a sufficiently strong oxidant to dissolve gold. Mixtures of either hydrochloric acid and nitric acid (aqua regia), or hydrochloric acid and chlorine, are required to provide both strongly oxidizing and complexing components.

Species that form stable gold complexes are listed in Table 4.1. Their stability constants show that some complexing ligands form more stable complexes with Au(I) and others with Au(III). This preferred oxidation state is related to the electron configuration of the donor (or complexing) ligand. The so-called soft electron donor ligands, such as cyanide, thiourea, thiocyanate, and thiosulfate, prefer metal ions of low valency (e.g., Au(I)), whereas hard electron donor ligands, such as chlorine and the other halides, prefer high-valency metal ions, for example, Au(III).

The preferred coordination numbers of 2 and 4 for Au(I) and Au(III) complexes result in linear and square planar structures, respectively. In water, Au^+ and Au^{3+} ions exist in a hydrated state, as illustrated in Figure 4.2. The geometry of the other complexes formed are similar, with ligands replacing the water molecules. Complexes and compounds with fewer, or excess, ligands may dimerize to satisfy the preferred coordination number, as shown in Figure 4.3.

4.2 CHEMICAL EQUILIBRIA

4.2.1 Definition of Equilibrium

Chemical equilibrium is the point in a chemical reaction at which there is no further change in the concentration of the relevant ionic and molecular species. The solution

114 | THE CHEMISTRY OF GOLD EXTRACTION

TABLE 4.1   Stability constants for selected Au(I) and Au(III) complexes [2]

Ligand	Au(I), β_2	Au(III), β_3
CN^-	2×10^{38}	10^{56}
SCN^-	1.3×10^{17}	10^{42}
$S_2O_3^{2-}$	5×10^{28}	—
Cl^-	10^9	10^{26}
Br^-	10^{12}	10^{32}
I^-	4×10^{19}	5×10^{47}
$CS(NH_2)_2^+$	2×10^{23}	—

NOTE: Dashes = not applicable.

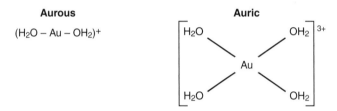

Aurous

$(H_2O - Au - OH_2)^+$

Auric

$\left[\begin{array}{c} H_2O \quad\quad OH_2 \\ Au \\ H_2O \quad\quad OH_2 \end{array} \right]^{3+}$

Preferred Gold–Cyano Complex:

$(N \equiv C - Au - C \equiv N)^-$

$Au(CN)_2^-$

Preferred Gold–Chloro Complex:

$\left[\begin{array}{c} Cl \quad\quad Cl \\ Au \\ Cl \quad\quad Cl \end{array} \right]^-$

$AuCl_4^-$

FIGURE 4.2   Geometry of hydrated and complexed aurous and auric ions in aqueous solution [2]

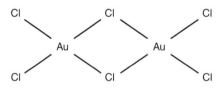

FIGURE 4.3   Geometry of solid Au(III) chloride, dimerized to satisfy the preferred coordination number 4

composition under these conditions can be described using an equilibrium constant (K), which is a ratio relating the activities of the reaction products and reactants. For example, the equilibrium established between two reactants (A and B) and two products (C and D) in aqueous solution can be expressed as follows:

$$wA + xB \rightleftharpoons yC + zD \qquad (EQ\ 4.5)$$

$$K = \frac{a_C^y a_D^z}{a_A^w a_B^x} \qquad (EQ\ 4.6)$$

where
$w, x, y,$ and z = the reaction stoichiometry
a_{A-D} = the activities of the species A–D, respectively

The larger the value of the equilibrium constant, the further the reaction will proceed toward completion, forming C and D. The factors that affect the absolute and relative concentrations of each species are:

- Value of the equilibrium constant (K)
- Total concentration of metal species
- Ligand concentration
- Metal–ligand concentration ratio
- Activity coefficients of all species

The energy change associated with a reaction determines whether, and in which direction, it will proceed. This can be expressed as the Gibbs free energy change (ΔG^0 kJ/mol) for the reaction, derived from standard enthalpy and entropy data:

$$\Delta G = \Delta H - T\Delta S \qquad (EQ\ 4.7)$$

where
ΔH = the enthalpy change (kJ/mol)
T = the temperature (K)
ΔS = the entropy change (kJ/mol K)

At equilibrium, the Gibbs free energy is minimized and it can be shown that [3]:

$$\Delta G = \Delta G^0 + RT \ln K \qquad (EQ\ 4.8)$$

This equation is known as the Van't Hoff isotherm. For a particular reaction, for example, Equation (4.5), ΔG^0 is constant and can be calculated from:

$$\Delta G^0 = yG^0 + zG^0 - wG^0 - xG^0 \qquad (EQ\ 4.9)$$

If ΔG is negative, the reaction proceeds spontaneously from left to right, whereas if it is positive, the reverse reaction occurs. Equilibrium is achieved when $\Delta G = 0$, and from Equation (4.8) this occurs when:

$$\Delta G^0 = -RT \ln K \qquad (EQ\ 4.10)$$

Therefore, the equilibrium constant for a given reaction can be derived from free energy calculations using actual or approximated activity data for the various species involved.

Thermodynamic data are available for standard conditions at 25°C for most species encountered in gold hydrometallurgy (see Table 4.2 for gold species). However, various

TABLE 4.2 Selected thermodynamic data for gold species [5]

Formula	State	ΔH^0 (kJ/mol)	ΔG^0 (kJ/mol)	S^0 (J mol^{-1} K^{-1})
Au	g*	366	326	180.39
Au	c†	0	0	47.40
Au$^+$	g	1,262.4	—	—
Au$^+$	aq‡	—	176	—
Au^{2+}	g	3.247×10^3	—	—
Au^{3+}	Aq	—	440	—
AuO$_3^{3-}$	Aq	—	51.9	—
HAu$_3^{2-}$	Aq	—	−142	—
H$_2$AuO$_3^-$	Aq	—	−218	—
Au(OH)$_3$	c	−424.7	−317	190
Au(OH)$_3$	aq	—	−283.5	—
AuCl	c	−35	—	—
AuCl$_2^-$	aq	—	−151.0	—
AuCl$_3$	c	−118	—	—
AuCl$_3$·2H$_2$O	c	−715.0	—	—
AuCl$_4^-$	Aq	−322	−234.6	267
AuBr	c	−14.0	—	—
AuBr$_2^-$	aq	−128	−115.0	220
AuBr$_3$	c	−53.26	—	—
AuBr$_3$	Aq	−39.3	—	—
AuBr$_4^-$	Aq	−192	−167	336
HAuBr$_4$·5H$_2$O	c	−1,668	—	—
AuI	c	0	−0.5	—
AuI$_2^-$	aq	—	−47.6	—
AuI$_4^-$	Aq	—	−45	—
Au(CN)$_2^-$	Aq	242	286	1.7×10^2
Au(SCN)$_2^-$	Aq	—	252	—
Au(SCN)$_4^-$	Aq	—	561.5	—

* g = gas.
† c = solid compound.
‡ aq = aqueous species.

reactions, for example, pressure leaching, are performed at elevated temperatures for which experimental enthalpy, entropy, or free energy values have not been determined. An approximation technique has been developed by, and named after, Criss and Cobble, which allows high-temperature entropy values to be estimated for various classes of ions in solution [4]. This allows free energy data to be calculated and applied to practical situations, for example, for the derivation of high-temperature E_h–pH diagrams.

4.2.2 Electrochemical Considerations

For electrochemical reactions, that is, those involving oxidation or reduction by electron transfer, the Equation (4.10) can be expressed in terms of electrode potentials. For example, the generalized electrochemical reduction is as follows:

$$M^{n+} + e \rightleftharpoons M^{(n-1)+} \qquad \text{(EQ 4.11)}$$

As for Equation (4.5), the equilibrium constant is the ratio of the activities of products divided by the activities of reacting species to the appropriate powers and can be expressed as follows:

$$K = \frac{a_M^{(n-1)+}}{a_M^{n+}}$$ (EQ 4.12)

The free energy is related to the electrode potential by:

$$\Delta G = -nFE$$ (EQ 4.13)

and if all species are in their standard states, then:

$$\Delta G^0 = -nFE^0$$ (EQ 4.14)

where
  n = the number of electrons transferred
  F = the Faraday constant, a convenient unit of charge (96,487 coulombs)

Hence it follows from Equation (4.8) that:

$$E = E^0 - (RT/nF)\ln\frac{[a_{\text{reduced states}}]}{[a_{\text{oxidized states}}]}$$

or

$$E = E^0 - \frac{RT \ln a_M^{(n-1)+}}{nF \ln a_M^{n+}}$$ (EQ 4.15)

Equation (4.15) is known as the Nernst equation, where R is the universal gas constant (8.3141 mol^{-1} K). It follows that the value of the coefficient RT/nF is 0.0591 V for a 1-electron ($n = 1$) reduction at 298 K.

Values of electrode potentials (E^0) can be calculated from standard free energy data. Values for selected half-reactions are listed in Tables 4.3 and 4.4. The Nernst equation (4.15) can be used to calculate electrode potentials in aqueous solutions at equilibrium. For example, in a solution containing only iron cations, if the concentrations of Fe(II) and Fe(III) are 0.001 and 0.002 M, respectively, then the solution potential may be estimated as follows:

$$E = 0.771 - RT/F \ln (0.001/0.002) = 0.812 \text{ (V)}$$ (EQ 4.16)

Conversely, Equation (4.15) enables solution activities to be estimated for known solution pH and potentials. Values of E may be measured against various reference electrodes, although for consistency the hydrogen electrode has been adopted as a standard, hence the potential is known as E_h. An example of an electrochemical reaction is the deposition of gold onto zinc (Chapter 8), which can be represented by two half-reactions:

Cathodic: $2Au(CN)_2^- + 2e \rightleftharpoons 2Au + 4CN^-$ $E^0 = -0.67$ (V)   (EQ 4.17a)

Anodic: $Zn(CN)_4^{2-} + 2e \rightleftharpoons Zn + 4CN^-$ $E^0 = -1.26$ (V)   (EQ 4.17b)

Overall: $2Au(CN)_2^- + Zn \rightleftharpoons 2Au + Zn(CN)_4^{2-}$ (EQ 4.17c)

TABLE 4.3 Summary of standard potentials for selected gold couples in aqueous solutions [5, 22]

Half-Reaction	Standard Potential, E^0 (V)
Au(I)/Au(0)	
$Au^+ + e \rightarrow Au$	1.68 to 1.83
$Au(CN)_2^- + e \rightarrow Au + 2CN^-$	−0.65 to −0.57
$AuCl_2^- + e \rightarrow Au + 2Cl^-$	1.154
$AuBr_2^- + e \rightarrow Au + 2Br^-$	0.960
$AuI_2^- + e \rightarrow Au + 2I^-$	0.578
$AuI + e \rightarrow Au + I^-$	0.530
$Au(SCN)_2 + e \rightarrow Au + 2SCN^-$	0.604 to 0.662
Au(III)/Au(0)	
$Au^{3+} + 3e \rightarrow Au$	1.42 to 1.52
$AuCl_4^- + 3e \rightarrow Au + 4Cl^-$	0.994 to 1.002
$AuBr_4^- + 3e \rightarrow Au + 4Br^-$	0.854
$AuI_4^- + 3e \rightarrow Au + 4I^-$	0.56
$Au(SCN)_4^- + 3e \rightarrow Au + 4SCN^-$	0.636
Au(III)/Au(I)	
$Au^{3+} + 2e \rightarrow Au^+$	1.36
$AuCl_4^- + 2e \rightarrow AuCl_2^- + 2Cl^-$	0.926
$AuBr_4^- + 2e \rightarrow AuBr_2^- + 2Br^-$	0.802
$AuI_4^- + 2e \rightarrow AuI_2^- + 2I^-$	0.55
$Au(SCN)_4^- + 2e \rightarrow Au(SCN)_2^- + 2SCN^-$	0.604 to 0.623

For the overall cementation reaction:

$$E_{cell} = E_{cathodic} - E_{anodic} \quad \text{(EQ 4.18)}$$
$$E_{cell} = -0.67 - (-1.26) = 0.59 \text{ (V)}$$

From Equation (4.14):

$$\Delta G^0 = -nFE^0 = -2 \times 96{,}487 \times 0.59 = -114 \text{ kJ/mol}$$

The negative free energy change indicates that gold deposition using zinc is thermodynamically favorable. The reaction would be even more favorable if there was a metallic zinc concentration in excess of that required by the reaction stoichiometry, or if zinc ions were removed from the system.

4.2.3 Activities and Concentrations

The activity of a particular species provides the most accurate indication of the availability of that species to determine solution properties, to react with other species, and/or to influence the position of chemical equilibria. Activities are related to ionic concentrations by the relationship:

$$a = \alpha c \quad \text{(EQ 4.19)}$$

where
  a = activity
  α = activity coefficient
  c = molality

TABLE 4.4 Selected standard reduction potentials

Reaction	E^0 (V)
$Ag^+ + e \rightleftharpoons Ag$	0.799
$AgCl + e \rightleftharpoons Ag + Cl^-$	0.222
$Ag_2S + 2e \rightleftharpoons 2Ag + S^{2-}$	−0.705
$Al^{3+} + 3e \rightleftharpoons Al$	−1.706
$H_2BO_3^- + 5H_2O + 8e \rightleftharpoons BH_4^- + 8OH^-$	−1.24
$Br_2(aq) + 2e \rightleftharpoons 2Br^-$	1.087
$Cl_2(g) + 2e \rightleftharpoons 2Cl^-$	1.358
$ClO^- + H_2O + 2e \rightleftharpoons Cl^- + 2OH^-$	0.90
$(CNS)_2 + 2e \rightleftharpoons 2CNS^-$	0.77
$(CN)_2 + 2H^+ + 2e \rightleftharpoons 2HCN$	0.37
$Cu^{2+} + 2e \rightleftharpoons Cu$	0.340
$Fe^{3+} + e \rightleftharpoons Fe^{2+}$	0.771
$2H_2O + 2e \rightleftharpoons H_2 + 2OH^-$	−0.828
$H_2O_2 + 2H^+ + 2e \rightleftharpoons 2H_2O$	1.776
$Hg^{2+} + 2e \rightleftharpoons Hg$	0.851
$Hg_2^{2+} + 2e \rightleftharpoons Hg$	0.796
$Mg^{2+} + 2e \rightleftharpoons Mg$	−2.375
$O_2 + 4H^+ + 4e \rightleftharpoons 2H_2O$	1.229
$O_2 + 2H^+ + 2e \rightleftharpoons H_2O_2$	0.682
$O_2 + 2H_2O + 4e \rightleftharpoons 4OH^-$	0.401
$Pb^{2+} + 2e \rightleftharpoons Pb$	−0.126
$Pt^{2+} + 2e \rightleftharpoons Pt$	1.2
$S + 2e \rightleftharpoons S^{2-}$	−0.508
$S + 2H^+ + 2e \rightleftharpoons H_2S$	0.141
$Sb^{5+} + 2e \rightleftharpoons Sb^{3+}$	0.75
$Zn^{2+} + 2e \rightleftharpoons Zn$	−0.763
$ZnO_2^{2-} + 2H_2O + 2e \rightleftharpoons Zn + 4OH^-$	−1.216

NOTE: Data from [22].

Where possible, activity data should be used in equilibria calculations. Unfortunately, the activities of individual solution species cannot be measured directly, and approximations must be used to complete equilibria calculations [6]. Although the use of such approximations is not rigorously correct, this approach gives useful and meaningful results for many hydrometallurgical systems.

Generally accepted approximations for activity data are given in Table 4.5. These are based upon the assumption that the ionic and molecular species do not interact electrostatically in aqueous solution; that is, their behavior is close to ideal. This assumption is valid for dilute solutions with species concentrations below approximately 0.001 mol/L.

The Debye–Huckel theory predicts activity coefficients for ionic species as a function of ionic strength and is valid for relatively dilute solutions at temperatures up to about 100°C (Figure 4.4). This is based on the assumption that the degree of departure of an ion from ideal behavior in a solvent is determined by the ionic strength of the solution and the ionic valencies, and is independent of the chemical nature of the ions. Ionic strengths can be estimated quite accurately from the specific conductance of a solution using the relationship developed by Lind (Figure 4.5) [7].

These graphs enable the estimation of approximate activity coefficients for ions in specific solutions, based on simple measurement of solution conductivity. For example,

TABLE 4.5 Approximate activity values for ionic or molecular species in dilute aqueous solution

Species	Approximated Activity Value
Pure solid or liquid	1
Gas or mixture of gases	Partial pressure of gas/gases (under atmospheric pressure)
Solute in solution	Concentration of solute
Molecular species in solution	Mole fraction of species (for dilute solutions = 1)

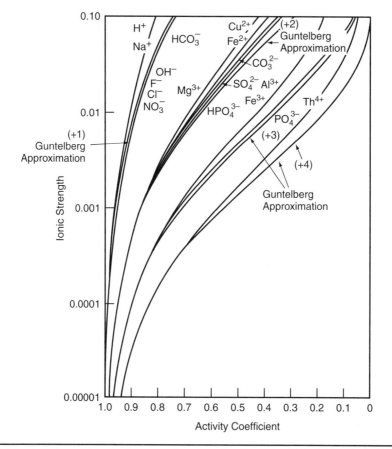

FIGURE 4.4 Activity coefficients of aqueous ions based on the extended Debye–Huckel equation [8]

Figures 4.4 and 4.5 indicate that, for a solution with a specific conductance of 700 μmhos/cm and estimated ionic strength of 0.01 M, the activity coefficient of monovalent ions (such as CN^-) is approximately 0.9. This value is significantly below unity, indicating that the use of concentrations in thermodynamic calculations is inaccurate but may nonetheless give a satisfactory estimate.

The approximation is useful when the ionic strength of the solution is low and individual chemical or ion concentrations are low; for example, gold leaching, precipitation from dilute solutions, and carbon or resin adsorption reactions from dilute solutions. In other cases, such as pressure stripping of carbon or resins, pressure leaching, precipitation from concentrated solutions, and pressure oxidation of sulfide minerals, the approximation

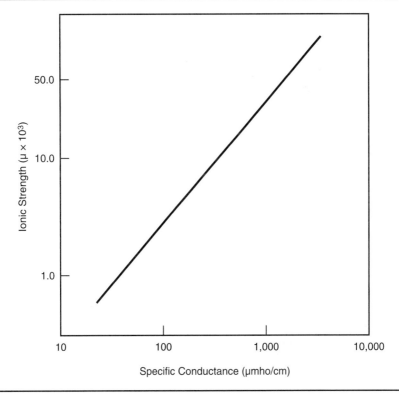

FIGURE 4.5 The use of specific conductance as a means of estimating ionic strength [7]

is less valid and ionic activities must be considered. The calculation of activity coefficients and their application to the interpretation of hydrometallurgical processes has been considered for several systems in the literature [6, 9].

4.2.4 pH Scale and pH Modification

Hydrometallurgical processes are performed in aqueous solutions containing hydrogen (H^+) and hydroxyl (OH^-) ions, and therefore reaction equilibria and kinetics are usually pH dependent. The effectiveness of leaching, carbon adsorption and desorption, surface chemical, aqueous sulfide oxidation, and zinc precipitation processes depends on the ability to measure and control pH. In particular, reagent consumptions, for example cyanide, are often strongly pH dependent. Consequently, it is important to consider pH and equilibria associated with pH modification.

For the complete dissociation of water to hydroxyl and hydrogen ions:

$$H_2O \rightleftharpoons H^+ + OH^- \qquad \text{(EQ 4.20)}$$

For which

$$K_w = \frac{a_{H^+} \cdot a_{OH^-}}{a_{H_2O}} = 1 \times 10^{-14} \qquad \text{(EQ 4.21)}$$

TABLE 4.6 Dissociation constants of acids and alkalis used in gold extraction

	Formula	pK_a
Acids		
Hydrochloric	HCl	(ª)*
Hydrocyanic	HCN	9.32–9.40
Hydrofluoric	HF	3.25
Hypochlorous	HOCl	7.43
Nitric	HNO_3	(ª)*
Nitrous	HNO_2	3.34
Sulfuric	H_2SO_4	(ª)*
Sulfurous	H_2SO_3	pK_1 = 1.92
		pK_2 = 7.91
Alkalis		
Calcium hydroxide	Ca(OH)$_2$†	k_s = 6.46 × 10^{-6}
Sodium hydroxide	NaOH	(ª)*

* (ª) Classified as strong acid/alkali that completely dissociates in solutions containing <1 mol/L of the acid/alkali.
† Value for Ca(OH)$_2$ is the solubility product, k_s.

The activity of water, a pure liquid, is assumed to be equal to 1 (Table 4.5), and the activities of H$^+$ and OH$^-$ for ionic concentrations between 10^{-3} and 10^{-11} (pH 3 to 11) are approximately equal to their concentrations, therefore:

$$K_w = [H^+] \cdot [OH^-] = 1 \times 10^{-14} \text{ mol}^2 \text{ dm}^{-6} \text{ (at 25°C)} \quad \text{(EQ 4.22)}$$

The value of K_w is temperature dependent, for example, 1.47 × 10^{-14} and 2.92 × 10^{-14} at 30°C and 40°C, respectively. The concentration of hydrogen ions is conveniently expressed as the negative logarithmic value (pH):

$$\text{pH} = -\log a_{H^+} \quad \text{(EQ 4.23)}$$

For example, a solution at pH 10 has a hydrogen ion activity of 1 × 10^{-10} mol/L.

A variety of acids and alkalis, for example, sulfuric acid (H_2SO_4), hydrochloric acid (HCl), sodium hydroxide (NaOH), and calcium hydroxide (Ca(OH)$_2$), are used for pH modification and control in industry. These reagents dissociate to varying degrees in aqueous solution, with acids increasing H$^+$ ion concentration and alkalis decreasing H$^+$ concentration. A strong acid will dissociate almost completely whereas a weak acid will only partially dissociate. The extent of dissociation is indicated by the equilibrium constant (K_a) for the reaction.

$$HA \rightleftharpoons H^+ + A^- \quad K_a = \frac{a_{H^+} \cdot a_{A^-}}{a_{HA}} \quad \text{(EQ 4.24)}$$

Dissociation constants for various pH modifiers used in gold extraction are listed in Table 4.6. More detailed reviews of acid–base theory, behavior, and associated equilibria calculations are available in the literature [8, 10].

4.2.5 Complexation

Complexes are formed by the association of two or more simple species and may be either cationic, anionic, or neutrally charged, depending on the number and charges of the components. For example, gold forms the anionic complex with cyanide ($Au(CN)_2^-$) and a cationic complex with thiourea ($Au[SC(NH_2)_2]_2^+$).

Metal ions in aqueous solution are generally solvated due to ion–dipole interactions between the metal and water. Complexation of metal ions results in the replacement of water molecules around the metal ions with complexing ions, or ligands, which bond chemically to the metallic species.

Complexation reactions are important in gold hydrometallurgy since they determine, *inter alia*, the extent of dissolution of gold and other metals or minerals. This affects the recovery of metal values, reagent consumptions, and the efficiency of subsequent processing stages. For example, the presence of soluble copper during cyanide leaching increases cyanide consumption and can severely interfere with carbon adsorption, precipitation, and electrowinning processes, thereby reducing their efficiency. Ultimately, such copper may contaminate the final gold bullion product.

The reaction of a metal cation (M^+) and an anionic ligand (L^-) in aqueous solution can be expressed in the general form:

$$M^{y+} + xL^- \rightleftharpoons ML_x^{(y-x)+} \qquad (EQ\ 4.25)$$

By convention, the water of hydration is omitted to simplify the expression. The number of ligands present in the complex (x) is the coordination number. This value, and the geometric properties of the complex, are determined by the electronic structures of the ionic species involved, principally their size, charge, and polarization. The equilibrium, or stability, constant for this reaction is defined by:

$$K = \frac{[ML_x]^{(y-x)+}}{[M^{y+}][L^-]^x} \qquad (EQ\ 4.26)$$

Polyvalent metal cations can react with ligands to form a series of complexes. The stability constant (K) can be defined for each of the metal complexes formed, giving a series of stability constants for a particular metallic species and ligand. A cumulative stability constant (β) can be calculated and is commonly quoted for the higher coordination complexes, as follows:

$$\beta_n = K_1 \times K_2 \times K_3 \ldots \times K_n \qquad (EQ\ 4.27)$$

An example of this is the complexation of copper with cyanide [11]:

$$Cu^+ + 2CN^- \rightleftharpoons Cu(CN)_2^- \qquad \log \beta_2 = 16.3 \qquad (EQ\ 4.28)$$

$$Cu^+ + 3CN^- \rightleftharpoons Cu(CN)_3^{2-} \qquad \log \beta_3 = 21.7 \qquad (EQ\ 4.29)$$

The ability to calculate the amount of each species present in a solution may be important for the optimization of a process; for example, in the Cu–CN system the various species are adsorbed onto activated carbon to differing extents depending on pH and free cyanide concentration (see Section 7.1.2.5).

4.2.6 Solubility of Solids

All ores contain a variety of gangue minerals, which may include silicates, alumino-silicates, sulfides, sulfates, carbonates, and oxides. Many of these minerals, for example, quartz and feldspars, have insignificant solubility in water and in most of the solutions encountered in gold extraction. These minerals generally play a passive role in gold dissolution and have little effect on solution chemistry. Others, such as the sulfides, sulfates, and carbonates may dissolve appreciably in water or process solutions. Indeed, all hydrometallurgical processes for the treatment of refractory sulfidic ores (see Chapter 5) rely on sulfide mineral dissolution, and it is important to be able to quantify the extent of mineral dissolution in various aqueous media.

When a solid (M_xL_y) dissolves in aqueous solution, an equilibrium is established, which is a function of the solubility of the solid and the extent of dissociation of dissolved species as follows:

$$M_xL_y(s) \rightleftharpoons M_xL_y(aq) \rightleftharpoons xM^{y+} + yL^{x-} \qquad \text{(EQ 4.30)}$$

The equilibrium constant or solubility product (K_s) is defined as [6, 10]:

$$K_s = \frac{(a_{M^{y+}})^x \cdot (a_{L^{x-}})^y}{a_{M_xL_y(s)}} \qquad \text{(EQ 4.31)}$$

For low concentrations of solution species and assuming that the activity of M_xL_y is unity, Equation (4.31) simplifies to yield an expression for the solubility product (K_s):

$$K_s = [M^{y+}]^x \cdot [L^{x-}]^y \qquad \text{(EQ 4.32)}$$

The solubility (S) of the solid M_xL_y is given by the concentration of either M or L:

$$S = [M^{y+}]^x = [L^{x-}]^y \qquad \text{(EQ 4.33)}$$

and by substitution it can be deduced that:

$$S = \frac{[K_s]}{[x^x y^y](x+y)} \qquad \text{(EQ 4.34)}$$

The solubilities of various ore constituents and reagents in water at 25°C are given in Table 4.7.

4.2.7 Solubility of Gases

The most important gas in gold hydrometallurgy is oxygen, which is used for the dissolution of gold in cyanidation and for the decomposition of sulfide minerals in oxidative pretreatment processes. Oxygen may be supplied to industrial systems as air, pure oxygen, or a combination of the two (i.e., enriched air). Chlorine gas has been used in the past in chlorine/chloride leaching and is employed in the chlorination pretreatment of mildly carbonaceous ores. Some other gases are produced during gold extraction processes, for example, hydrogen, which may be generated during gold precipitation with zinc. Sulfur dioxide is also used as a depressant in flotation or may be evolved during oxidation of sulfides. The solubilities of the oxides of nitrogen, bromine, and iodine are also of interest for use in potential gold extraction techniques.

The equilibrium for gas dissolution is expressed as a solubility product, for example, for oxygen dissolution in water:

$$O_2(g) \rightleftharpoons O_2(aq) \qquad \text{(EQ 4.35)}$$

$$K_{sol} = \frac{a_{O_2}}{pO_2} \qquad \text{(EQ 4.36)}$$

TABLE 4.7 Selected solubility products (log K_s) at 25°C [12]

Metal	Hydroxide	Sulfide	Arsenate	Phosphate	Carbonate	Sulfate	Chloride	Bromide	Iodide
Ag^+	−7.9	−49.2	−22.0	−16.0	−11.1	−4.8	−9.7	−12.1	−16.1
Al^{3+}	−32.0	—	−15.8	−17.0	—	—	—	—	—
Ba^{2+}	—	—	−50.1	−22.5	−8.3	−9.9	—	—	—
Be^{2+}	−21.3	—	—	−37.7	—	—	—	—	—
Bi^{3+}	−31.0	−98.8	−9.4	−23.0	—	—	—	—	—
Ca^{2+}	−5.3	—	−18.5	—	−8.2	−4.6	—	—	—
Cd^{2+}	−14.3	−28.9	−32.7	−32.6	−13.7	—	—	—	—
Ce^{3+}	−22.0	—	—	−22.0	—	—	—	—	—
Co^{2+}	−14.5	−22.1	−28.2	−34.7	−10.0	—	—	—	—
Cr^{3+}	−30.0	—	−20.1	—	—	—	—	—	—
Cu^+	—	−47.7	—	—	—	—	—	—	—
Cu^{2+}	−19.8	−35.9	−35.1	−36.9	−9.6	—	—	—	—
Fe^{2+}	−16.3	−18.8	—	—	−10.7	—	—	—	—
Fe^{3+}	−38.6	—	−20.2	−28.0	—	—	—	—	—
Hg^{2+}	−25.4	−52.2	—	—	—	—	—	—	—
La^{3+}	−19.0	—	—	−22.4	—	—	—	—	—
Mg^{2+}	−11.3	—	−19.7	−24.0	−7.5	—	—	—	—
Mn^{2+}	−12.7	−13.3	−28.7	—	−9.3	—	—	—	—
Ni^{2+}	−15.3	−21.0	−25.5	−31.3	−6.9	—	—	—	—
Pb^{2+}	−19.9	−28.1	−35.4	−42.0	−13.1	−7.8	−4.8	−4.1	−9.0
Sb^{3+}	—	−92.8	—	—	—	—	—	—	—
Sn^{2+}	−26.3	−27.5	—	—	—	—	—	—	—
Sr^{2+}	—	—	−18.4	—	−9.0	—	—	—	—
Th^{4+}	−44.0	—	—	—	—	—	—	—	—
Ti^{4+}	−53.0	—	—	—	—	—	—	—	—
UO_2^{2+}	−20.0	—	—	−48.0	—	—	—	—	—
Zn^{2+}	−16.1	−24.5	−27.6	−32.0	−10.0	—	—	—	—
Zr^{4+}	−52.0	—	—	—	—	—	—	—	—

NOTE: Dashes = not applicable.

The solubilities of gases in aqueous solution generally decrease with increasing temperature and rise with increasing gas partial pressure. The presence of ions in solution can decrease gas solubility, depending on the type and concentration, but for the majority of solutions in gold extraction processes the ionic strength is relatively low and this decrease is not significant. Table 4.8 shows the water solubility of some of the more important gases used in gold hydrometallurgy.

4.2.8 Deposition of Gold from Solution

Gold complexes in aqueous solution may be reduced to metallic gold by the addition of a species with a lower reduction potential (for example, metals such as zinc and aluminum), or by providing a potential difference from an external source (for example, during electrowinning). In either case, the driving force for the reduction reaction is provided by the difference between the two potentials, known as the electrochemical overpotential. The reduction potentials for selected gold half-reactions are given in Table 4.3, and selected values from the overall electrochemical series are listed in Table 4.4. These values show that the highly stable Au(I) cyanide complex requires a strong reducing agent, such as zinc, whereas gold chloride and gold thiourea can be reduced by more mild reductants, such as nitrite and sulfite ions.

TABLE 4.8 Solubility of gases in water [13]

Gas	Solubility (absorption coefficient)*
Oxygen	0.049
Chlorine	4.61
Hydrogen	0.021
Ammonia	1,300
Carbon dioxide	1.71
Sulfur dioxide	79.8
Hydrogen sulfide	4.7

* Absorption coefficient is volume of gas at 0°C and 1 atm, which will dissolve in unit volume of solvent under a partial pressure of 1 atm.

A number of metals (for example, zinc, magnesium, aluminum, and chromium) are thermodynamically suitable for reducing aurocyanide, and the commercial use of zinc and aluminum for gold cyanide solutions is discussed in Chapter 8. Of the other possible reductants, magnesium is violently reactive and chromium is passivated in alkaline media; consequently, neither can be considered for commercial use.

4.2.9 Graphical Representation of Equilibria

Complex equilibria may be summarized and presented in a graphical form to facilitate the understanding and interpretation of the thermodynamics of a chemical system. The most commonly used graphs are activity–activity diagrams, which show the relationship between activities or concentrations of related species, and potential–pH diagrams, which indicate the predominant species under a range of E_h and pH conditions. Simple diagrams may be calculated and drawn manually; however, the more complex diagrams are best generated by computer, by determining the most thermodynamically favorable reactions possible in a system. This is particularly important in mineral systems, as it allows many complex reactions to be assessed rapidly.

4.2.9.1 Activity–Activity Diagrams

Activity–activity diagrams display the activities of species present in solution under specified conditions. Activity–pH diagrams are probably the most common type, although metal–ligand diagrams are also useful. These diagrams are derived for a specified oxidation state (for example, Au(I) or Au(III)) and total concentration of the species, using literature values of the relevant equilibrium constants, and show the variation in concentration of the considered species.

4.2.9.2 Solubility Diagrams

Solubility diagrams are similar to activity–pH diagrams and show how the solubility of a solid varies with either pH or other solution species. An example for the zinc cyanide system is given in Figure 8.4. These are most useful for the examination of leaching or precipitation reactions, where a solid is passing into a liquid phase or vice versa. A similar figure can be derived showing the pH above which a given metal will precipitate from solution, as shown in Figures 4.6 and 4.7 for metal hydroxides and sulfides, respectively. Such diagrams clearly define the solution conditions under which a species precipitates or dissolves.

4.2.9.3 Potential–pH Diagrams

Potential–pH diagrams (also called E_h–pH or Pourbaix diagrams [15]) are commonly used to represent the thermodynamic stability of metals and other species in aqueous solution. The E_h and pH coordinates of the lines and the size of the predominance areas

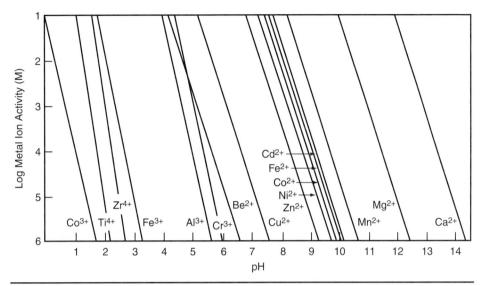

FIGURE 4.6 Precipitation diagram for metal hydroxides [14]

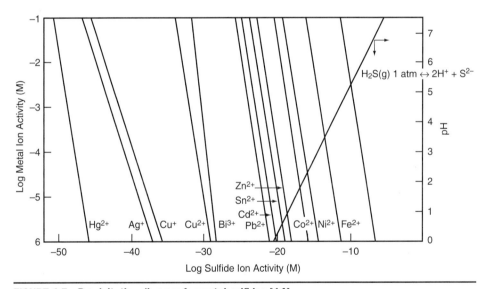

FIGURE 4.7 Precipitation diagram for metal sulfides [14]

provide a summary of the chemistry of a metal in an aqueous system. Figure 4.8 illustrates the conditions under which redox and acid–base reactions occur in such a system and shows the likely reaction products. For example, in the case of the generalized metal–water system shown:

- M^{2+} is oxidized at pH 1 to form M^{3+}, whereas at pH 7, the solid metal oxide, M_2O_3, forms as a precipitate in solution.
- M_2O_3 only dissolves in very acidic conditions and at high potentials (for example, in well-oxygenated solutions) to form M^{3+}, whereas, under reducing conditions, it

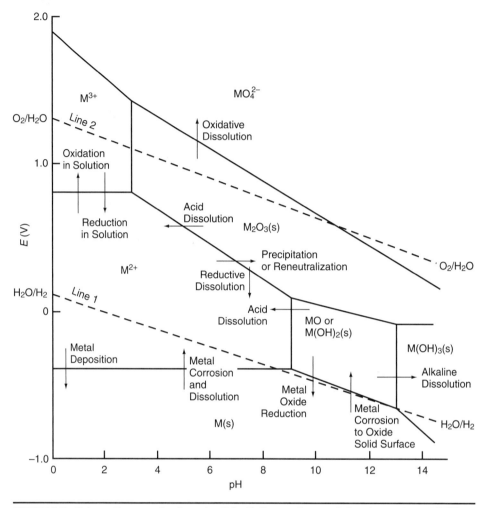

FIGURE 4.8 Schematic example of a potential–pH diagram for a metal–water system at 25°C

dissolves as M^{2+}, at pH values as high as 8. In a practical process this factor could lead to reduced acid consumption.

E_h–pH diagrams are constructed by applying the Nernst equation (4.15) to each reaction in a given system, using values of ΔG^0, K, and/or E^0 obtained from the literature [11, 15, 16, 17, 18]. Each line on the E_h–pH diagram represents the condition where the activities of reactants and products of the considered reaction are at equilibrium, that is, the reaction is 50% complete. On either side of a line one set of species predominates.

Lines 1 and 2 in Figure 4.8 indicate the equilibrium condition for the reduction of water to hydrogen and oxidation of water to oxygen, respectively, as given by the equations:

$$2H_2O + 2e \rightleftharpoons 2OH^- + H_2; \quad E = -0.0591\,pH - 0.0295 \log pH_2 \quad \text{(EQ 4.37)}$$

where
pH_2 = partial pressure of hydrogen gas

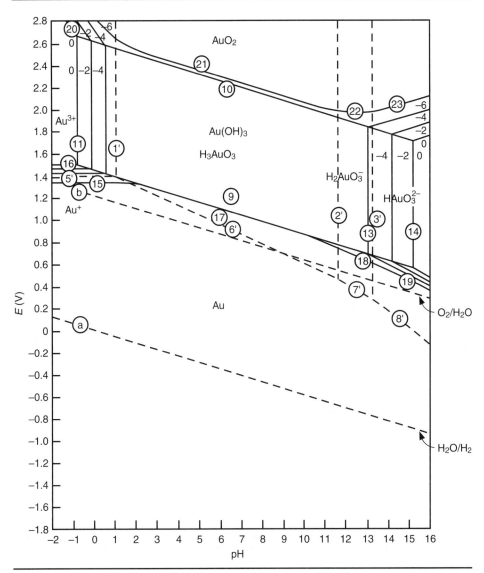

FIGURE 4.9 Potential–pH equilibrium diagram for the system gold–water, at 25°C [15]

$$O_2 + 4H^+ + 4e \rightleftharpoons 2H_2O; \quad E = 1.228 - 0.0591 \text{ pH} + 0.0147 \log pO_2 \quad \text{(EQ 4.38)}$$

where

pO_2 = partial pressure of oxygen gas

In the region between lines 1 and 2, oxidation reactions proceed with parallel reduction of oxygen. Below line 1, reactions are accompanied by reduction of water to hydrogen. Above line 2, species are stable in water, because there is no cathodic (reduction) reaction, which can be coupled to the anodic metal corrosion. In the case of gold (Figure 4.9), the stability area of the metal extends above the oxygen reduction line (line b), and it is apparent that gold is stable in pure water in the absence of complexants. The E_h–pH conditions that are

130 | THE CHEMISTRY OF GOLD EXTRACTION

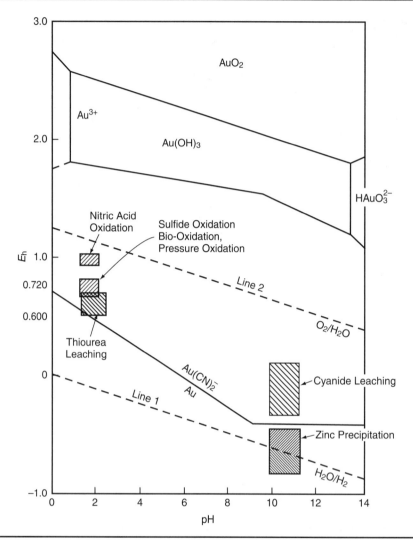

FIGURE 4.10 Indicative E_h–pH ranges employed in industrial gold-extraction processes

applied in industrial gold extraction processes are indicated in Figure 4.10. Several reviews are available on the generation and use of potential–pH diagrams [6, 17, 18, 19].

However, there are some limitations associated with the use of these diagrams in practical mineral systems:

- Only the predominant (>50% relative concentration) species are shown, and potentially important species are not displayed, for example, even at 40% relative concentration.
- Reaction kinetics are not considered, and often reactions in mineral systems do not reach equilibrium within the process residence time available.
- The chemical composition and concentrations of solid and solution species may be uncertain, making appropriate diagram derivation impossible.
- The formation of intermediate product layers, for example, sulfur or hydroxides, may restrict further reaction.

- Multi-mineral particles may produce catalyzing or galvanic effects (e.g., the presence of galena or pyrite, respectively, in gold leaching).
- Non-stoichiometric phases may be formed for which thermodynamic data are not considered or are unavailable.

Despite these limitations, potential–pH diagrams are extremely useful in practice. In special cases, E_h–pH diagrams describing complex process conditions that are not at equilibrium may be approximated either by manipulating the diagram derivation to disregard species that are unlikely to form (for example, SO_4^{2-} in the S–H_2O system or ClO_4^- in the Cl–H_2O system), or by superimposing measured values of species concentrations, pH, temperature, and solution potential. Such systems are considered in the chapters that follow for specific cases of relevance in gold extraction.

4.3 REACTION KINETICS

The thermodynamic prediction that a reaction can proceed is not sufficient to determine whether it will occur to any significant extent within a practical time scale. This depends on the reaction kinetics—an important consideration in the design and economics of all hydrometallurgical processes. Also, for existing process operations, maximizing kinetics often has the result of maximizing process efficiency.

Heterogeneous reactions are controlled either by the inherent chemical reaction kinetics or by the rate of mass transport of the individual reacting species. The major steps in a reaction are the following:

- Mass transport of gaseous reactants (where relevant) into the solution phase and subsequent dissolution
- Mass transport of reacting species through the solution–solid boundary layer, to the solid surface
- Chemical (or electrochemical) reaction at the solid surface, including adsorption and desorption at the solid surface and across the electrical double layer (see Section 9.1.1)
- Mass transport of reacted species through the boundary layer into bulk solution

A simplified reaction model is given in Figure 4.11. If the rate of overall reaction is controlled by stages 1, 2, or 4, the reaction is said to be mass transport controlled. If stage 3 controls the rate, the reaction is chemically controlled. The rate-determining stage may change for a reaction if conditions change; for example, if a rate-limiting reactant concentration is increased, a reaction may change from mass transport to chemically controlled as more reagent is available to be transported to the reaction surface. A reaction for which both mass transport and chemical factors affect the rate is said to proceed under mixed control.

Homogeneous reactions are generally faster than heterogeneous reactions, because they involve only mass transport through a single phase, and solution species react quickly because there is no requirement for adsorption onto, or desorption from, a solid surface. By contrast, heterogeneous reactions involve mass transport of species across a phase boundary, which can often be the rate-determining step in a chemical reaction, and if that particular step of the reaction can be accelerated, then the overall reaction rate is increased. The most important reactions in hydrometallurgical gold extraction processes are heterogeneous, involving the transfer of metals and minerals between solid and liquid phases.

Stage 3 in Figure 4.11 may involve reactions of two types: (1) those involving only a chemical reaction, for example, dissolution of an oxide in acid, or (2) those which are electrochemical, that is, an oxidation or reduction involving electron transfer. Stage 3 usually involves several of the following substages:

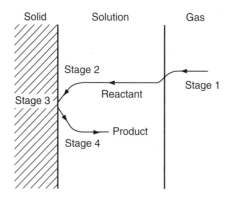

FIGURE 4.11 Schematic representation of the stages in a heterogeneous chemical reaction [19]

- Surface hydroxylation–hydration (at favorable sites)
- Reaction of surface species (e.g., protonation)
- Adsorption of reacting species onto the solid surface
- Desorption of product species from the solid surface
- Reaction of product in solution (e.g., aquation)

The chemical reaction step is rarely rate-determining in most industrial gold-extraction processes, which are designed to proceed at rates sufficient to provide an economic process, and hence mass transport is often reaction limiting. However, there are exceptions to this, including sulfide pressure oxidation and intensive cyanide leaching.

4.3.1 Modeling of Kinetics

The rate constant for any reaction may be expressed in the form of the Arrhenius equation, that is,

$$k = A\, e^{-E_a/RT} \qquad \text{(EQ 4.39)}$$

where
 A = constant related to collision frequency of solution species
 E_a = activation energy for the reaction
 R = gas constant
 T = absolute temperature

According to Equation (4.39) the reaction rate is found to increase exponentially with temperature. The value of E_a for a reaction can be derived by plotting log k versus $1/T$ and calculating the slope of the line (E_a/R). The term E_a is defined as follows:

$$E_a = RT^2 \frac{d \ln k}{dT} \qquad \text{(EQ 4.40)}$$

For chemically controlled reactions, the value of E_a is large, typically greater than 40 kJ/mol, whereas for mass transport controlled reactions E_a is small, between 5 and 20 kJ/mol. A useful general guide is that the rate of a chemically controlled reaction increases by a factor of two for a 10°C increase in temperature, other factors remaining constant. Whether a process is mass transport-, chemical- or pore diffusion-controlled may be

TABLE 4.9 Experimental tests for rate-controlling steps in fluid–solid reactions [20]

	Result Expected for:		
Factor	Surface Chemical Control	Rate Diffusion Control	External Mass Transfer Control
Increased fluid flow	No effect	No effect	Reaction rate increases
Activation energy on temperature change	40–400 kJ/mol	5–20 kJ/mol	5–20 kJ/mol
Change of solid particle size	For solids with zero initial porosity, kinetics proportional to size; for solids which are initially porous (and porosity develops during reaction), size has little or no effect on kinetics.	Reaction time proportional to size^2.	Reaction time proportional to size^n where 1<n<2.

TABLE 4.10 Activation energy data for selected processes in gold extraction [2, 19, 21]

Process	Activation Energy (kJ/mol)	Rate-Limiting Step
Gold dissolution		
—Atmospheric	8–20	Mass transport of either O_2 or CN^- depending on concentration and temperature
—Intensive (high CN, O_2)	60	Chemical reaction
Gold adsorption onto activated carbon	11–16	Pore diffusion control and mass transport of $Au(CN)_2^-$
Zinc precipitation	13–16	Mass transport of $Au(CN)_2^-$
Sulfide oxidation (by O_2)	30–70	Chemical reaction (low T)
		Mass transport of O_2 (high T)

determined experimentally by monitoring the effect of varying reaction conditions, as given in Table 4.9.

Activation energy data for a number of reactions of interest in gold hydrometallurgy are summarized in Table 4.10. It can be seen that the most important gold extraction reactions have activation energies within the range indicating mass transport control. The general expression used to describe the kinetics of a reaction is:

$$dC/dt = -k_m C^n \quad \text{(EQ 4.41)}$$

where
  C = the concentration of reacting species
  t = time
  k_m = a rate constant
  n = the reaction order

This expression can be integrated to produce an equation that shows how concentration changes with time for a given rate constant and reaction order. If the reaction is first order (n = 1), then the concentration after a certain time, t, is given by C_t, defined as follows:

$$C_t = C_0 e^{-kt} \quad \text{(EQ 4.42)}$$

where
  C_0 = the initial reactant concentration

The majority of reactions in gold hydrometallurgy can be approximated to first-order reactions. Notable exceptions are intensive cyanidation and pressure oxidation of sulfide minerals. For higher-order reactions, integration of Equation (4.41) results in more complex rate equations, which have been listed in the literature [22].

4.3.2 Mass Transport

4.3.2.1 Mass Transport Through the Boundary Layer

In heterogeneous systems a boundary layer is established at the solid surface, arising from the condition that the velocity of solution at the solid surface must be zero. A velocity profile is developed whereby the solution velocity increases with increasing distance from the solid surface until a maximum, steady state velocity in bulk solution is attained.

Mass transport of ionic and molecular species through a boundary layer occurs by diffusion. The driving force for this is the concentration gradient which results from the inherent difference in concentration between a reactant in bulk solution and a lower concentration at the reacting solid surface. Brownian motion also contributes to the transport process by increasing the net transfer of ions through the boundary layer.

When the concentration of a reactant at the surface is zero (Figure 4.12), then the rate of chemical reaction is sufficient to consume all of the reacting species, and the reaction is mass transport controlled, that is, dependent on the mass transport rate of species to the surface. If the reactant is not fully consumed at the surface, then the reaction is limited by the reaction rate and is chemically controlled.

The boundary layer thickness (δ) is the distance from the solid surface to the closest point at which the velocity equals the steady state bulk solution velocity (Figure 4.12). The concentration gradient across the boundary or interfacial layer has been described by Fick's first law, which is represented by Equation (4.43):

$$\frac{dn}{dt} = -D_i A \frac{dc}{dx} \qquad \text{(EQ 4.43)}$$

where dn is the number of moles flowing through a plane of surface area (A) during time dt. The ratio dn/dt is the diffusion driving force (j_d). The value of the diffusion coefficient (D_i) depends on the size of the species, the solution temperature, and viscosity. Nernst proposed the following approximation:

$$j_d = D_i A (C_0 - C_i)/N \qquad \text{(EQ 4.44)}$$

where

N = Nernst diffusion layer thickness
C_i = concentration of species i in solution
C_0 = concentration of species i at the solid surface

When the mass transport rate of species through the diffusion layer is the rate-determining step, then the concentration of reactant at the solid surface is zero. In this situation the expression reduces to:

$$j = -D_i \cdot C_b / N \qquad \text{(EQ 4.45)}$$

and then

$$j = -K_m \cdot C_b \qquad \text{(EQ 4.46)}$$

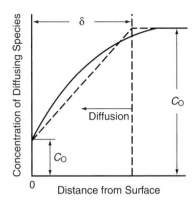

FIGURE 4.12 Concentration gradient of a reactant at an interface [19]

where
C_b = the concentration of species i in bulk solution
K_m ($= D_i/N$) = the mass transport coefficient, equivalent to a first-order chemical rate constant

This is generally valid for reactions where concentrations of reacting and produced species are low.

Mass transport rates through the diffusion layer may be increased by the following:

- Reducing the diffusion layer thickness
- Increasing the concentration gradient (e.g., increased solution concentration)
- Increasing solution temperature

The diffusion layer thickness is a function of the roughness of the solid surface; the viscosity of the liquid; solution and solid flow rates; stirring or mixing rates; and the degree of turbulence in bulk solution. The diffusion layer thickness is minimized by increasing solution flow rates (good mixing) and by increasing shear rate, for example, by use of baffles in tanks and high-shear impellers. This approach is of particular interest in the case of heap and dump leaching of gold ores where there is often little control over localized solution flow rates and conditions of turbulence because of solution channeling and the existence of dead zones (see Section 6.1.5). Systems with poor mixing and agitation are characterized by boundary layers of 0.5 mm and above. Good mixing produces boundary layers typically of about 0.01 mm and below.

Although increasing temperature is typically beneficial, it does not always increase the overall reaction rate because other factors, such as decreasing gas solubility with increasing temperature, may have an overriding effect. This must be considered carefully in each case.

4.3.2.2 Mass Transport in Bulk Solution

Bulk solution mass transport in well-agitated leaching systems is usually very effective and does not limit the reaction rate. However, such mass transport may be important in solution and slurry systems where the quality of mixing is poor, due to inherent particulate and/or engineering factors.

Processes such as heap leaching, vat leaching, and carbon-in-pulp (CIP) cannot always be operated to maintain a homogeneous solution–slurry phase. Examples of this are (1) solutions applied to heap and vat leaching systems, which may rapidly lose their solution concentration homogeneity (e.g., pH, sodium cyanide), and (2) CIP systems, which are prone to stratification of carbon concentration and slurry density in CIP tanks.

Good mixing in the bulk solution or slurry is important to maintain homogeneity and prevent dead spaces in a reactor, which might lead to some material having only limited exposure to other reactants. It is also important that agitation maximizes mass transport through the boundary layer, because this is often the rate-determining step.

4.3.3 Absorption of Gases in Liquids

The rate of absorption of a gas must be maintained in excess of its rate of consumption to prevent the absorption step from becoming rate controlling. For many gas–liquid reactions in gold extraction this can be achieved using simple gas addition techniques with little attention to the absorption kinetics required. Examples of such systems are cyanide leaching of gold, using air sparging, and chlorination of carbonaceous ores, using simple chlorine sparging methods. However, for some mass transport controlled processes, the rate of gas absorption, rather than diffusion of solution species, can be the rate-controlling step, for example:

- Gold dissolution in the presence of cyanicides (e.g., pyrrhotite), which may deplete dissolved oxygen concentration
- Chlorination in the presence of sulfide minerals, which may deplete hypochlorous acid concentration
- Sulfide oxidation reactions using gaseous oxygen, where the solid–liquid reaction kinetics are relatively fast, for example, pressure and biological oxidation of sulfide minerals

In these circumstances the gas absorption rate must be maximized to maximize the reaction rate. This is a function of the following:

1. Surface area of the liquid
2. Superficial velocity of gas bubbles
3. Partial pressure of the gas
4. Degree of agitation
5. Temperature
6. Concentration of dissolved gas and other species
7. Liquid viscosity

The absorption rate is increased by increasing factors 1 to 4 above and decreased by increasing factors 5 to 7. In the case of sulfide mineral oxidation by oxygen, high temperatures are used to increase the mineral reaction rate. This decreases the oxygen solubility, requiring that elevated oxygen overpressure be used to achieve adequate oxygen concentrations. The solubility of oxygen at various temperatures and pressures is illustrated in Figures 4.13A and 4.13B.

The surface area of the liquid is a function of the size and number of gas bubbles in the system and is highest for large numbers of small bubbles. This is promoted by vigorous mixing and the use of efficient sparging systems. Different sparging systems produce a variety of bubble sizes, and this is often an important factor in equipment selection during plant design. The superficial velocity of the gas is also a function of the sparging rate and the sparging method. Although high superficial velocities promote high absorption rates, the energy requirement to achieve this increases greatly as the velocity increases. In commercial pressure oxidation systems, superficial velocities of 0.10 to 0.30 cm/s are used, depending on the agitation method employed. The highest gas solubility is favored

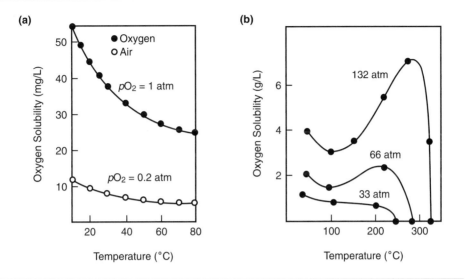

FIGURE 4.13 (a) Effect of temperature and gas composition on oxygen solubility in water (b) Effect of temperature and pressure on pure oxygen solubility in water [19]

at low temperatures, but the effect of temperature is much less important than factors 1 to 3 in the previous list. The operating temperature is selected based on overall reaction kinetics and not simply to maximize gas solubility.

The partial pressure of the gas and the concentration of dissolved gas species provide the overall driving force for absorption, and the absorption rate increases with increasing partial pressure. However, the closer the liquid–gas system comes to equilibrium (i.e., gas saturation of the liquid), then the weaker the driving force and the slower the absorption rate. This is usually not a problem because the rate-controlling step of such a system is unlikely to be controlled by gas absorption; rather, it is probably controlled by solid–liquid mass transport factors.

4.3.3.1 Gas Utilization

The efficiency of gas usage depends on the absorption kinetics, the efficiency of the mixing system into which the gas is being introduced, and the venting (gas removal) requirements. If it is assumed that absorption kinetics are maximized, then the gas utilization depends on the effectiveness of bubble dispersion (to avoid bubble coalescence) and bubble retention time in the liquid. In order to minimize process costs, it is generally desirable to maximize gas utilization.

4.3.4 Electrochemical Reactions

Electrochemical reactions are caused by, and result in, electron transfer. The electrochemical driving force is a potential (voltage) difference which results in a flow of electrons (i.e., an electrical current). Because this can be accurately measured directly, the rate of electrochemical reactions can be quantified very easily. Many sophisticated methods are available for studying reaction mechanisms and for quantifying and modeling reaction kinetics [23, 24]. The reaction model (Figure 4.11) also applies to electrochemical reactions, with the chemical reaction step replaced by an electrochemical one.

The Nernst equation (4.15) applies to a reaction at equilibrium. If there is a change in potential, the chemical system will no longer be at equilibrium, and the activities of reduced and oxidized species will be forced to change also. This driving force is called the overpotential (η) where:

$$\eta = E - E_{eqn} \qquad (EQ\ 4.47)$$

As well as determining the extent of a reaction, the overpotential can dictate the rate at which the reaction proceeds.

At equilibrium there is no net current, because the anodic and cathodic currents, called the exchange currents, are opposite and equal, with a value dependent on the standard reaction rate constant (k^0). The rate constant for a given half-reaction is expressed as follows:

$$k = k^0 \exp{(\alpha nF(E - E^0)/RT)} \qquad (EQ\ 4.48)$$

where
  α = transfer coefficient

This equation shows that the rate depends exponentially on the overpotential, as part of the generalized Butler–Volmer equation:

$$I_{net} = I_0 \left[\exp{(-\alpha nF\eta/RT)} - \exp{((1-\alpha)nF\eta/RT)}\right] \qquad (EQ\ 4.49)$$

This expression relates the main factors in electrochemical reactions and forms the basis for more specific and similarly useful reaction rate equations [23, 24, 25].

For simple solution redox reactions, the reaction rate is usually mass transport controlled when the overpotential is >0.36 V [21]. For overpotentials between 0.06 and 0.36 V, reactions proceed under mixed mass transport and electrochemical control, whereas <0.06 V reactions may be fully electrochemically controlled. For solution species with complicated structures (e.g., $Cr_2O_7^{2-}$), reaction kinetics are much slower and require much larger overpotentials to achieve mass transport control.

4.3.5 Particulate Factors in Solid–Liquid Systems

Particle characteristics, such as particle size, shape, and porosity, play an important role in both the kinetics and extent of completion of leaching reactions, because they control the surface area available for reaction with a solution phase reactant. These factors are particularly important in carbon adsorption, carbon elution, and precipitation processes.

4.3.5.1 Particle Size
The rate of reaction at a solid surface in aqueous solution is proportional to the surface area of the reacting species. This may be defined by the general first-order expression:

$$dn/dt = -A\ C\ k_0\ k' \qquad (EQ\ 4.50)$$

where
  A = solid surface area
  C = solution concentration
  k_0 = concentration of potential reactive surface sites (mol/cm^{-2})
  k' = rate constant

The particle size and shape are the main factors affecting the surface area of a mineral particle. In the case of gold leaching, little can be done to modify the surface area of gold available for dissolution other than to ensure optimum liberation of gold from the host rock, although flattening of gold particles during grinding may have an effect (see

Section 4.3.5.2). Liberation of gold locked within gangue minerals is, of course, an important factor in gold leaching, but this is considered in more detail in Sections 6.1.5 and 3.2.1.

In the case of oxidation processes to release locked gold in sulfide minerals, particle size may become an important reaction variable, because the host mineral is being dissolved and particle size decreases during the reaction. The rate of dissolution of spherical particles may be described by the shrinking particle and core models, derived from Equation (4.50) [21].

4.3.5.2 Particle Shape and Texture

Irregular particle shapes and rough surfaces lead to increased disturbance of fluid flow around particles and, for diffusion controlled reactions, may affect the reaction rate, although this effect is usually small. Because most mineral systems consist of a wide range of shapes, efforts to describe shape effects are usually highly complex and have limited value in industrial systems. Particle shape also affects surface area, but in most mineral systems the effect on reaction kinetics is negligible. However, the smearing of gold particles during fine grinding increases the surface area and may substantially reduce leaching time.

4.3.5.3 Mineralogical Factors

The reactivity of mineral grains is affected by a number of factors such as crystal orientation, polycrystallinity, and the presence of inclusions, dislocations, and impurities. In general, a mineral reacts more quickly if there are imperfections and many smaller crystals within a mineral grain.

The orientation of an exposed crystal surface is dependent on the mineral's crystal habit and fracture properties, for example, brittleness and cleavage. The presentation of different crystal faces can result in anisotropic dissolution or electrodeposition [6]. Because the crystal boundaries are regions of high reactivity, there can be preferential dissolution at these locations, which increases porosity and possibly results in breakup of a mineral grain. All these factors lead to an increase in the reaction rate with increased polycrystallinity.

The presence of impurities in a mineral lattice, particularly semiconducting sulfide minerals (e.g., Fe in sphalerite, As in chalcopyrite), can greatly affect the performance by altering the resistivity and "band gap" of the mineral and hence influencing the kinetics of electrochemical reactions. These factors may be particularly important during the oxidation of sulfide minerals (Chapter 5). Galvanic effects are discussed in Section 4.3.5.5.

4.3.5.4 Porosity

The rate of reaction in heterogeneous systems is dependent on the accessibility of solution to the mineral surface (boundary layer or pore diffusion controlled) or on the chemical reaction rate (dependent on the available surface area). In both cases the reaction rate will be increased if the rock or mineral is porous. The porosity of a reactive solid species or the host mineral(s) plays a critical role in the following gold extraction processes:

- Heap leaching of gold in a porous host mineral, where gold is incompletely liberated
- Oxidation of sulfide minerals along preferential sites, that is, fractures, cracks, and lattice defects. This is particularly significant for biological oxidation.
- Carbon and resin adsorption and desorption reactions. The well-developed pore structure of activated carbon accounts for its fast adsorption kinetics and high loading capacity.

When the chemical reaction is slow, it is possible for some of the reactants to diffuse into the pores before reaction takes place, and hence the active surface area is that of external surface plus the utilized internal pores, often with the pores contributing the greater proportion of surface area. Under these conditions, reaction rates are often observed to be independent of particle size. When chemical reaction rates are fast, the concentration gradient is steep and the reagent is consumed before penetrating the particle. Under these conditions the internal surface area is less significant, and the reaction rate becomes more dependent on particle size.

The rate of diffusion of a species (A) through a porous solid is described by Fick's law in the following form [21]:

$$n_A = -D_c \Delta C_A \qquad \text{(EQ 4.51)}$$

where

n_A = diffusion rate
D_c = effective diffusivity of A
ΔC_A = concentration gradient between the bulk solution and the pore

D_c is a constant that is related to the cross-sectional area occupied by the solid. The irregularity of pores is described by the Bosanquet formula [21]:

$$1/D_{Ac} = \frac{t}{E}(1/D_{AK} + 1/D_{AB}) \qquad \text{(EQ 4.52)}$$

where

D_{Ac} = effective diffusivity of A
D_{AK} = Knudsen diffusivity
D_{AB} = molecular diffusivity of A
t = tortuosity of the solid
E = porosity of the solid

Considering a porous solid (as shown in Figure 4.14), with fixed E and t involved in a reaction in which pore diffusion is rate controlling, then the rate can be increased either by increasing the concentration gradient, ΔC_A (i.e., by increasing the bulk solution reagent concentration), or by reducing the mean length of pores, that is, by reducing particle size.

For example, this may mean finer crushing of porous heap leach feed materials, finer grinding of sulfide-bearing materials prior to oxidation, or the use of a smaller-size range of activated carbon particles in adsorption circuits. However, these changes to a process require that other factors, for example, process economics and downstream impacts, also be taken into consideration.

4.3.5.5 Galvanic Effects

When two conducting minerals are in electrical contact in a slurry system, the one with the more negative rest potential will dissolve preferentially. The cathodic reduction reaction takes place at the surface of the mineral with the more positive rest potential, and there is a net flow of electrons between the minerals. This is referred to as galvanic corrosion, in the case of the dissolving mineral, or galvanic protection for the nondissolving mineral. In gold extraction systems, galvanic interactions can affect both metal and sulfide dissolution rates. In oxidative pretreatment processes, for example, accelerated oxidation of arsenopyrite occurs when in contact with pyrite, which acts cathodically, as shown in Figure 4.15.

The likelihood of galvanic interaction may be judged from the reduction potentials of the reactions that predominate under given solution conditions, values of mineral and metal rest potentials, and textural mineralogical information.

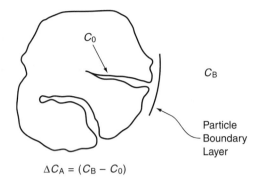

FIGURE 4.14 Schematic illustration of pore diffusion

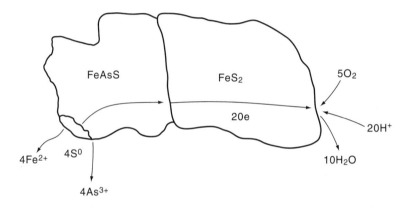

FIGURE 4.15 Schematic representation of galvanic interaction in mineral dissolution

4.3.5.6 The Effect of Competing Species

Minerals other than the valuable mineral(s) in ores and concentrates may react with reagents in a chemical system. This depletes the concentration of the reacting species and may retard the rate of the main reaction, for example:

- Reactive sulfides present during the cyanide leaching of gold, for example, pyrrhotite and marcasite, which consume cyanide and dissolved oxygen
- Barren sulfides present during sulfide oxidation, which consume oxygen

These undesirable side reactions can often be avoided, or at least reduced, by adjusting the system conditions and/or by passivating or removing the competing mineral species.

4.4 EXPERIMENTAL METHODS

In order to adequately understand factors affecting the reaction mechanisms and kinetics, experimental reaction information is required. A number of the most important techniques for this are introduced in the next sections, and extensive further reading is available on the subject [23, 24, 25]. General solution chemistry and metal extraction test work methods have not been covered, because these are often complex and highly specific to each chemical extraction system.

4.4.1 Measurement of Solution Potentials

Actual electrode potentials in aqueous solutions can be measured using suitable "indicator" and "reference" electrodes. A variety of reference electrodes are available, and care should be taken to ensure that potential measurements are compared against similar reference electrodes or that appropriate adjustments are made to measurements. The most commonly used reference electrodes are the following:

- Platinum/calomel ($HgCl_2$), E^0 = 0.2415 V vs. SHE (standard hydrogen electrode) in saturated potassium chloride
- Platinum/mercury sulfate, E^0 = 0.618 V vs. SHE
- Silver/silver chloride, E^0 = 0.222 V vs. SHE

The electrodes most suitable for measuring potentials are those that are unreactive in the conditions used. For example, platinum is effective under oxidizing conditions because the overpotential for oxygen evolution is high and Pt does not form oxides to any significant extent. However, under reducing conditions the hydrogen evolution side reaction is most likely to interfere with the measurement. The electrode material best suited to these conditions is mercury, either as a mercury drop, mercury pool, or mercury amalgamated on the surface of lead. A more practical material for wider application, for example, for use in slurries, is vitreous carbon or lead.

4.4.2 Rotating Disc Electrodes

The rotating disc electrode (RDE) has been used widely and successfully for the investigation of kinetics of homogeneous solution reactions and heterogeneous reactions, such as gold or silver dissolution and cementation. The advantage of using an RDE, rather than a stationary electrode, is that the hydrodynamics of solution flow caused by rotation have been modeled, and mass transfer rates can be calculated accurately from the measurements obtained [21, 26]. The electrode potential can also be controlled, for example, by a potentiostat, and consequently both the reaction thermodynamics and kinetics can be closely controlled for mass transport or chemically controlled reactions.

The RDE consists of a flat disc of an electronic conductor immersed in a solution. The disc is made of an unreactive material, such as carbon or Pt, or the metal or mineral to be studied, for example, Au or semiconducting sulfide minerals.

The rotation of the electrode causes a uniform boundary layer to be set up over the whole area of the disc and laminar flow displacement away from the disc. This causes solution flow perpendicularly toward the disc at a rate proportional to the rotation speed. The reaction rate is measured as an electrical current, which is dependent on the hydrodynamics, and the rate is described by the Levich equation [25], which relates the mass transport to the rotation rate:

$$i_d = 0.62\, n\, F\, D^{2/3}\, v^{-1/6}\, C\, \omega^{1/2} \qquad \text{(EQ 4.53)}$$

where

i_d = diffusion current for a mass transport controlled reaction
n = number of electrons in the reaction step
D = diffusion coefficient (cm^2/s)
v = kinematic viscosity (viscosity/density)
ω = rotation rate (rads/s)

Plots of current against the other terms in Equation (4.53) yield information about the rate-determining step in a reaction and the reaction mechanism [23, 24].

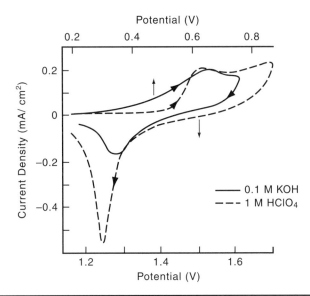

FIGURE 4.16 Cyclic voltammograms for the oxidation of gold in 0.1 M potassium hydroxide and 1 M perchloric acid. The abscissa scales are of different sensitivities as well as laterally displaced to allow the curves to overlap and to demonstrate their general similarity [26].

A useful development from the RDE is the rotating ring disc electrode, which has a ring electrode concentric to a disc electrode. The disc electrode is often constructed of a material to be studied, and the ring electrode is made from a relatively unreactive material, which is used to detect solution species resulting from a reaction at the disc. For example, if a gold RDE is being oxidized, then the ring would detect the solution species formed and, by the magnitude of the current under well-defined mass transport conditions, the oxidation state of the solution species formed. This information can be used to select and optimize possible leaching process conditions; for example, if a solution species is not detected at the ring electrode at low potential and there is an oxidation current at the disc, this may mean that the disc reaction product is a solid, such as oxide formation on a metal or sulfate formation on a sulfide mineral surface.

4.4.3 Potential Sweep Methods

Potential sweep methods are among the most powerful methods of investigating electrochemical reaction rates and mechanisms. The most important and easy to use is cyclic voltammetry, in which a linear potential sweep is applied to an electrode using a potentiostat, and the current that results from any reaction at the electrode is recorded.

This is usually plotted as a cyclic voltammogram, as shown in Figure 4.16 for the example of gold in two different solutions. This method can yield much useful data, for example:

- The potential at which reactions take place can be compared to thermodynamics (e.g., E^0) to identify reactions that may occur.
- The reversibility of reactions can be determined.
- Solid or solution reaction products can be determined from the effects of sweep rate.
- The number of electrons in a reaction step may be determined (e.g., gold dissolution as Au(I) or Au(III)).

- The formation of solid surface species, which may cause passivation, can be determined.

In addition, reaction rate constants may be determined for reversible, irreversible, and reactions involving intermediate chemical reaction steps [24].

These experimental techniques, when properly applied in conjunction with other techniques and an understanding of chemical thermodynamics, provide a powerful insight into the factors affecting hydrometallurgical processes and are invariably useful in modeling industrial systems.

REFERENCES

[1] Cotton, F.A., and G. Wilkinson. 1972. *Advanced Inorganic Chemistry*. 3rd edition. New York: Interscience.

[2] Adamson, R.J. 1972. *Gold Metallurgy in South Africa*. Johannesburg: Chamber of Mines of South Africa.

[3] Warn, J.R.W. 1969. *Concise Chemical Thermodynamics*. London: Van Nostrand.

[4] Criss, C.M., and J.W. Cobble. 1964. *Journal of American Chemical Society* 86: 5385–5390.

[5] Schmid, G.M. 1985. Gold. Pages 313–320 in *Standard Potentials in Aqueous Solutions*. Edited by A.J. Bard, R. Parsons, and J.K. Jordan. New York: Marcel Dekker.

[6] Burkin, A.R. 1966. *The Chemistry of Hydrometallurgical Processes*. London: E. & F.N. Spon.

[7] Lind, C.J. 1970. Specific conductance as a means of estimating ionic strength. Professional Paper No. 700-D. Reston, VA: U.S. Geological Survey.

[8] Snoeyink, V.L., and D. Jenkins. 1976. *Water Chemistry*. New York: John Wiley & Sons.

[9] Kusik, C., and H. Meissner. 1975. Calculating activity coefficients in hydrometallurgy—a review. *International Journal of Mineral Processing* 2:105–115.

[10] Butler, J.N. 1973. *Solubility and pH Calculations*. 2nd edition. Reading, MA: Addison-Wesley.

[11] Smith, R.M., and A.E. Martell. 1976. *Critical Stability Constants*. Volume 4. New York: Plenum Press.

[12] Wadsworth, M.E. 1984. Precipitation reactions in hydrometallurgical systems. Pages 481–494 in *Proceedings of the Advances in Mineral Processing*. Volume 5. Amsterdam: Elsevier.

[13] Stark, J.G., and H.G. Wallace. 1973. *Chemistry Data Book*. Revised edition. London: John Murray.

[14] Monhemius, A.J. 1977. Precipitation diagrams for metal hydroxides, sulphides, arsenates and phosphates. *Transactions of the Institution of Mining and Metallurgy* 86:C202.

[15] Pourbaix, M. 1974. *Atlas of Electrochemical Equilibria in Aqueous Solution*. 2nd edition. Houston, TX: National Association of Corrosion Engineers.

[16] Sillen, L.G., and A.E. Martell. 1964. *Stability Constants of Metal-Ion Complexes*. Publication No. 17. London: The Chemical Society.

[17] Garrels, R.M., and C.L. Christ. 1965. *Solutions, Minerals and Equilibria*. New York: Harper & Row.

[18] House, C.I. 1987. Potential–pH diagrams and their application to hydrometallurgical systems. Pages 3–19 in *Separation Processes in Hydrometallurgy*. Edited by G.A. Davies. Chichester, England: Ellis Horwood.

[19] Jackson, E. 1986. *Hydrometallurgical Extraction and Reclamation*. Chichester, England: Ellis Horwood.

[20] Evans, J.W. 1979. Mass transport with chemical reaction. *Minerals Science & Engineering*. 11(4):207–222.

[21] Sohn, H.Y., and M.E. Wadsworth. 1979. *Rate Processes in Extractive Metallurgy*. New York: Plenum Press.

[22] Weast, R.C. 1981. *Handbook of Chemistry and Physics*. 62nd edition. Boca Raton, FL: CRC Press.

[23] Bockris, J. O'M., and A.K.N. Reddy. 1977. *Modern Electrochemistry*. Volumes 1 and 2. 3rd edition. New York: Plenum.

[24] Bard, A.J., and L.R. Faulkner. 1980. *Electrochemical Methods: Fundamentals and Applications*. New York: John Wiley & Sons.

[25] Levich, V.G. 1962. *Physico-chemical Hydrodynamics*. Englewood Cliffs, NJ: Prentice Hall.

[26] Nicol, M.J. 1980. The anodic behaviour of gold: Part II. Oxidation in alkaline solution. *Gold Bulletin* 13(2):105–111.

CHAPTER 5

Oxidative Pretreatment

Oxidative processes may be used as a pretreatment for sulfide, carbonaceous, and telluride ores and concentrates to increase the extraction of gold by standard hydrometallurgical processing techniques, usually cyanidation. These methods are applied when direct treatment by cyanidation gives unacceptably low gold recovery or is uneconomic, for one of the following reasons:

1. Gold is locked in reactive gangue minerals, often sulfides, and cannot be adequately liberated, even by fine grinding.
2. Gold occurs with minerals that consume unacceptable quantities of reagents, for example, pyrrhotite, marcasite, and arsenopyrite.
3. Gold occurs with carbonaceous materials that adsorb gold during leaching.
4. Any combination of 1 to 3.

Such materials are commonly termed "refractory," literally meaning "difficult to treat." This is a misleading term because all gold ores exhibit some refractory properties, and no treatment process achieves 100% gold recovery. Each ore has a unique composition of minerals in different textural associations and hence an individual degree of refractoriness that must be considered to achieve optimum gold extraction.

The most important mineralogical associations of gold that may require oxidation prior to conventional treatment are illustrated in Figure 2.13. In the case of sulfide ores, oxidation may be necessary to dissolve some, or all, of the sulfide components in order to expose gold values and/or to passivate their surfaces, thereby preventing excessive consumption of reagents. In the treatment of ores containing deleterious carbonaceous components, oxidation is often required either to passivate the active surface of the carbonaceous matter to prevent adsorption of gold or to destroy it entirely. The need for oxidative pretreatment depends on the type, quantity, and properties of the refractory constituents in the ore.

The methods available for oxidation fall into two main categories: hydrometallurgical and pyrometallurgical. The options within these categories are summarized in Table 5.1. Pyrometallurgical oxidation of sulfide and carbonaceous ores by roasting has been practiced around the world for decades and is thoroughly proven. However, increasingly stringent legislation aimed at roaster emissions control for environmental protection worldwide has increased the complexity and cost of roasting processes. Hydrometallurgical methods, other than simple preaeration techniques, have been developed during the past 25 years, and these present attractive alternatives to roasting for many refractory ores and concentrates. The chemistry of both categories of oxidation methods are considered in this chapter.

5.1 HYDROMETALLURGICAL SULFIDE OXIDATION

In the absence of an oxidant, most sulfide minerals decompose very slowly in aqueous solution over a wide pH range and under atmospheric conditions, and are stable for all

TABLE 5.1 Summary of oxidative pretreatment processes

Process Type	Oxidation Method	State of Development of Technology	Ore Types Treated	Application Examples
Hydrometallurgical	Low-pressure oxygen preaeration	Proven commercially	Mildly refractory—contain small quantities of reactive sulfides (e.g., pyrrhotite and marcasite).	East Driefontein (South Africa) Lupin (Canada) Lead (South Dakota, United States)
	High-pressure oxygen—acidic media	Proven commercially	Refractory sulfidic and arsenical ores—low carbonate, high sulfur.	McLaughlin (California, United States) Sao Bento (Brazil) Goldstrike, Lone Tree, and Twin Creeks, (Nevada, United States) Lihir and Porgera (Papua New Guinea)
	High-pressure oxygen—nonacidic media	Proven commercially	Refractory sulfidic and arsenical ores—low sulfur, high carbonate.	Mercur (Utah, United States)
	Nitric acid	Proven commercially for silver concentrates; unproven for gold	Refractory concentrates containing silver, copper, and antimony.	Sunshine (Idaho, United States)
	Chlorine/chlorination	Proven commercially	Carbonaceous ores, low sulfur telluride ores.	Carlin and Jerritt Canyon (Nevada), Emperor (Fiji)
	Biological	Proven commercially for flotation concentrates; unproven for whole-ore treatment	Refractory arsenical and sulfidic ores. Gold preferably associated with arsenopyrite, marcasite.	Fairview (South Africa) Sao Bento (Brazil) Wiluna and Youanmi (Australia) Ashanti Sansu (Ghana)
Pyrometallurgical	Roasting	Proven commercially	Refractory sulfidic, arsenical, carbonaceous, and telluride ores.	Campbell Red Lake and Giant Yellowknife (Canada) Kalgoorlie Consolidated—Gidji (Australia) New Consort (South Africa) Big Springs, Carlin, Cortez, and Jerritt Canyon (Nevada)

TABLE 5.2 Standard electrode potentials for selected redox reactions for use as oxidants in oxidative pretreatment

Reaction	E^0(V)
$O_2 + 4H^+ + 4e \rightleftharpoons 2H_2O$	+1.29
$O_2 + 2H^+ + 2e \rightleftharpoons H_2O_2$	+0.682
$H_2O_2 + 2H^+ + 2e \rightleftharpoons 2H_2O$	+1.776
$O_2 + 2H_2O^+ + 4e \rightleftharpoons 4OH^-$	+0.401
$Fe^{3+} + e \rightleftharpoons Fe^{2+}$	+0.77
$NO_3^- + 4H^+ + 3e \rightleftharpoons NO + 2H_2O$	+0.96
$Cl_2 + 2e \rightleftharpoons 2Cl^-$	+1.36
$2HOCl + 2H^+ + 2e \rightleftharpoons Cl_2 + H_2O$	+1.64

practical purposes. They can be made to decompose rapidly by increasing the oxidizing potential of the solution, which may be achieved by the addition of a suitable oxidant, such as oxygen, chlorine, or nitric acid and, where necessary, by elevating temperature and pressure. Under optimized E_h–pH conditions, most sulfide mineral particles of 45 to 75 µm in diameter can be completely oxidized in a matter of hours—and even minutes in some cases. Standard electrode potentials for selected redox (i.e., reduction-oxidation) systems considered for the oxidation of sulfides are given in Table 5.2. The oxidation method and type of oxidant used depends on the mineralogy of material to be oxidized, the severity of oxidation required, the cost of equipment and reagents, downstream processing requirements, and safety, health, and environmental considerations.

In some cases catalysts, such as copper ions, activated carbon, or bacteria, are used to accelerate oxidation rates, although only the latter of these has been applied commercially in gold extraction [1]. Sulfide mineral oxidation rates can also be enhanced by mechanical activation of the mineral surfaces, which has been observed to occur during grinding operations. The effect can be particularly significant when ultrafine grinding is applied to reduce material down to sizes of less than about 15 µm, and this represents an important option for oxidative pretreatment of refractory concentrates.

The susceptibility of the different sulfide minerals to aqueous oxidation is dependent on their electrical and chemical properties, which include the mineral resistivity, standard electrode potential, and solubility in the oxidation media. In addition, galvanic interactions (Section 4.3.5.5) and the morphology of the oxidation reaction products can also significantly affect the oxidation rate. Oxidation reactions that yield sulfur involve a volume change, which may result in a buildup of sulfur at the mineral surface, potentially hindering oxidation and occluding exposed gold particles. Molar volume changes for selected sulfide mineral oxidation reactions are given in Table 5.3.

To understand the capabilities of the various treatment options discussed later, it is necessary to first consider the chemistry of aqueous oxidation of the most important sulfides that occur in gold ores. In the sections that follow, oxidation reaction products have generally been expressed as ions, unless a solid product is known to be formed under prevailing oxidation conditions. However, some of the ionic products may precipitate rapidly by reactions that are discussed in Section 5.1.6. In addition, the behavior of elemental sulfur formed by some of the reactions in acidic media is an important aspect of sulfide mineral oxidation, and this is considered separately in Section 5.1.5.

5.1.1 Iron Sulfides

Pyrite (FeS_2) is relatively stable in water, as illustrated by the large predominance area of FeS_2 in the E_h–pH diagram for the Fe–S–H_2O system, shown in Figure 5.1. In acidic solutions,

TABLE 5.3 Volume changes associated with selected sulfide mineral oxidation reactions [2]

Mineral	Molar Volume (cm³)	Sulfur Volume (cm³)	Volume Change (%)
Sphalerite (ZnS)	23.75	15.5	−35
Galena (PbS)	31.9	15.5	−51
Pyrrhotite (FeS)	18.5	15.5	−14
Pyrite (FeS$_2$)	24.0	31.0	+29
Chalcocite (Cu$_2$S)	28.4	15.5	−45
Covellite (CuS)	20.8	15.5	−25
Bornite (Cu$_5$FeS$_4$)	92.9	62.0	−33
Chalcopyrite (CuFeS$_2$)	42.65	31.0	−27
Molybdenite (MoS$_2$)	33.3	31.0	−7
Violarite (Ni$_2$FeS$_4$)	64.0	62.0	−3
Stibnite (Sb$_2$S$_3$)	82.5	46.5	−44
Orpiment (As$_2$S$_3$)	71.7	46.5	−35
Realgar (As$_2$S$_2$)	61.0	31.0	−49
Arsenopyrite* (FeAsS)	26.3	15.5 (30.5)*	−41 (+16)*
Enargite* (Cu$_3$AsS$_4$)	88.45	62.0 (74.6)*	−30 (+16)*
Tetrahedrite* (Cu$_3$SbS$_3$)	83.3	46.5 (64.5)*	−44 (+23)*
Tennantite* (CuAsS)	80.5	46.5 (59.1)*	−42 (+27)*

* $M_1M_2S_x \rightarrow M_1 + M_2S_x + ze$ (partial oxidation reaction).

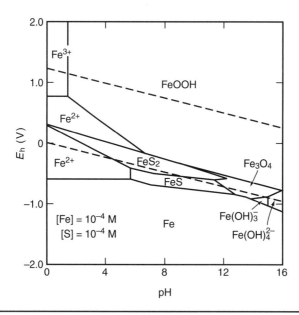

FIGURE 5.1 E_h–pH diagram for the Fe–S–H$_2$O system at 25°C [3]

below approximately pH 6 for the conditions given in the diagram, pyrite is oxidized to form Fe(II) and elemental sulfur:

$$Fe^{2+} + 2S + 2e \rightleftharpoons FeS_2; \quad E^0 = 0.34 \text{ (V)} \tag{EQ 5.1}$$

In sufficiently oxidizing solution, and below pH 1.5, the Fe(II) species are further oxidized to Fe(III), which is given for the case of the simple cations:

$$Fe^{3+} + e \rightleftharpoons Fe^{2+}; \quad E^0 = 0.77 \text{ (V)} \tag{EQ 5.2}$$

The Fe(III) species formed are themselves strong oxidizing agents, which can take part in further oxidation reactions, although their effect on iron sulfides, and pyrite in particular, is in doubt [1]. Pyrite is considered to be virtually immune to the action of Fe(III) species alone, but the reaction is catalyzed by several other species, such as Cu(II) ions and activated carbon [1, 4].

In less acidic and alkaline solutions, that is, above pH 6, pyrite is oxidized to Fe(III) hydroxide:

$$Fe(OH)_3 + 2SO_4^{2-} + 19H^+ + 15e \rightleftharpoons FeS_2 + 11H_2O; \quad E^0 = 0.38 \text{ (V)} \tag{EQ 5.3}$$

The Fe(III) hydroxide may form goethite (FeOOH) and then hematite (Fe_2O_3), as the water of hydration is removed.

Pyrite occurs in two distinct crystal habits: cubic and framboidal, which respond quite differently to oxidation. The framboidal form decomposes even in mildly oxidizing conditions, whereas the cubic structure is essentially stable under these conditions. Consequently, ores containing different forms of pyrite will respond differently to oxidation processes.

Marcasite (FeS_2) is dimorphic with pyrite; that is, it has the same chemical composition but a different structure. However, it is more reactive and oxidizes approximately twice as fast under similar conditions. Marcasite is therefore a greater reagent consumer than pyrite but has the advantage that it is easier to liberate encapsulated gold or passivate the surface by preaeration. The stoichiometry of oxidation reactions for marcasite are identical to those of pyrite.

The oxidation reactions of both pyrite and marcasite to sulfur and metal ions involve large positive volume changes; that is, the volume of sulfur produced is greater than the original mineral volume. This can cause passivation of the mineral surface to further oxidation if the reaction products are not removed faster than they are formed, as discussed in Section 5.1.5.

Pyrrhotite ($Fe_{(1-x)}S$, where $x = 0$ to 0.5) is much less stable than pyrite and, like marcasite, small amounts of the mineral can be oxidized using mild preaeration techniques (see Section 5.2). In acidic solution, below about pH 6, for the conditions given in Figure 5.1, the stoichiometry is as follows:

$$7Fe^{2+} + 8S + 14e \rightleftharpoons Fe_7S_8 \tag{EQ 5.4}$$

and in less acidic and alkaline solutions (i.e., above pH 6):

$$7Fe(OH)_3 + 8SO_4^{2-} + 85H^+ + 69e \rightleftharpoons Fe_7S_8 + 53H_2O \tag{EQ 5.5}$$

However, under conditions typically applied for sulfide oxidation, pyrrhotite has a restricted region of stability and does not react to form elemental sulfur at equilibrium [5].

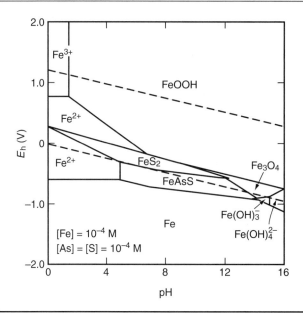

FIGURE 5.2 E_h–pH diagram for the Fe–S–As–H$_2$O system at 25°C [3]

5.1.2 Arsenic Sulfides

The E_h–pH diagram for arsenopyrite (Figure 5.2) indicates that arsenopyrite is less stable than pyrite by approximately 0.3 V and that the mineral is thermodynamically unstable in water. The kinetics of oxidation in water are slow, and arsenopyrite is relatively stable under ambient conditions. However, arsenopyrite is easier to oxidize than pyrite, a factor that is exploited in the hydrometallurgical treatment of some arsenical gold ores. In acidic solutions, the stoichiometry of arsenopyrite oxidation is as follows:

$$Fe^{2+} + AsO_2^- + 4H^+ + S + 5e \rightleftharpoons FeAsS + 2H_2O \qquad \text{(EQ 5.6)}$$

and in less acidic and alkaline solutions:

$$Fe^{2+} + AsO_2^- + 12H^+ + SO_4^{2-} + 11e \rightleftharpoons FeAsS + 6H_2O \qquad \text{(EQ 5.7)}$$

The As(III) species formed may be further oxidized to As(V), depending on solution conditions:

$$AsO_4^{3-} + 4H^+ + 2e \rightleftharpoons AsO_2^- + 2H_2O; \quad E^0 = -0.56 \text{ (V)} \qquad \text{(EQ 5.8)}$$

Realgar (AsS) and orpiment (As$_2$S$_3$) are more stable than arsenopyrite under acidic conditions but are very unstable in alkaline solution, as illustrated in Figure 5.3, with the driving force for oxidation increasing with increasing pH. In strongly acidic solutions, elemental sulfur is formed.

5.1.3 Copper Sulfides

The E_h–pH diagram for the Cu–Fe–S–H$_2$O system (Figure 5.4) shows that under acidic conditions (below pH 5) chalcopyrite oxidizes as follows [1]:

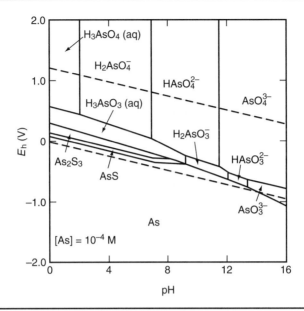

FIGURE 5.3 E_h–pH diagram for the S–As–H$_2$O system at 25°C [3]

$$Cu^{2+} + Fe^{2+} + 2S + 4e \rightleftharpoons CuFeS_2 \qquad (EQ\ 5.9)$$

The principal reaction above pH 5 is:

$$Cu(OH)_2 + Fe^{2+} + 2SO_4^{2-} + 18H^+ + 16e \rightleftharpoons CuFeS_2 + 10H_2O \qquad (EQ\ 5.10)$$

The ranges of stability of other copper sulfides appear on the E_h–pH diagram. These form similar oxidation products in acid and alkaline solutions, that is, copper ions and elemental sulfur, and copper hydroxide and sulfate, respectively. It should be noted that neither hydrogen sulfide nor pyrite is likely to be formed in actual oxidation systems [1].

5.1.4 Other Sulfides

In gold ores, sphalerite and galena are generally less common than the sulfides of iron, arsenic, and copper. The E_h–pH diagram for the Zn–S–H$_2$O system (Figure 5.5) shows that zinc forms Zn(II) in acidic solution and insoluble zinc hydroxide in alkaline solution. The behavior of zinc in aqueous solutions, and its effect on subsequent gold extraction, is considered further in Chapters 6, 7, and 8.

The lead diagram (Figure 5.6) is dominated by the formation of insoluble species: lead hydroxide above pH 7 and lead sulfate in acidic conditions. From a practical point of view this may be advantageous, because these species may form a passivating layer on the surfaces of more reactive sulfides, such as pyrrhotite and marcasite. This can be achieved by the addition of a soluble lead salt under mildly oxidizing solutions, for example, during preaeration (see Section 5.2).

For the conditions shown in the diagrams, the relevant reactions for zinc and lead in acidic media are:

$$Zn^{2+} + S + 2e \rightleftharpoons ZnS \qquad (EQ\ 5.11)$$

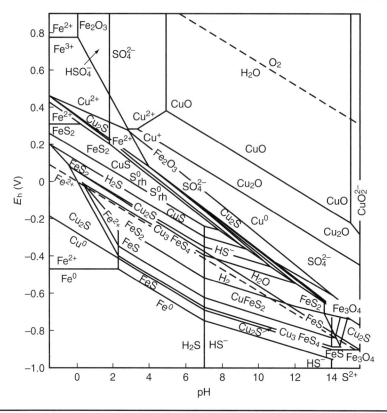

FIGURE 5.4 E_h–pH diagram for the Cu–Fe–S–H_2O system (conditions: 0.1 M Fe, S species; 0.01 M Cu species) [6]

$$Pb^{2+} + S + 2e \rightleftharpoons PbS \quad \text{(EQ 5.12)}$$

$$PbSO_4 + 8H^+ + 8e \rightleftharpoons PbS + 4H_2O \quad \text{(EQ 5.13)}$$

and in alkaline media:

$$Zn(OH)_2 + SO_4^{2-} + 10H^+ + 8e \rightleftharpoons ZnS + 6H_2O \quad \text{(EQ 5.14)}$$

$$Pb(OH)_2 + SO_4^{2-} + 10H^+ + 8e \rightleftharpoons PbS + 6H_2O \quad \text{(EQ 5.15)}$$

Stibnite (Sb_2S_3) behaves similarly to the arsenic sulfide minerals (i.e., As_xS_y), although antimony minerals are generally less soluble in acidic media.

5.1.5 Sulfur

The formation of elemental sulfur during sulfide mineral oxidation, as described in Section 5.1 and identified as a reaction product in Sections 5.1.1 to 5.1.4, can cause several problems since the elemental sulfur may cause the following effects [8]:

- Coating of sulfide particles, effectively occluding the mineral and preventing complete oxidation, as well as agglomerating unreacted sulfide particles

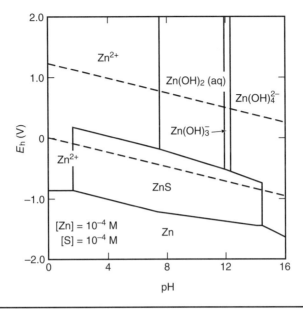

FIGURE 5.5 E_h–pH diagram for the Zn–S–H$_2$O system at 25°C [3]

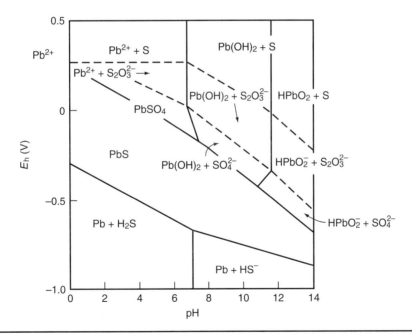

FIGURE 5.6 E_h–pH diagram for the Pb–S–H$_2$O system showing the region of stability (solid lines) and of metastability (dashed lines) for the mineral. Equilibrium lines correspond to dissolved species where [Pb] = 10^{-3} M. [7]

- Coating of exposed gold surfaces, inhibiting subsequent gold recovery processes
- Consumption of cyanide and oxygen through oxidation of sulfur to thiocyanate and thiosulfate in aqueous alkaline cyanide solutions

Consequently, it is generally desirable to avoid formation of elemental sulfur. If elemental sulfur is formed, then it is preferable to oxidize this further to the soluble sulfate species, which is not as detrimental to gold extraction processes.

$$SO_4^{2-} + 8H^+ + 6e \rightleftharpoons S + 4H_2O \qquad (EQ\ 5.16)$$

Figure 5.7 shows the E_h–pH diagram for the S–H_2O system. Thermodynamics predict that sulfur should be relatively easily oxidized to sulfate over a wide pH range, although the bisulfate species (HSO_4^-) is formed below pH 1.5, as indicated by the small region of sulfur stability in Figure 5.7 [2].

In practice, the range of sulfur stability is considerably larger than that predicted by thermodynamics because of kinetic constraints. Elemental sulfur is not oxidized to any appreciable extent in acidic solutions containing Fe(III) ions, despite a thermodynamic driving force for the reaction. A practical range of stability of sulfur, which is more consistent with observations over realistic time scales for metallurgical processes, is shown in Figure 5.8. The reduction of sulfate species in solution is also limited by slow kinetics, and the formation of sulfate ions is considered to be irreversible for all practical purposes. Figure 5.8 shows the E_h–pH diagram for the S–H_2O system with the sulfate species omitted. Although this diagram does not accurately represent the system under acidic conditions, it does reflect observed behavior in alkaline solutions, that is, the formation of thiosulfate and certain thionates in place of, or in addition to, sulfate ions:

$$S_2O_3^{2-} + S^{2-} + 3H_2O + 2e \rightleftharpoons 3S + 6OH^- \qquad (EQ\ 5.17)$$

In actual oxidation systems, some thiosulfate and a variety of other thionates are formed over a wide range of E_h–pH conditions.

The possible formation of thiosulfate is significant because it forms a stable complex with gold, resulting in some gold dissolution during oxidation. This is borne out in practice by the detection of low concentrations of gold in pressure oxidation process solutions in the absence of any other complexing agent. In principle, the simultaneous oxidation of sulfides and dissolution of gold using thiosulfate species generated by the oxidation reactions is a potential extraction route, although solution conditions would have to be carefully controlled to maintain adequate thiosulfate concentration and to prevent further oxidation to sulfate.

Equation (5.16) is strongly temperature dependent and proceeds rapidly above about 170°C, above which elemental sulfur formation can be avoided (Figure 5.9). Typically, pressure oxidation systems are operated in the range 180°C to 225°C, to allow for localized temperature variations and to minimize elemental sulfur formation. It is important to note that the viscosity of elemental sulfur is problematic between about 170°C and 180°C.

Another aspect of aqueous sulfur chemistry is that, in certain acidic systems and under locally reducing conditions, hydrogen sulfide may be formed as a decomposition product of sulfide minerals:

$$FeS_2 + 2H^+ + 2e \rightleftharpoons Fe^{2+} + H_2S + S \qquad (EQ\ 5.18)$$

$$S + 2H^+ + 2e \rightleftharpoons H_2S \qquad (EQ\ 5.19)$$

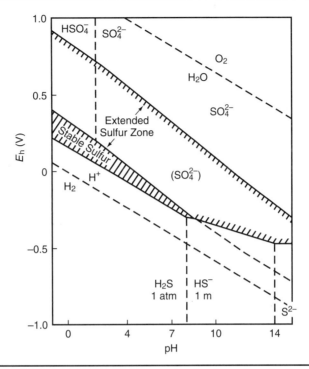

FIGURE 5.7 E_h–pH diagram for the S–H_2O system, showing the region of sulfur stability and the extended stability that is realized by a 300-kJ/mol barrier in the formation of SO_4^{2-} [2]

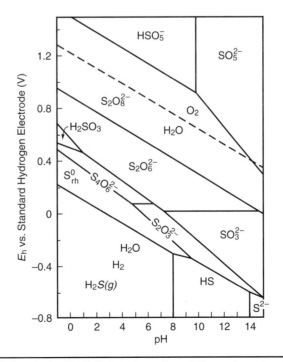

FIGURE 5.8 E_h–pH diagram for the S–H_2O system when all thermodynamic reference to SO_4^{2-} is deleted [2]

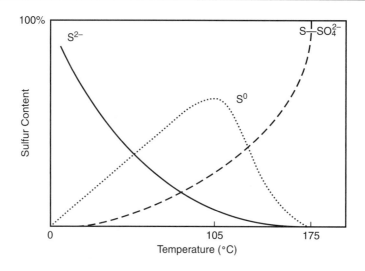

FIGURE 5.9 Oxidation of sulfides below pH 3 [9]

Any hydrogen sulfide evolved is rapidly oxidized in an oxidizing environment, as follows:

$$H_2S + 4H_2O \rightleftharpoons H_2SO_4 + 8H^+ + 8e \qquad \text{(EQ 5.20)}$$

$$2H_2S + 4H_2O \rightleftharpoons H_2SO_4 + S + 10H^+ + 10e \qquad \text{(EQ 5.21)}$$

5.1.6 Precipitation Reactions

Metal ions produced by the decomposition of sulfides may either remain in solution as simple cations or complex ions, or may be precipitated as other species such as hydroxides, oxides, sulfates, and sulfides. These precipitation reactions are important for the following reasons:

- Potentially hazardous species dissolved during oxidation may need to be precipitated as a stable solid product suitable for safe disposal, for example, arsenic.
- Precipitation reactions that occur during or after oxidation may need to be controlled so that any detrimental effect on subsequent gold extraction processes is minimized, for example, jarosite and hydroxide formation.

The precipitation reactions that occur depend on the slurry or solution conditions, specifically the temperature, pH, pulp density, and species concentrations. The latter of these in turn depends on the composition of the ore or concentrate, and the effect of oxidation on each of the constituents. To a large degree these conditions can be controlled to ensure that the most suitable species are produced.

Precipitation reactions of iron, which is generally present in the Fe(III) state in sulfide oxidation processes, are important because a number of undesirable products may be formed, including colloidal hydroxides, jarosites, and other complex salts. These products can severely interfere with subsequent treatment of the oxidized product as a result of poor filtration and settling properties, coating of exposed gold, and unreacted sulfide surfaces.

Precipitation of arsenic (and antimony) is important for environmental reasons as it is necessary to produce an arsenic (or antimony) product that has sufficient stability to allow satisfactory disposal, usually in tailings impoundments.

The precipitation of iron and arsenic is considered in more detail in the next section (antimony behaves similarly to arsenic). The precipitation of other species, such as mercury and selenium, may also be of importance in some systems and are considered in Chapter 11, but these are not considered further here.

5.1.6.1 Iron

The speciation of iron in solution depends on the pH, temperature, the sulfate concentration, and other solution conditions. Figure 5.10 shows the relative proportions of the major species that are formed as a function of pH at 25°C, and for Fe(II), Fe(III), and SO_4^{2-} concentrations of 0.1 M, 0.1 M, and 1 M, respectively [10].

Fe(II) ions are generally precipitated as Fe(II) sulfate, as indicated by Figure 5.10. However, for most of the oxidation techniques considered in this chapter, the majority of the dissolved iron is present in the higher, Fe(III), oxidation state. The precipitation chemistry of this species is highly complex and the reactions that occur can vary greatly for different ore types and oxidation methods.

At low temperatures, below approximately 150°C, and at sulfate concentrations typical of those achieved in oxidation circuits, Fe(III) is precipitated as goethite [2, 11]:

$$Fe^{3+} + 2H_2O \rightleftharpoons FeO(OH) + 3H^+ \quad \text{(EQ 5.22)}$$

At higher temperatures (i.e., >150°C) the precipitate formed depends on solution acidity. At high acidity, that is, >70 to 100 g/L H_2SO_4, and depending on temperature, a basic Fe(III) sulfate is formed [11]:

$$Fe_2(SO_4)_3 + 2H_2O \rightleftharpoons 2FeOHSO_4 + 2H^+ + SO_4^{2-} \quad \text{(EQ 5.23)}$$

This product can cause severe problems in downstream processes because it is hard to filter and has poor settling properties. In addition, the precipitate decomposes to hematite at higher pH—a reaction that releases acid and increases consumptions of alkali and cyanide in gold recovery circuits:

$$2FeOHSO_4 + H_2O \rightleftharpoons Fe_2O_3 + 4H^+ + SO_4^{2-} \quad \text{(EQ 5.24)}$$

The formation of the basic Fe(III) sulfate can be prevented to a large extent by keeping the temperature <190°C. Unfortunately, this is not practically possible in some oxidation systems (i.e., oxygen pressure oxidation) because of the need to maintain high temperatures to avoid elemental sulfur formation.

At acid concentrations below 30 to 50 g/L H_2SO_4, hematite is formed:

$$Fe_2(SO_4)_3 + 3H_2O \rightleftharpoons Fe_2O_3 + 6H^+ + SO_4^{2-} \quad \text{(EQ 5.25)}$$

Hematite is typically the preferred product of oxidation reactions, because it is relatively easy to handle in subsequent processing. It forms a porous precipitate which does not significantly interfere with gold extraction and is generally amenable to solid–liquid separation processes.

Over a wide range of solution conditions, a variety of complex jarosites can form, as follows:

$$3Fe_2(SO_4)_3 + X_2SO_4 + 12H_2O \rightleftharpoons 2XFe_3(SO_4)_2(OH)_6\ 12H^+ + SO_4^{2-} \quad \text{(EQ 5.26)}$$

where
    X = H_3O, Na, K, Ag, Hg, Pb, depending on the composition of the ore, the relative solubility of the species, and the solution conditions

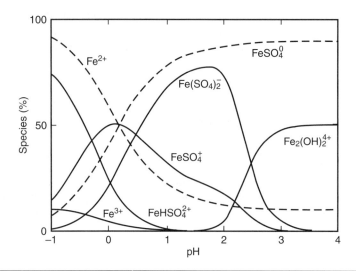

FIGURE 5.10 Effect of pH on species distribution of Fe(III) and Fe(II) in sulfate solution at 25°C (conditions: total SO_4^{2-} = 1 M; Fe(III) = Fe(II) = 0.1 M) (constants from Table 1 of [10])

Jarosites may further react with sulfuric acid to form basic Fe(III) sulfate, as well as Fe(III) sulfate:

$$2XFe_3(SO_4)_2(OH)_6 + 3H_2SO_4 \rightleftharpoons FeOHSO_4 + Fe_2(SO_4)_3 + 2X^+ + SO_4^{2-} + 10H_2O \quad \text{(EQ 5.27)}$$

The properties of jarosites are quite varied, and many are known to present problems in solid–liquid separation systems, which may have an impact on slurry viscosities in subsequent processes. These problems can usually be overcome by good oxidation circuit design.

5.1.6.2 Arsenic

Arsenic solution species can be precipitated as a variety of As(III) and As(V) compounds, depending on solution composition and conditions. In the presence of Fe(III), As(V) can be precipitated as basic Fe(III) arsenate, $FeAsO_4$:

$$Fe_2(SO_4)_3 + 2H_3AsO_4 \rightleftharpoons 2FeAsO_4 + 6H^+ + 3SO_4^{2-} \quad \text{(EQ 5.28)}$$

However, it is necessary for the arsenic to be present in the pentavalent state for effective removal from solution. This is generally the case for sulfide oxidation products, such as those produced by pressure oxidation.

Precipitation as amorphous Fe(III) arsenate, with an Fe–As ratio of 1:1, produces a relatively unstable product which may be considered unsuitable for direct disposal, although this depends on the location and the disposal method. The stability of the precipitate increases as the amount of excess Fe, and consequently the Fe–As ratio, is increased. For example, a product with an Fe–As ratio of 4:1 has a solubility 100 to 1,000 times lower than a product with a 1:1 ratio. Products with ratios greater than approximately 4:1 are essentially stable over the pH range 3 to 7, and the stability is not reduced even in the presence of acid-forming species such as carbon dioxide (from the atmosphere) or reactive sulfide minerals (in the product). Accelerated aging of the product also has no significant effect on the stability of arsenic in these products. Consequently, the more stable products, with Fe–As >4:1, are usually considered to be suitable for disposal [12, 13, 14].

There is increasing evidence that lower Fe–As ratios of 3:1, and potentially lower, can provide stable reaction products suitable for disposal.

If insufficient iron is available to form ferric arsenate, then the arsenic precipitates as a less stable alkali metal arsenate, such as calcium arsenate:

$$2H_3AsO_4 + 3Ca(OH)_2 \rightleftharpoons Ca_3(AsO_4)_2 + 6H_2O \quad \text{(EQ 5.29)}$$

Calcium arsenate is reasonably stable over the pH range 8 to 11, but readily decomposes at lower pH values, such as those produced by acid mine drainage or even rainwater [13].

5.2 OXYGEN: LOW-PRESSURE OXIDATION

Dissolved oxygen in solution under ambient conditions is capable of oxidizing some sulfide minerals. This can be applied as a simple, low-cost, preaeration step before cyanide leaching to oxidize and/or passivate the surfaces of some of the more reactive, reagent-consuming sulfides such as pyrrhotite and marcasite. This treatment is often only capable of partial (surface) oxidation of sulfides and is usually unsuitable for the treatment of ores where gold is intimately associated with sulfides.

5.2.1 Reaction Chemistry

The reduction of oxygen to water can proceed via several paths, as follows:

$$O_2 + 4H^+ + 4e \rightleftharpoons H_2O; \quad E^0 = +1.229 \text{ (V)} \quad \text{(EQ 5.30)}$$

$$O_2 + 2H^+ + 2e \rightleftharpoons H_2O_2; \quad E^0 = +0.682 \text{ (V)} \quad \text{(EQ 5.31)}$$

$$O_2 + 2H_2O + 4e \rightleftharpoons 4OH^-; \quad E^0 = +0.401 \text{ (V)} \quad \text{(EQ 5.32)}$$

$$H_2O_2 + 2H^+ + 2e \rightleftharpoons 2H_2O; \quad E^0 = +1.776 \text{ (V)} \quad \text{(EQ 5.33)}$$

Equation (5.30) represents the major reduction reaction at sulfide mineral surfaces and proceeds slowly under atmospheric conditions [2]. Pyrite, arsenopyrite, and chalcopyrite are relatively stable in oxygenated solutions over a wide pH range. Pyrrhotite and marcasite are less stable and are oxidized to Fe(III) hydroxide above approximately pH 2:

$$4Fe_7S_8 + 69O_2 + 74H_2O \rightleftharpoons 28Fe(OH)_3 + 64H^+ + 32SO_4^{2-} \quad \text{(EQ 5.34)}$$

$$4FeS_2 + 15O_2 + 14H_2O \rightleftharpoons 4Fe(OH)_3 + 16H^+ + 8SO_4^{2-} \quad \text{(EQ 5.35)}$$

The acid generated reacts with available alkali metal salts in the ore to precipitate gypsum or other sulfate species. Alternatively, if the reactive sulfide content is high or in the absence of neutralizing salts, a suitable material such as limestone, lime, dolomite, or sodium carbonate may be added to neutralize acid as it is formed.

Oxidation is usually performed in the pH range 8 to 11, although the pH does not appear to be critical. Most operations that use a preaeration step adjust the pH close to that required for subsequent cyanidation prior to treatment (i.e., 10.0–11.0).

The addition of small quantities of a soluble lead salt, such as lead nitrate or acetate, can assist in the passivation of reactive sulfide mineral surfaces [15]. The lead forms an insoluble hydroxide (Figure 5.6) and possibly some lead sulfate depending on the pH and sulfate concentration of solution. Both of these are largely insoluble in dilute alkaline cyanide solutions. The reactions are as follows:

$$Pb^{2+} + 2OH^- \rightleftharpoons Pb(OH)_2(s) \tag{EQ 5.36}$$

$$Pb^{2+} + SO_4^{2-} \rightleftharpoons PbSO_4(s) \tag{EQ 5.37}$$

In solutions with high carbonate content, insoluble lead carbonate may be formed [6]. Lead ions may also catalyze sulfide oxidation reactions, although this is not well understood. Lead ions can also increase gold leaching rates, as discussed further in Section 6.1.3.6.

Other reagents such as lime and cement have been added to ores during preaeration for the combined effect of pH modification and sulfide surface passivation by neutralizing any acid as it is formed at the mineral surface.

5.2.2 Reaction Kinetics

The oxidation rate depends on the mass transfer of oxygen to the mineral surface, which is principally a function of the dissolved oxygen concentration, degree of mixing (mass transfer rate), and temperature. The low solubility of oxygen in water under atmospheric conditions limits the oxidation rates that can be achieved in practice. Dissolved oxygen concentrations can be increased by raising the partial pressure of oxygen in the gas phase. For example, the use of pure oxygen rather than air increases the oxygen partial pressure by a factor of about 5. Table 5.4 indicates the saturated dissolved oxygen concentrations that can be obtained using air (21% oxygen) and pure oxygen at various temperatures.

Increasing temperature decreases the solubility of oxygen but increases sulfide mineral solubility in most media. Elevated temperatures cannot normally be economically justified for preaeration, which is usually intended to passivate mineral surfaces to reduce cyanide consumption during subsequent leaching, rather than completely oxidize sulfide mineral grains. However, when the process is used for more extensive oxidation of sulfides, such as prior to chlorination, higher temperatures may be used effectively to accelerate kinetics (see Section 12.2.9.1 on Jerritt Canyon).

The degree of agitation may also be important in optimizing oxygen mass transport through the gas–liquid interface. The requirements here are no different than the role of agitation in leaching processes where oxygen is used as the oxidant. This is discussed further in Chapter 6.

5.2.3 Process Considerations

Low-pressure oxidative pretreatment, or preaeration, is most commonly applied prior to cyanidation. Air or oxygen is sparged into agitated tanks, and sufficient retention time, typically 4 to 24 hr, is provided to allow adequate oxidation and/or passivation of cyanide-consuming minerals.

Air is a considerably cheaper source of dissolved oxygen than pure oxygen. However, in some cases pure oxygen can substantially increase oxidation rates and improve the degree of sulfide oxidation obtained with a corresponding further decrease in cyanide consumption. Hydrogen peroxide could also be used for low-pressure oxidation, but the economics are quite unfavorable for most ores, although one notable exception is Pine Creek (Australia), where hydrogen peroxide has been used during gold leaching, principally for sulfide mineral oxidation (see Section 12.2.2.8).

Soluble lead salts and pH modifiers, such as lime and cement, may be added to the first stage of preaeration or, more typically, in the grinding circuit.

TABLE 5.4 Saturated dissolved oxygen concentrations provided from air and pure oxygen as a function of temperature at sea level [16]

Temperature (°C)	Oxygen Concentration (mg/L)	
	21% O_2 (air)	100% O_2
0	14.58	69.45
5	12.75	60.72
10	11.27	53.68
15	10.12	48.02
20	9.11	43.39
25	8.25	39.31
30	7.53	35.88
35	6.96	33.15
37	6.75	32.22
40	6.47	30.82

Low-pressure (atmospheric) oxidative pretreatment has been applied at many operations for the treatment of ores, for example, Homestake Lead (United States), Pamour Porcupine (Canada), and East Driefontein (South Africa); and for concentrates, for example, Agnico Eagle (Canada). Some of these process flowsheets are described in Chapter 12 (see Sections 12.2.2.1, 12.2.2.4, and 12.2.4.1).

Mechanical activation of sulfide mineral surfaces can occur during milling operations and particularly during ultrafine grinding down to particle sizes <15 μm, and in some cases as low as 5 to 10 μm. Such mechanical activation can greatly increase oxidation rates of some sulfide minerals under atmospheric conditions. This can be enhanced by milling in an oxygen-rich atmosphere [17]. Different minerals exhibit different responses to such treatment. As expected, pyrite responds less favorably to this treatment than, for example, arsenopyrite. However, the degree of activation depends on the properties of the mineral of interest (grain size, presence of impurities, lattice defects, and deformities) and the conditions applied (grind size, grinding media, oxygen content of gas, and solution phase). It is expected that ultrafine grinding of refractory sulfide concentrates will be increasingly applied industrially with the associated benefit of mechanical activation of sulfide mineral surfaces exploited in some cases.

5.3 OXYGEN: HIGH-PRESSURE ACIDIC OXIDATION

Sulfide minerals can be made to decompose rapidly in acidic media at elevated temperature and pressure, using oxygen as the principal oxidant. The Fe(III) species that are formed also play an important role in many of the oxidation reactions. The reaction is performed in suitable pressure vessels, called autoclaves, which are capable of withstanding the high temperatures and pressures required. Refractory sulfide ores and concentrates containing greater than approximately 4% sulfide sulfur can be treated autogenously (i.e., using the sulfur content of the feed material as the total heat source at steady state operation and by employing efficient heat recovery systems) to liberate gold and render the ore amenable to cyanide leaching [18, 19, 20]. Materials containing less than about 4% sulfide sulfur may require additional heat to be added in the form of steam for effective pressure oxidation.

5.3.1 Reaction Chemistry

In strongly acidic conditions (below pH 2) at temperatures above approximately 100°C and below about 170°C, and in the presence of dissolved oxygen, the major oxidation reactions for pyrite, pyrrhotite, arsenopyrite, and chalcopyrite are as follows:

$$2FeS_2 + O_2 + 4H^+ \rightleftharpoons 2Fe^{2+} + 4S + 2H_2O \qquad (EQ\ 5.38)$$

$$2Fe_7S_8 + 7O_2 + 28H^+ \rightleftharpoons 14Fe^{2+} + 16S + 14H_2O \qquad (EQ\ 5.39)$$

$$4FeAsS + 5O_2 + 8H^+ \rightleftharpoons 4Fe^{2+} + 4HAsO_2 + 4S + 2H_2O \qquad (EQ\ 5.40)$$

$$4CuFeS_2 + 3O_2 + 12H^+ \rightleftharpoons 4Cu^+ + 4Fe^{2+} + 8S + 6H_2O \qquad (EQ\ 5.41)$$

In addition, Fe(II) species may be oxidized to Fe(III) under these conditions:

$$4Fe^{2+} + O_2 + 4H^+ \rightleftharpoons 4Fe^{3+} + 2H_2O \qquad (EQ\ 5.42)$$

The Fe(III) species are also strong oxidizing agents, which can assist in sulfide oxidation, for example, as follows:

$$FeAsS + 7Fe^{3+} + 4H_2O \rightleftharpoons 8Fe^{2+} + AsO_4^{3-} + 8H^+ + S \qquad (EQ\ 5.43)$$

$$CuFeS_2 + 10Fe^{3+} + 4H_2O \rightleftharpoons 11Fe^{2+} + Cu^{2+} + 8H^+ + S + SO_4^{2-} \qquad (EQ\ 5.44)$$

All of the sulfide oxidation reactions listed (Equations 5.38 to 5.44) yield elemental sulfur, which must be removed to avoid coating unreacted sulfide particles and agglomeration of the sulfur, because these may reduce oxidation efficiency, decrease gold extraction, and increase cyanide consumption. This is achieved by operating at temperatures above about 170°C at which point the sulfur is irreversibly oxidized to sulfate (see Section 5.1.5):

$$2S + 3O_2 + 2H_2O \rightleftharpoons 4H^+ + 2SO_4^{2-} \qquad (EQ\ 5.45)$$

In practice, temperatures of 180°C to 225°C are applied and the overall oxidation reactions are:

$$2FeS_2 + 7O_2 + 2H_2O \rightleftharpoons 2FeSO_4 + 2H_2SO_4 \qquad (EQ\ 5.46)$$

$$2Fe_7S_8 + 31O_2 + 2H_2O \rightleftharpoons 14FeSO_4 + 2H_2SO_4 \qquad (EQ\ 5.47)$$

$$4FeAsS + 11O_2 + 2H_2O \rightleftharpoons 4HAsO_2 + 4FeSO_4 \qquad (EQ\ 5.48)$$

$$4CuFeS_2 + 15O_2 + 2H_2O \rightleftharpoons 2Cu_2SO_4 + 4FeSO_4 + 2H_2SO_4 \qquad (EQ\ 5.49)$$

The iron, arsenic, and copper species are further oxidized to their higher oxidation states; that is, Fe(III), As(V), and Cu(II), respectively. Any carbonates present in the ore (e.g., limestone, dolomite) react with sulfuric acid as follows:

$$CaCO_3 + H_2SO_4 \rightleftharpoons CaSO_4 + CO_2 + H_2O \qquad (EQ\ 5.50)$$

The carbon dioxide evolved reduces the overall efficiency of oxidation by reducing oxygen partial pressure and oxygen utilization, and it is often beneficial to decompose at least a portion of the carbonate minerals beforehand (see Section 5.3.4.1).

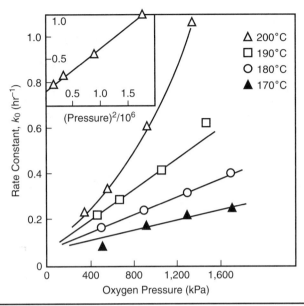

FIGURE 5.11 Pressure oxidation: Pressure dependence of the rate constant k_0 at a range of temperatures [25]

5.3.2 Reaction Kinetics

5.3.2.1 Oxygen Mass Transfer

The rate of sulfide oxidation in pressurized systems is generally dependent on the mass transfer of oxygen to the mineral surface, although this may not be the case for low-sulfur content materials (i.e., <1% to 2% S). The rate-controlling step is usually the transfer of oxygen across the gas–liquid interface, which depends on the partial pressure of oxygen, slurry temperature, gas sparging method, and the degree of agitation or mixing efficiency [21].

The effects of oxygen partial pressure and temperature on oxidation rate are shown in Figure 5.11. The oxygen partial pressure is a function of the oxygen sparging rate, the composition of sparged gas, the amount of carbonate in the feed (which determines how much carbon dioxide is produced in the autoclave), and the total system pressure. Typically, oxygen partial pressures of 150 to 700 kPa are applied with total pressures between 1,500 and 3,200 kPa. Operating conditions for several commercial pressure oxidation systems are summarized in Table 5.5.

The mass transfer rate of oxygen across the gas–liquid interface increases with increasing surface area of liquid and with increasing superficial velocity of gas bubbles. Consequently, mass transfer is maximized by minimizing bubble size and by maximizing gas sparging rate and turbulence, that is, the degree of agitation. However, the efficiency of oxygen utilization decreases as the superficial velocity, or sparging rate, is increased, and a compromise is necessary to achieve satisfactory oxidation rates with acceptable oxygen consumption (see also Section 4.3.3).

The stoichiometric oxygen requirement for sulfide minerals is approximately 20 kg/t per 1% sulfur. Additional oxygen is required for any arsenic and antimony in the feed, that is, the stoichiometric requirement to form AsO_4^{3-} and SbO_4^{3-}, which is 8.5 and 5.3 kg/t per 1% of As and Sb, respectively. In practice, oxygen utilization efficiency may vary from 50% to >90%, depending on the sparging method and oxygen partial pressure applied. An important limitation on oxygen utilization arises from the need to remove

TABLE 5.5 Operating conditions for selected commercial pressure oxidation plants

Plant	McLaughlin	Sao Bento	Mercur	Getchell	Goldstrike	Porgera	Campbell Red Lake	Lone Tree	Twin Creeks	Lihir
Country	U.S.	Brazil	U.S.	U.S.	U.S.	PNG[*]	Canada	U.S.	U.S.	PNG[*]
Start-up date	1985	1986	1988	1989	1990	1991	1991	1994	1997	1997
Media type	Acid	Acid	Alkali	Acid	Acid	Acid	Acid	Acid	Acid	Acid
Feed type	Ore	Concentrate	Ore	Ore	Ore	Concentrate	Concentrate	Ore + concentrate	Ore	Ore
Feed rate (tpd)	2,450	240	720	2,500	~18,000	2,150	70	2,270	3,600	~10,000
Particle size	80% −75 μm	90% −44 μm	75%–80% −75 μm	80% −75 μm	80% −75 μm	80% −38 μm	80% −75 μm[†]	80% −75 μm	80% −20 μm	80% −105 μm
Sulfide sulfur (%)	3.0	18.0	0.95	2.0–4.0	1.7	14.0	13.0–15.0	2.0–4.0	3.0–8.0	7.2
CO_3 (%)	—	8.0	16.0	1.5–7.5	3.5	1.0–3.0	~7.5[†]	—	0.5–12.0	—
Organic carbon (%)	—	—	0.3	0.4	0.75	—	—	—	0.1–1.0	—
Total pressure (kPa)	1,700–2,200	1,600	3,200	3,200	2,500–3,000	1,725	2,100	1,860	3,170	2,400–2,700
Oxygen pressure (kPa)	140–280	—	380	700	340	325–350[†]	300–400[†]	520	690	300–500[†]
Temperature (°C)	180	190	220	210	225	190–200	190–195	195	225	210
Retention time (min)	90	120	112	90	75	110	120	48	50	65
Sulfide oxidation (%)	>85	—	>70	—	86.0–97.0	99.0–99.5	—	75.0	97.0	98.0
Reference(s)	[28]	[27]	[22]	[28, 32]	[26]	[35]	[104, 105]	[106]	[34]	[36]

NOTE: Dashes = not available.
* PNG Papua New Guinea.
† Inferred data.

other gaseous products of the oxidation reactions. Consequently, unreacted oxygen is vented to the atmosphere with other exhaust gases (principally carbon dioxide) and associated steam [23]. A method for recycling some of the oxygen from the exhaust gases to improve overall oxygen utilization and reduce oxygen consumption has been used at Getchell (United States) [24]. The process consists of cooling the vent gases, removing excess moisture and acid vapors, and scrubbing carbon dioxide from the vent gas with hot potassium carbonate solution under pressure, as shown in Equation (5.51).

$$K_2CO_3 + CO_2 + H_2O = 2KHCO_3 \qquad \text{(EQ 5.51)}$$

The reaction is reversed when the pressure is removed and the carbon dioxide is stripped from the carbonate solution, allowing the solution to be recycled. The reaction is performed in a separate vessel, and steam is used to enhance the efficiency of CO_2 stripping. The cleaned, oxygen-rich, gas is cooled and recompressed, with a carbon dioxide content of approximately 2%. Oxygen utilization can be increased to well over 95% by this method, but there is additional cost associated with this practice.

5.3.2.2 Temperature and Pressure

The oxidation rate increases with increasing temperature, as shown in Figure 5.11. However, the pressure requirement also increases with rising temperature, and it becomes progressively more difficult (and costly) to design and operate oxidation systems as the temperature is increased. For example, oxidation at a temperature of 250°C requires an operating pressure of approximately 6,200 kPa. The major issues associated with operation at high temperature and pressure are the following:

- Limitations of mechanical agitator seals on the pressure vessels
- Increased material corrosion rates
- Increased capital and operating costs due to higher pressure ratings on equipment

Operating temperature and pressure are usually maintained at minimally sufficient levels to avoid elemental sulfur formation and to provide the desired oxygen partial pressure for effective sulfide mineral oxidation, as discussed further in Section 5.3.2.1.

5.3.2.3 Acid Concentration

It is generally desirable to maintain sufficient free acid to keep iron species in solution, avoiding excessive precipitation in the autoclave, and to maintain satisfactory oxidizing potential. On the other hand, excessively high acid concentrations result in additional, and unnecessary, neutralization requirements following oxidation, which is costly. The effect of acid concentration on the sulfide oxidation rate is illustrated in Figure 5.12, which shows the results of batch and continuous laboratory testing for one example. Typically, acid concentration is maintained above 10 g/L H_2SO_4 [19]. Systems that treat low-sulfur materials (i.e., less than approximately 2% S) and/or materials that contain acid consumers may require large additions of fresh acid to maintain the desired acid concentration.

High-sulfur materials (i.e., >3% S) generate sufficient acid by their decomposition, and the need for supplemental acid is usually limited to any feed preparation requirements, as dictated by the presence of carbonates. When oxidizing these materials, acid concentrations in the pressure leach vessel discharge slurry can be as high as 60 to 80 g/L.

5.3.2.4 Solution Potential

The electrical potential of the oxidized slurry discharged from an autoclave is a measure of the driving force for the reaction and provides a good indication of the extent of oxidation. Several commercial pressure oxidation plants use this value for process control. A slurry

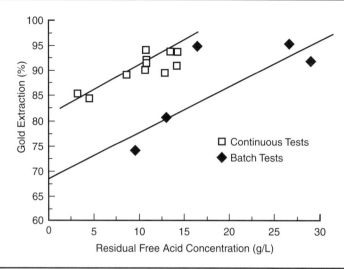

FIGURE 5.12 Example of gold extraction vs. residual free acid concentration during oxidation [19]

or solution potential of 480 mV (vs. SHE) is equivalent to an Fe(III)–Fe(II) ratio of 10:1, which indicates that a high degree of oxidation has been achieved.

5.3.2.5 Degree of Agitation
Sufficient agitation is required to ensure that the slurry in the autoclave has good heat and mass transfer properties. The mixing conditions also affect the absorption rate of oxygen into the liquid phase; increased agitation intensity increases the dispersion and retention of oxygen bubbles in the slurry. The degree of agitation achieved depends on reactor and impeller design, slurry properties (i.e., density and viscosity), mixing power input, and impeller tip speed [21]. Typically, radial flow multibladed impellers (e.g., Rushton turbines) are used, but other impeller designs have been employed effectively.

5.3.2.6 Pulp Density
The optimum pulp density for pressure oxidation is primarily a compromise between minimizing the reactor size (i.e., by maximizing pulp density) and maximizing oxygen mass transfer. However, sulfur product formation (i.e., elemental sulfur) and ore characteristics also influence the choice of operating slurry density. For low-sulfide sulfur ores (<5% S), the formation of sulfur is typically not a problem, and slurry densities of 45% to 55% solids can be used. Materials with higher sulfide contents (e.g., flotation concentrates) must either be treated at lower slurry density, typically 30% to 40% solids and sometimes as low as 10% to 15% solids, or a portion of the product must be recycled to disperse the elemental sulfur formed. Similarly, ores containing carbonates may also need to be treated at lower densities to counteract the effect of gypsum formation on slurry viscosity. Ores with high clay content may require operation at a lower density to maintain acceptable slurry viscosity.

5.3.2.7 Particle Size
Oxidation rates increase with increasing sulfide surface area, as discussed in Section 4.3.5, leading to reduced reaction residence time requirements. An example of this relationship for one particular concentrate is given in Figure 5.13. The optimum particle size is a function of comminution costs, which increase significantly at finer sizes, the cost of incremental oxidation retention time, and the degree of oxidation required. This economic

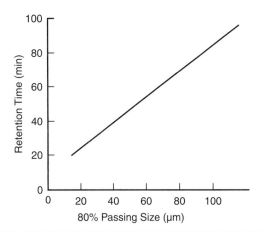

FIGURE 5.13 Effect of particle size on pressure oxidation retention time (schematic only) (adapted from [28])

optimum usually allows flotation concentrates to be ground finer than whole ores. Whole-ore pressure oxidation circuits in operation typically grind to 70% to 80% <75 µm, whereas pressure oxidation of concentrates typically treats a finer material between 70% to 80% <37 µm [26, 27, 28].

An important exception to this is the process developed for the treatment of carbonaceous refractory whole ore at Twin Creeks, Nevada, where the ore was ground to 80% <22 µm to ensure effective oxidation and/or passivation of the preg-robbing organic constituents in the ore (see also Section 3.3.6). In this case, pressure oxidation at 225°C was applied successfully to treat highly preg-robbing material containing 0.4% to 1.5% organic carbon and 2% to 6% sulfide sulfur. At the finer grind size, pressure oxidation achieved 97% to 98% sulfide sulfur oxidation and 30% carbon oxidation, compared with about 95% and 14%, respectively, at 80% <74 µm. Subsequent gold recovery by cyanidation was increased from 65% at 80% <74 µm to 85% to 90% at the 80% <22 µm size [29]. The operating conditions for this application are shown in Table 5.5.

5.3.3 Behavior of Other Species

Many base metal sulfides (e.g., copper, zinc, nickel) dissolve in the severely oxidizing media, and the metals remain in solution following oxidation (a notable exception is lead). Depending on the dissolution efficiency, this may permit them to be partially removed from the solid phase prior to cyanidation by solid–liquid separation, with separate neutralization (or other treatment) of the solution. This is important because many of the dissolved species can interfere with subsequent gold recovery processes.

Arsenic and antimony are precipitated from solution, provided there is sufficient iron present. An Fe–As/Sb ratio of 1:1 may be sufficient to precipitate most of these ions, although the stability of the products usually requires a higher Fe–As/Sb ratio (Section 5.1.6.2). Alternatively, if there is insufficient iron present, the species remain in solution (along with base metal ions) until the solution is neutralized.

Mercury and lead form partially soluble sulfates and jarosites during oxidation, a portion of which report with the solid oxidation products for gold recovery treatment.

Gold is stable under conditions employed for pressure oxidation in the absence of any complexing species. Low concentrations of thiosulfate ions may be formed in oxidation slurries as intermediate products of elemental sulfur oxidation, which may cause some

170 | THE CHEMISTRY OF GOLD EXTRACTION

gold to be dissolved as the gold–thiosulfate complex. Also, there is evidence that low chloride concentrations (from process water or ore constituents) can solubilize some gold. Gold-telluride minerals are believed to decompose under typical pressure oxidation conditions and form tellurites or tellurates, depending on the oxidizing potential [25]:

$$AuTe_2 + 2O_2 + 2H_2O \rightleftharpoons Au + 2H_2TeO_3 \quad (EQ\ 5.52)$$

Silver may be precipitated as an insoluble jarosite ($AgFe_3(SO_4)_2(OH)_6$) under acidic pressure oxidation conditions, depending on the acid concentration and temperature (as discussed in Section 5.3.2), and is essentially unrecoverable during subsequent cyanide leaching. Oxidized slurry can be treated with a hot lime or caustic wash to dissolve the jarosite, making the silver available for cyanide extraction [28]. The important reactions are thought to be as follows:

$$2KFe_3(SO_4)_2(OH)_6 + 3Ca(OH)_2 + 6H_2O \rightleftharpoons 6Fe(OH)_3 + K_2SO_4 + 3CaSO_4 \cdot 2H_2O \quad (EQ\ 5.53)$$

(Note that Na, Mg, and Ag can be substituted for K in Equation 5.53)

$$Fe(OH)SO_4 + Ca(OH)_2 + 2H_2O \rightleftharpoons Fe(OH)_3 + CaSO_4 \cdot 2H_2O \quad (EQ\ 5.54)$$

This is achieved by agitating the slurry for 1 to 2 hr at 80°C to 90°C and above pH 11, using lime for pH modification. In some cases this procedure also increases gold extraction due to the release of gold occluded in precipitates.

There is evidence that organic carbon naturally occurring in pressure oxidation feed material can become more activated through an oxidation circuit; that is, it may adsorb gold cyanide species more severely after oxidation [22]. Where applicable, this can be counteracted by using the carbon-in-leach (CIL) process after oxidation (see Section 12.2.5.4).

An example of the distribution of various elements in the pressure oxidation circuit feed and discharge components, as well as their deportment between solid and liquid phases, is given in Table 5.6.

5.3.4 Process Considerations

Acidic pressure oxidation processes consist of three main steps:

1. Feed preparation
2. Oxidation
3. Product neutralization

The manner in which these steps are applied and combined depends on the characteristics of the material to be treated and is discussed in the following sections. Table 5.5 summarizes operating conditions for selected commercial pressure oxidation plants. Flowsheets for eight commercial pressure oxidation circuits—Mercur, McLaughlin, Goldstrike, Campbell Red Lake, Sao Bento, Lone Tree, Lihir, and Porgera—are included in Chapter 12, as each of these differs slightly as a result of variations in ore mineralogy. Additional information on the Twin Creeks application is provided in Section 5.3.2.7.

5.3.4.1 Feed Preparation

Ground sulfide ore or concentrate may either be fed directly to the oxidation circuit or may first be pretreated with acid to remove some or all of any reactive carbonate material present. In either case sufficient free acid must be available in the first stage of the autoclave to promote rapid initial oxidation (Section 5.3.2.3).

Many sulfide materials contain carbonates which, if not removed, decompose during oxidation and evolve carbon dioxide, thereby reducing the partial pressure of oxygen in

TABLE 5.6 Example of distribution of elements in acidic pressure oxidation circuit feed and discharge materials [30]

Element	Concentrate Feed Range of Assay	Autoclave Discharge Liquid Phase Assay	Autoclave Discharge Solids Residue Assay	Distribution of Element (%)	
				Liquid Phase	Solid Phase
Aluminum	5% to 7%	1.0 to 2.5 g/L	5% to 7%	5	95
Antimony	80 to 100 g/t	<0.1 mg/L	80 to 150 g/t	<1	>99
Arsenic	0.1% to 0.4%	50 to 500 mg/L	0.07% to 0.35%	15	85
Copper	0.03% to 0.14%	120 to 180 mg/L	0.03% to 0.05%	85	15
Gold	6 to 30 g/t	ND*	6 to 30 g/t	0	100
Iron	11% to 13%	2.7 to 5.7 g/L	10 to 12.5%	5	95
Lead	0.14% to 0.64%	<5.0 mg/L	0.14% to 0.64%	<1	>99
Magnesium	0.4% to 1.5%	2.0 to 8.0 g/L	0.08% to 0.3%	80	20
Manganese	0.25%	1.3 g/L	0.025%	90	10
Mercury	14 to 36 g/t	0 to 1.0 mg/L	14 to 36 g/t	4	96
Nickel	70 to 160 g/t	40 to 90 mg/L	7 to 16 g/t	90	10
Selenium	2 to 18 g/t	<0.5 mg/L	2 to 18 g/t	<10	>90
Silver	30 to 80 g/t	Trace	30 to 80 g/t	<0.1	>99.9
Sulfur	9% to 12%	70 to 125 g/L	4% to 7%	50	50
Tellurium	0.5 to 2.0 g/t	ND	0.5 to 2.0 g/t	0	100
Zinc	0.3% to 1.8%	1.5 to 12.0 g/L	60 to 1,200 g/t	95.0	5

* ND = not detected.

the system. This can be countered by increasing the total system pressure and by increasing oxygen addition rates, but this adds to the costs of oxidation.

Either fresh acid or recycled acidic solution from the autoclave discharge can be used for the pretreatment step. Alternatively, a portion of the acidic oxidized slurry itself may be recycled. This not only provides acid to react with, and partially neutralize, the carbonates in the feed, but it also acts as a heat sink and helps to maintain the desired slurry density when treating high-sulfur materials, which would otherwise require the addition of cooling water during oxidation. This latter factor is important since the increased solids content dilutes the sulfur in the feed and helps to disperse any elemental sulfur that is formed, thereby helping to avoid the formation of agglomerates. This is particularly useful in the treatment of flotation concentrates with high sulfur content. Pretreatment is performed in agitated tanks, open to the atmosphere, typically.

The need for pretreatment depends on the sulfide and carbonate (and other acid consumers) content of the feed material, that is, the acid-generating and acid-consuming potential, and each ore must be considered individually. The reason for this is that acidic conditions are required in the first compartment of the autoclave, at which point only a portion of the sulfides have been oxidized; that is, only a portion of the acid-generating potential has been realized at this stage. As a general guideline, an ore containing 3% sulfide sulfur and more than approximately 3% carbonate (CO_3) will probably benefit from pretreatment. Materials with low sulfur (<1%) and high carbonate (>10%) content definitely require pretreatment, and may be better treated by nonacidic oxidation (see Section 5.4). Materials with high sulfide (>6%) and low carbonate (<3% CO_3) content are unlikely to need pretreatment.

The feed slurry may be preheated before it is fed into the autoclave, depending on the feed sulfur content. Steam that is produced during the pressure "let-down" of slurry from the autoclave is generally used for this purpose in two or three counter-current heat-up stages to achieve a slurry temperature of 150°C to 180°C. Feed materials containing as little as 3% sulfide sulfur can be treated autogenously if preheating is used.

5.3.4.2 Oxidation

A variety of pressure vessel designs have been developed for sulfide oxidation [21]; however, four- and five-compartment horizontal units have been used exclusively in gold ore oxidation systems. The slurry is preheated, if necessary, and delivered to the autoclave pressure oxidation vessel by positive displacement pump. The autoclave itself is usually a lead-lined closed tank with an inner lining of acid-resistant refractory brick. The slurry is introduced into the first compartment and cascades through subsequent compartments. Each compartment is fitted with an agitator(s), and oxygen is sparged into the slurry to provide effective gas dispersion and bubble retention time. The autoclave slurry level is controlled in the final compartment by regulating the slurry discharge through a choke valve. A series of pressure let-down, or "flash," vessels sequentially reduce the slurry pressure, allowing the slurry to be cooled and released at atmospheric pressure. In some cases a single flash let-down system has been used (e.g., at Lihir).

Materials that contain >3% sulfide sulfur can usually be oxidized autogenously [19]. The heat generated by the exothermic oxidation reactions is used to maintain the autoclave operating temperature and to heat the slurry prior to oxidation. Temperature control is achieved by steam and/or cooling water addition into each, or some, of the compartments [27]. Steam is used for heating during autoclave start-up and during periods of reduced sulfide content in the feed. Cooling is required when treating high-sulfur materials, especially if none of the oxidized slurry is recycled. These methods of temperature control are the most cost-effective, but they have the disadvantage of diluting the slurry, which reduces the retention time.

Ores that have insufficient sulfide sulfur for autogenous operation must be heated by steam, but the high cost associated with this generally makes this unattractive. A preconcentration step, such as flotation, should be considered for this type of material to increase the sulfide content.

Pressure oxidation of high-sulfur materials, that is, >10%, can lead to a buildup of sulfur on sulfide surfaces and may cause agglomeration of unreacted sulfides. This can be reduced by operation at higher temperatures (i.e., >220°C), or, if necessary, by the use of oxidation additives (e.g., calcium lignosulfonate), which help to disperse the sulfur, and/or by recycling a portion of the oxidized product to provide a greater surface area onto which sulfur deposition can occur. Both of the latter methods have been used at Sao Bento.

Residence times required within an autoclave depend on the oxidation kinetics, which is a function of the type and amount of sulfide minerals present and the operating conditions, the particle size of the material, and the degree of oxidation required [31]. The relationship between retention time and percent sulfide oxidation for three different sulfide materials is shown in Figure 5.14. Residence times of between 1 and 2 hr are usually necessary. Longer residence times are less feasible because of the potentially higher capital and operating costs of the process.

The materials of construction for equipment used in pressure oxidation circuits are extremely important because of the severe corrosion and erosion that occurs in the high-temperature, high-pressure, and acidic conditions in a slurry environment. A variety of metals, metal alloys, and ceramic materials have been tested and evaluated for different applications, and valuable information is available in the literature [27, 32].

5.3.4.3 Neutralization of Products and Precipitation Reactions

Following oxidation and let-down to atmospheric pressure, the oxidized slurry must be neutralized and adjusted to a pH of 10.0 to 11.0 in preparation for cyanide leaching. The two main options available to accomplish this are as follows:

1. Direct neutralization
2. One or more solid–liquid separation steps; two products neutralized independently

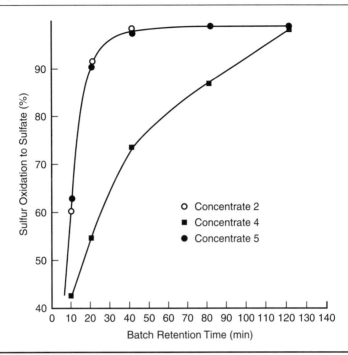

FIGURE 5.14 Example of sulfur oxidation rate by pressure oxidation [18]

Option 1 is the simplest; however, precipitation of dissolved species can occur onto and around exposed gold particles as the pH is increased. This may impair subsequent cyanide leaching, and the salts formed may significantly affect slurry density and viscosity. In addition, this option does not permit recycling of any of the acid generated during oxidation for use in conditioning the feed material, although a portion of the autoclave discharge slurry may be returned.

Option 2 allows for a large proportion of the solution to be separated from the gold-bearing solids prior to neutralization and precipitation. This reduces the amount of acid to be neutralized prior to cyanidation and allows acidic solution to be recycled for pre-conditioning of the feed before oxidation. Separation is achieved using one or more thickening stages, which allows the slurry to be washed. Species that are dissolved in the autoclave, such as iron, aluminum, magnesium, arsenic, antimony, and many base metals, can be recovered from the solution phase and disposed of independently, avoiding the formation of precipitates on and around ore particles. This can significantly improve the response of the oxidized solids to subsequent processing as a result of reduced slurry viscosity, less occlusion of gold particles by precipitated salts, and reduced quantities of cyanide and oxygen consumers. The advantages of separating and washing the oxidized solids are greater for high-sulfur materials, because these tend to produce larger amounts of deleterious precipitates.

Neutralization requirements are also less for option 2 than option 1, because of the ability to recycle a portion of the acidic liquid phase without neutralizing.

Neutralization is typically achieved by the addition of ground limestone (or other carbonate source) and/or milk of lime to the acidic slurry or solution. It may also be possible to use flotation plant tailings and/or cyanidation plant tailings for partial neutralization, depending on the properties of the materials and overall flowsheet considerations.

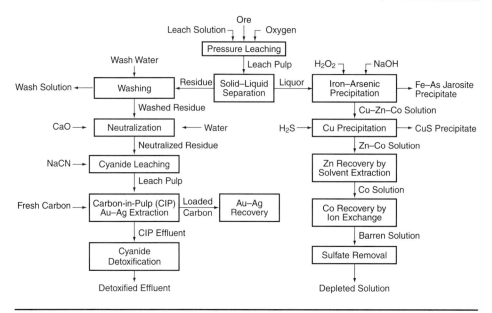

FIGURE 5.15 Scheme for base metals recovery from acidic pressure oxidation products [33]

Oxidation products may also be used to assist in cyanide destruction prior to tailings disposal.

Several schemes have been proposed for the recovery of base metals from oxidized slurries, and one such scheme is shown in Figure 5.15.

5.3.4.4 Whole Ore vs. Concentrate Treatment
The decision to treat a whole ore or a concentrate depends on a number of related factors:

- The response of the ore to flotation, that is, gold and sulfur recovery and concentrate grade
- Sulfide sulfur and carbonate content of the ore, and their effect on the heat and acid balances of the circuit
- The uniformity of the orebody, that is, the variability of sulfur content (because flotation may produce a more uniform feed to the oxidation circuit)

Excellent examples of the trade-off between treatment of whole ore and concentrate can be found in the literature [28, 30, 34, 35, 36]. In some cases, flotation concentrate may be blended with whole ore to optimize the sulfide sulfur content of the feed.

The processing of whole ore versus concentrate does not necessarily have a major impact on autoclave size, because this depends primarily on the total amount of contained sulfur in the feed. However, treatment of a smaller volume of concentrate may have a significant impact on the sizing of other unit process operations. In addition, flotation may serve to reduce the carbonate content of oxidation feed by rejecting carbonate minerals.

5.3.4.5 Gold Recovery
Of all the available oxidation processes, acidic pressure oxidation generally yields the best gold recoveries. A wide variety of ores and concentrates have been tested, with gold recoveries typically ranging from 90% to 95%. Some carbonaceous ores do not respond

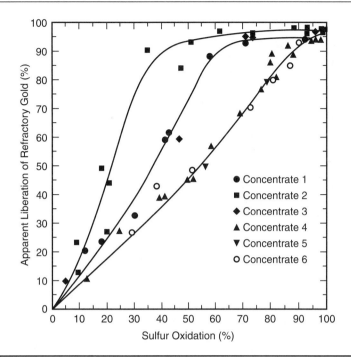

FIGURE 5.16 Examples of the effect of sulfur oxidation on refractory gold liberation [18]

as well because of adsorption of dissolved gold onto ore constituents. However, in some cases pressure oxidation has been applied successfully to treat ores containing carbonaceous, preg-robbing constituents [29].

The relationship between the degree of oxidation and gold recovery for six different sulfide materials is shown in Figure 5.16, which highlights the need for individual consideration of each ore or concentrate through detailed test work.

5.4 OXYGEN: HIGH-PRESSURE NONACIDIC OXIDATION

Nonacidic pressure oxidation uses similar conditions of temperature, pressure, and oxygenation to the acidic process, described in Section 5.3, but is operated under neutral or slightly alkaline pH. The process is applicable to the treatment of refractory ores which contain large amounts of acid-consuming carbonates but have low sulfide sulfur content and are consequently less suitable to acidic oxidation processes. The main difference is that no acid is added to the process, and any acid generated is rapidly neutralized by carbonates in the feed. Many of the factors that affect the efficiency of nonacidic pressure oxidation are similar to those for acidic systems (Section 5.3) and are not considered further.

5.4.1 Reaction Chemistry and Conditions

In neutral and alkaline solutions, and in the presence of dissolved oxygen, pyrite, pyrrhotite, arsenopyrite, and chalcopyrite are oxidized as follows:

$$4FeS_2 + 15O_2 + 14H_2O \rightleftharpoons 4Fe(OH)_3 + 16H^+ + 8SO_4^{2-} \quad \text{(EQ 5.55)}$$

$$4Fe_7S_8 + 69O_2 + 74H_2O \rightleftharpoons 28Fe(OH)_3 + 64H^+ + 32SO_4^{2-} \quad \text{(EQ 5.56)}$$

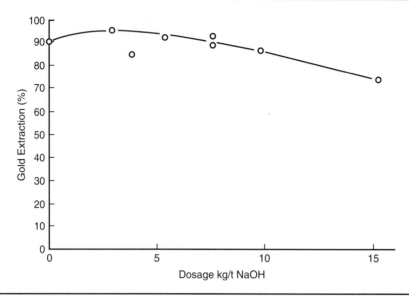

FIGURE 5.17 An example of the effect of sodium hydroxide dosage on nonacidic pressure oxidation: 30-min autoclave oxidation at 225°C [38]

$$2FeAsS + 7O_2 + 8H_2O \rightleftharpoons 2Fe(OH)_3 + 2H_3AsO_4 + 4H^+ + 2SO_4^{2-} \quad \text{(EQ 5.57)}$$

$$4CuFeS_2 + 17O_2 + 18H_2O \rightleftharpoons 4Cu(OH)_2 + 4Fe(OH)_3 + 16H^+ + 8SO_4^{2-} \quad \text{(EQ 5.58)}$$

The insoluble metal oxides or hydroxides formed may coat gold surfaces and sulfide minerals, reducing both gold solubility and the extent of sulfide oxidation. These problems are exacerbated as the sulfide content increases, and consequently this treatment is best suited to low-sulfur feeds.

Carbonates are essentially unreactive in neutral or alkaline media, although they may assist in neutralizing acid that is generated by sulfide oxidation reactions. Any carbon dioxide that is evolved reduces the efficiency of oxidation in a similar manner to that described for the acidic process (Section 5.3.4.1).

One advantage of nonacidic oxidation is that silver jarosite is not formed under these conditions, and high silver recoveries can generally be obtained directly by cyanide leaching following oxidation [37]. Some gold may also be dissolved by the complexing action of thiosulfate species formed during the oxidation process. An example of the high gold extractions possible by cyanide leaching following nonacidic pressure oxidation is given in Figure 5.17, which shows the effect of sodium hydroxide dosage on subsequent gold extraction by cyanidation.

In contrast to acidic pressure oxidation, mercury (and thallium) is dissolved and remains in solution in the autoclave discharge. If a solid–liquid separation step is used prior to cyanide leaching, then much of the mercury can be removed and precipitated from the solution, reducing the amount of mercury reporting to the gold recovery circuit. However, this may create a new problem because the precipitate formed must be disposed of in an acceptable manner.

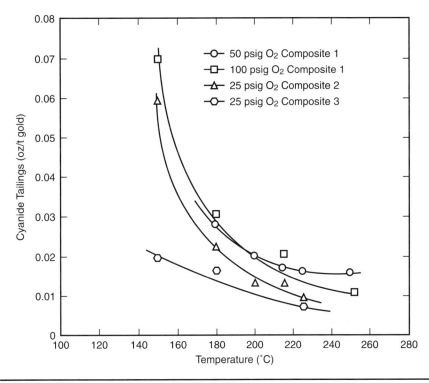

FIGURE 5.18 The effect of oxygen partial pressure and temperature in pressure oxidation on cyanidation extraction for Mercur ore [39]

5.4.2 Reaction Kinetics

The factors affecting oxidation kinetics are very similar to those affecting acidic pressure oxidation, as summarized in Table 5.5. Higher total pressures may be necessary to achieve desired oxygen partial pressures, because ores with high carbonate content may evolve large amounts of carbon dioxide, which dilutes the oxygen content of the gas phase in the autoclave. The effects of oxygen partial pressure and temperature on final solids tailings gold grade at Mercur appear in Figure 5.18.

5.4.3 Process Considerations

5.4.3.1 Applications

Ores that contain large amounts of acid-consuming carbonates (i.e., >10% CO_3^{2-}) and that have relatively low sulfide sulfur content (i.e. approximately <2%), and consequently low acid-generating potential, are most likely to be suitable for high-pressure nonacidic oxidation. In such cases, the cost of sodium hydroxide to maintain neutral or alkaline conditions is less than the cost of acid and neutralizing reagents for oxidation under acidic conditions. Nonacidic oxidation is only suitable for treatment of this specific type of material, because materials with lower acid-consuming potential and higher acid-generating potential result in prohibitively high sodium hydroxide consumption. The process was applied commercially at Mercur from 1988 into the 1990s, where oxidation was performed at 220°C and at 140 to 180 kPa oxygen partial pressure (3,300 kPa total pressure), with sodium hydroxide consumption of 2 to 2.5 kg/t [22].

5.5 NITRIC ACID OXIDATION

The use of nitric acid as an oxidant for the treatment of refractory pyritic and arsenopyritic ores and concentrates has been investigated since the early 1980s. Nitric acid oxidation has the potential to provide the fastest kinetics of the hydrometallurgical oxidation routes considered in this chapter (see Table 5.7). Several processes have been patented and developed to varying extents, including the Arseno, Nitrox, NSC, and Redox processes. The Sunshine mine in Idaho (United States) has used a nitric–sulfuric acid pressure oxidation (NSC, or nitrogen species-catalyzed) process since 1985, principally for silver and copper recovery from refractory concentrates [40]. The NSC process has potential application for the treatment of refractory gold-bearing materials. None of the other nitric acid–based oxidation processes (Arseno, Nitrox, Redox) have been applied commercially due to their relative complexity and high cost.

5.5.1 Reaction Chemistry

The chemical reactions involved in the proposed nitric acid oxidation processes are generally similar. Nitric acid completely dissociates in water, as follows:

$$HNO_3 \rightleftharpoons H^+ + NO_3^-; \quad K_a = 1 \tag{EQ 5.59}$$

The nitrate ion produced is an oxidizing agent, which can undergo a variety of oxidation–reduction reactions, for example:

$$NO_3^- + 2H^+ + e \rightleftharpoons NO_2 + H_2O; \quad E^0 = +0.81 \text{ (V)} \tag{EQ 5.60}$$

$$NO_3^- + 3H^+ + 2e \rightleftharpoons HNO_2 + H_2O; \quad E^0 = +0.94 \text{ (V)} \tag{EQ 5.61}$$

$$NO_3^- + 4H^+ + 3e \rightleftharpoons NO + 2H_2O; \quad E^0 = +0.96 \text{ (V)} \tag{EQ 5.62}$$

$$2NO_3^- + 10H^+ + 8e \rightleftharpoons N_2O + 5H_2O; \quad E^0 = +1.11 \text{ (V)} \tag{EQ 5.63}$$

$$2NO_3^- + 12H^+ + 10e \rightleftharpoons N_2 + 6H_2O; \quad E^0 = +1.24 \text{ (V)} \tag{EQ 5.64}$$

$$NO_3^- + 10H^+ + 8e \rightleftharpoons NH_4^+ + 3H_2O; \quad E^0 = +0.87 \text{ (V)} \tag{EQ 5.65}$$

The nitrite species produced in Equation (5.61) can be further reduced:

$$HNO_2 + H^+ + e \rightleftharpoons NO + H_2O; \quad E^0 = +0.99 \text{ (V)} \tag{EQ 5.66}$$

Both the nitrate and nitrite species are capable of oxidizing sulfides in a series of complex reactions. The nitrite species is the stronger oxidant; however, there is evidence that the oxidation mechanism takes place through an intermediate species, NO^+, which is thought to be a highly reactive and effective oxidant for sulfide minerals, as follows [41]:

$$NO^+ + e \rightleftharpoons NO(g); \quad E^0 = 1.450 \text{ (V)} \tag{EQ 5.67}$$

The main reduction product is nitrous oxide (NO). Lesser quantities of nitrogen dioxide (NO_2), dinitrogen oxide (N_2O), dinitrogen tetroxide (N_2O_4), and nitrogen (N_2) are also

TABLE 5.7 Operating conditions for various nitric acid oxidation processes

Process	Temperature (°C)	Oxygen Partial Pressure (kPa)	[HNO$_3$]*	References
Arseno	60 to 80	100 to 700	140 to 180 g/L	[43, 46]
Nitrox	85 to 95	atm	10 wt %	[37, 48]
Redox	195 to 210	345	70 to 110 g/L	[44]
Sunshine	50 to 170	200 to 300	2 g/L	[40, 47]

* HNO$_3$ concentration.

produced. Nitrous oxide is evolved in the gaseous phase and reacts rapidly in an oxygen-rich atmosphere to produce nitrogen dioxide:

$$2NO + O_2 \rightleftharpoons 2NO_2 \quad (EQ\ 5.68)$$

The nitrogen dioxide and its dimer (N$_2$O$_4$) are both highly soluble in water and can be scrubbed out of the gas to produce a mixture of nitrous and nitric acids, thereby regenerating the reactants, as follows:

$$2NO_2 + 2H_2O \rightleftharpoons HNO_2 + HNO_3 \quad (EQ\ 5.69)$$

$$N_2O_4 + H_2O \rightleftharpoons HNO_2 + HNO_3 \quad (EQ\ 5.70)$$

$$6NO_2 + 2H_2O \rightleftharpoons 4HNO_3 + 2NO \quad (EQ\ 5.71)$$

The regenerated reagents can be returned to the oxidation process, resulting in very efficient use of the oxidant. Approximately 99% of the NO$_x$ gases evolved in the process can be recovered and recycled in this manner. The remaining 1% is either lost as inert dinitrogen oxide, produced in Equation (5.63), or as a very small amount of unscrubbed nitrous oxide and nitrogen dioxide (see Section 5.5.3.4).

The use of nitrate–nitrite ions for sulfide oxidation has several advantages over oxygen, including:

- High solubility of nitrogen dioxide in water
- Ability to regenerate the oxidant in the gaseous phase
- Higher reduction potential resulting in faster oxidation kinetics

The simplified oxidation reaction for pyrite oxidation is as follows:

$$2FeS_2 + 10HNO_3 \rightleftharpoons 2Fe^{3+} + 2H^+ + 4SO_4^{2-} + 10NO + 4H_2O \quad (EQ\ 5.72)$$

This reaction proceeds rapidly above approximately 60°C and below pH 1.7 [42]. At high nitrate–nitrite concentrations (i.e., >50 g/L HNO$_3$), very little elemental sulfur is formed, even at low temperatures [43, 44]. At lower nitrate–nitrite concentrations, the amount of elemental sulfur formed increases, and at low temperatures substantial elemental sulfur is formed [40]. Investigations using linear potential sweep voltammetry have shown that the extent of sulfide sulfur conversion to elemental sulfur is strongly dependent on the potential, with the amount varying from 70% at 0.82 V to a negligible amount at 1.5 V (vs. SHE) [45].

Arsenopyrite reacts similarly to pyrite but is considerably more reactive, and oxidation is possible at ambient temperatures, depending on the acid concentration. The decomposition of arsenopyrite in nitric acid is known to generate elemental sulfur, as shown in the following simplified expression:

$$3FeAsS + 12HNO_3 \rightleftharpoons 3FeAsO_4 + 4H^+ + 2SO_4^{2-} + 4H_2O + S + 12NO \qquad (EQ\ 5.73)$$

Up to 70% of the contained sulfide sulfur may form elemental sulfur rather than the more soluble sulfate species [42]. This is exacerbated at low temperatures because sulfur is not oxidized to any appreciable extent below 160°C. Sulfur formation can be reduced at low temperatures by increasing the free acid concentration, that is, by the addition of sulfuric acid, either produced naturally by pyrite decomposition, or added as a reagent. The simplified expression for the modified reaction is as follows [44]:

$$3FeAsS + 3H^+ + 14HNO_3 \rightleftharpoons 3Fe^{3+} + 3SO_4^{2-} + 3H_3AsO_4 + 14NO + 4H_2O \qquad (EQ\ 5.74)$$

Although this helps to reduce sulfur formation, some sulfur is still formed, which can severely hinder subsequent gold recovery by coating previously exposed gold surfaces and by consuming cyanide. Several methods for removing sulfur have been investigated with mixed success [44]. Pyrrhotite, chalcopyrite, and sphalerite decompose rapidly in nitric acid but also produce large amounts of elemental sulfur at low temperatures, resulting in similar processing problems.

Gold is unreactive in pure nitrate–nitrite solutions, but some dissolution may occur in sulfide oxidation systems due to the formation of thiosulfates. Gold dissolution values ranging from 3% to 54% have been reported for two different nitric acid oxidation processes in the absence of cyanide [44, 46]. Cyanide leaching of neutralized nitric acid oxidation products yields highly variable gold extraction depending on the ore mineralogy. Typically, extractions of 85% to 95% have been obtained.

Both silver and copper are highly soluble in nitrate solutions and rapidly dissolve to form the respective nitrates. Very little, if any, silver jarosites are formed during the oxidation step, and, unlike high-pressure acidic oxidation processes, high silver recoveries should be obtained in subsequent extraction processes.

5.5.2 Reaction Kinetics

The high reduction potential of nitrate–nitrite species in solution and the high solubility of nitrogen dioxide means that high oxidation rates can be achieved at much lower temperatures and pressures than those required for pressure oxidation with oxygen alone (Sections 5.3 and 5.4). For example, Figure 5.19 shows the oxidation rates of pyrite and arsenopyrite in a solution containing 12 wt % nitric acid, corresponding to a solution potential of 750 mV (vs. SCE [standard calomel electrode]), at 80°C and with 100 kPa oxygen partial pressure. Almost complete oxidation of a coarse ground (15% <75 μm) pyrite sample was achieved in approximately 45 min and for a slightly finer arsenopyrite sample (48% <75 μm) in less than 10 min [42].

The kinetics of oxidation by nitric acid are strongly dependent on temperature and nitric acid concentration. The oxidation rate of finely ground sulfides at 200°C is approximately double that at 100°C. The oxidation times required for different materials obviously depend on many factors, including particle size distributions and sulfur content. The main advantage of operating at the higher temperature is the elimination of elemental sulfur and its associated problems; however, the choice of temperature depends on the mineralogy of the material to be oxidized.

The effect of nitric acid concentration on oxidation rates for pyrite and arsenopyrite is shown in Figures 5.20 and 5.21, respectively. Oxidation rates of both minerals are increased significantly as the acid strength is increased from 20 to 140 g/L HNO_3. Most of the nitric acid processes that have been developed use acid concentrations between 70 and 180 g/L to achieve fast oxidation. An exception to this is the Sunshine process,

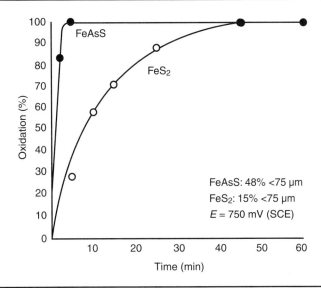

FIGURE 5.19 Rates of pyrite and arsenopyrite oxidation in nitric acid at 80°C, 10% solids [42]

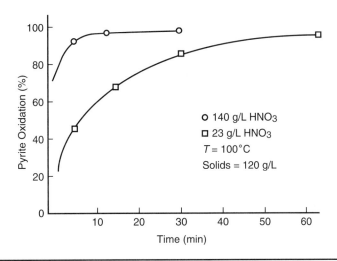

FIGURE 5.20 Effect of HNO_3 concentration on oxidation of pyrite [43]

where nitric acid concentration of 2 g/L is applied in conjunction with 200 g/L sulfuric acid. In this case, the process is applied to concentrate ground to 80% <25 μm to achieve acceptable kinetics [40].

The rate of oxidation of nitrous oxide to nitrogen dioxide is dependent on the partial pressure of the two reacting gases. Either air or oxygen can theoretically be used; however, the use of pure oxygen results in faster kinetics, with the added benefit of producing lower levels of the less soluble nitrogen oxides in the process off-gases. The rate of oxidation of nitrous oxide is inversely proportional to temperature, which means that in practice a compromise must be reached between sulfide mineral oxidation kinetics and oxidant regeneration [47].

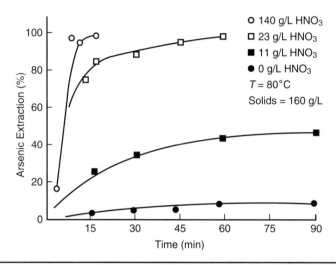

FIGURE 5.21 Effect of HNO$_3$ concentration on leaching of arsenic from arsenopyrite [43]

5.5.3 Process Considerations

5.5.3.1 Applications and Circuit Configurations

A number of process flowsheets have been proposed for nitric acid oxidation of sulfide minerals. Many of these are highly complex and have been specifically designed for oxidation of particular ores and concentrates. Although the major objective of nitric acid oxidation is to decompose the refractory sulfides in the material of interest, any nitric acid process developed must also meet several other important criteria:

- The process tailings must comply with environmental control regulations with respect to nitrate and dissolved arsenic (and other) species.
- Any gaseous discharge to the atmosphere must comply with environmental control regulations regarding NO$_x$ (and other gases) emissions.

All of the processes use quite similar steps, arranged in slightly different configurations and operated under different conditions. A simplified generic process scheme appears in Figure 5.22.

The operating conditions of four patented nitric acid processes are summarized in Table 5.7. The low-temperature processes (Arseno and Nitrox) rely on the oxidation of elemental sulfur by nitrate–nitrite species, which is not very effective, and a number of alternative methods for removing the elemental sulfur formed in these systems has been investigated [48]. The Redox process, which is effectively high-temperature pressure oxidation in the presence of nitric acid, operates under conditions that prevent elemental sulfur formation [44]. The NSC process operates at low nitrate–nitrite concentration (2 g/L) over a range of temperatures, depending on the material being treated, and forms elemental sulfur [40]. In the cases of the Arseno, Nitrox, and NSC processes, the formation of elemental sulfur may affect downstream gold extraction and recovery processes. If a significant amount of sulfur is present in the oxidation product, it must either be removed to avoid excessively high cyanide consumption or an alternative leaching scheme must be considered (see Chapter 6).

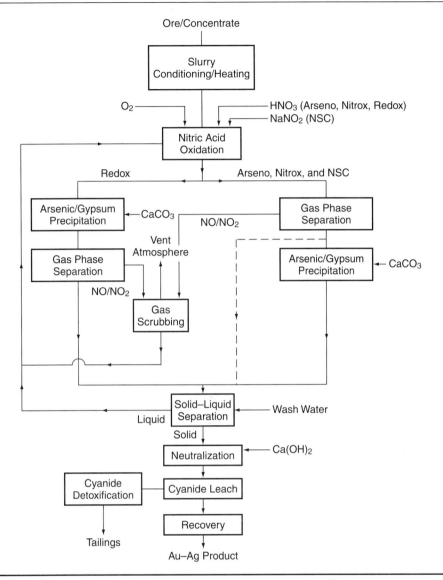

FIGURE 5.22 Flowsheet options for nitric acid sulfide oxidation (simplified)

5.5.3.2 Control of Nitrate Concentrations in Effluent Slurry

The concentration of nitrate ions in the effluent slurry is a function of the nitrate concentration at the completion of oxidation and the subsequent washing efficiency. Washing efficiencies are determined by the solid–liquid separation method and efficiency, and the process water balance. Several methods of reducing nitrate levels in the oxidized slurry have been investigated, including:

- Starvation of nitric acid so that most of the nitrate species are consumed during oxidation
- Addition of barren iron sulfides (e.g., pyrrhotite) or scrap iron to consume excess acid and nitrate species following oxidation

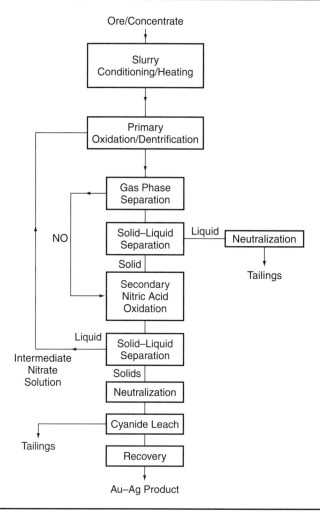

FIGURE 5.23 Flowsheet option for nitric acid sulfide oxidation, including denitrification step

- Inclusion of a denitrification step consisting of reacting fresh ore or concentrate feed with an intermediate nitrate solution (Figure 5.23).

All of these methods are sensitive to variations in feed mineralogy and to fluctuations in oxidation conditions. Consequently, circuits using this technology are expected to be difficult to consistently keep nitrate concentrations within acceptable limits. It is claimed that 99.5% recovery of nitrates can be achieved from the oxidized slurry to generate a product (i.e., cyanidation feed) containing less than 20 mg/L nitrate ions using any or a combination of the methods listed, coupled with an efficient slurry washing step.

5.5.3.3 Neutralization

Some of the proposed processes include a precipitation step, following oxidation, for the removal of dissolved arsenic species from solution prior to recycling it back to oxidation. Precipitation is carried out by raising the pH to between 3 and 4 with limestone or lime slurry. Stable Fe(III) arsenate is formed, provided that the Fe(III)–As(V) ratio is kept above approximately 4:1 (Section 5.1.6.2).

Prior to cyanide leaching, the oxidized slurry must be neutralized and the pH adjusted to approximately 10.5. This is achieved by the addition of lime. During neutralization the common iron precipitation reactions occur (Section 5.1.6.1), as well as precipitation of silver jarosite. It has been suggested that flowsheets for ores containing significant quantities of silver should probably include a step to recover silver directly from the nitrate solution used for oxidation. The inclusion of a hot lime treatment prior to cyanide leaching may help to prepare oxidized material for leaching and improve silver (and gold) recovery. This treatment consists of adjusting the pH of the slurry to 11 with lime and agitating for at least 3 hr at 80°C [49].

5.5.3.4 Control of NO_x in Gaseous Discharges

All of the nitrate-based processes need to discharge gases to the atmosphere. Although nitrogen dioxide is highly soluble in water, nitrous oxide is produced as a by-product of its dissolution (Equation [5.71]). Further reaction with oxygen takes this by-product into solution; however, several stages of adsorption are required to adequately scrub the exhaust gases. Commercial nitric acid plants use more than 20 stages of adsorption for NO_x recovery.

Alternatively, a combination of gas-recovery processes can be used whereby the majority of the nitrous oxide and nitrogen dioxide is adsorbed and recycled while the remainder is scrubbed out in alkaline solution. Efficient scrubbing can be performed on gases containing a nitrogen dioxide–nitrous oxide ratio of 1:1 or above. Several different alkaline media are available for scrubbing. The reactions for the different media are listed in order of decreasing effectiveness [42]:

$$NO + NO_2 + 2NaOH \rightleftharpoons 2NaNO_2 + H_2O \tag{EQ 5.75}$$

$$NO + NO_2 + Ca(OH)_2 \rightleftharpoons Ca(NO_2)_2 + H_2O \tag{EQ 5.76}$$

$$NO + NO_2 + CaCO_3 \rightleftharpoons Ca(NO_2)_2 + CO_2 \tag{EQ 5.77}$$

Sodium hydroxide is the most efficient scrubbing medium, followed by lime and then limestone. All of the alkaline scrubbing reactions listed are most effective at low temperatures, that is, 25°C rather than 80°C. The most cost-effective scrubbing scheme depends on many factors but may involve two stages using low-cost limestone in the first stage, followed by a second stage of caustic or lime scrubbing to produce a clean gas containing environmentally acceptable levels of NO_x species [42].

5.5.3.5 Equipment

The fast kinetics of the nitric acid processes, and the correspondingly short retention times theoretically required for oxidation, suggest that pipe reactors may be appropriate, especially for the treatment of small volumes of concentrates [44]. Pipe reactors provide close to plug flow mixing, good pressure and temperature control, and have excellent gas–liquid mass transfer characteristics. Alternatively, other more conventional, closed agitated tank or pressure leaching systems may be used. The Sunshine (NSC process) facility in Idaho utilized batch autoclaves for the process [40].

5.6 CHLORINE OXIDATION

Aqueous solutions of chlorine have strong oxidizing capabilities and have been used widely as oxidants in water and waste treatment. Chlorination of carbonaceous gold ores was developed in the late 1960s and has been applied successfully since 1971 at two operations in Nevada (see Section 12.2.9.1). In this case, chlorine was used to deactivate preg-robbing or preg-borrowing carbonaceous matter. Chlorine is also used in gold refining (Chapter 10) and cyanide detoxification processes (Chapter 11), and has been applied historically for leaching of gold ores and concentrates (Chapter 6). The application of chlorine for gold leaching is considered in detail in Section 6.2. The following sections focus on the use of chlorine–chloride media for sulfide mineral oxidation and deactivation of carbonaceous matter.

5.6.1 Chlorine Chemistry

Chlorine gas is highly soluble in water and dissolves to form hydrochloric and hypochlorous acids [50]:

$$Cl_2(g) \rightleftharpoons Cl_2(aq); \quad K = 6.2 \times 10^{-2} \qquad (EQ\ 5.78)$$

$$Cl_2(aq) + H_2O \rightleftharpoons HCl + HOCl; \quad K = 4 \times 10^{-4} \qquad (EQ\ 5.79)$$

Hydrochloric acid is a strong acid which completely dissociates in dilute aqueous solution:

$$HCl \rightleftharpoons H^+ + Cl^-; \quad K_a = 1 \qquad (EQ\ 5.80)$$

Hypochlorous acid is a weak acid:

$$HOCl \rightleftharpoons H^+ + OCl^-; \quad pK_a = 7.5 \qquad (EQ\ 5.81)$$

At pH values below 7.5 the hypochlorous species predominates, as shown in the distribution diagram for chlorine species in aqueous solution (Figure 5.24).

The two half-reactions for Equation (5.79) can be written as follows:

$$2HOCl + 2H^+ + 2e \rightleftharpoons Cl_2 + 2H_2O; \quad E^0 = +1.64\ (V) \qquad (EQ\ 5.82)$$

$$Cl_2(aq) + 2e \rightleftharpoons 2Cl^-; \quad E^0 = +1.358\ (V) \qquad (EQ\ 5.83)$$

It can be seen that both the hypochlorous (HOCl) and aqueous chlorine (Cl_2(aq)) species are strong oxidizing agents, with HOCl the preferred species for oxidation of sulfides and for deactivation of carbonaceous matter.

Above 50°C, chlorine dissolves to form the strongly oxidizing chlorate species (ClO_3^-):

$$3Cl_2 + 3H_2O \rightleftharpoons ClO_3^- + 5Cl^- + 6H^+ \qquad (EQ\ 5.84)$$

and

$$2ClO_3^- + 12H^+ + 10e \rightleftharpoons Cl_2 + 6H_2O; \quad E^0 = +1.47\ (V) \qquad (EQ\ 5.85)$$

Alternatively, the strongly oxidizing hypochlorite species can be generated by the addition of inorganic hypochlorite salts to a solution:

$$NaOCl \rightleftharpoons Na^+ + OCl^- \qquad (EQ\ 5.86)$$

$$Ca(OCl)_2 \rightleftharpoons Ca^{2+} + 2OCl^- \qquad (EQ\ 5.87)$$

The hypochlorite species further reacts with water to form hypochlorous acid with a corresponding increase in pH:

$$OCl^- + H_2O \rightleftharpoons HOCl + OH^- \qquad (EQ\ 5.88)$$

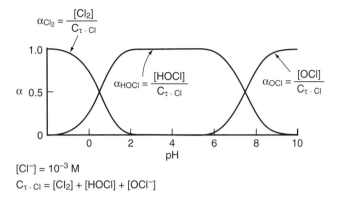

FIGURE 5.24 Distribution diagram for chlorine species in water at 25°C [50]

In addition to the application for oxidation of sulfide minerals and deactivation of carbonaceous matter, chlorine–chloride media is also capable of dissolving gold, silver, and, importantly, Au–Ag–Te alloys that are refractory to cyanidation, as considered in Chapter 6.

5.6.2 Deactivation of Carbonaceous Material

Since 1970, the primary application of chlorination has been to deactivate gold-adsorbing organic carbon, which occurs naturally in certain ores and may significantly reduce gold recovery by cyanidation. The mineralogical characteristics of these carbonaceous ores have been described in detail in Section 2.12. Ores containing minor amounts of gold-adsorbing carbonaceous material, or materials that have weak preg-robbing characteristics, may be processed effectively by adding a suitable surfactant to the ore prior to cyanide leaching (Section 3.3.6). This process is sometimes termed "blanking" or "blinding" of preg-robbing ores. Commonly used surfactants for this process include kerosene, diesel oil, sodium lauryl sulfate, and petroleum sulfate [51]. Although this technique is generally less costly than chlorination, it is not always effective and, in such cases, chlorination (or roasting) may be required.

The mechanism of deactivation, or passivation, with chlorine is not well understood. Direct oxidation of organic carbon to carbon dioxide is theoretically possible:

$$2HOCl + C \rightleftharpoons CO_2 + 2HCl \qquad \text{(EQ 5.89)}$$

However, the efficiency of deactivation, as quantified by the response of the ore to cyanide leaching following treatment, has been found to be unrelated to any change in the organic carbon content of the material being oxidized. This would indicate that Equation (5.89) is not responsible for deactivation. Changes in organic carbon content during chlorination have been recorded both experimentally and in practice [52, 53]. Such changes are probably related to the severity of oxidation, and carbon destruction does not appear to be necessary for effective deactivation.

During chlorination, the surfaces of the organic carbon are modified by chlorine, either by the formation of a chlorhydrocarbon layer or by the formation of carbonyl structures, composed mainly of carboxyl groups (COOH). These surface groups passivate the carbon by blocking active adsorption sites. It has been suggested that ionization of surface groups (such as COOH) in alkaline solution results in a negative charge at the surface of the organic material, thus repelling the negatively charged Au(I) cyanide ions [52].

188 | THE CHEMISTRY OF GOLD EXTRACTION

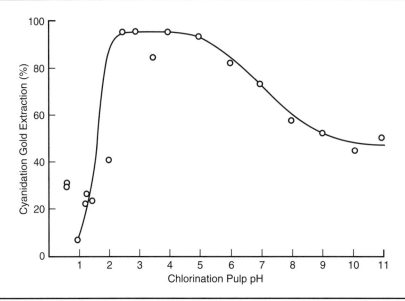

FIGURE 5.25 Cyanidation gold extraction as a function of the chlorination pulp pH [54]

Chlorine oxidation is most effectively achieved between pH 3 and 5, in the pH range where the hypochlorous species predominates (Figure 5.24). This is further evidence of the strong oxidizing properties of hypochlorous species over other chlorine solution species. Deactivation efficiency is reduced below pH 2 and above pH 6 as a result of reduced hypochlorous activity. The effect of pH on deactivation efficiency is shown in Figure 5.25, which indicates gold extraction for carbonaceous Carlin ore (Nevada) as a function of pH, following chlorination treatment.

Hypochlorous concentrations of approximately 1 g/L are required for effective deactivation of carbonaceous matter, although the exact level depends on the amount of competing reactive species in the ore, for example, sulfides and carbonates, and the solution conditions. Lower concentrations may be perfectly satisfactory with extended oxidation times.

The optimum temperature for oxidation is a compromise between chlorine gas absorption, favored at low temperatures, and the diffusion rate of hypochlorous species and chemical kinetics, both more favorable at higher temperatures. The effects of temperature on chlorine solubility and the diffusion coefficient of hypochlorous species are shown in Table 5.8 and Figure 5.26, respectively, from which it can be seen that the solubility decreases and the diffusion coefficient increases with increasing temperature. A temperature of 50°C has been found to be optimal in practice [54, 55].

5.6.3 Sulfide Oxidation

Aqueous chlorine will readily oxidize all sulfides commonly associated with gold. The reaction for pyrite oxidation within the pH range of operation for treatment of carbonaceous ores is given by:

$$2FeS_2 + 15HOCl + 7H_2O \rightleftharpoons 2Fe(OH)_3 + 23H^+ + 4SO_4^{2-} + 15Cl^- \quad \text{(EQ 5.90)}$$

The other sulfides react similarly, and sulfur is readily oxidized to sulfate in the strongly oxidizing conditions. Arsenic minerals decompose to form arsenic trichloride, which can

TABLE 5.8 The effect of temperature on chlorine solubility in water

Temperature (°C)	Cl_2 Solubility (mol Cl_2/mol H_2O)
20	7×10^{-3}
40	4.1×10^{-3}
60	2.5×10^{-3}

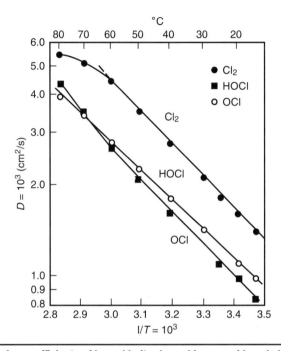

FIGURE 5.26 Diffusion coefficients of hypochlorite, hypochlorous acid, and chlorine [56]

be precipitated and recovered as a stable product [57]. The factors affecting the rate and extent of sulfide oxidation are similar to those affecting other aqueous sulfide oxidation processes and include particle size and sulfide mineralogy. Small spheroidal pyrite (0.5 to 2 μm) particles with well-developed pore structures are much more rapidly oxidized than large euhedral, nonporous, grains [58, 59].

Chlorination is not an economic method of treating refractory sulfide ores due to excessive chlorine consumption. For example, an ore containing 1% sulfide sulfur as pyrite will consume 82 kg/t of Cl_2, if the pyrite is completely oxidized. If the same amount of sulfur is present as pyrrhotite, which is also completely oxidized, then >100 kg/t Cl_2 is required. For this reason chlorination is restricted to treatment of ores with low sulfide content, typically below about 1%. Ores and concentrates with higher sulfide contents may need to be treated by one of the alternative options, for example, preaeration, pressure oxidation, or roasting.

5.6.4 Effect of Other Ore Constituents

Carbonate minerals react with both hydrochloric and hypochlorous acids to evolve carbon dioxide, as follows:

$$CaCO_3 + 2HCl \rightleftharpoons CaCl_2 + H_2O + CO_2 \qquad (EQ\ 5.91)$$

$$CaCO_3 + 2HOCl \rightleftharpoons Ca(OCl)_2 + H_2O + CO_2 \qquad (EQ\ 5.92)$$

Both reactions increase the pH of the slurry and consume chlorine (1% carbonate in an ore will consume 11.7 kg/t Cl_2, if the carbonate completely reacts). Carbonates may also react readily with the acid generated from sulfide oxidation reactions to form gypsum, as follows:

$$CaCO_3 + H_2SO_4 \rightleftharpoons CaSO_4 + H_2O + CO_2 \qquad (EQ\ 5.93)$$

which can adversely affect subsequent processing by increasing slurry viscosity and by coating gold surfaces, thereby inhibiting dissolution.

5.6.5 Process Considerations

Two commercial operations used chlorination for treatment of carbonaceous ores in the 1980s and 1990s: Carlin and Jerritt Canyon (see Section 12.2.9.1). A summary of ore compositions and chlorine consumptions for these operations is given in Table 5.9.

The chlorination step is performed in well-agitated tanks at 35°C to 50°C. Chlorine gas is sparged into the slurry beneath radial impellers for optimum gas dispersion and to achieve effective chlorine utilization (i.e., >90%). The tanks are closed and operated under a slightly negative pressure (5 cm H_2O) to control fugitive chlorine gas. The exhaust gases are scrubbed with a carbonate–bicarbonate solution, which is returned to the chlorination circuit as bleach. Chlorination circuit retention times vary from 10 to 20 hr. In 1987, Carlin introduced a "flash" chlorination system whereby chlorine was added over a short time (15 min) into small tanks, and the hypochlorous species were allowed to decay through the remainder of the oxidation circuit. A soak period of at least 1 hr at 1.0 g/L HOCl is required for satisfactory carbon deactivation [53, 55].

Carryover of chlorine to the cyanide leaching circuit is highly undesirable because it reacts with cyanide to form cyanogen chloride, resulting in high cyanide consumption. The flash chlorination system allows the hypochlorous species concentration to decay to <0.05 g/L in the final stage, which is acceptable as feed to the cyanidation circuit [55]. Residual hypochlorous ions in excess of this value are destroyed by the addition of sodium bisulfite.

A hot sodium carbonate pretreatment step may be required for ores with significant sulfide sulfur content (i.e., >0.5%), which would otherwise consume excessive chlorine. This process is most effectively performed at 60°C to 80°C with sodium carbonate additions of 25 kg/t in a vigorously air-agitated system. The cost of reagents, air, and heat required for this must offset the incremental cost of chlorine without the pretreatment step. The combination of hot sodium carbonate pretreatment and chlorination is sometimes referred to as "double oxidation." It was used at both Jerritt Canyon and Carlin in the late 1970s and early 1980s, but was subsequently discontinued in favor of simple chlorination as a result of ore-type changes and unfavorable economics. Both Jerritt Canyon and Carlin ceased the use of chlorination in the 1990s and subsequently have processed carbonaceous ores by roasting (see Section 5.8).

5.7 BIOLOGICAL OXIDATION

Biological oxidation has been used, sometimes unwittingly, on a commercial scale since the early 1900s for heap and dump leaching of low-grade copper ores. It was not until the

TABLE 5.9 Summary of carbonaceous ore types treated by chlorination

Ore Types	Organic Carbon Content (%)	Sulfide Sulfur Content (%)	Actual Chlorine Consumption (kg/t)	Theoretical* Chlorine Consumption (kg/t)	References
Carlin					
Carbonaceous ore	0.3 to 0.6	0.2 to 0.3	30 to 50	20 to 32	[52, 53]
Sulfide ore	0.1 to 0.4	0.2 to 1.0	>50†	18 to 87	[52, 53]
Jerritt Canyon					
Mildly carbonaceous ore	0.5	0.25 to 0.5	12 to 23	26 to 47	[55]
Highly carbonaceous ore	0.75 to 1.5	1 to 2	>40	90 to 170	[55]

* Assumes that all sulfide sulfur reacts.
† Inferred data.

TABLE 5.10 Historical development of biological oxidation [11, 60, 61]

Year	Event
1670	Copper recovered from acid mine drainage at Rio Tinto (Spain).
1900s	Dump, heap, and in situ leaching of copper ores, Rio Tinto (Spain).
1920s	Several copper ore stockpile and heap leach operations started in the United States.
1950s	Bacterial involvement in copper dump leaching demonstrated (United States). Bacterial action identified at Denison mine (Canada) and West Rand Consolidated (South Africa) as a result of acid mine drainage.
1951	Colmer and Hinkle identified a bacteria in coal mine drainage and named it *Thiobacillus ferro-oxidans*.
1960s to 1970s	Bacterial oxidation research initiated in South Africa, Canada, England, and United States.
1984	0.75-tpd pilot-scale bio-oxidation plant to treat a gold-bearing sulfide concentrate started at Fairview (South Africa) using Gencor BIOX technology.
1984	2-tpd pilot plant started at Equity Silver (Canada).
1986	10-tpd BIOX plant commissioned at Fairview to treat concentrates.
1991	Fairview plant expanded to 35-tpd concentrate capacity.
1991	300-tpd plant commissioned at Sao Bento to partially oxidize flotation concentrates ahead of acid pressure oxidation.
1993	Harbour Lights (Australia) BIOX plant commissioned (only operated 2 years).
1993	Wiluna (Australia) BIOX plant commissioned: 115-tpd concentrate capacity.
1994	Ashanti (Ghana) BIOX plant commissioned: 720 tpd concentrate capacity (three modules).
1994	Youanmi (Australia) BacTech plant commissioned: 120-tpd concentrate capacity.
1995	Ashanti (Ghana) BIOX plant expanded to 960-tpd concentrate capacity.
1995	Newmont Carlin (United States) commissioned 10,000-tpd biological heap leach to process refractory gold ore as a pretreatment ahead of milling and cyanidation.
1998	Tamboraque BIOX plant commissioned in Peru.
2000	Laizhou Gold Enterprise (Shandong, China) BacTech/Mintek (BACOX) plant commissioned: 100-tpd concentrates.

mid-1950s that the catalyzing effect of bacteria on the mineral oxidation reactions was recognized. Since then there has been extensive research into the field of bacterial oxidation of sulfides, including its use for enhancing gold extraction from refractory sulfide ores. Table 5.10 summarizes this historical development. The agitated tank biological oxidation process has emerged as a commercially proven, cost-effective, and environmentally attractive alternative for the treatment of some refractory gold-bearing concentrates, with more than 10 plants commissioned between 1990 and 2005 on five continents. Biological heap oxidation technology for treatment of low-grade refractory gold ores has been in development during this same time period, with a few small-scale operations.

5.7.1 Reaction Chemistry and Mechanism

Many types of naturally occurring bacteria are capable of catalyzing mineral oxidation reactions. *Thiobacillus thio-oxidans* and *Thiobacillus ferro-oxidans* (mesophiles) are broadly suitable for the oxidation of gold-bearing sulfide ores and concentrates, because they thrive under close to ambient temperatures (35°C to 45°C) and feed on unpathogenic (nondisease causing) inorganic compounds [61]. However, bacteria that thrive under higher-temperature conditions, including moderate thermophiles (e.g., *Sulfobacillus acidophilus*), which operate at intermediate temperatures (45°C to 65°C), and extreme thermophiles (e.g., *Sulfolobus*), which can thrive at higher temperatures (65°C to 80°C), are being applied increasingly to achieve faster oxidation kinetics in commercial systems [62, 63].

The bacteria derive energy from the oxidation of sulfur and iron species, but they also require oxygen, carbon, and nitrogen to support oxidation reactions and for cell growth. These chemicals must be supplied either from the ore or from nutrient reagent and/or air additions. The bacteria operate best within a pH range of 1.0 to 1.8.

The oxidation reactions for a generic iron sulfide mineral (FeS_x) in acidic media are given as follows:

$$FeS_x + xO_2 + 2xH_2O \rightleftharpoons FeSO_4 + 4xH^+ + (x-1)SO_4^{2-} + (2x+2)e \quad \text{(EQ 5.94)}$$

$$4FeSO_4 + O_2 + 2H_2SO_4 \rightleftharpoons 2Fe_2(SO_4)_3 + 2H_2O \quad \text{(EQ 5.95)}$$

$$FeS_x + Fe_2(SO_4)_3 \rightleftharpoons 3FeSO_4 + xS \quad \text{(EQ 5.96)}$$

$$2S + 3O_2 + 2H_2O \rightleftharpoons 2H_2SO_4 \quad \text{(EQ 5.97)}$$

Equations (5.95) and (5.97) rely entirely on bacterial catalysis and will not proceed to any appreciable degree in the absence of active bacteria, under ambient conditions. The reaction shown in Equation (5.96) is essentially chemical, with little or no bacterial involvement. There is evidence to suggest that Reaction (5.94) proceeds much faster in the presence of suitable bacteria [64]. The role of bacteria in the oxidation and removal of elemental sulfur produced by Reaction (5.96) is most important, because the sulfur, if not removed, builds up on the sulfide surface, hindering further oxidation of the mineral, possibly occluding gold values and increasing cyanide consumptions in leaching (Section 5.1.5).

The general form of Equations (5.94) to (5.97) can be applied for all the iron sulfide minerals, principally pyrite, marcasite, and pyrrhotite. The reaction chemistry for arsenopyrite and chalcopyrite is as follows:

Arsenopyrite:

$$4FeAsS + 13O_2 + 6H_2O \rightleftharpoons 4H_3AsO_4 + 4FeSO_4$$
$$\text{bacteria-catalyzed} \quad \text{(EQ 5.98)}$$

$$2FeAsS + 7O_2 + H_2SO_4 + 2H_2O \rightleftharpoons Fe_2(SO_4)_3 + 2H_3AsO_4$$
$$\text{bacteria-catalyzed} \quad \text{(EQ 5.99)}$$

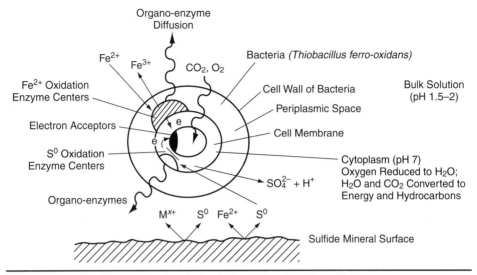

FIGURE 5.27 Simplified mechanism for bacteria-catalyzed sulfide oxidation

$$2FeAsS + Fe_2(SO_4)_3 + 4H_2O + 6O_2 \rightleftharpoons 4FeSO_4 + H_2SO_4 + 2H_3AsO_4$$
$$\text{chemical} \quad \text{(EQ 5.100)}$$

Chalcopyrite:

$$CuFeS_2 + O_2 + 2H_2SO_4 \rightleftharpoons CuSO_4 + FeSO_4 + 2S + 2H_2O$$
$$\text{bacteria-catalyzed} \quad \text{(EQ 5.101)}$$

$$4CuFeS_2 + 17O_2 + 2H_2SO_4 \rightleftharpoons 4CuSO_4 + 2Fe_2(SO_4)_3 + 2H_2O$$
$$\text{bacteria-catalyzed} \quad \text{(EQ 5.102)}$$

$$CuFeS_2 + 2Fe_2(SO_4)_3 \rightleftharpoons CuSO_4 + 5FeSO_4 + 2S$$
$$\text{chemical} \quad \text{(EQ 5.103)}$$

A number of mechanisms have been proposed for bacterial oxidation of sulfides, but the exact mechanism is the subject of debate and controversy. Figure 5.27 shows a simplified, generalized mechanism for bacteria-catalyzed sulfide oxidation. In a slurry suspension containing sulfide and gangue minerals, *Thiobacillii* bacteria are known to attach to the sulfide surfaces, although the exact mechanism of attachment is not well understood [65]. It has been proposed that the attached bacteria serve to oxidize elemental sulfur that is formed at the surface of the sulfide mineral as a result of chemical oxidation reactions involving Fe(III) and the sulfide mineral itself, rather than playing a "direct" role in the mineral oxidation [66]. This infers that the bacteria play an "indirect" role in sulfide mineral oxidation by converting sulfur to sulfate. In addition, in the bulk solution phase, the bacteria also oxidize Fe(II) to Fe(III), deriving energy from this reaction and creating valuable oxidant to supplement the direct action of bacteria on the sulfide minerals [60]. Regardless of the exact mechanism of bacterial action (i.e., indirect or direct), the bacteria play a critical role in accelerating sulfide mineral oxidation compared with the action of Fe(III) species alone.

From a strictly chemical point of view, the oxidation reactions described are exothermic. However, the exothermic nature of the reactions is reduced when bacterial catalysis is used, albeit by a small amount relative to the overall heat of reaction. This is thought to be due to the utilization of at least a portion of the available energy by the bacteria for cellular functions.

Bacterial and chemical oxidation take place at a faster rate along planes of weakness within the crystal structure, that is, cracks, fissures, and grain boundaries, which is often where the major gold mineralization occurs [65, 67]. As oxidation proceeds, the porosity of the sulfide mineral increases, with the development of penetrating pores within the sulfide mineral grains. Such pore systems have been observed in pyrite (and other mineral) grains with measured pore diameters ranging from 1 to 10 µm. Evidence of such differential rates of oxidation has been provided by the development of the pore system along three perpendicular planes which are parallel to the pyrite crystal axes [68]. For many ores this means that good gold recoveries may be achieved with only partial oxidation of the sulfide minerals with which the gold is associated [69]. This effect is a function of the mode of occurrence of gold (Chapter 2), and is illustrated in Figure 5.28 for two ore types with different responses to biological oxidation.

Bacterial oxidation is the only process that can realistically be operated to achieve controlled partial oxidation of sulfide minerals. This is not practically possible with roasting and is more difficult for aqueous pressure oxidation, although in the latter case it is possible that the feed rate might be controlled to maintain partial oxidation for a feed material that has a consistent composition. Partial oxidation not only reduces the oxygen and heat transfer requirements of the system but also has the added benefit of producing less acid with correspondingly lower neutralization requirements, all of which can have a significant beneficial effect on capital and operating costs.

5.7.2 Reaction Kinetics and Operating Conditions

The major factors affecting the kinetics of biological oxidation are as follows:

- Ore mineralogy (especially sulfide sulfur content, acid-consuming minerals, and constituents that may be toxic to bacteria)
- Solution chemistry and conditions (temperature, pH, potential, pulp density)
- Dissolved oxygen concentration (and oxygen mass transfer)
- Temperature
- Particle size

All these variables can have a controlling effect on bacterially assisted oxidation reactions and must be optimized and controlled to maximize bacterial activity and, consequently, the oxidation rate.

5.7.2.1 Ore Mineralogy

Because each sulfide mineral has different susceptibility to biological oxidation, the oxidation rate of the different minerals varies considerably. Absolute determinations of the oxidation rates of different minerals are impossible because of the variability of crystals of single mineral species due to lattice structure imperfections (e.g., dislocations and impurities), which weaken the crystal structure. Mineral grains with poorly ordered crystal lattices are more susceptible to bacterially assisted corrosion. For example, pyrite containing 1% As, substituted isomorphously for iron or sulfur in the lattice, oxidizes faster than pyrite containing 0.1% As [65]. This is because the arsenic–sulfur bond is weaker than the iron–sulfur and sulfur–sulfur bonds and, as a result, arsenopyrite is oxidized more rapidly than pyrite [65]. An indication of the relative susceptibility of various sulfide minerals to biological oxidation is given in Table 5.11.

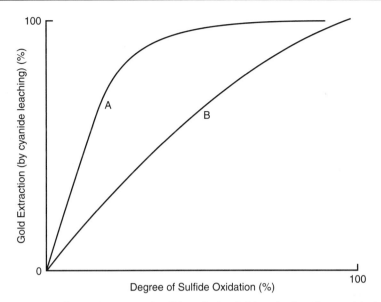

A: 82% pyrite, 15% arsenopyrite, 3% pyrrhotite. Gold predominantly associated with arsenopyrite.

B: 91% pyrite, 6.5% arsenopyrite, 0.5% pyrrhotite, 2% spharerite. Gold associated with pyrite and arsenopyrite.

FIGURE 5.28 Relationship between percent sulfide oxidation and gold extraction for biological oxidation of two different ore types (adapted from [68])

TABLE 5.11 Relative susceptibility of various sulfide minerals to bacterial oxidation (based on actual observations and electrochemical activity in 0.1 N ferric sulfate) [65, 70]

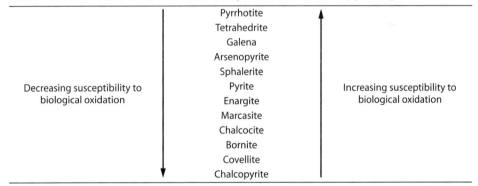

An important extension of this lattice imperfection concept is that biological corrosion also occurs more rapidly at gold inclusions in sulfide particles, because this represents a point of weakness in the lattice. Although there is considerable controversy over the nature of these inclusions—that is, whether the gold occurs as minute, discrete particles within the sulfide lattice or in solid solution—the fact that this preferential corrosion occurs explains the favorable relationship between percent gold recovery and percent sulfide oxidation that is obtained for some ores and concentrates (Figure 5.28). The other contributing factor is the frequent association of gold along grain boundaries of sulfide minerals, which is a result of the original mineralization sequence and mechanism during ore formation. As such, gold can be liberated by sulfide oxidation along grain boundaries, which can occur at a faster rate compared with lattice oxidation.

Galvanic interactions between different sulfide minerals in contact with each other may also result in preferential dissolution of one mineral over another, as considered in Chapter 4.

5.7.2.2 Factors Affecting Bacterial Activity

The oxidation rate is directly related to the activity of the bacteria or "biomass," which is determined and driven by the solution conditions. These factors are summarized as follows.

Temperature. Bacterial activity and growth rate are strongly temperature dependent, and the optimum temperature is a function of the bacteria culture and its degree of adaptation to particular temperature conditions as illustrated in Figure 5.29. For *Thiobacillii* bacteria cultures, the optimum temperature range is 35°C to 37°C. The bacterial activity decreases sharply outside the optimum temperature range; however, in practice the temperature can usually be controlled within 0.5°C in agitated tank systems. Some *Thiobacillii* cultures have been adapted to operate at 42°C to 45°C [69], which has the important advantage of reducing slurry cooling requirements when treating high-sulfur content materials (i.e., flotation concentrates). The use of moderate thermophiles (e.g., *Sulfobacillus acidophilus*), which operate at intermediate temperatures (45°C to 65°C) and extreme thermophiles (e.g., *Sulfolobus*), which can thrive at higher temperatures (65°C to 80°C), are being applied increasingly to achieve faster oxidation kinetics in commercial systems and to reduce cooling requirements and costs. In addition, the higher-temperature systems have been shown to oxidize essentially all of the elemental sulfur and intermediate sulfur species fully to sulfate, greatly reducing cyanide consumption in subsequent leaching operations compared with lower-temperature biological operations (using mesophiles). However, the use of higher temperatures creates additional problems for materials of construction in acidic, oxidizing slurries [62].

pH. The optimum pH range for bacterial growth has been demonstrated to be between 2.3 and 2.5. However, a pH of 1.0 to 1.8 is usually maintained to maximize oxidation rates and to prevent the formation of obstructive precipitates, such as jarosites, during oxidation [60, 65]. Solution pH is used to control the predominant species in a mixed bacteria culture. This must be optimized for gold extraction and depends on the feed material mineralogy and solution chemistry.

Pulp density. Tank volume requirements for bio-oxidation, and consequently the process economics, are favored by high pulp density. However, the maximum pulp density is generally constrained by oxygen mass transfer rate limitations, because the oxygen mass transfer requirements increase as the solids and sulfur content increases. Bio-oxidation systems that treat concentrates typically operate with pulp densities of 10% to 20% solids. Whole ores may be operated at higher slurry densities, typically 25% to 35% solids and possibly as high as 40%, as a result of the lower sulfide sulfur content and correspondingly lower oxygen requirements. Figure 5.30 provides an indication of the effects of pulp density (and aeration rate) on process economics, given for the example of oxidation of flotation concentrates at Equity Silver Mine (Canada).

Solution potential and Fe(II)–Fe(III) ratio. The solution potential and/or Fe(II)–Fe(III) ratio provide a useful indication of the metabolic energy, or activity, of the bacteria, and of overall bio-oxidation efficiency. Typically, solution potentials of 670 mV (vs. SHE) are obtained for solutions with poorly adapted bacteria and/or when high Fe(II)–Fe(III) ratios are present (i.e., early in the circuit). Solution potentials of 950 to 980 mV are typical for highly active bacteria in solution (indicative of maximum bacteria population growth) with low Fe(II)–Fe(III) ratios (as desired at the end of a bio-oxidation circuit) [11].

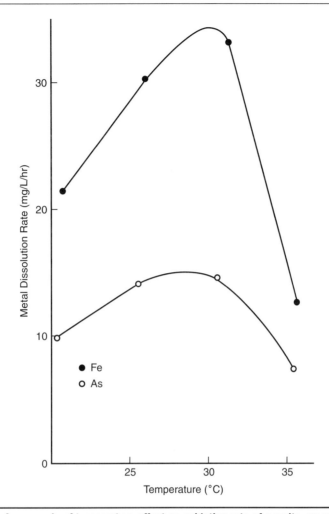

FIGURE 5.29 An example of temperature effect on oxidation rate of a pyrite–arsenopyrite concentrate: pulp density = 1.5%, pH = 2 [71]

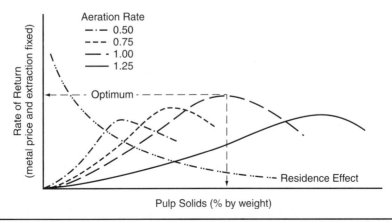

FIGURE 5.30 An example of bio-oxidation circuit design optimization showing effects of pulp density and aeration rate on process economics [64]

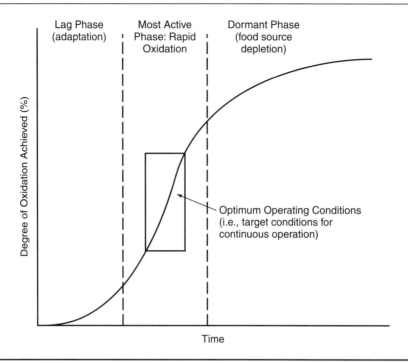

FIGURE 5.31 Schematic representation of bio-oxidation cycle for batch treatment of a sulfide feedstock

Dissolved oxygen concentration.   See Section 5.7.2.3.

Adaptation and the effect of other species in solution.   Bacteria have the ability to adapt to changing conditions. When a bacteria culture is introduced to a new type of feed, for example, a fresh, unoxidized sulfide ore or concentrate, the bacteria need time to adapt to the material. This process takes from 2 or 3 days to several months, depending on the mineralogy of the material and solution conditions [60]. In batch experiments, this period is called the lag phase, as shown in Figure 5.31. Fully adapted bacteria (the biomass) that are provided with adequate sulfur feed material will achieve oxidation rates close to the maximum attainable under the prevailing conditions. As the sulfur source is depleted, the bacteria go into a dormant phase and the oxidation rate declines. It is obviously desirable to operate a continuous oxidation circuit in the range of maximum rate of mineral oxidation, as indicated in Figure 5.31, and this is achieved by careful control of feed sulfur grades and the bio-oxidation conditions.

Changes in solid or solution phase composition can reduce the bacterial activity, in some cases very severely, but the bacteria can usually adapt relatively quickly to the new conditions. It is desirable to minimize the frequency and severity of chemical and mineralogical changes to maintain acceptable oxidation kinetics. *Thiobacillii* can tolerate relatively high concentrations of most metal species in aqueous solution. Approximate tolerance ranges for a variety of metals are given in Table 5.12. The most important aspect of bacterial tolerance is that the bacteria be allowed to adapt gradually to given conditions and that they are not subjected to "toxic shock," which results from a sudden increase in the concentration of a potentially toxic metal species. Commercial design of bacterial oxidation systems must consider feed variations and a control strategy to accommodate such variations, that is, ore-blending techniques, sulfur and toxic metal grade control, and so forth.

TABLE 5.12 Tolerance limits of *Thiobacillus ferro-oxidans* for various metal ions in solution [64, 65, 67 to 74]

Metal	Approximate Tolerance Limit (mg/L, unless specified)
As	>20 g/L
Sb	80 to 300
Fe	>50 g/L
Zn	3 to 80 g/L
Cu	>100
Pb	20
Se	80
Mg	>200
Ca	>200
Au	~5 µg/L*
Ag	~10 µg/L*

* Inconclusive literature data.

A number of other chemicals are toxic to *Thiobacillus ferro-oxidans*, including cyanide, thiocyanate, and a number of organic species that might be used for equipment lubrication or as process reagents, for example, hydraulic fluids, oils, greases, flotation collectors, and frothers. It is necessary to screen all chemicals that may come into contact with the bacterial process solutions prior to their introduction into the process. Materials of construction, notably rubber and synthetic rubber compounds, should also be rigorously tested for any adverse effect on the bacteria.

Effect of shear. Shear is important in bio-oxidation mixing systems because of the need to efficiently disperse air bubbles and to maximize mass transport rates in the slurry phase. However, there is some controversy over the effect of shear on bacteria activity and overall bio-oxidation efficiency. Some researchers claim that bio-oxidation systems must be gently agitated to avoid shearing bacteria off solid surfaces and even shearing the bacteria themselves [73]. The majority view is that shear, from a reasonable process standpoint, does not significantly affect bacterial activity. This conclusion is supported by the fact that a number of successful commercial bio-oxidation operations have used high-shear agitation to promote oxygen mass transfer; for example, Fairview (South Africa), Ashanti (Ghana), Wiluna (Australia), and others.

Nutrient requirements. Nutrient requirements vary depending on the feed material composition, the water source, and solution conditions. Sufficient carbon for energy conversion and cell growth is generally supplied from carbon dioxide in the air that is used to supply oxygen for oxidation. In addition to carbon, the bacteria require nitrogen and phosphorous. A fairly simple nutrient scheme, such as ammonium sulfate (0.5 to 1 kg/t) and potassium phosphate (0.1 to 0.2 kg/t), is usually adequate to maintain the biomass.

5.7.2.3 Dissolved Oxygen Concentration and Oxygen Mass Transfer

Two important dissolved oxygen limits are associated with bacterial oxidation of sulfides. The first considers the oxygen requirements for sulfide oxidation, where the oxidation rate increases with increasing dissolved oxygen concentration, up to a value above which the mass transfer of oxygen is no longer rate determining. The exact value of this limit depends on many factors, including the ore mineralogy, pulp density, mass transfer of oxygen (both through the liquid phase and from the gas to the liquid), bacterial activity, and other solution conditions. Generally, little benefit is gained at dissolved oxygen concentrations >4 mg/L.

The second limit is associated with bacterial activity. Below a critical value, which large-scale operating experience has shown to be in the range of 0.5 to 1.0 mg/L, the bacteria enter a dormant phase from which they may be slow to recover when higher dissolved oxygen concentrations are restored [64].

Because the saturated dissolved oxygen concentration in water at sea level and at 35°C is approximately 7 mg/L, the required dissolved oxygen concentration is theoretically readily achievable under bio-oxidation conditions. Little advantage is gained by increasing dissolved oxygen concentrations above this limit, but at lower concentrations the oxidation rate is dependent on the mass transfer of oxygen from the gas into the liquid phase. Dissolved oxygen depletion, especially at the mineral surface, can also cause buildup of elemental sulfur at the sulfide surface, as discussed in Section 5.7.1.

The rate at which oxygen is consumed by the bacteria-catalyzed oxidation reaction at the mineral surface is called the oxygen uptake rate and is equal to the rate at which oxygen must be absorbed into the solution phase. This uptake rate depends on the sulfur content of the material treated, the slurry density, and the bacterial activity. Typically, oxygen uptake rates of between 0.6 and 1.0 g/L/hr are observed. In order to satisfy the oxygen uptake requirements of a particular system, oxygen must be introduced to and absorbed into the solution phase, be transported through the bulk solution to the mineral surface, and finally be transferred across the mineral–solution interface. This requires:

- Efficient introduction of oxygen to the solution, as finely dispersed air or oxygen-enriched air bubbles, which allows adequate retention time for oxygen absorption
- Efficient mixing of the three-phase slurry system to maximize mass transport rates

Considerable effort has been directed at designing reactor systems that meet these requirements. Typically, air is used as the oxygen source, because this also provides the necessary carbon dioxide for bacterial cell growth. For the oxidation of concentrates, the air needs to be enriched with carbon dioxide. Oxygen-enriched air can be used to enhance oxygen mass transfer rates, but air containing >40% oxygen may reduce bacterial activity, as a result of reduced carbon dioxide content. Alternatively, a mixture of oxygen and carbon dioxide may be used, tailored specifically to the process requirements.

5.7.2.4 Particle Size

The effect of particle size on oxidation rate is discussed in Chapter 4. The minimum particle size at which bio-oxidation can be performed is generally limited by the settling properties of the oxidized product, since it is usually desirable to separate the slurry solid and liquid phases for the most effective neutralization of the acidic products, and to prepare the slurry for cyanide leaching. In practice, particle size ranges similar to those applied for cyanide leaching processes are used, that is, 70% to 80% <75 μm in agitated tank systems and 80% <12 to <25 mm in heap systems.

5.7.3 Process Considerations

5.7.3.1 Applications

Biological oxidation can theoretically be applied to both whole ores and flotation concentrates, but in practice it has found greatest success with the latter. A number of agitated tank leaching plants treating flotation concentrates have been operated successfully, including: Fairview (South Africa), Sao Bento (Brazil), Harbour Lights, Wiluna, Youanmi, Beaconsfield, and Fosterville (all in Australia), Ashanti Sansu (Ghana), Tamboraque (Peru), Laizhou (China), and Suzdal (Kazakhstan) [62, 63, 75]. By 2005, a number of other operations were using agitated tank leach biological oxidation technology. Biological heap leaching of low-grade refractory and mildly refractory ores also has promise and has been applied by Newmont at Carlin since the mid-1990s. Several commercial biological oxidation application flowsheets are provided in Chapter 12.

The economic viability of the process is strongly dependent on the relationship between gold recovery and the degree of sulfur oxidation, which ultimately determines the retention time. Oxidation times of 48 to 72 hr are typically required for materials that show a good response to biological oxidation, whereas materials with poorer response may require oxidation times in excess of 120 hr. The latter is unlikely to be economic as a result of the high capital costs associated with such long retention times (given the high cost of large biological reactor tanks) and large power requirements for agitation and air and/or oxygen addition.

5.7.3.2 Circuit Configurations and Design

A schematic flowsheet for a typical biological oxidation plant is given in Figure 5.32. The major factors to be considered in bio-oxidation circuit design include the following:

- Reactor design, size, and configuration
- Method and rate of aeration
- Heat balance and method of control
- Tank configuration
- Materials of construction
- Waste treatment and disposal
- Process control

Commercial circuits must provide sufficient retention time in the first stage of oxidation to allow the bacteria population to double and to ensure that the biomass is sustained. If this is not the case, or if the sulfide sulfur content of the feed is insufficient to support the biomass, then the bacteria are unable to duplicate sufficiently fast to maintain the biomass. This can result in a "wash-out" of the bacteria, after which oxidation ceases.

Although the risk of wash-out of the biomass is a real possibility, this can be minimized, and probably eliminated completely, by good plant design (i.e., correct equipment sizes, configuration, and heat transfer systems) and by appropriate and effective process control to accommodate sulfur mass flow variations. Also, good control of the sulfur feed grade helps to smooth out the operation. Some proposed bio-oxidation schemes utilize flotation as a preconcentration step to optimize the sulfur content of the feed. Other methods for managing the risk of wash-outs have been proposed, such as maintaining active batches of bacteria in separate tanks for such emergencies. This latter method is probably impractical because of the batch phase lag associated with reintroducing the bacteria to fresh ore.

The sulfide sulfur content of the feed material is extremely important. Systems treating low-sulfur material, that is, <1.0% sulfide sulfur, will require additional heating or the addition of extra sulfur (e.g., pyrite) to the feed to enable the optimum operating temperature to be maintained, although this also depends on other factors, such as climate, reactor design, and operating slurry density. Also, sufficient sulfur is required to support biological growth. High-sulfur materials (>2%) may require slurry cooling to maintain optimum temperature. Recycling of oxidized slurry has been suggested as a means of temperature control; however, this is generally not recommended because the concentrations of undesirable solution species are increased, in some cases to unacceptable levels, resulting in a decrease in bacterial activity. In addition, the operating pH range for effective bio-oxidation is quite narrow; thus the potential for recycling a significant amount of acid is less than that for pressure oxidation processes. Consequently, for feed materials containing carbonates, the acid consumption will be higher.

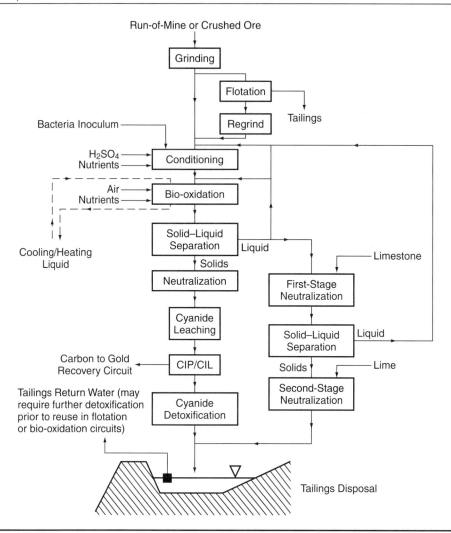

FIGURE 5.32 Schematic flowsheet for biological oxidation of sulfide gold ore. NOTE: Some process steps may not be required (adapted from [11, 64, 72]).

Following oxidation, the product must be neutralized prior to cyanide leaching. Two-stage neutralization is preferred to avoid the formation of excessive jarosites and other undesirable precipitates, which are hard to settle, difficult to filter, and may coat exposed gold surfaces. For example, a first stage using ground limestone ($CaCO_3$) may be used to increase the slurry pH to between 3.5 and 4.0 followed by a second stage where lime ($Ca(OH)_2$) is added to take the slurry pH up to approximately 10.5, suitable for feed to cyanidation.

5.7.3.3 Behavior of Gangue Minerals

Quartz and many silicate minerals are essentially inert but can dissolve to some extent in strongly acidic conditions, such as those encountered in bio-oxidation, and form a gelatinous silicate, which may coat exposed gold surfaces and is hard to filter. Chlorite is soluble in sulfuric acid and forms products that are typically difficult to filter. Carbonates react readily with sulfuric acid to form the respective sulfates, for example, gypsum and magnesium sulfate. These reactions often represent a significant source of acid consumption; for example, 1% limestone in an ore requires 10 kg/t H_2SO_4 ore. The reaction products of carbonate decomposition may also cause problems downstream with increased slurry viscosities and the formation of low permeability coatings on exposed gold surfaces.

5.7.3.4 Response to Cyanidation

The response of bio-oxidation products to cyanide leaching and subsequent recovery processes depends on the relationship between gold liberation and the degree of sulfide oxidation, the actual degree of oxidation achieved, and the effect of other oxidation products. Ores and concentrates that have a good response to bio-oxidation will generally yield 90% to 95% gold recovery in well-optimized circuits, as achieved at Fairview and Vaal Reefs (both in South Africa) in the 1990s. Cyanide and alkali consumptions vary, depending on the nature of the material, the degree of oxidation and the oxidation products, but are usually higher than those obtained following more complete oxidation, such as that typically achieved by pressure oxidation (Sections 5.3 and 5.4) and roasting (Section 5.8).

Bio-oxidation products have the potential to foul activated carbon by organic matter becoming entrained in the carbon pores, and therefore removal by thermal reactivation may be required.

5.7.3.5 Tailings Stability

The most important aspect of tailings produced from biological oxidation and cyanide leaching processes is the stability of arsenic. This is not well understood because the technology is relatively new. Limited work that has been performed indicates that arsenic in the tailings has very low solubility, even for measured iron–arsenic ratios of 2.2:1 (well below the 4:1 ratio recommended for pressure oxidation products). The solubility can be further decreased by detoxifying the cyanide content and by reducing the pH to 7 prior to disposal [74].

5.7.3.6 Biological Heap Oxidation

There is significant potential for biological oxidation of sulfide ores on heaps, in a manner similar to that used for bacterially-assisted dump and heap leaching of sulfide copper ores, for example, at Cerro Colorado, Quebrada Blanca, and Zaldivar (all in Chile), Cerro Verde (Peru), Girilambone (Australia), and Morenci (United States). However, a number of factors make this application to gold ores difficult to accomplish in practice. These include the following:

- Gold liberation requirements (more complex than sulfide mineral liberation requirements for copper leaching)
- Need for neutralization prior to cyanidation
- Likely need for double-handling of the ore
- Difficulty in controlling precipitation reactions
- High costs associated with providing extended leaching periods that may be required for effective gold liberation

A conceptual scheme, based on biological heap oxidation followed by cyanidation, is shown in Figure 5.33.

Biological heap oxidation systems may require the addition of bacteria, supplied from an external source, to the heap feed material (i.e., by "inoculation"). The external source could be a series of agitated tanks used to grow bacteria in high concentrations.

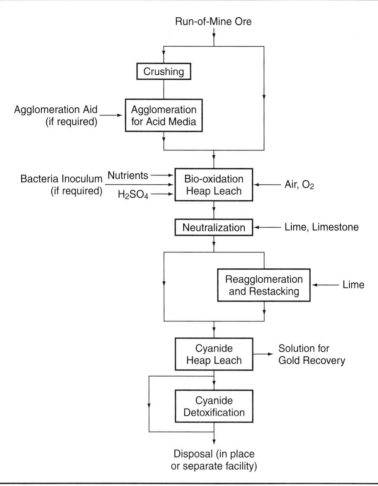

FIGURE 5.33 Conceptual flowsheet for biological heap leaching of low-grade sulfide ores (adapted from [72])

Alternatively, some systems may not require inoculation because the bacteria are present in the ore and/or leach solutions, and rapidly adapt to the leaching conditions applied in the heap. The addition of nutrients (e.g., NH_4^+ and PO_4^{3-}) are generally not required for heap systems because the ore generally contains sufficient nutrients to sustain adequate bacteria populations [76]. Effective biological heap oxidation requires the combination of sufficient flow of oxygen (generally supplied by air injected by blowers through a suitable distribution system under the heap [77]) and controlled solution application to maintain effective operating temperature and conditions throughout the heap.

In 1995, Newmont commenced operation of a demonstration-scale (approximately 700,000 t) biological heap leach at Carlin, applied originally as an oxidation step preceding thiosulfate heap leaching for treatment of a low-grade, carbonaceous refractory gold ore in a scheme similar to that shown in Figure 5.33 (see also Section 6.3.2.2) [78]. Further development of the technology in this configuration was suspended due to a period of low gold price. In 1999, the process flowsheet was modified to use the biological heap leaching process as a partial pre-oxidation step ahead of milling, agitated cyanide leaching, and CIL to treat siliceous sulfide refractory ore at a throughput rate of approximately

10,000 tpd. Prior to loading onto the leach pad, crushed ore (80% passing 19 mm) containing 1.4% to 2.0% sulfide sulfur was inoculated with a mixed bacteria culture containing *Thiobacillus ferro-oxidans, Leptospirillum ferro-oxidans, Sulfobacillus thermosulfido-oxidans,* and extreme thermophilic *Archaea.* Crushed ore was loaded onto a specially prepared pad with a crushed rock base up to a height of 12.8 m. Acidic solution (pH 1.75 to 1.90) was applied intermittently to the heap over a period of 150 days prior, and air was injected into the heap to provide the oxygen required to support the oxidation reactions. Temperatures ranging from 11°C to >80°C were observed in the heap, depending on sulfide content, carbonate content, solution conditions, and bacteria activity. Approximately 40% of the sulfide sulfur was oxidized during the biological oxidation step, allowing gold recoveries of 50% to 60% to be achieved during subsequent processing through the Newmont Mill No. 5 cyanidation circuit. These recoveries compared favorably to recoveries achieved without the biological oxidation step, which were in the range of 25%. However, the 60% gold recovery achieved in 2003 was less than the design recovery of 71%. Sulfuric acid consumption for biological oxidation was 0.9 kg/t. However, the process was complicated by lower than expected ore permeability, less time allowed for biological oxidation to occur than the original design (150 vs. 270 days), high carbonate content in the ore resulting in high solution pH (2.0 to 4.0) off the heap, and the need to aerate the solution pond to increase the Fe(II) to Fe(III) conversion rate [79, 80].

Biological heap oxidation processes will receive increasing attention as reserves of lower-grade (1 to 2 g/t) sulfide gold ores, which cannot be treated economically by higher-cost grinding and oxidation processes, are established. Practical application will probably be restricted to a few ore types, that is, porous ores with close to optimal sulfide content that are capable of sustaining bacteria through the oxidation cycle while avoiding excessive heat generation, and those that respond well to partial oxidation of sulfide minerals to achieve significantly improved gold recovery.

5.7.3.7 Future Applications

Biological oxidation will find increasing application as a cost-effective and environmentally acceptable process for the treatment of gold-bearing sulfide ores in both agitated tank and heap systems. The application of biological oxidation in an integrated agitated (tank) leaching and heap leaching process has been proposed whereby bacteria-inoculated sulfide concentrate is agglomerated with heap leach feed material. The combined mixture of concentrate and ore is heap leached using the methods described in Section 5.7.3.6.

Biological oxidation processes also have potential for treatment of carbonaceous ores, and this technology is being developed. Bacteria continue to be adapted to operate at increasingly high temperatures (i.e., moderate and extreme thermophiles) to increase oxidation kinetics and reduce temperature control requirements (particularly important when treating high-sulfur concentrates) and consequently to reduce cooling costs.

5.8 PYROMETALLURGICAL OXIDATION

Oxidation of refractory sulfide and carbonaceous constituents of ores and concentrates can be achieved pyrometallurgically by roasting in the presence of an oxidizing gas, such as air or oxygen. The objective is to produce a porous iron oxide calcine in which the gold is largely liberated, allowing access to cyanide leach solution and minimizing any reagent-consuming or gold-adsorbing potential of ore constituents.

Roasting can be performed using single- or two-stage processes, with the selection usually dependent on ore type. The single-stage method consists of direct roasting of the material in an oxidizing atmosphere. The two-stage process employs a first stage which operates under reducing conditions, creating a porous intermediate product, followed by a second-stage roast in an oxidizing atmosphere to complete oxidation.

The methods of handling the roaster products are also important for maximizing subsequent gold recovery, minimizing reagent consumption, and in satisfying environmental regulatory requirements.

Other roasting processes, such as sulfatizing and chloridizing techniques, which have been used in the past for gold extraction or may be considered as options for the treatment of some ores and concentrates in the future, are not discussed here, but further information is available in the literature [81, 82, 83]. The use of microwave energy as a pretreatment step for processing of refractory sulfide and carbonaceous materials has received considerable attention in the 1990s and 2000s and is considered briefly in this section.

5.8.1 Roasting Reaction Chemistry

5.8.1.1 Iron Sulfides

The phase–stability (or Evans) diagram for the Fe–S–O system at 602°C (Figure 5.34) shows that, under oxidizing conditions (i.e., low sulfur dioxide content in the gas phase), pyrite, marcasite, and pyrrhotite are directly oxidized to magnetite and then further to hematite [84, 85], as follows:

$$3FeS_2 + 8O_2 \rightleftharpoons Fe_3O_4 + 6SO_2 \quad \text{(EQ 5.104)}$$

$$3FeS + 5O_2 \rightleftharpoons Fe_3O_4 + 3SO_2 \quad \text{(EQ 5.105)}$$

$$4Fe_3O_4 + O_2 \rightleftharpoons 6Fe_2O_3 \quad \text{(EQ 5.106)}$$

These reactions are exothermic. Under reducing conditions (i.e., a sulfur and sulfur dioxide-rich atmosphere), pyrite decomposes to pyrrhotite and sulfur in a process termed desulfurization, represented by the following equation:

$$FeS_2 \rightleftharpoons FeS(s) + S(g) \quad \text{(EQ 5.107)}$$

This reaction is endothermic and requires an external source of heat to proceed, or heat supplied from other reactions [86]. The sulfur migrates to the surface of the mineral grain where it volatilizes, leaving a porous pyrrhotite structure. This migration of sulfur can be violent (particularly at high temperatures, >550°C), resulting in rupturing of particles in some cases. The porous, spongy, pyrrhotite structure is highly desirable as it leads to the formation of a highly porous product, which is favorable for subsequent cyanide leaching.

The volatilized sulfur is rapidly oxidized (exothermically) to sulfur dioxide in the presence of oxygen:

$$S(g) + O_2(g) \rightleftharpoons SO_2(g) \quad \text{(EQ 5.108)}$$

If a second-stage oxidizing roast is subsequently applied, then the porous pyrrhotite will be sequentially oxidized to magnetite and hematite, respectively, as described by Equations (5.105) and (5.106). This two-stage mechanism has been established by mineralogical investigation of various intermediate products of full-scale roasting operations and is illustrated schematically in Figure 5.35 [86]. Figure 5.36 shows an example of the mineralogical transformations involved in the two-stage roasting of pyrite (and arsenopyrite), clearly demonstrating the intermediate formation of pyrrhotite.

The two-stage process, if correctly applied, is known to produce a spongy, more porous product with a higher surface area [86, 87]. Typically, this enhances subsequent

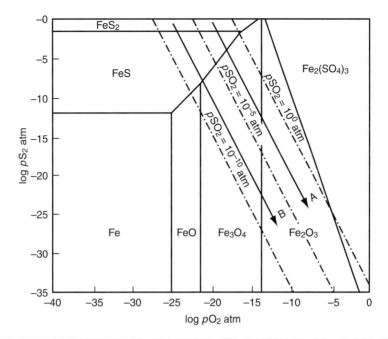

FIGURE 5.34 Phase stability diagram for Fe–S–O system at 602°C. Route A represents the sequence of mineralogical changes for a small grain of sulfide. Route B represents the sequence of mineralogical changes for a larger sulfide grain (adapted from [84]).

gold recovery, but the extent of this enhancement depends on the ore mineralogy, and two-stage roasting of pyrite is not always necessary, as discussed further in Section 5.8.2.

5.8.1.2 Arsenic Sulfides

The phase–stability diagram for the Fe–As–S–O system at 402°C is shown in Figure 5.37. Under oxidizing conditions, that is, low sulfur dioxide content, and even at such low temperatures, arsenopyrite reacts to form magnetite, as follows:

$$12FeAsS + 29O_2 \rightleftharpoons 4Fe_3O_4 + 6As_2O_3 + 12SO_2 \qquad \text{(EQ 5.109)}$$

The magnetite formed is further oxidized to hematite, as shown in Equation (5.106). The direct formation of magnetite is undesirable, because this may prevent adequate diffusion of arsenic to the surface, resulting in the retention of arsenic in the mineral matrix. Also, the calcine produced in this way typically has low porosity with poor accessibility of solution to contained gold values. Consequently, single-stage oxidizing, or oxidative, roasting is rarely applied to materials containing more than 2% to 3% arsenic [87, 89].

Under reducing conditions, in a sulfur dioxide–rich atmosphere, arsenopyrite decomposes to pyrrhotite and arsenic, in what is termed a "de-arsenification" process, analogous to the desulfurization of pyrite under similar reducing conditions. The arsenic diffuses through the thermally expanded lattice and is volatilized at the surface, leaving porous pyrrhotite:

$$FeAsS(s) \rightleftharpoons FeS(s) + As(g) \qquad \text{(EQ 5.110)}$$

Pyrrhotite is then oxidized to magnetite and subsequently to hematite, in a second stage of roasting, as is the case for pyrite oxidation (Section 5.8.1.1).

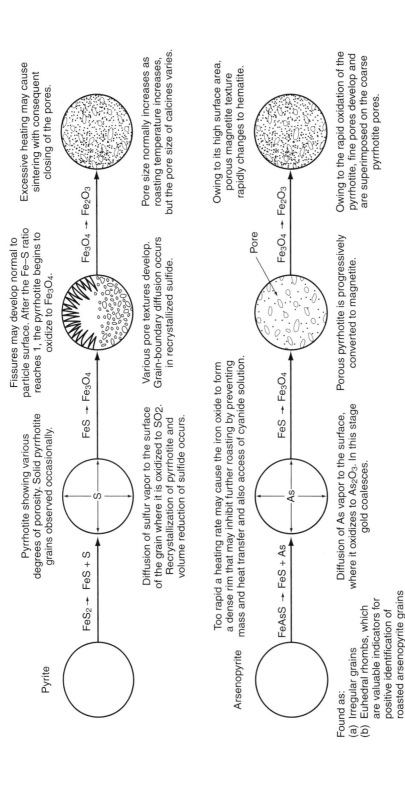

FIGURE 5.35 Mechanism of combustion of arsenopyrite and pyrite grains deduced from the results of small-scale tests and X-ray diffraction, scanning electron microscopy (SEM), optical microscopy, and chemical analysis of roaster feed and products [88]

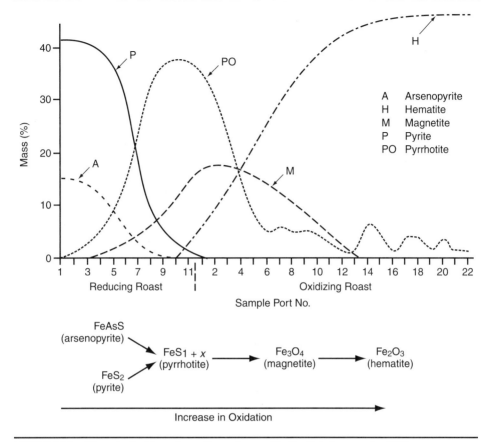

FIGURE 5.36 Mineralogical transformations observed in the Fairview roaster [88]

The volatilized arsenic is rapidly oxidized to arsenic trioxide in the presence of oxygen:

$$4As(g) + 3O_2(g) \rightleftharpoons 2As_2O_3(s) \tag{EQ 5.111}$$

Depending on conditions in the roaster, the arsenic trioxide may be oxidized to arsenic pentoxide:

$$As_2O_3(s) + O_2(g) \rightleftharpoons As_2O_5(s) \tag{EQ 5.112}$$

This reaction is significant because it may lead to a further undesirable reaction between hematite and arsenic pentoxide to form ferric arsenate, typically a nonporous and stable solid which tends to occlude gold and reduce subsequent gold extraction:

$$Fe_2O_3(s) + As_2O_5(s) \rightleftharpoons 2FeAsO_4(s) \tag{EQ 5.113}$$

The formation of arsenic pentoxide is minimized by roasting in a slightly oxygen-deficient atmosphere, although sufficient oxygen must be supplied to allow primary arsenic (and sulfur) oxidation to occur.

Other arsenic minerals, such as realgar and orpiment, are readily oxidized during roasting to arsenic trioxide (or pentoxide) and sulfur dioxide:

$$4AsS + 7O_2 \rightleftharpoons 2As_2O_3 + 4SO_2 \tag{EQ 5.114}$$

210 THE CHEMISTRY OF GOLD EXTRACTION

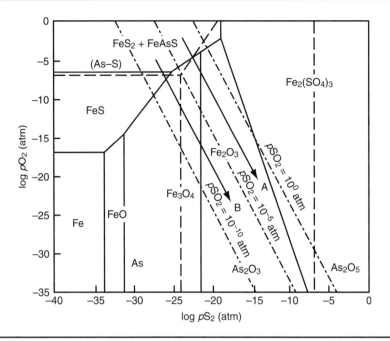

FIGURE 5.37 Phase–stability diagram for As-Fe–S–O system at 402°C [84]

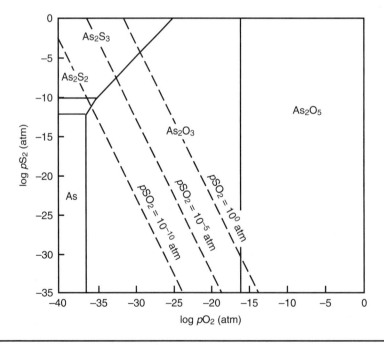

FIGURE 5.38 Phase–stability diagram for As–S–O system at 230°C [84]

$$2As_2S_3 + 9O_2 \rightleftharpoons 2As_2O_3 + 6SO_2 \qquad \text{(EQ 5.115)}$$

The phase-stability diagram for the As–S–O system at 230°C is given in Figure 5.38.

5.8.1.3 Antimony Minerals

Ores containing antimony may present special problems during oxidative roasting. Stibnite readily decomposes to form antimony trioxide and sulfur dioxide, as follows:

$$2Sb_2S_3 + 9O_2 \rightleftharpoons 2Sb_2O_3 + 6SO_2 \qquad \text{(EQ 5.116)}$$

The antimony trioxide thus formed can cause agglomeration of particles in the roaster bed, a process sometimes referred to as "clinkering." This may result in a buildup of material within the roaster as well as plugging of tuyeres and associated equipment, which may result in excessively high maintenance requirements. In addition, the mobilized antimony can form alloys with precious metals, which are insoluble in cyanide solution, for example, silver-containing ores where an Ag–Sb alloy ("antimony glass") can be formed. This alloy can coat gold surfaces and prevent access to leach solutions, especially when the silver occurs with the gold. The calcine may be treated with a sodium hydroxide wash to partially dissolve the alloy, if necessary (see Section 5.8.3.7) [57].

The collection of antimony fumes from the exhaust gases is important because of the high vapor pressure of antimony, even at low roasting temperatures, and the adverse health and environmental aspects associated with this metal. Treatment of roaster off-gases is considered in more detail in Section 5.8.3.8.

5.8.1.4 Other Sulfides

Simple copper, zinc, and lead sulfides oxidize to form their respective metal sulfides, as follows:

$$2MS + 3O_2 \rightleftharpoons 2MO + 2SO_2 \qquad \text{(EQ 5.117)}$$

where
  M = Cu, Zn, or Pb

Mixed metal sulfide minerals, for example, chalcopyrite ($CuFeS_2$), form the metal oxides and sulfur dioxide, although various intermediate oxidation products may be formed.

Because gold is less commonly associated with these minerals than with iron and arsenic sulfides, the reaction in Equation (5.117) rarely has a significant effect on gold extraction. However, the overall efficiency of oxidation of these sulfides is very important, because the metal sulfides are soluble in cyanide solution to varying degrees and may consume large amounts of reagent by the formation of metal cyanide complexes and thiocyanate if incompletely oxidized. The metal oxides may also be somewhat cyanide soluble (i.e., ZnO and Cu_2O), but the stoichiometric cyanide requirement of the metal oxides is significantly less than that of the sulfides.

5.8.1.5 Carbon and Carbonaceous Material

Sources of organic carbon, such as humic acid, graphitic carbon, and coals, are oxidized to carbon dioxide during roasting:

$$C + O_2 \rightleftharpoons CO_2 \qquad \text{(EQ 5.118)}$$

The rate of oxidation of some carbonaceous material may be slow under the roasting conditions applied for the treatment of sulfidic materials, possibly resulting in incomplete oxidation. However, the calcine materials produced rarely display any gold-adsorbing tendency, and it can be concluded that the type of materials responsible for this behavior (i.e., the surface functional groups) are oxidized rapidly during roasting.

Carbonates decompose to metal oxide and carbon dioxide during roasting, as follows:

$$CaCO_3 \rightleftharpoons CaO + CO_2 \qquad (EQ\ 5.119)$$

$$MgCO_3 \rightleftharpoons MgO + CO_2 \qquad (EQ\ 5.120)$$

If carbonates are present in sufficient quantity (i.e. >2% $CaCO_3$), the evolved carbon dioxide can form an unreactive blanket over the roaster bed, preventing mineral oxidation.

In the presence of sulfur dioxide, metal oxides may further react to form sulfates. This is the basis for a lime-fixing technique that has been developed to reduce levels of sulfur dioxide in roaster off-gases, as discussed in Section 5.8.3.8.

5.8.1.6 Gold

During the oxidation of gold-bearing sulfide minerals, the gold, which may be present either as fine particles or in solid solution, migrates toward pores or grain boundaries in the direction of diffusion of sulfur or arsenic. As migration occurs, the gold coalesces in the liquid phase. Further coalescence is thought to occur at the mineral surface, as arsenic and sulfur volatilize, and gold particles up to 1 μm diameter can be formed. A mechanism illustrating this phenomenon is given in Figure 5.39.

Further oxidation to magnetite and hematite, with associated recrystallization, renders these gold particles accessible to leach solutions. Significant coalescence of gold may also occur during this oxidation reaction [89]. These gold particles are, theoretically, highly amenable to leaching by virtue of their location and size [88]; however, the migration and coalescence process is known to be sensitive to roasting temperature and oxidation rate, and two-stage roasting is thought to provide the most favorable conditions for this mechanism to occur.

Some roasted calcine products exhibit slow leaching characteristics, although the final recovery achieved may be quite good, that is, >90%. This may be due to partial locking of gold within a porous calcine, where the leaching rate is controlled by pore diffusion. Alternatively, the slow leaching rate may be the result of dissolution of coarse gold particles, either present in the original ore or concentrate prior to leaching or, less likely, formed by excessive coalescence during roasting.

5.8.1.7 Gold Tellurides

All the common gold-telluride minerals can be decomposed by roasting to produce gaseous tellurium dioxide and metallic gold [90, 91]:

$$AuTe_2 + 2O_2 \rightleftharpoons Au + 2TeO_2 \qquad (EQ\ 5.121)$$

In the case of calaverite ($AuTe_2$), separation of the two metals starts at temperatures as low as 464°C, the melting point of the alloy. Oxidation of tellurium commences at approximately 600°C, above which the mineral decomposes rapidly. Gold-telluride ores have been successfully treated by roasting at Kalgoorlie (Australia) and Emperor (Fiji). In some cases, roasting is the only method for successfully treating such ores and concentrates.

5.8.2 Roasting Kinetics and Efficiency

The efficiency of roasting, as determined by the response of the calcine product to cyanide leaching, is strongly dependent on the roasting kinetics, which is principally a function of temperature, partial pressure of oxidizing gas, and particle size. Unlike many other

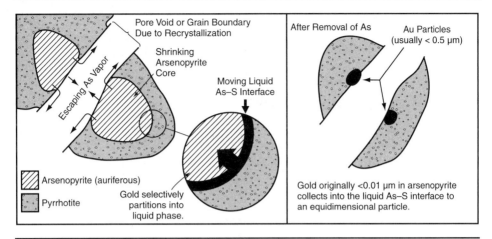

FIGURE 5.39 Mechanism for aggregation of gold particles during the roasting of auriferous arsenopyrite [88]

processes in gold extraction (i.e., leaching and precipitation), maximizing the reaction kinetics does not necessarily maximize gold recovery. This is discussed further in the following sections.

5.8.2.1 Temperature and Gas Phase Composition

The rate of volatilization and oxidation of sulfur and arsenic minerals increases with increasing temperature and increasing partial pressure of oxygen in the gas phase. At low temperatures (<400°C to 450°C), the rates of these reactions for pyrite and arsenopyrite are very slow (see Figures 5.40 and 5.41). Acceptable rates can only be achieved by increasing temperatures >450°C to 500°C. On the other hand, at very high temperatures, >700°C to 750°C, the porous iron oxide structure that is developed during oxidation may collapse, encapsulating gold within the particle and reducing subsequent gold recovery, possibly by as much as 50%. This process is often referred to as "sintering."

Sintering may occur as a result of high overall temperature conditions in the roaster bed, the development of hot spots in the bed due to locally high sulfur content, or poor temperature control within different regions of the bed. Similar effects may be produced by roasting for too long at high, or even marginally high, temperatures, commonly referred to as "overroasting."

The temperature ranges given for each of the effects described may vary, depending on the mineralogy, the roasting method, and specific operating conditions, such as particle size and gas flow rate. A general indication of the effect of temperature on subsequent gold recovery, which is a measure of roasting efficiency, is shown in Figure 5.42.

The porosity of the calcine product plays an important role in determining the accessibility of leach solution to gold particles contained in the calcine during subsequent cyanidation. This porosity is highly dependent on the rate of oxidation, and consequently the rates of diffusion of sulfur and arsenic, with the formation of a higher porosity product favored at slower roasting kinetics. At high temperatures and under strongly oxidizing conditions, sulfur and arsenic may have insufficient time to diffuse to the mineral surface before extensive oxidation and recrystallization of the mineral structure occurs, prohibiting further pore development.

The effect of temperature on product surface area, an indicator of porosity, is illustrated in Figure 5.43. This shows that, for the example given, a more porous product was

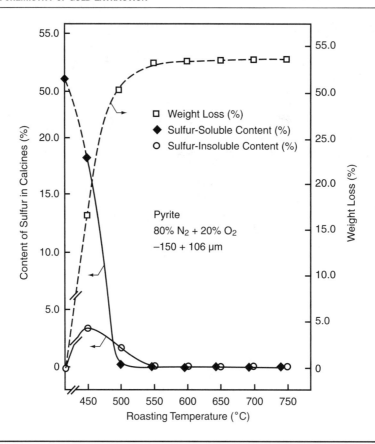

FIGURE 5.40 Content of sulfur (soluble and insoluble) in the calcines and weight losses in the roasting of pyrite at different temperatures [92]

generated at 550°C than at 650°C. In the case of both pyrite and arsenopyrite, the most porous product is known to be formed through the intermediate formation of pyrrhotite, as discussed in Sections 5.8.1.1 and 5.8.1.2.

The use of pure oxygen to upgrade the gas phase may be beneficial in some instances; the increased oxygen content can produce more complete oxidation of refractory minerals, improving subsequent gold recovery and reducing reagent consumptions. This may also allow roasting equipment to be reduced in size.

Single-stage roasting. The temperature and gas phase composition are controlled to optimize subsequent gold extraction by achieving satisfactory oxidation kinetics, while guarding against possible sintering. Typically, temperatures of 550°C to 700°C and a gas phase containing a slight excess of oxygen (6% to 8% O_2) in the roaster discharge are applied to ensure an oxygen supply in excess of the stoichiometric requirement [85]. The exact gas composition depends on the relative gas–solids flow rates through the roaster, feed sulfide content, and the quantities of other oxygen consumers in the feed. Higher oxygen concentrations may be used, for example, by adding pure oxygen into the feed gas, but the effectiveness of this depends on the sulfur content of the feed, the heat balance, and overall economics.

Two-stage roasting. The first stage of the two-stage roasting process is maintained under slightly reducing conditions by applying a sulfur dioxide-rich atmosphere to promote the formation of an intermediate porous pyrrhotite, as described in Sections 5.8.1.1

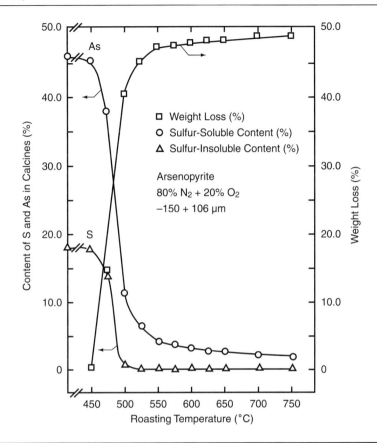

FIGURE 5.41 Content of sulfur and arsenic in calcines and weight losses in the roasting of arsenopyrite at different temperatures [92]

and 5.8.1.2. Typically, the temperature is controlled between 500°C and 575°C to encourage the formation of a highly porous product. Between 80% and 85% of the stoichiometric oxygen requirement may be supplied in the gas phase, ensuring that the roaster gases are deficient of oxygen, to maintain a reducing environment.

The second stage is operated under oxidizing conditions and has similar objectives to the single-stage process. The major difference is that any pyrite and arsenopyrite in the original feed should have a well-developed porous pyrrhotite structure prior to the second stage of roasting. Temperatures of 450°C to 700°C are used in the second stage, with gas phase compositions similar to those used for single-stage roasting [85, 88].

The exact operating conditions required for optimum roasting of a specific material depend on a number of related and nonrelated factors, such as the mineralogy of the material, the roasting method employed, and particle size distribution. These conditions must be determined by rigorous test work and evaluation.

5.8.2.2 Particle Size Distribution

The kinetics of sulfide roasting improve with decreasing particle size. Consequently, roaster circuits must be designed to treat a certain particle size range of material to ensure that satisfactory oxidation is achieved through the roaster, under optimized conditions, as described in Section 5.8.2.1. For a given roaster system, particles coarser than

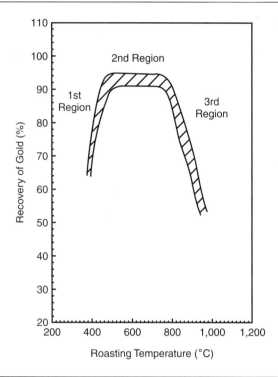

FIGURE 5.42 Recovery of gold as a function of roasting temperature for several gold ores and concentrates [93]

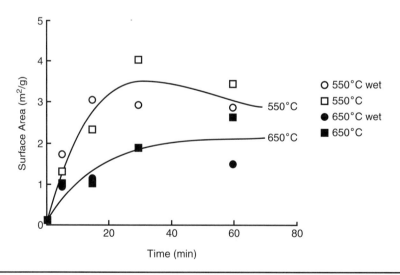

FIGURE 5.43 Effect of temperature on calcine surface area [89]

TABLE 5.13 Mineralogical analyses of concentrates from the Barberton area, South Africa (all values are expressed as percentages by mass) [88]

Constituent	Sheba Flotation Concentrate	Sheba Gravity Concentrate	Agnes Flotation Concentrate	Agnes Gravity Concentrate	New Consort Gravity Concentrate*	Fairview Flotation Concentrate
Pyrite	34.2	62.4	56.0	86.0	—	45.0
Arsenopyrite	14.0	21.4	2.2	<2.2	32.2	15.2
Pyrrhotite	<0.1	<0.1	2.0	<0.2	21.8	<0.1
Accessory minerals†	<2.0	<2.0	0.5	<0.9	<1.0	<1.0
Quartz	35.0	<10.0	25.0	<8	<25.0	<15.0
Micaceous minerals‡	<10.0	<3.0	8.0	<2	<20.0	<23.0
Iron carbonates	<1.0	<1.0	5.0	<1	<1.0	<6.0
Graphite carbon	1.0	0.4	0.6	0.2	0.1	0.3

NOTE: Dash = not available.

* High levels of platy minerals and poorly floating pyrrhotite do not allow for flotation at the New Consort Mine.

† These minerals vary somewhat but usually include some of the following: chalcopyrite, sphalerite, galena, bornite, chromite, bismuth sulfides, native silver, stibnite, jamesonite, tetrahedrite, and pentlandite.

‡ Mainly chlorite and muscovite (sericite), with lesser amounts of phlogopite.

the optimal particle size may be incompletely oxidized, whereas finer particles may be overroasted and possibly sintered. Sintering is a potentially serious problem because it can cause gold to become occluded and inaccessible to leach solutions. The magnitude of these effects varies depending on the roasting method and conditions, discussed further in Section 5.8.3.

The roaster feed particle size distribution should be kept as narrow as possible, minimizing the amount of fines and oversized material. Agglomeration or pelletization processes may be used to pretreat the roaster feed to produce a narrower and more even size distribution.

Particle sizes that are employed in practice depend on the material to be roasted and the roasting method. Sulfide ores can be successfully roasted at sizes as coarse as 12 mm, although this is rarely done. Typically ores and concentrates are ground to a size that gives optimum gold extraction in the recovery process that follows roasting, for example, 80% passing 75 to 150 μm. Gravity concentrates may be roasted at relatively coarse sizes (i.e., between 1 mm and 0.25 mm), with regrinding of the roasted calcine if necessary. Alternatively, they may be ground to a finer size at the outset for faster roasting kinetics.

5.8.3 Roasting Process Considerations

5.8.3.1 Applications

Roasting can be applied to a wide variety of sulfide, carbonaceous, and telluride ores and concentrates. Roasting has been employed by a number of major gold producers, including: Kalgoorlie Consolidated Gidji (Australia), Fairview and New Consort (South Africa), Campbell Red Lake and Con (Canada), Emperor (Fiji), and, more recently, at Jerritt Canyon, Big Springs, Cortez, Carlin, Goldstrike (all in the United States), and Minahasa (Indonesia). Mineralogical analyses for some materials that are treated by roasting are shown in Table 5.13.

Single-stage roasting is applied to feed materials that have low arsenic content, typically <3% As, and which show good response to the direct oxidation process. Two-stage roasting is preferred for high arsenic materials and for some iron sulfide feeds, even in the absence of arsenic. In these cases, the increased gold recovery that results from the more efficient two-stage roast more than compensates for increased capital and operating costs. This route is not always justified, and each refractory material must be considered individually.

Several roasting circuit flowsheet applications are presented in Chapter 12.

5.8.3.2 Feed Preparation

The feed to the roaster may either be in the form of a slurry (fluidized bed roasters) or dry (multiple hearth and fluidized bed roasters). This choice depends on the heat balance for the roaster circuit and other process factors, such as grinding requirements, optimum roaster feed size, and the need for any feed blending and/or stockpiling. The higher the moisture content of the roaster feed, the higher the sulfur content necessary to sustain autogenous roasting. In cases where the heat balance is insufficient to maintain autogenous operation, then supplemental fuel must be added, either as an additional fuel source with the feed (i.e., barren pyrite or coal) or supplied directly to the roaster burner (i.e., fuel oil or natural gas). As a general guideline, a dry feed containing at least 8% sulfur or a slurry (75% to 80% solids) containing >16% to 20% sulfur may be treated autogenously, depending on the mineralogy and conditions applied [94, 95].

When the roaster is given a dry or semi-dry feed, then the material may be preheated to drive off retained moisture, in order to reduce energy requirements in the roaster and improve efficiency. This may also assist with temperature control.

Where a dry feed is required the material must either be dry ground or wet ground, filtered, and dried prior to roasting. If a slurry feed system is used, then the ore may be wet ground, filtered, and repulped, as necessary. Agglomeration or pelletizing of dry, or semi-dry, feed can be used to ensure a consistent retention time for all particles in the roaster and hence to obtain a uniformly roasted product.

5.8.3.3 Roasting Equipment

A number of different types of roasting equipment are available for oxidation of refractory materials, for example:

- Multiple hearth (or Edwards) roaster
- Fluidized bed roaster
- Circulating fluidized bed roaster

The type of roaster used has an effect on the reaction chemistry and performance (summarized in Table 5.14) and is discussed further in the following sections.

Multiple hearth roaster. This equipment provides reasonably uniform retention time for all particle sizes. Although this is good for a narrow size range of particles, it has the disadvantage that fines may be overroasted and that sintering may occur. The temperature must be controlled by the flow rate of the gas phase, which tends to dilute the sulfur dioxide of the roaster discharge gas, increasing the difficulty (and cost) of gas cleaning processes. Also, because this technique uses a stationary bed of particles, the lack of mixing in the roaster bed makes temperature control within the bed very difficult and may result in uneven roasting of material. Despite such drawbacks, these units have been used reasonably successfully for almost a century for the treatment of a variety of refractory ores and concentrates. This technology has been employed for more than 100 years, with examples at Fairview (South Africa), Mount Morgan (Australia), and elsewhere.

Fluidized bed roaster. This technology emerged in the 1980s as a favorable alternative for treatment of ores and concentrates. The residence time of material in these units is dependent on particle size: Fine particles are removed quickly, avoiding overroasting (but with the possibility of underroasting), whereas coarse material is retained

TABLE 5.14 Roaster performance comparison [95]

Parameter	Multiple Hearth Roaster	Conventional Fluidized Bed Roaster	Circulating Fluidized Bed Roaster
Relative throughput per unit area	Very low	Moderate	High
Gas–solids mixing	Poor	Good	Very good
Bed temperature control	Poor	Good	Very good
Control of gas composition	Very limited	Good	Very good
Solids retention time	Fixed	Fixed	Variable
Slurry or dry feed	Dry only	Both	Both
Ability to roast whole ore	Not possible	Possible but heat input efficient only via fluidizing air. Fluidizing air temperature limited by bed plate arch design.	Possible. Several options available for heat input.
Ability to treat high arsenic content feed	Possible only with concentrate. Calcine quality inferior to two-stage conventional fluid bed roasting.	Good results using two-stage roasting	Not yet attempted at commercial scale. More development required.

longer. The gas phase is used for fluidization of the roaster bed, which provides good mixing of the gas–solid mixture. As a result, the heat and mass transfer properties of the bed are very good, and accurate temperature control, typically within 10°C, can be achieved. The occurrence of local hot spots is greatly reduced because process water can be used for cooling where necessary, avoiding the need to vary gas flow rates, with the associated dilution of the gas phase. Examples of fluidized bed roasters were Giant Yellowknife, Campbell Red Lake, and Jerritt Canyon [96, 97].

Circulating fluidized bed roaster.   This equipment uses similar operating principles to the fluidized bed roaster, except that the product is recirculated rapidly through the roaster, giving low single pass residence time and a relatively high circulating load. This has the following advantages:

- Improved mixing within the bed
- Closer control of residence time for a given particle size
- Possible recycling of roaster gases, which improves the efficiency of oxygen usage

Examples of circulating fluidized bed roasters are Kalgoorlie Consolidated, installed in 1990; Newmont Carlin, commissioned in 1994; and Barrick Goldstrike, commissioned in 2000 [98, 99, 100]. Design and operating characteristics for the Carlin and Goldstrike operations are summarized in Table 5.15. Both of these operations feature dry grinding prior to roasting, thereby avoiding the need for costly solid–liquid separation and drying steps that are required with wet grinding followed by roasting.

5.8.3.4 Retention Time

The retention time of each particle in the roaster determines the efficiency of oxidation of that particle. If the retention time is too short, either due to overfeeding of the roaster or due to bypassing within the roaster, then incomplete oxidation can occur. This may result in high cyanide consumptions downstream, and the calcine produced may still be refractory in nature, causing high gold losses. Long retention times can cause overroasting or sintering of calcined material, especially of fine particles.

TABLE 5.15 Design and operating parameters for selected circulating fluidized bed roaster operations [99, 100]

Operation	Newmont Carlin	Barrick Goldstrike
Design throughput (tpd)	7,800	10,800
Feed material		
Gold grade (g/t)	8.7	7.5
Organic carbon (%)	0.4	0.5 to 4.0
Carbonate (%)	0.05	5
Sulfide sulfur (%)	1.75	1.9
Fluorine (ppm)	1,000	—
Arsenic (ppm)	1,200	—
Feed particle size	100% –208 µm	80% –74 µm
Roaster type	Single-stage circulating fluidized bed (CFB)	Two-stage CFB
First stage temp (°C)	550 (1,020°F)	540 (1,000°F)
First stage residence time (min)	10	22
Second stage temp (°C)	NA	570 (1,060°F)
Second stage residence time (min)	NA	18
% Sulfide sulfur oxidation	~100	99
% Carbon oxidation	30	90
Gold recovery (%)	91	90 to 91
Roaster off-gas treatment	Some SO_2 is fixed in the roaster bed by addition of lime to roaster feed. Roaster off-gas cleaning includes particulates removal, fluorine removal, mercury removal, and acid plant for SO_2 recovery via three-stage catalytic conversion of $SO_2 \rightarrow SO_3$, followed by H_2O_2 scrubbing of acid plant tail gas and regenerative thermal oxidizer for CO conversion to CO_2 prior to release.	SO_2 is fixed in roaster bed by carbonate constituents of ore and by addition of lime (20% of the stoichiometric SO_2 requirement) to the first stage CFB.

NOTE: NA = not applicable. Dashes in blank cells = not available.

5.8.3.5 Oxidation Efficiency

Between 80% and 100% of the sulfide components and 60% to 70% of the carbonaceous components of materials treated by roasting are typically oxidized. The production of a calcine containing 75% to 85% hematite and 15% to 25% magnetite has been found to be optimal. This indicates that complete, or close to complete, sulfide oxidation has been achieved, but that secondary oxidation of magnetite to hematite is incomplete. This minimizes overroasting of hematite particles that remain in the roaster after oxidation has been completed. The color of the calcine provides a useful indication of the degree of oxidation achieved in the roaster: a bright red color indicates a product that is essentially hematite (i.e., very little magnetite), caused by overroasting. This may be due to high temperatures, high oxygen content, or excessively long retention time in the roaster. On the other hand, a very dark brown or black coloration is characteristic of a magnetite-rich product, resulting from too-low temperatures in the roaster, insufficient oxygen, or short particle-retention time. The efficiency of conversion of magnetite to hematite is important because magnetite tends to retain arsenic, sulfur, and gold in the matrix. Recrystallization to hematite helps to remove these products and to expose contained gold (Section 5.8.1.6). Ideally, the product has a chocolate brown color; however, variations in mineralogy may have an effect on this also [85]. In practice, actual performance must be assessed by effective chemical and mineralogical analyses.

5.8.3.6 Gold Losses During Roasting

Most roasting circuits have an unfavorable metallurgical balance of gold production between the roaster feed and discharge material (i.e., a loss of gold occurs through the roaster). This loss varies considerably for different operations, but typically ranges from 2% to 5%. The losses are attributed to either, or a combination, of two factors:

- Volatilization of gold, enhanced by the presence of chlorides (and fluorides) in the feed material or introduced in the process water supply
- Gold deposition within the roaster, resulting in hold-up of gold values

Gold deposition often necessitates that roasting equipment be shut down periodically and cleaned to recover gold values. Gold chloride volatilization losses can be minimized by the use of chloride-free water in roasting systems.

5.8.3.7 Treatment of Calcine and Gold Recovery

In preparation for gold leaching, the roaster calcine product must be cooled and mixed with water to form a slurry. This is usually performed in a single quenching–washing step. The water wash step helps to remove coatings that might otherwise impede gold dissolution. For many sulfide ores a combined washing and regrinding process is preferred. Regrinding not only helps to expose physically occluded gold in the oxidized material but also improves wetting of the calcine, particularly fine particles, and provides more efficient washing. Regrinding is applied when concentrates are roasted at a size coarser than that required for optimum gold extraction.

The quality of wash water can have a significant effect on the washing efficiency: If the solution is recycled around the quenching–washing circuit, there may be a buildup of gypsum and other dissolved salts which, if reprecipitated, can adversely affect the surface properties of the quenched calcine. In such cases, freshwater may need to be used and a bleed from the washing circuit removed for use elsewhere in the extraction–recovery process.

Other types of washing processes have been tested on a variety of ores. These have included the use of sulfuric acid, sodium carbonate, sodium silicate, potassium permanganate, and ammonia solutions. In the majority of cases, freshwater has been found to be the most economically effective.

The presence of significant quantities of lead minerals in the ore or concentrate to be roasted can lead to the formation of an insoluble lead oxide coating on precious metal surfaces. This can generally be removed with a dilute sodium chloride wash.

Roaster products show a wide range of responses to gold leaching and recovery processes, depending on roasting conditions and the mineralogy of the feed material. Gold recoveries obtained from calcine materials are generally lower than those from the products of suitable hydrometallurgical oxidation processes due to one, or more, of the following:

- The presence of incompletely roasted sulfides, either as discrete particles or particles with a partially oxidized core, which contain gold and which consume cyanide
- The presence of dense iron oxide products with low porosity, which contain gold that is inaccessible to leach solution
- Sintering of oxidized particles which encapsulates gold
- The presence of iron sulfate coatings around oxidized particles

These problems can be identified by diagnostic process mineralogy (Section 2.18) and can often be rectified by subtle changes in roasting conditions and calcine treatment processes described previously. Other problems have been noted with the treatment of roaster calcines, including fouling or blinding of carbon in adsorption circuits (i.e., carbon-in-pulp, -leach, or -solution), fouling of zinc in the precipitation process, and high reagent consumptions during leaching. In some cases, these factors may cause roasting to be rejected in favor of a hydrometallurgical oxidation route.

5.8.3.8 Treatment of Off-Gases

Because it is necessary to meet increasingly stringent environmental legislation worldwide, the quality and quantity of roaster off-gases is of primary importance in roasting. Also, in locations where such legislation is limited or nonexistent, it is prudent practice to ensure that appropriate gas-cleaning technology is employed.

The two most important gas components to be considered are sulfur dioxide and arsenic trioxide, but other emissions of potential concern include carbon dioxide, carbon monoxide, mercury, tellurium oxides, selenium, and antimony oxides. Other particulate emissions may also be regulated or may contain hazardous materials that are subject to legislative controls.

Particulate material. It is important to efficiently recover particulate matter from roaster gases in order to:

- Meet specifications for environmental emissions
- Recover contained gold values
- Eliminate any adverse effects on downstream gas treatment

Cyclones, dust chambers, and baghouses are typically used to recover the bulk of particles from the roaster gas. The underflow from such systems is quenched and combined with the solid roaster product. The gases overflowing the cyclone are cooled, to approximately 340°C to 360°C, and residual particulates are recovered by various methods, such as sedimentation and electrostatic precipitation. Wet scrubbing methods may also be used; however, such treatment is usually combined with scrubbing systems for arsenic and sulfur dioxide, where these are applied.

Sulfur dioxide. This can be recovered from gas streams by catalytic conversion into sulfur trioxide, which is then adsorbed into sulfuric acid to form oleum ($H_2S_2O_7$):

$$H_2SO_4 + SO_3 \rightleftharpoons H_2S_2O_7 \qquad \text{(EQ 5.122)}$$

Dilution of this product generates a commercial-grade sulfuric acid. Efficiently operated double-contact, double-adsorption systems can produce an effluent gas containing less than 100 ppm sulfur dioxide.

The two most important factors determining the viability of this process are:

- The availability and proximity of a market for sulfuric acid
- Production of a minimum of 4% SO_2 in the roaster discharge gas, to produce sulfuric acid economically

The sulfur content of the roaster off-gas is a function of the sulfide sulfur content of the feed material, the roasting method, and type or configuration of equipment used.

Roaster off-gases that contain lower sulfur dioxide levels than required for sulfuric acid production, or if sulfuric acid production is not considered to be a viable option for other reasons, must be treated differently. The alternatives available for this are reaction of the gas phase with lime or limestone slurry in a scrubbing system. For example, sulfur dioxide reacts with lime to form gypsum, which is stable and relatively inert:

$$2Ca(OH)_2 + 2SO_2 + O_2 \rightleftharpoons 2CaSO_4 + 2H_2O \qquad \text{(EQ 5.123)}$$

This reaction can be applied in two ways, either separately or in combination as follows:

- Roaster exhaust gases are pumped through a scrubber, using lime slurry as the scrubbing medium.
- Solid lime or ground limestone can be added to the roaster feed. Sulfur dioxide is "fixed" in the roaster bed as it is evolved during oxidation. Up to approximately 75% of the sulfur dioxide produced can be removed in this manner [85]. In addition, the lime acts as a heat sink for the highly exothermic reaction, allowing for better temperature control in the bed.

The reaction of sulfur dioxide with limestone produces carbon dioxide:

$$2CaCO_3 + 2SO_2 + O_2 \rightleftharpoons 2CaSO_4 + 2CO_2 \qquad (EQ\ 5.124)$$

Consequently, materials that contain large amounts of carbonates, or where limestone is used to fix sulfur dioxide, may experience reduced roasting efficiency by the formation of a blanket of carbon dioxide over the roaster bed. This may be countered by increasing gas flow rates through the roaster or by providing additional oxygen.

Lime fixing within the roaster bed can also cause other problems, such as excessive gypsum buildup in the roaster and associated equipment, and/or calcium sulfide formation (a high reagent consumer during cyanidation). Any gypsum produced may contain significant quantities of hazardous species, such as arsenic, cadmium, mercury, and lead, which may require special disposal or possibly further treatment prior to disposal.

Low concentrations of sulfur dioxide can also be removed from gas streams by scrubbing in solutions of sodium hydroxide, sodium hydroxide and sodium carbonate, or magnesium hydroxide; the latter two of these are regenerative processes, in that limestone can be used to precipitate the sulfate species out of solution as gypsum, allowing the functional alkali to be returned to the scrubbing circuit. All of these processes are similar in principle to the lime-scrubbing system described previously but have different levels of efficiency and economy. These processes must be considered on a case-by-case basis depending on specific gas cleaning requirements and economic factors.

Arsenic trioxide. When materials with high arsenic content (i.e., >2% or 3% As) are roasted, arsenic trioxide can be condensed and recovered in a relatively pure form by reducing the off-gas temperature from 360°C to below 120°C. The product is collected by standard baghouse gas-filtering technology. This method of arsenic recovery is typically unhygienic, messy, and maintenance intensive. As a result of the corrosive nature of the gas treated, the system is susceptible to leakage and failure of the bags. A product containing at least 98% As_2O_3 is required for sale. Alternatively, the material must be disposed of in an environmentally acceptable manner, for example, in underground vaults [96].

Gases with lower arsenic content can be treated by wet scrubbing with lime solution:

$$As_2O_3 + H_2O \rightleftharpoons 2HAsO_2 \qquad (EQ\ 5.125)$$

Solid arsenic trioxide can be produced from the solution by crystallization. Alternatively, the relatively stable Fe(III) arsenate can be produced by precipitation with suitable Fe(III) species.

With any of these methods, complete removal of arsenic is difficult to achieve. This results in a buildup of arsenic trioxide in ductwork and associated process equipment throughout the roaster gas-handling discharge system, requiring periodic cleaning.

Mercury. Small amounts of mercury can be recovered by alkaline scrubbing systems; however, materials with higher mercury content may require additional gas treatment. A number of methods are commercially available for this:

- Mercury chloride scrubbing at 25°C to 35°C, to produce solid mercury chloride

- Sulfuric acid scrubbing at 140°C to 180°C
- Sodium thiocyanate scrubbing followed by precipitation with sodium sulfide
- Filtration through a packed bed of sulfide-impregnated carbon

These processes are considered in detail in the literature [95].

5.8.4 Microwave Energy

The use of microwave energy as a pretreatment step for processing of refractory sulfide and carbonaceous materials has received considerable attention in the 1990s and 2000s, and several processes have been proposed. Sulfide minerals have dielectric (semiconductor) properties, and these materials act as energy receptors when irradiated with microwaves, whereas the gangue minerals (e.g., quartz, feldspar, other silicates) are insulators and are transparent to the microwaves. This results in rapid heating of the sulfide ore constituents in situ, with direct oxidation of sulfide sulfur to sulfur dioxide, provided sufficient oxygen is available. Depending on the conditions applied, sulfur dioxide may be released in violent eruptions of the rock or mineral surface, thereby propagating cracks and fissures in the rock matrix. Carbonaceous material (i.e., organic and graphitic carbon) probably behaves as conductive material, reflecting microwaves; however, this material is likely to be oxidized under the temperatures generated by the application of microwave energy. Microwave treatment can produce the following results [101, 102]:

- Direct oxidation of sulfide minerals and oxidation of carbonaceous materials (passivating the preg-robbing properties)
- Fracturing of the rock (due to gas formation or expansion and/or due to differential heating and expansion of adjacent minerals)
- Decrease in the crushing and/or Bond Work Index of the material
- Improvement in the leaching response of the material due to sulfide and carbonaceous mineral oxidation
- Changes in the surface properties of some mineral constituents, potentially affecting surface chemical processes such as flotation and coal–gold agglomeration

One investigation indicated that a carbonaceous refractory gold ore containing 3% sulfide sulfur, 1.42% total carbon, and 0.26% organic carbon responded favorably to microwave treatment [103]. In this case, gold extraction by cyanide leaching was increased from about 20% to between 85% to 95% after microwave energy irradiation with temperatures in the reactor measured at about 400°C. Approximately 15 kWh/t of electrical energy were consumed in the pretreatment step. The application of microwave energy for oxidative pretreatment of ores and concentrates is a complex process, and factors such as microwave frequency (Hz), specific microwave power (kWh/t), application method, irradiation duration, and material particle size all have an effect on the efficiency of treatment. Other investigations have indicated that the total energy consumption for effective microwave oxidative pretreatment is likely to be in the range of 10 to 60 kWh/t.

Although this technology shows promise for oxidative pretreatment of refractory sulfide and carbonaceous constituents, practical application of the technology has some significant challenges, and there are no commercial applications.

REFERENCES

[1] Peters, E. 1976. Direct leaching of sulphides: Chemistry and applications. *Metallurgical Transactions B* 7:505–517.

[2]   Peters, E. 1986. Leaching of sulphides. Pages 445–462 in *Advances in Mineral Processing*. Edited by P. Somasundaran. Littleton, CO: SME.

[3]   Osseo-Asare, K., T. Xue, and V.S.T. Ciminelli. 1984. Solution chemistry of cyanide leaching systems. Pages 173–197 in *Precious Metals: Mining, Extraction & Processing*. Edited by V. Kudryk, D.A. Corrigan, and W.W. Liang. Warrendale, PA: TMS.

[4]   Wadsworth, M.E. 1972. Advances in the leaching of sulphide minerals. *Minerals Science & Engineering* 4(4):36–47.

[5]   Burkin, A.R. 1966. *The Chemistry of Hydrometallurgical Processes*. London: E. & F.N. Spon.

[6]   Garrels, R.M., and C.L. Christ. 1965. *Solutions, Minerals and Equilibria*. New York: Harper & Row.

[7]   Woods, R. 1981. Mineral flotation. Pages 571–595 in *Comprehensive Treatise on Electrochemistry*. Volume 2, Electrochemical Processing. Edited by J. O'M. Bockris, B.E. Conway, E.Yeager, and R.E. White. New York: Plenum Press.

[8]   Berezowsky, R.M.G.S, A.K Haines, and D.R. Weir. 1988. The Sao Bento gold project pressure oxidation process development. Projects 1988, 18th Annual Meeting of the Hydrometallurgical Section of Canadian Institute of Mining, Metallurgy and Petroleum, Edmonton, AB: Canadian Institute of Mining, Metallurgy, and Petroleum.

[9]   Mackiw, V.N., T.W. Benz, and D.J.I. Evans. 1966. A review of recent developments in pressure hydrometallurgy. *CIM Metallurgical Review* 11(109):143–158.

[10] Senanayake, G., and D.M. Muir. 1988. Speciation and reduction potentials of metal ions in concentrated chloride and sulphate solutions relevant to processing base metal sulphides. *Metallurgical Transactions B* 19:37–45.

[11] Adam, K., M. Stefanakis, and A. Kontopoulos 1989. Pages 131–138 in *Biotechnology in Mineral & Metallurgical Processing*. Littleton, CO: SME.

[12] Krause, E., and V.A. Ettel. 1987. Solubilities and stabilities of ferric arsenates. Pages 3/1–3/18 in *Proceedings 17th Annual Hydrometallurgical Meeting of CIM*. New York: Pergamon Press.

[13] Krause, E., and V.A. Ettel. 1985. Ferric arsenate compounds: Are they environmentally safe? Solubilities of basic ferric arsenates. Pages 5/1–5/20 in *Proceedings of 15th Annual Hydrometallurgical Meeting of CIM*. Ottawa: Canadian Institute of Mining, Metallurgy, and Petroleum.

[14] Harris, G.B., and S. Monette. 1987. The stability of arsenic-bearing residues. Pages 469–488 in *Proceedings of Arsenic Metallurgy Fundamentals & Applications*. Edited by R.G. Reddy. Warrendale, PA: TMS.

[15] Liddell, K.S. 1989. Applications of ultra-fine milling and attrition reactor in the treatment of refractory gold ores. Pages 109–113 in *Proceedings of Randol Gold Forum*. Golden, CO: Randol International Ltd.

[16] Dean, J.A. 1985. *Lange's Handbook of Chemistry*. 13th edition. New York: McGraw-Hill.

[17] Welham, N.J. 2001. Mechanochemical processing of gold-bearing sulphides. *Minerals Engineering* 14(3):341–347.

[18] Weir, D.R., and R.M.G.S. Berezowsky. 1984. Gold extraction from refractory concentrates, *Proceedings of 14th Annual Hydrometallurgy Meeting of CIM*. Ottawa: Canadian Institute of Mining, Metallurgy, and Petroleum.

[19] Berezowsky, R.M.G.S., and D.R. Weir. 1989. Factors affecting the selection of pressure oxidation for the pretreatment of refractory gold ores. *Proceedings of International Gold Expo.* Chicago, IL: *Engineering & Mining Journal.*

[20] Weir, D.R., and R.M.G.S. Berezowsky. 1987. Aqueous pressure oxidation of refractory gold feedstocks. *Proceedings of the International Symposium on Gold Metallurgy.* Volume 1. Edited by R.S. Salter, D.M. Wyslouzil, and G.W. McDonald. New York: Pergamon Press.

[21] Sarkar, K.M. 1985. Selection of autoclaves in hydrometallurgical operations. *Transactions of the Institution of Mining and Metallurgy* C 94:184–194.

[22] St. Louis, R.M., and J.M. Edgecombe. 1990. Recovery enhancement in the Mercur Autoclave circuit. Pages 443–450 in *Proceedings of Gold 1990 Symposium.* Edited by D.M. Hausen, D.N. Halbe, E.U. Petersen, and W.J. Tafuri. Littleton, CO: SME.

[23] Hayden, A.S., P.G. Mason, and W.T. Yen. 1987. Refractory gold ore oxidation–Simulation of continuous flow. Pages 249-258 in *Proceedings of the International Symposium on Gold Metallurgy.* Volume 1. Edited by. R.S. Salter, D.M. Wyslouzil, and G.W. McDonald. New York: Pergamon Press.

[24] Mason, P.G. 1990. Energy requirements for the pressure oxidation of gold-bearing sulphides. *Journal of Metals* (September):15–18.

[25] Dziurdzak, G., J.H. Kyle, and R.C. Dunne. 1989. Pressure aqueous pre-oxidation of a refractory gold ore from the Golden Mile, Kalgoorlie, Australia. Pages 315–321 in *Proceedings of World Gold '89 Symposium.* New York: American Institute of Mining, Metallurgical, and Petroleum Engineers.

[26] Thomas, K.G., H. Pieterse, R. Williams, K.S. Fraser, and J.R. Goode. 1990. The Goldstrike autoclave plant design, commissioning and operation. Pages 181–191 in *Advances in Gold and Silver Processing—Proceedings of GOLDTech 4 Symposium.* Edited by M.C. Fuerstenau and J.L. Hendrix. New York: American Institute of Mining, Metallurgical, and Petroleum Engineers.

[27] Carvalho, T.M., A.K. Haines, E.J. da Silva, and B.N. Doyle. 1988. Start-up of the Sherritt pressure oxidation process at Sao Bento. Pages 152–156 in *Proceedings Randol Perth International Gold Conference.* Golden, CO: Randol International Ltd.

[28] Berezowsky, R.M.G.S., and D.R. Weir. 1989. Refractory gold: The role of pressure oxidation. Pages 295–304 in *Proceedings of World Gold '89 Symposium.* Edited by R.B. Bhappu and R.J. Harden. New York: American Institute of Mining, Metallurgical, and Petroleum Engineers.

[29] Simmons, G. 1996. Pressure oxidation process development for treating carbonaceous ores at Twin Creeks. Pages 199–208 in *Proceedings Randol Gold Forum '96.* Golden, CO: Randol International Ltd.

[30] Weir, D.R., J.A. King, and P.C. Robinson. 1986. Pre-concentration and pressure oxidation of Porgera refractory gold ore. *Minerals & Metallurgical Processing Journal* (November):201–207.

[31] Mason, P.G. 1990. Energy requirements for the pressure oxidation of gold-bearing sulphides. *Journal of Metals* (September):15–18.

[32] Wicker, G.R., and J.A. Cole. 1990. The development and implementation of a pressure oxidation flowsheet for the Getchell Mine. Pages 437–442 in *Proceedings of Gold 1990 Symposium.* Edited by D.M. Hausen, D.N. Halbe, E.U. Petersen, and W.J. Tafuri. Littleton, CO: SME.

[33] Simpson, W.W., and S.A.N. Sheya. 1990. Gold and base metal recovery from a massive sulphide ore. Pages 161–178 in *Advances in Gold and Silver Processing, the Proceedings of GOLDTech 4 Symposium*. Edited by M.C. Fuerstenau and J.L. Hendrix. Littleton, CO: SME.

[34] Simmons, G.L. 1996. Pressure oxidation process development for treating carbonaceous ores at Twin Creeks. Pages 199–208 in *Proceedings Randol Gold Forum '96*. Golden, CO: Randol International Ltd.

[35] Hille, S., and R. Raudsepp. 1998. Pressure oxidation performance at Porgera: 1995 and 1996. Pages 209–216 in *Proceedings Randol Gold & Silver Forum '98*. Golden, CO: Randol International Ltd.

[36] Ketcham, V.J., J.F. O'Reilly, and W.D. Vardill. 1993. The Lihir Gold Project: Process plant design. *Minerals Engineering* 6(8–10):1037–1065.

[37] Taylor, P.R., Z. Jin, and M. Spangler. 1989. Metallurgy of refractory gold ores—an overview. *Proceedings of International Gold Expo*. 1989. Chicago, IL: *Engineering & Mining Journal*.

[38] Mason, P.G., R. Pendreigh, F.D. Wicks, and L.D. Kornze. 1984. Selection of the process flowsheet for the Mercur gold plant. Pages 435–448 in *Precious Metals: Mining, Extraction & Processing*. Edited by V. Kudryk, D.A. Corrigan, and W.W. Liang. Warrendale, PA: TMS.

[39] Mason, P.G., F.D. Wicks, and J.C. Gathje. 1985. Process for the recovery of gold from refractory ores by pressure oxidation. U.S. Patent 4,552,589. November 12.

[40] Anderson, C.G., and K.D. Harrison. 1990. Optimization of nitric-sulfuric acid pressure leaching of silver from refractory sulfide concentrates. In *Precious Metals 1990: Proceedings of 14th International Precious Metals Conference*. Warrendale, PA: TMS.

[41] Peters, E. 1992. Hydrometallurgical process innovation. *Hydrometallurgy* 29:431–459.

[42] Fair, K.J., J.C. Schneider, and G. Van Weert. 1986. Options in the Nitrox process. Pages 279–291 in *Proceedings of the International Symposium on Gold Metallurgy*. Volume 1. Edited by. R.S. Salter, D.M. Wyslouzil, and G.W. McDonald. New York: Pergamon Press.

[43] Beattie, M.J.V., and R. Raudsepp. 1988. The Arseno process–an update. Paper presented at 90th Annual Meeting of Canadian Institute of Mining, Metallurgy and Petroleum, Edmonton, AB, Canada.

[44] Beattie, M.J.V., and A. Ismay. 1990. Applying the Redox process to arsenical concentrates. *Journal of Metals* (January):31–35.

[45] Flatt, J.R., and R. Woods. 2000. Oxidation of pyrite in nitric acid solutions: Relation to treatment of refractory gold ores. Pages 152–163 in *Proceedings of Electrochemistry in Mineral & Metallurgical Processing Journal*. Edited by R. Woods and J.M. Doyle. Pennington, NJ: Electrochemical Society.

[46] Foo, K.A., and M.D. Bath. 1989. Trends in the treatment of refractory ores. Paper presented at the Colorado Mining Association 92nd Western Mining Conference, Denver, CO, February 8–10.

[47] Anderson, C. 1996. Theoretical considerations of sodium nitrite oxidation and fine grinding in refractory precious metal concentrate pressure leaching. *Mineral & Metallurgical Processing Journal* (February):4–11.

[48] Van Weert, G., K.J. Fair, and V.H. Aprahamian. 1988. Design and operating results of the Nitrox process. Paper presented at 2nd International Gold Conference. Vancouver, BC, Canada.

[49] Van Weert, G., K.J. Fair, and J.C. Schneider. 1987. The Nitrox process for treating gold bearing arsenopyrites. Paper presented at 116th Annual Meeting of The Metallurgical Society of AIME, Denver, CO, February 23–26.

[50] Snoeyink, P.L., and D. Jenkins. 1979. *Water Chemistry*. New York: John Wiley.

[51] Adams, M.D., and A.M. Burger. 1998. Characterization and blinding of carbonaceous preg-robbers in gold ores. *Minerals Engineering* 11(10):912–927.

[52] Sibrell, P.L., R.Y. Wan, and J.D. Miller. 1990. Spectroscopic analysis of passivation reactions for carbonaceous matter from Carlin Trend ores. Pages 355–364 in *Proceedings of Gold 1990 Symposium*. Edited by D.M. Hausen, D.N. Halbe, E.U. Petersen, and W.J. Tafuri. Littleton, CO: SME.

[53] Brunk, K.A., G. Ramadorai, D. Seymour, and F.P. Traczyk. 1988. Flash chlorination—a new process for treatment of refractory sulphide and carbonceous gold ores. Pages 127–129 in *Proceedings of Randol Gold Forum 1988* Golden, CO: Randol International Ltd.

[54] Brunk, K.A., and R.L. Atwood. 1987. Practical aspects of cyanidation of carbonaceous gold ores. SME Preprint. Littleton, CO: SME.

[55] Birak, D., and K. Deter. 1987. Changes in the Jerritt Canyon metallurgical process as a result of geologic characteristics of the ores. Pages 135–139 in *Proceedings of the International Symposium on Gold Metallurgy*. Volume 1. Edited by R.S. Salter, D.M. Wyslouzil, and G.W. McDonald. New York: Pergamon Press.

[56] Chao, M.S. 1968. Diffusion coefficients of hypochlorite, hypochlorous acid and chlorine in aqueous media by potentiometry. *Journal of the Electrochemical Society* 115(11):1172–1174.

[57] Nagy, I., P. McKusic, and H.W. McCulloch. 1968. Chemical treatment of refractory gold ores—a literature survey. Report No. 38. Johannesburg, South Africa: National Institute for Metallurgy, South Africa.

[58] Guay, W.J., and M.A. Gross. 1981. The treatment of refractory gold ores containing carbonaceous material and sulfides. Paper presented at Annual Meeting of The Metallurgical Society of AIME, Chicago, IL, February.

[59] Guay, W.J. 1980. How Carlin treats gold ores by double oxidation. *World Mining* (March):47–49.

[60] Pooley, F.D., G.N. Shrestha, M.T. Errington, and H.E. Gibbs. 1987. The separate generator concept applied to the bacterial leaching of auriferous minerals. Pages 58–67 in *Separation Processes in Hydrometallurgy*. Edited by G.A. Davies. Chichester, England: Ellis Horwood.

[61] Van Aswegen, P.C., and A.K. Haines. 1988. *International Mining Magazine* 19–23.

[62] Van Aswegen, P.C., and J. van Niekerk. 2004. New developments in the bacterial oxidation technology to enhance the efficiency of the BIOX® process. Pages 181–189 in *Proceedings of Bac-Min Conference*. Carlton, Victoria, Australia: Australasian Institute of Mining and Metallurgy.

[63] Miller, P.C., F. Jiao, and J. Wang. 2004. The bacterial oxidation (BACOX) plant at Laizhou Shandong Province China—the first three years of operation. Pages 167–171 in *Proceedings of Bac-Min Conference*. Carlton, Victoria, Australia: Australasian Institute of Mining and Metallurgy.

[64] Marchant, P.B. 1985. Plant scale design and economic considerations for biooxidation of an arsenical sulphide concentrate. Paper presented at The International Symposium on Complex Sulfides, San Diego, CA, November.

[65] Southwood, M.J. 1986. *Mineralogical Aspects of the Bacterial Leaching of Auriferous Sulphide Concentrates, and a Mathematical Model for the Release of Gold.* Report No. M274. Council for Mineral Technology. Randburg, South Africa: Mintek.

[66] Sand, W., T. Gehrke, P-G. Jozsa, and A. Schippers. 2001. *Hydrometallurgy* 59:159–175.

[67] Southwood, M.J., and A.J. Southwood. 1985. Mineralogical observations on the bacterial leaching of puriferous pyrites. Pages 98–114 in *Fundamental and Applied Biohydrometallurgy: Proceedings of 6th International Symposium on Biohydrometallurgy.* Edited by R.W. Lawrence, R.M.R. Branion, and H.G. Ebner. New York: Elsevier.

[68] Lazer, M.J., M.J. Southwood, and A.J. Southwood. 1986. The release of refractory gold from sulphide minerals during bacterial leaching. Pages 287–297 in *Gold 100: Proceedings of International Conference on Gold.* Volume 2, Extractive Metallurgy of Gold. Johannesburg: South African Institute of Mining and Metallurgy.

[69] Marchant, P.B. 1985. Commercial piloting and the economic feasibility of plant scale continuous biological tank leaching at Equity Silver Mines Ltd. Pages 53–76 in *Fundamental and Applied Biohydrometallurgy: Proceedings of 6th International Symposium on Biohydrometallurgy.* Edited by R.W. Lawrence, R.M.R. Branion, and H.G. Ebner. New York: Elsevier.

[70] Karavainko, G.I., et al. 1977. The bacterial leaching of metals from ores. *Trans. Burns W. Technology Ltd.* (England).

[71] Pinches, A. 1970. Bacterial leaching of an arsenic-bearing sulphide concentrate. Ph.D. thesis, Cardiff, Wales: University College.

[72] Lawrence, R.W., and A. Bruynesteyn. 1983. Biological pre-oxidation to enhance gold and silver recovery from refractory pyritic ores and concentrates. *CIM Bulletin* (September):107–110.

[73] Gibbs, H.E., M.T. Errington, and F.D. Pooley. 1985. Economics of bacterial leaching. *Canadian Metallurgical Quarterly* 24(2):121–125.

[74] Hackl, R.P., F.R Wright, and L.S. Gormely. 1989. Bioleaching of refractory gold ores: Out of the lab and into the plant. Pages 533–549 in *Proceedings of International Symposium on Biohydrometallurgy.* Ottawa, ON, Canada: Canmet.

[75] Whincup, P., and M. Binks. 2004. A case study of the development, engineering and construction of a pyrite–arsenopyrite gold ore. Pages 191–202 in *Proceedings of Bac-Min Conference.* Carlton, Victoria, Australia: Australasian Institute of Mining and Metallurgy.

[76] Brierly, C.L. 2001. Bacterial succession in bioheap leaching. *Hydrometallurgy* 59: 249–255.

[77] Bartlett, R.W. 1996. Aeration requirements for heap bio-oxidation of refractory gold ores. Pages 273–274 in *Proceedings Randol Gold Forum '96.* Golden, CO: Randol International Ltd.

[78] Shutey-McCann, M.L., F.P. Sawyer, T. Logan, A.J. Schindler, and R.M. Perry. 1997. Operation of Newmont's biooxidation demonstration facility. Pages 75–82 in *Global Exploitation of Heap Leachable Gold Deposits.* Edited by D.M. Hausen. Warrendale, PA: TMS.

[79] Bhakta, P., and B. Arthur. 2002. Heap bio-oxidation and gold recovery at Newmont Mining: First year results. *Journal of Metals* (October):31–34.

[80] Temple, T. 2003. Commercial bio-oxidation challenges at Newmont's Nevada operations. SME Preprint No. 03-067. Littleton, CO: SME.

[81] Habashi, F. 1986. *Principles of Extractive Metallurgy*. Volume 3, Pyrometallurgy. New York: Gordon & Breach.

[82] Palmer, B.R. 1990. High-temperature gold-chlorination technology. Pages 127–142 in *Gold: Advances in Precious Metals Recovery*. Edited by N. Arbiter and K.N. Han. New York: Gordon & Breach.

[83] Kontopoulos, A., and M. Stefanakis. 1989. Process selection for the Olympias refractory gold concentration. Pages 179–210 in *Precious Metals '89*. Edited by M.C. Jha and S.D. Hill. Warrendale, PA: TMS.

[84] Jha, M.C., and M.J. Kramer. 1985. Recovery of gold from arsenical ores. Pages 337–366 in *Precious Metals: Mining, Extraction & Processing*. Edited by V. Kudryk, D.A. Corrigan, and W.W. Liang. Warrendale, PA: TMS.

[85] Coleman, R.B. 1990. Roasting of refractory gold ores and concentrates. Pages 381–388 in *Proceedings of Gold 1990 Symposium*. Edited by D.M. Hausen, D.N. Halbe, E.U. Petersen, and W.J. Tafuri. Littleton, CO: SME.

[86] Just, J., J. Graham, J. Dunn, J. Avraamides, and G. Nguyen. 1991. Mineralogy and textures of roasted pyrite. Presented at Southeast Asia Gold (SEAGOLD '91) Symposium, Seagold, Vietnam, December 5–7.

[87] Dunn, J.G., and A.C. Chamberlain. 1997. The recovery of gold from refractory arsenopyrite concentrates by pyrolysis-oxidation. *Minerals Engineering* 10(9):919–928.

[88] Swash, P.M., and P. Ellis. 1986. The roasting of arsenical gold ores—a mineralogical perspective. Pages 235–257 in *Gold 100: Proceedings of International Conference on Gold*. Volume 2, Extractive Metallurgy of Gold. Johannesburg: South African Institute of Mining and Metallurgy.

[89] Grimsey, E.J., and M.G. Aylmore. 1990. Roasting of arsenopyrite. Pages 397–410 in *Proceedings of Gold 1990 Symposium*. Edited by D.M. Hausen, D.N. Halbe, E.U. Petersen, and W.J. Tafuri. Littleton, CO: SME.

[90] Andon, R.G., J.F. Martion, and K.C. Mills. 1971. Thermodynamic properties of gold telluride. *Journal Chemical Society* (A)11:1788–1791.

[91] Johnstone, W.E. 1933. Tellurides, problem or alibi? *Engineering and Mining Journal* (August):333–334.

[92] Arriagada, F.J., and K. Osseo-Asare. 1984. Gold extraction from refractory ores: Roasting behaviour of pyrite and arsenopyrite. Pages 367–385 in *Precious Metals: Mining, Extraction & Processing*. Edited by V. Kudryk, D.A. Corrigan, and W.W. Liang. Warrendale, PA: TMS.

[93] Chen, B., and R.G. Reddy. 1990. Roasting characteristics of refractory gold ores. Pages 201–214 in *Advances in Gold and Silver Processing: Proceedings of GOLDTech 4 Symposium*. Edited by M.C. Fuerstenau and J.L. Hendrix. Littleton, CO: SME.

[94] Major, K.W., and P.G. Semple. 1990. Design and operating considerations for the application of a dry grinding and roasting circuit for refractory gold ore. Pages 175–184 in *Randol Gold Forum '90*. Golden, CO: Randol International Ltd.

[95] Maycock, A.R., W. Nahas, and T.C. Watson. 1990. Review of the design and operation of roasters for refractory gold bearing materials. Pages 389–396 in *Proceedings of Gold 1990 Symposium*. Edited by D.M. Hausen, D.N. Halbe, E.U. Petersen, and W.J. Tafuri. Littleton, CO: SME.

[96] Halverson, G.B. 1990. Fluosolids roasting practice at Giant Yellowknife Mines Ltd. Paper presented at 96th Annual Northwest Mining Association Meeting, Spokane, WA.

[97] Lahti, P.A. 1996. Refractory gold ore processing methods at Jerritt Canyon. Pages 225–231 in *Proceedings Randol Gold Forum '96*. Golden, CO: Randol International Ltd.

[98] Bunn, S. 1991. Process description—Fimiston plant and Gidji roaster. *Proceedings of Australasian Institute of Mining & Metallurgy Conference*. Melbourne: Australasian Institute of Mining & Metallurgy.

[99] DeSomber, R.K., R. Fernandez, L. McAnany, E. Stolarski, and G. Schmidt. 1996. Refractory ore treatment plant at Newmont Gold Company. Pages 239–247 in *Proceedings Randol Gold Forum '96*. Golden, CO: Randol International Ltd.

[100] Cole, A., S. Dunn, S. Bunk, and T. McCord. 1999. Refractory gold ore treatment by fluidized bed roasting for Barrick Goldstrike. Pages 79–84 in *Proceedings Randol Gold & Silver Forum*. Golden, CO: Randol International Ltd.

[101] Kruesi, P.R., and V.H. Frahm. 1982. Process for the recovery of copper from its ores. U.S. Patent No. 4,324,582. April 13.

[102] Tranquilla, J.M. 1997. Method and apparatus for microwave treatment of metal ores and concentrates in a fluidized bed reactor. International Patent Application PCT/C A97/00556.

[103] Tranquilla, J.M. 2000. Microwave for carbonaceous ores. Pages 45–47 in *Proceedings Randol Gold & Silver Forum*. Golden, CO: Randol International Ltd.

[104] Frostiak, J., and B. Haugrud. 1992. Start-up and operation of the Campbell Red Lake gold pressure oxidation plant. SME Preprint No. 92-14. Littleton, CO: SME.

[105] O'Rourke, J., B. Bissonnette, and T.Y. Chong. 2000. Placer Dome Inc.–Campbell Mine. Pages 41–47 in *Canadian Milling Practice*. CIM Special Volume 49. Edited by B. Damjanovic. Edmonton, AB, Canada: Canadian Institute of Metallurgy.

[106] Cole, J.A., W.J. Janhunen, and J.C. Lenz. 1995. Santa Fe Pacific Gold's first pressure oxidation circuit: Year one at Lone Tree. SME Preprint No. 95-200. Littleton, CO: SME.

CHAPTER 6

Leaching

In the context of gold extraction, leaching is the dissolution of a metal or mineral in a liquid. The reaction of primary concern is the dissolution of gold in an aqueous solution, which requires both a complexant and an oxidant to achieve acceptable leaching rates. Only a limited number of ligands form complexes of sufficient stability for use in gold extraction processes. Cyanide is universally used because of its relatively low cost, its great effectiveness for gold (and silver) dissolution, and its selectivity for gold and silver over other metals. Also, despite some concerns over the toxicity of cyanide, it can be applied with little risk to health and the environment. The oxidant most commonly used in cyanide leaching is oxygen, supplied from air, which contributes to the attractiveness of the process.

The different processes developed for leaching with cyanide, including agitation leaching, heap leaching, and intensive cyanidation, and their applications are reviewed in this chapter.

Noncyanide reagent schemes have the following potential advantages over cyanide:

- Environmental pressures, and in some cases restrictions, may make the application of cyanide difficult in certain locations.
- Some have faster gold leaching kinetics.
- Several can be applied in acidic media, which may be more suitable for refractory ore treatment.
- Some are more selective than cyanide for gold and silver over other metals, for example, copper.

However, they also have some significant disadvantages, and none appear widely applicable, at least without significant further advances.

Chlorine–chloride leaching was applied commercially in the 19th century, but its use diminished following the introduction of the cyanide process in 1889 (Chapter 1). Thiosulfate, thiourea, thiocyanate, ammonia, alkaline sulfide, and other halide (e.g., bromide, iodide) solutions have been investigated extensively in the laboratory, and several potential processes have been developed, but none commercially applied on a large scale. The chemistry of these alternative reagent schemes contributes to their lack of commercial success and is also discussed.

6.1 CYANIDATION

6.1.1 Chemistry of Cyanide Solutions

Simple cyanide salts, such as sodium, potassium, and calcium cyanide, dissolve and ionize in water to form their respective metal cation and free cyanide ions, as follows:

$$NaCN \rightleftharpoons Na^+ + CN^- \qquad (EQ\ 6.1)$$

TABLE 6.1 Properties of simple cyanide compounds [1]

Compound	Available Cyanide (%)	Solubility in Water at 25°C (g/100 cc)
NaCN	53.1	48
KCN	40.0	50
Ca(CN)$_2$	56.5	Decomposes

The solubility and relative cyanide content of the different cyanide salts are given in Table 6.1. All three salts have been used effectively on a commercial scale as sources of cyanide for leaching. Sodium and potassium cyanide are more readily soluble than calcium cyanide and are generally available in purer form, which has advantages for the handling and distribution of the reagent in leaching systems. Liquid sodium cyanide (i.e., in aqueous solution), which is widely available in some regions of the world, avoids the need to dissolve the reagent on site, reducing process requirements. Thus, the choice of cyanide type depends on the method of application, cost, and availability.

Cyanide ions hydrolyze in water to form molecular hydrogen cyanide (HCN) and hydroxyl (OH$^-$) ions, with a corresponding increase in pH:

$$CN^- + H_2O \rightleftharpoons HCN + OH^- \quad \text{(EQ 6.2)}$$

Hydrogen cyanide is a weak acid, which incompletely dissociates in water as follows [2]:

$$HCN \rightleftharpoons H^+ + CN^- \quad \text{(EQ 6.3)}$$

where
$K_a(25°C) = 6.2 \times 10^{-10}$
$pK_a = 9.31$

Figure 6.1 shows the extent of this dissociation reaction at equilibrium as a function of pH. At approximately pH 9.3, half of the total cyanide exists as hydrogen cyanide and half as free cyanide ions. At pH 10.2, >90% of the total cyanide is present as free cyanide (CN$^-$), while at pH 8.4, >90% exists as hydrogen cyanide. This is important because hydrogen cyanide has a relatively high vapor pressure (100 kPa at 26°C [3]) and consequently volatilizes readily at the liquid surface under ambient conditions, causing a loss of cyanide from the solution. The rate of volatilization depends on the hydrogen cyanide concentration (a function of total cyanide concentration and pH); the surface area and depth of the liquid; temperature; and transport phenomena associated with mixing [3]. As a result, most cyanide leaching systems are operated at a pH that minimizes cyanide loss, typically above pH 10, although adverse effects may be caused by excessively high pH, as discussed in Section 6.1.4.

Both hydrogen cyanide and free cyanide can be oxidized to cyanate in the presence of oxygen and under suitably oxidizing conditions, as illustrated in the E_h–pH diagram for the CN–H$_2$O system, given in Figure 6.2. The important reactions are as follows:

$$4HCN + 3O_2 \rightleftharpoons 4CNO^- + 2H_2O \quad \text{(EQ 6.4)}$$

$$3CN^- + 2O_2 + H_2O \rightleftharpoons 3CNO^- + 2OH^- \quad \text{(EQ 6.5)}$$

These reactions are undesirable during leaching, because they reduce the free cyanide concentration, and the cyanate species formed does not dissolve gold.

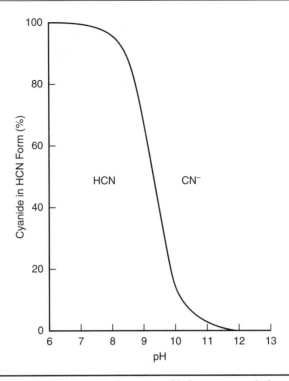

FIGURE 6.1 Speciation of cyanide and hydrogen cyanide in aqueous solution as a function of pH

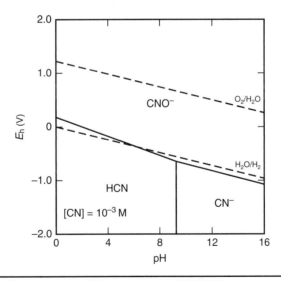

FIGURE 6.2 E_h–pH diagram for the CN–H_2O system at 25°C [43]

Figure 6.2 indicates that oxidation of cyanide to cyanate should occur spontaneously with oxygen, but the reaction is very slow and, in practice, strong oxidizing agents, such as ozone (O_3), hydrogen peroxide (H_2O_2), or hypochlorous acid (HOCl), are required for the reaction to proceed at a significant rate. In aerated cyanide solutions, the reaction is extremely slow, but can be accelerated by the action of ultraviolet light, heat, bacteria, and catalysts such as titanium dioxide, zinc oxide, and cadmium sulfide [3]. Some of these oxidation reactions are important for the destruction or degradation of cyanide and are considered in more detail in Chapter 11.

Free cyanide forms complexes with many metal species, principally the transition metals, which vary widely in stability and solubility:

$$M^{x+} + yCN^- \rightleftharpoons M(CN)_y^{(y-x)-} \quad \text{(EQ 6.6)}$$

where

$$K \rightleftharpoons [M^{x+}] \cdot [CN^-]^y / M(CN)_y^{(y-x)-}$$

The stability constants of some of the more important metal cyanide complexes are given in Table 6.2. The complexes may be grouped into three main categories, based on their stability [3]:

- Free cyanide (HCN, CN^-)
- Weak acid dissociable (WAD) cyanide complexes (for which $\log K \leq$ approximately 30)
- Strong cyanide complexes (for which $\log K >$ approximately 30)

These categories are used widely in the analysis of process solutions because they help to describe the behavior of the cyanide species present, while avoiding the need to provide detailed analytical information of every cyanide complex present, greatly simplifying analytical procedures. This general grouping is often useful for the evaluation and optimization of metallurgical performance or when working in environmental chemistry. Further information on the speciation of cyanide in aqueous solutions used for leaching is available in the literature [4, 5, 6, 7].

Metal cyanide complexes can form double salts with a variety of cations, for example, sodium, potassium, calcium, ammonium, and many other metal ions. For example, the iron(II) cyanide complex, $Fe(CN)_6^{4-}$, which is common to all gold leaching circuits, forms a large number of salts of varying solubility [8]. The solubility products for a number of these salts are listed in Table 6.3. Of these, the Fe(III) salt, $Fe_4(Fe(CN)_6)_3$, is commonly encountered in process effluents, appearing as "Prussian Blue" precipitate. The formation and solubility of these salts is an important consideration in effluent disposal and treatment, and is discussed further in Chapter 11.

6.1.2 Gold Dissolution

6.1.2.1 Anodic Reactions

In aqueous, alkaline cyanide solution, gold is oxidized and dissolves to form the Au(I) cyanide complex, $Au(CN)_2^-$, as shown in the E_h–pH diagram, Figure 6.3. The Au(III) cyanide complex, $Au(CN)_4^-$, is also formed, but the Au(I) complex is more stable than the Au(III) species by 0.5 V (Section 4.2.2) [16, 6]. For practical purposes, the stoichiometry of the dissolution reaction can be assumed to be:

$$Au(CN)_2^- + e \rightleftharpoons Au + 2CN^- \quad \text{(EQ 6.7)}$$

TABLE 6.2 Stability constants for selected metal cyanide complexes (adapted from [8])

Chemical Formula	Quantity	Remarks	Reference
CN (aq)	$G^0 = 41.2$ kcal/mol	25°C; $I = 0$	[9]
HCN (aq)	$G^0 = 48.6$ kcal/mol		[9]
$(CN)_2$(g)	$G^0 = 71.03$ kcal/mol		[9]
HOCN (aq)	$G^0 = 28.0$ kcal/mol		[9]
OCN (aq)	$G^0 = -23.3$ kcal/mol		[9]
HNCO (g) iso–	$G^0 = -25.66$ kcal/mol		[9]
HSCN (aq)	$G^0 = 23.3$ kcal/mol		[9]
SCN (aq)	$G^0 = 22.15$ kcal/mol		[9]
HCN	$\log K_1 = 9.21$	$I = 0$	[2]
$Fe(CN)_6^{4-}$	$\log \beta_6 = 35.4$	$I = 0$	[2]
$Fe(CN)_6^{3-}$	$\log \beta_6 = 43.6$	$I = 0$	[2]
$NiCN^-$	$\log K_1 = 7.03$	$I = 0$	[2]
$Ni(CN)_4^{2-}$	$\log \beta_4 = 30.22$	$I = 0.5$	[2]
$NiHCN_4$	$\log K_{11} = 5.4$	$I = 0.5$	[2]
NiH_2CN_4	$\log K_{12} = 4.5$	$I = 0.5$	[2]
NiH_3CN_4	$\log K_{14} = 2.6$	$I = 0.5$	[2]
$Ni(CN)_6^{4-}$	$\log = \beta_6 = 26\text{–}27.5$		[10, 11]
$CuCN_2^-$	$\log \beta_2 = 16.26$	$I = 0$	[2]
$CuCN_3^{2-}$	$\log \beta_3 = 21.66$	$I = 0$	[2]
$CuCN_4^{3-}$	$\log \beta_4 = 23.1$	$I = 0$	[2]
CuCN (s)	$pK_{sp} = 19.49$		[10, 11]
	$G^0 = 25.9$ kcal/mol		[9]
$ZnCN^+$	$\log K_1 = 5.3$	$I = 1.0$	[2]
$Zn(CN)_2$ (aq)	$\log \beta_2 = 11.07$		[2]
$Zn(CN)_3^-$	$\log \beta_3 = 16.05$		[2]
$Zn(CN)_4^{2-}$	$\log \beta_4 = 19.62$		[2]
$Zn(CN)_5^{3-}$	$\log \beta_5 = 20.17$		[10, 11]
$AgCN_2^-$	$\log \beta_2 = 20.48$		[2]
$AgCN_3^{2-}$	$\log \beta_3 = 21.40$		[2]
$AgCN_4^{3-}$	$\log \beta_4 = 20.80$		[2]
$Ag(OH)CN^-$	$\log \beta_{11} = 13.2$		[2]
Ag(CN) (s)	$pK_{sp} = 15.66$		[2]
$Au(CN)_2^-$	$\log \beta_2 = 39.3$		[10, 11]
$Au(CN)_4^-$	$\log \beta_4 = 56$		[10, 11]
	$G^0 = 68.3$ kcal/mol		[9]
$CdCN^+$	$\log \beta_1 = 6.01$		[2]
$CdCN_2$ (aq)	$\log \beta_2 = 11.12$		[2]
$CdCN_3^-$	$\log \beta_3 = 16.65$		[2]
$CdCN_4^{2-}$	$\log \beta_4 = 17.92$		[2]
$Co(CN)_6^{4-}$	$\log \beta_6 = 19.1$		[10, 11]
$Co(CN)_6^{3-}$	$\log \beta_6 = 64$		[10, 11]
	$\log \beta_6 = >50$		[12]
$Co(CN)_5^{3-}$	$H^0 = 257$ kJ/mol		[12]
$Co(CN)_6^{3-}/Co(CN)_6^{4-}$	$E^0 = -0.96$ V		[13]
$Co(CN)_5^{3-}/Co(CN)_5^{4-}$	$E^0 = -0.96$ V		[13]
$K[Fe(CN)_6]^{3-}$	$\log K_1 = 2.49$		[14]
$K_2[Fe(CN)_6]^{2-}$	$\log K_2 = 3.35$		[14]
$KH[Fe(CN)_6]^{2-}$	$\log K_2 = 5.84$		[14]
$H[Fe(CN)_6]^{3-}$	$\log K_1 = 4.25$		[14]
$H_2[Fe(CN)_6]^{2-}$	$\log K_2 = 6.8$		[15]

TABLE 6.3 Solubility products for selected metal cyanide compounds (adapted from [8])

Chemical formula	Quantity	Reference
$Zn(CN)_2$ (s)	$pK_{sp} = 15.50$	[2]
$Ag(CN)$ (s)	$pK_{sp} = 15.66$	[2]
$Cu_2[Fe(CN)_6]$ (s)	$pK_{sp} = 17.00$	[4]
$Cd_2[Fe(CN)_6]$ (s)	$pK_{sp} = 13.40$	[15]
$K_2Cd[Fe(CN)_6]$ (s)	$pK_{sp} = 17.10$	[15]
$K_2Cu_3[Fe(CN)_6]$ (s)	$pK_{sp} = 34.30$	[4]
$Zn_2[Fe(CN)_6]$ (s)	$pK_{sp} = 16.80$	[4]
$K_2Zn_3[Fe(CN)_6]$ (s)	$pK_{sp} = 38.50$	[4]
$KAg_3[Fe(CN)_6]$ (s)	$pK_{sp} = 19.00$	[4]
$Ag_4[Fe(CN)_6]$ (s)	$pK_{sp} = 19.00$	[4]
$Ni_2[Fe(CN)_6]$ (s)	$pK_{sp} = 15.89$	[10, 11]
$Co_2[Fe(CN)_6]$ (s)	$pK_{sp} = 37.32, 15.00$	[10, 11]
$K_4Ni_4[Fe(CN)_6]_3$ (s)	$pK_{sp} = 47.50$	[15]
$K_{12}Ni_8[Fe(CN)_6]_7$ (s)	$pK_{sp} = 113.60$	[15]
$K_{12}Cd_8[Fe(CN)_6]_7$ (s)	$pK_{sp} = 121.30$	[15]
$K_2Ni_3[Fe(CN)_6]_2$ (s)	$pK_{sp} = 32.56$	[15]
$K_2Co_3[Fe(CN)_6]_2$ (s)	$pK_{sp} = 27.86$	[15]
$K_4Co_4[Fe(CN)_6]_3$ (s)	$pK_{sp} = 45.50$	[15]
$Co_2[Fe(CN)_6]$ (s)	$pK_{sp} = 14.70$	[15]
$Fe_2[Fe(CN)_6]$ (s)	$pK_{sp} = 14.14$	[15]
$Cu_3[Fe(CN)_6]_2$ (s)	$pK_{sp} = 24.50$	[15]
$Cd_3[Fe(CN)_6]_2$ (s)	$pK_{sp} = 17.50$	[15]
$Ag_3[Fe(CN)_6]$ (s)	$pK_{sp} = 18.30$	[15]
$KCu_{10}[Fe(CN)_6]_7$(s)	$pK_{sp} = 74.00$	[15]
$KCd_{10}[Fe(CN)_6]_7$ (s)	$pK_{sp} = 69.00$	[15]
$Fe[Fe(CN)_6]$ (aq)	$logK = 1.30$	[10, 11]
$Fe_4[Fe(CN)_6]_3$ (s)	$pK_{sp} = 40.52$	[10, 11]
$N:Ni(CN)_4]$ (s)	$pK_{sp} = 8.77$	[10, 11]
$Co(CN)_2$ (s)	$pK_{sp} = 10.30$	[12]

for which the Nernst equation is:

$$E = -0.60 + 0.118 pCN + 0.059 \log a_{Au(CN)_2^-} \quad (V) \qquad (EQ\ 6.8)$$

Cyclic voltammetry has been used to study the mechanism of gold dissolution [16, 17]. Oxidation proceeds in three stages, as indicated by the three peaks shown in Figure 6.4. The peak at approximately −0.4 V probably represents the formation of an adsorbed intermediate species, AuCN, which causes temporary passivation of the gold surface:

$$AuCN_{(ads)} + e \rightleftharpoons Au + CN^- \qquad (EQ\ 6.9)$$

The second peak at approximately 0.3 V is attributed to the complexation reaction between free cyanide and the adsorbed intermediate $AuCN_{(ads)}$ species [16]:

$$AuCN_{(ads)} + CN^- \rightleftharpoons Au(CN)_2^- \qquad (EQ\ 6.10)$$

In the treatment of ores and concentrates with alkaline cyanide solutions, passivation of gold rarely occurs, even at low cyanide concentrations, probably due to the presence of low

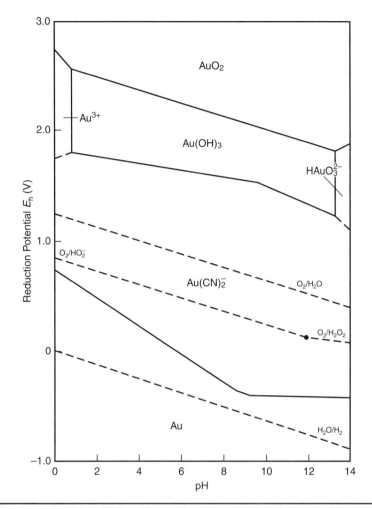

FIGURE 6.3 Potential–pH equilibrium diagram for the system Au–H_2O–CN^- at 25°C. Concentrations of all soluble gold species = 10^{-4} M [6]. $[CN^-]_{total}$ = 10^{-3} M, pO_2 = pH_2 = 1 atm, log ($[H_2O_2]$ – $[HO_2^-]/pO_2$)

concentrations of heavy metals ions (e.g., lead and mercury), dissolved from the feed material or introduced with reagents, which disrupt the formation of such a passivating layer [17, 18]. This is discussed further in Section 6.1.3.6.

The final peak at 0.6 to 0.7 V is thought to be due to the formation of an Au(III) oxide (Au_2O_3) layer which passivates the gold surface. However, such passivation is unlikely to be a problem in practice because of the highly positive potentials required for this to occur [6, 16, 17].

The formation of the adsorbed intermediate species (and the associated passivation), the beneficial effect of lead (and other divalent cations), and the passivation due to formation of the oxide layer have been confirmed using surface-enhanced raman scattering (SERS) spectroscopy to study the gold surface during cyanidation [19].

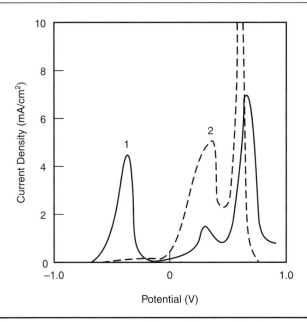

FIGURE 6.4 Current vs. potential curves for the oxidation of gold in alkaline cyanide solutions: (1) 0.077 M CN⁻, pH 12; (2) 0.1 M CN, 0.1 M OH (adapted from [17])

6.1.2.2 Cathodic Reactions

In aerated, alkaline cyanide solutions, the anodic dissolution reaction shown in Equation (6.7) is accompanied by the cathodic reduction of oxygen. The mechanism of this reaction has long been controversial and involves several parallel and series cathodic reactions. Experimental investigation of the stoichiometry of gold dissolution has shown the major reaction to be as follows [20]:

$$O_2 + 2H^+ + 2e = H_2O_2; \quad E^0 = +0.682 \qquad (EQ\ 6.11)$$

where
$$E = 0.682 - 0.059\ pH - 0.0295\ pO_2\ (V)$$

The hydrogen peroxide formed is a strong oxidizing agent, which may take part in further oxidation reactions:

$$H_2O_2 + 2e = 2OH^-; \quad E^0 = +0.88\ (V) \qquad (EQ\ 6.12)$$

The effect of hydrogen peroxide on gold leaching rates in alkaline cyanide solution is a matter of controversy and debate. Early studies indicated that the reduction of hydrogen peroxide on gold surfaces is kinetically hindered, and the dissolution rate of gold in oxygen-free solutions containing hydrogen peroxide is very slow [21, 22]. This work provided evidence of passivation of the gold surface by oxide layer formation, inhibiting gold leaching. On the other hand, hydrogen peroxide decomposes to oxygen and water, thus providing dissolved oxygen in solution as follows:

$$2H_2O_2 \rightarrow O_2 + 2H_2O \qquad (EQ\ 6.13)$$

It has been demonstrated that as much as 85% of the hydrogen peroxide formed by oxygen reduction (Equation [6.11]) diffuses away from the reaction site, with only a small proportion reduced directly to hydroxyl ions [21].

Other studies have indicated that hydrogen peroxide can play a direct role [23, 24]. One such investigation has demonstrated that for pure gold the leaching rate could be increased significantly using a concentration of 0.015 M H_2O_2 in a solution containing 0.01 M NaCN at pH 10. However, this study showed that smaller quantities of hydrogen peroxide (i.e., <0.0025 M) inhibited the gold dissolution rate, and the use of hydrogen peroxide at higher pH (i.e., >11) also reduced the gold dissolution rate, due to increased cyanide oxidation and decreased cyanide concentration [24].

Consequently, in general, hydrogen peroxide alone is not considered to be a very effective oxidant for use in gold leaching, except under conditions where dissolved oxygen is limited (e.g., at high altitude or in the presence of significant oxygen consumers in the ore).

Hydrogen peroxide is very effective for the oxidation of sulfides, although expensive for this purpose. However, it has been used successfully in some cases to accelerate low-pressure (atmospheric) oxidation kinetics, as discussed further in Section 5.2.3 [25].

Finally, oxygen may be directly reduced to hydroxide ions, rather than to H_2O_2 (Equation [6.11]), as follows:

$$O_2 + 2H_2O + 4e \rightleftharpoons 4OH^-; \quad E^0 = 0.401 \text{ (V)} \quad \text{(EQ 6.14)}$$

This reaction requires a large overpotential and is very slow, but occurs in parallel with (6.11) to a limited extent.

6.1.2.3 Overall Dissolution Reaction

The overall dissolution of gold in aerated, alkaline cyanide solutions, considering both the anodic and cathodic half-reactions, is most accurately described by the following reaction equations, which proceed in parallel:

$$2Au + 4CN^- + O_2 + 2H_2O \rightleftharpoons 2Au(CN)_2^- + H_2O_2 + 2OH^- \quad \text{(EQ 6.15)}$$

$$2Au + 4CN^- + H_2O_2 \rightleftharpoons 2Au(CN)_2^- + 2OH^- \quad \text{(EQ 6.16)}$$

The major reactions are illustrated schematically in Figure 6.5. The equation proposed by Elsner (Chapter 1):

$$4Au + 8CN^- + O_2 + 2H_2O \rightleftharpoons 4Au(CN)_2^- + 4OH^- \quad \text{(EQ 6.17)}$$

is stoichiometrically correct but does not completely describe the cathodic reactions associated with the dissolution.

6.1.3 Reaction Kinetics

The major factors affecting the dissolution rate of gold, namely cyanide and oxygen concentrations, temperature, pH, surface area of gold exposed, degree of agitation and mass transport, gold purity and the presence of other ions in solution, are discussed in detail in the following sections.

6.1.3.1 Cyanide and Dissolved Oxygen Concentration

Considering the general gold dissolution reaction in Equation (6.15), it is apparent that one mole of gold requires half a mole of oxygen and two moles of cyanide for dissolution,

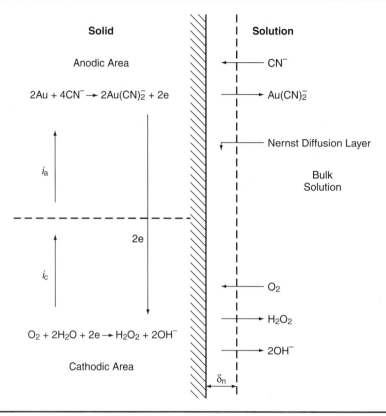

FIGURE 6.5 Schematic representation of the local corrosion cell at a gold surface in contact with an oxygen-containing cyanide solution: i_a = the anodic current; i_c = the cathodic current [6]

depending on the effectiveness of hydrogen peroxide reduction, and the major reaction is a two-electron process (Figure 6.5). The rate-limiting condition occurs when the diffusion rates of cyanide and oxygen are equal; that is, whichever species has the slower diffusion rate will provide the rate-limiting factor. At the mixed potential (E_m), the current arising from the cathodic reactions (i_c) is equal and opposite to the anodic current (i_a), as shown in Figure 6.6.

If Equation (4.41) is applied to this system at the rate-limiting condition, where $j(CN^-)$ is equal to $j(O_2)$, this yields the following:

$$(0.5 D_{CN^-} \cdot [CN^-])/\delta = (2 D_{O_2} \cdot [O_2])/\delta \qquad (EQ\ 6.18)$$

which simplifies to:

$$D_{CN^-} \cdot [CN^-] = 4 D_{O_2} \cdot [O_2] \qquad (EQ\ 6.19)$$

Values of the diffusion coefficients, D_{CN^-} and D_{O_2}, have been estimated at 1.83×10^{-9} and 2.76×10^{-9}, respectively [20]. This gives the ratio:

$$[CN^-]/[O_2] = 6 \qquad (EQ\ 6.20)$$

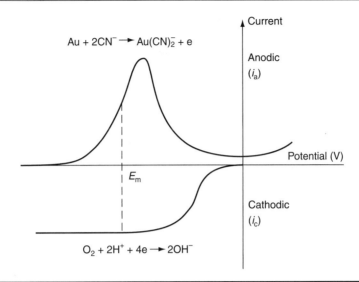

FIGURE 6.6 Simplified schematic diagram of the mixed-potential model for the dissolution of gold in cyanide solutions [17]

which has been found to agree closely with observed experimental and practical values. Investigations using rotating disc gold and gold–silver electrodes have indicated a range of observed optimal molar CN^-–O_2 ratios from about 4:1 to over 7:1 [20, 26, 27]. In practice, CN^-–O_2 ratios >6:1 are typically employed to ensure that cyanide concentration is not the rate-limiting factor (see Section 6.1.5).

Figure 6.7 shows anodic and cathodic current–potential curves, originally given for the simple case in Figure 6.6, superimposed on a common current scale (i.e., by using absolute values for the cathodic and anodic currents), for gold and silver at various cyanide concentrations. This shows the conditions under which the dissolution rate is limited by mass transfer of cyanide, oxygen, and a combination of the two. In aerated, alkaline cyanide solution, the dissolution rate is normally mass transport controlled, with activation energy values of 8 to 20 kJ/mol recorded [16, 20], and depends on the diffusion rate of cyanide, oxygen, or both, to the gold surface [28].

Cyanide concentration is relatively easy to control by the addition of concentrated cyanide solution or a solid cyanide compound. Control of the oxidant concentration (i.e., dissolved oxygen) is not as easy because of the low solubility of oxygen in water under atmospheric conditions. Consequently, the maximum rate of gold dissolution for processes that use air to provide oxygen in solution is determined by the conditions of temperature and pressure that the process operates under. At sea level and at 25°C, the saturated concentration of dissolved oxygen in solution is 8.2 mg/L. This value decreases with increasing altitude and increasing temperature, as shown in Table 6.4 (see also Figure 4.13). The corresponding cyanide concentration that gives maximum dissolution rate of gold at this oxygen concentration is approximately 0.005%, or 0.002 M CN^-, equivalent to 0.01% or 0.05 g/L NaCN. This is supported by practical observations, as illustrated in Figure 6.8, where close to maximum gold dissolution rate (i.e., 3 mg/in.^{2}/hr) is achieved at 0.02% or 0.10 g/L NaCN [30].

In mineral leaching systems, higher cyanide levels may be required because of the competition of other species for cyanide. An example of this for the leaching of calcine is given in Figure 6.9, which shows that increased dissolution rates are achieved at 0.25%

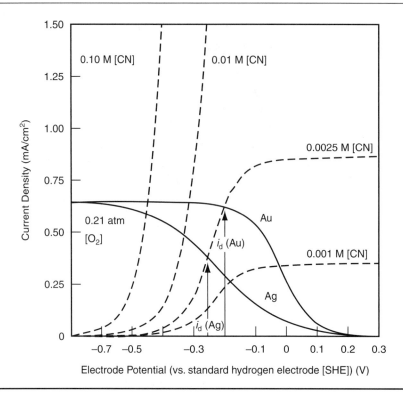

FIGURE 6.7 Electrochemical data for the anodic oxidation of silver and gold and the cathodic reduction of O_2 on silver and gold, calculated according to mixed kinetic models [29]

TABLE 6.4 Equilibrium concentration of dissolved oxygen (from air containing 21% O_2) in water at various temperatures and altitudes (values in mg/L)

Temperature (°C)	Sea Level (760 mm Hg)	914 m Altitude (680 mm Hg)	1,828 m Altitude (610 mm Hg)
0	14.6	13.1	11.7
5	12.8	11.4	10.3
10	11.3	10.1	9.1
15	10.1	9.0	8.1
20	9.1	8.2	7.3
25	8.3	7.3	6.6
30	7.5	6.7	6.1
35	7.0	6.2	5.6
40	6.5	5.7	5.2
45	6.0	5.3	4.8
50	5.6	4.9	4.5
60	4.8	4.2	3.8
70	3.9	3.4	3.0
80	2.9	2.4	2.0
90	1.7	1.1	0.7
100	0	0	0

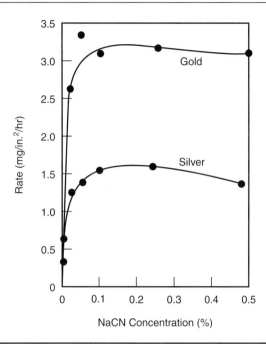

FIGURE 6.8 Effect of cyanide concentration on dissolution rate of gold and silver [30]

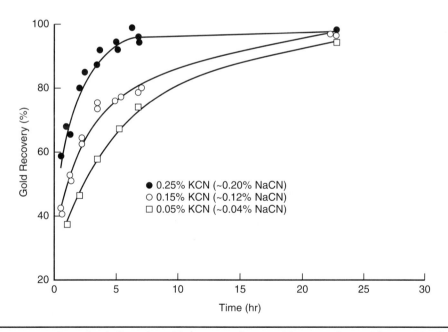

FIGURE 6.9 Example of effect of cyanide concentration on gold recovery from calcine material [31]

246 | THE CHEMISTRY OF GOLD EXTRACTION

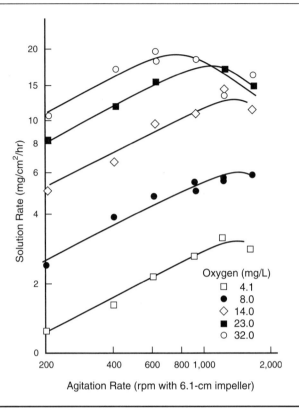

FIGURE 6.10 Dissolution rate of gold disc—effect of oxygen concentration and agitation [32]

potassium cyanide (KCN) (0.10% or 0.04 M CN⁻) compared with lower concentrations. In practice, cyanide concentrations are usually maintained above those at which a decrease in extraction is observed, although environmental concerns often dictate that cyanide concentrations be kept as low as practically possible.

The effect of dissolved oxygen concentration on the dissolution rate of gold discs is illustrated in Figure 6.10. In this case, increasing dissolved oxygen concentration increased the rate of dissolution, up to the maximum of 32 mg/L applied in the tests, in the presence of excess free cyanide. This effect is further illustrated in Figure 6.11 for leaching of a gold-bearing calcine. Although this shows the beneficial effect of dissolved oxygen concentration on dissolution rate, it also indicates that similar final gold extractions are achieved if leaching times are extended at the lower oxygen concentrations. This is an important practical consideration, and the cost of increasing dissolved oxygen concentration must always be weighed against the cost of providing extra leaching time [31, 32].

The dissolved oxygen concentration depends principally on the oxygen content of the gas phase in contact with the leach slurry or solution, temperature, and altitude. Table 6.4 provides the equilibrium concentrations of dissolved oxygen in water at various temperatures and at three different altitudes (sea level, 914 m, and 1,828 m). This shows that oxygen concentrations achievable in most gold leaching operations range from about 5 mg/L at high elevation (1,828 m) in a hot climate to more than 10 mg/L at sea level in cool conditions (i.e., 15°C).

In practice there are two methods for increasing the dissolved oxygen concentration above the equilibrium saturated condition, as follows:

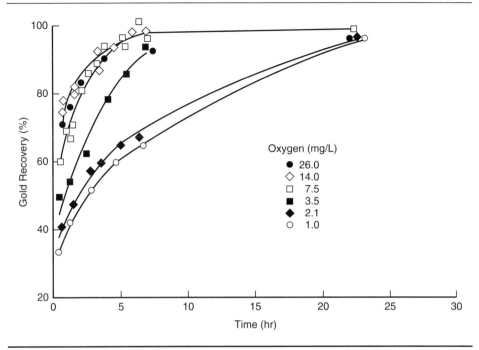

FIGURE 6.11 Example of cyanide extraction of gold from calcine showing effect of oxygen concentration: 0.25% KCN [31]

- Operation under pressure, using air as the oxidant
- Use of pure oxygen, oxygen-enriched air, hydrogen peroxide, or another oxygen source to supplement or replace air in the gas phase

Both methods are expensive and can usually only be justified for treatment of ores containing significant amounts of oxygen-consuming species.

It has been suggested that high dissolved oxygen concentrations, for example >20 mg/L, may cause passivation of the gold surface due to oxide layer formation [32]. In certain conditions, it has been shown that passivation can occur at dissolved oxygen concentrations as low as 7 mg/L in poorly agitated systems [33]. However, this is considered to be highly unlikely in practice because of the high solution potential that would be required, and there is little evidence of this from the majority of research and practical experience in this area.

A number of alternative oxidants have been proposed to increase gold dissolution kinetics in alkaline cyanide solution, including solid oxidants such as peroxides of barium, sodium, potassium, calcium, and manganese (each has different solubility and oxygen content), potassium chlorate, potassium permanganate, potassium bichromate, and potassium ferricyanide. In general, the high cost of these reagents prohibits their use, and none of these have been applied commercially to any significant extent [23, 34].

6.1.3.2 Temperature

As a result of increased activities and diffusion rates of reacting species, the gold dissolution rate increases with temperature, up to a maximum at approximately 85°C (Figure 6.12). Above this temperature the decrease in oxygen solubility outweighs the benefits of increased ionic activity and diffusion rates. Considering the example shown in Figure 6.12, it can

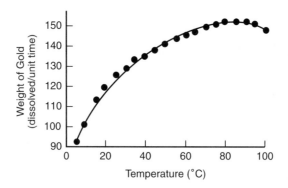

FIGURE 6.12 Effect of temperature on gold dissolution rate in aerated 0.25% KCN solution [35]

be seen that only a 20% to 25% increase in dissolution rate is achieved by elevating the temperature from 25 °C to 85 °C. This is supported by Figure 6.13, which indicates gold extraction versus time at temperatures between 21 °C and 45 °C. The high cost associated with such a temperature increase can rarely be justified for the treatment of low-grade materials, and ambient temperatures are usually applied. However, elevated temperatures have been applied to leaching of high-grade materials (e.g., gravity concentrates). This practice requires that additional free cyanide and oxygen be supplied, usually in pressurized systems. This forms the basis for an intensive cyanidation process, discussed in Section 6.1.5.3.

6.1.3.3 pH

The E_h–pH diagram for the gold–cyanide system (Figure 6.3) indicates that the electrochemical driving force for dissolution, that is, the potential difference between the lines representing gold oxidation and oxygen reduction reactions, is maximized at pH values between approximately 9.0 and 9.5. Typically, cyanide leaching is performed at pH values >9.4 to prevent excessive loss of cyanide by hydrolysis, as discussed in Section 6.1.1. Low pH cyanide leaching has been investigated as a means to reduce lime (or alkali) consumption and reduce scaling. However, as the pH decreases, the proportion of cyanide present in solution as hydrogen cyanide increases (Figure 6.1), and a closed leaching system must be used to prevent excessive loss of cyanide by volatilization of HCN. Consideration of the thermodynamics indicates that HCN should be capable of leaching gold, but investigations have shown that HCN does not leach gold at a sufficiently fast rate to compete with the kinetics of leaching with CN^- [36]. In some cases, low pH may be used either to reduce cyanide concentrations in leach system effluents or to reduce the rate and extent of other undesirable side reactions, for example, dissolution of antimony and arsenic minerals. In these cases, optimum conditions must be determined by test work on the material of interest.

The effect of pH on gold dissolution rate above pH 9.5 is small and depends on the presence of other solution species and ore constituents, as well as the type of alkali used for pH modification. In some cases, the rate may decrease markedly with increasing pH, due to an increase in the rate of interfering reactions, such as the dissolution of sulfides and other reactive species (see Section 6.1.4). These effects are generally more severe with calcium hydroxide than with either sodium or potassium hydroxide because of the lower solubility of many of the salts formed.

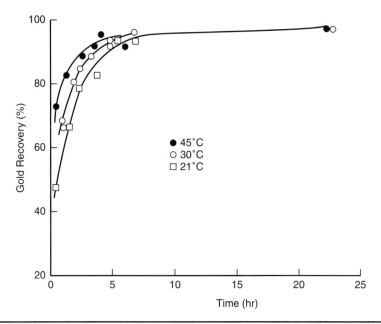

FIGURE 6.13 Effect of temperature on gold recovery from calcine material [31]

In practice, other process factors usually dictate the actual pH conditions applied, for example:

- Dissolution rate of other ore constituents, for example, copper, iron, tellurium, antimony, and arsenic minerals, which can adversely affect gold leaching
- Settling properties of the slurry
- Slurry viscosity
- Cost of pH modification
- Precipitation of solution species, for example, calcium and iron

Consequently, the optimum pH for leaching depends on a number of related factors and must be derived independently for each ore type and leaching system. Methods of pH modification for leaching are considered further in Section 6.1.5.

6.1.3.4 Surface Area

The dissolution rate is directly proportional to the exposed surface area of gold and other particulate factors (Section 4.3.5). The exposed surface area is related to the particle size distribution and liberation characteristics of the feed material, and is affected by the efficiency of the comminution processes preceding leaching. The rate generally increases with decreasing particle size, due to an increase in gold liberation and/or surface area of gold particles (due to flattening or physical breakage during grinding). However, this is not always the case and the rate of dissolution from ores containing cyanicides may decrease with decreasing particle size, due to the increased rate of competing, reagent-consuming side reactions. In such cases, the optimum particle size is a compromise between gold extraction and cyanide consumption. Alternatively, oxidative pretreatment may present the most attractive processing route for these materials. Particle sizes typically employed in leaching systems are discussed in Section 6.1.5.

6.1.3.5 Agitation

Gold dissolution is usually mass transport controlled under conditions normally applied for cyanide leaching, and therefore the rate depends on the diffusion layer thickness and mixing characteristics of the bulk solution. Increasing agitation increases the dissolution rate up to a maximum, above which agitation has little or no further benefit. The diffusion layer thickness is minimized by maximizing solution flow rates past solid particles. In slurry leaching systems this is achieved by mixing solids and solution with air or by mechanical agitation. Increasing the degree of agitation in poorly mixed systems may significantly enhance the gold dissolution rate, as a result of reduced diffusion layer thickness and improved homogeneity in the bulk solution. In well-mixed systems the effect is less significant because the bulk solution or slurry is more homogeneous; it becomes increasingly difficult to reduce the diffusion layer thickness by agitation alone, and increased agitation is often unjustifiable. This is certainly true of most systems that use modern mechanical mixing technology and probably also applies to most air-agitated systems.

In heap, dump, vat, or in situ systems, the diffusion layer thickness around stationary solid particles is determined by solution flow rates. On the other hand, where coarse particles are involved, pore diffusion plays a key (and sometimes dominant) role in getting reactants and products to and from the mineral (e.g., gold) surfaces. Although increasing the solution flow rate may have a similar effect to that produced by increasing the rate of agitation in slurry systems, that is, by increasing the mass transport rates of reacting species, the gold concentration of the leach solution decreases, which may reduce the efficiency of downstream processes. Consequently, an economic optimum must be established, discussed further in Section 6.1.5. In addition, sufficient time must be allowed for reactants and products to diffuse to and from mineral surfaces.

6.1.3.6 Effect of Lead and Other Metal Ions

It is well known that pure gold dissolves much more slowly than gold alloyed with silver, or gold containing minor quantities of other metals. Certain divalent cations can have a significant beneficial effect on the gold dissolution rate. Trace amounts of lead, mercury, thallium, and bismuth are known to depolarize gold surfaces and prevent or reduce the passivation effect that is observed at -0.4 V on the gold polarization curve, thereby accelerating gold leaching rates. This effect, shown in Figure 6.14 for mercury and lead ions, is thought to be due to the deposition of small amounts of metallic mercury or lead onto a portion of the gold surface, as follows:

$$Pb^{2+} + 2Au + 4CN^- \rightleftharpoons Pb + 2Au(CN)_2^- \qquad \text{(EQ 6.21)}$$

Many ores and concentrates contain minor amounts of soluble mercury and lead, and less commonly bismuth and thallium, which may naturally assist in the leaching process. When this is not the case, soluble lead (or mercury, thallium, bismuth) salts may be added to leaching systems to enhance gold dissolution. The potential disadvantage of such practice is that any metal species added may report to the final gold product or the tailings. Although the addition of lead ions to achieve concentrations of 1 to 10 mg/L have been found to be beneficial, lead concentrations in excess of 20 mg/L have been found to retard gold dissolution. Additional reading on the effect of lead on gold dissolution is available in the literature [37, 38, 39, 40]. The addition of as little as 20 mg/L thallium has been shown to have a significant beneficial effect on gold leaching, which is most pronounced under conditions where passivation may be occurring [33].

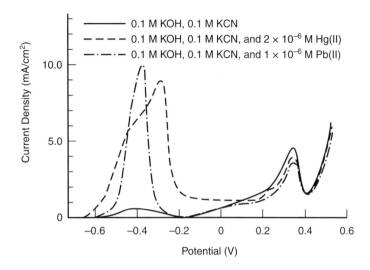

FIGURE 6.14 Effect of the addition of lead and mercury ions on the anodic behavior of gold in an alkaline cyanide solution. Before each anodic sweep, the electrode was left under open-circuit conditions for 5 min [16].

Although the presence of small amounts (5 to 10 mg/L) of sulfide ions has been demonstrated to have a significant retarding effect on gold dissolution rates, this can be alleviated by the addition of lead, as described in Section 6.1.4.6 [40, 41].

6.1.3.7 Galvanic Interactions with Sulfide Minerals

Many sulfide minerals have sufficient conductivity to allow electron transfer reactions at their surface. When such sulfide minerals are in direct electrical contact with gold or gold alloys, galvanic interaction may occur during leaching, which can affect the gold leaching rate. One investigation of this effect has indicated that the effect of such galvanic interaction increases the gold leaching rate when gold is in electrical contact with the following minerals (in order of the magnitude of the beneficial effect):

Pyrrhotite = galena > pyrite >> gold (not in contact with sulfide mineral)

On the other hand, when gold is in electrical contact with chalcopyrite, the galvanic interaction decreases the dissolution rate significantly, while chalcocite acts as the anode and stops dissolution entirely [42].

6.1.4 Behavior of Other Minerals in Alkaline Cyanide Solutions

Besides gold, many other metals and minerals also dissolve in dilute alkaline cyanide solutions. These reactions may consume cyanide and oxygen, as well as produce a variety of solution species which can reduce the efficiency of gold leaching and subsequent recovery processes.

Most metal sulfides decompose quite readily in aerated, alkaline cyanide solution to form metal ions, metal oxides, or metal cyanide complexes and various sulfur-containing species, including thiocyanate, sulfide, and thiosulfate ions. The general reaction for a sulfide containing a divalent metal cation is given as follows:

$$2MS + 2(x + 1)CN^- + O_2 + 2H_2O \rightleftharpoons 2M(CN)_x^{(x-2)-} + 2SCN^- + 4OH^- \qquad (EQ\ 6.22)$$

An example of this behavior is the dissolution of pyrite: Using the stoichiometry given in Equation (6.22), it can be shown that an ore containing 1% pyrite would consume 32.6 kg/t NaCN and 2.7 kg/t oxygen, if the pyrite dissolved completely, and if none of the iron nor sulfur were consumed by species other than cyanide. This cyanide consumption is a direct reagent cost and may deplete the leach solution of cyanide necessary for gold dissolution. The oxygen consumption given previously is equivalent to an oxygen mass transfer requirement of 1 g O_2/min/t solution in a leaching plant with a 24-hour retention time and a 2:1 liquid–solid ratio.

When sulfide mineral dissolution is significant, several pretreatment methods are available to improve the response to cyanidation. Preaeration may be considered for ores containing the more reactive sulfides, such as pyrrhotite and marcasite, and is considered further in Section 5.2. Ores containing significant amounts of sulfides that cannot be passivated adequately by preaeration, and which result in unacceptable cyanide and/or oxygen consumption, must be pretreated by alternative processes, for example, pressure oxidation, roasting, or biological oxidation (Sections 5.3 to 5.8).

Many metal oxides, carbonates, sulfates, and other compounds are soluble in alkaline cyanide solutions, with the solubility dependent on specific solution conditions. These minerals generally consume smaller quantities of cyanide and oxygen than do sulfides, because the anions produced in solution do not react with cyanide to any appreciable degree, unlike the sulfide species. Consequently, decomposition reactions of these compounds rarely have a large impact on gold dissolution, although the species formed may affect precipitation reactions and overall process efficiency.

6.1.4.1 Silver Minerals

Silver frequently occurs with gold in economically significant quantities, and therefore its behavior in cyanide solutions is most important. Metallic silver behaves similarly to gold in aqueous cyanide solution and anodically dissolves as follows:

$$Ag(CN)_2^- + e \rightleftharpoons Ag + 2CN^- \tag{EQ 6.23}$$

where
$$E^0 = -0.31 \text{ (V)}$$

The E_h–pH diagram for the Ag–CN–H_2O system (Figure 6.15) indicates the region of predominance of the Ag(I) cyanide complex and shows that insoluble silver cyanide, AgCN, is formed at low pH (<3.5). The area of predominance of AgCN(s) increases significantly as CN^- is reduced, for example, from 10^{-3} to 10^{-4} M, with important consequences for leaching systems. However, optimum cyanide concentrations for gold extraction are well in excess of those where such an insoluble species could form [35].

At very high cyanide concentrations, higher-order complexes, such as $Ag(CN)_3^{2-}$ and $Ag(CN)_4^{3-}$, may be formed, but these are of little practical importance.

The dissolution of silver has been shown to proceed via a four-electron mechanism, with direct reduction of oxygen to hydroxyl ions (Equation [6.14]), compared to the two-electron path for gold [29]. Oxygen reduction at the silver surface is thought to involve a mixed diffusion plus charge transfer process, incorporating the adsorption of oxygen onto the surface. The transfer coefficient for oxygen reduction on silver is 0.25, compared with 0.5 on gold. Under conditions typically applied for optimal gold dissolution, the dissolution rate of silver is significantly slower than gold, as indicated by the current–potential curves in Figure 6.7. For example, at a cyanide concentration of 0.0025 M NaCN and using air to provide oxygen, the dissolution current density for silver is approximately half that for gold. However, the difference in dissolution rates are reduced

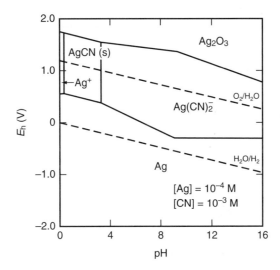

FIGURE 6.15 E_h–pH diagram for the Ag–CN–H_2O system at 25°C [43]

as the cyanide concentration is increased, and Figure 6.7 shows that the currents for gold and silver approach each other closely at concentrations in excess of 0.1 M NaCN (i.e., 5 g/L).

Consequently, operations that treat ores containing a significant amount of silver may employ elevated cyanide concentrations to improve silver extractions and to overcome any retarding effect on gold dissolution caused by competition for cyanide. An example of this may be found at Coeur–Rochester (Nevada, United States), where 0.06% to 0.08% NaCN (i.e., 6 to 8 g/L) has been used to treat an ore containing 0.3 g/t gold and 45 g/t silver. The recoveries of the two metals are 80% and 50%, respectively [44]. A further example of the treatment of high-silver ores may be found in Section 12.2.3.1.

6.1.4.2 Copper Minerals

Copper minerals dissolve to varying degrees in alkaline cyanide solutions, as summarized for some of the more important mineral species in Table 6.5. This indicates the solubility of various minerals ground to 100% <150 µm and leached in 0.1% sodium cyanide solution, expressed as the percentage of the total weight of mineral dissolved in solution. Copper dissolution is generally undesirable during leaching because it can consume cyanide and dissolved oxygen, retard gold dissolution rates, interfere with subsequent recovery processes, and contaminate the final product. In addition, some copper minerals (e.g., chalcopyrite) are capable of removing gold from solution by reduction at the mineral surface, exhibiting reversible preg-borrowing characteristics in cyanide-deficient solutions.

Chalcopyrite is the least soluble of the listed minerals and the least soluble sulfide mineral that is commonly encountered in gold extraction systems. Chalcocite, bornite, enargite, covellite (although not listed in Table 6.5), and copper oxides and carbonates are highly soluble, and dissolution of these minerals can severely affect leaching. Although native copper apparently dissolves quite readily in cyanide solutions, the rate of dissolution is much slower than that of both gold and silver.

The copper minerals dissolve to form a variety of Cu(I) cyanide complexes, $Cu(CN)_2^-$, $Cu(CN)_3^{2-}$, $Cu(CN)_4^{3-}$, as illustrated by the E_h–pH diagrams for the Cu–CN–H_2O system (Figure 6.16), and for the Cu–S–CN–H_2O system (Figure 6.17). These show the large

TABLE 6.5 Solubility of copper minerals in ~0.1% NaCN solutions [45]

		Percent Total Copper Dissolved	
Mineral		23°C	45°C
Azurite	$2CuCO_3 \cdot Cu(OH)_2$	94.5	100.0
Malachite	$CuCO_3 \cdot Cu(OH)_2$	90.2	100.0
Chalcocite	Cu_2S	90.2	100.0
Copper metal	Cu	90.0	100.0
Cuprite	Cu_2O	85.5	100.0
Bornite	Cu_5FeS_4	70.0	100.0
Enargite	Cu_3AsS_4	65.8	75.1
Tetrahedrite	$4Cu_2S \cdot Sb_2S_3$	21.9	43.7
Chrysocolla	$CuSiO_3$	11.8	15.7
Chalcopyrite	$CuFeS_2$	5.6	8.2

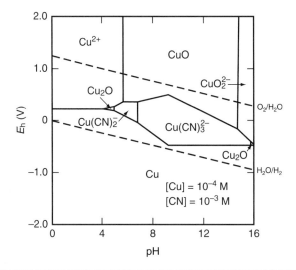

FIGURE 6.16 E_h–pH diagram for the Cu–CN–H_2O system at 25°C [43]

area of predominance of $Cu(CN)_3^{2-}$ under gold dissolution conditions. However, because the stability constants of the three complexes are quite close in value, all are present to some extent. The proportions of the different species present in a solution of known pH, temperature, copper, and cyanide concentrations can be calculated using stability constants for the various species formed. Figure 6.18 indicates the speciation of copper cyanide complexes as a function of pH. The diagram shows only the predominant species present and does not indicate proportions of minor species. Alternatively, an indication of the relative proportions of the different species present can be obtained by measurement of the molecular ratio of sodium cyanide to copper in solution, taking into account free cyanide and cyanide tied up with other species. Typically, this ratio varies between 2.5:1 and 3.5:1, which reflects the higher stability of the $Cu(CN)_3^{2-}$ complex under typical cyanide leaching conditions [45].

The formation of $Cu(CN)_2^-$ is favored at low pH (<6), under the conditions specified in Figure 6.18 and at very low cyanide concentrations, whereas $Cu(CN)_4^{3-}$ is the preferred species at high pH and at high cyanide concentrations. This has important practical

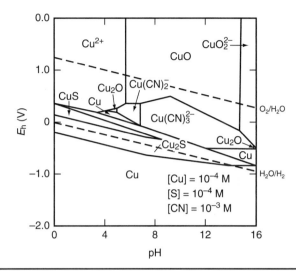

FIGURE 6.17 E_h–pH diagram for the Cu–S–CN–H$_2$O system at 25°C [43]

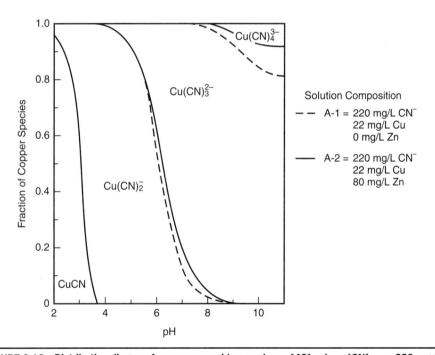

FIGURE 6.18 Distribution diagram for copper cyanide complexes [46], where [CN]$_{total}$ = 220 mg/L; [Cu]$_{total}$ = 22 mg/L

significance since the Cu(CN)$_4^{3-}$ ion is less readily adsorbed onto activated carbon. Consequently, high cyanide concentrations can be used to increase the proportion of this complex and improve the selectivity of gold and silver recovery from solutions.

The detrimental effect of copper ions on gold dissolution can, in some industrial systems, be attributed to the complications in analyzing solutions containing copper for free cyanide content. During titration with silver nitrate (the standard method of free cyanide analysis in most plants), a portion of the cyanide complexed with copper is released and complexes with silver to establish a new equilibrium. The extent of dissociation of the copper complexes depends on the specific solution conditions and properties. Other interactions with chemical indicators used in the analysis may further complicate the results obtained. This obviously gives a false indication of free cyanide concentration available for gold dissolution, and copper concentration must be measured and allowed for in free cyanide determinations [47].

Copper cyanide complexes have highly variable ability to dissolve gold. Solutions containing copper with little or no actual free cyanide will favor the formation of the $Cu(CN)_2^-$ complex over the higher-order complexes. Such solutions have very little ability to dissolve gold because of the low availability of free cyanide, and possibly because of the speculated formation of a passivating layer $AuCN \cdot CuCN_{ads}$ on the gold surface [48]. However, the $Cu(CN)_3^{2-}$ and $Cu(CN)_4^{3-}$ complexes do have the ability to dissolve gold, although at slower rates than free cyanide. Therefore, the detrimental effect of copper on gold (and silver) dissolution can be avoided by providing adequate free cyanide in solution to ensure that the gold dissolution rate is maximized. A molecular ratio of sodium cyanide–copper greater than 4.5:1 must be maintained for this purpose, which is equivalent to a mass ratio >3:1 [45, 47].

The use of ammonia–cyanide mixtures for leaching of copper–gold bearing materials is discussed in Section 6.6.

6.1.4.3 Iron Minerals

Hematite (Fe_2O_3), magnetite (Fe_3O_4), goethite (FeOOH), siderite ($FeCO_3$), and iron silicates are virtually insoluble in alkaline cyanide solutions. Similarly, metallic iron, which may be introduced to the process as grinding media or used in process equipment construction, corrodes very slowly, and the reaction accounts for insignificant cyanide and steel consumption in most leaching systems. Some iron carbonates, and other complex carbonate minerals, decompose in low-alkalinity solutions (<pH 10) to some extent, but are unreactive at the higher pH values usually applied for leaching [45].

Oxide minerals that do dissolve produce the Fe(II) cyanide complex, $Fe(CN)_6^{4-}$, as shown in Figure 6.19. This may be further oxidized to Fe(III) cyanide, $Fe(CN)_6^{3-}$, depending on solution conditions, but the rate of oxidation with dissolved oxygen is slow, and stronger oxidants, such as ozone or hydrogen peroxide, are required for this to proceed rapidly. The preferred complex coordination number is 6 and both the Fe(II) and Fe(III) complexes are very stable, as indicated in Table 6.2. The region of predominance of the Fe(III) complex is strongly dependent on free cyanide concentration and pH, with goethite formed outside this region.

Iron sulfides are much more reactive than the oxides and silicates, and most decompose in alkaline cyanide solutions to form iron cyanide complexes and various sulfur species. The E_h–pH diagram for the Fe–S–CN–H_2O system at 25°C is given in Figure 6.20 and for the Fe–As–S–CN–H_2O system in Figure 6.21. The thermodynamic prediction that pyrrhotite is the most reactive iron sulfide in alkaline cyanide solution is confirmed in practice. Pyrrhotite readily gives up one sulfur atom to equalize the stoichiometry and reacts with cyanide as follows:

$$Fe_7S_8 + CN^- \rightleftharpoons 7FeS + CNS^- \quad \text{(EQ 6.24)}$$

Further reaction produces Fe(II) cyanide and various aqueous sulfur species, considered in more detail in Section 6.1.4.6, but shown here for the case of sulfate:

$$2FeS + 12CN^- + 5O_2 + 2H_2O \rightleftharpoons 2Fe(CN)_6^{4-} + 2SO_4^{2-} + 4OH^- \quad \text{(EQ 6.25)}$$

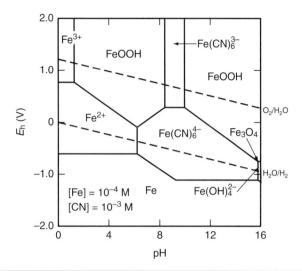

FIGURE 6.19 E_h–pH diagram for the Fe–CN–H_2O system at 25°C [43]

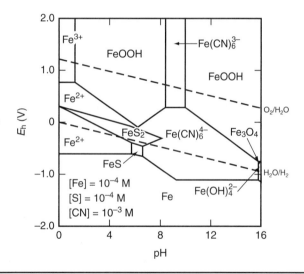

FIGURE 6.20 E_h–pH diagram for the Fe–S–CN–H_2O system at 25°C [43]

The other iron sulfides react similarly, albeit with different reaction stoichiometry. The order of decomposition rate of the most important sulfide minerals in cyanide solution is generally considered to be as follows [45]:

pyrrhotite >>> marcasite > arsenopyrite > pyrite

However, other factors can affect the decomposition rate significantly, including: the presence of lattice defects (i.e., impurity and foreign ion inclusions, lattice dislocations, etc.) in the mineral crystal structure, the association of the mineral with other reactive and/or conductive minerals that can result in galvanic interactions, and other factors. The decomposition of iron sulfide minerals can adversely affect the rate of gold dissolution and result in elevated consumptions of cyanide and lime. The adverse effect of iron

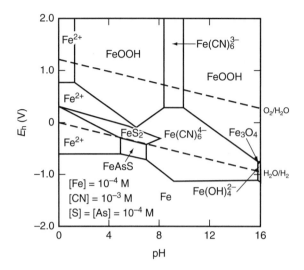

FIGURE 6.21 E_h–pH diagram for the Fe–S–As–CN–H_2O system at 25°C [43]

sulfide minerals on gold dissolution rates can be reduced by the addition of a suitable lead salt (e.g., 100 g/t lead nitrate $Pb(NO_3)_2$) [49].

Iron sulfides can be passivated to varying extents in alkaline cyanide solutions by the formation of a layer of Fe(III) hydroxide on the surface. This is achieved by conditioning in aerated, alkaline solution before cyanide leaching to partially dissolve the sulfide mineral and deposit an insoluble hydroxide layer on the surface (Section 5.2). However, the formation of insoluble iron oxides and/or hydroxides is not always desirable, either prior to or during cyanide leaching, because these products may coat gold particles and can reduce leaching efficiency, sometimes significantly.

Pyrite, when present in significant quantities (i.e., >20%), has been shown to exhibit preg-borrowing characteristics in cyanide-deficient solutions with gold reduction occurring on the mineral surface [50].

6.1.4.4 Arsenic and Antimony Minerals

Arsenic and antimony do not form stable complexes with cyanide, and consequently the presence of cyanide in solution does not appreciably affect the stability of the metal species formed. Under the conditions applied for gold leaching, arsenic and antimony sulfides decompose to arsenite (AsO_2^-) and arsenate (AsO_3^-), as shown in Figure 5.3, and stibnite (SbO_2^-), and stibnate (SbO_3^-), respectively, with the proportion of each depending on the solution composition and pH, shown for the lower oxidation states in the following equations:

Below approximately pH 11.5:

$$AsS + 3H_2O \rightleftharpoons H_2AsO_3^- + S + 4H^+ + 3e \quad \text{(EQ 6.26)}$$

and above this value:

$$AsS + 3OH^- \rightleftharpoons HAsO_3^{2-} + S + 2H^+ + 3e \quad \text{(EQ 6.27)}$$

The reaction for stibnite is similar, as follows:

$$SbS + 3OH^- \rightleftharpoons HSbO_3^{2-} + S + 2H^+ + 3e \quad \text{(EQ 6.28)}$$

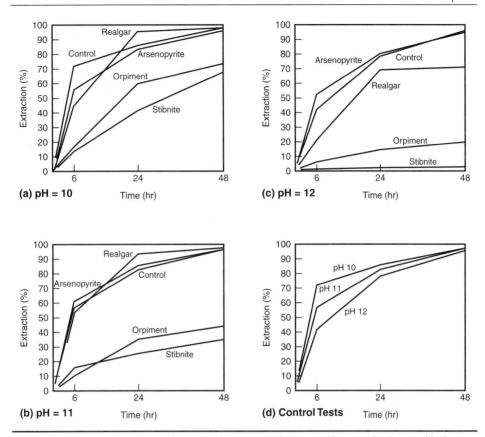

FIGURE 6.22 The effect of arsenic and antimony materials on gold extraction by cyanidation at (a) pH 10, (b) pH 11, (c) pH 12, and (d) control tests showing effects of lime alkalinity on gold extraction in absence of arsenic and antimony minerals [45]

Dissolution of these minerals has a detrimental effect on gold (and silver) extraction, and in some cases this effect is severe. This effect is thought to be due to the formation of a passivating layer of arsenic or antimony oxide layer on the gold surface. Decomposition of these minerals is strongly related to pH, with their solubility increasing with increasing pH. For example, at pH 12 orpiment, stibnite and realgar dissolve appreciably, while at pH 10 only orpiment dissolves to any significant degree [45].

Consequently, the detrimental effect of these minerals is strongly related to pH, as illustrated in Figure 6.22, and pH control is critical when leaching such ores. Under pH conditions normally applied for gold leaching, the detrimental effect of some of the more common arsenic and antimony minerals on gold extraction generally decreases in the order:

$$\text{stibnite} \ggg \text{orpiment} > \text{arsenopyrite} > \text{realgar}$$

The detrimental effect of stibnite (and probably also other antimony and arsenic sulfide minerals) can largely be alleviated by conducting leaching at low pH (i.e., 10) and by the addition of a suitable lead salt (e.g., lead nitrate) in sufficient quantity. In some cases, lead nitrate additions of 250 to 500 g/t may be required [49, 51].

A process for the dissolution and removal of antimony from refractory gold-bearing material prior to cyanide leaching has been proposed to reduce the adverse impact of stibnite on cyanidation. This process considered leaching with a solution containing sodium sulfide and sodium hydroxide, followed by direct electro-deposition of the antimony. The residue would be cyanide leached for gold extraction, taking into consideration the effects described [52]. This approach is only likely to be viable for materials containing very high antimony concentrations.

6.1.4.5 Zinc Minerals

Because zinc minerals occur infrequently and usually in small quantities in gold ores, their solubility in cyanide leaching systems is generally of limited importance. The solubility of various zinc minerals in cyanide solutions is available in the literature [45] and is not considered further here.

Metallic zinc is used for the recovery of gold from cyanide leach solutions (Chapter 8) and the behavior of zinc and its solution species are of considerable interest. Zinc metal dissolves readily in aerated alkaline cyanide solution to form the Zn(II) cyanide complex or zinc hydroxide, depending on the solution conditions. The reaction chemistry of the Zn–CN–H_2O system is considered in more detail in Chapter 8.

The zinc cyanide complex also interferes with the silver nitrate titrimetric technique for free cyanide analysis in a similar manner to copper (see Section 6.1.4.2), but the effect is more severe for the zinc species. This is because the stability constants of the zinc and silver cyanide complexes are much closer together than those of the copper and silver complexes (Table 6.2). On the other hand, zinc cyanide more readily gives up its cyanide for complexation with gold, if inadequate free cyanide is available. Consequently, zinc species generally present far less of a problem in cyanidation than copper.

6.1.4.6 Elemental Sulfur and Other Sulfur Species

Elemental sulfur, which may be formed during oxidative pretreatment or during sulfide decomposition in dilute cyanide solution, reacts rapidly with cyanide to form thiocyanate, sulfate, and other aqueous sulfur species, including sulfide, sulfite, and polysulfide ions. Some of these reactions are listed as follows:

$$S^0 + CN^- \rightleftharpoons SCN^- \quad \text{(EQ 6.29)}$$

$$xS^{2-} + CN^- \rightleftharpoons (x-1)S^{2-} + SCN^- + 2e \quad \text{(EQ 6.30)}$$

$$S_2O_3^{2-} + CN^- \rightleftharpoons SO_3^{2-} + SCN^- \quad \text{(EQ 6.31)}$$

The dissolution of elemental sulfur with cyanide may be beneficial for the removal of sulfur coatings on the surface of gold particles, but the reaction consumes cyanide (1% sulfur consumes 15.3 kg/t NaCN), and the formation of elemental sulfur should be avoided, if possible. Unfortunately, it is a natural product of some oxidation reactions; however, the economic impact of elemental sulfur production can be quantified and planned for in the extraction process.

Sulfide ions are oxidized to thiosulfate and more slowly to sulfate, in alkaline cyanide solution. However, sulfide ions are strongly adsorbed onto gold surfaces and can significantly retard gold dissolution by the formation of an inhibiting surface layer [53]. Sulfide ion concentrations of 1 to 10 mg/L may adversely affect gold dissolution, and concentrations as low as 0.5 mg/L have been shown to halve gold leaching rates in one instance [54]. The adverse effect of sulfide ions can be countered by the addition of small amounts of a soluble lead salt, such as lead nitrate. Lead sulfide is precipitated and then may be oxidized to the sulfate, generating Pb^{2+} for further reaction with sulfide [40, 41]. The effect of lead on gold leaching is discussed further in Section 6.1.3.6.

Polythionate ($S_2O_6^{2-}$) and polysulfide (S_n^{y-}) ions have also been detected in cyanide solutions, although generally in lower concentrations, and their effect on gold leaching is poorly understood.

Thiosulfate species may react further in cyanide solution to form thiocyanate as follows:

$$2S_2O_3^{2-} + O_2 + 2CN^- \rightleftharpoons 2SCN^- + 2SO_4^{2-} \qquad (EQ\ 6.32)$$

Both thiosulfate and thiocyanate species are capable of dissolving gold, and these reagent schemes are discussed in more detail in Sections 6.3 and 6.5.

6.1.4.7 Tellurium Minerals

Gold–tellurium minerals (Au_xTe_y) dissolve slowly in alkaline cyanide solution, although the mechanism is poorly researched and is not well understood. Tellurium does not form stable complexes with cyanide, as indicated in the E_h–pH diagram for the Au–CN–Te–H_2O system in Figure 6.23. In sufficiently oxidizing alkaline cyanide solution, gold tellurides decompose to Au(I) cyanide and tellurate species (TeO_3^{2-}), for example [55]:

$$AuTe_2 + 2CN^- + 6H_2O \rightleftharpoons Au(CN)_2^- + 2TeO_3^{2-} + 12H^+ + 9e \qquad (EQ\ 6.33)$$

The dissolution rate of gold tellurides is significantly slower than that for native gold and gold–silver alloys in cyanide solutions (Figure 6.24), and leaching rates vary for the different minerals, with leaching time requirements in excess of 14 days in some cases (depending on mineral type and particle size) [55, 57]. For example, hessite and petzite are known to leach much more readily than calaverite in alkaline cyanide solutions.

Leaching rates can be increased, often substantially, by fine grinding and by modifying solution conditions to optimize gold-telluride dissolution, that is, high pH (>12), addition of lead nitrate, and high dissolved oxygen concentrations. For example, it has been shown that ultrafine grinding of one telluride flotation concentrate (i.e., to 90% <10 μm) increased gold extraction by cyanide leaching from 80%, with no grinding, to 94% [58]. In another case, leaching of a telluride concentrate (220 g/t Au, 0.04% Te) ground to 98% <75 μm in a solution containing 2 g/L CaO, 2 g/L NaCN, and 1.5 kg/t $Pb(NO_3)_2$ yielded 92% gold extraction in 144 hr. The dissolution rate was increased significantly by changing the leach solution several times during leaching and introducing fresh solution to the material, which increased gold extraction to 98% in 72 hr. The use of oxygen, rather than air, as the oxidant further enhanced the leaching rate [55]. In some cases, tellurium oxide or hydroxide may be formed at the mineral surface, but this does not appear to significantly affect leaching kinetics.

Unfortunately, many telluride ores, in particular those of the well-known Colorado (United States), Kalgoorlie (Australia), and Emperor (Fiji) deposits, do not respond as favorably to the treatment described above, and pretreatment of the ore or concentrate by chlorination or roasting is required (see Chapter 5) [57, 59, 117].

6.1.4.8 Carbonaceous Materials

A number of naturally occurring carbonaceous materials can reduce gold extraction during cyanide leaching by the following [60, 61]:

- Physical locking of gold within carbonaceous constituents
- Retaining contained gold values as a result of the adsorptive properties of the material
- Adsorbing dissolved gold from the leach solution (reversible and nonreversible effects)

262 | THE CHEMISTRY OF GOLD EXTRACTION

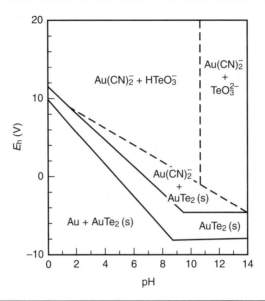

FIGURE 6.23 E_h–pH stability diagram for the Au–CN–Te–H$_2$O system: [Au] = [Te] = 1.0 × 10^{-3} M, and [CN] = 5.0 × 10^{-3} M [59]

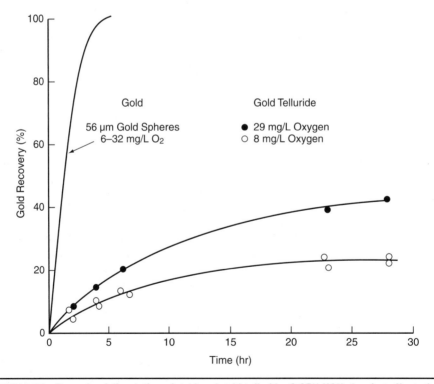

FIGURE 6.24 Example of dissolution of gold and gold telluride: 0.25% KCN showing effect of dissolved oxygen concentration [32]

Options for treating these ores are discussed in Chapter 3, and specific processes are considered in Sections 3.3.6.1 (passivation), 5.6 (chlorination), 5.8 (roasting), and 7.1.5.3 (carbon-in-leach).

6.1.5 Process Considerations

There are several methods available for cyanide leaching of gold-bearing materials, summarized as follows, in order of decreasing commercial importance:
- Agitation leaching
- Heap or dump leaching
- Intensive cyanidation leaching
- Vat leaching
- In situ leaching

The choice between these methods depends primarily on the relationship between particle size and recovery, capital and operating costs, and the dissolution rate in each case. An example of the relationship between particle size and gold recovery for two different ore types is given in Figure 6.25. Sometimes other factors, such as the recovery of other metals of value (i.e., silver, platinum group metals [PGMs]), environmental considerations, and availability of capital financing may play an important role in this selection.

6.1.5.1 Agitation Leaching

Commonly applied to a wide range of ore types, agitation leaching has been in use for well over 200 years. Leaching is typically performed in steel tanks, and the solids are kept in suspension by air or mechanical agitation. Air agitation in conical-bottomed leach tanks (Browns or Pachuca tanks) was widely practiced in the early years of cyanidation but has largely been superseded by more efficient mechanical agitation with reduced energy requirements and improved mixing efficiency. Well-designed systems can approach perfectly mixed flow conditions in a single reactor, which help to optimize reaction kinetics and make the most of available leaching equipment.

Particle size. The material to be leached is ground to a size that optimizes gold recovery and comminution costs, typically between 80% <150 μm and 80% <45 μm. In a few cases, whole ore is being ground to 80% <20 to 25 μm for optimal processing, either by oxidative pretreatment and/or leaching. Agitation leaching is rarely applied to material at sizes coarser than approximately 150 μm because it becomes increasingly difficult to keep coarse solids in suspension, and abrasion rates increase.

Increasingly, agitation leaching is being considered to treat very finely ground materials and, with the advances in ultrafine milling equipment (e.g., the Xstrata IsaMill and Metso SMD Detritor), concentrates have been ground to 80% <7 to 10 μm to liberate gold contained in refractory and nonrefractory sulfide mineral matrices prior to processing by agitation leaching and/or oxidative pretreatment.

Slurry density. Leaching is usually performed at slurry densities of between 35% and 50% solids, depending on the solids' specific gravity, particle size, and the presence of minerals that affect slurry viscosity (e.g., clays). Mass transport phenomena are maximized at low slurry densities; however, solids retention time in a fixed volume of leaching equipment increases as the density increases. In addition, reagent consumptions are minimized by maximizing slurry density, since optimal concentrations can be achieved at lower dosages, because of the smaller volume of solution per unit mass of material.

pH modification. Alkali, required for slurry pH modification and control, must always be added before cyanide addition to provide protective alkalinity, which prevents excessive loss of cyanide by hydrolysis (Section 6.1.1). Most leaching systems operate

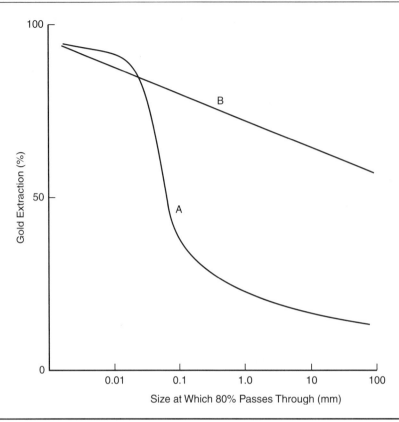

FIGURE 6.25 Schematic illustration of relationship between particle size (P_{80}) and gold extraction. A: Gold fine dispersed in a nonporous matrix—grinding required (Witwatersrand type). B: Gold located in cracks and fissures of rock structure—some degree of liberation achieved at coarse sizes (western United States oxidized ore type)

between pH 10 and 11. Staged addition of alkali may be required throughout the leaching circuit to maintain the desired operating pH, particularly when treating ores containing alkali-consuming materials. pH control is achieved by manual or automatic (on-line) measurement at various stages in the process. Calcium hydroxide (slaked lime, $Ca(OH)_2$), or sodium hydroxide can be used for pH modification. Calcium hydroxide (slaked lime) is the cheaper of the two but is less soluble and produces solutions that are much more susceptible to salt precipitation and scale formation. Unslaked lime (CaO) is sometimes used because it is less costly than slaked lime, but it is less effective for pH modification. For nonacidic- or nonalkali-consuming ores, calcium hydroxide concentrations of 0.15 to 0.25 g/L are typically required to achieve the desired pH range for leaching (i.e., pH 10 to 11). This represents typical lime consumptions of 0.15 to 0.5 kg/t for nonacidic ores. Sodium hydroxide is known to be more effective than calcium hydroxide at dissolving a variety of minerals, particularly at high alkalinities, and it is a highly effective dispersant. This may result in the dissolution of ore constituents, such as silicates, to produce various solution species, which can subsequently precipitate in a number of undesirable forms, potentially affecting downstream processes, including filtration, gold precipitation, or carbon adsorption. Consequently, calcium hydroxide is generally the preferred method of pH control in agitated leaching systems.

Cyanide. Cyanide may be added to agitated leaching systems either prior to the leaching circuit, that is, during grinding, or in the first stage of leaching. Subsequent reagent additions can be made into later stages of leaching to maintain or boost cyanide concentrations to maximize gold dissolution. In the absence of cyanide-consuming minerals in the ore or concentrate to be leached, cyanide concentrations used in practice range from 0.05 to 0.5 g/L NaCN, and typically between 0.15 to 0.30 g/L NaCN. Typical cyanide consumptions observed in agitated leaching systems for free-milling ores vary from about 0.25 to 0.75 kg/t. In cases where the feed material contains significant amounts of cyanide consumers and/or high silver content (i.e., >20 g/t), higher cyanide concentrations may be applied, that is, 2 to 10 g/L NaCN. In such cases, cyanide consumptions may vary from 1 to 2 kg/t, and in some cases much higher, depending on the nature and amount of cyanide-consuming minerals. Cyanide concentrations are usually monitored by manual titration techniques or less commonly by on-line cyanide analyzers, based on titrimetric, colorimetric, potentiometric, and ion-specific electrode techniques.

Oxygen. Oxygen is typically introduced into leaching systems as air, either sparged into tanks as the primary method of agitation, or supplied purely for aeration. In either case, crude sparging systems are usually sufficient to provide satisfactory bubble dispersion and to ensure that adequate dissolved oxygen concentrations are maintained. Typically, dissolved oxygen concentrations can be maintained at, or even slightly above, calculated saturation levels with air sparging (i.e., 8.2 mg/L O_2 at sea level at 25°C). Opinions vary on the best method of introducing air into leach tanks, which include sparging of air:

- Into the bottom of the tank (single or multiple addition points)
- Into the top of the tank using draught tube systems for dispersion
- Down the agitator shaft

The optimum sparging system depends on the geometry of the leach tanks. For example, conical-bottomed Pachuca tanks with single sparging points (common South African practice prior to about 1980) and flat-bottomed leach tanks with multiple sparging points, or simple down-the-agitator-shaft addition, have all been used.

In a few cases, particularly when treating ores that contain oxygen-consuming minerals, pure oxygen [62] or hydrogen peroxide [25] have been added to increase dissolved oxygen concentrations above those attainable with simple air sparging systems.

Residence time. Residence time requirements vary depending on the leaching characteristics of the material treated and must be determined by test work. Leaching times applied in practice vary from a few hours to several days. Leaching is usually performed in 4 to 10 stages, with the individual stage volume and number of stages dependent on the slurry flow rate, required residence time, and efficiency of mixing equipment used.

Counter-current leaching. Leaching efficiency can be enhanced by the application of Le Chatelier's principle. In summary, the lower the concentration of gold in solution, the greater the driving force for gold dissolution to occur, although in a mass transport controlled reaction it is debatable what role this plays in gold leaching. An alternative explanation for this phenomenon is the reversible adsorption of Au(I) cyanide onto ore constituents. The gold adsorption is reversed when the solution is exchanged for a lower-grade solution or when a material (such as activated carbon or suitable ion exchange resin) is introduced into the slurry, which actively competes for the Au(I) cyanide species. This effect can be exploited in practice by performing intermediate solid–liquid separation steps during leaching to remove high-grade gold solutions, and rediluting the solids in the remaining slurry with lower-grade leach solution and/or with freshwater plus reagents. Successful applications of this principle have been used at the Pinson and Chimney Creek, Nevada (United States), and East Driefontein (South Africa) plants, and at other operations [63, 64] (see Sections 12.2.2.1 and 12.2.2.5).

At many operating gold plants, an increase in gold dissolution is observed when a leach slurry is transferred from one type of process equipment to another (i.e., between leach tanks, thickeners, filters, pumps, and pipelines), which is illustrated in Figure 6.26. This is explained by the different mixing mechanisms in the different equipment, coupled with other factors, such as changes in slurry percent solids, changes in solution composition, and the effects of pumping transfer (i.e., plug flow mixing).

Likewise, the benefits of the carbon-in-leach (CIL) process compared with leaching and carbon-in-pulp (CIP) have been clearly demonstrated both experimentally and in practice, even without the presence of interfering preg-robbing constituents in the ore [65]. The CIL process results in improved conditions for gold dissolution as a result of the lower gold tenor, albeit at a cost of lower gold-on-carbon loading (see Section 7.1.5.3).

6.1.5.2 Heap Leaching

Heap leaching is a low-cost method that is most suitable for treatment of low-grade materials that do not justify the higher costs of grinding and agitation leaching. Ores can be treated either at a run-of-mine size or as crushed material, with the optimum size determined as a trade-off between gold recovery and crushing costs. Material handling requirements may also play a role in optimizing particle size for heap leaching because large particles may be difficult to transport by conveyor systems. This is an important factor in situations where conveying presents the most efficient method of material transport [66].

The process was first applied on a commercial scale in 1971 at the Carlin mine (Nevada) and was based on development work by the U.S. Bureau of Mines in Salt Lake City in the late 1960s and at the Carlin site from about 1968. At Carlin, crushed ore (nominally <18 mm) was placed on the heap at a rate of about 350 tpd during the six warmest months of the year. Gold recovery of approximately 65% was achieved with cyanide and lime consumptions of 0.05 and 0.50 kg/t, respectively [67]. Today, crushed ore heap and run-of-mine leaching operations around the world are processing large tonnages of ore (>200,000 tpd in some cases) on a year-round basis, at high altitudes and in severe climates (see Section 12.2.2.11 and [68]).

Heap leaching is usually performed on lined leach pads, which not only contain the gold-bearing leach solutions produced but also protect the ground and groundwater beneath and around the leach pad. In some rare cases, liners may not be required because of the particular topography and geology of the location, coupled with appropriate regulatory approval. A variety of effective leach pad liner systems have been developed, and much information is available in the literature [69, 70].

Materials handling. Heap leach feed material is placed on the lined leach pad by one of several possible methods, including conveyor stacking, truck dumping, loader-assisted stacking, or combinations of these methods. The method selected depends on the ore properties and specific project conditions, such as the processing rate, location, climate, and mining method. Material is typically stacked in lift heights ranging from 5 m to >15 m, depending on ore permeability, production schedule requirements, and liner area availability. Multiple lifts may be used to achieve the ultimate pad height (i.e., up to about 100 m). In some cases, ultimate heap heights have reached 200 m (e.g., Zortman, Montana, United States). This practice allows large amounts of ore to be leached in relatively small lined areas. The maximum pad height depends principally on the ore properties, such as permeability and the structural stability of the stacked material, but also on other local factors, for example, topography, climate, and regulatory requirements [68].

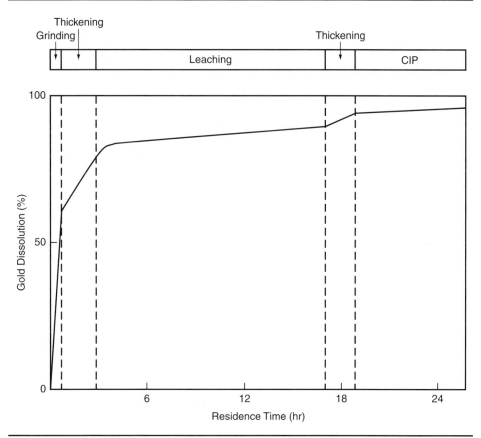

FIGURE 6.26 Schematic illustration of gold leaching profile through a grinding, partial countercurrent decantation (CCD)–CIP plant (Chimney Creek, Nevada, 1988)

Ores with high proportions of fines, particularly clays and fine silts, may require special stacking methods, such as loader-assisted stacking, to ensure that any detrimental segregation or compaction of fine particles is avoided. This could otherwise cause blinding of the heap, preventing solution percolation through the heap and reducing gold recovery. During leaching, fine particles (especially clay minerals) tend to migrate toward the bottom of the heap, which may also reduce pad permeability. In cases where these effects are particularly bad, the ore may need to be agglomerated prior to stacking on the leach pad [66, 71, 72]. Agglomeration is achieved by the addition of water, lime and/or cement to the ore followed by physical agglomeration, using either a series of short conveyor belts and drop points or a drum agglomerator. During this procedure, the fine particles adhere and bind to larger particles as agglomerates, which reduces the potential for migration in the heap. Agglomeration is particularly important for ores containing significant quantities of clay minerals, such as pyrophyllite, talc, kaolinite, montmorillonite, and other clay-forming minerals (i.e., sericite, muscovite, etc). The porosity, long-term stability, and mechanical strength of the agglomerates should be optimized by test work.

Solution application. Alkaline cyanide solution is applied to the top of the heap by a suitable distribution system, such as agricultural sprinkling or drip irrigation. Application rates vary depending on ore properties and the gold dissolution rate but typically range from 0.1 to 1.0 $l/m^2/min$ (or 0.001 to 0.01 gpm/ft^2). Solution is usually applied at a slower rate than the permeability of the leach pad allows, to prevent buildup of solution on the surface of the pad. This is often important to minimize any threat to wildlife (due to ponding on the heap surface) and to maintain the stability of stacked ore.

Agricultural-type sprays or sprinkler systems, and drip irrigation systems have all been used with good success in heap leaching operations worldwide. Sprays or sprinklers have the advantage of easy installation but the disadvantage of relatively high solution evaporation rates (typically 5% to 8% solution loss, depending on the climate). It is difficult to provide even solution coverage over the heap with sprays or sprinklers, and this is critical for effective leaching. On the other hand, drip irrigation is slightly more expensive to install but has the advantage of lower solution evaporation (typically 2% to 4% solution loss, again depending on climate). The use of tortuous-path agricultural drip emitters was proposed for application to gold (and other metals) heap leaching in 1984 [73] and was applied commercially at Coeur–Rochester (Nevada) in 1987 [44].

pH modification. Alkali, which is required for pH modification and control, may either be added to the leach solution and/or introduced to the ore prior to leaching. The latter practice is essential when treating ores that have high alkali consumption, and the reagent may be added directly to haulage trucks, onto conveyors during crushing or agglomeration, or onto the surface of the leach pad. Calcium hydroxide (slaked lime) is the most economic alkali to use for this purpose. When pH modification and control is required in leach solutions, either calcium hydroxide or sodium hydroxide may be used. Calcium hydroxide is the cheaper of the two but is less soluble and produces solutions that are much more susceptible to salt precipitation and scale formation. Unslaked lime (CaO) is sometimes used because it is less costly than slaked lime, but it is less effective for pH modification. For nonacidic- or nonalkali-consuming ores, calcium hydroxide concentrations of 0.15 to 0.25 g/L are typically required to achieve the desired pH range for leaching (i.e., pH 10 to 11). Sodium hydroxide is known to be more effective than calcium hydroxide at dissolving a variety of minerals, particularly at high alkalinities, and is a highly effective dispersant. This may result in the dissolution of ore constituents, such as silicates, to produce various solution species, which can subsequently precipitate in a number of undesirable forms, potentially affecting downstream processes including drip irrigation, filtration, gold precipitation, or carbon adsorption. Consequently, calcium hydroxide is generally the preferred method of pH control in heap leach systems.

Dissolution rate. Gold leaching rates in heap leaching environments are highly dependent on the mass transport of cyanide and oxygen to the exposed gold surfaces. The concentration of cyanide is depleted as the leach solution percolates through the heap and cyanide reacts with precious metals and other ore constituents, and as other cyanide-consuming reactions take place. However, in the absence of significant amounts of oxygen-consuming minerals, there is little evidence to suggest that dissolved oxygen concentrations are depleted significantly, and it is apparent that in most heaps, sufficient oxygen is drawn into the heap by the flow of solution to keep the solution at, or close to, oxygen saturation. Consequently, gold dissolution rates can usually be maximized by maintaining the cyanide concentration in leach pad run-off solutions above the minimum discussed in Section 6.1.3.1.

In the extreme case, when treating materials containing large amounts of reagent-consuming minerals, it is theoretically possible that both cyanide and oxygen may be completely depleted from the solution in the heap, but this is rarely experienced in practice. Such an ore would likely be considered to be refractory (see Chapter 5) and is unlikely to be suitable for heap leaching because of high cyanide consumption and poor gold (and silver) extractions, rendering such processing uneconomic.

During heap leaching of porous materials (e.g., sandstones and siltstones) and ores with cracks and fissures within single particles, pore diffusion may be important for the leaching process. Reacting species must be allowed sufficient time to diffuse to gold surfaces located within these materials, and for dissolved species to diffuse away to the bulk solution to permit subsequent recovery. Occasionally, the application of solution to the heap is suspended to allow time for these diffusion processes to occur. This resting period permits the leach solution to be used more efficiently elsewhere and may help to reduce reagent consumptions.

Leaching efficiency.   Gold extractions obtained by heap leaching are generally in the range of 50% to 80% and depend on the following:

- Degree of gold liberation achieved at the heap feed particle size
- Efficiency of solution contact with ore, which is a function of the uniformity of solution application and the homogeneity of material on the heap
- Relationship between dissolution rate and the time allowed for leaching

In heap leaching, many of these factors are difficult to control, and gold extractions achieved are often quite variable, even within a single ore type.

Solution management is particularly important in heap leaching, because it is usually desirable to produce a constant volume of pregnant, gold-bearing solution to the subsequent recovery process, for example, carbon-in-columns (see Chapter 7) or Merrill–Crowe zinc precipitation (see Chapter 8). Many innovative leaching schemes have been devised to achieve this. Commonly, low-grade ("intermediate") run-off solutions are used to leach fresh ore, increasing the gold concentration of the solution and enhancing subsequent recovery operations [74]. Still lower-grade barren solution is then applied to partially leached material that contains less gold. This is sometimes referred to as solution stacking. This method not only allows the production of a constant amount of high-grade pregnant solution independent of the amount of material being leached, but also assists gold dissolution according to the phenomenon described in Section 6.1.5.1 ("Counter-current leaching"), in that the lowest-grade solution is applied to material containing the least amount of gold. An example of such a scheme is given in Section 12.2.2.6 for the Round Mountain operation in Nevada. Other examples include Yanacocha (Peru), Coeur–Rochester (Nevada), and Mesquite (California, United States).

Careful monitoring of process solutions is critical to successful heap leach operation, particularly early in the leaching cycle when reactions involving the most reactive ore constituents occur. In addition to gold (and silver) concentrations, monitoring of pH, E_h, cyanide concentration, dissolved oxygen concentration, base metal ion concentrations, and temperature may need to be considered [75].

Cyanide consumptions for crushed ore heap and run-of-mine ore stockpile leaching operations are generally lower than those experienced in agitated leaching systems and, for nonrefractory ores, typically range from 0.1 to 0.5 kg/t. Lime consumptions are variable, depending on the ore type and properties, but typically range from 0.15 to 0.75 kg/t.

6.1.5.3 Intensive Cyanidation Leaching

Intensive cyanidation processes use high reagent concentrations, principally cyanide and oxygen, and often elevated temperature and/or pressure, to increase the dissolution rate of gold. They are applied to higher-grade materials, which can justify the higher treatment cost to achieve higher gold recovery. Such materials include flotation and gravity concentrates, which are not amenable to treatment by conventional cyanidation for any of the following reasons:

- Material contains coarse gold, which requires unacceptably long leaching times under standard cyanidation conditions.
- Some or all of the gold is locked in cyanide-soluble minerals (sulfides), whose dissolution rate is increased at high cyanide and oxygen concentrations.

- Gold occurs with other minerals that interfere with standard cyanidation practice, for example, tellurium and mercury.

Various advantages have been cited for the use of intensive cyanidation over other processes, such as the use of amalgamation for treatment of gravity concentrates, or oxidation and cyanide leaching for the treatment of flotation concentrates. These advantages include faster gold dissolution, reduced security risk, and less health hazard [76].

It has been shown in Section 6.1.3.1 that dissolved oxygen concentration is usually the rate-limiting factor in conventional cyanidation. The use of pressurized systems allows elevated concentrations of dissolved oxygen to be maintained which, when coupled with increased cyanide concentration, results in enhanced gold dissolution rates. Oxygen may be introduced either as air, pure oxygen, or mixtures of the two to achieve elevated oxygen partial pressures. Under these conditions the temperature of the system can be increased without the sharp reduction in dissolved oxygen concentration that occurs at atmospheric pressure, although some systems operate at ambient temperature. Cyanide concentrations of 0.5% to 2.5% NaCN have been applied in practice. Sodium hydroxide is often used for pH control to avoid scaling and other problems associated with calcium hydroxide use, and concentrations of 0.05% to 0.4% sodium hydroxide (NaOH) have been applied in practice. Occasionally leach additives (including alkali metal peroxides and other solid oxidants; chloride salts, lead salts, and other reagents) may be added to further enhance leaching rates [77, 78].

High gold (and silver) extractions are generally accomplished with intensive cyanidation processes, typically >97% and sometimes as high as 99%. Leach residence time requirements are usually 24 hr or less.

The increased severity of intensive cyanidation systems increases the dissolution rate of all cyanide-soluble minerals, as well as the rate of cyanide consumption, and consequently the consumption of reagents is typically high compared with agitation leaching. Cyanide consumptions typically range from 5 to 25 kg/t, depending on the mineralogy of the material and the conditions applied [77]. Sodium hydroxide (or less commonly lime) consumptions are also higher than those experienced in conventional cyanidation, ranging from 1 to 5 kg/t.

The process has been applied commercially for the treatment of gravity concentrates (see Section 12.2.2.3), especially in South Africa and Australia [76, 77]. For example, the ACACIA reactor comprises both equipment and a process developed specifically (by AngloGold Ashanti, Johannesburg, South Africa) for intensive leaching of gravity concentrates and has been applied at Union Reefs (South Africa), Sunrise Dam (Australia), and Porgera (Papua New Guinea). The ACACIA process consists of a prewashing step to remove fines, an intensive leaching step, and a direct electrowinning step for gold (and silver) recovery. The intensive leaching step utilizes cyanide concentrations of 15 to 25 g/L NaCN, 3 to 4 g/L NaOH, and 2 to 10 g/L of a suitable solid oxidant (such as an alkali metal peroxide [Section 6.1.3.1]) and is carried out at 50°C to 65°C. The application of this technology to treat gravity concentrates produced in conventional gold extraction circuits using leaching and CIP or CIL has the following potential advantages [79]:

- Faster dissolution of gold reporting to the leaching, and CIP/CIL circuits
- Improved recovery from the CIP/CIL circuit
- Reduced gold-in-circuit inventory
- Improved overall gold recovery
- Reduced overall cyanide consumption
- Improved oxygen demand in CIP/CIL

Gold recoveries of 97% to 99% have been achieved routinely from a range of gravity concentrates within 8 to 16 hr of leach residence time using the ACACIA process technology [80].

6.1.5.4 Vat Leaching

Vat leaching is usually performed in large wooden or concrete structures (vats) or steel tanks or may be accomplished by heap leaching in a valley-fill configuration, where the heap can be flooded. In either case, the ore is completely immersed in the leach solution either throughout the leach cycle or for portions of it. This has the advantage of efficiently wetting all surfaces of the ore to be leached and, to a degree, aiding mass transport. The proposed advantages are that channeling of solutions and the development of dead zones, which may be experienced during heap leaching, are avoided. The process is little used, because of the high capital cost associated with vat construction and the high operating cost, with only minor (if any) recovery benefits over standard heap leaching practice. However, the method was employed at Homestake Lead (South Dakota, United States) for about 100 years until the operation closed in 2000 (see Section 12.2.2.4).

6.1.5.5 In Situ Leaching

In situ leaching is the application of dilute cyanide solution to an ore in the location in which it is found [81]. This requires that the permeability of the ore be such that solution can gain access to an economically significant portion of the contained gold values. The required permeability may either be an inherent property of the orebody (i.e., a porous or highly fractured and/or fissured material) or may be induced by blasting to create sufficient fragmentation.

In practice, the process has largely been avoided for the following reasons:

1. The efficiency of solution contact with gold values is usually very poor, resulting in low recoveries.
2. The recovery of solution from in situ leaching operations is typically low, resulting in high reagent costs and low recoveries.
3. Environmental concerns and regulations may restrict the use of this process, particularly in view of No. 2.

There are no known commercial applications of in situ leaching with cyanide, and the authors see little potential for the future application of cyanide leaching in situ. In the 1980s, in situ leaching using thiosulfate solution was tested underground on the Witwatersrand in South Africa, but no commercial application emerged from this work (see also Section 6.3.2.2).

6.2 CHLORINATION

Chlorination was applied extensively in the 1800s, prior to the introduction of cyanidation, for the treatment of ores containing fine gold and gold occurring with sulfides, which were not amenable to gravity concentration and amalgamation. Chloride media have also been applied in electroplating processes since the early 19th century. Although chlorine–chloride media is no longer used for leaching of primary ores, several processes have been proposed for the treatment of refractory, or semirefractory, ores as an alternative to cyanide. Chlorination has also been applied for oxidative pretreatment of some carbonaceous refractory ores (Section 5.6), and significant proportions of the gold in the feed material may be dissolved during this process.

The general chemistry of chlorine solutions has been reviewed in detail in Section 5.6 and is not considered further in this chapter.

6.2.1 Mechanism of Gold Dissolution

Gold dissolves in aqueous chloride solution to form both the Au(I) and Au(III) chloride complexes, as follows:

$$AuCl_2^- + e \rightleftharpoons Au + 2Cl^-; \quad E^0 = +1.113 \text{ (V)} \quad \text{(EQ 6.34)}$$

$$AuCl_4^- + 3e \rightleftharpoons Au + 4Cl^-; \quad E^0 = +0.994 \text{ (V)} \quad \text{(EQ 6.35)}$$

The corresponding solution equilibria are shown on the E_h–pH diagram for the Au–Cl–H$_2$O system (Figure 6.27). The Au(III) complex is more stable than the Au(I) species by approximately 0.12 V. The gold chloride complexes are not as stable as the Au(I) cyanide complex, but one advantage of this is that they are more easily reduced to gold metal. Oxidation will only occur above approximately 1.2 V, and therefore a strong oxidant, such as chlorine, chlorate ions, or ozone, is required to dissolve gold at a reasonable rate. Chlorine is the most suitable oxidant for this purpose, because it also supplies chloride ions for gold dissolution in addition to the strongly oxidizing hypochlorous species (Section 5.6.1); however, bromine and iodine may also be used to achieve fast dissolution rates. Nitric acid is a sufficiently strong oxidant to dissolve gold in chloride media, and this is applied as aqua regia (a mixture of 33% HNO$_3$ and 66% HCl) in a well-established technique for gold analysis.

Several mechanisms have been proposed for gold dissolution in aqueous chloride solutions. The dissolution probably proceeds in two stages with the formation of an intermediate Au(I) chloride on the gold surface during the first stage [82]:

$$2Au + 2Cl^- \rightleftharpoons 2AuCl + 2e \quad \text{(EQ 6.36)}$$

The most likely theory for the second stage suggests that $AuCl_2^-$ is formed as a secondary intermediate, which then is either oxidized further to Au(III), as follows:

$$AuCl_2^- + 2Cl^- \rightleftharpoons AuCl_4^- + 2e \quad \text{(EQ 6.37)}$$

or diffuses into solution as $AuCl_2^-$, depending on the oxidizing potential of the solution [83]. At solution potentials above approximately 1.4 V, the gold surface becomes passivated with an oxide layer.

6.2.2 Reaction Kinetics

Cyclic voltammetry has been used to investigate the kinetics of gold dissolution in chloride solutions and has shown that the dissolution rate is proportional to the chloride ion concentration [83]. In practice, the mass transport of chloride ions to the gold surface is the rate-determining step, and the chemical reaction rate is fast. Chloride ion mass transport rates are maximized by high chlorine–chloride concentrations and by increased temperatures.

The dissolution rate of gold in aqueous chloride solutions is much faster than that achievable in aqueous alkaline cyanide solutions. Gold leaf tests have indicated dissolution rates of 0.008 g/m^2/s in cyanide solution compared with 0.3 g/m^2/s in chloride solution [82]. The main reason for the faster reaction rate is the high solubility of chlorine in water compared with oxygen, the oxidant used in cyanidation. It was also noted from this work that the presence of sodium chloride (3%) in the chlorine solution had a

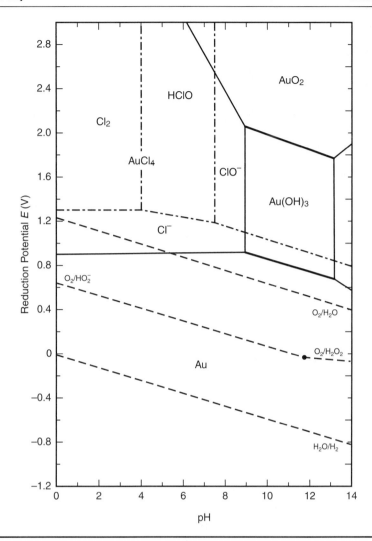

FIGURE 6.27 Potential–pH equilibrium diagram for the system Au–Cl⁻–H₂O at 25°C including some equilibria between chlorine and water: [Au(III)] = 10^{-2} M; [Cl⁻] = 2 M; pCl_2 = 0.1 atm; [HClO⁻] = [Cl⁻] = 6 × 10^{-3} M; pO_2 = pH_2 = 1 atm [6]

considerable accelerating effect on the dissolution rate, possibly due to the retarding effect of chloride ions on chlorine dissociation [82].

6.2.3 Behavior of Other Minerals in Chloride Solution

Both silver and lead react to form insoluble chlorides in chlorine–chloride solutions. This is significant because either, or both, of these insoluble products can reduce the solubility of gold due to the formation of an insoluble passivating layer and, in the case of silver, results in loss of metal recovery. However, passivation only occurs to any significant degree when the combined silver–lead content of the gold alloy exceeds approximately 13% [83].

Below pH 3, pyrite is attacked and dissolved in aqueous chloride solution using chlorine as the oxidant. However, between about pH 3 and 6, with hypochlorous species as the oxidant, the dissolution rate of pyrite is greatly reduced. One investigation has shown that the dissolution rate is reduced by a factor of 4 as the pH is increased from 2 to 4, and by a similar factor again when the pH is increased from 4 to 6 [84]. This presents an interesting process option for ores containing free gold with barren pyrite gangue, with the potential for selective gold dissolution above about pH 3.

Copper, zinc, and most of the other transition metal series form relatively unstable chloro-complexes, as indicated in Table 6.6. All of these chloro-complexes are less stable than Au(III) chloride. Of particular importance are the large relative differences in the stability constants of the copper cyanide–chloride and gold cyanide–chloride complexes. This indicates that for a particular ore, less copper will be dissolved in chloride media than in cyanide media, a feature that is potentially attractive for the treatment of some copper–gold ores.

Gold-telluride minerals are soluble in acidic chloride media in the presence of a sufficiently strong oxidant, such as ferric ion (Fe^{3+}) or chlorine (Cl_2), and dissolve to form gold and tellurium chloride complexes: $AuCl_4^-$ and $TeCl_6^{2-}$, respectively [59]. This method is applied at the Emperor mine for the recovery of gold and tellurium from an ore that is not amenable to direct cyanide leaching (see Section 12.2.8.1).

Carbonates and other acid-soluble minerals dissolve to varying extents in acidic chlorine–chloride media. The decomposition of these minerals may enhance gold extraction by exposing locked gold—a potential advantage over leaching with alkaline cyanide media.

The oxidation of sulfide minerals by chlorine is considered in Section 5.6.3.

6.2.4 Process Considerations

Chlorine leaching (or chlorination) is more difficult to apply commercially than cyanide leaching for a number of reasons. First, the leaching media are highly corrosive and require special materials of construction, for example, stainless steel and rubber-lined equipment, to withstand the acidic and strongly oxidizing conditions. Second, in cases where chlorine gas is used, the reactants must be contained in a closed system to allow optimum utilization of the gas and to avoid any health risk. Chlorine utilization is important in view of the high cost of the gas, compared with air and/or oxygen.

As a result, chlorine leaching is usually restricted to agitated or intensive agitated systems, where closed tanks can be used in a closely controlled environment. The process has typically been carried out at a pH of 2.0 to 2.5. The chlorine gas absorption rate is maximized by decreasing pH, but the solution becomes increasingly difficult to handle as pH decreases. Alternatively, a solid oxidant, such as sodium hypochlorite (NaOCl) may be used.

Consumption of the oxidant (i.e., either Cl_2 or NaOCl, which dissolve in solution to form the reactive HOCl/OCl⁻ species [see Equation 5.81]) is an important factor that has a large impact on the process cost. (HOCl is hypochlorous acid, OCl⁻ is hypochlorite species, produced when hypochlorous acid dissociates to H^+ and OCl⁻.) It has been reported that the addition of isocyanuric acid to the leaching system can significantly reduce the decomposition rate of HOCl/OCl⁻ species and thereby reduce the oxidant consumption [85].

Gold extractions achieved by chlorination vary widely, depending on ore type. Unfortunately, limited process data are available because chlorination of whole ores was discontinued for economic reasons after the advent of the cyanide process in 1888.

Chlorination has been used for pretreatment of carbonaceous ore at Newmont Carlin and at Jerritt Canyon (both in Nevada), and up to 85% gold dissolution is achieved by chlorination alone [61]. The treated slurry is subsequently neutralized, the pH adjusted to 10 11 and then cyanide leached, enabling high subsequent gold recovery to be achieved (see Sections 5.6 and 12.2.9.1).

TABLE 6.6 The stability constants of selected metal–chloride complexes [2, 10, 11]

Metal Ion	Stability Constant of Complex (β)			
	Cl⁻	2Cl⁻	3Cl⁻	4Cl⁻
Au^+	—	—	—	—
Au^{3+}	8.5	16.2	23.6	29.6
Ag^+	3.7	13.0*	6.2	6.0
Cu^+	2.7	6.1	5.9	5.6
Cu^{2+}	0.4	0.9	0.9	0.9
Sn^{2+}	1.2	1.8	1.5	1.3
Sn^{4+}	—	—	—	—
Pb^{2+}	1.5	2.0	1.8	1.3
Pb^{4+}	—	—	—	—
Zn^{2+}	−0.2	0.1	1.0	—
Ni^{2+}	—	−0.04	—	3.0

NOTES: Dashes = not applicable.
* Range of values from literature: β = 8.9 to 17.2.

A commercial process based on hot (75°C) agitated chloride leaching of a silver-bearing material (also containing lead, copper, and antimony) at the Itos mine in Bolivia was in operation in the late 1990s. This process used near-saturated solution of sodium chloride with hydrochloric acid (pH 0.3) and 15 g/L ferric ion as the oxidant [86].

An atmospheric leaching process using NaCl and NaOCl at pH 7 was developed by ISL Ventures in the late 1980s as an option for gold ores containing cyanide-soluble copper, because base metals are not leached under these conditions in this system. The use of isocyanuric acid (2,4,6 trihydroxy-s-triazine) was proposed to reduce the rate of consumption of the oxidant (i.e., ClO⁻ species, supplied by NaOCl). Gold recoveries in excess of 80% were reported for several gold ores when NaOCl concentrations of 1 to 2 g/L were employed; however, high NaOCl consumptions (0.5 to 1.0 kg/t ore) hamper the commercialization of the process [87].

It is well known that the presence of chlorides in pressure oxidation slurries can result in gold being dissolved during pressure oxidation based on operating experience and test work at Goldstrike and Twin Creeks in Nevada [88]. As an extension of this concept, during the 1990s, a chlorination process was developed to treat complex platinum–palladium–nickel–copper–cobalt ores and concentrates, referred to as the Platsol process. Although this process has not been applied commercially, it has been proposed for the extraction of gold from a variety of gold-bearing materials but particularly those containing by-products such as copper, nickel, and PGMs that present problems for conventional cyanidation. The process consists of high temperature (200°C to 225°C) pressure oxidation in sulfate media containing 5 to 20 g/L NaCl. In this case, the chloride provides the complexant, and oxygen and/or Fe(III) the oxidant. Gold extractions ranging from 80% to 99% have been achieved during testing of various gold-bearing materials [89].

However, the main reasons for the lack of commercial success of chlorine and chloride-based gold recovery processes are the following:

- Relatively high reagent (oxidant) consumptions
- Challenges with corrosion-resistance of construction materials for use in chloride media
- Well-known difficulties in recovering gold from chloride media (see Section 8.4)
- Poor overall economics

6.3 THIOSULFATE

In locations where environmental regulations and concerns prohibit the use of cyanide, thiosulfate has been proposed as an alternative to cyanide for the recovery of gold from ores and concentrates. The ability of thiosulfate species to form stable complexes with gold has been known for more than 100 years, but serious research into its use as an alternative to cyanide only started in the late 1970s. Extensive, worldwide research into the development of thiosulfate-based leaching processes escalated dramatically in the 1990s as a result of increased concerns over the use of cyanide and other factors discussed in Chapter 11.

Until recently, the use of thiosulfate media was proposed only for difficult ores and concentrates, such as those containing large amounts of cyanide-consuming copper, carbonaceous preg-robbing materials, refractory sulfides, and the products of partial oxidative pretreatment (e.g., biological oxidation) of refractory sulfides which contain cyanide-consuming sulfur species. One of the drivers for these potential applications is that reactive sulfide minerals and sulfur species react only to a limited extent with thiosulfate and generally do not consume large quantities of the reagent, unlike cyanide.

Much of the research and development work conducted since about 1995 focused on the development of an effective thiosulfate process to replace cyanidation for a broader range of material types. Thiosulfate media have been demonstrated to be capable of effectively dissolving gold, but complex systems are required that include a suitable oxidant (such as Cu(II) or Fe(III) ions) and an effective oxidant stabilizer (such as ammonia, or open chain polyamine ligands for copper, and oxalate for iron) to achieve an acceptable gold dissolution rate, while minimizing the thiosulfate oxidation rate. Despite showing considerable promise, the development of an effective thiosulfate-based process remains elusive because of high reagent consumption and costs (due to thiosulfate oxidation) and difficulties with metal recovery from the leach solution. However, this process continues to be researched aggressively by several of the major gold producers worldwide.

6.3.1 Reaction Chemistry and Kinetics

Gold forms a stable complex with thiosulfate species ($S_2O_3^{2-}$) in aqueous solution, as follows:

$$Au + 2S_2O_3^{2-} \rightleftharpoons Au(S_2O_3)_2^{3-} + e; \quad E^0 = +0.153 \text{ (V)} \qquad \text{(EQ 6.38)}$$

Gold dissolves in alkaline thiosulfate solution, using dissolved oxygen as the oxidant, to form the Au(I) complex, as follows:

$$4Au + 8S_2O_3^{2-} + O_2 + 2H_2O \rightleftharpoons 4Au(S_2O_3)_2^{3-} + 4OH^- \qquad \text{(EQ 6.39)}$$

Theoretically, based on E_h–pH data, dissolution should occur over a wide range of pH conditions. However, the reaction proceeds very slowly with oxygen in the absence of a suitable catalyst. Cu(II) is a very effective catalyst for this reaction, provided that it is used in combination with ammonia to stabilize the copper species in solution as Cu(II) tetramine ($Cu(NH_3)_4^{2+}$) [90].

$$Cu^{2+} + 4NH_3 \rightleftharpoons Cu(NH_3)_4^{2+} \qquad \text{(EQ 6.40)}$$

Gold forms stable complexes in solutions containing ammonia (as described in Section 6.6). Therefore, in ammoniacal thiosulfate solution, both the thiosulfate and ammonia species compete to form complexes with gold, according to the following equation:

$$Au(S_2O_3)_2^{3-} + 2NH_3 \rightleftharpoons Au(NH_3)_2^+ + 2S_2O_3^{2-} \qquad \text{(EQ 6.41)}$$

However, the exact values of the stability constants for the preferred thiosulfate and ammonia complexes of gold are in some doubt, and the two values are relatively close, resulting in a lack of clarity and certainty over which species predominates under the conditions applied for gold leaching with thiosulfate. The general consensus is that the thiosulfate complex prevails under the E_h–pH conditions applied for optimal leaching (i.e., a compromise between maximizing gold dissolution rate and minimizing thiosulfate oxidation rate). The E_h–pH diagram that best represents the most likely equilibria for the Au–NH_3–$S_2O_3^{2-}$–H_2O system (Figure 6.28) shows that dissolution can occur over a wide range of pH values [91].

Typically, leaching is carried out between pH 9 and 11. Below pH 9, the Cu(I) triamine ($Cu(NH_3)_3^+$) complex becomes prevalant, making the copper species less effective as a catalyst. Commonly, thiosulfate research is conducted using initial solution pH of 10.5 to 11 because the pH decreases during the reaction in batch testing. However, continuously operating thiosulfate systems are likely to be optimized between pH 9.0 and 10.0.

Fe(III) species are less suitable as the oxidant in the thiosulfate system because they are unstable above approximately pH 8, depending on the exact solution conditions. The use of Fe(III) with a suitable stabilizer such as oxalate has been proposed [19]. Hydrogen peroxide is also unsuitable because it oxidizes thiosulfate very rapidly and is difficult to stabilize.

The chemistry of the ammoniacal copper–thiosulfate system is highly complex, and, despite considerable research and investigation, the exact mechanism of gold dissolution and the catalytic action of Cu(II) species are still not completely understood [91, 92, 93]. The major reduction reaction is thought to be as follows:

$$Cu(NH_3)_4^{2+} + 5S_2O_3^{2-} \rightleftharpoons Cu(S_2O_3)_3^{5-} + 4NH_3 + S_4O_6^{2-} + e \qquad \text{(EQ 6.42)}$$

where the $Cu(S_2O_3)_3^{5-}$ species is more stable than the $Cu(NH_3)_2^+$ complex, under the conditions applied for effective gold dissolution. The reduction of the Cu(II) tetramine species is accompanied by oxidation of thiosulfate to tetrathionate.

The oxidant species, Cu(II), is then regenerated by oxygen reduction, according to the following equation:

$$4Cu(S_2O_3)_3^{5-} + 16NH_3 + O_2 + 2H_2O \rightleftharpoons 4Cu(NH_3)_4^{2+} + 8S_2O_3^{2-} + 4OH^- \qquad \text{(EQ 6.43)}$$

Therefore, the overall dissolution reaction for gold in ammoniacal copper–thiosulfate solutions has been proposed as follows [94]:

$$Au + 5S_2O_3^{2-} + Cu(NH_3)_4^{2+} \rightleftharpoons Au(S_2O_3)_2^{3-} + 4NH_3 + Cu(S_2O_3)_3^{5-} \qquad \text{(EQ 6.44)}$$

The presence of ammonia is critical to stabilize the Cu(II) species in solution as the Cu(II) tetramine ion, to prevent the formation of $Cu(OH)_2$ and to prevent passivation of the gold surface by preferential adsorption (i.e., avoiding coating with sulfur species). The mechanism of gold leaching in the ammoniacal copper–thiosulfate system is illustrated schematically in Figure 6.29.

The thiosulfate species is consumed by several possible oxidation and association reactions. These reactions include the formation of various intermediate sulfur species such as trithionate ($S_3O_6^{2-}$), tetrathionate ($S_4O_6^{2-}$), and other polythionates (e.g., $S_5O_6^{2-}$), and finally oxidation to sulfate (SO_4^{2-}) and, in some cases, elemental sulfur [88, 93, 95, 91].

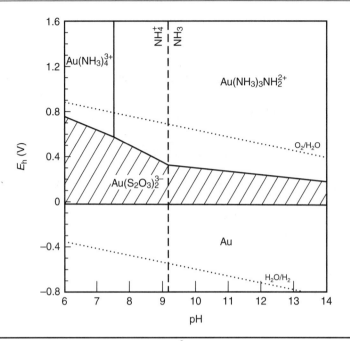

FIGURE 6.28 E_h–pH diagram for the Au–NH_3–$S_2O_3^{2-}$–H_2O system at 25°C: [Au(I)] = 10^{-5} M, [$Na_2S_2O_3$] = 0.1 M, [NH_3 + NH_4^+] = 1 M [91]

Some examples of the possible oxidation and association reactions are provided as follows:

$$2S_2O_3^{2-} + 2Cu^{2+} \rightleftharpoons S_4O_6^{2-} + 2Cu^+ \qquad \text{(EQ 6.45)}$$

$$4S_2O_3^{2-} + O_2 + 2H_2O \rightleftharpoons 2S_4O_6^{2-} + 4OH^- \text{ (slow at 25°C)} \qquad \text{(EQ 6.46)}$$

$$10S_2O_3^{2-} + 13O_2 + 4e \rightleftharpoons 4S_3O_6^{2-} + 8SO_4^{2-} \qquad \text{(EQ 6.47)}$$

$$3S_2O_3^{2-} + 8Cu^{2+} + 6OH^- \rightleftharpoons 2S_3O_6^{2-} + 8Cu^+ + 3H_2O \qquad \text{(EQ 6.48)}$$

$$5S_2O_3^{2-} + 3H_2O \rightleftharpoons 2S_5O_6^{2-} + 6OH^- \qquad \text{(EQ 6.49)}$$

The polythionates, and in particular the trithionate and tetrathionate species, are very detrimental to downstream gold recovery processes because they significantly reduce the loading of gold thiosulfate onto activated carbon and anionic ion exchange resins. Consequently, the formation and effects of these species must be mitigated, or an alternative recovery method used (see Section 8.4). The thiosulfate species can be stabilized to some extent by the addition of small amounts of sulfite ions (SO_3^{2-}), which react with other polythionate and sulfide species in the ore, to regenerate thiosulfate as follows:

$$S_5O_6^{2-} + SO_3^{2-} \rightleftharpoons S_4O_6^{2-} + S_2O_3^{2-} \qquad \text{(EQ 6.50)}$$

$$S_4O_6^{2-} + SO_3^{2-} \rightleftharpoons S_3O_6^{2-} + S_2O_3^{2-} \qquad \text{(EQ 6.51)}$$

$$3SO_3^{2-} + 2S^{2-} + 3H_2O \rightleftharpoons 2S_2O_3^{2-} + 6OH^- + S \qquad \text{(EQ 6.52)}$$

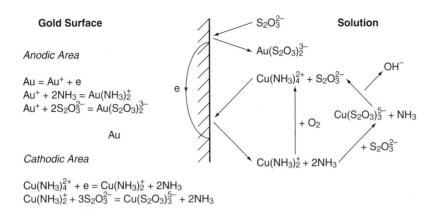

FIGURE 6.29 Schematic representation of the mechanism of gold leaching with ammoniacal copper–thiosulfate [88]

Other methods of stabilizing thiosulfate have been investigated with varying degrees of success. These methods include the use of multidentate ligands, such as open chain polyamine ligands with 2 to 5 nitrogen donors (e.g., tris[2-aminoethyl]amine) and thiourea to stabilize Cu(II), and oxalate to stabilize Fe(III). The chemistry of these systems is complex and, in the case of thiourea, is further complicated by the fact that thiourea forms stable complexes with gold (see Section 6.4) [96].

As thiosulfate is oxidized and the thiosulfate concentration in solution decreases, there is a risk that gold will reprecipitate out of solution onto ore constituents. There is evidence to suggest that gold will remain in solution as the thiosulfate complex down to 2 to 3 g/L $S_2O_3^{2-}$, however, this depends on specific solution conditions (e.g., temperature, pH, E_h, [NH_3], etc.) [97].

Silver, silver chloride, and silver sulfide all dissolve readily in thiosulfate media. The addition of sulfite species helps to prevent the precipitation of silver as the insoluble sulfide.

The rate of dissolution is dependent on thiosulfate and Cu(II) concentrations, up to a certain point. Thiosulfate concentrations ranging from 0.05 to 2.0 M $S_2O_3^{2-}$ have been used for investigations of gold leaching systems, but most researchers have focused on a thiosulfate concentration range of 0.1 to 0.2 M $S_2O_3^{2-}$ (i.e., about 11 to 22 g/L). Both sodium thiosulfate ($Na_2S_2O_3$) and ammonium thiosulfate ((NH_4)$_2S_2O_3$) have been used as the primary reagent. There is the obvious benefit to using ammonium thiosulfate in cases where the leaching system is ammoniacal copper–thiosulfate in that it provides some of the ammonia required. Sodium thiosulfate may be preferable to use in applications where copper is not used as the catalyst.

Copper concentrations applied in these investigations range from 0.0001 to 0.02 M; however, most commonly, Cu(II) concentration of 0.0005 to 0.002 M (i.e., 30 to 120 mg/L Cu^{2+}) is targeted. Unfortunately, the rate of thiosulfate oxidation increases with increasing Cu(II) concentration when all other parameters are held constant, and therefore from a practical standpoint this limits the maximum Cu(II) concentration that can be applied [93].

Ammonia concentration does not have any impact on gold dissolution rates, but it does have a big impact on the overall system effectiveness by stabilizing copper and reducing the thiosulfate oxidation rate. Ammonia concentrations between 0.2 to 0.4 M NH_3 have typically been applied, but up to 2 M NH_3 has been used for investigative work.

Ammonia is lost by volatilization from solution, which is exacerbated by sparging of air or oxygen into the solution phase, and by oxidation reactions.

Increased temperature does have a beneficial effect on the rate of gold extraction but also increases the rate of thiosulfate oxidation and the rate of ammonia loss from solution by volatilization. Most investigative work has been done at or slightly above ambient temperatures (i.e., 15°C to 30°C). Elevated temperature operation has been investigated up to 80°C, and a process using thiosulfate at an elevated temperature of 60°C has been proposed, in conjunction with the use of sulfite to stabilize the thiosulfate species at the higher temperature [97].

It has been reported that the presence of carbonates and bicarbonates in the ore can result in excessive oxidation of thiosulfate.

6.3.2 Process Considerations

Although thiosulfate leaching can be conducted in agitated systems, closed tanks are probably required to control ammonia loss, for industrial hygiene purposes, and where necessary to allow operation with an oxygen overpressure applied (i.e., operate system above atmospheric pressure). Agitated systems are applied to slurries typically containing 30% to 40% solids. Thiosulfate can also be applied to heap (or potentially vat) leaching systems, and this application is discussed further in this section.

6.3.2.1 Agitation Leaching

The thiosulfate system requires the addition of oxygen to regenerate the oxidant species ($Cu(NH_3)_4^{2+}$) in the solution phase. To accomplish this, either air or oxygen must be introduced into the system. In some cases, oxygen can be applied under pressure using an oxygen overpressure of 10 to 100 psi. However, there appears to be no direct correlation between the amount of oxygen supplied into the leach solution or slurry and the gold dissolution rate, likely due to competing reactions, including thiosulfate oxidation and oxidation of other sulfur species in addition to oxidation of Cu(I). Some investigators have reported a decrease in Cu(II) concentration when sparging the leach slurry with air and/or oxygen, and this was attributed to an increase in the oxidation rate of thiosulfate and the accompanying decrease in the free thiosulfate concentration [90]. However, it is likely that the decrease in Cu(II) is due to increased loss of ammonia by volatilization (and a corresponding decrease in free ammonia concentration) and possibly the precipitation of copper species from solution (as $Cu(OH)_2$). At steady state operation, the rate of oxygen supply to the leach solution or slurry should be just sufficient to maintain the oxidant concentration (i.e., Cu^{2+}) and to meet the needs of thiosulfate and sulfur species oxidation.

The use of Cu(II) as a catalyst for thiosulfate leaching of gold is potentially problematic from several different perspectives: First, the copper (and ammonia) will be present in the residue material and can present environmental concerns for the final treatment and disposal of the residue. Second, the copper can cause the precipitation of gold when the dissolved oxygen content is low. Finally, the copper can cause complications for the downstream recovery of gold (and silver) because copper tends to load onto activated carbon or ion exchange resins (with the amount and rate of loading dependent on solution conditions), and copper may be co-recovered with gold in the final recovery step (i.e., precipitation or electrowinning). As a result, attempts have been made to develop a process that does not require copper or ammonia. One such proposed process uses elevated temperature (90°C) and 10 to 100 psi oxygen in a closed system to achieve high gold dissolution rates. Sulfite is added to stabilize the thiosulfate species and to reduce trithionate and tetrathionate species prior to the gold recovery step. This process reportedly achieved >80% gold recovery at reasonable sodium thiosulfate consumptions of 7 to 10 kg/t and sodium sulfite consumptions of about 1 to 2 kg/t [97].

Agitated tank leaching systems (with and without ammonia and copper) are expected to be able to achieve gold extractions that are comparable to cyanidation (i.e., 80% to 95% recovery) within reasonable time frames (i.e., 6 to 24 hr), although this is, of course, dependent on the ore type, grind size, and the specific leaching conditions applied. Commercial thiosulfate consumptions are expected to be in the range of 5 to 15 kg/t which, depending on the type of thiosulfate used, translates into costs of between US$2.50/t and US$15.00/t ore (2004 constant dollar basis). To achieve consumptions at the lower end of this range would require that thiosulfate be regenerated and recycled in process solutions.

6.3.2.2 Heap, Vat, and In Situ Leaching

A major advantage of the thiosulfate-based process is that it can potentially be applied in a heap or vat leaching process, unlike other processes that use more aggressive oxidants such as chloride, bromine, and iodine. In the mid-1990s, Newmont commenced heap leaching of approximately three quarters of a million tons of previously biologically oxidized material (also on a heap) using ammonium thiosulfate as the lixiviant [56, 98]. The combined biological oxidation and thiosulfate leaching process was designed to treat low-grade carbonaceous refractory ore from Carlin (Section 5.7.3.6). An ammonium thiosulfate concentration of 10 to 13 g/L (0.06 to 0.08 M) was maintained, with 2 to 5 g/L free ammonia, 30 to 60 mg/L copper, at a pH of 8.8 to 9.2. The oxygen required for regeneration of the oxidant (Cu^{2+}) is provided from the air, and no special air or oxygen addition was required to the heap. The solution E_h was maintained in the range of 50 to 100 mV vs. Ag/AgCl. The material was leached for 176 days and achieved gold recovery of almost 53%, with ammonium thiosulfate consumption of 10 kg/t ore, ammonium hydroxide consumption of 0.75 kg/t, and minor consumptions of copper sulfate and other reagents. Gold was precipitated directly from the leach solution using metallic copper. Although the commercial demonstration was reportedly successful, by 2005 the technology had yet to be applied on a large commercial scale [56].

In the 1980s, an in situ thiosulfate-based process was tested on the Witwatersrand in an underground mine for leaching within underground stoping areas. However, this technology was not successfully commercialized. A process for the treatment of sulfidic gold ores has been proposed whereby thiosulfate ions are generated in situ under alkaline oxidation conditions. Likewise, this has not been applied commercially.

6.4 THIOUREA

Thiourea has been proposed as an alternative to cyanide for the treatment of sulfidic, cyanide-consuming ores, and for use in locations where environmental concerns make the use of cyanide difficult. Thiourea is a relatively nontoxic reagent, which behaves as a plant fertilizer in the environment, giving the impression that it might be an attractive alternative to cyanide in environmentally sensitive areas. On the other hand, the reagent is suspected to be carcinogenic and is capable of dissolving heavy metals in addition to gold and silver, which presents many similar environmental problems to cyanidation for the handling and disposal of effluents.

In addition, thiourea is oxidized and consumed very rapidly under the conditions required for leaching, resulting in prohibitively high reagent costs in most applications, particularly when compared with cyanidation.

Thiourea leaching has been used to treat an antimony-rich concentrate in New South Wales (Australia) and has been investigated as a process option for the treatment

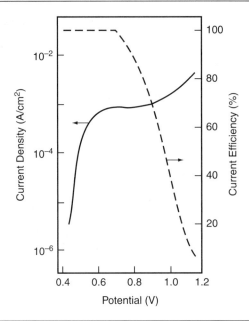

FIGURE 6.30 Effect of potential on the anodic dissolution of gold and the current efficiency of its dissolution into a solution containing 0.1 M sulfuric acid and 0.1 M thiourea at 30°C [103]

of several other ores, but no large-scale commercial processes have been developed [99, 100, 101].

6.4.1 Reaction Chemistry and Kinetics

Thiourea (NH_2CSNH_2) is an organic compound which dissolves readily in acidic solution in a stable molecular form. Gold dissolves in acidic thiourea solution to form a stable complex:

$$Au(CS(NH_2)_2)_2^+ + e \rightleftharpoons Au + 2CS(NH_2)_2; \quad E^0 = -0.38 \text{ (V)} \quad \text{(EQ 6.53)}$$

where the pK value for the gold–thiourea complex is 21.75. In this system, no oxide layer is formed as a result of the relatively low solution potentials that can be applied, and no passivation of gold occurs.

Figure 6.30 shows current–potential curves for leaching in thiourea solution. This indicates that gold dissolution will only occur at an acceptable rate if the solution potential is greater than approximately 0.5 V. Consequently, air and oxygen are unsuitable oxidants, and stronger oxidants, such as Fe(III), hydrogen peroxide, or ozone, are required to enable the reaction to proceed at a sufficient rate to achieve adequate gold extraction within practical time scales [102].

The overall equation for the dissolution reaction is:

$$2Au + 4CS(NH_2)_2 + 2Fe^{3+} \rightleftharpoons 2Au(CS(NH_2)_2)_2^+ + 2Fe^{2+} \quad \text{(EQ 6.54)}$$

Silver dissolves in aqueous thiourea solutions similarly to gold, as follows:

$$2Ag + 4CS(NH_2)_2 + 2Fe^{3+} \rightleftharpoons 2Ag(CS(NH_2)_2)_2^+ + 2Fe^{2+} \quad \text{(EQ 6.55)}$$

TABLE 6.7 Stabilities of metal–thiourea (Tu) complexes

Complex	pK
$Au(Tu)_2^+$	21.75
$Ag(Tu)_3^+$	13.10
$Cu(Tu)_4^{2+}$	15.4
$Zn(Tu)_2^{2+}$	1.77
$FeSO_4(Tu)^+$	6.64
$Cd(Tu)_4^{2+}$	3.55
$Pb(Tu)_4^{2+}$	2.04

for which pK = 13.1. Other metal species, some of which are listed in Table 6.7, dissolve to varying extents in acidic thiourea solution.

Thiourea is readily oxidized to formamidine disulfide, as follows [104]:

$$2CS(NH_2)_2 \rightleftharpoons (NH_2)_2CSSC(NH_2)_2^{2+} + 2e; \quad E^0 = -0.42 \text{ (V)} \tag{EQ 6.56}$$

This reaction proceeds slowly in the presence of oxygen but is accelerated by stronger oxidizing agents, such as ferric ion, hydrogen peroxide, or ozone that must be used for gold leaching. This oxidation is reversible, and the formation of formamidine disulfide can be controlled by controlling solution potential. This is important because formamidine disulfide disproportionates irreversibly to thiourea and a sulfinic compound, which decomposes (also irreversibly) to elemental sulfur and cyanamide ($CN(NH_2)$). The overall reaction is as follows:

$$NH_2(NH)CSSC(NH)NH_2 \rightleftharpoons CS(NH_2)_2 + S + CN(NH_2) \tag{EQ 6.57}$$

These irreversible reactions are undesirable because they consume thiourea (via formamidine disulfide) and are responsible for the high thiourea consumptions that are typically experienced in actual leaching systems.

Thiourea can be stabilized to some extent by the introduction of sulfur dioxide to the solution, added in the form of sodium bisulfite ($Na_2S_2O_5$). This partially reverses the thiourea oxidation reaction before the formamidine disulfide is oxidized further to elemental sulfur [104]:

$$((NH_2)_2CSSC(NH_2)_2)^{2+} + SO_2 + 2H_2O \rightleftharpoons 2CS(NH_2)_2 + SO_4^{2-} + 4H^+ \tag{EQ 6.58}$$

Other reactions that consume thiourea in aqueous solutions include the following:

$$CS(NH_2)_2 + H_2O \rightleftharpoons CO(NH_2)_2 + H_2S \text{ (hydrolysis)} \tag{EQ 6.59}$$

$$CS(NH_2)_2 \rightleftharpoons NH_4^+ + SCN^- \text{ (dissociation)} \tag{EQ 6.60}$$

These reactions proceed slowly in comparison with the other thiourea-consuming reactions described and probably do not account for a large proportion of the thiourea consumption.

However, formamidine disulfide is also capable of oxidizing gold in thiourea solutions [94, 105] and is considered to be a more efficient oxidizing agent than Fe(III), that is:

$$2Au + 2CS(NH_2)_2 + NH_2(NH)CSSC(NH)NH_2 + 2H^+ \rightleftharpoons 2Au(CS(NH_2)_2)_2^+ \tag{EQ 6.61}$$

The pH at which thiourea leaching may be carried out is limited at the upper end by two factors:

- Precipitation of Fe(III) hydroxide above approximately pH 3
- Sharp increase in the kinetics of thiourea oxidation above pH 3.5–4

Consequently, leaching is generally carried out at a pH of 1.4 to 1.8 [100, 101].

Leaching rates in thiourea solutions are very fast—much faster than cyanide dissolution and at least comparable to the other leaching methods available. The rate of gold dissolution in acidic thiourea solutions is usually controlled by the diffusion of reactants to the gold surface and is consequently related to the concentrations of Fe(III), formamidine disulfide, and thiourea species. Formamidine disulfide plays an important role in the kinetics of thiourea leaching, because optimum kinetics are achieved when approximately half the thiourea present in solution is converted to this species [100].

The addition of sodium sulfite (2.5 g/t Na_2SO_3) to thiourea leach solutions has been shown to be beneficial by reducing thiourea consumption and increasing the gold dissolution rate [106].

6.4.2 Process Considerations

The economics of gold leaching with thiourea are principally determined by thiourea consumption, which is related to thiourea and oxidant concentrations, solution pH, and the solution potential. High gold extractions can typically be obtained using thiourea, that is, >95% for ores in which the gold is well liberated. Thiourea concentrations between 5 and 50 g/L have been applied in laboratory and pilot-scale test work. Sufficient oxidant (i.e., Fe(III) or H_2O_2) is required to oxidize approximately 50% of the thiourea to formamidine disulfide, for optimal leaching conditions; however, the presence of excess oxidant increases thiourea consumption significantly. For this reason, close control of solution potential would be required through all stages of leaching in any commercial process. Thiourea consumptions of 1 to 4 kg/t have been projected for optimized thiourea leaching systems based on currently available technology, although estimates as high as 10 to 12 kg/t have been made. Such high consumptions, coupled with the requirements of H_2SO_4 for pH control, and H_2O_2 and SO_2 for potential control, make the overall cost of the process very high, probably at least twice the cost of cyanidation for direct leaching of the same material [94, 107]. However, other methods for stabilization of thiourea in solution under conditions suitable for gold dissolution continue to be investigated.

6.5 THIOCYANATE

Gold dissolves in aqueous, acidified thiocyanate (SCN^-) solutions to form both the Au(I) and Au(III) complexes, depending on the solution potential [103]:

$$Au(SCN)_2^- + e \rightleftharpoons Au + 2SCN^-; \quad E^0 = +0.662 \text{ (V)} \quad \text{(EQ 6.62)}$$

$$Au(SCN)_4^- + 3e \rightleftharpoons Au + 4SCN^-; \quad E^0 = +0.636 \text{ (V)} \quad \text{(EQ 6.63)}$$

The stability constants for the two complexes, $Au(SCN)_2^-$ and $Au(SCN)_4^-$, are approximately 10^{17} and 10^{42} repsectively, and $Au(SCN)_4^-$ is by far the most stable.

Fe(III) is the most suitable oxidant for the reaction since the kinetics of dissolution are prohibitively slow if oxygen is used and thiocyanate is oxidized very rapidly by hydrogen peroxide. In addition, the stability of the thiocyanate ion is increased in the

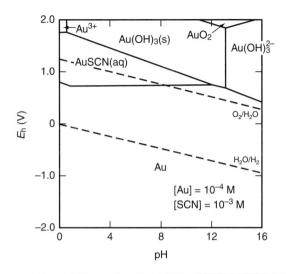

FIGURE 6.31 E_h–pH diagram for the Au–SCN–H_2O system at 25°C [43]

presence of Fe(III), presumably due to the many complexes it forms with these ions, as follows:

$$Fe^{3+} + nSCN^- \rightleftharpoons Fe(SCN)_n^{(3-n)} \quad \text{(EQ 6.64)}$$

where
$\quad n = 1$ to 5

Figure 6.31 shows the E_h–pH diagram for the Au–SCN–H_2O system under atmospheric conditions. The pH range for the reaction is limited at the low end (below about pH 1) by the reaction:

$$SCN^- + H^+ \rightleftharpoons HSCN; \quad pK = 0.85 \quad \text{(EQ 6.65)}$$

Above about pH 3, Fe(III) is precipitated by hydrolysis. Consequently, the optimum pH for gold leaching is in the range of 1.5 to 2.5.

The stability of thiocyanate species is strongly potential dependent, with stability achieved below approximately 0.64 V (vs. SHE). On the other hand, a potential >0.64 V is required to achieve satisfactory gold leaching rates at practical thiocyanate concentrations. In practice, a compromise between these two requirements is required to achieve acceptable gold leaching rates and to avoid excessive oxidation of thiocyanate, according to the following reactions:

$$SCN^- + 4H_2O \rightleftharpoons SO_4^{2-} + CN^- + 8H^+ + 6e \quad \text{(EQ 6.66)}$$

$$SCN^- + 5H_2O \rightleftharpoons SO_4^{2-} + CNO^- + 10H^+ + 8e \quad \text{(EQ 6.67)}$$

$$SCN^- + 7H_2O \rightleftharpoons NH_3 + CO_3^{2-} + SO_4^{2-} + 11H^+ + 8e \quad \text{(EQ 6.68)}$$

A number of other intermediate species also may be formed, including trithiocyanate $(SCN)_3^-$ and thiocyanogen $(SCN)_2$.

The rate of gold dissolution increases with increasing thiocyanate concentration and, to a lesser extent, Fe(III) concentration. Thiocyanide concentrations of 0.5 to 5 g/L (0.01 to 0.10 M) and Fe(III) concentrations of 6 to 12 g/L (0.1 to 0.2 M) have been used in laboratory column leach testing and small-scale pilot work. Because the thiocyanate consumption rises with increasing thiocyanate concentration, it is important to maintain sufficient thiocyanate concentration for effective gold dissolution but no excess. The oxidant, Fe(III), must be regenerated, and this can be accomplished using either air and/or oxygen or by some other means [94].

Increasing temperature increases gold dissolution rate but also significantly increases the rate of thiocyanate consumption, and, in view of the high reagent consumptions even at ambient temperatures, elevating leaching temperature is not likely to be a viable option. A significant drop in gold extraction has been reported at temperatures >40°C when using thiocyanate, probably due to the increased oxidation of thiocyanate.

Silver forms a relatively insoluble product, silver thiocyanate, in thiocyanate solutions and is essentially unrecoverable by this process.

In the late 1990s, Newmont Gold investigated the potential application of acidic thiocyanate leaching as an alternative to cyanide leaching to treat the product from a biological oxidation pretreatment on low-grade, refractory Carlin ore (Section 5.7). During this investigation, laboratory testing of ground products using 0.05 M SCN^- and 0.2 M Fe(III) concentrations yielded gold extractions ranging from 55% to 65%, compared to about 70% for conventional cyanidation. However, column leach testing (on coarser material) using thiocyanate resulted in >50% gold recovery compared to 40% for cyanidation, possibly due to continued oxidation of refractory sulfides in the acidic thiocyanate media. Sodium thiocyanate consumption varied from 0.6 to 0.8 kg/t, compared with 0.3 kg/t sodium cyanide [108].

Although acidic thiocyanate solution shows promise as an alternative lixiviant for gold, more work is required to optimize conditions, minimize thiocyanate consumption, and develop effective methods for downstream recovery of gold from leach solution.

6.6 AMMONIA

Gold is soluble in aqueous ammoniacal solution in the presence of a suitable oxidant, such as oxygen, hypochlorite, hydrogen peroxide, or bromine. The dissolution proceeds through the formation of Au(I) species, $(Au(NH_3)_2^+)$, to potentially form the stable Au(III) amine complex (depending on the solution potential), as follows:

$$Au + 2NH_3 \rightleftharpoons Au(NH_3)_4^+ + e \qquad \text{(EQ 6.69)}$$

In the case where oxygen is used as the oxidant, high ammonia concentrations (2 to 8 M) are required, and elevated temperatures (i.e., >100°C) must be applied for the reaction to proceed at an acceptable rate [109, 110].

The use of aqueous ammonia–cyanide mixtures has been proposed for leaching of gold from copper–gold ores, with Cu(II) and oxygen (air) as the oxidant. It has been suggested that the ammonia disrupts the formation of a passivating layer on the gold surface as well as catalyzing the oxidation of Cu(I) species to Cu(II). Typical conditions proposed for leaching of gold in ammonia–cyanide are pH 10.5, 0.5 to 1.5 kg/t NaCN, 1.0 to 3.0 kg/t NH_3, 20 to 50 mg/L Cu(II), and 100 to 500 mg/L Cu(I) [111, 112]. The effectiveness of gold leaching is very sensitive to solution conditions. The concentrations of cyanide, ammonia, and copper, and solution pH all have an impact on the gold leaching rate and recovery, and the concentrations of each must be optimized (not necessarily maximized) to maximize gold extraction. Unfortunately, the copper–ammonia–cyanide system is rather complex, and changes in the concentration of one species affects the concentration of the other species and pH. A key feature of this system is that, as copper dissolves from the ore, copper concentration increases to a point of equilibrium at which copper precipitates as $Cu(OH)_2$ (and also potentially as CuO/Cu_2O and CuCN, depending on the solution conditions). This results in a leach solution containing less copper than that obtained with cyanide leaching in the absence of ammonia and consequently less adverse impact on gold dissolution (Section 6.1.4.2). In one study, two copper–gold materials were leached with cyanide only (1.65 kg/t NaCN) and with a cyanide–ammonia mixture (0.55 kg/t NaCN and 2 to 3 kg/t NH_3). Gold extractions were increased from 17% and 70%, respectively, to 89% and 85%, for the two materials. Residual copper concentrations in the final leach solutions were 200 and 100 mg/L, respectively [112]. (The behavior of copper in cyanide solutions, and its effect on gold dissolution and recovery, are discussed in more detail in Sections 6.1.4.2 and 7.1.2.5.) Another key issue associated with this process is ammonia loss, which affects solution pH (and complicates solution pH control), resulting in significant ammonia consumption [48, 111]. The ammonia-cyanide leaching system has been applied at small commercial scale at the Paris Dump at Kalgoorlie (Western Australia) and at Ajkjoujt (Mauritania) [112]. The ammonia–cyanide system appears to have potential for leaching certain copper–gold bearing materials.

The use of ammonia to stabilize copper in thiosulfate media has been discussed in detail in Section 6.3, and is not considered further here.

The potential use of halides (i.e., Cl_2, Br_2, or I_2) as the oxidant in ammoniacal leach solution has been investigated since the mid-1990s. The use of halides is complicated by the fact that the halide anions (Cl^-, Br^- and I^-) form stable complexes with gold (see Section 6.7). However, it has been found that gold dissolution rates are 100 to 500 times faster in ammoniacal iodine–iodide and ammoniacal bromine–bromide solutions compared with those achieved in the absence of ammonia [92, 113]. However, such systems appear to have limited potential for commercial application for the reasons discussed in Section 6.7.

6.7 OTHER LIXIVIANTS

Other halide systems, such as bromine–bromide, iodine–iodide, and bromine–chloride, are capable of dissolving gold at very fast rates, as predicted by the electrode potentials of the relevant reduction reactions, for example:

$$Au + 2Br^- + Br_2 \rightleftharpoons AuBr_4^- + e; \quad E^0 = +0.95 \text{ (V)} \quad \text{(EQ 6.70)}$$

$$Au + 2I^- + I_2 \rightleftharpoons AuI_4^- + e; \quad E^0 = +0.69 \text{ (V)} \quad \text{(EQ 6.71)}$$

These systems are strongly oxidizing, and dissolution rates are typically several orders of magnitude faster than those achieved with cyanide and oxygen under ambient conditions [94]. In addition, they are capable of dissolving many sulfide minerals, and bromine-chloride solution has been used commercially on a small scale for leaching of refractory gold-bearing materials. The dissolutions rate of gold in halide media is strongly dependent on the concentration of complexant and oxidant, and can be increased significantly at elevated temperatures (e.g., 150°C to 180°C) [111].

Unfortunately, the commercial application of bromine and iodine solutions for gold leaching is restricted by the high cost of the reagents, the high cost of materials of construction to withstand the severe process conditions (as with chlorine [Section 6.2]), and industrial hygiene and health issues associated with their use. Any process developed must be able to effectively regenerate the reagent to make it economic. One possible means of achieving this is by electrolytically regenerating the bromine or iodine from solution, with the potential for simultaneous recovery of gold and other metals [114, 115].

Other leaching systems using cyanamide, cyanoform, organic nitrile, and malononitrile-related compounds for gold dissolution have been proposed; however, despite some potential advantages, there is little prospect for commercial development, and they are currently of academic interest only [94, 116].

REFERENCES

[1] Weast, R.C. 1981. *Handbook of Chemistry and Physics.* 62nd edition. Boca Raton, FL: CRC Press.

[2] Smith, R.M., and A.E. Martell. 1976. *Critical Stability Constants.* Volume 4. New York: Plenum Press.

[3] Huiatt, J.L., J.E. Kerrigan, F.A. Olson, and G.L. Potter. 1983. Cyanide from mineral processing. *Proceedings of Workshop Sponsored by National Science Foundation, USBM & Industry.* Salt Lake City, UT: Utah Mining and Mineral Resources Research Institute.

[4] Bellomo, A. 1970. Formation of copper(II), zinc(II), silver(II) and lead(II) ferrocyanide. *Talanta* 7:1109–1114.

[5] Adams, M.D. 2001. A methodology for determining the deportment of cyanide losses in gold plants. *Minerals Engineering* 14(4):383–390.

[6] Finkelstein, N.P. 1972. The chemistry of the extraction of gold from its ores. Pages 284–351 in *Gold Metallurgy on the Witwatersrand.* Edited by R.J. Adamson. Cape Town, South Africa: Cape & Transvaal Printers Ltd.

[7] Adams, M.D. 1990. The chemical behaviour of cyanide in the extraction of gold. 1. Kinetics of cyanide loss in the presence and absence of activated carbon. *Journal of South African Institute of Mining and Metallurgy* 90(2):37–44.

[8] Wang, X., and K.S.E. Forssberg. 1990. The chemistry of cyanide–metal complexes in relation to hydrometallurgical processes of precious metals. *Mineral Processing & Extractive Metallurgy Reviews* 6:81–125.

[9] Dean, J.A., editor. 1985. *Lange's Handbook of Chemistry.* 13th edition. New York: McGraw-Hill.

[10] Sillen, L.G., and A.E. Martell. 1964. *Stability Constants of Metal-Ion Complexes.* Publication No. 17. London: The Chemical Society.

[11] Sillen, L.G., and A.E. Martell. 1970. *Stability Constants of Metal-Ion Complexes.* Supplement No. 1. Publication No. 25. London. The Chemical Society.

[12] Sharpe, A.G. 1976. *The Chemistry of Cyano-Complexes of the Transition Metals.* London: Academic Press.

[13] Perrin, D.D. 1964. Pages 56–66 in *Organic Complexing Reagents: Structures, Behaviour and Application to Inorganic Analysis.* New York: Wiley Interscience.

[14] Capone, S., A. Robertis, and S. Sammartano. 1986. Studies on hexacyano-ferrate(II) complexes: Formation constants for alkali metals. *Thermochemica Acta* 112(3):1–14.

[15] Hogfeldt, E. 1982. *Stability Constants of Metal-Ion Complexes. Part A: Inorganic Ligands.* No. 21. New York: Pergamon Press.

[16] Nicol, M.J. 1980. The anodic behaviour of gold. Part II. Oxidation in alkaline solutions. *Gold Bulletin* 13:105–111.

[17] Nicol, M.J., C.A. Fleming, and R.L. Paul. 1987. The chemistry of the extraction of gold. Pages 831–905 in *The Extractive Metallurgy of Gold*. Monograph M7. Edited by G.G. Stanley. Johannesburg: South African Institute of Mining and Metallurgy.

[18] Jeffrey, M., and I. Ritchie. 2000. Electrochemical aspects of gold cyanidation. Pages 176–186 in *Proceedings of Electrochemistry in Mineral & Metal Processing V*. Edited by R. Woods and F.M. Doyle. Pennington, NJ: The Electrochemical Society.

[19] Jeffrey, M.I., I. Chandra, I.M. Ritchie, G.A. Hope, K. Waling, and R. Woods. 2005. Innovations in gold leaching research and development. Pages 207–221 in *Innovations in Natural Resource Processing: Proceedings of Jan D. Miller Symposium*. Edited by C.A. Young, J.J. Kellar, M.L. Free, J. Drelich, and R.P. King. Littleton, CO: SME.

[20] Habashi, F. 1966. The theory of cyanidation. *Transactions of the Mineralogical Society of AIME* 235:236–239.

[21] Zurilla, R.W., R.K. Sen, and E.Yeager. 1978. The kinetics of oxygen reduction reaction on gold in alkaline solution. *Journal of Electrochemical Society* 125:1103–1109.

[22] Kirk, D.W., F.R. Foulkes, and W.F. Graydon. 1978. A study of anodic dissolution of gold in aqueous alkaline cyanide. *Journal of Electrochemical Society* 125:1436–1443.

[23] Ball, S.P., A.J. Monhemius, and P.J. Wyborn. 1989. The use of inorganic peroxides as accelerators for gold heap leaching. Pages 149–164 in *Precious Metals '89*. Edited by M.C. Jha and S.D. Hill. Warrendale, PA: TMS.

[24] Guzman, L., M. Segarra, J.M. Chimenos, M.A. Fernandez, and F. Espiell. 1999. Gold cyanidation using hydrogen peroxide. *Hydrometallurgy* 52:21–35.

[25] Lee, V., P. Robinson, and F. Merz. 1989. Peroxide addition improves gold recovery and saves reagents at Pine Creek Gold Mine. Pages 170–182 in *Proceedings Randol Gold Conference*. Golden, CO: Randol International Ltd.

[26] Heath, A.R., and J.A. Rumball. 1998. Optimizing cyanide:oxygen ratios in gold CIP/CIL circuits. *Minerals Engineering* 11(11):999–1010.

[27] Lorenzen, L., and J.S.J. van Deventer. 1992. Electrochemical interactions between gold and its associated minerals during cyanidation. *Hydrometallurgy* 30:177–104.

[28] Kudryk, V., and H.H. Kellogg. 1954. The mechanism and rate controlling factors in the dissolution of gold in cyanide solution. *Journal of Metals* 6(5):541–548.

[29] Hiskey, J.B., and V.M. Sanchez. 1990. Mechanistic and kinetic aspects of silver dissolution in cyanide solutions. *Journal of Applied Electrochemistry* 20:479–487.

[30] Barsky, G., S.J. Swainson, and N. Hedley. 1934. Dissolution of gold and silver in cyanide solution. *Transactions A of the Institution of Mining and Metallurgy* 112:660–677.

[31] Cathro, K.J., and D.F.A. Koch. 1964. The anodic dissolution of gold in cyanide solutions. *Journal of Electrochemical Society* 111:1416–1420.

[32] Cathro, K.J., and A. Walkley. 1961. *The Cyanidation of Gold*. CSIRO Publication Melbourne, Australia: CSIRO.

[33] Cathro, K.J. 1963. The effect of oxygen in the cyanide process for gold recovery. Pages 181–205 in *Proceedings Australasian Institute of Mining & Metallurgy*. Carlton, Victoria, Australia: Australasian Institute of Mining & Metallurgy.

[34] Norris, R.D., R.A. Brown, and F.E. Carpreso. 1983. The use of peroxygen chemicals in the heap leaching of gold and silver ores. AIME Preprint No. 83-26. Littleton, CO: SME-AIME.

[35] Caruso, S.G. 1975. *The Chemistry of Cyanide Compounds and Their Behaviour in the Aquatic Environment.* Pittsburgh, PA: Carnegie Mellon Institute of Research.

[36] Perry, R., R.E. Browner, R. Dunne, and N. Stoitis. 1999. Low pH cyanidation of gold. *Minerals Engineering* 12(12):1431–1440.

[37] Wadsworth, M.E., and X. Zhu. 2005. Kinetics of enhanced gold dissolution: Activation by dissolved lead. Pages 3–20 in *Innovations in Natural Resource Processing: Proceedings of Jan D. Miller Symposium.* Edited by C.A. Young, J.J. Kellar, M.L. Free, J. Drelich, and R.P. King. Littleton, CO: SME.

[38] Sandenbergh, R.E., H. Hutchison, T. von Molke, and K.P. Cerovic. 2003. The influence of lead additions on the anodic behaviour of gold in alkaline aqueous cyanide. Pages 1222–1230 in *Proceedings of XXII International Mineral Processing Congress.* Edited by L. Lorenzen and D.J. Bradshaw. Johannesburg, South Africa: South African Institute of Mining and Metallurgy.

[39] Mussati, D., J. Mager, and G.P. Martins. 1997 Electrochemical aspects of the dissolution of gold in cyanide electrolytes containing lead. Pages 245–265 in *Aqueous Electrotechnologies: Progress in Theory and Practice.* Edited by D.H. Dreisinger. Warrendale, PA: TMS.

[40] Lapidus, G. 1995. Unsteady state model for gold cyanidation on a rotating disc electrode. *Hydrometallurgy* 39:251–263.

[41] Tavani, S., R.E. Browner, and R. Dunne. 2000. Gold cyanidation rates in the presence of sulfide and lead ions. Pages 309–314 in *Proceedings Randol Gold & Silver Forum 2000.* Golden, CO: Randol International Ltd.

[42] Aghamirian, M.M., and W.T. Yen. 2005. Mechanisms of galvanic interactions between gold and sulfide minerals in cyanide solution. *Minerals Engineering* 18: 393–407.

[43] Osseo-Asare, K., T. Xue, and V.S.T. Ciminelli. 1984. Solution chemistry of cyanide leaching systems. Pages 173–197 in *Precious Metals: Mining, Extraction and Processing.* Edited by V. Kudryk, D.A. Corrigan, and W.W. Liang. Warrendale, PA: TMS.

[44] Wilder, A.L., and S.N. Dixon. 1988. Heap leach solution application at Coeur-Rochester. AIME Preprint No. 88-19. Littleton, CO: SME.

[45] Hedley, N., and H. Tabachnick. 1958. Chemistry of cyanidation. *Mineral Dressing Notes* 23. New York: American Cyanamid Company.

[46] Byerley, J.J., K. Enns, C.V. Trang, and V.T. Lee. 1988. A treatment strategy for mixed cyanide effluents-precipitation of copper and nickel. Pages 331–340 in *Proceedings Randol Gold Forum.* Golden, CO: Randol International Ltd.

[47] Hamilton, E.M. 1920. *Manual of Cyanidation.* New York: McGraw-Hill.

[48] Muir, D., and S. LaBrooy. 1993. Why ammonia–cyanide mixtures are better than either cyanide or ammonia for leaching copper–gold ores. Pages 255–260 in *Proceedings Randol Gold Forum Beaver Creek '93.* Golden, CO: Randol International Ltd.

[49] Guo, H., G. Deschenes, A. Pratt, M. Fulton, and R. Lastra. 2005. Leaching kinetics and mechanisms during cyanidation of gold in the presence of pyrite or stibnite. *Mineral & Metallurgical Processing Journal* 22(2):89–95.

[50] Rees, K.L., and J.S.J. van Deventer. 2000. Preg-robbing phenomena in the cyanidation of sulfide gold ores. *Hydrometallurgy* 58:61–80.

[51] Guo, H., G. Deschenes, A. Pratt, M. Fulton, and R. Lastra. 2005. Leaching kinetics and mechanisms of surface reactions during cyanidation of gold in the presence of pyrite or stibnite. *Mineral and Metallurgical Processing Journal* 22(2):89–95.

[52] Ubaldini, S., F. Veglio, P. Fornari, and C. Abbruzzese. 2000. Process flow sheet for gold and antimony recovery from stibnite. *Hydrometallurgy* 57:187–199.

[53] Plaskin, I.N., and M.D. Ivanovsky. 1958. A study of the effect of some components of the liquid phase on the rate of solution of gold and silver in cyanide solution. *Met. i. Metalloved* (Moscow) 182–188.

[54] Fink, C.G., and G.L. Putnam. 1950. The action of sulphide ion and of metal salts on the dissolution of gold in cyanide solutions. *Transactions of American Institute of Mining & Metallurgical Engineers* 187:952–955.

[55] Jackman, I., and K. Sarbutt. 1990. The recovery of gold from a telluride concentrate. Pages 55–58 in *Proceedings Randol Gold Forum*. Golden, CO: Randol International Ltd.

[56] Bhakta, P. 2003. Ammonium thiosulfate heap leaching. Pages 259–268 in *Hydrometallurgy 2003: Proceedings of 5th International Symposium*. Edited by C.A. Young, A. Alfantazi, C. Anderson, A. James, D. Dreisinger, and B. Harris. Warrendale, PA: TMS.

[57] Cornwall, W.G., and R.J. Hisshion. 1976. Leaching of telluride concentrates for gold, silver and tellurium—Emperor process. *Transactions of the Metallurgical Society of AIME* 260:108–112.

[58] Liddell, K.S., and R.C. Dunne. 1989. The recovery of gold–refractory telluride concentrates by the Metprotech fine milling process. Pages. 349–358 in *Proceedings Randol Gold Conference*. Golden, CO: Randol International Ltd.

[59] Jayasekera, S., I.M. Ritchie, and J. Avraamides. 1988. Electrochemical aspects of the leaching of gold telluride. Pages 187–189 in *Proceedings Randol Perth International Gold Conference*. Golden, CO: Randol International Ltd.

[60] Urban, M.R., J. Urban, and P.J.D. Lloyd. 1973. The adsorption of gold from cyanide solutions onto constituents of the reef, and its role in reducing the efficiency process. Research Report No. 32/73. Johannesburg, South Africa: Chamber of Mines of South Africa.

[61] Hausen, D.M., and C.H. Bucknam. 1985. Study of preg-robbing in the cyanidation of carbonaceous gold ores from Carlin, Nevada. Pages 833–856 in *Applied Mineralogy: Proceedings of 2nd International Congress on Applied Mineralogy*. Warrendale, PA: TMS.

[62] Bosch, D.W. 1987. Retreatment of residues and waste rock. Pages 707–744 in *The Extractive Metallurgy of Gold in South Africa*. Edited by G.G. Stanley. Monograph M7. Johannesburg: South African Institute of Mining and Metallurgy.

[63] Mansanti, J.G., J.R. Arnold, J.H. Gourdie, and J.O. Marsden. 1989. Double thickener circuit at Gold Fields' Chimney Creek. *Mineral and Metallurgical Processing Journal* (November):179–185.

[64] Thorndycraft, R.B. 1982. Pinson Mining Company—Mill Design. AIME Preprint No. 82-162. Littleton, CO: SME-AIME.

[65] Rees, K.L., and J.S.J. van Deventer. 2000. The mechanism of enhanced gold extraction from ores in the presence of activated carbon. *Hydrometallurgy* 58:151–167.

[66] Arnold, J.R., J.M. Keane, V.G. Loftus, and J.K. Ahlness. 1990. Gold heap leach information exchange. Pages 283–311 in *Gold '90. Proceedings of Gold 1990 Symposium*. Edited by D.M. Hausen, D.N. Halbe, E.U. Petersen, and W.J. Tafuri. Littleton, CO: SME.

[67] Pizzaro, R., J.D. McBeth, and G.M Potter. 1974. Heap leaching practice at the Carlin Gold Mining Co., Carlin, Nevada. Paper presented at Annual American Institute of Mining, Metallurgical, and Petroleum Engineers Meeting, Dallas, TX, February 23–28.

[68] Marsden, J.O., L.C. Todd, and R. Moritz. 1995. Effect of lift height, overall heap height and climate on heap leaching efficiency. Pages 272–283 in *Proceedings Randol Gold Forum Perth '95*. Golden, CO: Randol International Ltd.

[69] East, D.R., J.P. Haile, and R.V. Beck 1987. Optimization technology for leach pad liner selection. *Geotechnical Aspects of Heap Leach Design*. Littleton, CO: SME.

[70] Strachan, C., and D. Van Zyl. 1988. Leach pads and liners. Pages 176–202 in *Introduction to Evaluation, Design and Aeration of Precious Metal Heap Leaching Projects*. Edited by D.J.A. Van Zyl, I.P.G. Hutchison, and J.E. Kiel. New York: American Institute of Mining, Metallurgical, and Petroleum Engineers.

[71] Heinen, H.J., G.E. McClelland, and R.E. Lindstrom. 1979. Enhancing percolation rates in heap leaching of gold–silver ores. U.S. Bureau of Mines Report of Investigations No. 8388. Salt Lake City, UT: U.S. Department of Interior.

[72] McClelland, G.E., and J.A. Eisele. 1982. Improvements in heap leaching to recover silver and gold from low-grade resources. U.S. Bureau of Mines Report of Investigations 8612. U.S. Department of Interior.

[73] Menne, D. 1984. Heap leaching. Gold mining, metallurgy and geology. Pages 229–244 in *Proceedings of Perth and Kalgoorlie Branches Regional Conference*. Carlton, Victoria, Australia: Australasian Institute of Mining and Metallurgy.

[74] Pennstrom, W.J., and J.R. Arnold. 1999. Optimizing heap leach solution balances for enhanced performance. *Mineral & Metallurgical Processing Journal* 16:(1):12–17.

[75] Irish, M.G., M.J. Jahraus, P.R. Taylor, and E.E. Vidal. 2004. On-line diagnostic of gold leaching parameters in a laboratory scale simulation of a heap leach system. SME Preprint No. 04–64. Littleton, CO: SME.

[76] Dewhirst, R.F., S.P. Moult, and J.A. Coetzee. 1983. Intensive cyanidation for coarse gold recovery. *Journal of South African Institute of Mining and Metallurgy* (May):111–117.

[77] Davidson, R.J., G.A. Brown, C.G. Schmidt, N.W. Hanf, D. Duncanson, and J.D. Taylor. 1978. The intensive cyanidation of gold plant gravity concentrates. *Journal of South African Institute of Mining and Metallurgy* (January):146–165.

[78] Lethlean, W. 2000. Leaching gravity concentrates using the Acacia reactor. Pages 93–100 in *Proceedings Randol Gold & Silver Forum*. Golden, CO: Randol International Ltd.

[79] Campbell, J., and B. Watson. 2003. Gravity leaching with the ConSep ACACIA Reactor—Results from AngloGold Union Reefs. Pages 167–175 in *Proceedings of 8th Mill Operators' Conference*. Melbourne: Australasian Institute of Mining and Metallurgy.

[80] Watson, B., and G. Steward. 2002. Gravity leaching—The ACACIA Reactor. Pages 383–390 in *Proceedings of Metallurgical Plant Design & Operating Strategies*. Melbourne: Australasian Institute of Mining and Metallurgy.

[81] Chamberlain, P.D. 1989. Status of heap, dump and in-situ leaching of gold and silver. Pages 225–232 in *Proceedings of World Gold '89: 1st Joint Meeting of Society of Mining Engineers and Australasian Institute of Mining & Metallurgy*. Edited by R.B. Bhappu and R.J. Harden. Littleton, CO: SME-AIME.

[82] Putnam, G.L. 1944. Chlorine as a solvent in gold hydrometallurgy. *Engineering and Mining Journal* 145(3):70–75.

[83] Nicol, M.J. 1980. The anodic dissolution of gold. Part I. Oxidation in acidic solutions. *Gold Bulletin* 13:46–55.

[84] Welham, N.J., and G.H. Kelsall. 2000. Recovery of gold from pyrite (FeS_2) by aqueous chlorination I. Oxidation rates of gold and pyrite in aqueous chlorine solutions. Pages 141–148 in *Proceedings of Electrochemistry in Mineral & Metal Processing V*. Edited by R. Woods and F.M. Doyle. Pennington, NJ: The Electrochemical Society.

[85] Huff, R.V., D.R. Baughman, and P.D. Chamberlain. 1999. NaCl–NaOCl lixiviant for precious metals. Pages 163–168 in *Proceedings Randol Gold & Silver Forum '99*. Golden, CO: Randol International Ltd.

[86] Kappes, D.W. 1998. The Itos chloride leach plant for silver—Two years into production. Pages 247–248 in *Proceedings Randol Gold & Silver Forum '98*. Golden, CO: Randol International Ltd.

[87] Huff, R.V., D.R. Baughman, and P.D. Chamberlain. 1999. NaCl–NaOCl lixiviant for precious metals. Pages 163–168 in *Proceedings Randol Gold & Silver Forum*. Golden, CO: Randol International Ltd.

[88] Aylmore, M.G., and D.M. Muir. 2001. Thermodynamic analysis of gold leaching by ammoniacal thiosulfate using E_h/pH and speciation diagrams. *Mineral & Metallurgical Processing Journal* 18(4):221–227.

[89] Ferron, C.J., C.A. Fleming, D. Dreisinger, and T. O'Kane. 2003. Chloride as an alternative to cyanide for the extraction of gold—going full circle? Pages 89–103 in *Hydrometallurgy 2003: Proceedings 5th International Symposium Honoring Prof. I.M. Ritchie*. Edited by C.A. Young, A. Alfantazi, C. Anderson, A. James, D. Dreisinger, and B. Harris. Warrendale, PA: TMS.

[90] Breuer, P.L., and M.I. Jeffrey. 2003. A review of the chemistry, electrochemistry and kinetics of the gold thiosulfate leaching process. Pages 139–154 in *Hydrometallurgy 2003: Proceedings 5th International Symposium Honoring Prof. I.M. Ritchie*. Edited by C.A. Young, A. Alfantazi, C. Anderson, A. James, D. Dreisinger, and B. Harris. Warrendale, PA: TMS.

[91] Senayake, G., W.N. Perera, and M.J. Nicol. 2003. Thermodynamic studies of the gold(III)/(I)/(0) redox system in ammonia-thiosulfate solutions at 25°C. Pages 155–168 in *Hydrometallurgy 2003: Proceedings 5th International Symposium Honoring Prof. I.M. Ritchie*. Edited by C.A. Young, A. Alfantazi, C. Anderson, A. James, D. Dreisinger, and B. Harris. Warrendale, PA: TMS.

[92] Peri, K., G. Yuan, and K.N. Han. 1999. Dissolution behavior of gold in ammoniacal solutions with iodine as an oxidant. SME Preprint No. 99–117. Littleton, CO: SME.

[93] Lam, A.E., and D.B. Dreisinger. 2003. The importance of the Cu(II) catalyst in the thiosulfate leaching of gold. Pages 195–211 in *Hydrometallurgy 2003: Proceedings of 5th International Symposium Honoring Prof. I.M. Ritchie*. Edited by C.A. Young, A. Alfantazi, C. Anderson, A. James, D. Dreisinger, and B. Harris. Warrendale, PA: TMS.

[94] Hiskey, J.B., and V.P. Atluri. 1988. Dissolution chemistry of gold and silver in different lixiviants. *Mineral Processing & Extractive Metallurgy Review* 4:95–134.

[95] Zipperian, D., S. Raghavan, and J.P. Wilson. 1986. Thiosulphate technology for precious metal recovery. Paper presented at 115th American Institute of Mining, Metallurgical, and Petroleum Engineers Convention, New Orleans, LA, March 2–6.

[96] Brown, T., A. Fischmann, L. Spiccia, and D.C. McPhail. 2003. Alternative copper(II) catalysts for gold leaching: Use of multidentate ligands to control thiosulfate oxidation. Pages 213–226 in *Hydrometallurgy 2003: Proceedings of 5th International Symposium Honoring Prof. I.M. Ritchie.* Edited by C.A. Young, A. Alfantazi, C. Anderson, A. James, D. Dreisinger, and B. Harris. Warrendale, PA: TMS.

[97] Ji, J., C.A. Fleming, P.G. West-Sells, and R.P. Hackl. 2003. A novel thiosulfate system for leaching gold without the use of copper and ammonia. Pages 227–244 in *Hydrometallurgy 2003: Proceedings 5th International Symposium Honoring Prof. I.M. Ritchie.* Edited by C.A. Young, A. Alfantazi, C. Anderson, A. James, D. Dreisinger, and B. Harris. Warrendale, PA: TMS.

[98] Wan, R.Y., and M. LeVier. 2003. Solution chemistry factors for gold thiosulfate heap leaching. *International Journal of Mineral Processing* 72(1–4):312–322.

[99] Hisshion, R.J., and C.G. Waller. 1984. Recovering gold with thiourea. *Mining Magazine* (September):237–243.

[100] Raudsepp, R., and R. Allgood. 1987. Thiourea leaching of gold in a continuous pilot plant. Pages 87–96 in *Proceedings of International Symposium on Gold Metallurgy.* Edited by R.S. Salter, D.M. Wyslouzil, and G.W. McDonald. New York: Pergamon Press.

[101] Marchant, P.B., L.M. Broughton, and M.J. Lake. 1988. Comparative analysis of cyanidation and acidothioureation. Pages 171–178 in *Proceedings Randol Gold Forum.* Golden, CO: Randol International Ltd.

[102] Groenewald, T. 1977. Potential applications of thiourea in the processing of gold. *Journal of South African Institute of Mining and Metallurgy* 77:217–223.

[103] Barbosa, O.F., and A.J. Monhemius. 1988. Thermochemistry of thiocyanate systems for leaching gold and silver ores. Pages 307–339 in *Precious Metals '89.* Edited by M.C. Jha and S.D. Hill. Warrendale, PA: TMS.

[104] Hiskey, J.B. 1984. Thiourea leaching of gold and silver—technology update and additional applications. *Mineral & Metallurgical Processing Journal* (November):173–178.

[105] Li, J., and J.D. Miller. 1999. Reaction kinetics for gold dissolution in acid thiourea solution using formamidine disulfide as oxidant. SME Preprint No. 99-62. Littleton, CO: SME.

[106] Deng, T., and M. Liao. 2002. Gold recovery enhancement from a refractory flotation concentrate by sequential bioleaching and thiourea leach. *Hydrometallurgy* 63:249–255.

[107] Zhang, Y.Z. 1996. Evaluation of thiourea consumption for gold extraction from complex and refractory gold ores. Pages 443–460 in *Proceedings EPD Congress 1996.* Edited by G.W. Warren. Warrendale, PA: TMS.

[108] Wan, R.Y., J.A Brierley, and K.M. LeVier. 2003. Using thiocyanate as lixiviant for gold recovery in an acidic environment. Pages 105–121 in *Hydrometallurgy 2003: Proceedings 5th International Symposium Honoring Prof. I.M. Ritchie*. Edited by C.A. Young, A. Alfantazi, C. Anderson, A. James, D. Dreisinger, and B. Harris. Warrendale, PA: TMS.

[109] Han, K.N., and X. Meng 1992. Ammonia extraction of gold and silver from ores and other materials. U.S. Patent No. 5,114,687. May 19.

[110] Han, K.N., and X. Meng. 1993. The dissolution behaviour of gold in ammoniacal solutions. Pages 205–221 in *Proceedings 4th International Symposium on Hydrometallurgy*. Littleton, CO: SME.

[111] Muir, D., S. Vukcevic, and J. Shuttleworth. 1995. Optimizing the ammonia–cyanide leaching process for copper–gold ores. Pages. 225–229 in *Proceedings Randol Gold Forum Perth '95*. Golden, CO: Randol International Ltd.

[112] Vukcevic, S. 1996. A comparison of alkali and acid methods for the extraction of gold from low grade ores. *Minerals Engineering* 9(10):1033–1047.

[113] Kim, P.N., and K.N. Han. 2005. Leaching behaviour of gold in ammoniacal solutions in the presence of bromine as an oxidant. Pages 261–280 in *Innovations in Natural Resource Processing: Proceedings of Jan D. Miller Symposium*. Edited by C.A. Young, J.J. Kellar, M.L. Free, J. Drelich, and R.P. King. Littleton, CO: SME.

[114] Dadgar, A., Sanders, B.M., J.A. McKeown, R.H. Sergent, and R.H. Jacobson. 1990. Leaching and recovery of gold from black sand concentrate and electrochemical regeneration of bromine. Pages 75–90 in *Advances in Gold and Silver Processing. Proceedings of GOLDTech 4 Symposium*. Edited by M.C. Fuerstenau. New York: American Institute of Mining, Metallurgical, and Petroleum Engineers.

[115] Hiskey, J.B., and P.H. Qi. 1990. Leaching and electrochemical behavior of gold in iodine solutions. Part I: Dissolution kinetics. SME Preprint No. 90-112. Littleton, CO: SME.

[116] Woodcock, J.T. 1988. Innovations and options in gold metallurgy. Pages 115–131 in *Proceedings of XVI International Mineral Processing Congress, Stockholm, Sweden*. Edited by E.K.S. Forssberg. Amsterdam: Elsevier.

[117] Schnabel, C. 1921. Gold. Page 1171 in *Handbook of Metallurgy*. 3rd edition. London: Macmillan and Co. Ltd.

CHAPTER 7

Solution Purification and Concentration

The processes considered in this chapter selectively concentrate dilute (e.g., 1 to 20 g Au/t) gold-bearing solutions to produce a higher-grade solution from which gold can be extracted most efficiently by the recovery methods described in Chapter 8. Gold is first adsorbed from leach solutions onto an extractant, such as activated carbon or a synthetic ion exchange resin. The loaded extractant is then separated from the process stream, and the gold values are desorbed into a smaller volume of solution suitable for metal recovery. The stripped extractant is regenerated, if necessary, and then reused in the process. Activated carbon is the most widely used extractant for this purpose. Alternatively, ion exchange resins have been used in some applications and continue to be developed. Both extractants can be used to treat leach slurries directly, called in-pulp processing, as well as unclarified and clarified solutions, thereby obviating the need for solid–liquid separation steps required in conventional flowsheets.

Liquid solvents have also been investigated for use in gold extraction and are considered briefly.

7.1 CARBON ADSORPTION

7.1.1 Properties of Activated Carbon

Activated carbon, or charcoal as it is now less commonly called, is an organic material which has an essentially graphitic structure. Due to a highly developed internal pore structure, it has an extremely large specific surface area, and values in excess of 1,000 m^2/g are not uncommon. As a result, activated carbon has found diverse industrial applications in both gas and liquid separation processes; however, its use in the gold recovery industry has only been widespread since about 1980 (Chapter 1).

The most important properties of activated carbon for use in gold extraction are the following:

- Adsorptive capacity
- Adsorption rate
- Mechanical strength and wear resistance
- Reactivation characteristics
- Particle size distribution

Other nontechnical considerations such as cost, availability, and service by supplier also affect carbon selection.

The properties listed here are considered in more detail in Section 7.1.2; however, it is first necessary to consider the factors that determine these properties, namely the type of source material and the methods of manufacture and activation.

7.1.1.1 Manufacture and Activation

If properly treated, virtually any carbonaceous material can be used to produce activated carbon. The most commonly used source materials are wood, peat, coconut shells, bituminous coal, anthracite, and fruit pips. The type of source material has a marked influence on the physical structure of the product, in particular, the pore volume and particle size distribution. For example, wood is used as the source material for decolorizing carbons, whereas coconut shells and coal-based carbons are generally used for gas phase adsorption and gold recovery applications. By the 1980s, some activated carbon was being manufactured by extruding powdered and agglomerated peat- or coal-based source materials to produce similar sized pellets.

Carbon is activated by removing hydrogen, or hydrogen-rich fractions, from a carbonaceous raw material to produce an open, porous residue. This process is called activation, and is typically achieved in two stages (Figure 7.1).

In the first stage, the material is heated to approximately 500°C in the presence of dehydrating agents, a process called carbonization. Many of the impurities are removed as gases (e.g., carbon monoxide, carbon dioxide, or acetic acid) or remain as a tar-like residue on the carbon. As a consequence, carbon atoms are freed to some extent and group together as crystallographic formations, known as elementary crystallites. This results in the development of a product with a specific surface area between 10 and 500 m^2/g (and sometimes as high as 1,000 m^2/g), the majority of which is due to micropore formation [2, 3].

The second stage consists of exposing the carbonized material to an oxidizing atmosphere of steam, carbon dioxide, and/or oxygen (air) at temperatures of 700 to 1,000°C to burn off the tar-like residues and to develop the internal pore structure. Further reaction results in partial or complete burnout of carbon layers, producing a widening of existing pores and exposing the surfaces of the elementary crystallites formed during carbonization. Carbon atoms at the edges and corners of the elementary crystallites, and at defects or discontinuities, are especially reactive due to their unsaturated valencies and are called active sites.

The reaction of steam with carbon is thought to first involve the adsorption of water vapor onto the carbon surface, followed by the evolution of hydrogen and carbon monoxide. The reactions that follow are postulated, where a square bracket (]) indicates bonding with the carbon surface [4, 5].

$$]C + H_2O \rightleftharpoons]C(H_2O) + CO \rightleftharpoons]C(O) + H_2$$
$$\parallel$$
$$CO \qquad (EQ\ 7.1)$$

This mechanism is represented by the overall reaction:

$$C + H_2O \rightleftharpoons H_2 + CO \qquad \Delta G^0 = -130\ kJ/mol \qquad (EQ\ 7.2)$$

This process is sometimes applied during carbon reactivation in gold extraction plants by the addition of steam into reactivation kilns (see Sections 7.1.4.4 and 7.1.5.7). The activation of carbon is catalyzed by iron, copper, and oxides and carbonates of the alkali metals.

Although coconut shell carbons are the most commonly used activated carbons in the gold extraction industry, extruded (peat-based) carbon has been used increasingly, and there is potential for other source materials to provide alternative carbons in the future, for example, peach and apricot pips and sugar cane residue [6, 7].

In the early 2000s, magnetic activated carbons (MACs) were produced by mixing a magnetic precursor material (e.g., iron citrate) with a suitable carbon source (such as

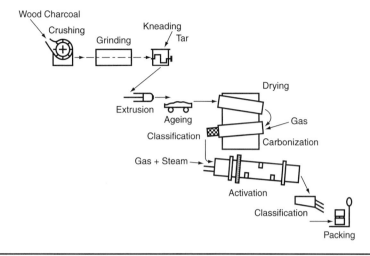

FIGURE 7.1 Flowsheet for production of active carbon by activation with steam [1]

pinewood) and heat treating the mixture under controlled conditions of temperature and gas phase composition. Investigation and evaluation of activated carbon produced in this way have indicated the presence of micropores as well as mesopores during the early stages of activation [8]. MACs show potential for future application and optimization of carbon adsorption systems, with the potential for faster adsorption kinetics and the ability to separate carbon from solution or slurry by wet, high-intensity magnetic separation.

7.1.1.2 Physical Properties

Activated carbon has a similar, though less well-ordered, structure to that of graphite. X-ray studies have suggested that activated carbon has two basic structures [4]:

- Small regions of elementary crystallites, composed of roughly parallel layers of hexagonally ordered atoms
- Disordered, cross-linked, spaced lattice of carbon hexagons, which is more pronounced in chars formed from materials of high oxygen content.

Commonly quoted dimensions for elementary crystallites are temperature dependent, but typically vary from 9 to 12 Å high and 20 to 23 Å wide. From these values it has been estimated that the crystallite structures are approximately three layers high, with widths equivalent to the diameter of nine carbon hexagons.

The activation process generates an extremely large internal surface area, which is practically infinite relative to the outer surface of a carbon granule, and a wide range of pore sizes and shapes. Unfortunately, because it is not possible to accurately determine the shape of pores, this leads to some difficulty in expressing pore size. The classification of pore size by Dubinin [5] is generally accepted and is based on changes in gas or vapor adsorption mechanisms with pore size:

- Macropores: $x > 100$ to 200 nm
- Transitional pores or mesopores: $1.6 < x < 100$ to 200 nm
- Micropores: $x < 1.6$ nm

where x is the characteristic size. The term "supermicropore" has been used to describe the range 0.6 to 1.6 nm [9].

The main distinction between gas-adsorbing and decolorizing carbons lies in their pore size distributions (Figure 7.2). Coal-based carbon has a large number of mesopores, which is important for adsorption kinetics as they allow access to micropores. This is significant because the gold cyanide anion is relatively large and may be inaccessible to as much as 90% of the carbon micropores. However, coal-based carbons are not as mechanically strong as coconut shell carbons and are less suitable for gold extraction systems.

Activation with carbon dioxide can produce carbon with smaller pore volumes and a greater proportion of micropores. On the other hand, activation in the presence of oxygen only develops porosity to a limited extent due to pore blockage by surface oxides [9]. In many other applications, including gold cyanide adsorption, transportation of the adsorbate within the pores can be rate determining. Consequently, decreasing the particle size of granular carbon has a large effect on the rate of adsorption, despite only a small increase in net surface area.

Typical coconut shell activated carbon has an ash content of 2% to 4%, an apparent density of 420 to 450 kg/m^3, BET surface area of 900 to 1,000 m^2/g, and an iodine number of 1,100 to 1,200 mg I_2/g.

7.1.1.3 Chemical Properties

The adsorptive properties of activated carbon are not only determined by surface area but also by its chemical properties. Although these characteristics are less well understood, the activity of carbon is attributed to the effects listed:

- Disturbances in the microcrystalline structure, such as edge and dislocation effects, which result in the presence of residual carbon valencies. This affects the adsorption of both polar and polarizable species.
- The presence of chemically bonded elements such as oxygen and hydrogen in the source material or chemical bonding between the carbon and species in the activating gas. The nature of chemically bound oxygen and hydrogen is dependent upon the type of source material and the activation conditions, such as atmosphere composition and temperature.
- The presence of inorganic matter, that is, ash components and impregnation agents, which may be detrimental to adsorption or may encourage specific adsorption

To some extent these effects are interactive. For example, inorganic impurities can create disorder within the carbon lattice, resulting in the formation of defects at locations where oxygen can be preferentially adsorbed during activation.

Activated carbons have been divided into two types: H-carbons and L-carbons [11]. H-carbons are formed at temperatures >700°C, typically around 1,000°C, and are characterized by their ability to adsorb hydrogen ions when immersed in water, thereby reducing the pH in bulk solution. L-carbons are activated at temperatures <700°C, usually between 300°C and 400°C, and preferentially adsorb hydroxyl ions. Steam-activated carbons are generally used for gold recovery and have predominantly H-carbon type characteristics.

Oxygen is chemisorbed onto carbon more readily than other elements, and the C–O complexes that are formed can influence surface reactions, wettability, and electrical and catalytic properties of the carbon. Approximately 90% of oxygen on the surface is thought to be present as functional groups. The remainder exists as neutral bonds in ether bridges [12, 13].

A variety of analytical techniques have identified carboxyl, phenol, quinone, and hydroxyl, as well as ester groups (such as lactones, carboxylic anhydride, and cyclic peroxides), at the carbon surface [14, 15, 16]. The structures of some of these groups are shown in Figure 7.3. In general, oxide groups formed at low temperature appear to be carboxylates, whereas those formed at higher temperatures tend to be phenolic. The

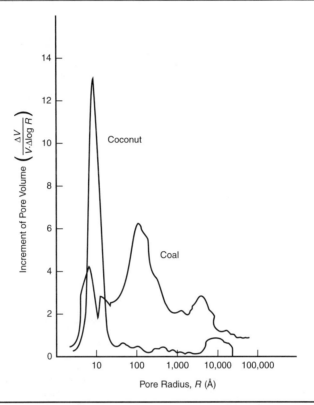

FIGURE 7.2 Pore-size distribution data for a typical thermally activated coal-based activated carbon and a coconut carbon [10]

importance of these groups lies in their ability to affect the acid–base characteristics of the surface. For example, Figure 7.4 illustrates the effect of pH on the zeta potential and acid–base adsorption characteristics of an extruded carbon. This behavior is consistent with the presence of carboxylic acid groups (pK_a = 4.8) and phenolic groups (pK_a = 9.8), that is, an oxidized surface.

Reduction of surface groups on activated carbon may be represented by the reactions [17]:

$$Q_2 + 4H_2O + 4e \rightleftharpoons 2H_2Q + 4OH^- \quad (EQ\ 7.3)$$

$$Q + 2H_2O + 2e \rightleftharpoons H_2Q + 2OH^- \quad (EQ\ 7.4)$$

where
  H_2Q = the quinhydrone group

The potential resulting from Equation (7.4) is given by:

$$E = E^0(H_2Q/Q) - 0.00259 \log [H_2Q]/[Q] - 0.059\ pH + 0.826 \quad (EQ\ 7.5)$$

where
  E^0 (H_2Q/Q) = 0.699 (V) [21]

Most commercial carbons have a reduction potential in the range of 0.1 to 0.4 V (see Figure 4.10).

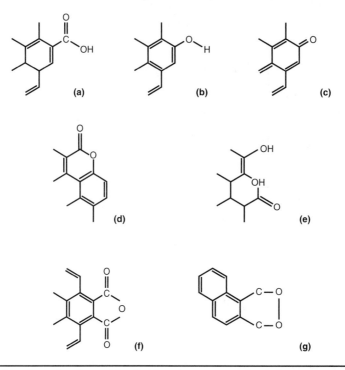

FIGURE 7.3 Structure of surface oxides that have been proposed as being present on the surface of activated carbon: (a) carboxylic acid, (b) phenolic hydroxyl, (c) quinone-type carbonyl groups, (d) normal lactones, (e) fluorescein-type lactones, (f) carboxylic acid anhydrides, and (g) cyclic peroxides [10]

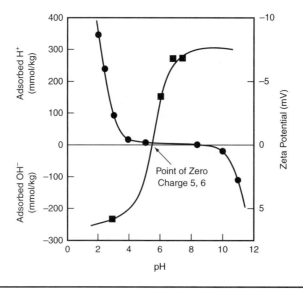

FIGURE 7.4 The adsorption of acid and base by Norit 2020 carbon and the effect of pH on charge and zeta potential [17]

7.1.2 Adsorption from Cyanide Solutions

7.1.2.1 Physical Factors Affecting Adsorption
The major physical factors affecting gold adsorption onto carbon are reviewed in the following sections.

Carbon type. Carbons produced using different methods or source materials (Section 7.1.1) have a range of chemical and physical properties, which affect the adsorption rate and loading capacity. In general, higher-activity carbons are softer, due to a more extensive pore structure that reduces the mechanical strength of the carbon. These carbons typically result in higher attrition losses in plants. Attrition losses are important, not only because they consume carbon but also because of the associated gold loss (see Section 7.1.5.8).

The type of carbon required for a particular process application depends on many factors, including the type of adsorption process (i.e., carbon-in-pulp [CIP], carbon-in-leach [CIL], or carbon-in-columns [CIC]), the gold concentration, solution or slurry flow rate, gold production rate, carbon attrition rate, and the severity of process conditions. High-activity carbons are used when high adsorption efficiency is required, either to prevent loss of soluble gold values or to improve overall circuit efficiency; that is, by achieving superior gold loadings or by improving solution equilibria to favor gold dissolution. Lower-activity carbons are used most effectively in circuits that are less susceptible to gold losses resulting from poor carbon adsorption performance and have the advantage of lower attrition losses.

The activity of carbons used for gold extraction decreases with plant usage, and reactivation techniques are commonly employed to limit the extent of this degradation (see Section 7.1.4).

Carbon particle size. Although the carbon particle size distribution has a significant effect on its external surface area, it has only a very small effect on the specific surface area because of the highly developed internal pore structure. As a result, the ultimate carbon-loading capacity is virtually independent of particle size. However, the size has a large effect on the mean pore length within the carbon particles, and the rate of adsorption increases with decreasing particle size, as illustrated in Figure 7.5. This is an important factor in industrial adsorption systems because the majority of these operate at gold loadings well below the true equilibrium loading capacity of the carbon. Particle size ranges of carbons used in industrial applications typically vary from 1.2×2.4 mm to 1.7×3.4 mm.

In practice, several other factors affect the choice of carbon particle size:

- The separation of carbon from the solution or slurry phase becomes increasingly difficult at finer sizes (typically, screening of carbon can be performed at 0.7 to 0.8 mm in most slurry applications).

- Finer carbon is more susceptible to attrition losses because of its higher ratio of surface area to mass, and generally it is reduced to a size where it can leave the plant more quickly than coarse carbon particles.

- Smaller carbon has a lower fluidization velocity than coarser carbon, which affects process equipment design (i.e., upflow carbon columns in CIC circuits, carbon elution systems, acid wash vessels, etc.).

Systems that contain carbon with a wide size distribution may experience less of a difference in gold loading with increasing size due to an effect called contact ion exchange [19]. In this readily measurable effect, gold is transferred from carbon of high gold loading to carbon of low loading—achieved through direct contact of the thin films surrounding the carbon particle, with negligible gold passing into bulk solution [20].

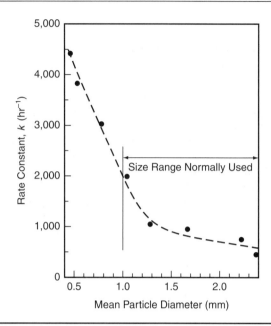

FIGURE 7.5 Rate of gold loading of carbon particles of various sizes [18]

Mixing efficiency.   Mixing conditions have an important effect on the gold adsorption rate, as illustrated in Figure 7.6. This effect is due to the fact that most carbon adsorption systems are operated at a pseudo-equilibrium, below the maximum equilibrium loading, where the adsorption rate is dependent to some extent on diffusion through the solid–liquid boundary layer (see Section 7.1.2.3). The pseudo-equilibrium is attributed to the proportion of the pores that are utilized within the operating residence time of the adsorption system. Consequently, the degree of mixing of carbon in a solution or slurry must be sufficient to:

- Keep the carbon, solution, and solids suspended and to keep the mixture as homogeneous as possible
- Maximize the mass transport rate of gold cyanide species to the surface of the carbon, preferably faster than the actual rate of adsorption at the surface

Effect of solids.   The rate of gold cyanide adsorption decreases with increasing slurry density, as illustrated in Figure 7.7 [18]. This effect is attributed to the following factors:

- Decreased mixing efficiency resulting from increased viscosity and decreased energy input per unit mass of slurry
- Physical blinding of the carbon surfaces and pores by fine ore particles
- Reduced solution–carbon ratio at higher slurry densities

The mixing efficiency can also be reduced by increased slurry viscosity caused by changes in ore type rather than as a result of a change in slurry density. Ore types that produce high viscosity slurries also have a greater tendency to impair carbon performance; for example, by blinding of carbon pores with very fine particles. This effect is particularly evident when treating clay-bearing ores.

Pulp density is affected by the density of solids, and changes in the type of material treated in adsorption systems must also be considered. For example, when considering

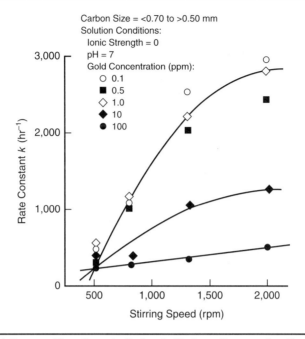

FIGURE 7.6 The influence of impeller velocity in a baffled reaction vessel on the rate of extraction of gold cyanide by Metsorb 101 activated carbon [20]

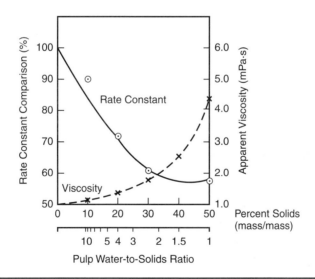

FIGURE 7.7 The effect of the pulp solid content on the rate of gold loading onto the carbon [18]

the treatment of calcine (essentially Fe_2O_3) which has a particle density of 5,000 kg/m^3 compared with a quartz-based material with a density of 2,700 kg/m^3, the following effects must be taken into account:

- Particle sedimentation rates are greater for the higher-density material, and therefore some settling and dead space may result if inadequate mixing is provided, which reduces solution and carbon mobility.

- The volume proportion of solids is lower for a specific slurry density, and therefore contact between carbon and solution is improved.

7.1.2.2 Mechanism of Gold Adsorption

The complex physical and chemical structure of activated carbons allows the adsorption of different species by various mechanisms. Consequently, the exact mechanism of adsorption of gold from cyanide solutions has been difficult to determine; however, during the 1980s a clearer picture emerged.

The adsorption mechanisms proposed prior to about 1978 can be split into four categories, as follows:

- Adsorption as the Au(I) cyanide ion
- Adsorption as molecular AuCN
- Reduction and adsorption as metallic gold
- Adsorption in association with a metal cation such as Ca^{2+}

Studies performed since about 1978 proposed a number of mechanisms that attempted to account for some well-established adsorption characteristics, most importantly [10, 21, 22, 23]:

- Extraction of $Au(CN)_2^-$ and $Ag(CN)_2^-$ is enhanced by the presence of electrolytes, such as calcium chloride and potassium chloride.
- Adsorption kinetics and equilibrium loading increase as the pH decreases.
- The adsorption of gold cyanide increases the pH of the bulk solution.
- Neutral cyanide complexes, for example, $Hg(CN)_2$, adsorb strongly and independently of ionic strength.
- Gold cyanide adsorption is a reversible process with generally faster kinetics for desorption under slightly modified conditions.
- There is some evidence that gold adsorption is dependent on the reduction potential of the system, for example:
  - Gold adsorption increases with increasing reducing power of the carbon.
  - Gold adsorption decreases if the carbon has been oxidized by chlorine or nitric acid.
- Under most conditions the molar ratio of loaded gold to nitrogen is 0.5:1.0, which is consistent with the presence of the $Au(CN)_2^-$ group.
- Gold adsorption decreases with increasing temperature.

Detailed investigations using Mossbauer spectroscopy, X-ray photoelectron spectroscopy (XPS or ESCA), and model extractants on high ionic strength solutions, typical of those obtained in actual gold leaching systems, have shown that the gold cyanide complex is adsorbed predominantly as an ion pair [22, 24]. Further evidence for this has been provided by surface chemical and other analyses, which have established that the oxidation state of gold on carbon is +1 [25]. The mechanism is best illustrated by the equation:

$$M^{n+} + nAu(CN)_2^- \rightleftharpoons M^{n+}[Au(CN)_2^-]_n \qquad (EQ\ 7.6)$$

where the ion pair, $M^{n+}[Au(CN)_2^-]_n$, is the adsorbed gold species. Detailed experimental evidence leading to this conclusion is available in the literature [10, 17, 22, 23, 24, 25].

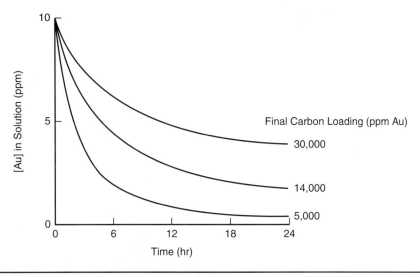

FIGURE 7.8 An example of kinetics of gold loading onto activated carbon [26]

7.1.2.3 Adsorption Kinetics and Loading Capacity

The adsorption of gold cyanide onto activated carbon is dependent upon many chemical and physical factors, which affect both the adsorption kinetics and the equilibrium gold-loading capacity. The initial rate of adsorption of gold cyanide is rapid, with adsorption occurring at the most accessible sites in macropores, and possibly mesopores, but the kinetics decrease as equilibrium is approached (Figure 7.8). Under these conditions the rate is controlled by the mass transport of gold cyanide species to the available activated carbon surfaces. However, once this adsorption capacity has been utilized, a pseudo-equilibrium is established beyond which adsorption must take place in the micropores. This requires diffusion of gold cyanide species along pores within the carbon structure, typically a much slower process than boundary layer diffusion, due to the length and tortuosity of the pores [20].

The activation energy for gold adsorption onto carbon has been estimated at 11 kJ/mol, which is well within the range expected for mass transport control [27].

The rate of gold adsorption onto carbon can be described by the first-order rate equation [28]:

$$\log C_t = mt + \log C_0 \tag{EQ 7.7}$$

where

C_t = gold concentration at time, t
C_0 = initial gold concentration
m = a rate constant which can be readily determined from a plot of log C versus time, using data obtained from simple laboratory tests

A typical equilibrium gold-loading isotherm is given in Figure 7.9. The loading capacity of carbon has traditionally been expressed as an iodine number (the mass of iodine adsorbed per gram of carbon in a 0.02 N iodine solution) or as a carbon tetrachloride number (weight percent CCl_4 loading on carbon exposed to air saturated with CCl_4 at 0°C). Both of these values provide a useful approximation of the available surface

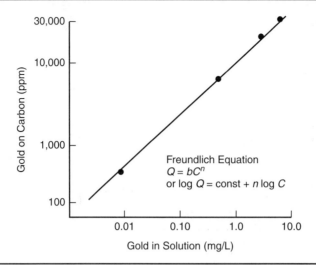

FIGURE 7.9 Equilibrium adsorption isotherm for loading of gold on carbon [26]

area for some vapor phase adsorbates, but it has been demonstrated that such estimates are poorly correlated with gold adsorption capacity [29], due to the complex combination of physical and chemical processes involved in gold adsorption from cyanide solutions. As a consequence, actual gold-loading rate data are generally of more practical use for optimizing industrial adsorption systems, particularly since true equilibrium between gold in solution and gold on carbon is never attained.

For the same reason, it is most appropriate to use an empirically developed equilibrium gold-loading capacity (K value) for the evaluation of carbons for use in gold adsorption systems. This is obtained by reacting various weights of carbon with a standard borate-buffered gold solution for a fixed time. The results are plotted as the Freundlich isotherm (Figure 7.9), and the K value is interpolated as the carbon loading in equilibrium with a residual gold solution concentration of 1 mg/L [29].

Several variations of these expressions for loading rate and capacity have been developed and applied for specific operations around the world, and further information is available in the literature [18, 28, 29, 30, 31].

7.1.2.4 Chemical Factors Affecting Adsorption Efficiency

The major chemical factors affecting the efficiency of gold adsorption onto carbon are reviewed in the following sections.

Temperature. The adsorption of gold onto carbon is exothermic, which accounts for the ability to reverse adsorption by increasing temperature [10]. Consequently, the loading capacity decreases as the temperature increases, as shown in Figure 7.10 and Table 7.1. This is exploited in the high-temperature elution of gold from loaded carbon, discussed in detail in Section 7.1.3. The adsorption rate increases slightly with increasing temperature (see Table 7.1 and Figure 7.11) due to the accelerated diffusion of gold cyanide species, following a behavior described by the Arrhenius equation (4.39).

Gold concentration in solution. The rate of gold adsorption and the equilibrium loading capacity both increase with increasing gold concentration in solution, as illustrated in Figure 7.9. Typically gold-loading rates of 10 to 100 g Au/hr/t carbon and loadings of 5 to 10 kg Au/t carbon are achieved in practice at gold concentrations produced by standard cyanide leaching processes (Chapter 6).

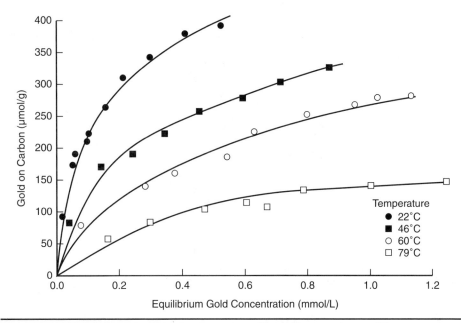

FIGURE 7.10 Equilibrium adsorption isotherms for gold cyanide on carbon at different temperatures (experimental conditions: volume of solution = 50 mL; mass of carbon = 0.25 g; adsorption medium contained 2.8 g/L $CaCl_2$ and 0.5 g/L KCN) [31]

TABLE 7.1 Effect of temperature and sodium cyanide concentration on gold loading: [Au] = 25 mg/L, pH = 10.4 to 10.8 [20]

Temperature (°C)	Free Cyanide (mg/L)	Rate Constant, k (per hr)	Gold-Loading Capacity (mg/L)
20	0	3,400	73,000
25	130	3,390	62,000
24	260	2,620	57,000
23	1,300	2,950	59,000
44	0	4,190	48,000
43	130	4,070	47,000
42	260	3,150	42,000
43	1,300	3,010	33,000
62	0	4,900	35,000
62	130	4,920	29,000
62	260	3,900	29,000
62	1,300	4,060	26,000
81.5	260	5,330	20,000

Cyanide concentration. Both the loading rate and capacity of gold on carbon decrease with increasing free cyanide concentration. This is illustrated in Table 7.1, which shows data for tests at constant ionic strength, and the effect is attributed to increased competition of free cyanide species for adsorption sites on the carbon [20]. However, the selectivity of activated carbon for gold over other metal cyanide species increases with increasing cyanide concentration, as exploited in the treatment of high copper ores (see Section 7.1.2.5).

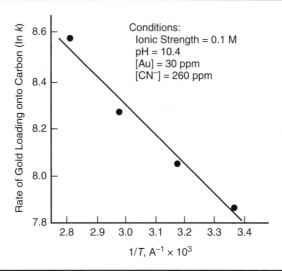

FIGURE 7.11 Effect of temperature on the rate of gold extraction [20]

In practice, the cyanide concentration used in adsorption systems is often determined by the requirements for optimal gold dissolution and by natural cyanide degradation rates within the extraction circuit (i.e., 0.1 to 0.3 g/L sodium cyanide [NaCN]).

Solution pH. A decrease in solution pH increases both the adsorption rate and loading capacity, as shown in Figure 7.12 and Table 7.2. The effect on the adsorption rate is quite small over the pH range 9 to 11, as applied in cyanidation circuits, with only a small advantage to be gained by reducing pH. The capacity is increased by approximately 10% as the pH is lowered from 11 to 9. In practice, the pH is usually maintained at >10 to avoid loss of cyanide by hydrolysis, or alternatively the pH may be allowed to decrease naturally through a CIP or CIL circuit to assist with cyanide degradation prior to tailings disposal.

Ionic strength. The effect of ionic strength on gold adsorption is shown in Table 7.2. This effect is also illustrated by the finding that the gold cyanide complex can be eluted off carbon with deionized water. Both the adsorption rate and loading capacity are increased with increasing ionic strength [32].

Concentration of other metals. Under laboratory conditions, gold loading capacity increases with increasing concentration of cations in solution in the following order:

$$Ca^{2+} > Mg^{2+} > H^+ > Li^+ > Na^+ > K^+$$

and decreases with anion concentration in the order:

$$CN^- > S^{2-} > SCN^- > S_2O_3^{2-} > OH^- > Cl^- > NO_3^-$$

These effects are compounded under industrial conditions by the adsorption of other metal cyanide species, which compete for active, available adsorption sites. This results in a slower adsorption kinetics and reduces the equilibrium capacity for gold. The adsorption of these metals is considered further in Section 7.1.2.5.

Dissolved oxygen. The beneficial effect of oxygen on the adsorption of gold from cyanide solutions has been reported [33]; however, the effect, as has been demonstrated, is most significant in low ionic strength solutions, which are atypical of most industrial leach solutions. Despite this, some benefit is observed in actual adsorption

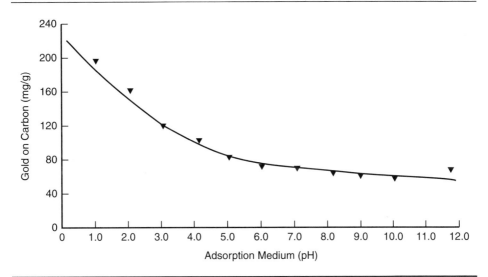

FIGURE 7.12 Effect of pH value of the adsorption medium on the gold capacity of the carbon; experimental conditions: volume of solution = 300 mL, mass of carbon = 0.25 g, nitrogen atmosphere, initial concentration of gold = 190 mg/L [23]

TABLE 7.2 Effect of pH and ionic strength on rate of loading and equilibrium capacity: ionic strength = 0.2 M; pH = 6.5 [20]

pH	Rate Constant, k (per hr)	Gold-Loading Capacity (mg/L)	Ionic Strength (M)	Rate Constant, k (per hr)	Gold-Loading Capacity (mg/L)
11.3	3,010	45,000	0.005	3,150	56,000
9.1	3,000	86,000	0.010	3,690	60,000
7.1	3,660	92,000	0.020	3,480	63,000
4.2	3,900	122,000	0.050	3,902	73,000
3.1	4,420	143,000	0.100	3,310	84,000
1.5	4,880	216,000	1.000	4,150	113,000

systems—attributed to the catalytic oxidation of cyanide. This results in a decrease in cyanide concentration which favors adsorption.

Carbon fouling. Carbon fouling, or poisoning, due to the adsorption, precipitation, or physical trapping of other solution species and ore constituents can have a severe adverse effect on gold adsorption efficiency, as considered in Sections 7.1.4.1 and 7.1.4.3.

7.1.2.5 Adsorption of Other Metals

Leach solutions usually contain a variety of metal ions and complexes, including silver, copper, nickel, zinc, iron, and mercury, which are adsorbed onto activated carbon to varying extents, depending on the concentration of each species, the properties of the carbon, and the solution conditions. The adsorption of silver, and in some cases mercury, may be important as they may be economic by-products of gold. In contrast, the adsorption of noneconomic metals is detrimental to gold extraction, because these species compete with gold (and silver) for active carbon sites. In addition, the adsorbed metals may be difficult to desorb under the conditions most suitable for gold desorption, resulting in a buildup of the metals on the carbon, which decreases carbon activity. Any metals adsorbed, for example, copper or mercury, may contaminate the final product, requiring additional treatment step(s).

Fortunately, activated carbon is highly selective for gold and silver over most other metal species, mercury being the most important exception. The general order of preference of adsorption for several commonly encountered metal complexes is given as follows:

$$Au(CN)_2^- > Hg(CN)_2 > Ag(CN)_2^- > Cu(CN)_3^{2-} > Zn(CN)_4^{2-} > Ni(CN)_4^{2-} \gg Fe(CN)_6^{4-}$$

High loadings of nonprecious metals can be achieved onto activated carbon in the absence of significant precious metal values, which makes it possible for heavy metal ions to be removed from water, a process commonly used for water purification.

Silver. The mechanism of Ag(I) cyanide adsorption is similar to that of Au(I) cyanide; however, the adsorption capacity of carbon for silver is substantially less than that for gold [10]. Also, the gold complex tends to displace silver from carbon. Both these factors are important in plant design and operation because a larger amount of carbon is required to recover an equivalent amount of silver from solution, and the silver loading rate is slower than gold. Many gold ores contain significant quantities of silver, often many multiples of the gold concentration, potentially making them less amenable to treatment by carbon adsorption (see also Sections 3.3.5 and 12.2.3.1).

Mercury. The neutral mercury cyanide complex, $Hg(CN)_2$, competes directly with $Au(CN)_2^-$ for adsorption sites and can even displace some of the adsorbed gold from carbon. Fortunately, mercury is usually present in leach solutions in relatively low concentrations, partly due to its low grade in most ores and its rather poor dissolution characteristics, and it rarely has a severely detrimental effect on gold adsorption. However, the highly effective adsorption and desorption of mercury under conditions applied for gold extraction require that, for treatment of ores containing significant quantities of mercury, a method of removing mercury must be provided downstream (see Chapter 10).

Copper. The adsorption of copper is strongly related to pH and cyanide concentration. The $Cu(CN)_2^-$ complex, which is favored at low pH and low cyanide concentrations, is most readily adsorbed, whereas at high pH and high free cyanide concentration, the $Cu(CN)_4^{3-}$ complex predominates, and adsorption is poor (see Figure 6.18). Therefore, the adsorption of copper species increases in the order:

$$Cu(CN)_4^{3-} < Cu(CN)_3^{2-} < Cu(CN)_2^-$$

The detrimental effect of copper on gold adsorption has been reported extensively, and copper concentrations as low as 100 mg/L can interfere severely with adsorption processes [34]. The effect of cyanide concentration on gold- and copper-loading capacities is shown in Figure 7.13. In order to minimize the adsorption of copper onto carbon (and to minimize the adverse impact on gold loading), the molar ratio of CN−Cu should be maintained at or above 4:1 in leach solutions prior to feeding the carbon adsorption process. Alternatively, copper can be allowed to co-adsorb onto carbon with gold and can be removed selectively using a cold desorption step (see Section 7.1.3.8).

Processes that treat materials containing high concentrations of cyanide-soluble copper, that is, yielding >200 mg/L Cu in solution, will require very careful control of pH and cyanide to allow satisfactory treatment. In the extreme case, these materials may be unsuitable for treatment by carbon adsorption [35].

7.1.3 Elution

Activated carbon that has been loaded with gold and other metals in adsorption processes must be treated by an elution step to desorb the metals from the carbon. This produces a

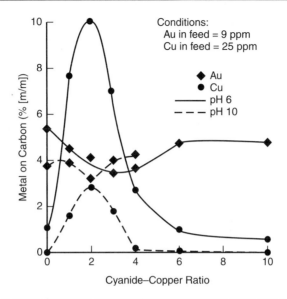

FIGURE 7.13 Effect of the ratio of cyanide to copper on the relative equilibrium loading capacities of copper cyanide and gold cyanide [20]

smaller volume of high-grade gold solution, suitable for final gold recovery by electrowinning or zinc precipitation (Chapter 8), and allows the carbon to be recycled to the adsorption circuit. Carbon is typically reused between 100 and 400 adsorption–elution cycles, depending on the carbon quality and the effectiveness of reactivation procedures applied (see Section 7.1.4).

The desorption process, commonly referred to as either elution or stripping, is a reversal of the adsorption process, and the chemical and physical factors that inhibit adsorption generally enhance desorption. For gold adsorbed from cyanide solutions, the desorption reaction is most simply represented by:

$$M^{n+}[Au(CN)_2^-]_n(ads) \rightleftharpoons n\, Au(CN)_2^- + M^{n+} \qquad \text{(EQ 7.8)}$$

although the exact mechanism of adsorption is considered to be more complicated.

7.1.3.1 Temperature and Pressure

Temperature is the most important factor in the elution of gold cyanide from carbon, with approximately an order of magnitude increase in the elution rate for a 100°C increase. For example, the elution rate at 180°C is eight times faster than at 90°C at atmospheric pressure. Figure 7.14 shows the effect of temperature on gold desorption efficiency, given for the example of the Zadra elution scheme. Although it is possible to reduce elution times substantially by operating at temperatures >100°C, this requires the use of elevated pressure to keep the eluting media in the liquid phase and enable practical application of the system. Consequently, elution systems have evolved into two classes:

- Processes that operate at atmospheric pressure and temperatures < 100°C
- Processes that operate at elevated pressures to allow operation at elevated temperatures, that is, >100°C, to achieve faster elution rates

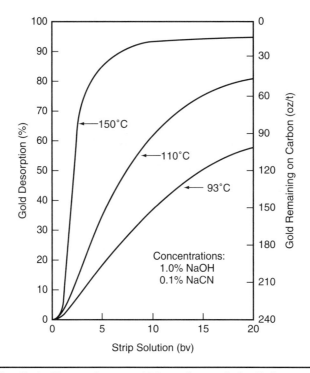

FIGURE 7.14 The effect of temperature on gold desorption employing Zadra process solution [36]

Unpressurized systems operate at temperatures just below the solution boiling point (95°C), whereas pressurized systems have been operated as high as 160°C and at 500 kPa. At temperatures >180°C, most metal cyanide complexes, including gold cyanide, decompose to the metallic species and free cyanide, which results in high residual gold concentrations on carbon, which are very hard to remove [37]. Cyanide decomposition, resulting in ammonia evolution, also increases at elevated temperatures (see Chapter 6).

7.1.3.2 Cyanide Concentration
The effect of cyanide on the rate of gold desorption is illustrated in Figure 7.15. Increasing cyanide concentration increases the competition of cyanide ions with gold cyanide species for adsorption sites on the carbon and assists with the displacement of gold cyanide species from the carbon. However, the presence of free cyanide throughout the desorption process is not a requirement for effective elution (as illustrated by the OH⁻ line in Figure 7.15) and several procedures have been developed that use a cyanide presoak step followed by deionized water elution. Consequently, elution systems can be divided into those using cyanide throughout the process and those using cyanide during a presoak only, as described in Section 7.1.5.6.

7.1.3.3 Ionic Strength
Ionic strength has a greater effect on elution rate than cyanide concentration, as shown in Figure 7.16. Gold may be desorbed quite effectively with low ionic strength solution, for example, deionized water, even in the absence of free cyanide [38]. The beneficial effect of divalent cations, such as calcium and magnesium, on gold adsorption onto activated carbon is reversed for elution processes.

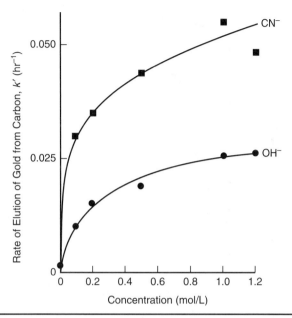

FIGURE 7.15 Variation of the rate of elution with eluant strength at a constant ionic strength of 1.2 at 95°C [38]

7.1.3.4 pH
Elution processes that have been developed since the original Zadra process (1950) have all used solutions containing 1% to 3% sodium hydroxide, either in a presoak step or for the main elution step [39, 40]. The hydroxide ions displace gold cyanide ions on the carbon in a manner similar to free cyanide ions, as discussed in Section 7.1.3.2, and as illustrated in Figure 7.15. In addition, sufficient alkalinity is required to avoid loss of cyanide by hydrolysis, that is, to typically maintain the pH between 10.0 and 12.0. The control of pH during elution is most important for processes that use zinc precipitation for subsequent gold recovery, because insoluble zinc hydroxide can be formed (Chapter 8).

7.1.3.5 Organic Solvents
The rate of elution can be significantly increased by the addition of organic solvents, such as alcohols and glycols, to the aqueous eluant. These increase the activity of other ionic species in solution, particularly smaller ions such as cyanide, in preference to larger ones, for example, gold cyanide. This effect increases the efficiency of displacement of gold cyanide from carbon.

A variety of solvents have been tested and used to enhance elution processes over a range of temperatures (Figures 7.17 and 7.18). Alcohols, such as ethanol, methanol, and isopropanol, at concentrations of 15% to 25% in the eluant, can be used to reduce elution times by a factor of 3 to 4 (e.g., from 48 to 12 hr for Zadra-type elution at 90°C). Unfortunately, these solutions are highly flammable and may constitute a fire hazard in industrial plants. Glycols, such as ethylene and propylene glycol, have been used in similar concentrations but with generally smaller reductions in elution times—approximately half the reductions that are achieved with alcohols. Glycols are less flammable than alcohols and therefore may be preferred in some cases.

The relative effectiveness of various organic solvents decreases in the order [41]:

acetonitrile > methyl ethyl ketone > acetone >>> demethyl formamide > ethanol

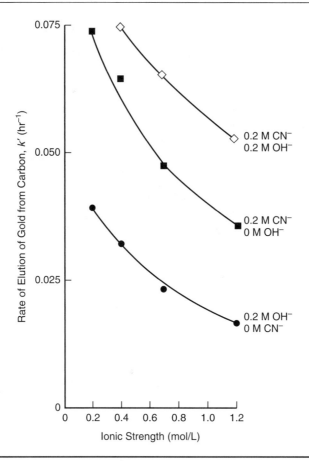

FIGURE 7.16 Effect of the ionic strength of the eluant on the rate of elution at 95°C [38]

In addition, the use of organic solvents in eluant solutions without cyanide has been proposed. Solutions containing 1% by weight of sodium hydroxide with 20% (by volume) of a suitable organic solvent, such as isopropyl alcohol, ethylene glycol, or ethanol, have been demonstrated to achieve effective elution at 80°C in about 8 hr [42]. The effectiveness of the various solvents decreases in the following order:

$$\text{isopropyl alcohol} > \text{ethylene glycol} > \text{ethanol}$$

The use of organic solvents at high temperatures may benefit the removal of adsorbed organic species (e.g., oils, humic acid, etc.) from carbon [43]. This does not appear to have any significant adverse effect on subsequent electrowinning or zinc precipitation recovery processes.

7.1.3.6 Solution Flow Rate

The solution flow rate applied for elution is usually expressed as carbon bed volumes per hour (bv/hr). The elution rate tends to be virtually independent of flow rate above about 1 bv/hr, but the residual gold loading on carbon decreases with increasing flow rate after a fixed time, as shown in Figure 7.19. Typically, flow rates of 2 to 4 bv/hr are used to produce a carbon with low residual gold loading, while avoiding the need to treat excessively large volumes of eluant.

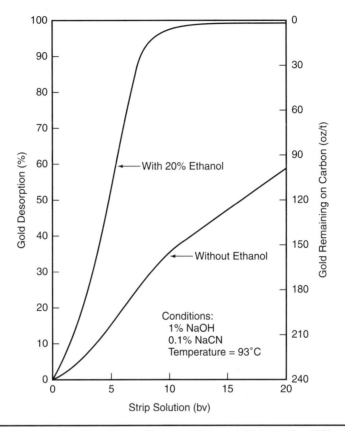

FIGURE 7.17 Effect of ethanol additive to Zadra solution on gold desorption [36]

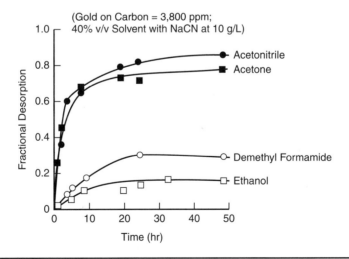

FIGURE 7.18 Desorption of gold by organic solvents at 25°C [23]

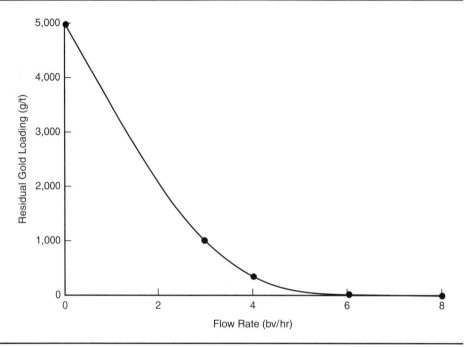

FIGURE 7.19 Effect of flow rate on average loading of residual gold after 20 hr [38]

7.1.3.7 Gold Concentration in Solution

The elution rate decreases, and the residual gold loading increases, with an increase in the gold concentration of the eluant (Figure 7.20). This reduces the rate of elution with time in a batch process—a most important factor for elution systems that recirculate solutions directly following metal recovery. An example of this is atmospheric Zadra elution coupled with electrowinning or zinc precipitation for metal recovery. In this case, the gold concentration in the solution used for elution, and consequently the efficiency of elution, depends on the efficiency of the associated electrowinning or zinc precipitation step.

7.1.3.8 Elution of Other Metals

Metal cyanides, other than gold, are also eluted from carbon under the conditions applied for gold desorption. Copper, silver, and mercury preferentially desorb before gold, as illustrated for copper and silver in Figure 7.21. The possibility of sequentially desorbing copper and then precious metals in a two-stage elution process has been applied by operations that experience high copper loadings on carbon, for example, El Indio (Chile) (see Section 12.2.6.3). In this case, copper (and other base metal cyanides) are eluted from loaded carbon using a cold (ambient temperature), alkaline cyanide solution. Depending on the metal loading on carbon and the eluant properties, >90% of the copper (and potentially other base metals) can be removed selectively over gold and silver. The gold and silver (and a portion of the remaining base metal complexes) are subsequently desorbed using one of the elution processes described in Section 7.1.5.6.

7.1.4 Carbon Fouling and Reactivation

Carbon fouling is the buildup of organic and inorganic substances on carbon, which detrimentally affect gold adsorption. This results in a decrease in the kinetics and equilibrium

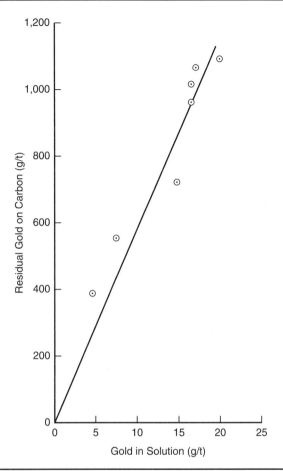

FIGURE 7.20 Equilibrium isotherm for the distribution of aurocyanide between activated carbon and a solution containing 0.2 M NaOH and 0.2 M NaCN at 95°C [38]

loading of gold adsorption onto carbon and may adversely affect the efficiency of desorption processes by reducing desorption kinetics and elevating residual (eluted) carbon gold loading. Fouling may also affect the eluate composition adversely, for example, by the formation of silica gel or release of particulate matter into the eluate solution.

The effectiveness of carbon adsorption as a commercial process relies on the ability of activated carbon to be reused many times, which depends on the degree of fouling and the efficiency of any reactivation processes used. Carbon fouling can occur by any or all of the following mechanisms:

- Undesirable organic or inorganic species are adsorbed onto the carbon surface, taking up active sites which would otherwise be available for gold adsorption
- Inorganic salts are precipitated onto the carbon surface, blocking active sites
- Solid particles or precipitates are physically trapped in carbon pores, restricting access to gold-bearing solution

In industrial systems this fouling can be counteracted in three ways:

- Reducing fouling during adsorption and desorption processes

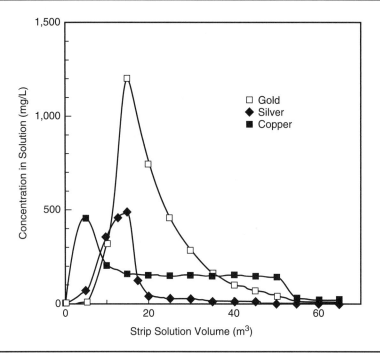

FIGURE 7.21 Typical elution curves from carbon stripping [44]

- Removing the fouling species, following adsorption or desorption
- Removing fouled carbon from the circuit and replacing with new carbon

Consequently, it is important to understand the sources and nature of carbon fouling, as well as the methods available for reactivation, in order to optimize carbon adsorption systems. These are reviewed in more detail in the following sections.

7.1.4.1 Inorganic Fouling

The most important forms of inorganic fouling are [45]:

- Calcium salts, primarily carbonate, but also to a lesser extent sulfate and other species
- Magnesium and sodium salts
- Fine ore minerals, such as silica, complex silicates, and aluminates (including clays and clay-forming minerals)
- Fine iron particles and associated products resulting from grinding media
- Base metal precipitates from the leach solution

The mechanisms by which inorganic salts are deposited onto activated carbon are largely unrelated to the adsorption of gold species.

Calcium carbonate is formed by carbon dioxide from the atmosphere dissolving in water to form CO_3^{2-} which reacts with available Ca^{2+} ions as follows:

$$CO_2 + H_2O \rightleftharpoons HCO_3^- + H^+ \quad \text{(EQ 7.9)}$$

$$HCO_3^- \rightleftharpoons H^+ + CO_3^{2-} \quad \text{(EQ 7.10)}$$

$$CO_3^{2-} + Ca^{2+} \rightleftharpoons CaCO_3 \quad \text{(EQ 7.11)}$$

The most common sources of calcium ions in gold extraction are from lime (CaO) and slaked lime (Ca(OH)$_2$), added to process slurries and solution for pH control, and soluble ore constituents (i.e., limestone or dolomite).

Alternatively, carbonate ions may be formed by the oxidation of cyanide at the carbon surface:

$$2CN^- + O_2 + 4H_2O \rightleftharpoons 2CO_3^{2-} + 2NH_4^+ \quad \text{(EQ 7.12)}$$

$$Ca^{2+} + CO_3^{2-} \rightleftharpoons CaCO_3 \quad \text{(EQ 7.13)}$$

At higher temperatures, the formation of methanoic and oxalic acids is favored. There is evidence that the carbon surface imposes a region of solution which is atypical of the bulk solution and in which the local solubility of metal compounds is reduced. This leads to a precipitation reaction on the carbon surface. This behavior is similar to the adsorption of polyvalent metal species onto oxide and silicate minerals [47]. The solution conditions for metal deposition are similar to those for precipitation in bulk solution; however, adsorption occurs at lower pH values and lower metal ion concentrations [46, 48]. Although hydroxides may be precipitated onto carbon, carbonate precipitation is usually the more important problem in industrial systems.

Studies of calcium deposits using scanning electron microscopy (SEM) have revealed the material to be crystalline [45]. Such precipitates fill the cracks and depressions on the external surfaces of the carbon, and are widespread within the pore structure, even after acid washing. This inhibits the diffusion of gold cyanide species through carbon pores, thus reducing both the active surface area and the adsorptive properties of the carbon.

The precipitation of calcium carbonate can largely be avoided by reducing the pH to <8.3, below which the more soluble calcium bicarbonate is formed. However, in most cyanide adsorption systems this is impractical, and possibly hazardous, because of the increased hydrogen cyanide evolution. On the other hand, a possible environmental advantage of such practice is the associated destruction of free cyanide in CIP and CIL circuits prior to discharging slurry tailings.

Descalant reagents can be added to process solutions to reduce calcium carbonate precipitation on carbon in CIC systems. These have been used successfully at many operations, for example, Chimney Creek, Carlin, and Pinson (Nevada, United States).

7.1.4.2 Inorganic Removal

Many inorganic foulants can be removed by acid washing, whereby the precipitated salts are dissolved in dilute mineral acid (HCl or HNO$_3$) and then rinsed from the carbon. Clearly, the success of this technique depends on the solubility of the salt deposited, and the type and concentration of acid solution used. Dilute mineral acids will readily dissolve calcium carbonate and many other metal salts, leaving the adsorbed gold species essentially unaffected. Consequently, acid washing may be performed either before or after the desorption (elution) of precious metals from the carbon. Several factors affect this decision, discussed in Section 7.1.5.6.

The general equation for the dissolution of a divalent metal carbonate in mineral acid is given as follows:

$$MCO_3 + 2H^+ \rightleftharpoons M^{2+} + CO_2 + H_2O \quad \text{(EQ 7.14)}$$

In practice, both hydrochloric and nitric acid have been used for acid washing.

Hydrochloric acid. The preferred reagent in industry, hydrochloric acid has been applied at both ambient and elevated temperatures, up to approximately 85°C. Volume concentrations of 1% to 5% HCl are used, depending on the loading of inorganic foulants on the carbon and the acid washing conditions applied. The efficiency of calcium removal is largely unaffected by acid concentration (3% to 10% HCl) and temperature (25°C to 90°C) for calcium loadings below approximately 1% Ca. However, significant benefits may be gained by increasing acid concentration (i.e., to 5%–7.5%) for carbon loaded with higher calcium levels, that is, >2% Ca (see Figure 7.22).

The efficiency of calcium removal is strongly related to the efficiency of carbon–acid contact (mixing) during acid washing, as is the case for other carbon adsorption and desorption processes. Because most carbon acid washing systems are operated as stationary, semifluidized, or fully fluidized beds, the calcium removal efficiency depends on the vessel geometry and the acid flow rate. Also, the dissolution rate of calcium salts within carbon pores is diffusion controlled. Consequently, the process residence time is likely to be important in optimizing removal of inorganic salts. An example of the effect of acid strength and reaction time on final calcium loading is shown in Figure 7.22.

In well-mixed systems, with relatively low calcium loadings (<1%, for example), residence times of 10 to 15 min may be adequate, whereas for poorly mixed systems or when high calcium loadings are experienced (i.e., >3%) over an hour may be required.

Dilute hydrochloric acid is usually capable of removing between 80% and 95% of calcium loaded on carbon. Although the efficiency of removal of sodium and magnesium salts may be somewhat lower (<80%), these species are usually present on carbon in lesser quantities than are calcium salts, and the efficiency of their removal is generally not as critical. In addition, up to 50% of nickel, zinc, iron, and silicon loaded on the carbon may also be removed, depending on the adsorbate and the acid washing conditions applied. Silver, mercury, and copper are not removed from carbon by dilute hydrochloric acid.

The major drawback associated with the use of hydrochloric acid is the presence of residual chloride ions in carbon pores, which are difficult to remove effectively following acid washing. This may cause severe problems when acid washing precedes thermal reactivation processes because the highly corrosive chloride ions are released and vaporized at elevated temperatures. Similarly, if acid washing precedes desorption (elution), residual chloride ions may be released, creating a corrosive environment in the elution circuit, although this is generally less severe and can often be controlled. This effect, and the possible reprecipitation of salts, is controlled by washing and neutralizing acid-washed carbon with sequential water and sodium hydroxide washes. Despite this, 100% removal of chloride species is rarely achieved. Alternatively, acid washing may be performed after desorption and thermal reactivation to allow chloride release to occur after the carbon has been returned to the adsorption circuit, as discussed in Section 7.1.5.6.

Nitric acid. Used in some operations, particularly in North America, nitric acid is applied in a similar manner to hydrochloric acid. The use of nitric acid avoids the corrosion problems associated with the use of hydrochloric acid but may cause other problems, such as oxidation and deactivation of carbon surfaces. This effect is thought to be small in very dilute solutions (<5% HNO_3) but increases with increasing concentration. Mercury, and to a lesser extent silver, are removed from carbon by dilute nitric acid, which may be beneficial or detrimental, depending on their concentrations and the specific process requirements. Finally, the use of nitric acid introduces nitrate ions into the process which may, depending on the nature of the process, present environmental problems.

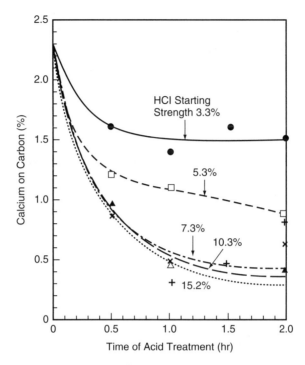

NOTE: Percentages indicate HCl starting concentration during treatment.

FIGURE 7.22 Effect of acid strength and time on the residual calcium content of carbon at Grootvlei (South Africa) [49]

7.1.4.3 Organic Fouling

Because activated carbon is a relatively nonpolar and hydrophobic material, it readily adsorbs most organic compounds from aqueous solutions. The organics that contribute most to fouling are as follows:

- Diesel oil, lubricating oils, and antifreeze chemicals from mining and processing equipment
- Humic acid and other vegetation decomposition products that are present in the ore
- Flotation reagents, such as collectors and frothers
- Flocculants and other surface-active reagents

Adsorption occurs by mechanisms involving hydrogen bonding and van der Waals attraction between organic species and the carbon surface. It is most favorable under conditions of limited solubility, for example, when the potential adsorbate has high molecular mass, low polarity, or low ionization potential.

Such fouling can lead to a significant proportion of the carbon surface area being inaccessible to gold cyanide species, resulting in decreased adsorption rate and decreased gold loading. This principle is exploited when blocking agents are employed to deactivate organic carbon components in carbonaceous ores (Section 5.6).

7.1.4.4 Organic Removal

Organic species that foul activated carbon fall into two categories with respect to their removal [50]:

- Adsorbates that are either highly volatile or easily thermally decomposed to gaseous products at normal kiln temperatures (500°C to 800°C).
- Nonvolatile species that leave a carbonaceous residue on pyrolysis. These residues can be removed using steam at temperatures >650°C, according to the following reaction:

$$(C)_n + nH_2O(\text{steam}) \rightarrow nCO + nH_2 \quad \text{(EQ 7.15)}$$

This reaction also causes some loss of the original carbon, particularly if inorganic salts of calcium, magnesium, and iron are present, which catalyze the carbon–steam reaction and should therefore be removed by acid washing prior to thermal reactivation (kilning).

Fouled carbon can be regenerated by heating to 650°C to 750°C in a nonoxidizing atmosphere. A steam atmosphere is often used for the reason previously discussed and as illustrated in Table 7.3. The most important variables during thermal reactivation are the following:

- Temperature
- Steam addition
- Residence time
- Initial moisture content of carbon
- Presence of extraneous mineral matter
- Reactivation equipment type

Figure 7.23 (a and b) shows the effects of temperature, retention time, and steam addition on relative carbon activity for a South African gold plant [50]. In this case, it was also observed that carbon loss was reduced and activity was increased by acid washing prior to the elution step.

If the temperature or residence time is too low, then removal of the organic material may be incomplete. If the temperature is too high, then further activation of the carbon may occur, resulting in increased material loss and decreased hardness. The latter is due to an excessive increase in pore structure, which reduces mechanical strength. To reduce energy consumption, the carbon should be dewatered prior to reactivation.

Hot, regenerated carbon should be cooled by quenching in water, which minimizes exposure to oxygen and maintains activity. Some operations use warm water for quenching to avoid excessive thermal shock on the carbon, which might otherwise result in particle fracturing and degradation.

The performance of kilning may also be affected by the presence of coarse mineral particles, wood chips, and plastics (e.g., electrical cable insulation). Mineral particles may be removed by acid washing, then by separation on a jig or shaking table if necessary. Kiln off-gases consist primarily of carbon monoxide, carbon dioxide, hydrogen, and steam. Small quantities of other substances (e.g., hydrogen sulfide, carbon disulfide, and ammonia) may result from decomposition of flotation reagents and adsorbed cyanide. Any residual mercury on the carbon is readily volatilized during thermal reactivation, and the effects of mercury should be carefully considered on a case-by-case basis.

7.1.5 Process Considerations

7.1.5.1 Carbon Preparation

Despite being manufactured to particular size specifications, fresh carbon always contains small quantities of fine carbon, which is unsuitable for use in carbon adsorption

TABLE 7.3 The effect of steam flow rate and contact time on the activity of carbon using Rintoul thermal reactivation kiln [18]

Steam Flow Rate (kg/hr)	Percentage Activity Compared to Fresh Carbon	
	After 1 hr	After 15 hr
125	77.9	83.5
150	91.5	94.0
175	93.9	99.4

systems. In addition, much of the carbon produced by manufacturers is angular in shape, containing points and sharp edges which are readily attrited during transportation and handling under normal process conditions. Similarly, flat (plate-like) particles and damaged (i.e., cracked or fractured) particles are susceptible to rapid degradation in adsorption systems, resulting in the loss of carbon fines and any contained gold.

Consequently, it is desirable to remove the carbon fines that are most rapidly generated from fresh carbon prior to introducing it into the adsorption circuit. This is usually achieved by vigorous mechanical agitation in water at 10% to 20% solids for 0.5 to 2 hr. Typically, between 1% and 3% of the total carbon weight is removed as fines, depending on the carbon type and the severity and duration of attritioning. The conditioned mixture (coarse carbon, carbon fines, and water) is then screened at a size slightly coarser than the screen size employed for interstage screening within the adsorption circuit. The coarse carbon produced is ready for use, while the fine carbon slurry is discarded from the process.

7.1.5.2 Carbon-in-Pulp Process

A well-established technology, the CIP process, is commonly applied for the extraction of gold from cyanide leach slurries. The process is usually configured with carbon flowing counter-current to the process slurry in mechanically agitated tanks (Figure 7.24). An alternative to this is a carousel-type system, which will be discussed later. The slurry is introduced to carbon following or during cyanide leaching and passes through several stages of carbon adsorption, the number depending principally on the tank sizes, the carbon concentration, and the amount of gold to be adsorbed. The number of stages ranges from four to ten, with five to six most typical. The gold concentration in solution is depleted as the gold is adsorbed onto carbon. Fresh or reactivated carbon is introduced at the tail end of the process and is transferred either continuously or in batches up the adsorption stages, in the opposite direction to the slurry although at a much slower rate. The carbon in each stage becomes loaded to the pseudo-equilibrium, which depends on the gold concentration in solution in each stage. The carbon in the first stage has the highest gold loading and is contacted with the highest-grade solution while the carbon in the last stage has the lowest loading and, consequently, the highest activity.

Carbon is retained in each adsorption stage by interstage screens with apertures slightly smaller than that of the activated carbon, which are large enough to allow free flow of slurry between the stages. The efficiency of interstage screening has proven to be one of the biggest challenges in the application of carbon adsorption and has received much attention. The major screen types and their approximate slurry-handling capacities are summarized as follows [52]:

- Horizontal screen with wiper (e.g., Kambalda): 40 to 60 $m^3/hr/m^2$
- Vibrating screen with inclined deck (e.g., Derrick): 90 to 110 $m^3/hr/m^2$
- Equal pressure, air-cleaned (EPAC): 10 to 50 $m^3/hr/m^2$
- Vertical, cylindrical, with wiper and/or bottom discharge (e.g., NKM, Kemix): 30 to 60 $m^3/hr/m^2$

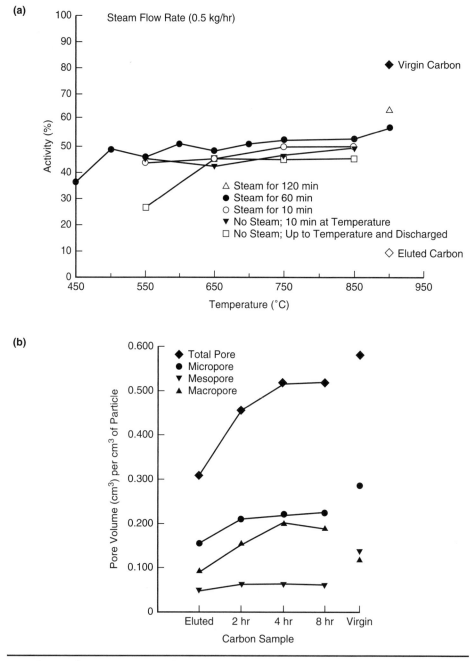

FIGURE 7.23 Results of regeneration tests on carbon from a South African plant [50]

However, the choice of screen depends on carbon size, ore particle size, slurry density and viscosity, screen size, and the capital cost for each screening system.

A typical CIP circuit configuration is shown in Figure 7.24. The design and configuration of CIP adsorption systems has been well studied, and several excellent design approaches are available based on empirical models of carbon adsorption processes [18, 53, 54]. In the conventional cascade-type system, slurry flows by gravity from one stage

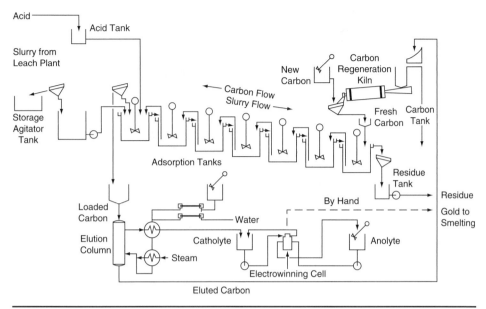

FIGURE 7.24 Example of a CIP cascade circuit [51]

to the next and carbon is transferred counter-current to slurry flow by pumping or elution. In the early 1990s, Anglo American Research Laboratories introduced the AARL "pump-cell," which combines the functions of agitation, screening, and interstage slurry transfer into one compact and highly efficient unit [55]. A unique feature of the pump-cell is that the pump impeller lifts the screened slurry from inside the cylindrical screen and deposits it into the launder to feed the next stage in the circuit. The pumping action generates a pressure differential across the screen surface, promoting slurry flow through the screen. This equipment lends itself to configuration in "carousel" mode, where the counter-current flow of slurry and carbon is accomplished without transferring the carbon from tank to tank, but rather by switching fresh slurry feed from tank to tank. The carbon is removed only for elution and regeneration.

This configuration has the advantages of reducing carbon attrition losses in the adsorption circuit, eliminating short circuiting of carbon during carbon transfer, and potentially improving the stage efficiency and carbon loading profile. Disadvantages include more complex piping design and the need to take a tank off-line to transfer carbon to elution (i.e., less efficient circuit operation during this time) [54]. However, a number of large gold plants have utilized the pump-cell CIP technology, including Vaal River No. 2, Hartebeesfontein, Consolidated Murchison, Blyvooruitzicht, West Driefontein, East Driefontein, Kloof, and Leeudoorn (all in South Africa), and El Indio (Chile) [54, 56]. Of these plants, the largest is Vaal River No. 2 with a throughput rate of approximately 13,000 tpd, indicating that the technology is truly mature.

Either granular or extruded carbons can be used in a variety of size ranges, typically 1.7×3.4 mm, 1.2×3.4 mm, and 1.2×2.4 mm. Carbon concentrations are usually maintained between 5 and 30 g/L, depending on the specific needs of each operation and the carbon properties. High carbon concentrations result in high gold and carbon inventories, and increased fine carbon particle losses (and consequently increased gold losses with the fine carbon). The most important factor is the amount of carbon required in the circuit to optimize gold recovery and reduce operating risk, that is, the potential of losing soluble gold to tailings. These requirements vary from operation to operation. Typical activated carbon consumptions are in the range of 20 to 40 g/t ore, depending on the specific conditions applied and the type and quality of the carbon.

Control of slurry density and viscosity are important in CIP systems, not only for the reasons outlined in Section 7.1.2.1, but also because of the potential for carbon concentration stratification within adsorption tanks, arising from the low apparent density of carbon (0.8 to 0.9 g/mL).

7.1.5.3 Carbon-in-Leach Process

CIL is a modification of the CIP process (Section 7.1.5.2) where the leaching and adsorption process steps are performed in the same tanks simultaneously. The process offers advantages of lower capital cost than separate leaching and carbon adsorption systems and can significantly improve gold extraction from ores containing constituents that adsorb gold from leach solutions. In the latter case, the carbon competes with the gold-adsorbing, preg-robbing, or preg-borrowing minerals, and preferentially adsorbs the gold values. However, CIL has some inherent disadvantages compared with CIP, summarized as follows:

1. Larger carbon inventory is required (due to lower loaded carbon concentrations).
2. As a result of item 1, the in-plant gold inventory, or "lockup," is higher.
3. Fine carbon particles due to carbon attrition and the associated gold losses are typically higher.
4. Carbon gold loadings are usually lower due to treatment of lower-grade solutions, which increases the carbon transfer frequency, as well as increasing elution and reactivation requirements.
5. Operating costs are typically higher.

Consequently, CIL must be evaluated in detail for each specific application.

AARL pump-cell, carousel-type circuits can be considered in some CIL installations, but generally are only effective if applied in conjunction with a preleach step to provide a higher gold grade to the first-stage CIL (i.e., effectively a partial-leach and CIP circuit rather than true CIL). Circuits that require the use of CIL (due to the presence of preg-robbing constituents in the ore) must generally utilize conventional carbon adsorption slurry tank systems with higher slurry retention times [54].

7.1.5.4 Use of Air or Oxygen in Carbon Adsorption Systems

Several carbon adsorption systems, and in particular CIL, sparge air or oxygen into the process tanks to provide dissolved oxygen for gold leaching. (The role of dissolved oxygen in the cyanide leaching reaction is discussed in detail in Chapter 6.) However, air or oxygen can affect carbon adsorption systems in a number of other ways:

- Oxygen enhances the adsorption of gold cyanide onto activated carbon (Section 7.1.2.4).
- The presence of air and oxygen bubbles in slurry systems can cause the carbon to float, thereby impairing homogeneous mixing.
- The oxidation of cyanide by oxygen is catalyzed in the presence of activated carbon (see Chapter 11).

These effects need to be carefully considered for each operation.

7.1.5.5 Carbon-in-Solution Process

Activated carbon can be used to extract gold from a wide range of gold cyanide solution streams, including heap or run-of-mine stockpile (dump) leaching solutions, thickener

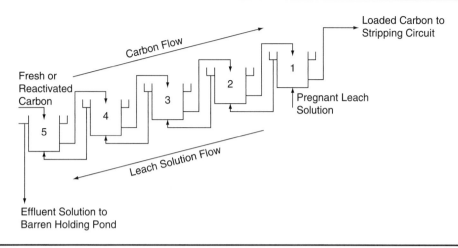

FIGURE 7.25 Five-stage carbon adsorption circuit for the recovery of gold onto activated carbon from solution (CIC)

overflow solutions, unclarified filtrates, and tailings reclaim solutions. Two methods are available for treating solutions:

- Fluidized or expanded bed systems
- Fixed, packed, or pinned bed systems

Several configurations of each have been designed and implemented to meet specific needs, including the following:

- Multiple-stage fluidized beds stacked in a single column (e.g., NIMCIX by Mintek)
- Multiple-stage fluidized beds as a series of cascading columns
- Single, deep fluidized bed column
- Up-flowing fixed bed (single or multiple stages)
- Down-flowing fixed bed (single or multiple stages)

The choice of system depends on the flow rate to be treated, the solution clarity, the gold concentration, and the desired mode of operation. Fluidized beds are preferred for unclarified solutions, where the solids will not rapidly plug the bed. Bed expansions of 10% to 100% are used in practice. Fluidized bed systems have better mass transport properties than packed beds and channeling is usually avoided. A schematic diagram of a multiple-stage, cascading, fluidized bed adsorption system is shown in Figure 7.25.

Packed bed systems are effective for treatment of clean solutions. A gold-loading profile develops across the bed and, because there is no carbon movement, the packed bed columns behave like a plug flow reactor, providing an effective mode of contact. However, channeling of solution through a packed bed is often a major problem in commercial applications, as it causes mass transport rates to be reduced in some regions of the bed, resulting in inefficient adsorption.

7.1.5.6 Elution

Several elution systems have been developed and applied on a commercial scale, and these are reviewed here. Typical operating conditions for the different elution systems are summarized in Table 7.4, together with an indication of the applicability of each in industry.

TABLE 7.4 Comparison of typical operating conditions for various carbon elution methods [26, 36, 39, 41, 43, 49, 51]

Elution type	Procedure	Reagent Scheme		Temperature (°C)	Pressure (kPa)	Time (hr)	Maximum Gold Concentration* (mg/L)	Application
		Presoak	Elution					
Atmospheric (no solvent)	Zadra	None	10 g/L NaOH 1–2 g/L NaCN	90–100	100	36–72	150	Limited use in the United States and Australia. Declining popularity.
Pressurized	Pressure Zadra	None	10 g/L NaOH 2 g/L NaCN	135–140	400–500	8–14	1,000	Widely applied in the United States and Australia, to a lesser extent in South Africa.
	AARL†	20–50 g/L NaCN 10–20 g/L NaOH	H$_2$O	110–120	170–200	8–14	1,500	Widely applied in South Africa and throughout the world. The generally preferred elution method.
Solvent-assisted (atmospheric)	Zadra/Duval	None	10%–20% ethanol 10 g/L NaOH 2 g/L NaCN	80	100	6–12	1,500	Very limited application.
	Anglo	20–50 g/L NaCN 10–20 g/L NaOH	90% acetone or ethanol in H$_2$O	70–90	100	6–8	1,000–2,000	Very limited application.
	Murdoch	80% acetonitrite in H$_2$O	20%–40% acetonitrite 10 g/L NaCN 2 g/L NaOH	25–70	100	8–14	1,500–6,000	Very limited application.
	Micron Research	20–50 g/L NaCN 50–100 g/L NaCN	60%–80% methanol in H$_2$O	60–80	100	8–80	3,000–10,000	Limited application in Australia and South Africa.

* Maximum gold concentrations in solution given for elution of carbon loaded to approximately 4,000–5,000 g/t Au.
† AARL = Anglo American Research Laboratories, South Africa.

Atmospheric elution with cyanide and caustic (Zadra process). A solution containing approximately 1% to 2% sodium hydroxide and 0.1% sodium cyanide is used at 95°C and at a flow rate of 2 bv/hr. The process takes between 36 and 72 hr to elute loaded carbon to a low residual loading (<100 g/t), equivalent to 100 to 150 bv of solution. Mild steel equipment can be used throughout.

Pressure elution with cyanide and caustic (pressure Zadra process). The system is used at elevated temperature (135°C–140°C) and pressure (400 to 500 kPa) to reduce the elution time to 8 to 14 hr at a flow rate of 2 bv/hr (15 to 30 total bv of solution). Stainless steel elution columns are required.

Deionized water elution with cyanide presoak (AARL process). The carbon is acid washed in dilute mineral acid, water-washed, then soaked in 3% sodium cyanide and 1% to 2% sodium hydroxide solution for approximately 30 min. The carbon is eluted with 6 to 10 bv (at 2 bv/hr) of deionized water at 110°C to 120°C and 70 to 100 kPa pressure [57]. Elution is completed in 8 to 14 hr. A butyl rubber-lined elution column is required to withstand acid and alkaline media, which has the drawback of a 113°C maximum operating temperature. In some cases special steel alloys (e.g., Hastelloy) may be preferred to enable operation at higher temperatures.

Solvent-assisted elution. Several processes use varying proportions of alcohols and glycols to assist in atmospheric elution (i.e., up to 95°C). Systems using 20% of a suitable alcohol (ethanol or methanol) can reduce the elution times of conventional Zadra systems below 12 to 16 hr. Alternatively, glycols (e.g., ethylene or propylene glycol) in proportions of 20% to 25% can reduce elution times to 24 to 36 hr, avoiding the use of alcohols, which are a fire hazard.

Solvent distillation elution. In this process the elution column is configured as a packed bed distillation tower with a solution heater at the base of the column, an overhead condenser, and a reflux pump to recirculate the solvent. The loaded carbon acts as the tower packing. The carbon is presoaked with 1% to 2% sodium hydroxide and 5% to 10% sodium cyanide at ambient temperature. Ethanol, methanol, or acetonitrile is used as the solvent (approximately 0.5 bv). The carbon is refluxed at 65°C to 80°C for 8 hr, with gold values eluted from the carbon by the downflowing condensate. A total of about 1 bv of solution is used to produce a very concentrated solution. The eluted carbon typically has high activity, which is thought to be due to the efficient removal of organic foulants by the hot solvent.

The benefits of using water of low ionic strength for elution have been discussed previously (Section 7.1.3). The presence of dissolved solids not only reduces elution efficiency but also increases the scaling tendency of the solution and may affect its filtration properties. Consequently, water to be used in elution circuits may need to be softened. Plant water containing chlorides should not be used for elution due to the corrosive nature of chlorine gas evolved at the anode during subsequent electrowinning, where applicable [58].

The concentrations of various solution species, which are eluted from carbon more efficiently than they are removed by the subsequent recovery process and operated in parallel with elution, tend to build up. This can have the effect of poisoning the eluate solution, reducing the efficiency of either elution, precious metals recovery, or both. The adverse effect of copper loaded on carbon has been discussed (Section 7.1.3.8), and copper can be removed from carbon prior to gold and silver using a cold elution step. Other examples of eluate poisoning are the following:

- Buildup of silica and nickel in the eluate, neither of which are deposited during electrowinning
- Buildup of zinc ions in the eluate, dissolved during zinc precipitation

This requires that a portion of the elution and recovery system solution be removed or bled off after each cycle or after a number of cycles. This may be delivered to the leach or CIP circuit with no detrimental effect [58]. The use of a slow, continuous bleed of solution may help to reduce the effects of this solution on the process stream to which it is returned, particularly with respect to cyanide concentration.

Overall gold elution efficiencies typically range from 97% to 99.5% for most carbon circuits.

7.1.5.7 Reactivation

Sequence of desorption, acid washing, and thermal reactivation. The optimum sequence for desorption and reactivation processes is a subject of controversy within the gold extraction industry. The options available are as follows:

1. Acid washing–desorption–thermal reactivation
2. Desorption–acid washing–thermal reactivation
3. Desorption–thermal reactivation–acid washing.

In the first option, inorganic fouling species are removed before desorption (and thermal activation), which may improve the efficiency of precious metals desorption. In addition, the amount of calcium and other metals that are introduced into eluate solutions is reduced, which may have some beneficial effect on subsequent recovery processes (i.e., electrowinning and zinc precipitation) and may reduce scaling in the desorption circuit. This is an advantage for ores where high lime addition is required to neutralize oxidized leach slurries.

The second option avoids any possible corrosion effects in the desorption circuit due to chloride ions introduced during acid washing (where HCl is used) but may result in corrosion of the thermal reactivation kiln and associated equipment.

Both options 1 and 2 have the advantage that inorganic foulants are removed prior to thermal reactivation. This is important because carbon losses during thermal reactivation are increased in the presence of calcium carbonate. For example, carbon containing 0.5% calcium results in a carbon mass loss that is 6% greater than that for carbon containing 0.1% Ca, when reactivated at a temperature of 750°C [18].

Finally, the third option prevents any corrosion of elution column and thermal reactivation process equipment resulting from residual chloride species, but the benefits of acid washing prior to both elution and thermal reactivation are lost.

Carbon quality. The quality of carbon that is returned to the adsorption circuit determines the efficiency of gold adsorption. Carbon quality, or activity, may be quantified by determining the loading rate and equilibrium loading factors, described earlier in Section 7.1.2.3. More commonly, quicker and easier tests are applied to provide a relative indication of carbon quality. These are summarized as follows:

- A fixed mass of carbon (usually a few grams) is mixed with a standard gold cyanide solution for a fixed time. The amount of gold adsorbed is expressed as a percentage of the total gold in the feed solution to provide a relative indication of carbon activity. Commonly, the amount of gold adsorbed is expressed as a percentage of gold adsorbed by the same weight of a fresh "virgin" sample of the same carbon type.
- The bulk density of the carbon is estimated to give an indication of the proportion of organic and inorganic contaminants on reactivated carbon. Virgin activated carbons typically used in gold extraction have bulk densities between 540 and 570 g/L. As a comparison, fouled carbon with a calcium content of, for example, 3% may have a bulk density of 640 g/L, depending on the amounts of other contaminants present.

- The calcium content of the carbon is determined by a wet analytical technique (i.e., acid digestion and atomic adsorption analysis). This is most useful where calcium carbonate is one of the major causes of carbon fouling. As a general guide, the presence of <0.5% calcium on carbon, in the absence of other significant contaminants, will typically have a minor effect on activity, whereas the activity may be severely reduced for calcium content >3%. The calcium content of carbon will increase by 0.1% to 1.0% for each cycle through an adsorption circuit, depending on the ore properties, solution conditions, and method of pH modification used.

In addition, carbon particle size distribution may be determined after each, or selected, reactivation cycle(s). This indicates the rate of carbon breakdown and may also provide useful information for optimizing carbon screening size.

7.1.5.8 Carbon Attrition Losses

Activated carbon breaks down in gold extraction systems, representing a loss of carbon itself, but also resulting in the loss of gold values. Carbon losses occur by a combination of the following mechanisms:

- Carbon–solid attrition in slurry systems (CIP, CIL)
- Carbon–carbon attrition in solution and slurry systems (CIC, CIP, CIL, elution, acid washing)
- Carbon–carbon, carbon–steel attrition in dry systems (during thermal reactivation, transfer in bins, etc.)
- Carbon breakage during transfer (in pumps, eductors, and on screens)
- Breakage due to thermal shock during thermal reactivation and quenching
- Chemical shock or during chemical removal of carbon–inorganic composites

Carbon attrition losses vary from plant to plant, depending on the process conditions applied (i.e., treatment rate, carbon transfer rate, type of adsorption system, etc.), the types of equipment used (especially carbon transfer pumps), carbon residence time in the circuit, and the type of carbon. Typically carbon consumption varies from 15 to 80 g/t, with an industry average of about 40 g/t. Carbon consumption is generally lower in CIC (solution) systems (20 to 40 g/t) compared with CIP and CIL (slurry) systems (40 to 60 g/t), but this depends on the configuration of the system and carbon cycle times.

Considerable effort has been directed at establishing where attrition losses occur, but quantification of these losses is difficult because of the small carbon weight loss achieved in each process unit. In one study, carbon losses were distributed approximately as follows [18, 49]:

- Adsorption circuit mixing: 40%
- Interstage carbon transfer (including transfer to elution): 6%
- Elution (including transfer to regeneration): 7%
- Regeneration (including quenching and final sizing): 47%

An estimated 41% of the carbon lost within the adsorption circuit (i.e., adsorption + interstage carbon transfer = 46%) was lost to the final tailings, with only 5% recovered on a tailings safety screen. The carbon lost to the tailings contains any gold that it adsorbed while in the adsorption system. There is some controversy as to whether this carbon desorbs any of the contained gold or transfers it to particles with lower gold loadings as it moves through the carbon adsorption circuit. It is unlikely that either of these mechanisms occurs to any significant extent because it must be assumed that the fine carbon, once generated, passes through the circuit rapidly with the slurry, and little if any equilibration of gold loading is likely.

7.1.6 Adsorption from Noncyanide Solutions

The recovery of gold from noncyanide solutions is an important consideration in the development of possible alternative gold leaching systems, for example, chloride, thiosulfate, thiourea, or thiocyanate leaching. Although no such processes are used commercially, the chemistry of adsorption differs from cyanide solutions and is worthy of consideration.

The ability of activated carbon to adsorb gold complexes follows the sequence [10, 31]:

$$AuCl_4^- > Au(CN)_2^- > Au(SCN)_2^- > Au(CS(NH_2)_2)_2^+ > Au(S_2O_3)_2^{3-}$$

The potentials for the reductions of gold chloride and other complexes suggest that the mechanism of adsorption from chloride and thiocyanate solutions involves the reduction of gold species, whereas alternative mechanisms are likely to apply from other media (e.g., thiosulfate), with thiourea being a borderline case.

7.1.6.1 Chloride

Activated carbon was used at the Youanmi mine (Western Australia) in the 1920s to recover gold from industrial chloride leaching solutions. The mechanism of adsorption of gold from acidic chloride solutions appears to be relatively straightforward and involves the reduction of Au(I) or Au(III) to metallic gold on the carbon surfaces. The relevant reactions are:

$$AuCl_4^- + 3e \rightleftharpoons Au^0 + 4Cl^-; \quad E^0 = 1.00 \text{ (V)} \qquad \text{(EQ 7.16)}$$

$$AuCl_2^- + e \rightleftharpoons Au^0 + 2Cl^-; \quad E^0 = 1.16 \text{ (V)} \qquad \text{(EQ 7.17)}$$

$$AuCl_4^- + 2e \rightleftharpoons AuCl_2^- + 2Cl^-; \quad E^0 = 0.92 \text{ (V)} \qquad \text{(EQ 7.18)}$$

where the reducing electrons are supplied by the activated carbon, according to the following anodic reaction:

$$C + 2H_2O \rightleftharpoons 4H^+ + CO_2 + 4e; \quad E^0 = 0.21 \text{ (V)} \qquad \text{(EQ 7.19)}$$

This reduction potential agrees with measured values, which vary between 0.1 and 0.4 V [17, 31].

Gold deposition from chloride solutions is thought to occur initially at suitable sites to form spherical particles of 0.5 to 35 μm and spreads over the surface to yield a metallic gold-colored surface [59]. The main factors affecting the reaction are carbon particle size (i.e., superficial surface area), temperature, and chloride ion activity, with HNO_3 and Fe(III) concentration having little effect [60]. The adsorption, or deposition, kinetics are increased with increased temperature, due to the decrease in stability of the $AuCl_4^-$ complex [59].

The rate of gold chloride reduction is controlled by boundary layer diffusion and is insensitive to the initial gold concentration. The rate is also independent of pH at values <7 but decreases markedly as the pH increases above this value. Also, sodium hypochlorite, which is commonly present in chloride leaching systems, has an adverse effect on the adsorption rate because of its oxidizing properties.

The effect of chloride concentration can be understood from the Nernst equation, as follows:

$$E = E^0 + (0.059/n) \log ([AuCl_4^-]/[Cl^-]) \qquad \text{(EQ 7.20)}$$

As [Cl⁻] increases, E decreases and therefore the difference between the reduction potential of activated carbon and the gold deposition reaction decreases. Hence, the adsorption driving force is decreased and with it the gold-loading rate.

Gold is not deposited throughout the carbon granule but only on the surface and in the larger pores open to bulk solution. In contrast to adsorption from cyanide solutions, carbon granule size has a major effect on the equilibrium loading. The size of the spherical gold particles varies in size in the range 0.5 to 35 μm and is directly dependent on the gold concentration in solution [60].

Elution of gold from carbon loaded from chloride solutions is difficult to achieve and must be approached as a dissolution, rather than a desorption, process.

7.1.6.2 Thiosulfate

The gold thiosulfate complex is not adsorbed onto activated carbon to any appreciable extent and the process is not suitable gold recovery from thiosulfate media [61, 62]. Ion exchange resins have been proposed for recovery of gold from thiosulfate and this is discussed further in Section 7.2.5.2.

7.1.6.3 Thiourea

The gold thiourea complex, $Au(CS(NH_2)_2)_2^+$, is thought to be adsorbed onto activated carbon without undergoing any chemical change [63]. This mechanism is based on observations that gold and sulfur are loaded onto carbon in the molar proportion 1:2, the same ratio that exists in the solution phase. However, it is also possible that adsorption of an ion pair involving anions such as ClO_4^-, Cl^-, or HNO_3 occurs [63]. An interesting aspect of thiourea adsorption is that the gold is present in solution as a cationic complex, unlike other systems where the gold complex anions are adsorbed.

The loading characteristics are very similar to those obtained from cyanide solutions: The adsorption kinetics are relatively slow, and a true equilibrium is not attained due to the very slow diffusion of the large thiourea molecule through the carbon micropores. Disadvantages of the thiourea system are the detrimental effect of the thiourea ion concentration in solution, which needs to be sufficiently high to ensure satisfactory gold leaching performance, and the possibility of depositing elemental sulfur within the carbon pores, thereby reducing adsorption capacity and kinetics.

Elution of gold from thiourea-loaded carbon is possible using cyanide or sulfide solutions. Simple acids (sulfuric and hydrochloric) and bases (sodium hydroxide) have been found to be ineffective eluants [63].

7.1.6.4 Ammonia–Cyanide

The recovery of gold from cyanide solutions containing ammonia is complicated by the fact that ammonia is adsorbed onto the carbon, competing for active sites and potentially reducing gold-loading efficiency. The ammonia–cyanide system has been proposed for the treatment of copper–gold ores, and leach solutions from this system contain copper cyanide complexes, which also load onto carbon. However, the effect of copper can be reduced significantly by adding free cyanide to clarified leach solutions prior to contact with carbon to ensure that the $Cu(CN)_3^{2-}$ and $Cu(CN)_4^{3-}$ complexes predominate (see Section 7.1.2.5) [35].

7.2   ION EXCHANGE RESINS

The use of ion exchange resins for the concentration and purification of gold from dilute cyanide solutions has been investigated since the late 1940s. Early research was performed in the United States, South Africa, Romania, and the former USSR, with the latter culminating in the installation of the first resin-in-pulp (RIP) plant at the large Muruntau operation (western Uzbekistan) in about 1970. In this case the decision to use resin was undoubtedly influenced by the mining company's experience in resin technology for uranium extraction. Other plants in Russia, Uzbekistan, Kazakhstan, and Kyrgyzstan (all former USSR states) have also reportedly used resin systems for gold extraction [64].

The application of resins outside the former USSR was preempted by the widespread use of carbon adsorption systems from about 1979. However, a 250-tpd RIP plant was commissioned at Golden Jubilee (Eastern Transvaal, South Africa) in 1988 [65].

Commercially available resins have been unable to compete with activated carbon in most mineral systems because of poor selectivity, mechanical breakdown of the beads, and the requirement for complex elution and regeneration processes. However, resins offer some chemical advantages over activated carbon and have excellent technical potential for application in gold extraction systems.

7.2.1 Properties of Resins

Ion exchange resins are synthetic materials which consist of an inert matrix (e.g., polystyrene-divinyl benzene co-polymers) and contain surface functional groups, such as amines and esters. Resins prepared without using a solvent diluent in the process have a gel-type matrix structure, whereas those prepared with solvents have a more open, macroporous structure. The latter are preferred for use in gold extraction systems because they provide high surface area for ion adsorption and have better mechanical strength than the gel-type resins.

The functional groups can exchange ions with other similarly charged ionic species in solution, depending on the preference of a particular functional group for a specific ion. This depends on the properties of the functional group and the charge, size, and polarizability of the ions in solution. Functional groups can be basic (anion exchangers) or acidic (cation exchangers) and can be further classified as having weak or strong properties, depending on their degree of dissociation in solution. Alternatively, highly selective chelating functional groups can be attached to a resin matrix, but these are of limited interest in gold extraction. A summary of resin types and their capacity for gold cyanide is given in Table 7.5.

Resins are usually produced as beads, generally ranging from 0.25 to 0.60 mm in diameter. Because the physical strength of resins depends strongly on the matrix structure, consequently different resin types show variable attrition resistance. Resin structures are also susceptible to thermal and osmotic shock. The maximum recommended operating temperature for many commercially available resins is between 60°C and 70°C, and osmotic shock resulting from repeated exposure to acid and alkaline solutions can lead to severe chemical and physical degradation. These factors present significant problems for some of the elution and regeneration processes, discussed in Section 7.2.3. More recently, the use of impregnated and/or interpenetrated polyurethane-based polymers that can be produced as foams, films, and fibers, as well as beads, have been proposed and shows promise for gold adsorption from cyanide and noncyanide solutions [67].

7.2.2 Adsorption from Cyanide Solutions

Adsorption of gold and silver from cyanide solutions can be achieved with both strong- and weak-base resins. Strong-base resins typically have high loading capacity and fast loading rates, but they have poor selectivity (due to base metals adsorption) and are difficult to elute. Weak-base resins are more selective and much easier to elute but have lower loading capacity (25% to 50% of strong-base resins capacity) and slower loading rates

TABLE 7.5 Selected ion exchange resins and their capacity for gold–cyanide [66]

Resin No.	Structure R=⟨ring⟩	Strong-Base Capacity (weak-base capacity) mol/L	Gold in Solution After Extraction* (mg/L) (amount of resin used in extraction)			
			(25 mg)	(30 mg)	(55 mg)	(80 mg)
1[†]	R–N⁺(CH₃)₃	1.40 (0)	0.45	0.31	0.15	0.12
2[‡]	R–N⁺(CH₃)₃	0.80 (0)	0.84	0.67	0.37	0.26
3	R–N⁺(CH₃)₂–R and R–N⁺(CH₃)₂	0.125 (0.23)	0.23	0.10	0.03	0
4	R–N-pyridine	0.3 (0)	22.40	9.30	0.46	0.17
5	R–N⁺(imidazole)N–CH₃	0.47 (0)	0.51	0.33	0.14	0.07
6	R–N⁺(imidazole)N–R and R–N(imidazole)N	0.21 (0.29)	1.46	0.65	0.02	0
7	Similar to Resin No. 6	0.17 (0.69)	11.80	1.29	0.06	0

* Initial solution: volume 10 mL, [Au] = 415 g/L, pH 10.
† Rohm & Haas IRA 400 commercial gel strong-base resin.
‡ Mintek macroporous strong-base resin.

[65]. Resins with predominantly weak-base properties but with some strong-base characteristics, resulting from the presence of a small percentage (10% to 15%) of quarternary amine functional groups, are likely to be optimal for the extraction of gold from cyanide solutions.

7.2.2.1 Adsorption onto Strong-Base Resins

The functional groups of strong-base resins are quarternary, and occasionally tertiary, amines which have a permanent positive charge in aqueous solution.

The adsorption of the Au(I) cyanide complex is represented as follows:

$$]-^+NR_3X^- + Au(CN)_2^- \rightleftharpoons]-^+NR_3Au(CN)_2^- + X^- \quad \text{(EQ 7.21)}$$

where

$]-$ = the inert portion of the resin
R = the CH_3 species
X^- = an anion such as sulfate or bisulfate, depending on the elution and regeneration methods used (Section 7.2.3)

The resin gold-loading capacity depends on the resin structure, the concentration of functional groups present, the concentrations of the various ions in solution, and their properties. Gold loadings >100 g/L resin (i.e., about 25% gold content) can be achieved when little or no competing ions are present in the solution; however, the capacity decreases with increasing ionic strength (i.e., more competing ions) and with increasing temperature. The effect of temperature on resin performance is similar, though less significant (particularly loading capacity), than its effect on metal adsorption by activated carbon, as shown in Figures 7.26 and 7.27.

Strong-base resins also readily adsorb silver, nickel, cobalt, copper, zinc, Fe(II), and Fe(III) cyanide complexes and consequently show poor selectivity for gold. It can be seen from Figure 7.28 that copper, nickel, and zinc are adsorbed more readily than gold and silver, and consequently must be extracted almost completely onto the resin before the precious metals can be effectively recovered. The selectivity of adsorption of the different metal cyanides is pH dependent, as illustrated in Figure 7.29. At low pH, metal cyanide salts are precipitated, which reduces competition for resin adsorption sites. Selective precipitation of base metals prior to precious metal extraction by strong-base resins has been suggested as a process option but has not been applied commercially [65].

The rate of gold adsorption is first order with respect to gold activity in solution at low resin loadings (less than approximately 50 g/L) and for solution gold concentrations up to approximately 40 mg/L. In addition, the adsorption rate:

- Decreases as the resin becomes loaded (similarly to carbon)
- Is largely unaffected by pH between 2 and 12
- Increases with increasing temperature
- Increases with increasing agitation up to a limiting level, above which pore diffusion is rate limiting

The activation energy for gold adsorption or to a strong-base resin has been estimated at 16.5 kJ/mol, which is within the range of mass transport control. The rate-controlling step is either diffusion across the resin-solution boundary layer or diffusion along resin pores. Pore diffusion is strongly affected by solution pH and ionic strength and is favored by high pH and low ionic strength. It has been suggested that boundary layer diffusion significantly affects the rate of gold extraction under most conditions encountered in commercial RIP and solution circuits, and hence the rate is maximized by good mixing [65]. Strong-base resins are capable of faster gold-loading rates than activated carbon (see Figure 7.26).

7.2.2.2 Adsorption onto Weak-Base Resins

Weak-base resins are characterized by primary, secondary, or tertiary amine functional groups, or mixtures of these. In aqueous solution the functional groups become protonated, that is:

$$]-NR_2 + HX \rightleftharpoons]-^+NR_2HX^- \tag{EQ 7.22}$$

where

$]-$ = the inert portion of the resin
R = the CH_3 species

A pK_a value can be used to define the condition at which 50% of the functional groups are protonated. In acidic solutions the equilibrium of Equation (7.22) is shifted to the right, and the resin behaves like a strong-base ion exchanger.

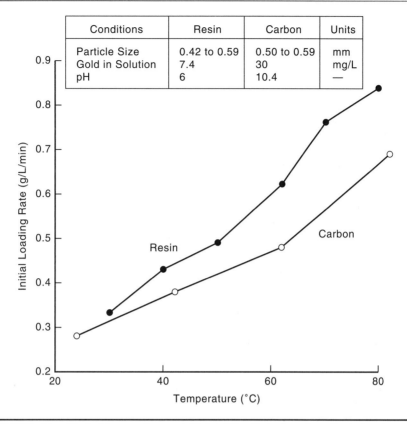

FIGURE 7.26 The effect of temperature on initial gold-loading rate onto activated carbon and a strong-base ion exchange resin (adapted from [65])

The adsorption of the Au(I) cyanide complex is given by:

$$]-{}^+NR_2HX^- + Au(CN)_2^- \rightleftharpoons]-{}^+NR_2HAu(CN)_2^- + X^- \quad (EQ\ 7.23)$$

The loading capacity of weak-base resins is approximately half that of strong-base resins under similar conditions. This capacity depends on the number of protonated functional groups per unit volume of resin and their degree of protonation, as well as the concentrations of gold and other competing species in solution. Consequently, the capacity is strongly dependent on the pK_a of the resin and solution pH, as shown in Figure 7.30. The higher the resin pK_a, the higher the pH at which optimum gold loading is achieved. For resins with pK_a values between 8 and 9, maximum loadings are typically achieved in the range of pH 6 to 8.

The selectivity of weak-base resins for gold and silver over other metals is illustrated in Figure 7.31. Although selectivity improves with increasing pH, both the capacity and adsorption rate are reduced markedly as the pH is increased >8. Weak-base resins, which have relatively low pK_a values (i.e., 6 to 8), are generally unsuitable for use in alkaline cyanide solutions, because the ideal pH conditions for cyanide leaching (10 to 11) tend to strip metal values off the resin. Development work during the late 1980s has produced weak-base resins with higher pK_a values (8 to 12), which are more suitable for use in cyanidation systems [66].

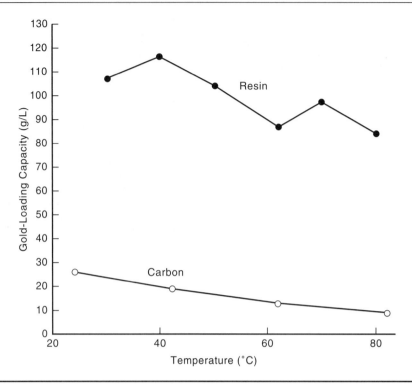

FIGURE 7.27 The effect of temperature on gold-loading capacity on activated carbon and a strong-base ion exchange resin (adapted from [65])

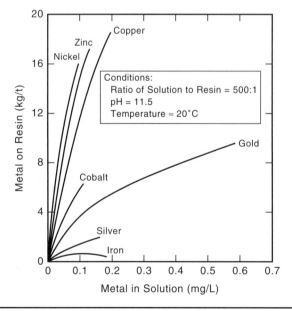

FIGURE 7.28 Equilibrium loading of various metal cyanide complexes from a Grootvlei (South Africa) pregnant solution onto a strong-base resin (A101 DU) [65]

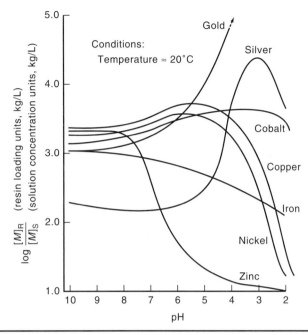

FIGURE 7.29 Effect of pH on the selectivity of a strong-base resin (A101 DU) for metal cyanide complexes [65]

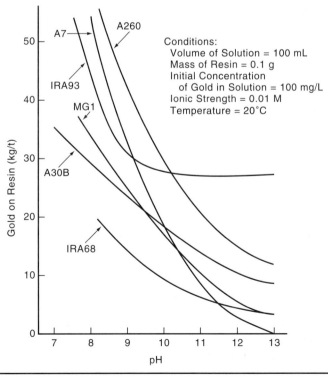

FIGURE 7.30 Effect of pH on the equilibrium loading of aurocyanide onto weak-base resins (0.1 g of each) from a solution containing potassium aurocyanide alone [65]

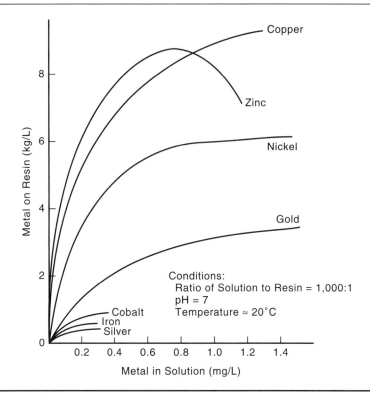

FIGURE 7.31 Equilibrium loading of various metal cyanide complexes from Durban Roodepoort Deep (South Africa) pregnant solution onto a weak-base resin (A7) [65]

Other factors affecting the rate of adsorption onto weak-base resins are similar to those discussed for strong-base resins and are not discussed further here.

7.2.2.3 Adsorption onto Mixed Weak- and Strong-Base Resins

Weak-base resins always show some strong-base properties due to the presence of a few quarternary amine functional groups, formed when some of the secondary or tertiary amine groups cross-link. Resins with a low, optimized, strong-base content, which is evenly distributed in the matrix, should give excellent selectivity for gold and silver. Resins with strong-base capacities of 12% to 16% have been shown to have good selectivity at pH 10.6, but selectivity decreases with increasing strong-base capacity. The optimum resin for gold and silver extraction should have predominantly weak-base characteristics with a high pKa, preferably between 10 and 12, and an optimized strong-base content [66].

7.2.2.4 Chelating Resins

Chelating resins have been examined, as they can be highly selective for gold, for example, a chelant synthesized from diallyamime, with 1,6-bis(diallyamino)-hexane as the preferred co-monomer. The gold may be recovered from the resin by combustion at about 800°C or by eluting with a solution of 0.5% thiourea in 2 M HCl [68]. However, this system is unattractive compared with anion exchange resins since lower gold loadings are achieved and elution is relatively difficult.

7.2.3 Elution and Regeneration

Loaded resins must be eluted to enable subsequent gold recovery and allow recycling of the resin to the adsorption circuit, for the process to be economic. Stripped resin should contain 25 to 50 g Au/t for effective reuse. In addition, to avoid buildup of poisoning elements that would otherwise reduce gold adsorption, other adsorbed species must be removed from the resin and eliminated from the system.

The advantages and disadvantages of various elution processes for strong- and weak-base resins are summarized in Table 7.6. These are considered further in the following sections.

7.2.3.1 Elution from Strong-Base Resins

The elution of metal species from strong-base resins is more difficult than from weak-base resins because of the greater strength of ion adsorption. Elution can be achieved either by shifting the equilibrium of Equation (7.21) to the left by increasing the concentration of a competing anion in solution, or by converting the adsorbed metal ions to a nonionic species (e.g., the gold–thiourea complex). In the former case, simple anions, such as thiocyanate, chloride, bisulfate, nitrate, or cyanide ions may be used. High concentrations of these ions are required because they are more weakly adsorbed than the gold cyanide complex, and a polar organic solvent is usually required to increase the activity of the eluate, for example, acetone or acetonitrile. Alternatively, a complex anion, such as Zn(II) cyanide, which is more strongly adsorbed than the Au(I) cyanide, may be used at relatively low concentrations to elute gold-loaded resin [65, 70].

Four elution processes have been developed from this theoretical basis, each of which may have application for elution of specific resin types in particular applications: zinc cyanide, thiourea, thiocyanate, and chloride.

The relative elution efficiency of the first three of these methods for gold removal is illustrated in Figure 7.32. However, the effectiveness for the removal of poisoning elements decreases in the following order (as indicated by Table 7.7):

$$\text{zinc cyanide} > \text{thiocyanate} \ggg \text{thiourea}$$

and the relative cost of the three methods has been estimated as follows [69, 71]:

$$\text{zinc cyanide} \ll \text{thiocyanate} < \text{thiourea}$$

Consequently, the most appropriate elution method depends on the specific chemistry of a particular ore treatment.

The zinc cyanide elution process has been used commercially at Golden Jubilee; a complex process using thiourea is applied on a large scale at Muruntau; and the thiocyanate method has been proposed for elution of mixed strong- and weak-base resins developed specifically for gold extraction [66, 69]. A sequential chloride elution process has also been developed and proposed for treatment of resin loaded with mercury, silver, and gold. Details of the four resin elution processes are provided in the following sections.

Zinc cyanide. The elution of gold cyanide from strong-base resin using zinc cyanide is given by the following equation:

$$2]\text{-}^+\!NR_3Au(CN)_2^- + Zn(CN)_4^{2-} \rightleftharpoons (]\text{-}^+\!NR_3)_2 Zn(CN)_4^{2-} + 2Au(CN)_2^- \qquad \text{(EQ 7.24)}$$

where
$\quad$]– = the inert portion of the resin
$\quad$ R = the CH_3 species

TABLE 7.6 Advantages and disadvantages of various processes for the elution of strong-base (S/B) and weak-base (W/B) resins [69]

	$Zn(CN)_4^{2-}$ (S/B)	SCN^- (S/B)	$CS(NH_2)_2$ (S/B)	NaOH (W/B)
Advantages				
	All anionic cyanide complexes eluted efficiently.	All anionic cyanide complexes stripped in elution–regeneration.	Resin does not have to be regenerated after elution.	All anionic cyanide complexes stripped in elution and regeneration.
	Resin can be regenerated to full capacity each cycle with efficient recycle of chemicals.	Resin can be regenerated to full capacity each cycle with efficient recycle of chemicals.		Resin can be regenerated to full capacity each cycle with inexpensive chemicals.
	No problems with materials of construction in elution and electrowinning.	Nontoxic chemicals for elution and regeneration.	Nontoxic chemicals for elution.	Nontoxic chemicals for elution and regeneration.
		Fast elution kinetics.		Fast elution and electrowinning kinetics.
				No problems with materials of construction in elution and electrowinning
Disadvantages				
	Resin has to be chemically regenerated after elution.	Resin has to be chemically regenerated after elution.	Elution kinetics slow.	Resin should be treated with acid after elution.
	Elution kinetics slow.	Slight corrosion of electrodes in electrowinning cell.	Corrosion of electrodes in acidic eluate.	
	Toxicity of hydrocyanic gas produced during regeneration.		Poisoning of resin by cobalticyanide.	
			Decomposition of thiourea in acidic solution.	

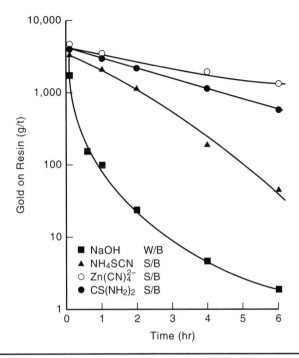

FIGURE 7.32 Relative rates of gold elution from strong-base (S/B) and weak-base (W/B) resins under the standard electro-elution conditions [69]

TABLE 7.7 The elution efficiency of various anionic metal cyanide complexes from anion exchange resins using zinc cyanide, thiocyanate, thiourea, and sodium hydroxide under standard conditions [68]

	Elution Efficiency (%)			
	Strong-Base Resin			Weak-Base Resin
Metal Ion	$Zn(CN)_4^{2-}$	SCN^-	$CS(NH_2)_2$	NaOH
Silver	~100	~100	~100	70
Copper	~100	~100	53	~100
Cobalt	87	63	0	78
Nickel	85	58	~100	66
Zinc	ND*	0	0	0
Iron	90	87	0	58

* ND = not determined.

The elution rate increases with increasing zinc cyanide concentration up to 0.15 M $Zn(CN)_4^{2-}$; however, the optimum concentration to minimize residual gold loadings is 0.5 M, taking into consideration that elevated temperatures (>50°C) are required to prevent crystallization of zinc cyanide salts.

The activation energy of the reaction is estimated at 26 kJ/mol, which indicates some degree of chemical control, favoring high temperature elution. In practice, temperatures of 50°C to 60°C are used. Other metal species, such as silver, cobalt, nickel, zinc, and iron, are also effectively stripped from the resin by this method.

The eluted resin must be regenerated to remove the adsorbed zinc cyanide species from the functional groups to allow reuse for gold adsorption. This is most effectively achieved with sulfuric acid <pH 2, as follows [65]:

$$(]-^+NR_3)_2Zn(CN)_4^{2-} + 3H_2SO_4 \rightleftharpoons 2]-^+NR_3HSO_4^- + ZnSO_4 + 4HCN \qquad (EQ\ 7.25)$$

This method reduces zinc loadings rapidly, that is, from 100 kg/t to <100 g/t in 4 to 8 hr. The hydrogen cyanide evolved is scrubbed into alkaline solution (caustic or lime) in a closed tank system as it is produced.

Thiourea. Thiourea reacts with adsorbed gold species to form a positively charged complex, which can be washed off the resin with water. The overall elution reaction is given as follows:

$$]-^+NR_3Au(CN)_2^- + 2CS(NH_2)_2 + 2H_2SO_4 \rightleftharpoons]-^+NR_3HSO_4^- + \\ Au(CS(NH_2)_2)_2HSO_4 + 2HCN \qquad (EQ\ 7.26)$$

where

]– = the inert portion of the resin
R = the CH$_3$ species

The rate of elution increases with increasing thiourea concentration, up to a maximum of 1 M; and with increasing sulfuric acid strength, also up to approximately 1 M. The activation energy, estimated at 19 kJ/mol, indicates that some benefit might be obtained from eluting at elevated temperatures. However, this is not possible, because thiourea decomposes at ambient temperature in acidic solution to form elemental sulfur, and the decomposition rate increases with increasing temperature. This reaction is thought to be catalyzed by the resin itself, resulting in high reagent consumptions.

Eluant solutions containing only thiourea are very inefficient at removing base metal ions, especially cobalt, iron, and to some extent zinc and nickel. Consequently, the process is unsuitable for treating solutions containing high concentrations of these species. For example, thiourea elution of a resin loaded from a solution containing >1 mg/L cobalt will result in rapid resin poisoning.

The use of thiourea in acidic solutions also presents some other problems:

- Steel wool cathodes in electrowinning cells and other metal process components are readily corroded in the acidic media required for thiourea elution.
- Osmotic shock produced by repeated elution in acidic media and adsorption in alkaline media degrades the resin structure, resulting in resin loss.

Despite this, the thiourea method found favor in the former USSR (Russia, Uzbekistan, and Kazakhstan) and has been used in combination with other elution steps to remove fouling elements (Section 7.2.4.6).

Thiocyanate. Elution by thiocyanate involves the displacement of the gold cyanide complex by the thiocyanate (SCN$^-$) ion, represented by the following reaction:

$$]-^+NR_3Au(CN)_2^- + SCN^- \rightleftharpoons]-^+NR_3\ SCN^- + Au(CN)_2^- \qquad (EQ\ 7.27)$$

where

]– = the inert portion of the resin
R = the CH$_3$ species

The elution rate increases with increasing thiocyanate concentration, up to a maximum of 2 M SCN$^-$, and within an optimum pH range of 7 to 8. The activation energy is estimated at 15 kJ/mol, which suggests that elevated temperatures are unlikely to improve

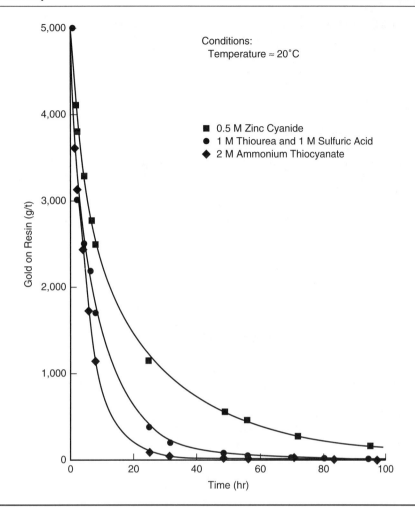

FIGURE 7.33 Rate of gold elution from a strong-base resin with various eluants [69]

elution kinetics significantly. Typically, faster elution rates are achieved with thiocyanate than either with zinc cyanide or thiourea, as indicated in Figure 7.33.

After elution, the resin must be regenerated to allow reuse for metal adsorption. Mineral acids (e.g., sulfuric and hydrochloric) can be used, but the thiocyanate ion decomposes rapidly to elemental sulfur in the presence of strong acids. An alternative is to treat the resin with an Fe(III) solution to form the cationic Fe(III) thiocyanate complex, which is then readily washed off the resin with water:

$$4]\text{-}^+NR_3\ SCN^- + Fe_2(SO_4)_3 \rightleftharpoons 2(]\text{-}^+NR_3)_2SO_4^{2-} + 2Fe(SCN)_2^+ + SO_4^{2-} \quad (EQ\ 7.28)$$

In this case an Fe(III) concentration of approximately 0.5 to -1 M is required for effective regeneration.

The thiocyanate is recovered by the addition of hydroxide to precipitate Fe(III) hydroxide:

$$Fe(SCN)_2^+ + 3OH^- \rightleftharpoons Fe(OH)_3 + 2SCN^- \quad (EQ\ 7.29)$$

The thiocyanate ions thus formed are available for recycling to the elution process.

Chloride. A sequential elution process using chloride ions for removal of gold and silver has been developed [68, 72]. The process uses 2 N HCl for elution of mercury, followed by treatment with 200 g/L sodium chloride (NaCl) in 1 N HCl for silver removal, and finally gold elution with 0.73% NaOCl in a solution containing 150 g/L NaCl and 5 g/L sodium hydroxide (NaOH). The kinetics and relative economics of this process are not known.

7.2.3.2 Elution from Weak-Base Resins

At pH values above the resin pK_a, the equilibrium of the ion exchange reaction given by Equation (7.23) lies to the left, and the functional groups are no longer able to hold the adsorbed anions, that is, the resin is converted to the free base form. As a result, weak-base resins can be eluted by simple hydrolysis with sodium hydroxide [69]:

$$]\text{-}^+\text{NR}_2\text{HAu(CN)}_2^- + \text{OH}^- \rightleftharpoons]\text{-NR}_2 + \text{Au(CN)}_2^- + \text{H}_2\text{O} \quad \text{(EQ 7.30)}$$

where

]– = the inert portion of the resin
R = the CH_3 species

The rate of elution increases with increasing sodium hydroxide concentration up to an optimum concentration of 0.5 M. At higher concentrations the rate is reduced, probably due to the precipitation of zinc hydroxide and/or other metal hydroxides. The rate is also increased at elevated temperatures, but fast elution is achieved under ambient conditions, typically much faster than carbon and strong-base resin elution processes (see Figure 7.32); for example, in one case loaded resin was stripped from 1,800 g/t to <10 g/t in <1 hr [69]. The effect of gold concentration on elution kinetics and efficiency is less pronounced than that observed for strong-base resin elution.

Sodium hydroxide elution removes gold and copper quite efficiently (Table 7.7); however, silver, nickel, zinc, and iron are relatively poorly eluted, particularly in acidic solutions, due to the formation of insoluble metal cyanides, for example:

$$\text{Zn(CN)}_4^{2-} + 2\text{H}^+ \rightleftharpoons \text{Zn(CN)}_2 + 2\text{HCN} \quad \text{(EQ 7.31)}$$

$$\text{Zn(CN)}_2(\text{aq}) \rightleftharpoons \text{Zn(CN)}_2(\text{s}) \quad \text{(EQ 7.32)}$$

The elution of silver, iron, and nickel during caustic treatment can be significantly improved by the addition of cyanide to the eluant. The removal of acid-dissociable cyanides, such as copper, zinc, and nickel, can be achieved effectively by treatment with dilute mineral acid (e.g., 5% sulfuric acid) following elution, which breaks down the adsorbed metal cyanide species to Cu^{2+}, Zn^{2+}, and Ni^{2+} ions, respectively, which are easily washed off the resin.

7.2.4 Process Considerations

7.2.4.1 Application of Resins

Resins have the potential to provide faster loading kinetics and higher loading capacity than activated carbon in many gold extraction systems, particularly leach solutions with low concentrations of competing ions. In addition, resins have clear advantages over the use of activated carbon for the treatment of some specific ores because they are less affected by the following:

- Fine particles that tend to coat carbon surfaces and block pores, for example, clay minerals and roaster calcine products

- Calcium carbonate and other inorganic foulants
- Operation at elevated temperature (e.g., roaster calcine pulps)
- Poisoning by naturally occurring organic compounds (i.e., humates and fulvates) or organic reagents introduced into the process (e.g., kerosene and flotation reagents)

Resins can be applied to the treatment of solution (RIS) or pulp (RIP/RIL), in a similar manner to the well-established methods for applying activated carbon (CIC and CIP/CIL).

A number of resin manufacturers have focused efforts on the development of resins for the gold extraction industry, including Cognis (Tucson, Arizona) with its guanidine-based extractant (AuRIX100), which is a weak-base ion exchange resin with intermediate basicity between that of simple amines and quarternary amines (i.e., it has some strong-base properties) [76, 77]. This appears to meet, or come close to, the requirements for an optimal gold extraction resin and has great potential for commercial application (see Section 7.2.2). An important feature of this resin is that the guanidine function is sufficiently basic to deprotonate water to form the guanidium cation. When the alkalinity of the solution is increased, neutral guanidine functionality is established, and negatively charged ions (i.e., Au(I) cyanide) are released. This provides an effective elution mechanism in alkaline media, which eliminates the need to expose the resin to acid (Sections 7.2.3.1 and 7.3.2.2). Mintek (South Africa) has worked on the development of resins for the gold industry, discussed further in Section 7.2.4.7.

In addition, research and development of other applications of ion exchange resins for metals and cyanide recovery continues (see Section 11.2.3 for discussion on the use of ion exchange for effluent treatment and recycling, and the AuGMENT process) [74, 75, 78].

7.2.4.2 Resin-in-Solution System

RIS systems can be operated in the "pinned" or "packed" bed mode (i.e., downflow or closed columns) or as fluidized beds (i.e., upflow columns). Downflow systems have proved notoriously difficult to operate unless well-clarified solutions are used, however schemes where resin is removed from pinned/packed bed adsorption systems for elution in an upflow column have been applied effectively. Resins have lower density than activated carbons, and lower flow rates (per unit cross-sectional area) than those used in carbon systems must be applied for fluidized bed applications.

7.2.4.3 Resin-in-Pulp System

Resin-in-pulp/resin-in-leach (RIP/RIL) systems are operated in similar configuration to CIP/CIL systems (Sections 7.1.5.2 and 7.1.5.3), with resin concentrations between 10 to 20 g/L maintained. The flowsheet for the Mintek–Grootvlei RIP pilot plant appears in Figure 7.34. The lower density of resins compared to carbon results in a greater tendency to develop a concentration profile in adsorption tanks with an increasing probability of resin floating on the slurry surface, as the slurry density goes higher. In-pulp adsorption systems may therefore need to be operated at lower slurry densities in general, although this depends on the characteristics of the slurry. Resin concentrations are typically higher than carbon in CIP and CIL systems, with volume concentrations ranging from 5% to 15% likely to be applied. The preferred configuration of adsorption tanks for RIP and RIL, based on developments at Mintek, and experience at Golden Jubilee and in Russia, Uzbekistan, and Kazakhstan, favors the use of air-agitated tanks (Pachuca-type) to keep resin and slurry well mixed; air-lifting for resin-slurry transfer into and out of the tanks; and static sieve bend screens (DSM type) to separate the resin from the slurry [78].

The resin particle size affects interstage screening which, due to the smaller size of commercially available resins compared with carbon, must separate at a smaller size.

350 | THE CHEMISTRY OF GOLD EXTRACTION

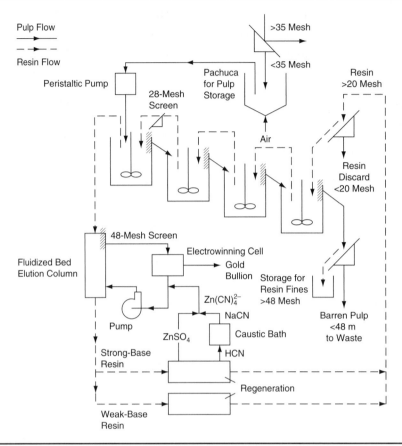

FIGURE 7.34 Schematic flowsheet for RIP gold extraction systems [69]

Also, advances in pulp preparation, such as the development of high-efficiency linear screens for slurry separation at fine sizes and interstage screening technology, have made it possible to screen efficiently at the finer sizes required for RIP/RIL applications. Typically, a screen aperture of 0.5 mm is likely to be required to separate the largest available resin beads (i.e., 0.8 to 1.2 mm) from slurry, although development of larger resin beads, foams, fibers, and other materials is ongoing.

7.2.4.4 Elution
All of the elution systems available for treatment of weak- and strong-base resins operate under relatively mild conditions, that is, low temperature and pressure. Temperatures above approximately 60°C may result in excessive resin degradation. Similar equipment to that used for carbon elution (Section 7.1.5.6) is required for resin elution. The downstream metal recovery process, usually electrowinning or precipitation, should be operated in series with elution to reduce the gold concentration in the eluant and allow recycling of the solution. Process solutions must be bled to remove base metal ions that would otherwise build up in the eluant.

7.2.4.5 Golden Jubilee Mine
At Golden Jubilee (South Africa), the high clay content of the ore made conventional solid–liquid separation and zinc cementation processes unattractive, and the ore gave

poor response to CIP due to high concentrations of organic acids (humic and fulvic acid), which poisoned the carbon [71, 79]. A comparison of the response of this ore to resin and carbon treatments is shown in Figure 7.35. In 1988 the CIP plant was replaced by RIP, using a strong-base resin. The increased kinetics and gold-loading capacity enabled plant throughput to be increased almost twofold to 390 tpd. The plant used four stages of adsorption with approximately 30 min retention time per stage, and achieved gold loadings of 5,000 to 7,000 g/t resin (Table 7.8).

Thiourea elution was used initially but was found to be uneconomic because of the need to discard the entire eluate after each elution cycle. This was necessary because solution fouling caused excessive corrosion in the electrowinning circuit [71]. The elution circuit was eventually replaced with a zinc cyanide elution system which operated more successfully. The resin was regenerated using sulfuric acid to remove zinc cyanide prior to returning the resin to the adsorption circuit (see Section 7.2.3.1).

The Golden Jubilee mine shut down in 1994 after the ore reserve was depleted [78].

7.2.4.6 Muruntau Mine

At Muruntau in Uzbekistan a mixed weak/strong-base resin has been used for adsorption of gold from cyanide leach solution since the early 1970s. The flowsheet reportedly incorporated a relatively complex three-stage elution process for sequential base metal and precious metal recovery. The flowsheet for this plant is discussed in more detail in Section 12.2.2.9. Although reports of the details of the complex elution process are conflicting [65, 80], the most likely scheme includes a first-stage treatment with dilute mineral acid to remove zinc, nickel, and cyanide; a second-stage elution with ammoniacal ammonium nitrate to remove copper; and, finally, simultaneous thiourea elution and electrowinning to recover gold and silver.

7.2.4.7 Mintek's MINRIP Process

Mintek (Randburg, South Africa) has continued to advance the understanding and commercialization of RIP technology with the development of the MINRIP process, based on the use of the DOWEX MINIX strong-base resin. The DOWEX MINIX resin and the MINRIP process have been used commercially at the Penjom mine in Malaysia for treatment of a highly carbonaceous preg-robbing ore (see Section 12.2.9.3). In this case, the RIP process was used in conjunction with a blanking agent (kerosene) to reduce the preg-robbing characteristics of the ore. Gold is eluted from the resin with acidic thiourea solution and recovered by electrowinning from thiourea media [73, 74, 75].

7.2.4.8 Other Resin Applications

The use of ion exchange resins for the recovery of base metals (i.e., copper and zinc) from cyanidation leach solutions and slurries, together with the recovery of cyanide for reuse in the leaching circuit, is considered in detail in Section 11.2.3.

7.2.5 Adsorption from Noncyanide Solutions

7.2.5.1 Halide Solutions

A weak-base cation exchanger (with ester functional groups) can be used to selectively recover gold chloride from solution, albeit at modest loadings, that is, <20 g/L. The gold is rapidly eluted with 4 M hydrochloric acid and acetone, and the gold can be recovered by precipitation with sulfur dioxide or formic acid. The resin must be regenerated by distillation [66].

Ion exchange resins have also reportedly been used successfully for gold recovery from bromide solutions [81].

352 | THE CHEMISTRY OF GOLD EXTRACTION

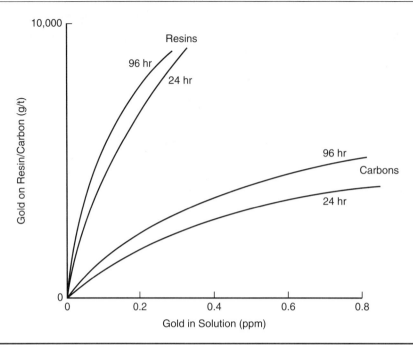

FIGURE 7.35 Equilibrium loading of gold cyanide from pulp from the Golden Jubilee mine onto strong-base resin and activated carbon [71]

TABLE 7.8 Concentrations of elements on loaded and eluted–regenerated resin at Golden Jubilee [79]

	Concentration (g/t)	
Element	**Loaded Resin**	**Eluted Resin**
Gold	6,730	462
Silver	101	<50
Cobalt	302	96
Copper	3,470	88
Zinc	21,900	1,470
Nickel	1,840	770
Iron	11,800	1,700
Calcium	644	89
Silicon	7,730	595

7.2.5.2 Thiosulfate Solutions

Strong-base resins can be used for the recovery of gold from thiosulfate leach solutions. The adsorption of the Au(I) thiosulfate complex is represented as follows:

$$3(]-^+NR_3)_2SO_4^{2-} + 2Au(S_2O_3)_2^{3-} \rightleftharpoons 2(]-^+NR_3)_3Au(S_2O_3)_2^{3-} + 3SO_4^{2-} \quad \text{(EQ 7.33)}$$

where
  $]-$ = the inert portion of the resin
  R = the CH_3 species

However, the recovery of gold is complicated by the presence of various sulfur species generated as a result of thiosulfate oxidation reactions (see Section 6.3). The major sulfur-species of concern are the polythionates, tetrathionate ($S_4O_6^{2-}$), and trithionate ($S_3O_6^{2-}$), because both load onto strong-base resins and compete with gold thiosulfate, potentially reducing gold loading significantly [61, 82]. Research has indicated typical concentrations of tetrathionate and trithionate in thiosulfate leach solutions of 0.5 to 2.0 g/L. Tetrathionate and trithionate concentrations of 420 and 350 mg/L, respectively, reduced gold loading onto Purolite A500C resin by an order of magnitude, that is, from 26 to about 2 kg Au/t resin from a solution containing 0.3 mg/L Au. The addition of sodium sulfite (Na_2SO_3, at 0.5 g/L concentration) to the leach solution in an oxygen-free atmosphere (i.e., using a nitrogen purge) has been shown to counteract the detrimental effect of tetrathionate and trithionate species to some extent by converting tetrathionate to thiosulfate and reducing the tetrathionate concentration [61]. This may represent a path toward a commercial recovery process for gold from thiosulfate leach solutions.

In addition, thiosulfate leaching systems proposed generally incorporate the use of Cu(II) and ammonia as a catalyst. Unfortunately, strong-base ion exchange resins are not very selective for gold (and silver) over other base metals, with the rate and extent of loading dependent on the concentration of each species in solution. Typical gold loadings on resins of 2 to 5 kg/t have been reported from solutions containing 2 to 5 mg/L of gold using four to six stages of RIP with a total residence time of 6 hr. However, copper loadings of 10 to 20 kg/t were obtained from solution containing as little as 20 to 25 mg/L copper [78]. A process incorporating the selective desorption of copper prior to gold, similar to the process used for removal of copper from activated carbon (see Section 7.1.3.8) has been proposed. This process is rapid (1 to 2 hr), efficient (>99%), and takes place at pH 7 (avoiding exposing the resin to osmotic shock). It also is possible that some other method of removing copper from solution prior to resin loading could be used, but this is a complex problem, because such methods could also affect the concentrations of other detrimental solution species (e.g., tetrathionate and trithionate).

The elution of gold from loaded resin is accomplished rapidly and efficiently with ammonium thiocyanate, ammonium nitrate, a sulfite-ammonia solution, or, less efficiently, with thiosulfate [82, 83]. It is likely that elution could be accomplished at ambient temperature and pressure, with expected elution times of 2 to 6 hr (less than for carbon systems) to achieve 97% to 99% desorption of gold. Elution can be conducted at pH 7 to 9 with the significant advantage of eliminating the effect of osmotic shock on the resin, which occurs when the resin is subjected to large changes in pH (see Section 7.2.1). Other elution schemes including two-stage processes to remove copper and gold–silver sequentially have been proposed and show some promise for commercialization. Resin regeneration requirements are unknown at this time but would likely be required in any commercial application.

7.2.5.3 Thiourea Solutions

A strong-base cation exchanger (sulfonated polystyrene) is capable of strongly adsorbing the gold–thiourea complex. The loaded resin can be eluted with ammonium thiosulfate at pH 6 and the gold removed as the thiosulfate complex [84]. Other resin systems have been researched and proposed for gold recovery from thiourea solutions but these are of academic interest only, given the limited potential of a thiourea leaching system for commercial application in gold extraction (see Section 6.4).

7.3 SOLVENT EXTRACTION

During precious metals refining, solvent extraction has been applied industrially for the separation of gold, platinum group metals, and base metals [85]. In this application, the process competed favorably with the well-established electrorefining technique, described in Chapter 10, and it is believed that similar technology has been applied in

Russia and other former USSR states (Uzbekistan, Kazakhstan, Kyrgyzstan, etc.) since the 1970s. The method also forms the basis for a well-established analytical technique to determine gold concentration in dilute solutions. Solvent extraction has been used with great success for the extraction of copper and other base metals.

The use of solvents for purification and concentration of gold in dilute leach solutions has been proposed but has not been applied commercially. Liquid extractants offer some potential advantages over activated carbon and ion exchange resins; namely, rapid extraction kinetics and high gold loadings. These factors have the potential to reduce process equipment requirements, reduce gold inventory, and possibly simplify refining requirements of the product. However, unlike carbon and resin materials, liquid extractants cannot easily be applied directly in-pulp, and their potential use is likely to be restricted to treatment of clarified solutions. Also, liquid extractants have some solubility in water, which results in solvent (and gold) losses to the aqueous phase.

7.3.1 General Principles

Solvent extraction uses suitable liquid organic extractants to selectively remove gold species from aqueous solution. The extractant is dissolved in a supporting matrix (or diluent), such as kerosene, to distribute the functional groups in an optimal concentration for metal extraction. Typically, solvent concentrations of 10% to 20% in the diluent are used, although this may vary depending on the type of solvent. Gold is recovered from the loaded extractant either directly, by precipitation or electrolysis, or indirectly by stripping the solute back into the aqueous phase to allow recovery by the methods described in Chapter 8.

The factors affecting the effectiveness of liquid extractants for gold extraction are the following:

- Extraction and stripping kinetics
- Loading capacity
- Selectivity
- Density
- Flash point
- Solubility in water

The kinetics of solvent extraction are typically much faster than carbon adsorption and ion exchange resin processes, and high levels of extraction can usually be achieved within a few minutes. This is attributed to superior mass transport properties of the liquid–liquid system. However, the kinetics of stripping are much slower, and 2 to 4 hr may be required to achieve satisfactory metal recovery.

Gold loading on organic solvents can exceed 200 g/L, but such high loadings increase the density of the organic phase. This may interfere with phase separation and can ultimately cause phase inversion, whereby the organic phase settles below the aqueous phase. Consequently, the efficiency of phase separation depends to a large extent on the density of the solvent itself, as well as the loading of gold and other species. For this reason gold loadings must be kept below 80 g/L. Even allowing for such restrictions, a conservative gold loading of 40 g/L is between 5 and 10 times higher than the equivalent loadings achieved on activated carbon or resin.

7.3.2 Extraction Systems

Solvent extraction systems of importance in gold extraction can be grouped into two categories, summarized as follows:

Extraction by ion exchange. Metal ions or metal complexes are adsorbed onto specific sites on the extractant molecule by virtue of their charge and size, in a process similar to the ion exchange mechanism described in Section 7.2.2. Primary, secondary, tertiary, and quarternary amine extractants, with similar ion exchange properties to those of resins with amine functional groups, fall into this category.

Extraction by ion solvation. Extractant molecules replace the water of solvation around the metal ions in solution, rendering them soluble in the organic phase. Suitable extractants are characteristically oxygen-containing species, such as ethers (e.g., dibutyl carbitol), ketones (e.g., methyl isobutyl ketone), and various phosphorous-containing molecules (e.g., tributyl phosphate). It is thought that metal ions and metal complexes are extracted as neutral ion pairs, that is, $M^{n+}[Au(CN)_2^-]_n$, in a similar mechanism to that of adsorption onto activated carbon [86].

The formulae and properties of selected solvents are given in Table 7.9.

7.3.2.1 Amines

A variety of amines are capable of extracting gold from aqueous alkaline cyanide solutions. Both weak- and strong-base amines can be applied in similar manner to the application of weak- and strong-base resins, as discussed in Section 7.2.2. Quaternary amines are the most selective for gold but are difficult to strip, whereas tertiary, secondary, and primary amines are progressively less selective but easier to strip. Tertiary amines, such as tridecylamine and trioctylamine, have shown the most promise for use in gold extraction systems. These solvents are reportedly selective for gold and silver over copper, iron, and other base metals, because of their greater affinity for univalent complex ions (e.g., $Au(CN)_2^-$) over polyvalent species (e.g., $Cu(CN)_3^{2-}$) [87].

The activity of weak-base amines is pH dependent, as is the case for weak-base resins; however, the basicity of these solvents is lower than that of most weak-based resins. The pK_a values of primary, secondary, and tertiary amines are approximately 6.5, 7.5, and 6.0, respectively, compared with 9.0 for weak-base resins [86]. If amines are to be applied directly to leach solutions, then the pH must be modified close to the relevant pK_a. Alternatively, the basicity can be increased by adding between 20% to 80% by volume of a second solvent, such as tributyl phosphate or di-*n*-butyl butyl phosphonate [88], which can increase the effective pK_a of the solvent close to 10. However, this also increases the solubility of the solvent in water (i.e., resulting in higher solvent losses) and is usually detrimental to the phase separation characteristics of the system [89].

Gold can be stripped from most weak-base amines by modifying the pH of the aqueous phase above the pK_a of the extractant. In the case of tridecylamine, this means increasing the pH to 12–13, which allows the extractant to be recycled.

Quaternary amines are difficult to strip effectively, but it has been suggested that gold could be recovered by burning the extractant, after prior distillation of the diluent. This may be attractive for some specific applications because of the excellent selectivity and very high loading capacity of this type of extractant; however, the process is unlikely to find widespread application due to its relatively high cost.

A guanidine-based extractant (LIX 79, Cognis Corp., Tucson, Arizona) that has predominantly weak-base properties but with some strong-base properties (based on the guanidine functionality) has been developed. As with the guanidine-based resin (see Section 7.2.4.1), the reagent is capable of deprotonating water to form the guanidium cation when the solution pH is increased above 10 to 11.5. The hydroxide ion is exchanged for the Au(I) cyanide ion, thereby providing a method for stripping the reagent without the need to use acidic conditions [76, 90].

A number of other amine-based solvent extractants have been proposed, including primary amines (e.g., Primene JMT), secondary amines (e.g., Amberlite LA2), tertiary

TABLE 7.9   Selected organic solvents for the extraction of gold from aqueous solutions

Solvent Type	Chemical Name	Chemical Formula	Density (kg/L)	Flash Point (°C)	Solubility (g/100 g H_2O)
Ketones	Methyl isobutyl ketone (MIBK)	$(CH_3)_2CHCH_2COCH_3$	0.801	13	1.7
	Di-isobutyl ketone (DIBK)	$[(CH_3)_2CHCH_2]_2C=O$	0.806	48	0.06
Phosphates and phosphonates	Tributyl phosphate (TBP)	$(C_4H_9O)_3P=O$	0.972	146	0.04
	Di-n-butyl butyl phosphonate (DBBP)	$(C_4H_9O)_2P\text{-}O\text{-}(C_4H_9O)$ $\mid$ H	0.995	121	slight
Ethers	Diethylene glycol dibutyl ether (Dibutyl carbitol, DBC)	$(C_4H_9OCH_2CH_2)_2O$	0.885	47	0.3
Amines	Tridecylamine	$H(CH_2)_{13}NH_2$	—	—	—
	Trioctylamine	$[H(CH_2)_8]_3N$	0.809	>112	—

amines (e.g., Hostarex A327), and tetradecldimethylbenzylammonium chloride with tributyl phosphate.

7.3.2.2   Ethers

Diethylene glycol dibutyl ether (dibutyl carbitol, or diethylene glycol dibutyl [DBC]) has been used for the separation of gold during refining of precious metals, principally platinum group metals, since 1971 [91]. DBC extracts gold very selectively over platinum group and base metals. The reagent also has the advantages of a low flash point and low vapor pressure but it has high solubility in water (3 g/L) which results in high solvent losses to the aqueous phase (Table 7.9).

A flowsheet for solvent extraction of gold during precious metals refining is given in Figure 7.36. Gold is extracted from a concentrated gold chloride leach solution, obtained by aqua regia leaching of an intermediate precious metal product. Efficient extraction can be achieved if sufficient acid concentration is maintained in the aqueous phase (i.e., 3 to 6 M HCl) and provided that Au(III) chloride is the predominant gold species present. Single-stage extraction efficiencies >90% have been reported [91], which compare favorably with carbon and resin systems.

An example of an equilibrium loading isotherm for the extraction of gold onto DBC is shown in Figure 7.37, which indicates the barren aqueous phase (raffinate) gold concentrations that can be achieved at given extractant gold loadings.

The concentrated leach solution may also contain other impurities, that is, palladium, iron, arsenic, and antimony, which are co-extracted with gold. Unfortunately, the extraction of these species is favored by the strongly acidic conditions required for the most efficient extraction of gold. Consequently, the contaminants must be removed prior to gold recovery by washing with dilute hydrochloric acid, that is, 1 to 2 M HCl.

Gold is recovered directly from the solvent by reduction with hot, aqueous oxalic acid:

$$2DBC \cdot HAuCl_4 + 3(COOH)_2 \rightleftharpoons 2DBC + 2Au + 8HCl + 6CO_2 \qquad (EQ\ 7.34)$$

Other stronger reductants, such as sulfur dioxide, have also been considered, but these are generally less suitable because they precipitate greater proportions of contaminants [91].

7.3.2.3   Phosphorus-Containing Species

Phosphorus containing organic molecules, such as tri-n-butyl phosphate (TBP) and di-n-butyl butyl phosphonate (DBBP), are capable of extracting gold from cyanide solutions [87, 88].

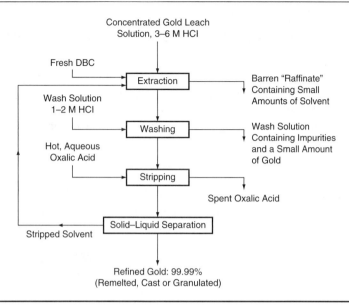

FIGURE 7.36 Process flowsheet for solvent extraction of gold for precious metals refining using DBC (adapted from [91, 68])

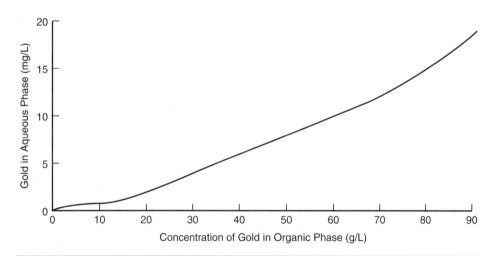

FIGURE 7.37 Equilibrium loading isotherm for gold in aqueous solution in contact with DBC [91]

These compounds have relatively high densities which result in poor phase separation properties for the extraction system if the solvents are used on their own. Consequently, they are usually applied in conjunction with a suitable amine, as discussed in Section 7.3.2.1. Also, stripping of gold from these extractants is particularly difficult because loading is generally independent of pH, but limited success has been achieved by using elevated temperatures and reduced solution ionic strength.

A possible process utilizing TBP to generate a 99.99% purity gold product has been proposed. The scheme has been demonstrated to be selective for gold over nickel, cobalt, iron, and zinc, but not copper or silver. However, the proposed process includes a scrubbing step to remove co-extracted copper and silver to produce a gold-bearing organic phase from which the gold could potentially be recovered by direct reduction or electrowinning [92].

7.3.2.4 Ketones

Methyl isobutyl ketone (MIBK) and di-isobutyl ketone (DIBK) are used for the extraction of gold from aqueous cyanide solution or aqua regia leach solution during the analysis of gold. Both have high affinity for gold and exhibit good selectivity, but neither is suitable for larger-scale applications because they are difficult to strip and have high solubility in water (resulting in large solvent losses).

7.3.3 Process Considerations

A major disadvantage of solvent extraction is that it must either be applied to clarified leach solutions or the solvent must be contained and separated from solid components in an unclarified leach solution or slurry. This is necessary because any solids present tend to form a crud layer at the aqueous–organic interface, reducing extraction efficiency. However, an in-pulp system was proposed in the former USSR in the 1980s whereby solvent would be placed in membrane pouches or bags and suspended in slurry tanks. The membrane bags would allow the passage of ions but not allow the flow of organic phase or solid particles. It is not known whether any such system has been used commercially.

A second significant disadvantage to the application of solvent extraction is that some solvent is always lost to the aqueous phase, either dissolved or entrained in it. The amount of solvent that is lost depends on its solubility in water, the ratio of aqueous–organic liquids in the process, and the degree of physical entrainment. For this reason it is desirable to minimize the aqueous–organic ratio, and this is favored by treating small volumes of concentrated gold solution. The amount of physical entrainment of solvent is a function of the type of mixing and settling equipment used, and, although substantial improvements have been made to this equipment over the past 20 to 30 years (especially in the copper industry), some losses are inevitable.

In practice, these factors result in overall solvent losses of up to 5% of the solvent flow rate, which makes the process rather unattractive economically for treatment of low-grade leach solutions, when compared with carbon and ion exchange resin processes.

REFERENCES

[1] Smisek, M., and S. Cerny. 1970. *Active Carbon.* New York: Elsevier.

[2] Czechowski, F., A. Jankowski, H. Marsh, T. Siemieniewska, and K. Tomkov. 1979. The development of porosity in group components of brown coals on carbonisation and activation. In *Characterisation of Porous Solids.* Edited by S.J. Gregg, K.S.W. Sing, and H.F. Stoeckli. London: Society of Chemical Industry.

[3] Marsh, H., M. Iley, J. Berger, and T. Siemieniewska. 1975. The adsorptive properties of activated plum stone chars. *Carbon* 13:103–109.

[4] Riley, H.L. 1947. *Quarterly Reviews* 1:59.

[5] Dubinin, M.M., and P.L. Walker. 1966. Porous structure and adsorption properties of active carbons. In *Chemistry and Physics of Carbon.* Volume 2. New York: Marcel Dekker.

[6] Yalcin, M., and A.I. Arol. 2002. Gold cyanide adsorption characteristics of activated carbon of non coconut shell origin. *Hydrometallurgy* 63:201–206.

[7]   Syna, N., and M. Valix. 2003. Modelling of gold(I) cyanide adsorption based on the properties of activated bagasse. *Minerals Engineering* 6:421–427.

[8]   Miller, J.D., G.A. Munoz, and S. Duyvesteyn. 2004. Design and synthesis of powdered magnetic activated carbon for aurodicyanide anion adsorption from alkaline cyanide leaching solutions. Pages 277–291 in *Fundamentals and Applications of Anion Separations*. Edited by B.A. Moyer and R.J. Singh. New York: Kluwer Academic/Plenum Publishers.

[9]   Dubinin, M.M. 1979. Microporous structures of carbonaceous adsorbents. In *Characterisation of Porous Solids*. Edited by S.J. Gregg, K.S.W. Sing, and H.F. Stoeckli. London: Society of Chemical Industry.

[10] MacDougall, G.J., and R.D. Hancock. 1981. Gold complexes and activated carbon: A literature review. *Gold Bulletin* 14:(4):138–153.

[11] Steenberg, B. 1944. *Adsorption and Exchange of Ions on Activated Charcoal.* Uppsala, Sweden: Almqvist and Wiksells.

[12] Van der Plaas, T.H. 1970. The texture and surface chemistry of carbons. *Physics and Chemical Aspects of Adsorbents and Catalysts*. Edited by B.G. Linden. London: Academic Press.

[13] Ryvin, D. 1963. *Rubber Chemistry & Technology Journal* 36:729.

[14] Puri, B.R. 1970. Pages 191–283 in *Surface Complexes on Carbon, Chemistry and Physics of Carbon*. Volume 6. Edited by P.L. Walker Jr. New York: Marcel Dekker.

[15] Donnet, J.B. 1967. The reactivity of carbons. *Carbon* 6:161–176.

[16] Mattson J.S., and H.B. Mark. 1971. *Activated Carbon.* New York: Marcel Dekker.

[17] Hughes, H.C., D.M. Muir, N. Tsuchida, and R. Dalton. 1984. Oxidation/reduction of activated carbon during anion loading. Pages 151–157 in *Proceedings of Regional Conference on Gold Mining, Metallurgy and Geology Conference,* Melbourne: Australasian Institute of Mining and Metallurgy (Perth and Kalgoorlie branches).

[18] Bailey, P.R. 1987. Application of activated carbon to gold recovery. Pages 379–614 in *The Extractive Metallurgy of South Africa*. Edited by G.G. Stanley. Monograph Series M7. Johannesburg: South African Institute of Mining and Metallurgy.

[19] Bunzyl, K., and W. Schultz. 1982. Kinetics of differentially small conversions in isotopic and contact ion exchange. *Analytical Chemistry* 54:272–277.

[20] Fleming, C.A., and M.J. Nicol. 1984. The adsorption of gold cyanide onto activated carbon, III: Factors influencing the rate of loading and equilibrium capacity. *Journal of South African Institute of Mining and Metallurgy* 84(4):85–93.

[21] Weast, R.C. 1981. *Handbook of Chemistry and Physics*. 62nd edition. Boca Raton, FL: CRC Press.

[22] McDougall, G.J., R.D. Hancock, M.J. Nicol, D.L. Wellington, and R.G. Copperthwaite. 1980. The mechanism of the adsorption of gold cyanide on activated carbon. *Journal of South African Institute of Mining and Metallurgy* 80:344–356.

[23] Tsuchida, N., M. Ruane, and D.M. Muir. 1984. Studies on the mechanism of gold adsorption on carbon. Pages 647–656 in *Proceedings of Mintek 50; International Conference on Mineral Science and Technology*. Randburg, South Africa: Council for Mineral Technology.

[24] Adams, M.D. 1989. The mechanism of adsorption of aurocyanide onto activated carbon—the latest developments and practical ramifications. Pages 166–168 in *Proceedings Randol Gold Conference*. Golden, CO: Randol International Ltd.

[25] Ibrado, A.S., and D.W. Fuerstenau. 1988. The adsorption of silver cyanide on activated carbon. SME Preprint No. 88-176. Littleton, CO: SME.

[26] Muir, D.M. 1982. Recovery of gold from cyanide solutions using activated carbon—a review. Pages 7-22 in *Proceedings of Carbon-in-Pulp Technology for the Extraction of Gold*. Parkville, Victoria: Australasian Institute of Mining and Metallurgy.

[27] Dixon, S., E.J. Cho, and C.H. Pitt. 1976. The interaction between gold cyanide, silver cyanide and high surface area charcoal, Paper presented at American Institute Chemical Engineers Meeting, Chicago, Illinois, November 26-December 2.

[28] Avraamides J. 1989. CIP carbons—selection, testing and plant monitoring. Pages 288-292 in *Proceedings of World Gold '89 Symposium*. Littleton, CO: SME.

[29] Davidson, R.J., W.D. Douglas, and J.A. Tumilty. 1982. The selection of granular activated carbon for use in a carbon-in-pulp operation. Pages 199-218 in *Proceedings of Carbon-in-Pulp Technology for the Extraction of Gold*. Parkville, Victoria: Australasian Institute of Mining and Metallurgy.

[30] van Deventer, J.S.J. 1984. Criteria for selection of activated carbons used in CIP plants. Pages 155-160 in *Reagents in the Mineral Industry*, Edited by M.J. Jones and R. Oblatt. London: The Institution of Mining and Metallurgy.

[31] McDougall, G.J. 1988. The mechanism of adsorption of gold cyanide onto activated carbon. Pages 249-252 in *Proceedings Randol International Gold Conference*. Golden, CO: Randol International Ltd.

[32] Davidson, R.J., and D. Duncanson. 1977. The elution of gold from activated carbon using deionized water. *Journal of South African Institute of Mining and Metallurgy*: 77:254-261.

[33] van Der Merwe, P.F., and J.S.J. Van Deventer. 1988. The influence of oxygen on the adsorption of metal cyanides on activated carbon. *Chemical Engineering Communications* 65:121-138.

[34] Boehme, W.R., and G.M. Potter. 1983. Carbon adsorption of gold. SME Preprint No. 83-422. Littleton, CO: SME.

[35] Muir, D., S. Vukcevic, and J. Shuttleworth. 1995. Optimizing the ammonia-cyanide leaching process for copper-gold ores. Pages 225-229 in *Proceedings Randol Gold Forum '95*. Golden, CO: Randol International Ltd.

[36] Heinen, H.J., P.G. Petersen, and R.E. Lindstrom. 1976. Gold desorption from activated carbon with alkaline alcohol solutions. Chapter 33 in *World Mining and Metals*. Volume 1. Edited by A. Weiss. New York: American Institute of Mining, Metallurgical, and Petroleum Engineers.

[37] Davidson, R.J. 1986. A pilot plant study on the effects of cyanide concentration on the CIP process. Pages 209-223 in *Gold 100: Proceedings International Conference on Gold*. Volume 2. Johannesburg: South African Institute of Mining and Metallurgy.

[38] Adams, M.D., and M.J. Nicol. 1986. The kinetics of the elution of gold from activated carbon. Pages 111-122 in *Gold 100: Proceedings of International Conference on Gold*. Volume 2. Johannesburg: South African Institute of Mining and Metallurgy.

[39] Zadra, J.B. 1950. A process for the recovery of gold from activated carbon by leaching and electrolysis. Report of Investigations No. 4672. Washington, DC: U.S. Bureau of Mines.

[40] Zadra, J.B., A.L. Engel, and H.J. Heinen. 1952. Process for recovering gold and silver from activated carbon by leaching and electrolysis. Report of Investigations No. 4843. Washington, DC: U.S. Bureau of Mines.

[41] Muir, D.M., W.D. Hinchcliffe, N. Tsuchida, and M. Ruane. 1985. Solvent elution of gold from CIP carbon. *Hydrometallurgy* 14:47–65.

[42] Ubaldini, S., R. Massidda, C. Abbruzzese, and F. Veglio. 2003. A new technology for gold extraction from activated carbon after cyanidation. Pages 1176–1184 in *Proceedings of XXII International Mineral Processing Congress*. Edited by L. Lorenzen and D. Bradshaw. Cape Town, South Africa: International Mineral Processing Congress.

[43] Muir, D.M., W.D. Hinchcliffe, and A. Griffen. 1985. Elution of gold from carbon by the Micron solvent distillation procedure. *Hydrometallurgy* 14:151–169.

[44] Rollwagen, D., P. Kresin, and C. Lam. 1987. Gold recovery at Detour Lake mine. Pages 41–58 in *Proceedings of International Symposium on Gold Metallurgy*. Edited by R.S. Salter, D.M. Wyslouzil, and G.W. McDonald. New York: Pergamon Press.

[45] Macrae, C.M., G.J. Sparrow, and J.T. Woodcock. 1988. Gold, calcite and other materials on activated carbons from CIP plants. Pages 53–59 in *Proceedings of Third Mill Operators Conference*. Melbourne: Australasian Institute of Mining and Metallurgy.

[46] House, C.I., and H.L. Shergold. 1984. Adsorption of polyvalent metal species onto activated carbon. *Transactions of the Institution of Mining and Metallurgy* 93:C19–C22.

[47] Fuerstenau, M.C., and B.R. Palmer. 1976. Anionic flotation of oxides and silicates. Pages 148–196 in *Flotation: A.M. Gaudin Memorial Volume*. Edited by M.C. Fuerstenau. New York: American Institute of Mining, Metallurgical, and Petroleum Engineers.

[48] James, R.O., and T.W. Healy. 1972. Adsorption of hydrolyzable metal ions at the oxide–water interface. *Journal Colloid and Interface Science* 40:43–81.

[49] Bailey, P.R. 1985. Ancilliary operations: Acid treatment, elutriation, carbon sizing, carbon breakage. Lecture No. 24 in *South African Institute of Mining and Metallurgy School on Use of Activated Carbon for Gold Recovery*. Johannesburg: South African Institute of Mining and Metallurgy.

[50] Cole, P.M., D.S. Von Broembsen, and P.A. Laxen. 1986. A novel process for regeneration of carbon. Pages 133–155 in *Gold 100: Proceedings International Conference on Gold*. Volume 2. Johannesburg: South African Institute of Mining and Metallurgy.

[51] Young, G.C., W.D. Douglas, and M.J. Hampshire. 1984. Carbon in pulp process for recovering gold from acid plant calcines at President Brand. *Mining Engineering*. (March):257–264.

[52] Ritson, G.D. 1998. World gold surveys—carbon technology. Pages 153–155 in *Proceedings Randol International Gold & Silver Forum*. Golden, CO: Randol International Ltd.

[53] Johns, M.W. 1995. The optimum configuration of a CIP circuit. Pages 333–340 in *Proceedings Randol Gold Forum '95*. Golden, CO: Randol International Ltd.

[54] Macintosh, A., D. McArthur, and R.M. Whyte. 2000. Process choices for carbon technology. Pages 327–332 in *Proceedings Randol International Gold & Silver Forum*. Golden, CO: Randol International Ltd.

[55] Whyte, R.M., P. Dempsey, and W. Stange. 1990. The development and testing of the Anglo American Corporation pump-cell at Vaal Reefs Exploration and Mining Company Limited. *Proceedings of International Deep Mining Conference: Innovations in Metallurgical Plants*. Johannesburg: South African Institute of Mining and Metallurgy.

[56] Taunyane, B.D., and A.B. Phillips. 2003. The phased metallurgical upgrade strategy at Gold Fields Limited South African operations. Pages 143–152 in *Proceedings of 8th Mill Operators' Conference*. Melbourne: Australasian Institute of Mining and Metallurgy.

[57] Brittan, M.I. 1988. The Bateman AARL carbon elution process. Pages 315–320 in *Proceedings Randol Gold Forum*. Golden, CO: Randol International Ltd.

[58] McDonald, N.W. 1982. The influence of the CIP process on the metallurgy of gold by reference to Australia. *Proceedings of 12th CMMI Congress*. Edited by H.W. Glen. Johannesburg: South African Institute of Mining and Metallurgy.

[59] Avraamides, J., G. Hefter, and C. Budiselic. 1985. The uptake of gold from chloride solutions by activated carbon. *Bulletin of Australasian Institute of Mining and Metallurgy* 240(7):59–62.

[60] de Siegel, E.A., and A.M. Soto. 1984. Microscopic observations on adsorption of metallic gold on activated carbon. *Transactions of the Institution of Mining and Metallurgy* 93:C90–C92.

[61] West-Sells, P.G., and R.P. Hackl. 2003. A process for counteracting the detrimental effect of tetrathionate on resin gold adsorption from thiosulfate leachates. Pages 245–256 in *Proceedings of Hydrometallurgy 2003*. Edited by C.A. Young, A. Alfantazi, C. Anderson, A. James, D. Dreisinger, and B. Harris. Warrendale, PA: TMS.

[62] Aylmore, M.G., and D.M. Muir. 2001. Thiosulfate leaching of gold—a review. *Minerals Engineering* 14(2):135–174.

[63] Fleming, C.A. 1987. The recovery of gold from thiourea leach liquors with activated carbon. Pages 259–277 in *Proceedings of International Symposium on Gold Metallurgy*. Edited by R.S. Salter, D.M. Wyslouzil, and G.W. McDonald. New York: Pergamon Press.

[64] McQuiston, F.W., and R.S. Shoemaker. 1980. *Gold and Silver Cyanidation Plant Practice*. Volume II. New York: American Institute of Mining, Metallurgical, and Petroleum Engineers.

[65] Fleming, C.A., and G. Cromberge. 1984. The extraction of gold from cyanide solutions by strong and weak base anion exchange resins. *Journal of South African Institute of Mining and Metallurgy* 84(5):125–137.

[66] Green, B.R, A.H. Schwellnus, and M.H. Kotze. 1986. Recent developments in resins for the extraction of gold. Pages 321–334 in *Gold 100: Proceedings of International Conference on Gold*. Volume 2. Johannesburg: South African Institute of Mining and Metallurgy.

[67] Jay, W.H., F. Lawson, and D.F.A. Koch. 1995. New ion exchange polymers for gold cyanide recovery. Pages 433–438 in *Proceedings Randol Gold Forum*. Golden, CO: Randol International Ltd.

[68] Lakshmanan, V.I., and P.D. Tackaberry. 1990. A review on the application of ion exchange technology for gold recovery. Pages 257–262 in *Advances in Gold and Silver Processing: Proceedings of GOLDTech 4 Symposium*. Edited by M.C. Fuerstenau. New York: American Institute of Mining, Metallurgical, and Petroleum Engineers.

[69] Fleming, C.A., and G. Cromberge. 1984. The elution of aurocyanide from strong and weak base resins. *Journal of South African Institute of Mining and Metallurgy* 84(9):269–280.

[70] Hazen, W.C. 1957. Method for eluting adsorbed complex cyanides of gold and silver. U.S. Patent 2,810,638.

[71] Fleming, C.A., and G. Cromberge. 1984. Small scale pilot plant tests on the resin-in-pulp extraction of gold from cyanide media. *Journal of South African Institute of Mining and Metallurgy* 84(11):369–378.

[72] Palmer, R.G. 1986. Ion exchange research in precious metals recovery. Pages 2–9 in *Proceedings of Precious Metal Recovery from Low Grade Resources*. Washington, D.C.: U.S. Bureau of Mines.

[73] Lewis, G.O., and W. Bouwer. 2000. Resin-in-leach: An effective option for gold recovery from carbonaceous ores. Pages 35–43 in *Proceedings Randol International Gold & Silver Forum*. Golden, CO: Randol International Ltd.

[74] Green, B.R., M.H. Kotze, and J.P. Wyethe. 2002. Developments in ion exchange: The Mintek perspective. *Journal of Metals* (October):37–44.

[75] Green, B.R., M.H. Kotze, and J.P. Engelbrecht. 1998. Resin-in-pulp—After gold, where next? Pages 119–136 in *Proceedings of EPD Congress 1998*. Edited by B. Mishra. Warrendale, PA: TMS.

[76] Virnig, M.J., J.M.W. Mackenzie, and C. Adamson. 1996. The use of guanidine-based extractants for the recovery of gold. Pages 151–156 in *Proceedings Hidden Wealth Conference*. Johannesburg: South African Institute of Mining and Metallurgy.

[77] Davis, M.R., M.W. Mackenzie, and M.J. Virnig. 1999. An engineering cost study: CIS vs. RIS with AuRIX100. Pages 189–195 in *Proceedings Randol Gold & Silver Forum '99*. Golden, CO: Randol International Ltd.

[78] Fleming, C.A. 1998. The potential role of anion exchange resins in the gold industry. Pages 95–117 in *Proceedings of EPD Congress 1998*. Edited by B. Mishra. Warrendale, PA: TMS.

[79] Seymore, D., and C.A. Fleming. 1989. Golden Jubilee resin-in-pulp plant for gold recovery. Pages 297–307 in *Proceedings Randol International Gold Conference* Golden, CO: Randol International Ltd.

[80] Riveros, P.A., and W.C. Cooper. 1987. Advances in the recovery of silver and gold from cyanide solutions by ion exchange. In *Separation Processes in Hydrometallurgy*. Edited by G.A. Davies. Chichester, England: Ellis Horwood.

[81] Read, K.J., and R. Mensah-Biney. 1988. Gold bromide loading characteristics on selected adsorbents. Pages 306–309 in *Proceedings Randol International Gold Conference*. Golden, CO: Randol International Ltd.

[82] Zhang, H., and D.B. Dreisinger. 2002. The adsorption of gold and copper onto ion-exchange resins from ammoniacal thiosulfate solutions. *Hydrometallurgy* 66:67–76.

[83] Nicol, M.J., and G. O'Malley. 2002. Recovering gold from thiosulfate leach pulps via ion exchange. *Journal of Metals* (October):44–46.

[84] Bjerre, A.B., E. Sorensen, and T.E. Chantson. 1989. Recovery of gold from a gold-thiourea loaded ion exchanger. *Transactions of the Institution of Mining and Metallurgy* 98C:84–87.

[85] Barnes, J., and J.D. Edwards. 1982. Solvent extraction at Inco's Acton precious metals refinery. *Chemistry and Industry* (March):151–155.

[86] Nicol, M.J., C.A. Fleming, and R.L. Paul. 1987. The chemistry of gold extraction. Pages 831–905 in *The Extractive Metallurgy of Gold*. Edited by G.G. Stanley. Johannesburg: South African Institute of Mining and Metallurgy.

[87] Wan, R.Y., and J.D. Miller. 1990. Research and development activities for the recovery of gold from alkaline cyanide solutions. *Mineral Processing and Extractive Metallurgy Review* 6:143–190.

[88] Mooiman, M.B., and J.D. Miller. 1986. The chemistry of gold solvent extraction using modified amines. *Hydrometallurgy* 16:245–261.

[89] Sibrell, P.L., and J.D. Miller. 1986. Soluble losses in the extraction of gold from alkaline cyanide solutions by modified amines. Pages 187–194 in *Proceedings of ISEC '86*. Volume 2. Munich, Germany: International Solvent Extraction Conference.

[90] Valenzuela, J.L., S. Aguayo, J.R. Parga, and R.G. Lewis. 2003. Gold solvent extraction from alkaline cyanide solutions using LIX 79 extractant. Pages 881–889 in *Proceedings of Hydrometallurgy 2003*. Volume 1, Leaching and Solution Purification. Edited by C.A. Young, A. Alfantazi, C. Anderson, A. James, D. Dreisinger, and B. Harris. Warrendale, PA: TMS.

[91] Thomas, J.A., W.A. Phillips, and A. Farais. 1984. The refining of gold by a leach-solvent extraction process. Paper presented at 1st International Symposium on Precious Metals Recovery, Reno, NV, June 10–14.

[92] Adams, M.D. 2003. On-site gold refining of cyanide liquors by solvent extraction. *Minerals Engineering* 16:369–373.

CHAPTER 8

Recovery

Recovery processes, as defined in this chapter, remove gold and other metal values from solution into a concentrated solid form. The high-grade solid product may then either be sold directly or treated further on-site by refining (Chapter 10) to improve the grade. Recovery may be achieved from a wide range of solutions, varying dramatically in gold concentration and purity, both of which can greatly affect the efficiency of the various recovery processes available.

Prior to the widespread adoption of carbon adsorption processes for treatment of dilute leach slurries and solutions in the late 1970s (Chapter 7), zinc precipitation was used almost exclusively for direct recovery of gold from clarified leach solutions. Subsequently, both electrowinning and zinc precipitation have been used to treat the more concentrated gold solutions produced by carbon elution. Direct recovery from dilute solutions by zinc precipitation is still preferred over carbon adsorption for treatment of some ores, for example:

- Ores with high silver content
- Ores containing species that interfere with carbon adsorption (e.g., clay minerals, organic material)
- Small orebodies that cannot justify the generally higher costs of carbon adsorption, elution, and regeneration systems at that scale of operation

The factors affecting this process selection are considered further in Section 3.2.8. The chemistry of both electrowinning and zinc precipitation processes are considered in detail in this chapter.

Although no commercial processes now use aluminum dust for gold precipitation, it has been used in the past, and the chemistry of this recovery system is briefly reviewed. Activated carbon has also been used in a few instances for final gold recovery, usually with incineration of the loaded carbon to produce a high-grade ash for refining (see Chapter 7).

Possible schemes for gold recovery from noncyanide leaching systems are also considered.

8.1 ZINC PRECIPITATION

Precipitation, or "cementation," of gold with zinc was introduced commercially for the treatment of cyanide leach solutions in 1890 [1], and has subsequently been applied widely in the industry. The process, which is commonly referred to as Merrill–Crowe precipitation after its pioneers (Chapter 1), has evolved to be highly efficient, with gold recoveries from solution >98% routinely achieved and sometimes as high as 99.5%. The Merrill–Crowe process has cost advantages over the carbon adsorption process for treatment of solutions containing high silver content (see Chapter 3) and has the advantage of being able to more easily handle fluctuations of gold grade in the feed. For the treatment of low silver content solutions and slurries, the use of carbon adsorption as an intermediate step has emerged as the preferred process route, mainly due to lower capital and operating costs.

Zinc precipitation was first used to treat hot, high-grade solutions produced by carbon elution in 1981 in the United States and South Africa, and has subsequently been applied widely around the world as an alternative to electrowinning.

8.1.1 Reaction Chemistry

8.1.1.1 Anodic Behavior of Zinc in Cyanide Solution
The anodic oxidation of zinc in aqueous solutions is given by:

$$Zn^{2+} + 2e \rightleftharpoons Zn; \quad E^0 = -0.763 \quad \text{(EQ 8.1)}$$

where
$$E = -0.763 + 0.0295 \log[Zn^{2+}] \text{ (V)}$$

In cyanide solutions zinc forms a stable cyanide complex:

$$Zn^{2+} + 4CN^- \rightleftharpoons Zn(CN)_4^{2-} \quad \text{(EQ 8.2)}$$

where
$$\beta = 3.98 \times 10^{19} \text{ [2]}$$

Combining this with Equation (8.1) gives:

$$Zn(CN)_4^{2-} + 2e \rightleftharpoons Zn + 4CN^- \quad \text{(EQ 8.3)}$$

where
$$E = -1.25 + 0.0295 \log[Zn(CN)_4^{2-}] + 0.118 \text{ pCN (V)}$$

In sufficiently oxidizing alkaline solution, and depending on the cyanide concentration, zinc may corrode to form other species as follows:

$$Zn(OH)_2 + 2e \rightleftharpoons Zn + 2OH^-; \quad E^0 = -1.34 \text{ (V)} \quad \text{(EQ 8.4)}$$

$$HZnO_2^- + H^+ + 2e \rightleftharpoons Zn + 2OH^-; \quad E^0 = -1.24 \text{ (V)} \quad \text{(EQ 8.5)}$$

$$ZnO_2^{2-} + 2H_2O + 2e \rightleftharpoons Zn + 4OH^-; \quad E^0 = -1.22 \text{ (V)} \quad \text{(EQ 8.6)}$$

The regions of stability of these solution species are illustrated in the E_h–pH diagrams for the Zn–CN–H_2O system at 25°C (Figure 8.1), given for three different zinc ion concentrations.
The oxidation products of zinc can also undergo a number of other reactions in aqueous solution, many of which are also considered as lines in Figure 8.1:

$$Zn^{2+} + OH^- \rightleftharpoons Zn(OH)^+; \quad K_1 = 1.41 \times 10^4 \quad \text{(EQ 8.7)}$$

$$Zn^{2+} + 2OH^- \rightleftharpoons Zn(OH)_{2(s)}; \quad K_{s0} = 7.08 \times 10^{-18} \quad \text{(EQ 8.8)}$$

$$Zn(OH)_{2(s)} \rightleftharpoons HZnO_2^- + H^+; \quad K_{s3} = 1.2 \times 10^{-3} \quad \text{(EQ 8.9)}$$

$$Zn(CN)_4^{2-} + 2OH^- \rightleftharpoons Zn(OH)_{2(s)} + 4CN^-;$$
$$(K \text{ obtained from Equations (8.2) and (8.7))} \quad \text{(EQ 8.10)}$$

$$HZnO_2^- \rightleftharpoons ZnO_2^{2-} + H^+; \quad K_4 = 18.2 \quad \text{(EQ 8.11)}$$

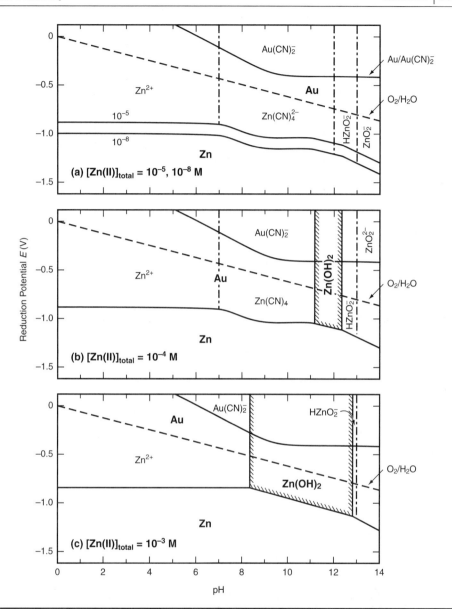

FIGURE 8.1 Potential–pH equilibrium diagram for the system Zn–CN–H_2O at 25°C including some equilibria between gold, cyanide, and water: $[CN^-]_{total} = 10^{-3}$ M, $[Au(CN)_2^-] = 10^{-4}$ M, $pH_2 = 1$ atm; (a) $[Zn(II)]_{total} = 10^{-5}$ M and 10^{-8} M; (b) $[Zn(II)]_{total} = 10^{-4}$ M; (c) $[Zn(II)]_{total} = 10^{-3}$ M [2]

The formation of zinc hydroxide precipitate is highly undesirable because this product may coat the zinc surface, causing passivation and inhibiting gold and silver precipitation. It is important, therefore, to consider the effect of pH, cyanide, and zinc concentration on the formation of zinc hydroxide, shown graphically in Figures 8.2 and 8.3 [3]. The combined effect of cyanide and zinc concentrations on the solubility of the zinc species formed is illustrated elegantly in Figure 8.4. It is apparent that zinc hydroxide formation is favored at high zinc and low cyanide concentrations under conditions

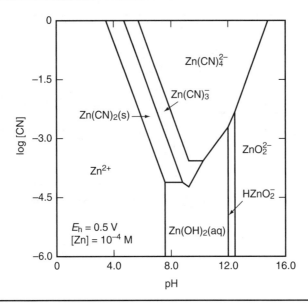

FIGURE 8.2 Log [CN⁻]–pH diagram for the Zn–CN–H$_2$O system at 25°C [3]

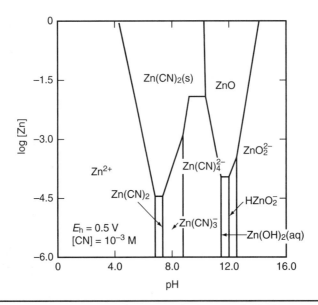

FIGURE 8.3 Log [Zn]–pH diagram for the Zn–CN–H$_2$O system at 25°C [3]

typically applied in commercial precipitation operations. The factors affecting zinc hydroxide formation are discussed further in Section 8.1.2.

8.1.1.2 Cathodic Reactions

The dissolution of zinc in cyanide solution forms the anodic half of a pair of coupled electrochemical reactions. The accompanying cathodic reduction is either the desirable precipitation of gold, and other metals, or one of several possible undesirable side reactions, including the reduction of water, oxygen, and other species in solution.

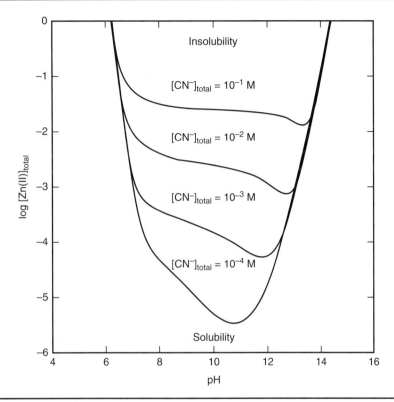

FIGURE 8.4 Domains of solubility and insolubility of Zn(II) in cyanide solutions at 25°C [2]

Gold reduction. The cathodic reduction of gold from alkaline cyanide solution is expressed as:

$$Au(CN)_2^- + e \rightleftharpoons Au + 2CN^- \qquad (EQ\ 8.12)$$

where
$$E = -0.60 + 0.118 \log[CN^-] + 0.0591 \log[Au(CN)_2^-]\ (V)$$

The line representing this equation has been superimposed on the E_h–pH diagram for the Zn–CN–H$_2$O system, given in Figure 8.1. It is apparent that between pH 9.5 and 11.0, over the range of free cyanide, zinc, and gold concentrations usually present in cyanide leaching and carbon elution solutions, the difference in potentials for the two half-reactions is >0.5 V, indicating a strong thermodynamic driving force for the precipitation reaction to proceed.

The precipitation reaction mechanism is illustrated schematically in Figure 8.5. Several expressions have been used to describe the overall zinc precipitation reaction, most commonly [2, 4]:

$$2Au(CN)_2^- + Zn \rightleftharpoons 2Au + Zn(CN)_4^{2-} \qquad (EQ\ 8.13)$$

Although this expression neatly combines the two half reactions and cancels out the free cyanide species associated with each reaction, it is probably misleading since the cathodic and anodic reactions may not occur in close physical proximity, a requirement

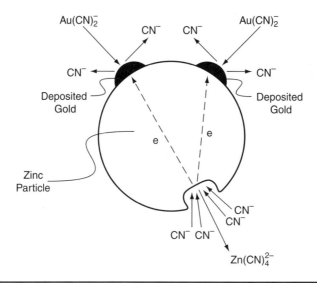

FIGURE 8.5 Schematic representation of mechanism of gold precipitation on zinc [5]

for the direct transfer of cyanide ions between two gold and one zinc species, implied by Equation (8.13). Such direct transfer is improbable, to any significant extent, due to the low solution gold concentrations (typically <0.0005 M) produced during leaching, and even carbon elution.

The expression that most accurately describes the stoichiometry of the reaction is [5, 6]:

$$2Au(CN)_2^- + Zn + 4CN^- \rightleftharpoons 2Au + 4CN^- + Zn(CN)_4^{2-} \quad (EQ\ 8.14)$$

This is further substantiated by the fact that the reaction rate is first order with respect to cyanide concentration, up to a certain limiting value of cyanide (see Section 8.1.2.2).

Reduction of other metals. Other metal cyanide complexes that are more positive than zinc in the electrochemical series, shown in Table 8.1, will also be reduced and possibly coprecipitated with gold into the solid product. The overpotentials for both silver and mercury are both considerably larger than for gold, and both are precipitated effectively. Copper, nickel, and cobalt are also displaced by zinc, but have smaller overpotentials, and their precipitation is less efficient.

Reduction of water and oxygen. Zinc dissolves at potentials more negative than both oxygen and water reduction reactions, as indicated in Figure 8.1. The relevant half-reactions are as follows:

$$O_2 + 4H^+ + 4e \rightleftharpoons 2H_2O \quad (EQ\ 8.15)$$

where
$$E = 1.229 - 0.0591\ pH + 0.0147 \log pO_2\ (V)$$

and

$$2H_2O + 2e \rightleftharpoons H_2 + 2OH^- \quad (EQ\ 8.16)$$

where
$$E = -0.828 - 0.0591\ pH - 0.0295 \log pH_2\ (V)$$

TABLE 8.1 Equilibrium potentials for the reduction of various metal cyanide ions: metal ion concentration = 10^{-4} mol/L, free CN^- concentration = 0.04 mol/L (0.2% NaCN) [7]

Reaction	No. Electrons Involved	E_{eq} (V)
$Hg(CN)_4^{2-} \rightarrow Hg$	2	–0.33
$Pb(CN)_4^{2-} \rightarrow Pb$	2	–0.38
$Ag(CN)_2^- \rightarrow Ag$	1	–0.45
$Au(CN)_2^- \rightarrow Au$	1	–0.63
$Cu(CN)_3^{2-} \rightarrow Cu$	1	–0.75
$Fe(CN)_6^{4-} \rightarrow Fe$	2	–0.99
$Ni(CN)_4^{2-} \rightarrow Ni$	2	–1.07
$Zn(CN)_4^{2-} \rightarrow Zn$	2	–1.22
$O_2 \rightarrow 2OH^-$	pH = 13	0.45
$2H_2O \rightarrow H_2 + 2OH^-$	pH = 13	–0.78

For the two Nernst equations described (Equations [8.15] and [8.16]), the partial pressures of both oxygen and hydrogen are generally taken as unity under atmospheric conditions.

Consequently, zinc may corrode by mechanisms involving reduction of both water and oxygen. The combined reaction with water under E_h–pH conditions favoring $Zn(CN)_4^{2-}$ formation is:

$$Zn + 4CN^- + 2H_2O \rightleftharpoons Zn(CN)_4^{2-} + 2OH^- + H_2 \qquad (EQ\ 8.17)$$

The overall reaction with oxygen under similar conditions is:

$$2Zn + O_2 + 8CN^- + 2H_2O \rightleftharpoons 2Zn(CN)_4^{2-} + 4OH^- \qquad (EQ\ 8.18)$$

These side reactions are important because they consume zinc, and in practice between 5 and 30 times the stoichiometric requirement is used.

The presence of dissolved oxygen can also result in redissolution of precipitated gold by the standard cyanide leaching mechanism, described in Chapter 6:

$$2Au + 4CN^- + O_2 + 2H_2O \rightleftharpoons 2Au(CN)_2^- + H_2O_2 + 2OH^- \qquad (EQ\ 8.19)$$

However, when the precipitated gold is in electrical contact with zinc, as is usually the case, the zinc will preferentially corrode. Despite this, redissolution can occur in operating systems, and it is not uncommon to detect higher gold concentrations in precipitation effluent solutions (compared with the feed) during upset conditions or when dissolved oxygen concentrations are elevated (i.e., >0.5 mg/L).

8.1.2 Reaction Kinetics and Factors Affecting Efficiency

Zinc precipitation of gold from cyanide solution proceeds via the following steps:
- Mass transport of Au(I) cyanide and free cyanide species to the zinc surface from bulk solution
- Adsorption of Au(I) cyanide species onto the surface of zinc, which may involve the formation of an intermediate adsorbed species, AuCN
- Electron transfer between adsorbed Au(I) cyanide species and zinc, and simultaneous dissociation of Au(I) cyanide species and formation of zinc cyanide complex
- Desorption of zinc cyanide species from zinc surface
- Mass transfer of zinc cyanide species into bulk solution

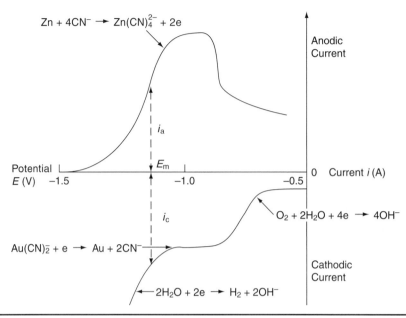

FIGURE 8.6 General form of the current–potential curves for zinc precipitation of gold [10]

Rotating disc electrode investigations of precipitation kinetics yield the general form of the current–potential curves shown in Figure 8.6. These investigations have shown that, under conditions typical of those applied industrially, the rate of gold precipitation is mass transport controlled, with film diffusion of Au(I) cyanide ions to the zinc surface as the rate-determining step [8, 9]. This conclusion is supported by measurements of the activation energy for the reaction of 13 kJ/mol, which is well within the range characterized by mass transport control.

Precipitation kinetics are generally first order, although deviation may occur depending on the morphology of the precipitate formed and any formation of passivating layers or coatings on the zinc surface, for example, $Zn(OH)_2$. These factors are considered in detail in the Sections 8.1.2.1 to 8.1.2.12.

The precipitation reaction proceeds at the mixed potential (E_m), at which point the anodic current is equal and opposite to the cathodic current. As discussed in Chapter 4, the rate of electrochemical reactions is directly related to the current, and at the mixed potential the rates of both anodic and cathodic half-reactions are equal. The reaction rate constant for gold precipitation has been estimated at 0.004 and 0.017 cm/s [1, 6, 10]. Interestingly, the precipitation rate of silver is between three and four times faster than gold because of the greater overpotential for the reaction (see Table 8.1) [6].

The rate of the anodic half-reaction only affects the overall kinetics if:

- The cyanide concentration falls below a critical value, as discussed in Section 8.1.2.2
- The surface of the zinc is blocked by an insoluble product or film, for example, colloidal silica, alumina, or zinc hydroxide

8.1.2.1 Gold Concentration

In dilute gold solutions, the rate of precipitation increases with increasing gold concentration, as expected for a reaction controlled by the mass transport of gold cyanide species to the zinc surface. This effect is illustrated in Figure 8.7, which shows experimental and

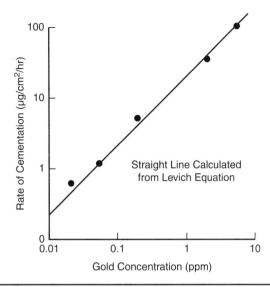

FIGURE 8.7 Effect of gold concentration on the rate of cementation [8]

theoretical (calculated from the Levich equation [61]) rate data, indicating close conformity of the reaction to the mass transport controlled mechanism previously proposed [8].

Other work has indicated a deviation from first-order behavior, with the reaction rate actually decreasing with increasing gold concentration. This effect is most prevalent in high-grade gold solutions and is attributed to the morphology of the precipitate formed: either a retarding effect of deposited gold, a particle aggregation phenomenon (which occurs at high gold concentrations and low temperatures), or a combination of the two [11, 12, 13]. All of these possible mechanisms restrict access of gold cyanide solution to the zinc surface. For example, one investigation has shown that precipitation from a solution containing 80 g/t Au yielded a porous product layer around the zinc (see Plate 11), while a higher-grade solution (640 g/t) produced a dense, nonporous deposit [13].

In the extreme, the deposition of a dense layer of gold precipitate around zinc particles may result in anodic closure, completely stopping the reaction.

8.1.2.2 Cyanide Concentration

The general effect of free cyanide concentration is illustrated by the cyclic voltammograms in Figure 8.8, which show that the current–potential curves for both the anodic and cathodic reactions are shifted negatively by approximately 0.12 V for an order of magnitude increase in cyanide concentration [8]. In practice, the free cyanide concentration affects the precipitation rate only if it is below a certain minimum value, which depends on the gold concentration (Figure 8.9) and pH. Below this value, the rate is either controlled by the diffusion of cyanide to the zinc surface or is retarded by the formation of a passivating layer of zinc hydroxide. For treatment of dilute (approximately <1 g/t Au) gold cyanide solutions at pH 10.5, the critical concentration is between 0.001 and 0.004 M free cyanide, equivalent to 0.05 and 0.20 g/L NaCN, respectively [8, 9, 14]. This is illustrated in Figure 8.10. The critical concentration is in good agreement with the value derived from the application of Fick's law to Equation (8.14).

During the treatment of hot, concentrated (20 to 100 g/t Au) solutions, the formation of insoluble zinc hydroxide usually has an overriding effect on the minimum cyanide concentration, as considered in Section 8.1.2.3. The critical (minimum) cyanide concentration

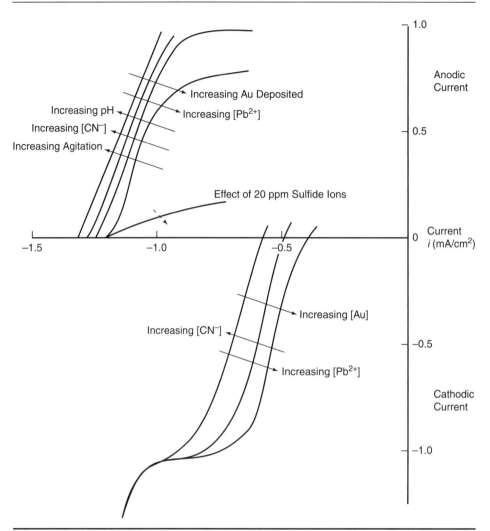

FIGURE 8.8 Simplified current–potential curves for rotating zinc and gold electrode investigations showing general effect of various parameters. (NOTE: The anodic and cathodic curves were obtained under slightly different solution conditions and are schematic only.) (adapted from [8])

required for treatment of solutions containing between 20 and 100 g/t Au has been found to be 2 to 5g/L NaCN [5, 13].

The rate of precipitation from both cold dilute and hot concentrated solutions is independent of cyanide concentration above the critical minimum value. However, the dissolution rate of zinc increases with increasing cyanide concentration, and hence it is undesirable to increase cyanide concentration significantly above the minimum required for effective precipitation.

8.1.2.3 Zinc Concentration

The dissolution rate of zinc decreases with increasing zinc ion concentration; however, it can be shown from Equations (8.3) and (8.12) that, even at relatively high zinc concentrations, metallic zinc is still sufficiently reducing to displace gold from cyanide solution.

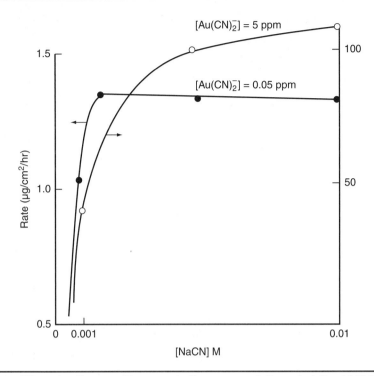

FIGURE 8.9 Effect of sodium cyanide concentration on cementation rate (pH = 10.5) (adapted from [8])

High zinc concentrations can result in the formation of insoluble zinc hydroxide, which can passivate the zinc surface and severely reduce the precipitation rate. However, the zinc concentrations produced by treatment of leach solutions (i.e., Merrill–Crowe precipitation) are usually low as a result of relatively low zinc addition rate required. Also, zinc is continuously removed from the system by precipitation of zinc salts in tailings facilities or in heap leach pads. Zinc concentrations in solutions produced by precipitation from carbon eluates are typically much higher because of higher zinc addition rates per unit volume of solution and as a result of closed-circuit operation; zinc concentrations as high as 2 g/L have been recorded [5]. The maximum recommended zinc concentration varies, depending on pH and cyanide concentration, for example, 0.75 g/L Zn^{2+} for a solution containing 3 g/L NaCN at pH 11. Zinc concentration can be controlled by routine bleeding of solution to the leaching process [5].

8.1.2.4 Zinc Particle Size

Because zinc precipitation is a mass transport controlled reaction, an increase in the available zinc surface area increases the kinetics (Figure 8.11). The fineness of zinc that can be used in industrial systems is usually limited by the filtration requirements, the filtration properties of the precipitate, and by the availability (and quality) of different sized zinc dust products. Increased surface area also increases the rate of side reactions, which increases zinc consumption. One other factor to be considered is that finer-sized zinc dust may be more prone to surface oxidation, and extra care must be taken to store in sealed containers in a cool, dry environment.

376 | THE CHEMISTRY OF GOLD EXTRACTION

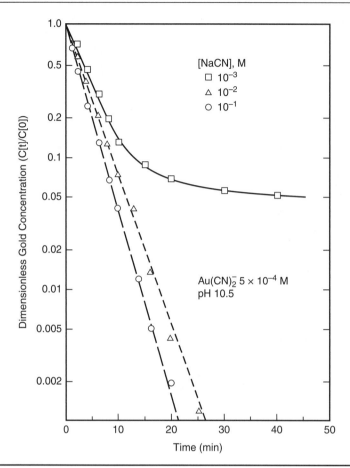

FIGURE 8.10 Effect of sodium cyanide concentration on the cementation of gold at pH = 10.5 and rpm = 1,000: C[0] = concentration when t = 0, C[t] = concentration after time, t [12]

A typical particle size distribution for commercially available zinc dust is given in Figure 8.12. Zinc dust has a surface area approximately 500 times that of zinc shavings, which illustrates the significance of the change from shavings to dust in the early 1900s (Chapter 1). Zinc shavings are still used in a few small plants because of the low cost and simplicity of the system, largely due to the fact that filtration is not required to recover the gold precipitate [14, 15].

8.1.2.5 Temperature

The accelerating effect of elevated temperature on precipitation kinetics was observed during early investigations into the reaction chemistry. In one instance it was noted that the same amount of gold was precipitated in 2 hr at 35°C as that precipitated in 24 hr at 20°C [14]. The effect is illustrated clearly in Figure 8.13, which shows the first-order reaction plots for a relatively high-grade gold solution containing 80g/t Au.

Elevated temperatures increase the rate of zinc dissolution and hydrogen evolution, with an associated decrease in precipitation efficiency [13]. Under these circumstances, the addition of Pb(II) may reduce zinc consumption and improve precipitation efficiency. In high-temperature systems treating carbon eluates, such adverse effects do not have a significant effect on overall precipitation efficiency, discussed further in Section 8.1.3.2 [5].

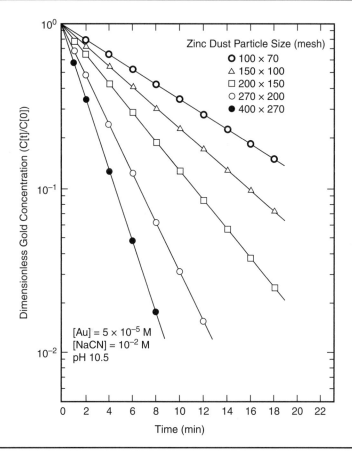

FIGURE 8.11 First-order plots showing the rate removal of gold from cyanide solution by suspended zinc particles of various size: C[0] = concentration when t = 0, C[t] = concentration after time, t [12]

8.1.2.6 Dissolved Oxygen Concentration

The presence of dissolved oxygen reduces precipitation kinetics and efficiency as the oxygen reduction reaction competes with gold reduction. The severity of this effect depends on the dissolved oxygen concentration and other precipitation conditions, principally the temperature and gold concentration. In dilute gold cyanide solutions at ambient temperatures, such as those applied in Merrill–Crowe precipitation, the effect becomes significant at dissolved oxygen concentrations above 0.5 to 1.0 mg/L. As a result, solutions must be deaerated prior to precipitation to reduce dissolved oxygen <1.0 mg/L, and preferably <0.5 mg/L.

There is some evidence that low concentrations of dissolved oxygen (i.e., 0.5 to 1.0 mg/L) enhance precipitation from dilute, cold gold cyanide solutions. This is attributed to the depolarizing action of oxygen at cathodic areas of the zinc surface by reaction with hydrogen as it is evolved.

The dissolved oxygen concentration has a less severe effect on precipitation from hot, high-grade gold solutions, such as those obtained from carbon elution systems [5, 13], and deaeration is unnecessary because of the low solubility of oxygen at normal operating temperatures and the consumption of residual oxygen during carbon elution (see Section 7.1.3).

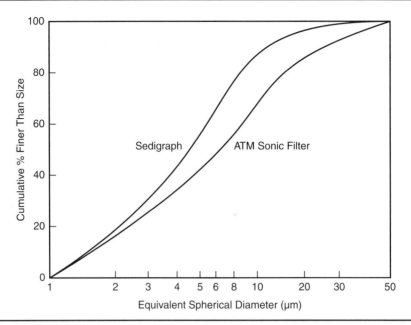

FIGURE 8.12 Typical size distribution of commercially available zinc dust [12]

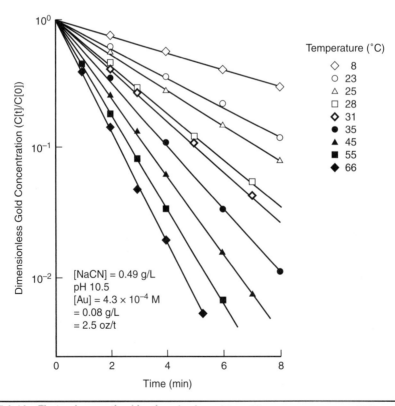

FIGURE 8.13 First-order reaction kinetics plot for gold at different temperatures [12]

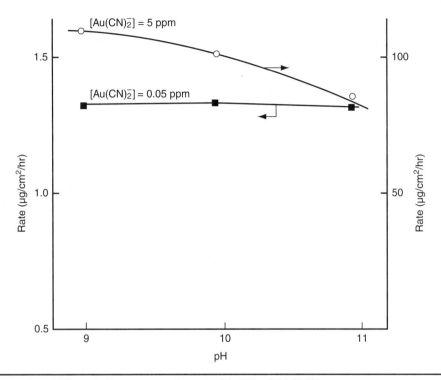

FIGURE 8.14 Effect of pH on cementation rate: [NaCN] = 0.01 M [8]

8.1.2.7 pH

Early investigations indicated that the optimum pH for precipitation from cold, dilute solutions was in the range of 11.5 to 11.9 [8, 11]. Subsequently, the effect of pH has been found to be small over the range 9 to 12, as shown in Figure 8.14. Precipitation is severely retarded below pH 8, largely due to the sharp reduction in free cyanide concentration. In practice, the minimum pH is limited by the loss of cyanide by hydrolysis which becomes significant below a pH of approximately 9.5. At pH values >12 the precipitation rate has been found to drop sharply due to excessive hydrogen evolution. Consequently, industrial systems usually operate within the pH range applied for cyanide leaching, that is, pH 10.5 to 11.5 (or approximately 0.10 to 0.20 g/L $Ca(OH)_2$).

Solution pH also has a marked effect on the formation of zinc hydroxide, which can inhibit precipitation by the formation of an insoluble layer on the zinc (see Figures 8.2, 8.3, and 8.4) [16].

8.1.2.8 Effect of Certain Polyvalent Heavy Metal Ions

The beneficial effect of lead ions on precipitation from dilute gold solutions at low temperatures is well known [8, 10, 13, 14]. Pb(II) species are reduced at the zinc surface, forming cathodically charged areas of metallic lead. The negatively charged gold cyanide species are reduced preferentially at these polarized regions. This helps to localize gold deposition, preventing the entire surface of the zinc from becoming coated with gold and allowing continued corrosion of zinc, which is coupled with further gold deposition. Lead also has a very large overpotential for hydrogen evolution and, when deposited on the zinc surface, greatly reduces hydrogen evolution, thereby improving efficiency. The general effect of lead on the polarization curve for zinc in aqueous solution is illustrated in Figure 8.8.

The lead addition rate is critical and is most effective at a Pb–Zn weight ratio of between 1:7 and 1:10. In practice, Pb^{2+} concentrations between 0.001 and 0.01 g/L have been found to be optimal for Merrill–Crowe precipitation from solutions containing 1 to 10 g/t Au. Excess addition is detrimental and can completely stop precipitation, probably due to complete coating of zinc particles by a lead film. This effect typically occurs at concentrations above 0.06 to 0.10 g/L Pb^{2+}, but concentrations as low as 0.01 g/L may be detrimental [8, 17].

Lead has also been shown to reduce the inhibiting effect of zinc hydroxide formation on the zinc surface at low cyanide concentrations, as illustrated in Figure 8.15.

Other divalent metal cations, such as mercury, thallium, bismuth, cadmium, and copper at very low concentrations, have been shown to have similar effects to lead, which possibly explains why some operations with solutions containing only low concentrations of one or more of such species find the benefits of lead addition to be insignificant [18].

Another advantage of lead addition is that it reacts with sulfide ions in solution, forming insoluble lead sulfide in preference to the highly detrimental zinc sulfide that would otherwise be formed directly on the zinc surface, inhibiting zinc dissolution.

8.1.2.9 Effect of Other Ions in Solution

A summary of the effects of a variety of ionic species on zinc precipitation, which are mainly based on laboratory testing, appear in Table 8.2. Although these results cannot be applied directly to plant solutions because of variations in solution conditions, they do give a relative indication of the critical concentrations of various species.

Sulfide ions have the most severe effect, and precipitation ceases completely at concentrations >14 mg/L (see also Figure 8.8), due to the formation of an insoluble zinc sulfide layer on the zinc surface. This effect may be reduced to some extent in the presence of lead, due to preferential precipitation of lead sulfide.

Arsenic, antimony, copper, nickel, and cobalt can all have a severe detrimental effect on precipitation, whereas sulfite, sulfate, thiosulfate, thiocyanate, and iron have relatively little effect, even at high concentrations.

Although beneficial to precipitation in low concentrations (Section 8.1.2.8), mercury may disrupt precipitation and reduce efficiency by forming a gelatinous deposit that is difficult to filter. The presence of high concentrations of mercury (>50 g/t) has been shown to reduce the efficiency of gold precipitation significantly [20]. Mercury is recovered effectively by zinc precipitation from cyanide solution, and typically close to 100% of the mercury in solution is coprecipitated with gold (and silver) in the Merrill–Crowe process [21]. This is important because carbon adsorption processes (e.g., carbon-in-columns [CIC], carbon-in-pulp [CIP], carbon-in-leach [CIL]) are less efficient at removing mercury from cyanide solution, and typically between 30% and 70% is recovered, allowing the mercury concentration in solution to build up if it is not removed by other means. This is important for ores containing mercury. Methods for the precipitation of mercury from cyanide solutions by means other than zinc precipitation are considered briefly in Section 11.3.3.1.

Calcium ions have a small beneficial effect on precipitation; however, in the presence of sulfate ions, insoluble gypsum may be formed and, in the presence of carbonate, calcite may precipitate as a scale. Both are detrimental to precipitation, not only because of possible coating and passivation of zinc, but also by degradation of the precipitate filtration properties and scale formation on filtration equipment.

In addition to the species listed in Table 8.2, chromate (CrO_4^{2-}) ions are known to be detrimental to precipitation by the deposition of metallic chromium on the zinc surface, which reduces available surface area.

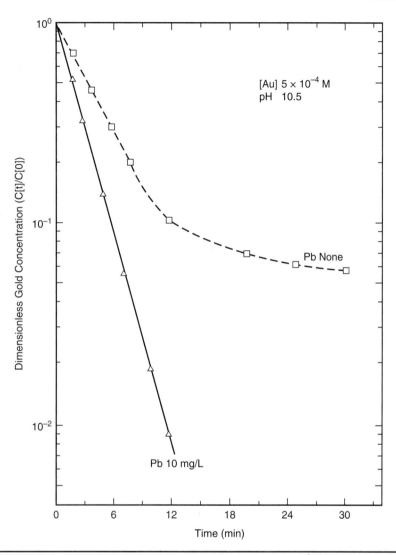

FIGURE 8.15 Effect of lead addition on the cementation of gold at 10^{-3} M sodium cyanide [12]

8.1.2.10 Organic Species

Various organic species, including humic and fulvic acids, which occur naturally in some ores and surface-active reagents, are suspected to have detrimental effects on zinc precipitation under certain conditions. Examples of these are flotation collectors and frothers, oils and hydraulic fluids from process equipment, antiscalant chemicals (phosphonates, acrylates, maleic acid, etc.), and flocculants. In all cases the adverse effects are poorly quantified and are not well understood, but are attributed to blinding of zinc surfaces and disruption of the filtration properties of the product.

In the extreme case where precipitation is severely impaired, solutions may need to be treated, for example by saponification to neutralize organic acids, or solution may be bled out of the circuit to remove the offending species. Care should be taken to avoid excessive addition of organic reagents/chemicals into the process upstream of zinc precipitation.

TABLE 8.2 Effect of various ionic species on zinc precipitation efficiency from dilute gold cyanide solutions (i.e., 10^{-4} to 10^{-5} M Au(CN)$_2^-$) [2, 8, 14, 17, 19]

Solution Species	Concentrations That Impair Zinc Precipitation (mg/L)		Concentration Above Which Zinc Precipitation Ceases (mg/L)
	Minor Effect	Major Effect	
Sulfide	0.01–0.60	>0.6	14
Copper cyanide	>0.2 or 25 (variable)	>0.2 or 25 (variable)	850
Antimony	—	>0.1	20
Arsenic	—	>0.1	17
Nickel	5–150	150–500	—
Cobalt	—	>5	—
Sulfite	>10	—	—
Sulfate	>2,000	—	—
Thiosulfate	>200	—	—
Thiocyanate	>150	—	—
Iron(II) cyanide	>100	—	—

NOTE: Dashes = not available.

On the other hand, zinc precipitation is far less likely to be adversely affected by flotation reagents and other organic species than by activated carbon (Chapter 7), which is sometimes an important factor in process selection.

8.1.2.11 Solution Clarity
The presence of fine particulate material suspended in solution, particularly clays and colloidal silicates, reduces precipitation efficiency, possibly by coating zinc surfaces or by interfering with oxygen removal, although neither of these effects are well understood or quantified. In addition to these possible effects, any solid matter present in solution is recovered with the final precipitate by filtration and contaminates the final product, increasing subsequent treatment requirements. This is discussed further in Sections 8.1.3.1 and 8.1.3.2.

8.1.2.12 Zinc Quality
The quality of zinc used for precipitation greatly affects the efficiency of gold recovery and the zinc consumption. Fresh zinc dust generally contains some zinc oxide (up to 6%), which is formed during the manufacturing process and is mainly present as a coating on the zinc surface. Further oxidation may occur if the dust is exposed to air for long periods, which is exacerbated in the presence of moisture. Oxidation forms a surface layer of zinc oxide and/or hydroxide around zinc particles, which can inhibit gold precipitation. The oxide or hydroxide coating is readily removed in cyanide solution [12], and some operations include a zinc polishing step prior to precipitation, whereby the zinc is wetted in alkaline cyanide solution to remove such coatings, as follows [10]:

$$ZnO + H_2O + 4CN^- \rightleftharpoons Zn(CN)_4^{2-} + 2OH^- \quad \text{(EQ 8.20)}$$

Alternatively, low-grade (barren) gold solution may be recirculated through the precipitate filtration system to establish a bed of "polished" zinc for effective precipitation. Although the polishing step consumes some zinc, the beneficial effects of surface preparation may be quite substantial.

In addition, zinc dust may contain small concentrations of other metals, for example, cadmium, which can have a variety of effects on precipitation (Sections 8.1.2.8 and 8.1.2.9) and downstream refining processes (Chapter 10), and may adversely affect ultimate bullion purity. These impurities should be determined by chemical analysis prior to application of zinc in a plant, and steps should be taken to routinely monitor the quality of zinc supply.

8.1.3 Process Considerations

The practical application of zinc precipitation for the treatment of relatively dilute gold leaching solutions (the Merrill–Crowe process) and for hot, concentrated carbon eluates are quite different, and the two systems are considered separately in detail in the following sections.

8.1.3.1 Precipitation from Cold, Low-Grade Solutions

The Merrill–Crowe process is used to treat solutions at temperatures close to ambient, ranging from just above freezing to 40°C, depending on climate and processing methods, and which vary in gold concentration between approximately 0.5 and 10 g/t.

Solution preparation. Precipitation from cold, dilute gold cyanide solutions can only be achieved effectively from clarified feed solution containing <10 mg/L suspended solids, and preferably <5 mg/L. Consequently, leached slurries must be treated by effective solid–liquid separation processes, such as thickening or filtration. The solutions produced by these methods, and gold-bearing solutions produced directly by other processes such as heap and vat leaching, usually require further clarification to produce an acceptable feed for precipitation. Clarification may be achieved in a variety of equipment, depending on the quantity and quality of solution. Such equipment includes plate-and-frame, leaf, candle and disc filters, sand-bed clarifiers, settling tanks, and "double-vee" (or "hopper") clarifiers [6, 62, 63].

The solution must also be deareated to a dissolved oxygen concentration <1.0 mg/L, and preferably <0.5 mg/L, to avoid undesirable side reactions which reduce precipitation efficiency and consume excess zinc (see Section 8.1.2.6). This is usually achieved mechanically by applying a vacuum to a deaeration vessel (e.g., Crowe tower) through which the solution is allowed to cascade. Various methods have been employed to increase the solution surface area within the vessel to assist in the release of oxygen, such as feeding the solution as a spray or allowing the solution to flow over high-surface area media. The vacuum required to sufficiently deaerate the solution varies, depending on the temperature and the altitude of the operation, as shown in Table 8.3. The vacuum requirement decreases with increasing temperature and with increasing altitude above sea level. For example, at 20°C at sea level, a vacuum of 27.4 in. Hg is required to reduce the dissolved oxygen concentration to 0.5 mg/L, compared with only 22.0 in. Hg at 1,640 m (5,400 ft) above sea level. If the temperature increased to 30°C, the vacuum requirements would be 26.7 and 21.3 in. Hg at the two altitudes, respectively.

The solution cyanide concentration and alkalinity must be adjusted to optimal levels prior to precipitation, as discussed in Sections 8.1.2.2 and 8.1.2.7.

Precipitation. Zinc is added at a rate between 5 and 30 times the stoichiometric precious metals (Au + Ag) requirement, depending on the solution composition and process operating efficiency. For example, a solution containing 5 g/t gold would require a zinc addition rate of 17 g/t solution, using 10 times the stoichiometric requirement (Equation [8.14]). The zinc dust is either added directly to the gold-bearing (pregnant) solution, sometimes called "emulsifying," or may be premixed with cyanide solution, to prepare the zinc surfaces for precipitation, and added as a slurry. Typical zinc consumptions vary from about 1 g/g precious metal precipitated to about 10 g/g, depending on solution conditions and operating efficiency [22].

A soluble lead salt, if required (see Section 8.1.2.8), should be added to the solution after clarification to avoid entrainment and loss of gold by any secondary lead salts (e.g.,

TABLE 8.3 Vacuum gauge requirements for deaeration of solutions prior to Merrill–Crowe zinc precipitation

Elevation Above Sea Level (ft)	Barometer (in. Hg)	Vacuum Gauge (in. Hg)						
0	29.9	29.4	28.4	27.4	26.4	25.4	23.4	21.4
1,300	28.5	28.0	27.0	26.0	25.0	24.0	22.0	20.0
3,300	26.5	26.0	25.0	24.0	23.0	22.0	20.0	18.0
5,400	24.5	24.0	23.0	22.0	21.0	20.0	18.0	16.0
7,600	22.5	22.0	21.0	20.0	19.0	18.0	16.0	14.0
Absolute Values	0	0.5	1.5	2.5	3.5	4.5	6.5	8.5
Vapor Pressure (in. Hg)	**Temp. (°C)**	**Oxygen Content (mg/L)**						
7	5	0.1	0.5	0.9	1.3	1.8	2.6	3.4
9	10	0.1	0.4	0.8	1.2	1.5	2.3	3.0
13	15	<0.1	0.3	0.7	1.0	1.3	2.0	2.7
18	20	<0.1	0.3	0.5	0.7	1.1	1.7	2.3
24	25	<0.1	0.2	0.4	0.7	1.0	1.5	2.1
31	30	<0.1	0.1	0.3	0.6	0.9	1.4	1.9
Absolute torr* (mm)	0	13	38	63	89	114.0	165	216

NOTE: Instructions for use of table: Identify desired oxygen content at specified temperature and vapor pressure conditions in lower half of table; then consult vacuum gauge column immediately above the desired oxygen content and locate value corresponding to altitude of operation (left-hand column).

* Torr = units of pressure. 1 torr = 133.3 N/m².

lead sulfide or sulfate) that may be precipitated. Lead nitrate is typically used for this purpose because of its high solubility, low cost, and worldwide availability. Lead acetate has been used on occasion but has been found to cause filtration problems in some instances due to the formation of crystalline (calcium, aluminum) acetate salts. Lead is usually added at a rate of approximately one twentieth to one tenth that of zinc [22].

The retention time allowed for precipitation varies greatly at different operations. The reaction kinetics are such that, in the absence of any interfering species (i.e., oxygen, sulfide ions), precipitation is complete within a few minutes of zinc addition. Reaction vessels with 2 to 5 min retention time are occasionally provided within circuits, but more usually the reaction is achieved in a length of piping (and possibly a pumping system) between the zinc addition point and the filtration system used for precipitate–solution separation. Further reaction may occur within the filter bed itself [10, 11].

A number of different filtration systems have been used to remove the precipitate from the zinc-solution mixture, including various types of plate-and-frame, leaf, disc, and candle filters operating over a range of pressures and flow rates. Often diatomaceous earth is used to aid filtration, either added to the solution during precipitation or used to precoat the filter to create a suitable filter medium, or a combination of the two. Filters may also be precoated with zinc to provide an initial bed on which precipitation can take place.

Process efficiency. The process is capable of producing a "barren" tailings solution containing <0.03 g/t Au, with some operations achieving <0.01 g/t Au. Consequently, depending on solution feed grade, precipitation efficiency is usually consistently >99.5%.

The barren tailing solution produced by filtration is either recycled to the process or bled to the final tailings disposal system. The solid products of precipitation vary greatly, depending on the original solution composition, the efficiency of precipitation, and the method of solids removal from solution. The most important components and their typical approximate ranges are:

- Gold and silver: 25% to 75%
- Zinc: 3% to 20%
- Diatomaceous earth: 0% to 25%
- Silica/silicates: 1% to 10%
- Other metals (e.g., lead, mercury, and copper): 1% to 10%

Treatment of these products is considered further in Chapter 10.

8.1.3.2 Precipitation from Hot, Concentrated Solutions (Modified Merrill Precipitation)

Solutions produced by carbon elution vary in temperature, gold concentration, and purity (i.e., concentrations of other metals, such as silver, mercury, and copper), depending on the elution method (Section 7.1.3) and the original concentrations of different species loaded on the carbon. Typically, the solution temperature is between 60°C and 90°C, depending on the elution temperature and the amount of time allowed for the solution to cool prior to precipitation. Gold concentrations vary from 1 to 1,000 g/t, with an average between 50 and 200 g/t produced from carbon loaded to between 5,000 and 10,000 g Au/t. The variability of gold concentration in solution depends on the volume of surge capacity, if any, that may need to be provided ahead of precipitation because of the nature of precious metal elution profiles (see Figure 7.20).

Under these conditions, highly efficient precipitation can be achieved. Deaeration of the solution is not necessary, because of the low solubility of oxygen at temperatures commonly employed for elution, coupled with the consumption of oxygen by cyanide oxidation reactions to produce cyanate, carbon dioxide, and ammonia during the carbon elution step. In addition, solution clarification is generally not necessary as carbon eluates predominantly contain only fine carbon and silica washed from the carbon during elution, which have a negligible effect on precipitation.

Zinc is prepared and added in similar manner to that applied in the Merrill–Crowe precipitation process, although at the higher zinc addition rates required, the use of a zinc slurry feed system may help to avoid agglomeration of zinc particles. Dry zinc feed systems work effectively in many applications. Turbulent mixing is beneficial to ensure that the zinc is dispersed effectively in the solution phase.

It has been suggested that lead nitrate may assist in zinc precipitation at high temperatures, and in particular by reducing the tendency for zinc hydroxide formation [13]. In practice, this is rarely necessary due to the presence of low concentrations of divalent heavy metal cations in most carbon eluates, which have similar beneficial effects to lead.

An important difference in the application of zinc precipitation to treat carbon eluates, rather than leach solutions, is that the tailings solution produced is typically reused for carbon elution, whereas tailings (barren) solution from the Merrill–Crowe process is reused in preceding solid–liquid separation stages, and a portion of the solution is lost to the final plant tailings. Although important, the absolute value of gold concentration in solution is far less critical for reuse in a carbon elution system, and barren solution containing <1 g/t Au is usually perfectly acceptable for elution systems to operate effectively. Consequently, a different approach can be taken for the operation of carbon eluate zinc precipitation systems, for example [23]:

- Zinc addition may be closer to the stoichiometric requirement, reducing both zinc consumption and the residual zinc content of the product.
- Precipitates can be "washed" with relatively high-grade gold solution without adding zinc, to displace residual zinc and to upgrade the gold content of the product.

Consequently, the final precipitate generally contains higher grades of precious metals (50% to 90%) and lower zinc concentrations (3% to 10%) than those obtained from dilute solutions (discussed in Section 8.1.3.1). High-grade gold solutions require correspondingly high zinc addition rates which, when coupled with the need to recirculate precipitation tailings to carbon elution, results in the buildup of high concentrations of zinc in solution. This increases the risk of zinc hydroxide formation, as discussed in Section 8.1.2.3. Consequently, a portion of the circulating precipitation–elution circuit barren solution must be routinely bled from the precipitation system to keep the zinc concentration below the level at which zinc hydroxide will form. For example, for a solution containing 5 g/L NaCN and 50 g/t Au at pH 11, a zinc concentration of 0.05 g/L has been found to be a safe operating level [5].

8.2 ALUMINUM PRECIPITATION

The use of aluminum for precipitation of gold from alkaline cyanide solutions was originally proposed and patented by Moldenhauer in 1893 [24]. Despite some advantages over zinc, the process has not been applied widely because of the more favorable economics of zinc precipitation. Aluminum was used commercially at Nipissing and at the Deloro smelter (both in Canada) and has recently been tested for the recovery of gold and silver from various leaching systems other than cyanide (see Section 8.4).

The oxidation of aluminum in aqueous solution is given by:

$$Al^{3+} + 3e \rightleftharpoons Al \qquad (EQ\ 8.21)$$

where

$$E = -1.66 + 0.0197 \log [Al^{3+}]\ (V)$$

Clearly aluminum has a sufficiently low reduction potential to reduce the Au(I) cyanide complex to gold. Aluminum does not form a stable complex with cyanide but dissolves in alkaline solution as follows:

$$AlO_2^- + 2H_2O + 3e \rightleftharpoons Al + 4OH^- \qquad (EQ\ 8.22)$$

where

$$E = -2.35 + pH + \log [AlO_2^-]\ (V)$$

The aluminate ion (AlO_2^-) can hydrolyze further to form a relatively insoluble hydroxide:

$$AlO_2^- + 2H_2O \rightleftharpoons Al(OH)_3 + OH^- \qquad (EQ\ 8.23)$$

The equilibrium of this reaction can be kept to the left, favoring the aluminate species, by keeping the pH >12 to avoid passivation of the aluminum surface by hydroxide layer formation.

The overall stoichiometry of the gold precipitation reaction is:

$$3Au(CN)_2^- + Al + 4OH^- \rightleftharpoons 3Au + 6CN^- + AlO_2^- + 2H_2O \qquad (EQ\ 8.24)$$

Several other stoichiometries have been proposed based on plant observations, which take into account some of the many side reactions that may occur, but these are usually dependent on specific solution conditions [15]. The stoichiometric requirement of aluminum is less than zinc, because the metal reduction is a three-electron reaction compared with the two-electron reaction for zinc. Also, it can be seen from Equation (8.24) that the gold cyanide reduction yields 2 moles of cyanide for every mole of gold, unlike the zinc precipitation reaction that actually consumes 2 additional moles of cyanide for every mole of gold precipitated. Consequently, the cyanide is effectively regenerated by aluminum precipitation.

An important disadvantage of aluminum is that lime cannot be used for pH control because highly insoluble calcium aluminate is formed, which tends to foul filters and contaminates the final gold precipitate, as follows:

$$2AlO_2^- + Ca(OH)_2 \rightleftharpoons CaAl_2O_4 + 2OH^- \qquad (EQ\ 8.25)$$

Therefore, solutions containing even moderate calcium content would have to be treated for calcium removal prior to aluminum precipitation. This could be achieved by the addition of sodium carbonate to precipitate calcium carbonate, with the added benefit of a net increase in pH.

Aluminum precipitates gold much more slowly than zinc, despite a larger electrochemical driving force, due to the faster dissolution rate of zinc in complexing cyanide solution. It is far less effective than zinc for recovery from solutions containing little or no silver but works quite well for the precipitation of silver, or both gold and silver, from solutions containing >50 g/t silver. Deaeration of solutions is required prior to precipitation because of the rapid oxidation of aluminum in the presence of oxygen; however, the metal is less affected by "poisoning" ions, such as sulfide, arsenic, and antimony, than is zinc.

8.3 ELECTROWINNING

Electrowinning is used for treatment of high-grade gold solutions, that is, carbon eluates, to produce loaded cathodes and cathodes cell sludges, which require relatively little further refining.

There is also some potential for the application of electrowinning to more dilute leach solutions; however, further developments in electrowinning technology are required for this to be economically viable.

8.3.1 Electrowinning Fundamentals

A cathodic reduction reaction in aqueous solution can be driven by applying a voltage across a pair of electrodes immersed in the solution. The voltage applied must exceed the reversible electrode potential for the desired reaction to occur, and must allow for the voltage drop due to the resistance of the solution—a function of the solution conductivity—and other cell losses [25]. The amount by which the applied voltage exceeds the reversible electrode potential (E_r) is referred to as the voltage overpotential, η, defined as follows:

$$\eta = E_{applied} - E_r \qquad (EQ\ 8.26)$$

The reduction at the cathode is accompanied by a parallel oxidation reaction at the anode, usually the oxidation of water to oxygen.

Considering the electro-reduction of a metal ion (M^{z+}) the total cell voltage is represented by:

$$V = E_{(O_2/H_2O)} + E_{(M^{z+}/M)} + \eta_a + \eta_c + iR \qquad (EQ\ 8.27)$$

where

η_a and η_c = the anodic and cathodic overpotentials, respectively
iR = the potential drop through the solution due to the resistance (or finite conductivity) of the solution

This is illustrated for a simple cell system in Figures 8.16 and 8.17.

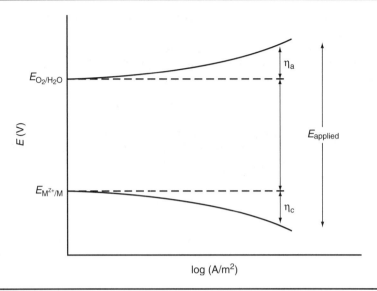

FIGURE 8.16 Schematic representation of potential requirements of a simple cell system

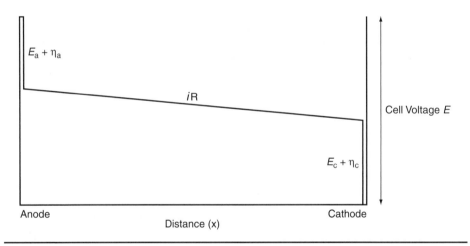

FIGURE 8.17 Schematic illustration of potential drop across an electrolytic cell

The rate of the electron transfer process is described by the Butler–Volmer equation:

$$i = i_0 e^{(-\alpha n F \eta / RT)} - e^{((1-\alpha)(nF\eta/RT))} \qquad \text{(EQ 8.28)}$$

At high overpotentials, above approximately 50 mV, the second term of Equation (8.28), which is due to the reverse or "back" reaction, may be neglected. The larger the overpotential, the faster the reaction proceeds. As the overpotential is increased, the concentration of reduced species close to the cathode surface becomes depleted, because the mass transport of the ionic species through the boundary layer cannot keep up with the electrochemical reaction rate. At this point, the reaction rate becomes mass transfer controlled. The current at which this occurs is known as the limiting current density (i_L)

which is expressed in units of amp/m² of cathode [25]. For the case where the concentration of species to be reduced at the surface of the cathode is zero, then:

$$j_T = -D_{M^{z+}}(c_b/N) \qquad \text{(EQ 8.29)}$$

Combining this expression with Faraday's second law, it can be shown that:

$$i_L = -nFk_m c_b \qquad \text{(EQ 8.30)}$$

The value of the limiting current density is important because this is the maximum current that can be passed through the solution before ionic species other than the desired metal ions are reduced. The value of the limiting current density, and consequently the deposition rate, increases with increasing concentration of the metal ion to be deposited.

8.3.2 Reaction Chemistry

8.3.2.1 Cathodic Reduction of Gold

Gold is electrolytically displaced from aqueous alkaline cyanide solution according to the reaction:

$$Au(CN)_2^- + e \rightleftharpoons Au + 2CN^- \qquad \text{(EQ 8.31)}$$

where

$$E_{rev} = -0.60 + 0.118 \log a_{CN^-} + 0.059 \log a_{Au(CN)_2^-} \text{ (V)}$$

Deposition of the metal occurs at potentials below approximately −0.7 V, although the exact potential at which reduction starts depends on the solution conditions, such as conductivity, the concentrations of ionic species present, and temperature. Reported values vary between −0.7 and −1.1 V [26].

The mechanism of electrolytic deposition of gold probably proceeds by the adsorption of aurocyanide at the cathode, followed by reduction of the adsorbed species, as follows:

$$Au(CN)_2^- \rightleftharpoons AuCN_{ads} + CN^- \qquad \text{(EQ 8.32)}$$

$$AuCN_{ads} + e \rightleftharpoons AuCN_{ads}^- \qquad \text{(EQ 8.33)}$$

The reduction step is then followed by dissociation of the reduced species:

$$AuCN_{ads}^- \rightleftharpoons Au + CN^- \qquad \text{(EQ 8.34)}$$

At high cathodic overpotentials, the electron transfer step is likely to be direct, and the intermediate adsorbed species is probably not formed. This distinction between deposition mechanisms under different conditions is important because the physical characteristics of gold deposited by the two are very different. Whereas gold that is deposited at relatively low cathodic overpotentials forms a dense, solid product on the cathode, at higher cathodic overpotentials, a fluffy, porous deposit is formed, which may end up as sludge on the cell floor. Either of these product types may be desirable in particular circumstances, for example:

- Sludge may be considered less of a security risk or health hazard (e.g., if the product contains mercury) because it can be stored under solution and can be handled by washing and pumping.

- A dense solid product on cathodes may be of higher purity, reducing subsequent refining requirements.
- Loaded electrodes may be transferred to replating cells for refining (i.e., sludge is undesirable).

These factors are discussed further in Section 8.3.4.

8.3.2.2 Cathodic Reduction of Other Metals

Several other metal cyanide complexes, for example, silver, mercury, and lead, will be reduced in preference to gold at the potentials applied for gold reduction (Table 8.1). The recovery of silver is usually desirable as an economic by-product of gold. However, the difference in standard electrode potentials for gold and silver cyanide reduction is in >0.2 V, which indicates some possibility for separate recovery. In situations where the silver concentration is greater than approximately 50% of the gold concentration, silver is deposited in preference to gold, as indicated in Figure 8.18. However, as the silver concentration is depleted, codeposition of gold occurs, and such a separation is not feasible in existing industrial electrowinning systems [27].

If present in significant concentrations, mercury may disrupt electrowinning of gold due to the need to deposit most of the metal from solution before gold deposition will start. This mercury deposition can affect the quality of the cathode deposit, encouraging sludge formation rather than a cohesive cathode deposit. In addition, mercury may present a health hazard above electrowinning cells and for the handling of cathodes as a result of the low vapor pressure of the metal. The chemistry and health aspects of mercury are considered further in Section 9.3.

Low concentrations of lead ions in the electrolyte, for example, up to 1 mg/L, have a beneficial effect on gold deposition. This is due to codeposition of lead, which acts as a catalyst in the electrical double layer and depolarizes the gold reduction reaction [27]. On the other hand, higher concentrations of lead may adversely affect gold recovery by depositing prior to and during gold deposition, thereby contaminating the final cathode product.

The reduction potentials for copper, nickel, iron, and zinc are more negative than that for gold reduction (Table 8.1), and their deposition at potentials applied for gold reduction is possible, depending on their concentration. The concentration above which deposition of each of these metals starts can be calculated from the Nernst equation:

$$E^0(Au(CN)_2^-/Au) + 0.0591 \log [Au(CN)_2^-]/[CN^-]^2$$
$$= E^0(M^{z+}(CN)_x^{(z-x)-}/M) + 0.0591 \log [M^{z+}(CN)_x^{(z-x)-}]/[CN^-]^x \qquad (EQ\ 8.35)$$

which can be expressed in terms of the concentration of the relevant metal ion, as follows:

$$\log[M^{z+}(CN)_x^{(z-x)-}] = 16.92(E^0(Au(CN)_2^-/Au) - E^0(M^{z+}(CN)_x^{(z-x)-}/M)) +$$
$$\log [Au(CN)_2^-]/[CN^-]^2 + (z - x) \log [CN^-] \qquad (EQ\ 8.36)$$

The presence of copper in moderate concentrations (i.e., up to 300 mg/L) has a beneficial effect on gold deposition, with a similar mechanism to that of lead, described earlier [27]. However, even at such low concentrations, some copper may be recovered, despite the fact that the concentration is below the threshold value for the onset of copper deposition. This is attributed to codeposition of copper as a result of the high copper–gold ratio in the electrolyte, illustrated in Figure 8.18, which shows the percentage of copper recovery increasing with increasing applied voltage, as expected. Copper deposition during

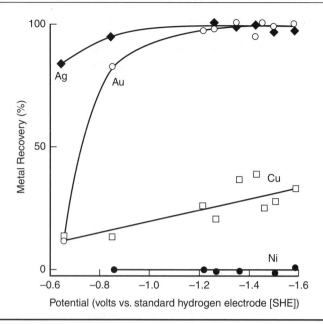

FIGURE 8.18 Percentage of metal recovery from industrial leach solutions at various potentials: flow rate = 15 mL/min, T = 22.5 ±1.5°C, pH = 11.4, Cu = 300 mg/L, Au = 17.5 mg/L, Ag = 2.3 mg/L, Ni = 250 mg/L [27]

gold electrowinning is usually undesirable as it contaminates the cathode and interferes with subsequent refining processes.

The presence of nickel has an adverse effect on gold deposition and may, depending on its concentration, negate the beneficial effect of any copper and lead present [27]. However, nickel recovery is typically poor in industrial electrolytes, as illustrated in Figure 8.18.

8.3.2.3 Reduction of Oxygen and Water

Oxygen that is evolved at the anode by oxidation of water (Section 8.3.2.4) may, in a membraneless cell, migrate or diffuse to the cathode where it is reduced back to water, by either the two- or four-electron path [27]:

$$O_2 + 2H^+ + 2e \rightleftharpoons H_2O_2; \quad E^0 = +0.682 \text{ (V)} \quad \text{(EQ 8.37)}$$

$$O_2 + 2H_2O + 4e \rightleftharpoons 4OH^-; \quad E^0 = +0.410 \text{ (V)} \quad \text{(EQ 8.38)}$$

It has been shown that oxygen reduction may account for >50% of the cathode current [28]. The use of an ion exchange membrane to separate the anode and cathode eliminates this problem, but there are disadvantages with these systems and these are discussed further in Section 8.3.4.2.

Water can be reduced to hydrogen, as given by the following equation:

$$2H_2O + 2e \rightleftharpoons H_2 + 2OH^-; \quad E^0 = -0.828 \text{ (V)} \quad \text{(EQ 8.39)}$$

It can be seen from Figure 8.19 that, under the conditions for which the deposition of gold is mass transport controlled (i.e., the lower portion of the curve), hydrogen is evolved. The evolution of hydrogen is a potential safety hazard that should be monitored

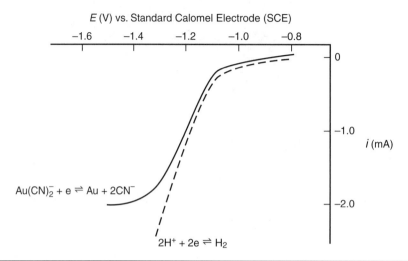

FIGURE 8.19 General form of current–potential curve for gold deposition from cyanide solution [29]

carefully and controlled (see Section 8.3.4.1). The reduction of both water and oxygen at the cathode tends to increase the pH of solution.

8.3.2.4 Anodic Reactions

In alkaline cyanide solution, the major reaction at the anode is the oxidation of water to oxygen:

$$O_2 + 4H^+ + 4e \rightleftharpoons 2H_2O \qquad (EQ\ 8.40)$$

where

$$E_{rev} = 1.228 + 0.015 \log pO_2 - 0.059\ \text{pH (V)}$$

This reaction tends to decrease the solution pH close to the anode—an important consideration in determining possible corrosion of the anode, as discussed in Section 8.3.3.8.

Cyanide may be oxidized to cyanate at the anode:

$$CN^- + 2OH^- \rightleftharpoons CNO^- + H_2O + 2e;\quad E^0 = -0.97\ (V) \qquad (EQ\ 8.41)$$

and may hydrolyze to form ammonia and carbon dioxide according to the following reaction, although this occurs very slowly at high pH:

$$CN^- + H_2O + OH^- \rightleftharpoons NH_3 + CO_2 + 2e \qquad (EQ\ 8.42)$$

This reaction may contribute to the strong ammonia smell above many electrowinning cells, although a significant, if not dominant, portion of the ammonia is probably carried over from the oxidation of cyanide in carbon elution systems.

8.3.3 Reaction Kinetics and Factors Affecting Efficiency

Gold deposition, Equation (8.31), is electrochemically controlled down to potentials in the order of −0.85 to −1.0 V, depending on solution conditions. The reaction rate in this region is described by the Butler–Volmer equation (8.28). At more negative potentials, the deposition rate is dependent on the mass transport rate of Au(I) cyanide species to the cathode.

For solutions containing low concentrations of gold, as is generally the case in gold extraction systems, cells must be operated under mass transport controlled conditions to maximize deposition rates and allow electrowinning to be performed economically. The optimum current for gold deposition is that at which the deposition rate is just mass transport controlled because this minimizes other side reactions at the cathode [28, 30].

Consequently, for an electrowinning system operating at or above the limiting current density (i_L), the rate is determined by factors that affect the mass transport of Au(I) cyanide to the cathode, namely gold concentration (c_b) and the mass transfer coefficient of the system (k_m). The mass transfer coefficient depends on the hydrodynamics of the cell–electrolyte system (i.e., cell geometry, electrolyte flow rate, cell mixing, etc.), solution temperature, and the available cathode surface area [28].

Other factors that affect the efficiency of electrowinning systems are the applied cell voltage and current, which are themselves determined by the solution properties (i.e., temperature, pH, cyanide concentration, concentrations of other ions, conductivity). These factors are discussed further in the following sections.

8.3.3.1 Gold Concentration
The effect of gold concentration on the deposition rate is illustrated by the current–potential curves for two gold concentrations, given in Figure 8.20. These show that, as predicted by Equation (8.30), the limiting current is approximately doubled for a 100% increase in gold concentration [29].

8.3.3.2 Electrolyte Hydrodynamics
The degree of mixing within the electrolyte has a large effect on the mass transport of solution species and consequently on the gold deposition rate. In electrowinning cells with fixed electrodes, such as those normally used for gold recovery (and refining), and with no mechanical agitation, the hydrodynamic conditions are determined by the structure of the electrode(s), the cell configuration, and the flow rate of solution. Electrode structures and cell configuration are considered briefly in Section 8.3.4.2.

8.3.3.3 Temperature
Elevated electrolyte temperature has the following advantages for the electrodeposition of gold:

1. Diffusion coefficient of Au(I) cyanide is increased.
2. Solution conductivity is increased.
3. Solubility of oxygen is decreased, reducing the amount of oxygen available for reduction at the cathode.

Item 1 increases the deposition rate, whereas items 2 and 3 improve the cell current efficiency. Consequently there is an overall advantage to operating electrowinning systems at elevated temperature. However, these effects are quite small, particularly when compared with the effects of gold concentration, the degree of mixing of the electrolyte, and the cathode surface area.

8.3.3.4 Cathode Surface Area
Within reason, electrowinning efficiency is maximized by maximizing the cathode surface area, that is, by the use of three-dimensional electrodes such as steel wool, as used in the Mintek cell and others. The presence of any particulate material in the electrolyte that builds up in the cathode will tend to reduce voidage, restrict flow, and reduce cell efficiency. In addition, as the amount of gold deposited onto the cathode increases, the surface area of the cathode is also increased, which in turn increases the rate of water

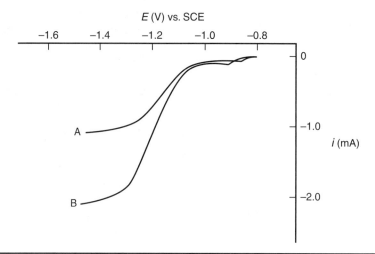

FIGURE 8.20 Current–potential curves showing effect of gold concentration: [KCN] = 3.14 × 10^{-2} M, sweep rate = 10 mV/s, rotation speed = 11.43 Hz; [Au] = (A) 1.0 × 10^{-2} M, (B) 5.0 × 10^{-3} M [29]

and oxygen reduction at the cathode. The increased hydrogen evolution increases the potential drop across the cathode, which may produce regions at the electrode surface where gold deposition cannot occur.

8.3.3.5 Cell Voltage

The voltage and current that must be applied to a cell for the most efficient gold recovery depends on the cell design and the solution conditions, including conductivity, pH, temperature, and concentrations of all ionic species present. In industrial cell systems, voltages of the order 2 to 4 V are typically applied to allow for solution losses and side reactions (see also Table 8.4) [30].

8.3.3.6 Cell Current and Current Efficiency

The rate of deposition increases with increasing current, up to the limiting current, at which point maximum current efficiency is obtained in the system. Above the limiting value the excess current is consumed by side reactions, principally the reduction of water to hydrogen, but also potentially the reduction of oxygen and other metals. Cell currents of between 30 and 100 amp/m^2 cathode are applied (see Table 8.4), with typical current efficiencies of 2% to 8% achieved depending on the cell configuration, gold concentration, electrolyte composition, and hydrodynamic properties of the electrolyte. Current efficiencies as high as 27% have been recorded for high-grade (>300 g/t) gold solutions [31].

8.3.3.7 Solution Conductivity

The solution conductivity determines the ohmic potential drop across the cell: the higher the conductivity, the lower the electrical loss in solution. Typically cells are operated with electrolyte conductivities in the order of 2 S/m. Electrolytes produced by carbon elution usually contain a sufficient concentration of sodium hydroxide for adequate solution conductivity, but conductivity may be increased further if necessary by the addition of a suitable electrolyte, such as sodium hydroxide [28]. Leach solutions generally have relatively low conductivity, which must be significantly increased if gold is to be electrodeposited efficiently. This factor, coupled with the extremely low limiting current densities achievable, prevents the use of electrowinning on solutions containing less than approximately 10 g/t Au.

TABLE 8.4 Selected electrowinning system operating parameters [34, 36, 37, 38]

Operation	Units	Mesquite (United States)	Beisa (South Africa)	Kambalda (Australia)	Williams (Canada)
Cell type	—	Custom-built	Mintek	Custom-built	Custom Equipment Corp.
No. of cells	—	2	2	2	4
Configuration	—	Parallel	Parallel	Series	2 × 2 series
No. of cathodes per cell	—	16	6	5	18
No. of anodes per cell	—	17	7	6	19
Weight steel wool per cathode	kg	1.46	0.50	0.50	2.80*
Feed solution, gold concentration	g/t	140	180	20–200	350–400
Barren solution, gold concentration	g/t	1.6	9	15–20	5–10
Single-pass efficiency	%	99	60†	60	97
Overall efficiency	%	99	95	99	>97
Steel wool gold loading	kg/kg	1.5	2–4	2–3*	4
NaCN concentration	%	1–1.5*	2*	0.5	0
NaOH concentration	%	1.0	0.5*	2.0	1.0
Temperature	°C	77	NA	80–90	50–60
Solution flow rate	L/min	38	62	65	150
Cell current	amp	800	500	170	650
Cell voltage	V	2–4	5	4	2.5–3.5

* Knitted stainless steel mesh cathodes.
† Estimated from available data.
NA = not available.

8.3.3.8 pH

In addition to its effect on solution conductivity (Section 8.3.3.7), pH is also important for electrode stability. The majority of commercial electrowinning cells use stainless steel anodes (and cathodes in many cases; see Section 8.3.4), which corrode in solutions below a pH of approximately 12.5. This corrosion increases the concentrations of chromium and iron species in solution, which may undergo side reactions and reduce cell efficiency. For example, chromite ions are reduced at the cathode to form an insoluble layer of chromium hydroxide:

$$CrO_4^{2-} + 4H_2O + 3e \rightleftharpoons Cr(OH)_3 + 5OH^-; \quad E^0 = -0.12 \text{ (V)} \qquad \text{(EQ 8.43)}$$

Chromium ion concentrations as low as 5 g/t can significantly affect the cell efficiency and reduce gold deposition down to low levels [28]. Chromium concentrations >100 g/t have been observed to completely prohibit gold electrowinning.

Fe(III) cyanide species are also reduced at the cathode to the Fe(II) cyanide complex, thereby setting up a redox system across the cell:

$$Fe(CN)_6^{4-} \rightleftharpoons Fe(CN)_6^{3-} + e; \quad E^0 = 0.46 \text{ (V)} \qquad \text{(EQ 8.44)}$$

These problems can be eliminated by the use of alternative anode materials, discussed in Section 8.3.4.2, although there are some disadvantages associated with this practice, including higher cost.

8.3.3.9 Cyanide Concentration

The effect of cyanide concentration on the current–potential curve for gold deposition is shown in Figure 8.21. This indicates a negative shift in potential of approximately 0.2 V

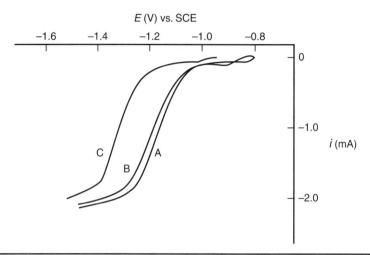

FIGURE 8.21 Current–potential curves showing effect of cyanide concentration: c_{Au} 1.00 × 10^{-2} M, sweep rate = mV/s, rotation speed = 11.43 Hz; c_{KCN}: (A) 3.14 × 10^{-2}, (B) 0.10, (C) 1.00 M [29]

for a tenfold increase in cyanide concentration (i.e., from 5 g/L NaCN to 50 g/L) [29]. A higher voltage must be applied at the higher concentration to ensure that the system is operated at, or slightly above, the limiting current.

8.3.3.10 Effect of Other Solution Species
Sulfide ions, which may be present in low concentrations in gold electrolytes, are oxidized to polysulfide species at the anode. These ions may then be transported to the cathode where they may react with deposited gold to form stable complexes (i.e., potentially causing redissolution of the deposited metal).

8.3.3.11 Solution Flow Rate
The solution flow rate is important because it determines the mass transport of species in the cell, for example, most importantly, the transport of gold cyanide to the cathode. Lower flow rates will decrease the limiting current density and promote the formation of powder and/or sludge under a given set of operating conditions; conversely, the use of higher flow rates promote the formation of solid, dense product at the cathode.

8.3.4 Process Considerations

8.3.4.1 Applications
Gold solutions produced by the elution of loaded carbon and the leaching of high-grade materials (e.g., intensive leaching of gravity concentrates) are too dilute for treatment in conventional electrowinning cells (i.e., copper recovery cells). However, the relatively small volumes of electrolytes containing between 20 and 100 g/t Au that are produced by these processes can be treated successfully in customized cell systems (Section 8.3.4.2), achieving current efficiencies between 1% and 10%. Application of existing electrowinning technology to more dilute solutions (i.e., 1 to 10 g/t Au), for example, solutions produced by direct leaching of ores, results in much lower current efficiencies, making the process uneconomic.

Electrowinning has several advantages over zinc precipitation for the treatment of high-grade solutions, for example:

- No new chemicals or metals are introduced into the process.

- The process is more selective for gold and silver over copper.
- The product is generally of higher purity.

However, electrowinning has the following disadvantages [32]:

- Low single-pass efficiencies are achieved at high flow rates per unit cell volume, and most systems rely on recirculating solutions to achieve acceptable gold extraction.
- Mercury may present more of a health hazard in electrowinning systems than in closed zinc precipitation circuits.
- The release of ammonia (produced by breakdown of cyanide during high-temperature carbon elution and carried over into electrowinning) and hydrogen gas generated during electrowinning of gold from alkaline cyanide solutions may pose safety, health, and environmental problems, requiring specialized venting and control.

8.3.4.2 Electrowinning Cell Configurations

Electrowinning is performed in cells, generally constructed of a suitable nonconducting material such as fiberglass or plastic, containing a number of cathodes and anodes spaced alternately in the cell. The spacing of the electrodes is set so that there is no danger of the cathode deposit reaching the anode while minimizing solution power losses. The gold-bearing solution is fed into one side of the cell and passes out the opposite end. The direction of solution flow may either be parallel or perpendicular to the orientation of the electrodes, depending on the specific cell design. The choice is a function of the flow characteristics of the cell, that is, the flow-through properties of the electrodes and the cell geometry. Several customized electrowinning cells have been developed to treat these gold-bearing electrolytes, including the Zadra, Anglo American Research Laboratories, NIM (graphite chip), Mintek, Kemix, and other cells. In addition, cells using membranes to separate the cathode and anode portions of the electrolyte have been proposed but have not gained acceptance in industry because of difficulties in maintaining the membrane. Information on the design of electrowinning systems is available in the literature [7].

A number of methods of increasing mass transport, and consequently for improving current efficiencies, have been investigated and applied in cell systems. These have included various methods for improving the hydrodynamics of the electrolyte by [33]:

- Recirculating the electrolyte through the cell (e.g., pumped cell arrangement)
- Use of moving (rotating) electrodes
- Agitating the electrolyte, for example, mechanically, with air, or by ultrasonics

In addition, various types of electrodes have been developed to increase the available cathode surface area and to improve solution flow characteristics through and around the electrodes. During the early development of cell systems for gold electrowinning, flat plate stainless steel cathodes were used but these were quickly replaced with punched plate and wire mesh (to allow better solution flow), and subsequently with three-dimensional electrodes comprised of steel wool in polypropylene baskets. The latter provided much higher cathode surface area, while having excellent flow-through characteristics, and these have become the most widely used electrodes in gold electrowinning systems. Porous graphite cathodes have also been used, but these were found to plug with extraneous material, and the evolved gases tended to impede solution flow through the cathode [34]. Standard practice is to use stainless steel wool cathodes, which are harvested manually. However, the unique Kemix cell design uses a rotating cylindrical stainless steel wool cathode, which is fully enclosed and can be harvested automatically using high-pressure water sprays (see also Section 8.3.4.4) [35]. The rotation of the cathode increases mass transport rates and increases single-pass extraction efficiency.

The original flat plate stainless steel anodes have also been replaced with punched-plate electrodes in many cases, and occasionally with graphite as the anode material. Graphite has the advantage that it does not corrode to produce deleterious Fe and Cr(VI) ions; however, graphite anodes tend to break relatively easily, particularly in industrial environments, and it is difficult to make punched-plate out of graphite, which increases their cost. Titanium plate and mesh anodes have also been used because of their superior chemical stability over stainless steel and their greater mechanical strength compared with graphite, but they are much more expensive than either of these alternatives.

8.3.4.3 Operating Conditions

Operating conditions for several industrial electrowinning systems are given in Table 8.4. Single-pass extraction efficiencies, defined as the percentage of gold in the cell feed removed during a single pass of solution through the cell, vary between 60% and 99%, depending primarily on the flow rate and total residence time in the cell. Gold extraction efficiency data, displayed as a function of cathode residence time, is shown for selected industrial applications in Figure 8.22. Some systems are designed to operate with relatively low single-pass extraction efficiency, that is, between 60% and 80%, which allows cell sizes to be kept small with associated capital cost savings. However, in such systems, the electrolyte must be recirculated through the cell to achieve high overall extraction efficiency (i.e., >99%), comparable with that achieved routinely by zinc precipitation. This results in an electrowinning cycle that is longer than the carbon elution cycle and may limit overall plant throughput in certain cases.

The single-pass extraction efficiency decreases with increasing solution flow rate. The operating solution flow rate is a balance between single-pass efficiency and the total volume of solution that must be treated to recover a fixed amount of gold. Usually solution flow rates in the range of 250 to 500 L/min/m^2 cathode (or 0.25 to 0.5 m/min equivalent linear velocity) are employed [39, 40].

Steel wool cathodes may be loaded up to 20 times their weight in precious metals. This is rarely done in practice because the single-pass efficiency decreases with increasing gold loading on the cathodes. Loadings of 1 to 2 kg Au/kg steel wool are typically achieved. The amount of steel wool that is charged into a unit volume of cathode space is important because it determines both the flow-through and loading characteristics of the electrode; steel wool packing densities of 5 to 15 g/L are typically applied [31], although densities as high as 30 g/L have been used [37]. Variations in steel wool wire diameter and unit surface area have been reported, ranging from 30 to 200 µm and 0.003 to 0.020 m^2/g, respectively [40].

The factors affecting the nature of the deposit formed (Section 8.3.3) may have an important impact on how the cathodes are treated for final metal recovery. Some electrowinning systems are deliberately operated at high current densities to encourage powder and/or sludge formation [38]. In such cases, gold (and silver) form a loose deposit on the cathode which can be washed off with high-pressure water jets, thus avoiding the need for treatment of the cathodes.

8.3.4.4 Product Handling and Treatment

The loaded cathodes are removed periodically from the electrowinning cell for final metal recovery and bullion production (Chapter 10). There are three main options for the handling of electrowinning products to produce gold bullion:

1. Loaded cathodes in their entirety are treated directly by refining, for example, by acid washing and smelting.

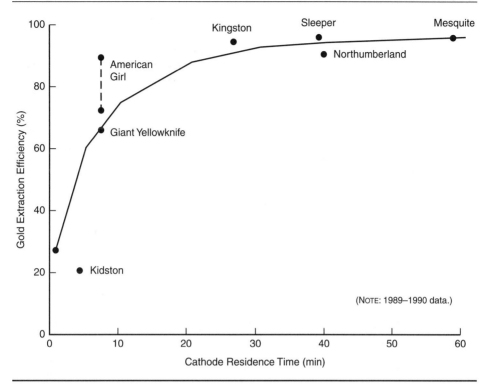

FIGURE 8.22 Gold extraction efficiency vs. cathode residence time for selected electrowinning operations [31]

2. Precious metal deposit is washed from the cathodes, using water sprays and the sludge.
3. Loaded cathodes are transferred to a secondary electrowinning system for electroplating of the gold onto stainless steel plates. This procedure is described in Section 8.3.4.5.

In the case of option 2, the removal of precious metal sludge from the cathodes can either be accomplished manually or automatically. The Kemix cell design is a fully enclosed system that uses high-pressure water sprays to wash the sludge off a cylindrical stainless steel wool cathode. This is an innovative configuration that significantly reduces the health and safety exposure (due to ammonia, hydrogen, etc.), and also helps to reduce the risk of theft since the precious metal sludge is kept in an enclosed and controlled environment. The process of harvesting the cathode can be automated, reducing the labor requirements for the process and further reducing the risks and exposure [35].

In addition, the sludge that collects in the bottom of the cells also typically contains high concentrations of precious metals and is collected and treated further to produce bullion. In some cases, it may be highly desirable to produce a sludge at the cathode for easier removal of precious metals from cathodes, as defined in option 2.

8.3.4.5 Crude Electrorefining

A few industrial operations have applied an electroplating process to treat loaded steel wool cathodes, yielding a product that requires little or no further treatment prior to shipment and sale. The cathodes from the primary electrowinning operation are loaded into a secondary electrolytic cell, where they become the anodes. Flat, stainless steel plates are used as the cathodes. High conductivity electrolyte (0.5% NaCN, 1% NaOH) is circulated through the cell to provide agitation and to aid mass transport. The cell is operated at relatively low voltage and current (i.e., 1.8 V and 150 to 200 amp, respectively), at least initially, to produce a dense product at the cathode. The product is subsequently scraped off the cathodes as a foil, recovered as a granular product off the surface of the foil or, less commonly, recovered as sludge from the bottom of the cell. The impurities pass into, and remain in, solution or deposit out with the cell sludge. The quality of the product obtained depends on the primary cathode composition and the conditions applied. More than 98% of the gold can be redeposited from the original steel wool electrodes [38].

This procedure allows the steel wool to be reused many times and significantly reduces subsequent refining and refinery by-products treatment requirements [38]. It has the disadvantage of being quite labor-intensive and is unsuitable for cathodes with high levels of certain impurities, for example, copper and nickel.

8.3.4.6 Electrowinning from Dilute Solutions

The use of novel electrowinning cells utilizing three-dimensional cathodes, including fluidized bed cathodes, has been investigated extensively for the application of electrowinning to treat increasingly dilute metal solutions [41, 42]. The potential application of such systems for gold recovery from dilute solutions is no exception. One investigation of a cylindrical electrowinning cell with a fixed three-dimensional steel wool cathode indicated that gold could be recovered effectively from a cyanide solution containing 5 mg/L Au, albeit at very low current efficiency of 0.33% and overall electrical energy consumption of 112 kWh/kg ($0.17/oz Au) [43]. The presence of low concentrations of impurities (i.e., 10 mg/L of Ag, Ni, Cu, Zn, and Fe) had a relatively small effect on electrowinning efficiency; higher concentrations (i.e., 100 mg/L) were severely detrimental in most cases. Although this technology is unlikely to compete with carbon adsorption and zinc precipitation in its present state of development, it does show promise for potential application if improvements in efficiency can be realized.

Fluidized bed systems offer a number of potential advantages over fixed cathode systems, including extremely high mass transfer rates within the bed. However, a significant problem associated with fluidized bed systems is the large potential drop across the fluidized bed and the inability to control the potential accurately within the bed, as illustrated in Figure 8.23. This can result in nonselective metal deposition, reduced current efficiency, and poor control of side reactions. On the other hand the potential benefit of greatly improved mass transport at the cathode is a significant incentive for the development of fluidized bed electrode systems. Such systems are likely to receive increasing attention from researchers in the quest for an effective method to recover gold directly from dilute leach solutions.

8.4 RECOVERY FROM NONCYANIDE SOLUTIONS

The use of noncyanide solution systems for gold leaching and recovery (i.e., using carbon adsorption, ion exchange [IX], and solvent extraction [SX] systems) is commercially undeveloped, and consequently there is no clear consensus on the best method of gold recovery from solutions produced by these processes. The economic and technical problems associated with such alternative media have tended to focus research and development on the primary processes, that is, leaching, adsorption, and elution, rather than ultimate metal recovery. However, Table 8.5 reveals a number of chemical species that should be capable of reducing, and precipitating, gold from chloride, thiourea, thiosulfate, and thiocyanate solutions.

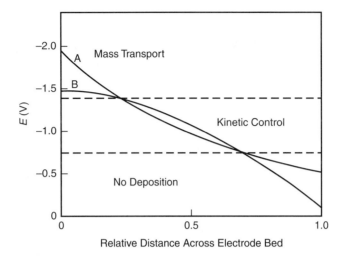

FIGURE 8.23 Schematic representation of potential distribution across a fluidized bed electrode: A = potential at solid electrode surface, B = potential in solution

TABLE 8.5 Electrode potentials for gold complexes and possible reductant systems for gold precipitation [10, 44, 45, 46, 47, 48]

Gold Complex	E^0	Reductant System
$Au^+ + e \rightleftharpoons Au$	1.69	
$AuCl_2^- + e \rightleftharpoons Au + 2Cl^-$	1.15	
$AuCl_4^- + 2e \rightleftharpoons AuCl_2^- + 2Cl^-$	0.93	
	0.77	$Fe^{3+} + e \rightleftharpoons Fe^{2+}$ (iron)
$Au(SCN)_2^- + e \rightleftharpoons Au + 2SCN^-$	0.66	
$Au(CS(NH_2)_2)_2^+ + e \rightleftharpoons Au + 2CS(NH_2)_2$	0.35	
	0.17	$SO_4^{2-} + 4H^+ + 2e \rightleftharpoons H_2SO_3 + H_2O$ (sulfur dioxide)
$Au(S_2O_3)_2^{3-} + e \rightleftharpoons Au + 2S_2O_3^{2-}$	0.15	$Sn^{4+} + 2e \rightleftharpoons Sn^{2+}$ (tin)
	0.14	$S + 2H^+ + 2e = H_2S$ (hydrogen sulfide)
	0	$2H^+ + 2e \rightleftharpoons H_2$ (hydrogen)
	−0.39	$2H_2CO_3 + 2H^+ + 2e \rightleftharpoons (COOH)_2 + 2H_2O$ (oxalic acid)
	−0.48	$S + 2e = S^{2-}$ (sulfide)
$Au(CN)_2^- + e \rightleftharpoons Au + 2CN^-$	−0.57	
	−0.75	$BO_3^{3-} + 7H_2O + 7e \rightleftharpoons BH_4 + 10OH^-$ (borohydride)
	−0.76	$Zn^{2+} + 2e \rightleftharpoons Zn$ (zinc)
	−1.66	$Al^{3+} + 3e \rightleftharpoons Al$ (aluminum)

Of these, zinc is unsuitable for cementation from acidic solutions because it is highly soluble in acid, evolving large quantities of hydrogen and resulting in prohibitively high consumptions of the metal [49]. Aluminum dust has been used effectively (with 99.5% efficiency) to precipitate gold and silver from thiourea solutions, with metal consumption in the order of 1 kg Al/kg precious metal [50]. Aluminum should also be effective at reducing the gold chloride, thiocyanate, and thiosulfate complexes, depending on the gold concentration and solution conditions. However, the purity of precious metals products obtained by this method is generally poor, and costly refining procedures are necessary.

The use of copper for the direct recovery of gold by cementation from dilute ammoniacal thiosulfate solution produced by heap leaching has been proposed. In this case, the use of copper is synergistic with the leaching process because Cu(II) ions are used in conjunction with ammonia as a catalyst for gold leaching (see Section 6.3) [40]. The major reaction is as follows:

$$2Au(S_2O_3)_2^{3-} + Cu + 4NH_3 \rightleftharpoons 2Au + 4S_2O_3^{2-} + Cu(NH_3)_4^{2+} \qquad \text{(EQ 8.45)}$$

The cementation reaction has been shown to be mass transport controlled (activation energy = 14 to 18 kJ/mol), with kinetics favored at elevated temperatures, high pH, and high ammonia concentration. The kinetics were negatively affected by sulfite and copper ions in solution [51]. This process would be expected to generate precipitate containing a high concentration of copper (i.e., 10% to 20% Cu) which would likely require additional process steps during refining, for example, acid leaching to remove copper (see Section 10.2).

The precipitation of gold from ammoniacal thiosulfate solution using ammonium sulfide, sodium sulfide or bisulfide, or hydrogen sulfide has been proposed [52]. At pH 9–10, the solution pH typical of those produced during ammoniacal thiosulfate leaching (see Section 6.3.1), the sulfide ions reportedly protonate to form bisulfide ions (HS^-). The bisulfide species reduce gold in thiosulfate solution, as follows:

$$Au(S_2O_3)_2^{3-} + HS^- \rightleftharpoons Au + 2S_2O_3^{2-} + S \qquad \text{(EQ 8.46)}$$

Copper and other transition metals are also efficiently precipitated from solution. This method of gold reduction from thiosulfate solution is rather elegant because the bisulfide will also react with polythionates (e.g., trithionate and tetrathionate species) to regenerate the thiosulfate reagent according to the following reactions:

$$S_3O_6^{2-} + HS^- + OH^- \rightleftharpoons 2S_2O_3^{2-} + H_2O \qquad \text{(EQ 8.47)}$$

$$4S_4O_6^{2-} + 2HS^- + 8OH^- \rightleftharpoons 9S_2O_3^{2-} + 5H_2O \qquad \text{(EQ 8.48)}$$

However, oxygen must be excluded from the process to avoid the formation of excess elemental sulfur. It has been proposed that the gold-bearing precipitate produced could be further upgraded by reacting the material with sulfite species (added as the sodium or ammonium salt) at 90°C to 100°C [52]. The sulfite ion reacts with the elemental sulfur to form thiosulfate, as follows:

$$S + SO_3^{2-} \rightleftharpoons S_2O_3^{2-} \qquad \text{(EQ 8.49)}$$

The thiosulfate that is generated in this manner can be used beneficially in the leaching circuit. This overall scheme for gold recovery from ammoniacal thiosulfate solution by direct precipitation from solution (after solid–liquid separation, if required) appears to hold considerable promise for commercial application, largely because it eliminates the problems associated with polythionate fouling of ion exchange resins that have been proposed as an alternative for gold recovery from thiosulfate solutions (see Section 7.2.5.2).

Sodium borohydride will effectively precipitate gold from acidic thiourea, thiosulfate, thiocyanate, and chloride solutions, and, as predicted by thermodynamics, the reaction is very fast for all the complexes. The process is quite selective for gold, with the possibility of separate recovery of gold and silver if a satisfactory two-stage process could be developed. A high-purity precious metal product is obtained at relatively low reagent cost, similar to that achieved by zinc precipitation from alkaline cyanide solution [49].

The use of hydrogen, under elevated pressure and temperature, has been investigated to a limited extent for gold recovery from thiourea solutions, but a catalyst such as nickel or platinum is required to make the reaction proceed sufficiently fast, and the pressure required is high (i.e., >2,000 kPa) [53]. This process is likely to be more expensive than some of the alternatives.

The electrolytic recovery of gold from chloride solutions is applied commercially for gold refining, as discussed in Chapter 10, and may be successfully applied to lower-grade solutions, depending on the gold grade and other solution properties (see Section 8.3). Chlorine is evolved at the anode, which is attractive from the point of view that the chlorine can potentially be reused for leaching or elution, but any chlorine released must be contained in a closed system to prevent health hazards and corrosion of surrounding equipment and structures. One such process has been investigated for the recovery of gold (and other metals) from electronic scrap [54]. In this case, selective leaching of copper using Cu^{2+} (to remove a significant portion of the copper) was proposed, followed by more aggressive chlorine–chloride leaching of gold, palladium, silver, and other metals. Selective recovery of the metal by direct electrowinning from the leach solution was then suggested, with a gold–palladium alloy deposited at 0 V (vs. SHE), a copper–silver alloy deposited at –0.2 to –0.3 V (vs. SHE), and a lead–tin alloy deposited at –0.7 V. Chlorine would be regenerated at the anode for reuse in the leaching step.

Gold can be recovered similarly by electrowinning from bromide and iodide solutions, with the possibility of electroregeneration of bromine and iodine oxidants, respectively [55, 56].

Gold can be electrodeposited from acidic thiourea (and potentially thiocyanate and thiosulfate) solutions; however, this probably requires the use of two-compartment electrowinning cells, with anolyte and catholyte solutions separated by a membrane to avoid excessive oxidation of the reagent at the anode and other deleterious reactions at the cathode (principally involving sulfur species), which may contaminate the product and reduce cell current efficiency [57, 58, 59]. One study has shown that gold can be recovered by electrowinning from solution containing 45 mg/L gold, 6 g/L thiourea, and 20% by volume of isopropanol (i.e., synthetic ion exchange resin eluate solution) with a current efficiency of 0.5% to 1.0% and overall energy consumption of 15 kWh/kg [60].

The potential for direct electrowinning of gold from noncyanide solutions remains an attractive target for research and development activity, similar to its potential and attractiveness for gold recovery from cyanide solutions. However, there are many technical and economic challenges to such technology, and commercial application remains elusive.

In a few rare cases, activated carbon has been used to recover gold from thiourea and chloride solutions up to high loadings (>6 kg/t), and the carbon is sold for further treatment by pyrometallurgical techniques (i.e., by ashing and/or direct smelting) [58, 59]. The potential for the recovery of gold from noncyanide solutions by activated carbon, and also ion exchange resins and solvents, is considered further in Sections 7.1.6, 7.2.4, and 7.3.

REFERENCES

[1] Rose, T.K., and W.A.C. Newman. 1937. *The Metallurgy of Gold.* 7th edition. London: Charles Griffin.

[2] Finkelstein, N.P. 1972. The chemistry of the extraction of gold from its ores. Pages 284–351 in *Gold Metallurgy on the Witwatersrand.* Edited by R.J. Adamson. Cape Town, South Africa: Cape and Transvaal Printers Ltd.

[3] Osseo-Asare, K., T. Xue, and V.S.T. Ciminelli. 1984. Solution chemistry of cyanide leaching systems. Pages 173–196 in *Precious Metals: Mining, Extraction and Processing*. Edited by V. Kudryk, D.A. Corrigan, and W.W. Liang. Warrendale, PA: TMS.

[4] Julian, H.F., and E. Smart. 1921. *Cyaniding Gold and Silver Ores*. 2nd edition. London: Charles Griffin.

[5] Marsden, J.O. 1990. Practical aspects of the precipitation of gold from high temperature carbon eluates. Pages 289–294 in *Proceedings Randol Gold Forum*. Golden, CO: Randol International Ltd.

[6] Dorr, J.V.N., and F.L. Bosqui. 1950. *Cyanidation and Concentration of Gold and Silver Ores*. 2nd edition. New York: McGraw-Hill.

[7] Bailey, P.R. 1987. Application of activated carbon to gold recovery. Pages 550–570 in *The Extractive Metallurgy of South Africa*. Edited by G.G. Stanley. Monograph Series M7. Johannesburg: South African Institute of Mining and Metallurgy.

[8] Nicol, M.J., E. Schalch, P. Balestra, and H. Hedegus. 1979. A modern study of the kinetics and mechanism of the cementation of gold. *Journal of South African Institute of Mining and Metallurgy* (February):191–198.

[9] Barin, I., H. Barth, and A. Yaman. 1980. Electrochemical investigation of the kinetics of gold cementation by zinc from cyanide solutions. *Erzmetall* 33(7–8):379–403.

[10] Nicol, M.J., C.A. Fleming, and R.L. Paul. 1987. The chemistry of gold extraction. Pages 831–905 in *The Extractive Metallurgy of Gold*. Edited by G.G. Stanley. Johannesburg: South African Institute of Mining and Metallurgy.

[11] Parga, J.R., R.Y. Wan, and J.D. Miller. 1988. Zinc dust cementation of silver from alkaline cyanide solutions—analysis of Merrill–Crowe plant data. *Minerals and Metallurgical Processing Journal* 5:170–176.

[12] Miller, J.D., R.Y. Wan, and J.R. Parga. 1989. Characterisation and electrochemical analysis of gold cementation from alkaline cyanide solution by suspended zinc particles. *Hydrometallurgy* 24:373–392.

[13] Paul, R.L., and D. Howarth. 1986. Cementation of gold onto zinc from concentrated aurocyanide electrolytes. Pages 157–172 in *Gold 100: Proceedings of the International Conference on Gold:* Volume 2. Edited by C.E. Fivaz. Johannesburg: South African Institute of Mining and Metallurgy.

[14] Leblanc, R. 1942. Precipitation of gold from cyanide solution by zinc dust. *Canadian Mining Journal* 63 (three parts: April, May, and June).

[15] Hutchings, W. 1949. Laboratory experiments on aluminum and zinc dusts as precipitants of gold in cyanidation practice. *Canadian Mining Journal* (October):74–79.

[16] Chi, G., M.C. Fuerstenau, and J.O. Marsden. 1997. Study of Merrill–Crowe processing. Part I: Solubility of zinc in alkaline cyanide solution. *International Journal of Mineral Processing* 49:171–183.

[17] Halbe, D. 1985. Recovery of gold and silver from leach solutions. *Heap and Dump Leaching Newsletter* May (Parts I and II). Denver, CO: DHL Company.

[18] McIntyre, J.D.E., and W.F. Peck. 1976. Electro deposition of gold—depolarisation effects of heavy metal ions. *Journal of the Electrochemical Society* 123:1800–1813.

[19] Kammel, R., and H.W. Lieber. 1984. Cell design for electrolytic silver recovery from various dilute aqueous solutions. Pages 261–280 in *Precious Metals: Mining, Extraction & Processing*. Edited by V. Kudryk, D.A. Corrigan, and W.W. Liang. Warrendale, PA: TMS.

[20] Sheerin, C.H., R.A. Smith, and M.G. Eiselein. 1990. Effect of mercury on the Merrill-Crowe process at FMC Gold's Paradise Peak mine. In *Proceedings Gold Symposium*. New York: American Institute of Mining, Metallurgical, and Petroleum Engineers.

[21] Harris, L., D. Malhotra, and J. McGregor. 2000. Recent developments in processing of oxide and refractory gold ores. Pages 321–325 in *Proceedings Randol Gold and Silver Forum 2000*. Golden, CO: Randol International Ltd.

[22] Marsden, J.O., and M.C. Fuerstenau. 1993. Comparison of Merrill-Crowe precipitation and carbon adsorption for precious metals recovery. Pages 1189–1194 in *Proceedings of XVIII International Mineral Processing Congress*. Carlton, Victoria: Australasian Institute of Mining and Metallurgy.

[23] Mansanti, J.G., and M.F. Gleason. 1989. Funda filters for zinc precipitation: Start-up and operation. Paper presented at 2nd Annual Intermountain Mining and Processing Operators Symposium. Elko, NV, October 25–28.

[24] Clennell, J.E. 1915. *The Cyanide Handbook*. New York: McGraw-Hill.

[25] Bockris, J.O.M., and A.K.N. Reddy. 1970. Pages 623–1432 in *Modern Electrochemistry*. Volume 2. New York: Plenum.

[26] Kuhn, A.T. 1978. Electrochemical recovery of metals from dilute solutions. *Chemistry and Industry* (July):447–464.

[27] Kirk, D.W., and F.R. Foulkes. 1984. A potentiometric study of metals affecting precious metal recovery from alkaline cyanide solutions. *Journal of the Electrochemical Society: Electrochemical Science and Technology* (April):760–769.

[28] Paul, R.L., A.O. Filmer, and M.J. Nicol. 1983. The recovery of gold from concentrated aurocyanide solutions. Pages 689–704 in *Proceedings of 3rd International Symposium on Hydrometallurgy: Hydrometallurgy Research, Development and Plant Practice*. Edited by K. Osseo-Asare and J.D. Miller. Warrendale, PA: TMS.

[29] Harrison, J.A., and J. Thompson. 1973. The electrodeposition of precious metals. *Electrochimica Acta* 18:829–834.

[30] Raub, J. 1984. Precious metal alloy electrodeposits. In *Precious Metals: Mining, Extraction & Processing*. Edited by V. Kudryk, D.A. Corrigan, and W.W. Liang. Warrendale, PA: TMS.

[31] Halbe, D. 1988. *Electrowinning Cell Efficiency*. Salt Lake City, UT: Custom Equipment Corporation.

[32] Diaz, X. 2004. A preliminary mathematical model for the prediction of the evolution of mercury, ammonia and hydrogen from electrowinning cells in a gold-cyanide solution system. Ph.D. dissertation. Salt Lake City, UT: University of Utah.

[33] Walker, R. 1997. Ultrasound improves electrolytic recovery of metals. *Ultrasonics Sonochemistry* 4:39–43.

[34] Briggs, A.P.W. 1983. Problems encountered during commissioning the CIP plant at Beisa mines. In *Proceedings of South African Institute of Mining and Metallurgy Colloquium*. South Africa, Johannesburg: South African Institute of Mining and Metallurgy.

[35] Tuanyane, B.D., and A.B. Phillips. 2003. The phased metallurgical upgrade strategy at Gold Fields Limited South African operations. Pages 143–152 in *Proceedings of 8th Mill Operators' Conference*. Melbourne: Australasian Institute of Mining and Metallurgy.

[36] Scerensini, B.J.S. 1982. The design, construction and operation of a 500,000 tonnes per annum carbon-in-pulp plant at Kambalda, Western Australia. Pages 237–278 in *Proceedings of CIP Technology for the Extraction of Gold*. Melbourne: Australasian Institute of Mining and Metallurgy.

[37] Barnes, D., and T.R. Raponi. 1991. Electrowinning and refining at the Williams mine. SME Preprint No. 90-401. Littleton, CO: SME.

[38] Arnold, J.R., and W.J. Pennstrom. 1986. The gold electrowinning-replate circuits at Gold Fields' Ortiz and Mesquite operations. SME Preprint 86-309. Littleton, CO: SME.

[39] Jha, M.C. 1984. Recovery of gold and silver from cyanide solutions: A comparative study of various processes. Paper presented at 1st International Symposium on Precious Metals Recovery, Reno, NV, June 10–14.

[40] Wan, R.Y., K.M. LeVier, and R.B. Clayton. 1994. Hydrometallurgical process for the recovery of precious metal values from precious metal ores with thiosulfate lixiviant. U.S. Patent No. 5,354,359. October 11.

[41] Huh, T., and J.W. Evans. 1987. Electrical and electrochemical behaviour of fluidised bed electrodes. Parts I and II. *Journal of the Electrochemical Society* 134(2): 308–321.

[42] Evans, J.W., and V. Jiricny. 1987. Fluidised bed electrodes. In *Proceedings CHISA Conference*. August 1987.

[43] Brandon, N.P., M.N. Mahmood, P.W. Page, and C.A. Roberts. 1987. The direct electrowinning of gold from dilute cyanide leach liquors. *Hydrometallurgy* 18: 305–319.

[44] Cotton, F.A., and G. Wilkinson. 1972. *Advanced Inorganic Chemistry*. 3rd edition. New York: Interscience.

[45] Schmid, G.M. 1985. Gold. Pages 313–320 in *Standard Potentials in Aqueous Solutions*. Edited by A.J. Bard, R. Parsons, and J.K. Jordan. New York: Marcel Dekker.

[46] Weast, R.C. 1981. *Handbook of Chemistry and Physics*. 62nd edition. Boca Raton, FL: CRC Press.

[47] Stark, J.G., and H.G. Wallace. 1973. *Chemistry Data Book*. Revised edition. London: John Murray.

[48] Dean, J.A. 1985. Chapter 5 in *Lange's Handbook of Chemistry*. 13th edition. New York: McGraw-Hill.

[49] Awadalla, F.T., and G.M. Ritcey. 1990. Recovery of gold from thiourea, thiocyanate or thiocyanate solutions by reduction-precipitation with a stabilised form of sodium borohydride. Pages 295–305 in *Proceedings Randol Gold Forum*. Golden, CO: Randol International Ltd.

[50] Raudsepp, R., and R. Allgood. 1987. Thiourea leaching of gold in a continuous pilot plant. Pages 87–96 in *Proceedings of the International Symposium on Gold Metallurgy*. Edited by R.S. Salter, D.M. Wyslouzil, and G.W. McDonald. New York: Pergamon Press.

[51] Guerra, E., and D.B. Dreisinger. 1999. A study of the factors affecting copper cementation of gold from ammoniacal thiosulfate solution. *Hydrometallurgy* 51: 155–172.

[52] West-Sells, P.G., and R.P. Hackl. 2005. A novel thiosulfate leach process for the treatment of carbonaceous gold ores. Pages 209–223 in *Proceedings of the International Symposium on the Treatment of Gold Ores*. Calgary, Alberta: Canadian Institute of Mining and Metallurgy.

[53] Deschenes, G. 1987. Investigation on the potential techniques to recover gold from thiourea solution. Pages 359–377 in *Proceedings of the International Symposium on Gold Metallurgy*. Edited by R.S. Salter, D.M. Wyslouzil, and G.W. McDonald. New York: Pergamon Press.

[54] Pilone, D., and G.H. Kelsall. 2003. Metal recovery from electronic scrap by leaching and electrowinning IV. Pages 1565–1576 in *Proceedings Hydrometallurgy 2003: 5th International Conference in Honor of Professor I. Ritchie*. Volume 2. Edited by C.A. Young, A. Alfantazi, C. Anderson, A. James, D. Dreisinger, and B. Harris. Warrendale, PA: TMS.

[55] Sergent, R.H., and A. Dadgar. 1990. Advancements in bromine chemistry for use in large-scale gold extraction. Pages 225–228 in *Proceedings Randol Gold Forum*. Golden, CO: Randol International Ltd.

[56] von Michaelis, H. 1987. The prospects for alternative leach reagents. *Engineering and Mining Journal* (June):42–46.

[57] Hiskey, J.B. 1981. Thiourea as a lixiviant for gold and silver. Pages 83–92 in *Gold and Silver Leaching, Recovery and Economics*. Edited by W.J. Schlitt, W.C. Larson, and J.B. Hiskey. New York: Society of Mining Engineers of the American Institute of Mining, Metallurgical, and Petroleum Engineers.

[58] Hisshion, R.J., and C.G. Waller. 1984. Recovering gold with thiourea. *Mining Magazine* (September):237–243.

[59] Fleming, C.A. 1987. The recovery of gold from thiourea leach liquors with activated carbon. Pages 259–278 in *Proceedings of the International Symposium on Gold Metallurgy*. Edited by R.S. Salter, D.M. Wyslouzil, and G.W. McDonald. New York: Pergamon Press.

[60] Urbanski, T.S., P. Fornari, and C. Abbruzzese. 2000. Gold electrowinning from aqueous-alcohol thiourea solutions. *Hydrometallurgy* 55:137–152.

[61] Levich, V.G. 1962. *Physico-chemical Hydrodynamics*. Englewood Cliffs, NJ: Prentice Hall.

[62] Marsden, J.O., and J.H.L. van der Westhuizen. 1986. Double-vee clarification of pregnant gold-bearing solution at the East Driefontein gold mine. Circular No. 2/86. Johannesburg, South Africa: Mine Metallurgical Managers' Association.

[63] Cox, C., and F. Traczyk. 2002. Design features and types of filtration equipment. Pages 1342–1357 in *Mineral Processing Plant Design, Practice, and Control*. Volume 2. Edited by A.L. Mular, D.N. Halbe, and D.J. Barratt. Littleton, CO: SME.

CHAPTER 9

Surface Chemical Methods

Because surface chemistry influences all heterogeneous reaction systems, it is important in gold ore treatment processes, particularly solid–liquid systems, such as cyanidation, electrowinning, and amalgamation, and those in which gases also play a primary role, for example, flotation. Used for gold recovery since the 1930s (Chapter 1), flotation is a cost-effective process option for many gold ores, either as a preconcentration step or to generate product for direct smelting, as discussed in Chapter 3. Amalgamation with mercury has been applied widely historically and, despite a general decline in its use for health and environmental reasons, the process continues to play an important role in some flowsheets.

This chapter covers the principles and practices of surface chemical methods of mineral separation for the selective recovery of free gold and gold-bearing sulfides, including flotation, amalgamation, and the emerging coal–gold agglomeration process. Knowledge of surface chemistry is also important for flowsheet selection, treatment of refractory ores, tailings retreatment, roasting, pressure leaching, and gravity concentrate processing.

9.1 PRINCIPLES OF SURFACE CHEMISTRY

9.1.1 Mineral–Water Interface

When a mineral is placed in water, the interfacial regions of the two phases alter to accommodate the new environment. An electrical double layer is established at the solid–liquid interface, which balances the overall electrical charge in the system. This affects the behavior of the mineral surface and its interaction with chemical reagents, a factor that is crucial to surface chemical separation processes.

The electrical double layer established at a solid–solution interface can be considered analogous to an electrical capacitor, or impedance, which can be investigated experimentally to characterize surfaces or reactions [1]. The surface charge on the mineral (γ_s) in a pure mineral–water system can be established by several mechanisms:

- Polarization of a conductor or semiconductor by an external source, for example, a charged electrode
- Dissolution of ions from, and/or adsorption of ions onto, the mineral surface (e.g., ionic solid such as argentite [Ag_2S]), which results in a net uneven charge distribution
- Rearrangement of the mineral lattice surface to present ions of different valency to the solid solution interface, for example, complex silicates

The magnitude of the surface charge is determined by the density of adsorbed potential-determining ions. In the case of oxide minerals such as quartz (SiO_2) the potential-determining ions are OH^- and H^+.

The structure of an idealized electrical double layer for a negatively charged surface is shown in Figure 9.1. In this case, positive ions in solution are attracted to the surface to

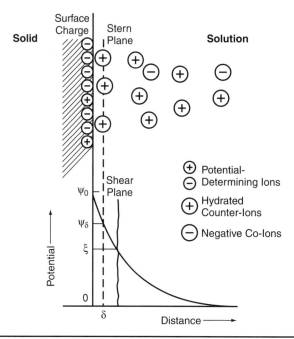

FIGURE 9.1 Schematic representation of the double layer according to Stern's model [2]

form a plane of positive charge, which is then balanced by a less structured collection of negative ions in solution. These are called counter-ions, which make up the diffuse part of the double layer to maintain electrical neutrality. Consequently, this results in a drop in potential across the double layer.

If adsorption of counter-ions occurs by coulombic attraction only, then the solution is called an indifferent electrolyte. The closest distance of approach of the counter-ions to the surface is called the Stern plane (or outer Helmholtz plane). The potential at this plane is called the Stern potential (γ_s). If the ions are not hydrated (i.e., not surrounded by water molecules), they can approach closer to a position called the inner Helmholtz plane. A solution structure is established, which is indicated by a reduction of the dielectric constant to a value of between 78 for bulk solution and 6 for fully oriented water molecules in a primary hydration sheath. This can, in turn, affect other important processing factors, such as ion mobilities, local solubilities, and dissolution or adsorption kinetics.

If the surface area of a mineral powder is known, then the surface charge can be determined by titration with acid or alkali until charge balance is achieved [3]. The actual surface potential (γ_s) cannot be measured directly; however, the potential at a hydrodynamic shear plane close to the Stern layer (Figure 9.1), known as the zeta potential, can be easily measured and has been extremely useful in interpreting surface effects in flotation. The activity of the point-determining ions at which the zeta potential is zero is called the point of zero charge (pzc). The pzc pH values for selected minerals are given in Table 9.1.

For sulfide minerals, the potential-determining ions are difficult to determine because the surface charge is many times less than that for oxides. This is due to limited hydration of mineral surfaces resulting from the inability of sulfides to form hydrogen bonds. Sulfide mineral surfaces are therefore less hydrophilic, and, because of the low charge, their surface properties are influenced by relatively minor surface changes such as the addition of surface-active reagents. This leads to the possibility of effective sulfide

TABLE 9.1 Point of zero charge (pH) values for selected minerals

Mineral	pzc pH Value
Augite	2.7–3.5
Calcite	8.5–10.5
Cassiterite	4.5
Chromite	5.6–7.0
Cuprite	9.5
Garnet	4.4
Geothite	6.7
Hematite	5.0–6.5
Kaolinite	3.4
Magnetite	6.5
Malachite	9.5–10.0
Muscovite	3.6
Pyrite	6–7
Quartz	1.5–2.5
Talc	1.5–3.6
Zircon	5.8

mineral separation from oxide and silicate minerals with only small reagent additions, as is common industrial practice, and in some cases enhanced by the use of inert gas (e.g., N_2) flotation to prevent sulfide mineral surface oxidation [4].

In certain cases (e.g., the addition of flotation collectors), some counter-ions are attracted to the surface much more strongly than by simple coulombic attraction. These adsorb by other forces such as hydrophobic bonding, hydrogen bonding, and covalent bond formation. Adsorption by these mechanisms can reverse the charge at the Stern layer, an effect called specific adsorption.

The electrical double layer properties of a mineral can affect the performance of particle separation processes, such as flotation, in the following ways [5]:

- The sign and magnitude of the surface charge controls the adsorption of physically adsorbing flotation reagents.
- A high surface charge can inhibit the chemisorption of chemically adsorbing collectors.
- The extent of flocculation and dispersion of mineral suspensions is controlled by the electrical double layer.
- The occurrence and magnitude of slime coatings are determined by electrical double-layer interaction.
- Flotation kinetics are dependent on the effect of double layers on the kinetics of film thinning, which also affects particle bubble attachment (see Section 9.1.4.5).

The magnitude of these effects is more pronounced for fine particles, which are represented in Figure 9.2.

If the solid is a semiconductor, there is a potential drop within the solid space charge region on the solid side of the interface. Most sulfides are semiconductors (a property that affects their behavior in flotation), have narrow energy gaps, and possess high surface concentrations of conducting electrons or positive holes. In some cases (e.g., pyrite), sulfides can have different properties, depending whether they are n- or p-type semiconductors.

Semiconductor surfaces can act as catalysts for anodic and cathodic reactions. The practical benefit of this is that collectors (such as xanthates), which undergo an electrochemical change, can act rapidly and selectively for sulfides.

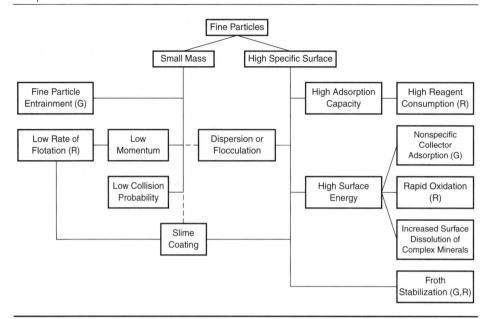

FIGURE 9.2 Schematic diagram showing the relationship between the physical and chemical properties of fine particles and their behavior in flotation. (G) and (R) refer to whether the phenomenon affects grade and/or recovery (adapted from [6]).

9.1.2 Hydrophobicity

Flotation, amalgamation, and coal–gold agglomeration processes aim to remove valuable minerals selectively from an aqueous slurry into an alternative medium. In these three cases, the separation media are air, mercury, and oil, respectively. The objective is to reduce the attraction of the valuable mineral(s) to water, that is, decrease their wettability and increase hydrophobicity.

A measure of wettability is given by the angle established when a drop of water is placed on a mineral surface. If the water spreads over the surface, then the contact angle (θ_{H_2O}) is high and the surface is hydrophilic. Similarly, for the case of flotation, if water doesn't spread and droplets are formed, (θ_{H_2O}) is low and the surface is considered to be hydrophobic. If an air bubble is attached to a surface in water with a significant θ angle (Figure 9.3), then it is relatively hydrophobic and aerophilic, suggesting that flotation may be successful. In an ideal system at equilibrium, this phenomenon is described by Young's equation, as follows:

$$\gamma_{SG} = \gamma_{SL} + \gamma_{LG} \cos \theta \qquad (EQ\ 9.1)$$

where

γ_{SG}, γ_{SL}, and γ_{LG} = the tensions of the solid–gas, solid–liquid, and liquid–gas interfaces, respectively
θ = the contact angle

Surface tension (γ_{SG}) is an effect of the forces of attraction existing between the molecules of a liquid.

In practice, whether a particle is actually floatable is far more complex and involves dynamic factors such as adsorption kinetics, particle collision frequency, hydrodynamics, and bubble adhesion.

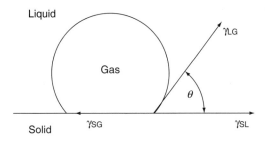

FIGURE 9.3 Schematic representation of the equilibrium contact between air bubble and solid immersed in liquid

9.1.3 Surface Chemistry of Gold

In noncomplexing aqueous media, gold is unreactive because gold ions and oxides are thermodynamically unstable, requiring strongly oxidizing conditions for their formation (see also Chapter 4).

$$Au_2O_3 + 6H^+ + 6e \rightleftharpoons Au + 3H_2O; \quad E^0 = 1.457 - 0.059\,pH\,(V) \qquad (EQ\ 9.2)$$

The chemisorption of oxygen onto gold begins at potentials above about 1.4 V with monolayer coverage approached at 2.0 V. This is attributed to the growth of an oxide layer (Au_2O_3) in acidic solutions and a hydroxide layer ($Au(OH)_3$) in alkaline solutions at high potentials [7]. Voltammetry (Chapter 4) has determined that these reactions occur over ranges that are dependent upon the crystal orientation of the gold and the presence of impurities. For example, if a small amount of metallic platinum is present, then oxygen adsorption can occur at potentials as low as 0.8 V. In gold ore slurries, potentials are generally lower than the 1.4 V required for Equation (9.2), unless an external potential is applied, as is the case in electrolysis. Similarly, gold is the only metal that does not react significantly on heating in air or oxygen and only forms a thin surface layer of oxide, even at a temperature of 900°C.

Pure, clean gold surfaces are naturally hydrophilic. However, less than a monolayer of carbonaceous contaminant, deposited from the air or solution, can be enough to render the surface hydrophobic [8]. As a result gold is one of the most naturally hydrophobic components found in industrial mineral extraction systems.

9.1.4 Reagents

The surface chemical reagents used for flotation processes (Section 9.2) and coal–gold agglomeration (Section 9.4) can be classified broadly as either (1) collectors, (2) activators, (3) depressants, (4) dispersants, or (5) frothers. These generic terms describe the primary action of a reagent, although some secondary effects may also occur. The selection of reagents is usually dependent on the ore characteristics, the concentrations of the various ore constituents, the target concentrate grade, and recovery.

9.1.4.1 Collectors
Collectors are heterogeneous compounds that contain an active polar group, which adsorbs onto the mineral, and a hydrocarbon chain, which presents a nonpolar and therefore hydrophobic surface to the bulk solution. These reagents are used to selectively render the valuable mineral hydrophobic. Examples of gold and sulfide mineral collectors are

given in Table 9.2. The most important of these is the thiol group of collectors, which are characterized by functional groups that contain a sulfur atom bonded to a carbon or phosphorus atom and can be oxidized to form the even more hydrophobic dithiolate dimers, as discussed next.

Xanthates. The general formula of the xanthate (alkyl dithiocarbonate) ion is $ROCS_2^-$, where R is a hydrocarbon chain, shown schematically for sodium ethyl xanthate:

$$H-\underset{\underset{H}{|}}{\overset{\overset{H}{|}}{C}}-\underset{\underset{H}{|}}{\overset{\overset{H}{|}}{C}}-O-\overset{\overset{S}{\|}}{C}-S-Na$$

Nonpolar group   Polar group

The solubilities of xanthates in water decrease from ethyl (C_2) to amyl or pentyl (C_5), which is the longest chain xanthate commonly used. As the hydrocarbon chain length increases, the collector strength and mineral recovery generally increase, whereas selectivity decreases because minerals of lower initial hydrophobicity may also adsorb the collector, under the process conditions applied.

Xanthates are prepared by reacting an alcohol with carbon disulfide and hydroxide, for example, as follows:

$$C_2H_5OH + NaOH + CS_2 \rightleftharpoons C_2H_5OCS_2Na + H_2O \qquad (EQ\ 9.3)$$

Xanthates (X^-) are readily oxidized to the dithiolate, dixanthogen (X_2 or $ROC(S)SS(S)COR$), a nonpolar covalent compound:

$$X_2 + 2e \rightleftharpoons 2X^- \qquad (EQ\ 9.4)$$

Dixanthogen is the active hydrophobic species in the flotation of gold, pyrite, and certain other sulfides commonly encountered in gold extraction (Table 9.3) [9, 10]. The potentials required to oxidize the xanthate series are given in Table 9.4.

Xanthate solutions are unstable due to several main decomposition reactions:

- Oxidation to form dixanthogen
- Acid decomposition to form carbon disulfide and an alcohol
- Alkali decomposition, to form CSP_2, monothiocarbonate, CO_3^{2-}, and CS_3^{2-}

The half-lives of xanthates are given in Table 9.5. The rapid decomposition reactions in the acidic region are much better understood than alkali decomposition, which is slower and has a complex mechanism [11, 12]. The general ranges of stability of thiol collectors are given in Table 9.6, indicating the ranges of practical application.

Dithiophosphates. Dithiophosphate (DTP) collectors are stable under more acidic conditions than xanthates and hence are used when flotation at pH <8 is desirable. These collectors will not float pyrite effectively under alkaline conditions and therefore have good selectivity for recovery of other nonferrous sulfides.

DTPs generally give a lower recovery (although possibly with higher selectivity) than the corresponding xanthate and are less readily oxidized, forming the neutral dimer dithiophosphatogen, for example, as follows:

$$[Et(DTP)]_2 + 2e \rightleftharpoons 2Et(DTP)_2^-; \quad E^0 = 0.225\ (V) \qquad (EQ\ 9.5)$$

Standard potentials for the DTP homologous series are also given in Table 9.4. DTPs can also react with metal sulfides in a similar manner to xanthates and have particularly

TABLE 9.2 Collectors used to increase precious metal and sulfide mineral hydrophobicity [14]

Collector Type	Formula
Alkyl xanthates (alkyl dithiocarbonates)	R—O—C(=S)—SNa
Dialkyl dithiophosphates	(R—O)$_2$P(=S)—SNa
Mercaptobenzothiozole	benzothiazole—C—SH
Dialkyl thionocarbamates	R—O—C(=S)—NHR′
Xanthic esters	R—O—C(=S)—S—R′
Thiocarbanilide	(R—NH)(R^1—NH)C=S
Diaryl dithiophosphoric acid	(R—O)(R^1—O)P(=S)—SNa

good selectivity for copper minerals (Table 9.7). The diethyl and sec-butyl derivatives are commonly used as collectors for gold, silver, and copper.

Mercaptobenzothiozole. Despite being used in many industrial plants, including those for flotation of oxidized copper and zinc minerals, mercaptobenzothiozole (MBT) is a less well-investigated collector. It is used widely for free gold and pyrite flotation under acidic conditions. MBT has a pK_a of about 7 and is ionized in alkaline solutions. It has a lower tendency to form the dithiolate, (MBT)$_2$, than other thiol collectors and instead forms metal–MBT species, which render the mineral surface hydrophobic (Table 9.7).

Dithiocarbamates. Dithiocarbamates (DTCs), for example, isopropyl ethyl thiocarbamate, are most commonly used for flotation of copper sulfide minerals. They are relatively stable collectors which oxidize to form a dithiolate, (DTC)$_2$, in a manner similar to xanthates, as follows:

$$(DTC)_2 + 2e \rightleftharpoons 2DTC^- \qquad (EQ\ 9.6)$$

where E^0 for the diethyl DTC is in the range −0.068 to −0.010 (V). The hydrophobic species formed on sulfide mineral surfaces may be the dithiolate, as for pyrite, or a metal–DTC compound, for example, with lead, molybdenum, and copper sulfides (Table 9.7) [9].

TABLE 9.3 Products extracted from xanthated mineral surfaces [9]

Mineral	Methyl	Ethyl	Propyl	Butyl	Amyl	Hexyl
Orpiment	NPI	NPI	NPI	NPI	NPI	MX
Realgar	NPI	NPI	NPI	NPI	NPI	MX
Sphalerite	NPI	NPI	NPI	NPI	NPI	MX
Stibnite	NPI	NPI	NPI	NPI	NPI	MX
Cinnabar	NPI	NPI	NPI	NPI	NPI	MX
Antimonite	NPI	MX	MX	MX	MX	MX
Galena	MX	MX	MX	MX	MX	MX
Bornite	NPI	NPI	MX	MX	MX	MX
Chalcocite	NPI	NPI	MX	MX	MX	MX
Covellite	X_2	X_2	X_2+ MX	X_2+ MX	X_2+ MX	X_2+ MX
Chalcopyrite	X_2	X_2	X_2	X_2	X_2	X_2
Pyrite	X_2	X_2	X_2	X_2	X_2	X_2
Pyrrhotite	NPI	X_2	X_2	X_2	X_2	X_2
Arsenopyrite	X_2	X_2	X_2	X_2	X_2	X_2
Alabandite	NPI	X_2	X_2	X_2	X_2	X_2
Molybdenite	X_2+?	X_2+?	X_2+?	X_2+?	X_2+?	X_2+?

NPI = no positive identification.
MX = metal xanthate.
X_2 = dixanthogen.
? = possible presence of other species.

TABLE 9.4 Standard reduction potentials for homologous series of dithiolate–thiol couples (V) [9]

Homologue	Xanthate	Monothio-carbonate	Dithiophosphate
Methyl	−0.004	0.020	0.315
Ethyl	−0.060	0.002	0.255
n-Butyl	−0.091	−0.022	0.187
Isopropyl	−0.096	ND	0.196
n-Butyl	−0.127	−0.038	0.122
Isobutyl	−0.127	ND	0.158
n-Amyl	−0.159	−0.080	0.050
Isoamyl	ND	ND	0.086
Hexyl	ND	−0.120	−0.015

ND = not determined.

Amines. Amines have generally found only limited use in gold flotation but have been used occasionally under alkaline conditions (pH 10 to 11) for the recovery of pyrite and gold [15].

9.1.4.2 pH Modifiers
Commonly used industrial reagents and their characteristics in flotation are given in Table 9.8.

9.1.4.3 Activators
Activators are chemicals that enhance the flotation performance of a specific mineral or minerals. These are usually inorganic compounds which react with or are adsorbed such that a stronger reaction is possible with a collector. Examples include the following [17]:
- Addition of Cu(II) ions, which activate minerals such as sphalerite (ZnS), pyrite (FeS_2), pyrrhotite (FeS), stibnite (Sb_2S_3), and arsenopyrite (FeAsS).

TABLE 9.5 Xanthate half-life at 25°C and 40°C, as a function of pH, under nitrogen [13]

		Xanthate Half-life (hr)					
pH	Temp (°C)	Ethyl	Isopropyl	n-Propyl	Isobutyl	n-Butyl	Amyl
10	25	709	2722	866	1187	1021	1079
9	25	720	2928	912	1175	1035	1195
8	25	707	2628	873	1172	1045	1124
7	25	402	728	479	475	492	469
6	25	63.2	72.8	61.9	63.5	51.4	57.0
5	25	6.66	6.81	7.20	6.49	5.86	5.72
4	25	0.78	0.79	0.81	0.90	0.84	0.83
10	40	124	517	158	209	182	194
9	40	128	512	163	214	188	211
8	40	123	395	150	184	161	177
7	40	62.2	96.1	72.0	88.9	71.6	74.5
6	40	14.5	14.8	15.4	16.9	15.2	15.5
5	40	1.66	1.70	1.78	1.90	1.75	1.70

TABLE 9.6 pH stability ranges for various thiol collectors [14]

Collector	pH Range
Dithiocarbamate	5–12
Dithiophosphate	4–12
Dixanthogen	1–11
Mercaptobenzothiazole	4–9
Thionocarbamate	4–9
Xanthate	8–13

TABLE 9.7 Products extracted from mineral surfaces after reaction [9]

	DTP	DTC	MBT
Pyrite	$(DTP)_2$	$(DTC)_2$	$(MBT)_2 + Fe(MBT)_3$
Pyrrhotite	NPI	NPI	NPI
Arsenopyrite	NPI	NPI	NPI
Galena	$Pb(DTP)_2$	$Pb(DTC)_2$	NPI
Molybdenite	$Mo(DTP)_x$	$Mo(DTC)_x$	$Mo(MBT)_x$
Covellite	$Cu(DTP)_2$	$Cu(DTC)_2$	$Cu(MBT)_2$
Chalcopyrite	$Cu(DTP)_2$	$Cu(DTC)_2$	$Cu(MBT)_2$
Chalcocite	$Cu(DTP)_2$	$Cu(DTC)$	$Cu(MBT)$
Bornite	$Cu(DTP)_2$	$Cu(DTC)_2$	$Cu(MBT)_2$
Sphalerite	NPI	NPI	NPI
Stibnite	NPI	NPI	NPI
Antimonite	NPI	NPI	NPI
Orpiment	NPI	NPI	NPI
Realgar	NPI	NPI	NPI

DTP = dithiophosphate.
DTC = diothiocarbamate.
MBT = mercaptobenzothiozole.
NPI = No positive identification

TABLE 9.8 pH modifiers used in flotation (adapted from [16])

Reagent	Usual Addition (g/t)	Characteristic Action Sulfides	Precious Metals
Lime	250–2,500	Depresses iron sulfides; also lead, marmatitic zinc, and certain copper minerals if excess used.	Depresses gold. Little effect on silver sulfides.
Soda ash	250–1,500	Assists separation of sulfides from each other by acting as gangue slime dispersant. Aids recovery of arsenopyrite when used with copper sulfate.	Assists flotation of precious metals and sulfides.
Alkaline silicates	250–1,500	Disperses gangue slimes, assists grade and recovery. Produces brittle-type froth. Depresses quartz and silicates.	In controlled amounts, aids selectivity and grade of concentrate.
Sodium hydroxide	250–2,000	Gangue slime regulator. With copper sulfate, activates arsenopyrite.	Some assistance to recovery of free gold.
Alkaline phosphates	250–1,000	Improves grade for some sulfide ores by dispersing gangue slimes. Particularly effective on ores containing iron–oxide slimes.	Assists recovery of precious metals from slimy ores.
Sulfuric acid	250–2,500	Assists recovery of iron sulfides, especially after depression by lime or cyanide.	Assists recovery of gold in dilute pulp. Less benefit in thick pulp.

TABLE 9.9 Response of sulfide minerals to collectors of xanthate type [12]

Mineral	Methyl Xanthate	Sodium Aerofloat	Ethyl Xanthate	Butyl Xanthate	Amyl Xanthate	Hexadecyl Xanthate
Sphalerite		Activator required				
Pyrrhotite						
Pyrite						
Galena						
Chalcopyrite				No activator required		
Bornite						
Covellite						
Chalcocite						

NOTE: Shaded area represents response to collectors only in the presence of activators. Nonshaded area represents response to collectors without activation.

- Addition of sulfide ions, which activate carbonate, and oxidized and partially oxidized minerals, in a process sometimes referred to as sulfidization

The relative requirement for activation of different sulfide minerals is shown in Table 9.9. Activation is widely practiced for oxidized or partially oxidized materials. In general, activators are added before collectors in order to establish mineral surface conditions most suitable for effective collector action.

9.1.4.4 Depressants

Depressants are either inorganic or organic chemicals (Table 9.10), which selectively decrease a specific mineral (usually gangue) recovery. The reaction mechanisms for inorganic depressants, which are important in precious metals flotation, are reasonably well understood and specific to each mineral system [14].

TABLE 9.10 Reagents used as depressants in flotation (other than by pH modification) (adapted from [16])

Reagent	Usual Addition (g/t)	Characteristic Action		
		Sulfides	Precious Metals	Others
Cyanide compounds	0.5–250	Depresses zinc, antimony, nickel, and iron sulfides, and copper sulfide if excess used.	Depresses gold due to solubility effect.	—
Ferri- and ferrocyanides	50–1,000	Depresses iron sulfides.	—	—
Sulfites, bisulfites, sulfur dioxide	250–2,000	Depresses zinc and iron sulfides.	—	—
$FeSO_4/Fe_2(SO_4)_3$	50–1,000	Depresses sulfides.	—	—
Chromates/ dichromates	100–2,500	Depresses galena. Excess depresses copper and iron sulfides.	—	—
Permanganates	50–1,000	Depresses pyrrhotite and arsenopyrite away from pyrite.	—	—
Quebracho, tannic acid	10–100	Excess depresses all sulfides.	Excess depresses gold and silver sulfides.	Excellent calcite, dolomite depressant.
Starch	50–500	Excess depresses sulfides, particularly lead and silver sulfides.	Aids recovery from clay ore. Excess depresses gold.	Talc depressant. Depresses carbon.

Dashes = action not significant.

9.1.4.5 Frothers

Frothers are usually nonionic surface active molecules which, in flotation, are used to produce a large, stable air–water interface to ensure that floated material remains in the froth to allow recovery to concentrate. This may be achieved by the following:

- Reducing the size of bubbles formed in the flotation pulp and stabilizing the froth formed
- Enhancing the approach of the bubble and the particle
- Thinning of the water film between the bubble and particle until rupture occurs
- Establishing equilibrium contact

The chemical formulas and structure of common frothers are shown in Table 9.11. The addition of a frother (e.g., an alcohol) reduces the surface tension of the solution from that of water (0.073 N/m) closer to that of the alcohol, although it is added only in small amounts. This indicates that a large fraction of the solution surface (i.e., bubble surfaces) is composed of the alcohol.

9.2 FLOTATION

9.2.1 Application of Flotation

Free metallic gold can generally be recovered very effectively by flotation (see Section 9.1.3), although more commonly it is recovered together with sulfide minerals, where gold is intimately associated with the sulfides as fine unliberated grains (in solid solution or as discrete inclusions), or occurs with barren, hydrophobic sulfides. The most common

TABLE 9.11 General types of frothing reagents used in flotation* [18]

Type	Formula	Actual Frothers			
Alcohols	R—OH	Terpineal	Fenchyl alcohol	Borneol	1
		Pine oil			
		Xylenol	Cresol	Phenol	2
		Cresylic acid			
		Methyl isobutyl carbinol (MIBC)			3
Hydroxylated polyglycol ethers	R′O(RO)$_x$H	CH$_3$(OC$_3$H$_6$)$_x$H Polypropylene glycol methyl ether			
Alkoxy substituted paraffins	(R′O)$_x$H	1,1,3-Triethoxy butane			

* In this table, C refers to CH$_x$ group.

gold-bearing sulfides are pyrite (FeS$_2$), arsenopyrite (FeAsS), and to a lesser extent pyrrhotite (Fe$_{1-x}$S).

The flotation of gold from sulfide-free ores containing very low concentrations of free gold is difficult because of the low mass of material reporting to the concentrate and the high density of gold (19,300 kg/m^3). For example, 0.005% (50 g/t) Au would be a very high gold ore grade, compared with grades of >0.5% for most copper, lead, or zinc ores treated by flotation. This results in very poor froth stability, decreased recovery, and/or concentrate grade. Despite these factors, close to 100% recovery of free gold has been achieved by flotation under optimized conditions in some applications, with concentration ratios of between 30:1 and 300:1 achieved [19, 20].

Gold flotation is inherently a slow rate process compared to the flotation of other naturally floating minerals, such as chalcopyrite, chalcocite, and sphalerite. Ore mineralogy has a profound effect on the flotation conditions employed. The selection of operating pH, type and amount of frother, collector system and, where necessary, activators and/or depressants, are all critical to achieve effective recovery (and concentrate grade) for a given ore type. The major classes of treatment for different mineral systems are considered in the following sections.

Free gold flotation with an oxide or silicate gangue. For the treatment of placer deposits or gravity concentrates, most of the gangue minerals are oxides or silicates, which are hydrophilic, and strong collectors may be used to maximize gold recovery with little concern for corecovery of sulfides. This type of flotation is rare but has been proposed for low-grade ores where the gold is too fine to be recovered effectively by gravity concentration [20]. Thus, conditions can be selected solely for gold recovery and not to optimize selectivity against sulfide minerals.

Free gold with a sulfide gangue. When the gangue mineral has no economic value, the objective is to selectively recover gold from the barren sulfide (and silicate) gangue. Optimum reagent selection is required and high gold recoveries are difficult to achieve without corecovery of some sulfide minerals. Commonly, free gold can be floated at neutral pH with little or no collector addition [21].

If sulfide components in the gangue are of value, then conditions should be selected to give the best economic return of all economic minerals, not necessarily the maximum recovery of gold or any other single mineral. For example, indicative smelter specifications for copper concentrates, including the value for gold credit, are given in Table 9.12.

Unliberated gold in a sulfide gangue. When gold is very fine (i.e., <10 μm) and intimately associated with sulfide minerals (either in solid solution or as fine inclusions), cyanidation performance is typically poor. Flotation may be used as a preconcentration step to allow more expensive refractory ore treatment to be performed on a smaller fraction of the material. Under these conditions, the flotation circuit is usually operated to maximize the recovery of all the minerals containing gold (i.e., bulk flotation). Conditions must be optimized carefully for the recovery of both free gold and gold associated with sulfide minerals, although some of the free gold is typically readily recovered with other floatable minerals [23]. However, interactions between the flotation of free gold and gold-bearing sulfide minerals, as well as conflicting chemical and physical conditions for optimal flotation, must be considered carefully (see Section 9.2.2.3) [24, 25].

Flotation of gangue minerals. Gangue minerals may be removed by flotation prior to the gold recovery stage to improve performance or reduce reagent consumption, for example, carbonates that affect pH control during oxidative pretreatment or carbonaceous material, which can adsorb gold during cyanidation. The use of flotation in this way also acts as a preconcentration step.

The characteristics of sulfide mineral flotation have been well documented [14, 26]; however, certain aspects are particularly relevant to gold recovery processes. The grade and composition of a gold concentrate will dictate the subsequent process steps, the main options being: (1) smelt directly, (2) oxidize and leach by intensive cyanidation, or (3) regrind and leach. These options are considered further in Chapter 3 and other process-specific chapters.

9.2.2 Gold

Free gold is naturally floatable in most industrial systems (see Section 9.1.3), which means that it can be recovered without collector addition [14, 27]. This is due to the adsorption of hydrocarbons (and other surface reactions), which are dependent on gold's metallic properties, particularly its high electrical conductivity, which allows surface electrochemical reactions to occur catalytically and selectively. However, the surface properties of gold and gold alloys (Au–Ag, Au–Te) can be significantly affected by preceding mineral processing steps such as crushing, grinding, oxidative pretreatment, leaching, and

TABLE 9.12 Typical smelter specifications for copper concentrates [22]

Metal	Concentrate Payment Basis
Copper	Payment = $(T-1)Q$
Gold	Payment = $0.95Q(t-1)$
Silver	Payment = $0.95Q(t-35)$
Zinc	None (penalties when $T > 2.5\%$)
Lead	None (penalties when $T > 2.5\%$)
Mercury	Penalties when $T > 0.0025$
Antimony, bismuth, arsenic	Penalties when $T > 0.1\%$

T = concentrate grade (%).
Q = price of metal per ton (or per gram for gold and silver).
t = concentrate grade (g/t).

solid–liquid separation. These process steps can lead to surface coatings and impregnation of foreign material on the gold surface, as well as slimes formation, which may interfere with flotation of both gold and gold-bearing minerals. In many cases, gold can be effectively recovered by flotation at neutral or near-neutral pH [15, 28]. Particle size distribution can also play an important role in free gold flotation efficiency. These factors are considered in more detail in the following sections.

9.2.2.1 Flotation Chemistry

Gold hydrophobicity is enhanced by the addition of flotation collectors such as xanthates, DTP, and MBT, as used in sulfide mineral flotation. Usually collector concentrations in the range of 25 to 75 g/t are used. The mechanism by which gold hydrophobicity is enhanced is similar to that of certain sulfides, for example, pyrite. Xanthate ions are oxidized at the gold surface to form the neutral dimer dixanthogen:

$$(EtX)_2 + 2e \rightleftharpoons 2EtX^-; \quad E^0 = 0.057 \text{ (V)} \tag{EQ 9.7}$$

$$(AmX)_2 + 2e \rightleftharpoons 2AmX^-; \quad E^0 = 0.159 \text{ (V)} \tag{EQ 9.8}$$

$$O_2 + 2H_2O + 4e \rightleftharpoons 4OH^-; \quad E^0 = 0.401 \text{ (V)} \tag{EQ 9.9}$$

where
  Et = ethyl (C2) alkyl group
  Am = amyl (C5) alkyl group

The nonpolar liquid species, dixanthogen, forms an oily surface coating on gold surfaces, rendering them hydrophobic. However, the oxidation of xanthate ions in bulk solution is very slow:

$$4ROCS_2^- + O_2 + 4H^+ \rightleftharpoons 4(ROCS_2) + 2H_2O \tag{EQ 9.10}$$

which demonstrates that the conducting surface of gold or semiconducting minerals is required for dixanthogen formation.

Oxide and silicate minerals are not electrical conductors and cannot sustain redox couples described by Equations (9.7), (9.8), and (9.9). Hence, the collecting action of the reagent is very selective for gold and sulfides, which means that reagent additions and concentrations are typically low.

Gold electrodes have been used as a model in mechanistic studies against which sulfide mineral flotation could be compared in order to understand collector and mineral interactions [29]. For example, linear potential sweep experiments on a particulate bed electrode of pure gold particles have shown that flotation with nitrogen bubbles occurs when approximately a monolayer of dixanthogen is formed, at 0.191 V and 0.132 V for EtX$^-$ concentrations of 10 and 100 mg/L, respectively. In the absence of oxygen and a collector, the gold surface was not sufficiently hydrophobic for flotation to occur. When pentyl xanthate was used, only 4% of a monolayer was required to render the gold floatable [30]. This indicates that increased chain length is beneficial, with the practical advantage of reducing reagent addition requirements.

Cyclic voltammetry has also indicated that gold can be rendered hydrophobic ($\theta_{air} \approx$ 50°) by deposition of a surface layer of sulfur from solutions containing S^{2-} or HS^- [31]. Charge measurements indicate that 20 monolayers of sulfur are required to produce a hydrophobic gold surface. This mechanism may be important in industrial systems where dissolved sulfide concentrations are in excess of this requirement. This process has been applied commercially for flotation of partially oxidized copper minerals (i.e., sulfidization; see Section 9.1.4.3).

The mechanisms for other collector types are less well understood. Cyclic voltammetry, impedance studies, and contact angle measurements have been used to develop a reaction model for DTP behavior at gold surfaces [32]. Reversible chemisorption of DTP anions occurs at potentials negative to the pzc (i.e., 0.14 V). Above 0.5 V the dithiolate forms, together with a metal–collector (Au–DTP) compound.

Similar investigations for the MBT system suggested that, below −0.2 V, MBT anions were adsorbed; between −0.2 and 0.4 V, the dithiolate was formed; and at potentials greater than 0.4, the dithiolate was either removed or oxidized, with a corresponding decrease in θ_{air} [32].

For most collector systems, the rate of flotation reaches a maximum as collector dosage is increased, and the rate decreases beyond this maximum. Obviously, this maximum addition varies depending on the collector type, but it is important not to overdose the collector since this not only decreases the flotation rate but may also increase the flotation on undesirable minerals [21]. Also, the overaddition of any surface-active reagents into a system where free gold is floated can have a significant negative impact on flotation efficiency (both gold recovery and concentrate grade).

It is common practice in free gold flotation, and for the flotation of free gold with gold-bearing sulfides, to use a dual-collector system. This is discussed further in Section 9.2.2.3. Also, dual-frother systems are sometimes used when maintaining a stable froth proves to be difficult.

9.2.2.2 Particulate Factors Affecting Gold Recovery

Liberation. The primary mineralogical requirement for the effective flotation of gold is that it is liberated (i.e., as free gold grains), or that it occurs in composite particles which are floatable. This latter condition usually requires gold to be a component of a particle with a predominantly gold and/or sulfide surface. If oxides or silicates form a composite with gold, then floatability would be lower and dependent on their respective surface areas and hydrophobicities.

Coatings. The floatability of free gold is dependent on the condition of the exposed surface. For example, coatings of hydrophilic metal salts may have been precipitated on the gold surface, reducing the overall surface hydrophobicity and consequently reducing gold recovery. Common coatings are Fe(III) oxides or hydroxides, which may form during ore formation, may be produced from iron-containing minerals present in the ore, or may be generated as a result of grinding media loss. Grinding media loss commonly accounts for between about 0.5 and 1 kg Fe/t ore [33]. In old tailings deposits, several decades of mineral dissolution and reprecipitation may have occurred, resulting in the formation of coatings of calcium, magnesium (Mg), manganese, aluminum, and iron salts, such as oxides and carbonates on gold and other mineral surfaces. In contrast, depending on mineralogical factors, the state of gold liberation may increase with prolonged oxidation of the associated gangue minerals, such as pyrite.

Some impurities in the gold (e.g., silver and copper) are more reactive than gold and can form hydrophilic surface phases. Molar volume increases associated with reactions of these impurities can form coatings on the gold, for example, Ag_2S formation, as follows:

$$Ag_2S + 2e \rightleftharpoons 2Ag + S^{2-}; \quad E^0 = -0.705 \text{ (V)} \tag{EQ 9.11}$$

which has an associated molar volume change of 183%. Conversely, the hydrophobicity of native gold increases if silver is leached from the surface regions, because this reduces the hydrophilic portion of the metal surface [20].

Particle size and shape. Because of the high density of gold (19,300 kg/m^3), particle size has a large effect on its recovery by flotation. Flotation is effective for gold particles in the size range 20 to 200 μm. Flotation kinetics are generally faster for finer-sized gold particles than for larger particles [21]. At finer sizes (i.e., < 20 μm), the selectivity for gold decreases due to coflotation of gangue components, although such gold particles can be recovered effectively in some cases, provided that slimes formation can be controlled [34]. In the coarser-size range, flotation should be performed at high slurry densities (i.e., >35% solids), as this helps to reduce gold particle sedimentation. For treatment of ores containing gold particle sizes coarser than about 200 μm, the more successful flowsheets have included gravity concentration and, historically, amalgamation, or more recently intensive cyanidation.

Because of its high density and malleability, gold tends to get flattened during grinding, and surfaces can become coated or embedded with gangue mineral particles and iron coatings. Although flattened particles present a larger surface area than spherical particles, the detrimental effects of surface degradation can have a significant impact on floatability. Such flattened gold particles have been found to have flat but very rough, surfaces, and the greater the roughness, the less hydrophobic the gold particles become, thereby hindering bubble attachment and floatability [34, 35, 36].

Several trials on the flotation of coarse gold from low-grade rougher gravity concentrates have produced concentrates of sufficiently high grade to be smelted directly [37, 38]. At Village Main (South Africa), a gravity concentrate was floated with DTP collector and achieved 98% gold recovery into a concentrate that was successfully smelted [39]. However, in another case, the flotation of gold from a belt gravity concentrate containing 2.5 kg/t Au was not economically viable, and intensive cyanidation was the favored alternative [38].

9.2.2.3 Chemical and Physical Factors Affecting Gold Recovery

An excellent conceptual model for free gold (and gold-bearing mineral) flotation has been proposed, providing a broad overview of the chemical and physical factors affecting the recovery of gold [24]. An adaptation of this conceptual model is shown in Figure 9.4. The factors affecting gold recovery are considered in more detail in the following sections.

Effect of sulfide mineral flotation on free gold flotation. Most investigative work on gold flotation has focused either on free gold or on the gold-bearing sulfide minerals (e.g., pyrite, pyrrhotite, and arsenopyrite) but not both together. Research conducted into the recovery of free gold and gold-bearing sulfide minerals has revealed important interactions that can significantly affect the recovery of each of these mineral classes and, consequently, the overall gold recovery [24, 25]. When free gold is floated at the same time as sulfide minerals, there is competition for bubble surface sites. Under conditions

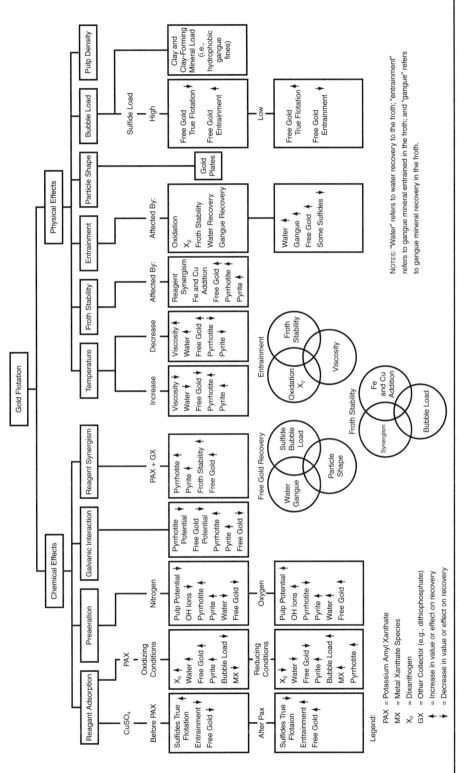

FIGURE 9.4 Conceptual model for gold flotation, showing effect of selected variables on individual component recovery or other floation property (adapted from [24])

where sulfide mineral flotation is optimized, evidence suggests that free gold particle attachment to bubble surfaces is hindered. This effect is exacerbated in finer size ranges (i.e., <10 μm) where differences in the relative degree of hydrophobicity between particles is reduced.

Effect of nonsulfide gangue minerals (and slimes) on free gold flotation. Various ore constituents can hinder the attachment of free gold to bubble surfaces, including fine clay or clay-forming minerals, humic acid and fulvates, and hydrated iron oxides. The adverse effects of these constituents is exacerbated in the fine size ranges (i.e., <10 μm) [15]. These fine particles, or slimes, are entrained in the froth, occupy bubble surfaces, and can coat otherwise floatable minerals. In severe cases, the addition of small amounts of dispersants, such as sodium polyacrylate (10 to 25 g/t) or sodium silicate (25 to 250 g/t), can have a positive impact on flotation, increasing both flotation rate and overall gold recovery. The addition of dispersants prior to flotation must be carefully tested to determine any interactive effects that may occur with other reagents and chemicals used in the flotation scheme.

Synergistic effects of dual-collector reagent schemes. The beneficial effect of applying dual-collector schemes for the recovery of free gold and gold-bearing sulfides is well known but not well understood. Examples of such dual-collector systems applied in practice include: DTP with DTC; xanthate and DTP; and (potassium ethyl) xanthate and trithiocarbonate. In such cases, the two collectors used in combination yield higher overall gold recovery, sometimes >20% higher, than is the case when each is used separately. Often, the combined collector addition can be optimized at a lower level than that required for either collector used separately. The addition sequence of the dual collectors can also have a significant effect on gold recovery.

Depressants. The depressing effect of excessive concentrations of $Ca(OH)_2$ (i.e., >pH 9.5) and NaOH in the slurry on free gold flotation by xanthates and other collectors is well known. In one case, gold recovery (as a coproduct with chalcopyrite) was improved by 10% by decreasing the pH from 10.7 to 9.5 [40]. Evidence also suggests that high concentrations of sulfide species can depress gold flotation [19].

Modifiers. The use of organic modifiers such as citric and oxalic acids (200 to 300 g/t) has been proposed and, when used in conjunction with xanthate and DTP collectors for flotation of a copper–gold ore, showed >10% benefit to gold recovery [34]. The addition of hydrocarbon oils, such as diesel oil or kerosene (200 to 300 g/t) have also been demonstrated to increase gold recovery, in some cases by more than 25%.

Activators. Copper sulfate is commonly used as an activator to promote free gold flotation, although the exact role and mechanism are not well understood.

Pulp density. The flotation response of free gold particles is generally optimized in slurries at high pulp density (i.e., >35% solids), which give the greatest opportunity for gold particle–bubble contact in slurries containing relatively low concentrations of gold (i.e., most gold ores). The use of higher pulp densities favors coarse gold flotation by helping to keep larger gold particles in the froth phase. However, there is a conflict between pulp density and viscosity, and high viscosity slurry can result in hindrance of gold attachment to bubble particles due to increased competition with other gangue minerals in the ore [41].

Galvanic effects. Galvanic interactions can occur between conductive ore minerals, such as pyrrhotite, pyrite, and other sulfides, and iron constituents in the flotation feed (e.g., grinding media fines or metallic iron added to grinding). According to one study, the presence of metallic iron reduces the rest potential of pyrrhotite, thereby reducing the extent of oxide and/or hydroxide formation on the mineral surface. This in turn increases the potential for activation with copper and hence increases floatability. A similar result is indicated for pyrite, although the effect is not nearly as strong. However, although the overall gold recovery was increased in the presence of metallic iron, the recovery of free gold was reduced, probably because of the increased competition for bubble surface sites due to improved floatability of sulfide minerals [24].

Nonoxidizing (or reducing) slurry preconditioning. Minerals such as pyrrhotite, arsenopyrite, and, to a much lesser extent, pyrite are susceptible to surface oxidation prior to and during preconditioning for flotation. Any oxide, hydroxide, or hydroxyl species present on the surface of the sulfide will inhibit activation (i.e., by copper adsorption) and subsequent collector adsorption, thereby adversely affecting flotation response. This can be addressed in two ways:

- Purging of nitrogen gas into the slurry can be used as a preconditioning step prior to activator and/or collector addition. This minimizes the dissolved oxygen content of the slurry and helps to prevent mineral surface oxidation. The effect on pyrite is much less pronounced than on pyrrhotite (but this certainly depends on the exact form of the pyrite—i.e., framboidal vs. cubic structure, grain size, presence of foreign inclusions, and lattice imperfections). When oxygen is present in the slurry and/or when copper is added as an activator in the presence of oxygen, it is generally agreed that dixanthogen is formed at the sulfide mineral surfaces. When nitrogen is added to the extent that oxygen is substantially depleted from the slurry, the presence of dixanthogen at the mineral surface is very low, and the majority of the hydrophobic species is metal–xanthate (e.g., Fe–X or Cu–X) [35].

- The addition of reductants such as sodium hydrosulfide (NaHS) can reduce the redox potential of the slurry, with similar effects on reactive sulfide mineral surface oxidation to that achieved with nitrogen.

It has been suggested that the metal–xanthate species formed by these methods are more selective for pyrrhotite and pyrite (and potentially other sulfide minerals), and result in less water and gangue entrainment in the froth [24]. The disadvantage of increased selectivity for sulfide minerals is that free gold recovery may be reduced, but overall gold recovery (i.e., free gold and gold-bearing sulfides) and subsequent processing should drive the optimal conditions. Where copper ions are used as an activator, it must be added before the collector for most effective action.

Froth stability. Froth stability is a critical factor for effective free gold flotation, and adequate froth stability must be established using frothers (if necessary), carefully controlling pulp density and viscosity, using nitrogen preaeration (where necessary), and considering separate flotation of different size distributions and/or slimes removal (if necessary). The presence of metallic iron in the slurry (i.e., grinding media) has also been shown to help froth stability. Where froth stability is problematic, the use of dual-frother systems (e.g., pine oil with polypropylene glycol methyl ether) may be beneficial.

Temperature. One study has shown that increasing temperature (from 15°C to 30°C) decreases the recovery of free gold but increases the recovery of sulfide minerals (e.g., pyrite and pyrrhotite). Other investigations have shown that low temperatures (<25°C) can adversely affect flotation due to increased slurry viscosity, and in some cases heating of the slurry may be beneficial. However, temperatures above about 50°C are clearly detrimental to flotation, presumably due to desorption of the collector or other surface-related phenomena [21].

Water quality. Free gold flotation is generally insensitive to moderate variations in water chemistry, and effective flotation can often be accomplished in highly saline water (e.g., in Western Australia) and even seawater. One exception to this is the effect of high concentrations of Ca^{2+} ions, which depress free gold flotation.

9.2.3 Gold Tellurides

Gold-telluride minerals are not as naturally hydrophobic as clean native gold but are at least as floatable as most sulfide minerals. This allows for some selectivity and the possibility of producing a gold-telluride concentrate ahead of sulfide flotation. At the Emperor mine in Fiji, gold tellurides have been floated selectively at pH 9 to depress iron sulfides, with the addition of a frother but no collector. The concentrate produced contains 3 to 4 kg/t Au from a feed containing 5 to 10 g/t. The gold-telluride flotation concentrates were subsequently treated by chlorination and cyanidation for tellurium and gold–silver extraction, respectively (see Section 12.2.8.1). In addition, the concentrate from a secondary sulfide float was roasted. An advantage of this two-stage process is that gold associated with tellurium is not lost by volatilization during roasting.

At the Kalgoorlie (Western Australia) operation, ores containing gold-telluride minerals (calaverite and petzite) are processed by flotation to recover free gold, gold tellurides, and gold-bearing sulfides, followed by roasting of the concentrate and subsequent cyanidation of the calcine product. The feed composition is approximately 3 to 6 g/t Au, 1% to 3% S, and 1% to 7% Te. Flotation is carried out on material ground to 80% −130 μm using highly saline water at the natural pH. Sodium ethyl xanthate is used as the collector (20 g/t) together with a frother, 1,1,3 triethoxybutane–"InterFroth 50" (20 g/t). Gold recovery to the flotation concentrate varies from 70% to 90%, depending on ore type [42].

9.2.4 Sulfide Minerals

9.2.4.1 Pyrite

Pyrite has a cubic and predominantly covalently bonded crystal structure in which Fe(II) ions are at the unit cell corners with the face centers occupied by S^{2-} ions. Pyrite tends to oxidize in air, which affects its surface properties and flotation behavior.

Pyrite flotation is better understood than any other common sulfide system. In the presence of xanthate collectors in oxygenated solutions, the pyrite surface is rendered hydrophobic by the formation of dixanthogen, through the same electrochemical mechanism as that for gold (Section 9.2.2.1). There is little evidence for the existence of a hydrophobic Fe–xanthate species in industrial systems. Therefore, flotation is performed at potentials above that required for dixanthogen formation. Industrial pyrite flotation circuits are usually controlled by modification of collector type, collector addition, pH, pH modifier type, and cyanide addition.

The pH range used for pyrite flotation depends on the ore components and the resulting natural pH of the slurried material. The collector is then chosen to operate close to this pH, for example, MBT at pH 3 to 4 or xanthate in the pH range 9 to 11. Flotation with xanthates is possible below pH 9; however, reagent decomposition and consumption become increasingly significant as pH is reduced. In the region pH 4 to 8, reduced recovery can be demonstrated at low collector additions (Figure 9.5).

The mechanism of pyrite flotation with DTP is similar to xanthate where, in the acidic region (below pH 4 to 6), DTP is oxidized to the dithiolate, dithiophosphatogen, the most important hydrophobic species. In base metal flotation, where metal–DTP species are formed, for example, in copper, lead, and zinc recovery, pyrite is depressed under alkaline conditions, that is, above pH 9.5 and preferably pH 10.5 to 11.5.

The depression of pyrite is a major subject in base metal flotation and can be achieved by the addition of reagents such as $Ca(OH)_2$, NaCN, or Na_2S [14, 26]. However, in gold systems it is extremely rare to attempt to depress pyrite (if it is present, almost always some gold is associated with it), and pyrite activation is almost always practiced.

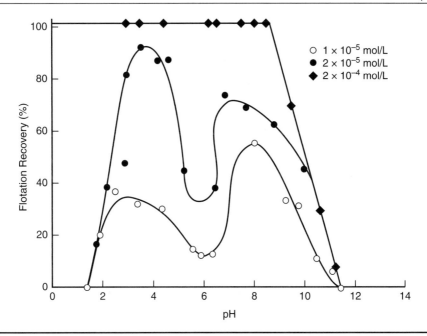

FIGURE 9.5 Recovery of pyrite as a function of flotation pH with various additions of potassium ethyl xanthate [14]

An exception would be if gold were present largely as free gold or occurred principally with arsenopyrite rather than pyrite.

Copper sulfate ($CuSO_4$) is commonly used as an activator in gold and pyrite flotation, particularly in the pH range 7 to 9 where silicate gangue flotation is reduced. Laboratory tests have shown that both pyrite grade and recovery can be increased markedly if copper sulfate is added prior to collector addition (Figures 9.6 and 9.7). The mechanisms of action are uncertain but the major effects include the following:

- Adsorption of Cu(II) onto pyrite, which enhances the rate and/or extent of collector adsorption, as follows:

$$FeS_2 + Cu^{2+} \rightleftharpoons Fe^{2+} + S + CuS \qquad (EQ\ 9.12)$$

- Depression of gangue particles
- Modification of froth structure
- Complexation with free cyanide, which would otherwise depress pyrite

More recent work has shown that copper sulfate can adsorb onto pyrite surfaces, but this has no effect on the rate of xanthate adsorption [43]. See also Section 9.2.2.3 for additional discussion on pyrite and gold flotation.

Other more exotic reagent schemes, such as propylene-trithiocarbonate and oxypropylated sulfides (PROCS) with ferric ions at pH 4 to 7, and dimethyl-dithiocarbamate (DMDC) with butyl xanthate, have been proposed for selective flotation of pyrite and arsenopyrite [44].

9.2.4.2 Arsenopyrite

Arsenopyrite behaves in a manner similar to other iron-containing sulfides. The E_h–pH diagram for the Fe–As–S–H_2O system (Figure 9.8) shows a region of stability for arsenopyrite,

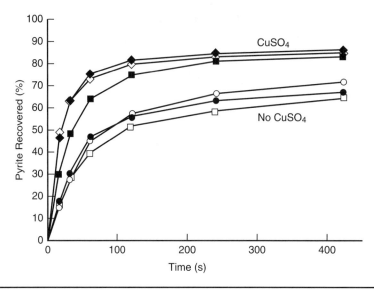

FIGURE 9.6 Effect of copper sulfate on recovery of pyrite [43]

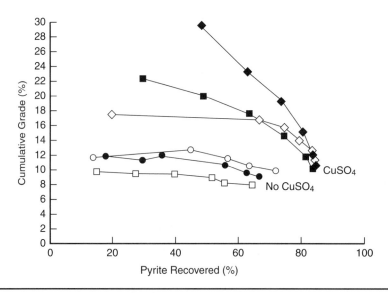

FIGURE 9.7 Effect of copper sulfate on pyrite grade and recovery [43]

which is similar to that of pyrite or pyrrhotite (see Figure 5.1, Section 5.1.1) but displaced to a lower potential, indicating that arsenopyrite is more prone to oxidation.

Arsenopyrite can be floated with xanthate collectors with best results achieved in the pH range 5 to 8 by activation with Cu(II). In general, the factors that enhance pyrite flotation also improve arsenopyrite flotation. The determination of the hydrophobic xanthate species that renders arsenopyrite floatable has received less attention than the other iron and base metal sulfides; however, it is likely dixanthogen. The formation of dixanthogen has been found to be decreased if arsenopyrite is oxidized to Fe(III) hydroxide and arsenate, as follows:

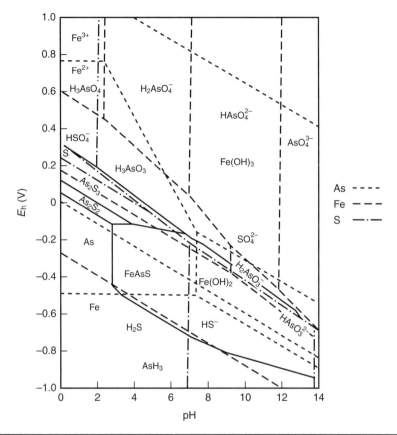

FIGURE 9.8 Potential–pH diagram for Fe–As–S–H$_2$O arsenopyrite (activities taken to be 10^{-3} M) [45]

$$Fe(OH)_3 + HAsO_4^{2-} + SO_4^{2-} + 18H^+ + 14e \rightleftharpoons FeAsS + 11H_2O; \quad E^0 = 0.569 \text{ (V)} \quad \text{(EQ 9.13)}$$

This reaction is apparent on a cyclic voltammogram at pH 11 (Figure 9.9) [46]. The potential required for arsenopyrite oxidation (Equation [9.13]) can easily be achieved with common oxidizing agents, such as hydrogen peroxide, hypochlorite, and permanganate in solution. Pyrite is only oxidized to Fe(OH)$_3$ at much higher potentials, and this forms a basis for separation of arsenopyrite and pyrite where pyrite is floated from arsenopyrite under moderately oxidizing conditions at high pH (Figure 9.10) [47]. If both arsenopyrite and pyrite need to be depressed, for example, to improve gold flotation selectivity, this can be achieved using Ca(OH)$_2$, NaCN, or SO$_2$-air.

9.2.4.3 Pyrrhotite

Pyrrhotite is stable at lower potentials than pyrite, which leads to a greater oxidation of the sulfide surface and generally lower recoveries, although this can be compensated for by Cu(II) activation. The mechanism of flotation with thiol collectors involves dixanthogen formation, as in the case of pyrite. In gold ore treatment, pyrrhotite flotation is generally undesirable unless it contains gold, or unless it is intimately associated with other sulfides containing gold. This is because concentrate grades would be reduced, and pyrrhotite is a cyanide oxygen consumer in the cyanidation process (Section 6.1), both of

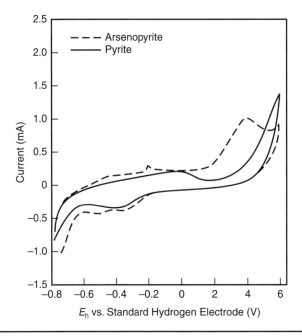

FIGURE 9.9 Cyclic voltammogram for arsenopyrite and pyrite at pH 11 (20 mV/s) [46]

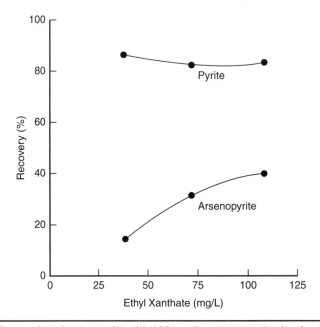

FIGURE 9.10 Depression of arsenopyrite with 100 mg/L permanganate showing the effect of xanthate concentration on selectivity [47]

which increase downstream processing costs. Section 9.2.2.3 contains additional information on pyrrhotite flotation in conjunction with free gold.

9.2.4.4 Stibnite
Stibnite is unusual because its surface properties render the mineral naturally hydrophobic and easily recoverable by flotation with thiol collectors. The hydrophobic cleaved surface of stibnite is formed by the rupture of weak antimony–sulfur bonds. This factor, coupled with hydrophilic sites at the end of antimony–sulfur chains, which may form surface groups of the structure $Sb_xO_yH_2^{n-}$, results in considerable anisotropic behavior and a lower contact angle than other naturally hydrophobic minerals, such as molybdenite (MoS_2).

Stibnite is generally floated using thiol collectors with activators, if the mineralogy allows this to be done without decreasing concentrate grade. At Consolidated Murchison (South Africa) lead nitrate is used as an activator with isobutyl xanthate and DTP collectors at pH 6 to 8. Cyanide is added to depress arsenopyrite. The flowsheet is particularly interesting (see Section 12.2.7.1) and produces a concentrate (25 to 35 g/t Au at 30% to 40% recovery), which is treated by pressure cyanidation. Elsewhere, stibnite concentrates are usually processed by roasting to produce a saleable antimony product (Sb_2O_3) and a calcine suitable for cyanidation.

9.2.4.5 Copper Sulfides
A considerable amount of gold is produced as a by-product of base metal production, principally from copper ores (see Section 12.2.6). The flotation of copper minerals, for example, chalcopyrite ($CuFeS_2$), bornite (Cu_5FeS_4), chalcocite (Cu_2S), and covellite (CuS), is relatively straightforward with thiol collectors, and differs from gold–pyrite–arsenopyrite flotation because a copper–xanthate compound is formed, which renders the copper minerals hydrophobic. In the case of chalcopyrite flotation, dixanthogen is also present.

Chalcopyrite can be floated with ethyl xanthate over a wide pH range (Figure 9.11) due to the stability of cuprous xanthate. Commonly, chalcopyrite ores are subjected to rougher flotation at pH 9 to 10 at a relatively coarse particle size, followed by regrinding of the rougher concentrate and cleaning at higher pH to reject gangue minerals, such as pyrite. Chalcopyrite can only be depressed at pH >11 to 12, unless cyanide or sulfide ions are added. Flotation is also possible with DTP, dithionocarbamates, and the whole range of xanthates (Tables 9.7 and 9.9) [40, 48]. When free gold is present together with copper sulfide minerals, it is common practice to use a dual-collector system, such as dithionocarbamate–DTP, or xanthate–DTP. In such cases, the use of two collectors provides higher gold recovery at reasonable collector addition.

9.2.4.6 Other Sulfides
Information on the flotation of galena (PbS), sphalerite, and other base metal sulfides, which are generally of limited importance in gold extraction, is available in many reviews [14, 26]. The Fachinal deposit in Chile contains free gold associated with sphalerite and galena, and the ore has been treated by flotation at natural pH (~7) using a potassium amyl xanthate collector and dual-frother system.

9.2.5 Carbonaceous Matter

9.2.5.1 Activated Carbon
Because it is an organic material, activated carbon can be floated with small additions of hydrocarbons, such as fuel oil or kerosene, at neutral or near-neutral pH. Although the activation process (Section 7.1.1) gives the carbon surface some polarity, its hydrophobicity is still greater than most ore minerals.

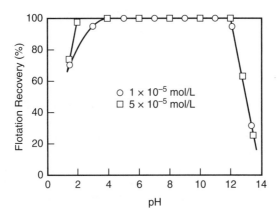

FIGURE 9.11 Flotation recovery of chalcopyrite as a function of pH and ethyl xanthate concentration [14]

A process incorporating the flotation of fine loaded carbon after gold adsorption was patented in 1939 (Chapter 1); however, the use of fine carbon as a primary means of gold recovery from leach slurries is not commercially utilized. One of its potential advantages is that, due to the fine carbon size, relatively short adsorption times are required. However, the major disadvantages are that the activated carbon recovered in a froth flotation concentrate cannot easily be recycled, and the plant cannot be operated in a counter-current manner as applied in the carbon-in-pulp (CIP) and carbon-in-leach (CIL) process. As a result, final gold loadings are low, and the consumption of activated carbon is uneconomically high.

Flotation of activated carbon has been used to scavenge dissolved gold in cyanidation plant residues by adding powdered activated carbon (at 100 g/t) to the tailings slurry and agitating for 6 hr to extract >80% of the dissolved gold [49]. The loaded carbon was recovered by flotation into a pyrite concentrate in the existing flotation circuit, and therefore no additional flotation plant was required. This fit with the existing flowsheet greatly increased the economic viability of the process.

The tailings of CIP and CIL plants contain some gold-loaded carbon resulting from attrition losses which occur in the process; however, the gold-on-carbon content is too low to justify recovery by flotation.

9.2.5.2 Ore Components

Carbonaceous materials, which occur naturally in many ores, tend to cause problems during cyanidation due to their ability to adsorb gold from solution, as discussed in Section 6.1. Occasionally, carbonaceous material is floated either before cyanidation, to remove the gold-adsorbing components, or after cyanidation, to increase gold recovery by further treatment of the gold-bearing carbonaceous concentrate. The success of the former process depends on the ability to remove a sufficient proportion of the carbonaceous material and to reduce preg-robbing to acceptable levels but without floating significant amounts of naturally hydrophobic gold. This must also be cost competitive with alternative treatment processes, such as chlorination (Section 5.6).

At the McIntyre mine (Canada) a flotation circuit was used in which carbon was floated using fuel oil, methyl isobutyl carbinol (MIBC) frother, and Quebracho (acting as a graphite dispersant) to produce a concentrate that was discarded [50]. The flotation tailings were refloated, in the presence of a dextrin–guar carbon depressant, to recover a gold concentrate, which was then leached with cyanide.

In the treatment of South African gold plant residues, a significant proportion of gold is associated with a uraniferous kerogen component called thucholite (Chapter 2). Gold is present in concentrations of up to 500 g/t and is refractory to direct cyanidation due to the kerogen's adsorptive properties. When retreating residues containing this material, it was found that overall gold recoveries by flotation could be increased by up to 5% by the addition of paraffin, which was thought to adsorb onto thucholite, causing passivation [51]. The flotation concentrate in this application was primarily pyrite (containing 30% to 32% S), which was subsequently roasted to liberate the gold for cyanidation.

In an alternative approach, a hydrocarbon such as kerosene may be added to carbonaceous ores to reduce preg-robbing. This is effective because the oil is selectively adsorbed onto the naturally hydrophobic carbonaceous components of the ore, thereby reducing access to the gold cyanide species in solution. This has been employed at Kerr-Addison (Canada) (see also Section 3.3.6.1).

9.2.6 Silicates

The recovery of silicate minerals is generally undesirable in gold–sulfide flotation because it lowers concentrate grades, decreases the efficiency of subsequent concentrate processing stages, and/or increases the transport costs of a saleable concentrate. Generally, silicate recovery is low, but some presence is inevitable because of unselective physical entrainment in froths. This problem is most significant when the feed material contains fine clays and silty material.

In certain cases, silicate depressants must be added, particularly for treatment of ores that contain clay minerals, naturally hydrophobic because of their three-layer silicate structures, which have electrically neutral crystal surfaces, for example, talc ($Mg_3Si_4O_{10}(OH)_2$) or pyrophyllite ($Al_2Si_4O_{10}(OH)_2$) [14].

The structure of talc is shown in Figure 9.12, where the outer layer consists of uncharged silicon–oxygen bonds and the inner portion is $Mg(OH)_2$. Stacked layers are only bound by weak Van der Waals forces, which creates the possibility of a cleavage plane. Talc therefore occurs naturally as platey particles, where the flat surfaces are hydrophobic and the edges are hydrophilic, and the flotation characteristics are strongly dependent upon shape factors.

Investigation of the surface chemistry of talc has shown that H^+ and OH^- ions are potential determining with the charge in the acidic region produced by the dissolution of Mg^{2+} ions and, under alkaline conditions, by the adsorption of OH^- ions onto Mg^{2+} sites.

Hydrophobic silicate minerals may be depressed using reagents such as starch, dextrin, and modified guar gums. Some South African ores, particularly in the Orange Free State, which contain significant amounts of hydrophobic pyrophyllite (up to 15%), are depressed using these reagents at pH >9 [43, 51].

9.2.7 Process Considerations

9.2.7.1 Flotation Circuits
The engineering and process design of flotation circuits is very important in achieving effective flotation performance and in producing a concentrate that is acceptable to downstream processes. This requires adequate feed conditioning, residence time, agitation, effective cell design, and optimum rougher, cleaner, and scavenger capacities, as reviewed previously [52, 53].

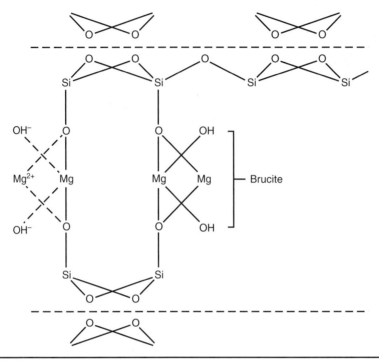

FIGURE 9.12 A schematic diagram of the structure of talc

Common gold and gold–sulfide mineral flotation circuit configurations include the following [17, 40, 51, 54, 55]:

- Rougher flotation, regrind rougher concentrate, and cleaner flotation
- Rougher flotation, then scavenger flotation with cleaner flotation of the scavenger concentrate (cleaner concentrate is combined with the rougher concentrate to produce a final concentrate)
- Rougher flotation, scavenger flotation (of rougher tails), regrind rougher–scavenger concentrate followed by cleaner flotation
- Rougher flotation, regrind and gravity concentration of rougher concentrate, followed by cleaner flotation

Many other innovative flotation circuit configurations are available for consideration. For example, for the treatment of ores containing both free gold and gold-bearing sulfides (e.g., pyrite, arsenopyrite), the use of two-stage rougher flotation has been proposed. This innovative flowsheet consists of a neutral or near-neutral pH first-stage rougher with a small dosage of weak (uncharged) collector, followed by a second-stage rougher flotation using a higher dosage of stronger collector [21]. A significant benefit of this flowsheet is that it allows greater retention time to be allocated for the slower floating material in the second stage. This scheme is commonly applied in gold recovery circuits through the use of flash flotation within the primary grinding circuit as the first stage, followed by conventional rougher flotation as the second stage. A number of other issues related to the design and operation of gold flotation circuits are reviewed in the next section.

A number of specific examples of the industrial application of flotation are provided in Chapter 12.

9.2.7.2 Flash Flotation

Flash flotation, or unit cell flotation, as it is sometimes called, can be applied for the recovery of valuable minerals at a coarse size early in a flowsheet. This usually requires flotation cells designed specifically for coarse particle treatment (i.e., to handle rapidly settling solids and rapidly floating components). Flash flotation of coarse gold within milling circuits has been used since the 1940s with several benefits to the overall gold recovery flowsheet:

- Reduced gold lockup in grinding circuits (i.e., behind mill liners)
- Faster gold flotation at coarse sizes
- Avoidance of overgrinding of sulfides
- Removal of large gold particles, which require long leach times (but potentially at the expense of recovery to a gravity concentrate)
- Higher recovery and decreased security risk compared with gravity concentration

Flash flotation gained popularity in the 1980s and has been employed at several gold plants in the United States, Australia, Canada, and elsewhere, for example, McCoy–Cove (Nevada, United States) and Kanowna Belle (Western Australia) [56].

9.2.7.3 Column Flotation

In several industry sectors, conventional, mechanically agitated, dispersed air flotation cells have been replaced by column flotation cells, particularly in cleaner circuits. These have the potential advantage of recovering finer particles and producing higher-grade concentrates, due to the highly effective cleaning action of a deep froth height and substantial froth washing capability. They have additional advantages of lower operating costs and smaller footprints (i.e., less surface area requirements) than conventional cells. The chemistry involved is generally similar to conventional flotation systems.

9.2.7.4 Inert Gas Flotation

The beneficial effect of inert gas (e.g., nitrogen) on the flotation of many sulfide minerals has been known for a long time [4, 57, 58]. When gold is intimately associated with fine-grained sulfide minerals that benefit from fine grinding prior to flotation, significant benefits may be obtained from conducting flotation under an inert nitrogen atmosphere, in addition to nonoxidizing slurry preconditioning (as described in Section 9.2.2.3). Newmont's N_2TEC process is one such technology that exploits the benefits of maintaining nonoxidizing conditions in flotation slurry to reduce dixanthogen formation [25, 59]. The benefits of an inert gas atmosphere on sulfide mineral flotation are attributable to the decrease in surface oxidation; however, evidence suggests that free gold is suppressed to some extent by flotation under nitrogen, so each ore must be tested carefully to accurately quantify the effects (see Section 9.2.2.1 regarding the role of dixanthogen in gold flotation).

9.2.7.5 Air-Sparged Hydrocyclones

Air-sparged hydrocyclones are similar to other hydrocyclones except that air is introduced into the slurry by injection into the feed stream or through a porous cyclone wall. The latter method has the advantage of producing a dispersion of very fine bubbles. Because the heavy (dense and/or large) solids are separated from the light particles, as in a hydrocyclone, the majority of air bubbles and their associated recovered particles report with the light fraction. Concentration ratios typically range from 3 to 10, depending on the particle size distribution. The potential advantages of air-sparged hydrocyclones for gold applications are that very fine gold may be recovered and that a stable froth does not need to be formed, which is sometimes a problem in the conventional flotation of gold ores with very low sulfide grades. This new development has potential for application to gold recovery and for the recovery of cyanide from effluent streams [60].

9.2.7.6 Cyanidation and Flotation

Cyanidation and flotation are often used in the same flowsheet, with flotation applied either before or after cyanidation (see Sections 3.3.3 and 3.3.4). Each process can have a detrimental effect on the other.

Gold metallurgists are sometimes reluctant to use flotation prior to cyanidation and CIP because of the deleterious effects of thiol collectors on leaching and carbon adsorption processes. In some cases this has led to the decision to float gold and/or pyrite after cyanidation, unless there is an acid leach step prior to cyanidation, as in the reverse leaching circuits employed at some South African gold–uranium plants (Section 3.3.2.4).

The effect of thiol interference can be avoided by the use of alternative collectors, such as amines; however, their flotation performance may be inferior and more limited in application than thiols.

Cyanide ion concentrations above approximately 5 mg/L effectively depress sulfide minerals. Consequently, natural degradation of cyanide with time or a cyanide destruction stage (see Chapter 11) is required before the flotation of cyanidation tailings, although sometimes the destruction products also depress flotation (e.g., $Fe(CN)_4^{2-}$). If the feed slurry to flotation is alkaline (e.g., recently cyanided tailings), a reasonably long acid conditioning stage may be used to: decrease the depressant effect of lime, destroy the free cyanide, and remove surface coatings. For the treatment of cyanidation tailings, the agitation intensity during conditioning also has an effect on the rate and extent of surface coating removal (Figure 9.13) [61, 62].

The pH of slurries produced by old residues are typically more acidic (pH 3 to 4), as a result of pyrite oxidation, and such materials may be floated using MBT collectors or neutralized and floated with xanthates.

9.3 AMALGAMATION

In the amalgamation process, relatively pure metal particles (such as gold or silver) are incorporated into liquid mercury and separated from gangue minerals. Gold is readily amalgamated, and throughout history this has been used as an effective means of separating gold from host rock. In the mid-1800s amalgamation accounted for 48% of South African gold production. For human health and environmental reasons, amalgamation of run-of-mine ore has gradually diminished. With the exception of its use by informal miners in Indonesia, Russia, China, Brazil, and some other Latin American countries, the process is now virtually obsolete. Amalgamation is still commonly employed for the treatment of gravity concentrates and has the advantages of being a simple and cost-effective process in applications where health and environmental aspects can be effectively controlled. However, the application of amalgamation for the treatment of gravity concentrates is rapidly being superseded by intensive cyanide leaching processes (see Chapter 6).

9.3.1 Properties of Mercury

Mercury is important in gold extraction chemistry, not only for use in amalgamation, but also due to its natural occurrence in ores as native mercury or cinnabar (HgS). Because it tends to be concentrated together with gold in a flowsheet, mercury can be a significant impurity in intermediate products of gold extraction processes (see Chapter 8).

Mercury is a shiny liquid at ambient temperatures with an appreciable vapor pressure (1.3×10^{-3} mm at 20°C) and significant solubility (6.4×10^{-7} g/L) in air-free water [63]. It has a tendency to combine readily with many metals, for example, sodium, zinc, potassium, gold, and silver, to produce amalgams. However, several important transition metals, for example, iron, do not amalgamate.

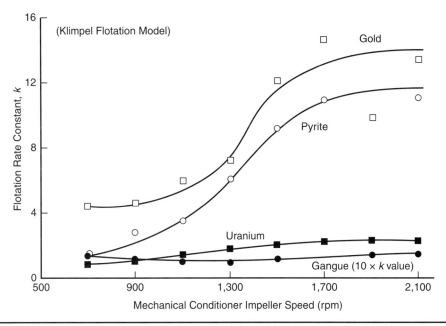

FIGURE 9.13 Effect of conditioner impeller speed on flotation rate constant: pH 11.3, 72% <75 μm [62]

Generally, mercury occurs as a very pure metal and one which, by virtue of its low boiling point, is easy to purify, for example, by distillation. Even the cheaper commercial grades of mercury contain total impurity concentrations of only 1 mg/L. Chemical and electrochemical methods are available to increase mercury purity and may need to be applied in industrial gold applications (i.e., amalgamation) to ensure that mercury surfaces are kept clean and highly surface active.

Mercury has a unique place in analytical and industrial electrochemistry as an electrode material which is highly conducting, has high hydrogen overpotential, is a liquid at ambient temperature, and is electro-inactive over a wide range of potentials.

Mercury oxidizes to form oxides and hydroxides in oxygenated solutions:

$$Hg_2^{2+} + 2e \rightleftharpoons 2Hg(liquid); \quad E^0 = 0.796 \text{ (V)} \quad \text{(EQ 9.14)}$$

$$Hg^{2+} + 2e \rightleftharpoons Hg(liquid); \quad E^0 = 0.8545 \text{ (V)} \quad \text{(EQ 9.15)}$$

The rate of mercury corrosion is very slow in pure solutions but is catalyzed by impurities to yield the overall reaction:

$$4Hg + O_2 + 2H_2O \rightleftharpoons 2Hg_2^{2+} + 4OH^- \quad \text{(EQ 9.16)}$$

In practice, mercury solubility is negligible, except in cyanide solutions, in which stable Hg–CN complexes can form (see Section 6.1.3.6).

9.3.2 Factors Affecting Amalgamation

The inclusion of gold into mercury is predominantly a surface chemical process, although gold also has a finite solubility in liquid mercury. The solubilities of various

TABLE 9.13 Solubilities of metals in mercury [64]

Metal	Solubility (% wt Hg)
Antimony	2.9×10^{-5}
Barium	0.33
Bismuth	1.4
Copper	2×10^{-3}
Gold	0.13
Indium	27.0
Iron	1.5×10^{-6}
Lead	1.3
Silver	0.042
Sodium	0.68
Thallium	42.8
Tin	0.62
Zinc	2.15

metals in mercury are given in Table 9.13. Generally the proportion of gold dissolved into mercury during industrial amalgamation is about 5% of the total present in the amalgam. The following criteria are necessary for efficient precious metal recovery by amalgamation:

- Gold and silver should be the only metallic components in the ore.
- Gold grains must be sufficiently clean to be wettable by mercury.
- The surface tension of the mercury–water interface must be sufficiently high for the mercury-wetted gold to be included in the mercury bulk.

The gold concentrate and mercury are usually contacted for several hours under mildly agitated conditions, for example, using a rotating drum with a gold–mercury ratio of 1:10 (batch process) or by passing a slurry over mercury-coated plates or mercury pools (continuous or semicontinuous processes). Amalgamation equipment, which has been described previously in the literature [41], has not changed significantly in the last 50 years. The gold-loaded amalgam may be separated from the waste minerals by elutriation or sedimentation prior to removal of excess mercury by filtration and retorting (Section 10.3.1).

As in the case of flotation, several mineralogical and surface chemical factors have an effect on amalgamation operation; however, the magnitude of these effects differs due to the difference in surface energy for gold adsorption at a mercury, rather than oil, surface. The main factors affecting amalgamation are the following:

- Recovery can be poor from ore feeds containing gold finer than 50 µm.
- Gold can occur in minerals such as tellurides and selenides, which do not amalgamate effectively.
- Mercury quality may be impaired by oxidation to yield oxides and hydroxides in oxygenated solutions.
- Locking of gold within gangue particles can mean that gold surfaces are not sufficiently exposed for efficient recovery.
- Tarnishing or coating of gold surfaces reduces gold hydrophobicity and wettability by mercury. Grinding procedures, for example, by the addition of steel balls to an amalgamation drum, are often employed to clean and liberate coated gold surfaces.

Dilute mineral acids may be added to dissolve light coatings of base metal and/or mercury oxides.

- Sulfide minerals and sulfide ions in solution can result in the formation of highly insoluble, black-colored, HgS which can reduce gold recovery:

$$Hg_2^{2+} + S^{2-} \rightleftharpoons HgS + Hg^0 \qquad \text{(EQ 9.17)}$$

This can be countered by the addition of oxidizing agents such as potassium dichromate ($K_2Cr_2O_7$) or potassium permanganate ($KMnO_4$), as follows:

$$Cr_2O_7^{2-} + 14H^+ + 6e \rightleftharpoons 2Cr^{3+} + 7H_2O; \quad E^0 = 1.33 \text{ (V)} \qquad \text{(EQ 9.18)}$$

$$MnO_4^- + 8H^+ + 5e \rightleftharpoons Mn^{2+} + 4H_2O; \quad E^0 = 1.49 \text{ (V)} \qquad \text{(EQ 9.19)}$$

Such reactions may help to reduce mercury "sickening" or "flouring" by oxidizing the deleterious sulfide ions.

- Contamination of the mercury–water interface with hydrocarbons, such as grease or diesel fuel, may result in poor mercury–gold contact and increase the probability of recovering hydrophobic ore components such as sulfides, carbonaceous matter, and some silicates (e.g., talc).

- Overgrinding of ores can result in hydrophilic gangue particles becoming embedded in the malleable gold surfaces, which reduces gold recovery by amalgamation.

9.3.3 Process Considerations

Despite the health and environmental concerns associated with the use of mercury, amalgamation has few serious competitors for the treatment of some placer ore concentrates and will likely continue to be used as part of the flowsheet for treating placer ores, particularly in developing countries, for some time to come. On the other hand, increasingly, many placer ores and concentrates can be effectively processed by modern centrifugal gravity concentration equipment, followed by direct smelting or intensive cyanidation. However, the use of amalgamation is likely to continue in some long-established plants treating free-milling ores and at small-scale operations in some lesser developed countries.

9.4 COAL–GOLD AGGLOMERATION

The affinity of gold for oil is well known, as evidenced by the addition of kerosene to gold flotation circuits to improve gold recovery or grade and the need to avoid oil contamination in amalgamation, as this reduces gold recovery. The principle of using oil to recover gold has been exploited in the coal–gold agglomeration (CGA) process, which uses (diesel or lubricating) oil in the form of spherical agglomerates of coal and oil into which gold grains are recovered. This has been found to be a very selective separation, and product grades of several kilograms per ton can be achieved by recycling the agglomerates. Testing and development work has shown that, by using suitably sized coal particles (i.e., 38 to 106 μm) and a heavy oil, free gold particles ranging in size from 1 to 500 μm can be recovered effectively [65, 66]. More recent work has indicated that some sulfide minerals (chalcopyrite, pyrite, and galena) can also be recovered by the CGA process [67].

A schematic CGA flowsheet is shown in Figure 9.14 [68]. The loaded agglomerates are burnt to yield a gold ash which increases the gold grade further by a factor of 10 to 20.

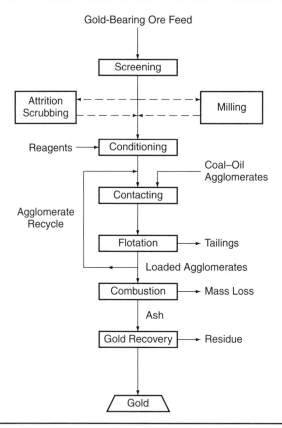

FIGURE 9.14 Schematic coal–gold agglomeration flowsheet [68]

Depending on the grade, the ash is either smelted directly or leached and the gold electrowon to produce gold bullion. The use of coal-based agglomerates overcomes the disadvantages of the uneconomic use of oil and the relatively low concentrate grades found in earlier, similar work [69]. Recycling of the agglomerates within the process to increase gold loading has been found to be beneficial.

A process scheme has also been developed by which the precious minerals are extracted from the agglomerate media by agitation with a suitable solvent to disintegrate the agglomerates. The slurry is then centrifuged to yield a heavy mineral concentrate and the solvent recovered by distillation for reuse [70, 71].

An alternative recovery scheme involves the use of flotation to separate the loaded agglomerates from the bulk slurry using potassium amyl xanthate as the collector [72].

The CGA process has the greatest potential for treating placer ores, gravity concentrates (and is a potential alternative to mercury amalgamation for finer-sized concentrates), free-milling ores, and gravity plant tailings [65, 73, 74]. The potential advantages of the process are high recovery of gold, which is largely independent of size (i.e., 1 to 1,000 μm particles recovered efficiently), very rapid recovery kinetics (i.e., <30 min), no use of cyanide, and relatively low capital and operating costs [68].

Several related process developments have occurred subsequent to the original CGA process developed by BP Minerals, including the development of the Bateman and Carbad processes [67].

REFERENCES

[1] Bard, A.L., and L.R. Faulkner. 1980. *Electrochemical Methods–Fundamentals and Applications.* New York: John Wiley & Sons.

[2] Shaw, D.J. 1980. *Introduction to Colloid and Surface Chemistry.* 3rd edition. London: Butterworths.

[3] De Bruyn, P., and G.E. Agar. 1962. Surface chemistry of flotation. Pages 91–138 in *Froth Flotation.* Edited by D.W. Fuerstenau. New York: American Institute of Mining, Metallurgical, and Petroleum Engineers.

[4] Martin, C.J., S.R. Rao, J.A. Finch, and M. Leroux. 1989. Complex ore processing with pyrite flotation by nitrogen. *International Journal of Mineral Processing* 26: 95–110.

[5] Fuerstenau, D.W. 1982. Mineral-water interfaces and the electrical double layer. Pages 17–30 in *Principles of Flotation.* Edited by R.P. King. Johannesburg: South African Institute of Mining and Metallurgy.

[6] Fuerstenau, D.W., S. Chander, and A.M. Abouzeid. 1979. The recovery of fine particles by physical separation methods. Pages 3–59 in *Beneficiation of Mineral Fines: Problems and Research Needs.* Edited by P. Somasundaran and N. Arbiter. New York: American Institute of Mining, Metallurgical, and Petroleum Engineers.

[7] Nicol, M.J. 1980. The anionic behaviour of gold. *Gold Bulletin* 13:46–55,105–111.

[8] Tennyson, S. 1980. The hydrophilic nature of a clean gold surface. *Journal of Colloid and Interface Science* 75(1):51–55.

[9] Finkelstein, N.P., and G.W. Poling. 1977. Role of dithiolates in the flotation of sulfide minerals. *Minerals Science and Engineering* 9(4):177–196.

[10] Fuerstenau, M.C., M.C. Kuhn, and D.A. Elgillani. 1968. The role of dixanthogen in xanthate flotation of pyrite. *Transactions AIME* 241:431–436.

[11] Harris, P.J. 1984. Influence of substituent group on the decomposition of xanthates in aqueous solutions. *South African Journal of Chemistry* 31(3):91–95.

[12] Jones, M.H., and J.T. Woodcock. 1978. Applications of pulp chemistry to regulation of chemical environment in sulphide mineral flotation. Pages 147–183 in *Principles of Mineral Flotation.* Parkville, Victoria, Australia: Australasian Institute of Mining and Metallurgy.

[13] Maillot, M., J.L. Cecile, and R. Bloise. 1984. Stability of ethyl xanthate ion in neutral and weakly acidic media. Part 1: Influence of pH. *International Journal of Mineral Processing* 13:193–210.

[14] King, R.P., editor. 1982. *Principles of Flotation.* Johannesburg: South African Institute of Mining and Metallurgy.

[15] Allison, S.A., R.C. Dunne, and S.A. De Waal. 1982. The flotation of gold and pyrite from South African gold-mine residues. Pages 9.1–9.18 in *14th International Minerals Processing Congress.* Volume 2. Edmonton, AB, Canada: Canadian Institute of Mining and Metallurgy.

[16] American Cyanamid. 1986. *Mining Chemicals Handbook.* Parsippany, NJ: American Cyanamid.

[17] Orwe, D., S.R. Grano, and D.W. Lauder. 1998. Increasing fine copper recovery at the Ok Tedi concentrador, Papua New Guinea. *Minerals Engineering* 11(2):171–187.

[18] Harris, P.J. 1982. Pages 237–250 in *Principles of Flotation*. Edited by R.P. King. Johannesburg: South African Institute of Mining and Metallurgy.

[19] Leaver, E.S., and J.A. Woolf. 1934. Pages 9–17 in *Flotation of Metallic Gold; Relation of Particle Size to Floatability*. Report of Investigation 3226. Washington, DC: U.S. Bureau of Mines.

[20] Wenqian, W., and G.W. Poling. 1983. Methods for recovering fine placer gold. *Canadian Institute of Mining and Metallurgy Bulletin* 76(860):47.

[21] Klimpel, R.R. 1999. Industrial experiences in the evaluation of various flotation reagent schemes for the recovery of gold. *Minerals and Metallurgical Processing Journal* 16(1):1–11.

[22] Barbery, G., A.W. Fletcher, and L. Sirois. 1980. Pages 135–150 in *Exploitation of Complex Sulphide Deposits: A Review of Processing Options from Ore to Metals*. London: Institution of Mining and Metallurgy.

[23] O'Connor, C.T., and R.C. Dunne. 1994. The flotation of gold-bearing ores—a review. *Minerals Engineering* 7(7):839–849.

[24] Teague, A.J., J.S.J. Van Deventer, and C. Swaminathan. 1999. A conceptual model for gold flotation. *Minerals Engineering* 12(9):1001–1019.

[25] Teague, A.J., C. Swaminathan, and J.S.J. Van Deventer. 1998. The behaviour of gold bearing minerals during froth flotation as determined by diagnostic leaching. *Minerals Engineering* 11(6):523–533.

[26] Forsberg, E. 1988. *Sulphide Mineral Flotation—Developments in Mineral Processing*. Volume 9. Amsterdam: Elsevier.

[27] Bath, M.D., A.J. Duncan, and E.R. Rudolph. 1973. Some factors affecting gold recovery by gravity concentration. *Journal of South African Institute of Mining and Metallurgy* 73(11):363.

[28] Botelho De Sousa, A.M.R., C.T. O'Connor, and R.C. Dunne. 1986. The influence of various chemical, physical and mineralogical factors on the flotation of gold-bearing pyrites. Pages 493–503 in *Gold 100: Proceedings of International Conference on Gold*. Edited by C.E. Fivaz. Johannesburg: South African Institute of Mining and Metallurgy.

[29] Gardner, J.R., and R. Woods. 1973. The use of a particulate bed electrode for the electrochemical investigation of metal and sulphide flotation. *Australian Journal of Chemistry* 26:1635–1644.

[30] Gardner, J.R., and R. Woods. 1977. The hydropohilic nature of gold and platinum. *Journal of Electroanalytical Chemistry* 81:285–290.

[31] Walker, G.W., C.P. Walters, and P.E. Richardson. 1984. Correlation of the electrosorption of sulphur and thiol collectors with contact angle and flotation. Pages 202–217 in *Electrochemistry in Mineral and Metals Processing I*. Edited by P.E. Richardson and R. Woods. Pennington, NJ: The Electrochemical Society.

[32] Groot, D.R. 1987. *The Reactions of Some Thiol Collectors at Noble Metal and Pyrite Electrodes*. Mintek Report M312. Randburg, South Africa: National Institute of Metallurgy.

[33] Viljoen, E.A., and P. Mihalik. 1968. *The Determination of the Nature of Coatings on Gold Recovery*. Mintek Report 385. Randburg, South Africa: National Institute of Metallurgy.

[34] Bulatovic, S.M. 1997. Flotation behaviour of gold during processing of porphyry copper–gold ores and refractory gold bearing sulfides. *Minerals Engineering* 10(9):895–908.

[35] Aksoy, B.S., and B. Yarar. 1989. Natural hydrophobicity of native gold flakes and their flotation under different conditions. *Processing of Complex Ores*. Edited by G.S. Dobby and S.R. Rao. New York: Pergamon Press.

[36] Knipe, S.W., S.L. Chryssoulis, and B. Clements. 2004. Flakey gold: Problems with recovery and mineralogical quantification. *Journal of the Minerals, Metals and Materials Society* (July):58–62.

[37] DeKok, S.K. 1973. Gold concentration by flotation. *Journal of the South African Institute of Mining and Metallurgy* (special issue):139–141.

[38] Davidson, J.D., G.A. Brown, C. Schmidt, N.W. Hanf, D. Duncanson, and J.D. Taylor. 1978. The intensive cyanidation of gold plant gravity concentrates. *Journal of the South African Institute of Mining and Metallurgy* 78:146–165.

[39] Lloyd, P.J.D. 1981. The flotation of gold, uranium and pyrite from Witwatersrand ores. *Journal of the South African Institute of Mining and Metallurgy* 81(2):41–47.

[40] Kendrick, M., W. Baum, P. Thompson, G. Wilkie, and P. Gottlieb. 2003. The use of the QemSCAN automated mineral analyzer in the Candelaria concentrator. Pages 415–430 in *Proceedings Coppe-Cobre 2003*. Volume III, Mineral Processing. Ottawa: Canadian Institute of Mining and Metallurgy.

[41] Taggart, A.F. 1950. *Handbook of Mineral Dressing*. New York: John Wiley & Sons.

[42] Weller, K.R., J.J. Campbell, G.J. Wilkie, M.R. Thornber, R. Bateman, and S. Ellis. 1998. The ores of the Golden Mile, Kalgoorlie: Coherent metallurgical testwork from comminution, through flotation, roasting and leaching, to mineralogical determination of key mineral deportments using QEMSEM. Pages 339–347 in *Proceedings AusIMM 1998–The Mining Cycle*. Melbourne: Australasian Institute of Mining and Metallurgy.

[43] O'Connor, C.T., C. Botha, M.J. Walls, and R.C. Dunne. 1988. The role of copper sulphate in pyrite flotation. *Proceedings Randol Gold Forum*. Golden, CO: Randol International Ltd.

[44] Chanturiya, V.A., A.A. Fedorov, T.N. Matveyeva, T.V. Nedosekina, and T.A. Ivanova. 2003. Theoretical aspects of gold-bearing sulfides selective flotation. Pages 753–763 in *Proceedings of 22nd International Minerals Processing Congress*. Edited by L. Lorenzen and D. Bradshaw. Johannesburg, South Africa: South African Institute of Mining and Metallurgy.

[45] Beattie, M.J.V., and G.W. Poling. 1987. A study of the surface oxidation of arsenopyrite using cyclic voltammetry. *International Journal of Mineral Processing* 20:87–108.

[46] Poling, G.W., and M.J.V. Beattie. 1988. Selective depression in complex sulphide flotation. Pages 137–146 in *Principles of Mineral Flotation*. Edited by M.H. Jones and J.T. Woodcock. Melbourne: Australasian Institute of Mining and Metallurgy.

[47] Vreudge, M.J.A. 1982. Flotation characteristics of arsenopyrite. Ph.D. thesis. Vancouver, BC, Canada: University of British Columbia.

[48] McMullen, J., P. Pelletier, Y. Breau, and F. Robichaud. 2002. Gold–copper ores processing-flotation process optimization by the development of an on-line expert system. SME Preprint 02-010. Littleton, CO: SME.

[49] Koekemoer, M.J.M., and R.W. Way. 1986. Pages 9–18 in *The Recovery of Dissolved Gold from Plant Residues onto Powdered Carbon and Its Subsequent Recovery by Flotation*. Circular 1-85. Johannesburg, South Africa: Mine Metallurgical Managers' Association of South Africa.

[50] Nice, R.W. 1971. Recovery of gold from active carbonaceous ores at McIntyre. Pages 41–49 in *Proceedings of Canadian Mineral Processors Conference*. Ottawa, ON, Canada: Canadian Institute of Mining and Metallurgy.

[51] Cabassi, P.A.J., B.K. Loveday, P.H. Radcliffe, and M.J. Wilkinson. 1983. The improved flotation of gold from the residues of Orange Free State ores. *Journal of South African Institute of Mining and Metallurgy* 83(11):270–275.

[52] Arbiter, N., P.B. Hobsbawn, J.F. Mahoney, and C.C. Harris. 1978. Conceptual design of flotation circuits. Pages 447–465 in *Mineral Processing Plant Design*. Edited by A.L. Mular and R.B. Bhappu. New York: American Institute of Mining, Metallurgical, and Petroleum Engineers.

[53] Weiss, N.L., editor. 1985. Pages 5.1–5.109 in *SME Mineral Processing Handbook*. Littleton, CO: SME.

[54] Wong, K., M. Nelson, M. Brooks, and R. Smith. 2000. Flotation characteristics of Kalgoorlie Super Pit sulfidic gold ore. Pages 239–242 in *Proceedings 7th Mill Operators' Conference*. Victoria, Australia: Australasian Institute of Mining and Metallurgy.

[55] Harbort, G., D. Lauder, J. Miranda, and A. Murphy. 2000. Size by size analysis of operating characteristics of Jameson cell cleaners at the Bajo de Alumbrera copper–gold concentrator. Pages 207–220 in *Proceedings 7th Mill Operators' Conference*. Victoria, Australia: The Australasian Institute of Mining and Metallurgy.

[56] Alexander, D.J., T. Bilney, and S. Schwarz. 2005. Flotation performance improvement at Placer Dome Kanowna Belle gold mine. Pages 171–201 in *Proceedings 37th Annual Canadian Mineral Processors Conference*. Calgary, Alberta: Canadian Institute of Mining and Metallurgy.

[57] Rao, S.R., and J.A. Finch. 1991. Adsorption of amyl xanthate at pyrrhotite in the presence of nitrogen and implications in flotation. *Canadian Metallurgical Quarterly* 30(1):1–6.

[58] Miller, J.D. 2002. The low potential hydrophobic state of pyrite in amyl xanthate flotation with nitrogen. SME Annual Meeting. Littleton, CO: SME.

[59] Simmons, G.L., J.C. Gathje, W. de Beer, and R. Moritz. 2003. Inert gas flotation technologies. Pages 1134–1142 in *Proceedings 22nd International Mineral Processors Congress*. Edited by L. Lorenzen and D. Bradshaw. Johannesburg: South African Institute of Mining and Metallurgy.

[60] Hupka, J., B. Dabrowski, J.D. Miller, and D. Halbe. 2004. Air-sparged hydrocyclone technology for cyanide recovery. *Minerals and Metallurgical Processing Journal* 22(2):135–139.

[61] Ferron, C.J. 1993. Process alternatives for the treatment of gold tailings. Pages 53–57 in *Proceedings Randol Gold Forum*. Golden, CO: Randol International Ltd.

[62] Duchen, R.B., and L.A.E. Carter. 1986. An investigation into the effects of various flotation parameters on the flotation behaviour of pyrite, gold and uranium, contained in Witwatersrand type ores and their practical exploitation. Pages 505–525 in *Gold 100: Proceedings of International Conference on Gold*. Edited by C.E. Fivaz. Johannesburg: South African Institute of Mining and Metallurgy.

[63] Sanemasa, I. 1975. The solubility of elemental mercury vapor in water. *Bulletin of the Chemical Society of Japan* 48:1795–1798.

[64] Wilkinson, M.C. 1972. The surface chemistry of mercury. *Chemical Reviews* 72(6):575.

[65] House, C.I., I.G. Townsend, and C.J. Veal. 1988. The coal gold agglomeration process. *International Mining (UK)* 5(9):17–19.

[66] Bonney, C.F. 1988. Coal gold agglomeration—a novel approach to gold recovery. Pages 125–128 in *Proceedings Randol Gold Forum*. Golden, CO: Randol International Ltd.

[67] Calvez, J.P.S., M.J. Kim, P.L.M. Wong, and T. Tran. 1998. Use of coal–gold agglomerates for particulate gold recovery. *Minerals Engineering* 11(9):803–812.

[68] Buckley, S.A., C.I. House, and I.G. Townsend. 1989. A techno-economic analysis of the coal gold agglomeration gold recovery process. *Canadian Mineral Processors Conference*. Ottawa: Canadian Institute of Mining and Metallurgy.

[69] Farnard, J.R., F.W. Meadus, F.W. Goodhue, and Puddington. 1969. *The Beneficiation of Gold by Oil Phase Agglomeration*. Ottawa: Canadian Institute of Mining and Metallurgy.

[70] Cadzow, M., and R. Lamb. 1989. Carbad gold recovery. Pages 375–379 in *World Gold '90*. New York: American Institute of Mining, Metallurgical, and Petroleum Engineers.

[71] Bonney, C.F. 1989. The BP-CGA and Carbad processes: Technical and cost differences. Pages 71–73 in *Proceedings Randol Gold Forum*. Golden, CO: Randol International Ltd.

[72] Moses, L.B., and F.W. Petersen. 2000. Flotation as a separation technique in the coal gold agglomeration process. *Minerals Engineering* 13(3):255–264.

[73] House, C.I., I.G. Townsend, and C.J. Veal. 1988. Coal gold agglomeration–pilot scale retreatment of tailings. Pages 111–116 in *Proceedings Randol International Gold Conference*. Golden, CO: Randol International Ltd.

[74] Bellamy, S.R., C.I. House, and C.J. Veal. 1989. Recovery of fine gold from a placer ore by coal gold agglomeration. Pages 347–352 in *World Gold '89*. New York: American Institute of Mining, Metallurgical, and Petroleum Engineers.

CHAPTER 10

Refining

Refining processes are used to upgrade the products of earlier recovery processes, principally zinc precipitate slimes, loaded cathodes, electrowinning sludges, and mercury–gold amalgam, all of which generally contain >10% gold. In addition, a number of by-products of gold extraction and refining processes are treated further for gold recovery; for example, loaded carbon fines, refinery slags, high-grade dusts, old crucibles and furnace liners, and refinery floor sweepings. The methods applied depend on the nature of the material, and the type and amount of impurities present.

Refining is usually performed in two stages:

1. Treatment at the point of production (e.g., at the mine site) to produce a crude bullion (typically 90% to 99% total precious metals)
2. Refining of crude bullion from the first stage to produce high-purity gold and silver for sale

The first stage can be applied at relatively low cost at the mine site, even for small quantities of precious metal production. This yields a low-volume product, which is easy to transport and which can be sampled representatively, either in the molten state or as homogeneous solid bullion, for accurate accounting purposes. Alternatively, lower-grade products of recovery processes (e.g., precipitates, cathodes, electrowinning sludges) may be shipped directly to an independent refinery, although this is rarely done.

The second stage is usually performed by dedicated refineries, often located close to major gold- (and other metal-) producing regions, which can process the doré bullion more economically on a bigger scale; for example, the Rand Refinery (Johannesburg, South Africa), Johnson Matthey (Utah, United States), and the Perth Mint (Perth, Western Australia). The purity of the final product depends on the ultimate use: typically 99.6% for jewelry and monetary bullion, and 99.99% for production of coins.

10.1 PYROMETALLURGY OF GOLD

Gold melts at 1,064°C and boils at 2,808°C (quoted values range from 2,530°C to 3,100°C) under atmospheric pressure; and at 1,800°C in a vacuum. The melting and boiling points for other metals and minerals of importance in pyrometallurgical refining processes are given in Table 10.1. At temperatures above the melting point, gold volatilizes as a red-colored vapor, with the rate of volatilization increasing with increasing temperature. For example, the volatilization rate at 3,000°C is sufficient to completely remove a gold film on porcelain within minutes. However, the volatility of pure gold is negligible below approximately 1,050°C, and is low below 1,250°C. The volatility increases with the presence of impurity metals, particularly tellurium; for example, an alloy containing 5% Te loses between 2% and 4% of the contained Au in 1 hr at 1,245°C. Alloys containing 5% Hg or Sb lose approximately 0.2% of the gold under similar conditions [1].

Gold does not oxidize in air or oxygen, even at red heat, but does have the ability to absorb gases; for example, hydrogen, carbon monoxide, carbon dioxide, and nitrogen

TABLE 10.1 Melting and boiling points of selected minerals and metals [1]

Mineral or Metal	Melting Point (°C)	Boiling Point (°C)
Au^0	1,064	2,808
Ag^0	961	2,210
Pt^0	1,769	4,530
Hg^0	−38.9	357
Zn^0	420	907
Pb^0	327	1,744
Cu^0	1,083	2,595
SiO_2	1,723	2,230
Al_2O_3	2,072	2,980
Fe_2O_3	1,565	—
PbO	886	—
ZnO	1,975	—
Ag_2O	230	—
$Na_2B_4O_7 \cdot 10H_2O$	75	—
Na_2CO_3	851	—
$AuTe_2$	472	—
FeS_2 (pyrite)	1,171	—
FeS_2 (marcasite)	450	—
Cu_2S	1,100	—
CuS	103	220
CaF_2	1,423	~2,500
MnO_2	535	—
$NaNO_3$	271	320

NOTE: Dashes = not available.

can be absorbed up to concentrations of 0.48%, 0.29%, 0.16%, and 0.20%, respectively. Gold chloride is formed when gold is heated above approximately 140°C in the presence of chlorine gas. This compound has a dark red color [1].

Gold forms alloys with most metals. Those most commonly encountered in refining processes are summarized as follows [1]:

- Gold and mercury form alloys over the complete range of proportions (i.e., 0% to 100% of each). This characteristic is exploited in the amalgamation process (Chapter 9). Mercury can be separated from gold by distillation, leaving a gold product containing 0.1% to 1.0% Hg.

- Gold and silver form alloys in all proportions. Alloys containing more than 66% Ag can be separated by acid parting. Alloys with less silver must be "inquarted" with additional silver to increase the silver content to allow parting. The effect of silver content on the alloy color, shown in Figure 2.2 (Chapter 2), can be a most useful indicator of metal composition in refinery operations. The melting point of gold–silver alloys decreases with increasing silver content.

- Gold and lead readily form a wide range of alloy compositions. Lead has a high affinity for gold (and other precious metals) and can be used as a collector for gold in pyrometallurgical processes. The lead can subsequently be removed by volatilization.

- Copper can be separated from gold by fusion with lead, by electrolysis, or by smelting to produce a copper matte. Copper imparts a red color to gold–silver–copper alloys, as illustrated in Figure 2.2—a property that is exploited in the jewelry industry. The melting point of gold–copper alloys reaches a minimum at about 50% Cu content.

10.2 ACID LEACHING

Acid leaching is used to remove acid-soluble impurities in refinery feed materials to allow more effective treatment by subsequent roasting and smelting processes.

10.2.1 Zinc Precipitates and High-Grade Sludges

The composition of gold-bearing slimes produced by zinc precipitation processes (Section 8.1) depends on the nature of the precipitation feed solution, the method of precipitation, and the type of solid–liquid separation used to recover the precipitate from the solution–slurry. For example, the grades of zinc precipitates from Merrill–Crowe precipitation are typically lower than those obtained from high-grade solutions, such as carbon eluates. Examples of precipitate composition are given in Table 10.2.

Roasting and smelting processes are most effectively performed when the zinc content in the feed material is minimized. The zinc content can be reduced by acid leaching to form zinc in solution and hydrogen gas.

$$Zn^0 + 2H^+ \rightleftharpoons H_2 + Zn^{2+} \qquad \text{(EQ 10.1)}$$

Dissolution may be achieved under mildly oxidizing conditions, or even under what would normally be considered a reducing environment, due to the low reduction potential (E = –0.76 V) of the Zn(II)–Zn couple. A 10% to 30% solution of sulfuric acid is normally used, although solutions of 70% are occasionally applied for materials with high zinc–iron content [3], and other types of acid (i.e., HCl, HNO_3) have been used in some circumstances. The dissolution reaction is exothermic and vigorous, often with quite violent gas evolution. For this reason, the zinc precipitate may need to be added slowly to the acidic solution, or the acid added slowly to the zinc slurry. A stoichiometric excess of acid must be used, possibly in stages, to ensure complete reaction. The efficiency of zinc dissolution is increased by agitating the mixture, either continuously or periodically.

The hydrogen gas evolved is highly flammable and must be vented, together with small quantities of hydrogen sulfide (H_2S) and hydrogen cyanide (HCN) that may be produced from any sulfide and/or residual cyanide content of the precipitate, respectively:

$$S^{2-} + 2H^+ \rightleftharpoons H_2S \qquad \text{(EQ 10.2)}$$

$$CN^- + H^+ \rightleftharpoons HCN \qquad \text{(EQ 10.3)}$$

If arsenic and antimony are present in the precipitate, these are reduced to the gaseous hydrides, arsine and stibnine respectively, as follows [4]:

$$As + 3H^+ + 3e \rightleftharpoons AsH_3(g); \quad E^0 = -0.54 \text{ (V)} \qquad \text{(EQ 10.4)}$$

$$Sb + 3H^+ + 3e \rightleftharpoons SbH_3(g); \quad E^0 = -0.51 \text{ (V)} \qquad \text{(EQ 10.5)}$$

These gases are highly toxic and must be efficiently vented to avoid any health risk. The addition of nitric acid to the solution up to a concentration of about 35% has been shown to reduce the production of arsine gas (AsH_3) by the preferred formation of arsenous acid.

TABLE 10.2 Typical ranges of composition of various refinery materials and products [adapted from 3 to 6] (all data in percent by weight)

Element or Material	Zinc Merrill–Crowe Precipitate*	Acid-Treated Zinc Precipitate*	Calcined and Acid-Treated Zinc Precipitate*	Retorted Zinc Precipitate*	Zinc Precipitate (from carbon eluate)†	Loaded Steel Wool Cathodes	Bullion	Final Slag (cleaned)
Gold	25–40	35–55	35–60	25–40	50–75	60–75	80–95	0–0.2
Silver‡	2–5	4–10	4–8	2–5	5–10	5–10	8–15	0–0.1
Mercury	0–5	0–10	<0.05	<0.05	0–10	0–5	<0.05	<0.05
Copper	0–3	0–2	0–2	0–3	0–5	0–10	0–3	0–1
Lead	5–20	5–25	5–25	5–20	<0.2	—	<0.5	0–15
Zinc	10–15	2–10	3–12	10–25	5–10	—	<0.2	0–15
PGMs§	←			Variable, depends on ore type		→	0–1	<0.01
Silica	0–15	0–25	0–25	0–15	10–25	—	—	20–40
Sulfur	0–8	0–12	<2	<2	<0.5	—	—	<0.2
Iron	0–2	0–3	<2	0–2	—	10–25	—	1–3
Calcium	0–2	0–1	<1	0–2	—	—	—	1–3
Nickel	<0.2	<0.2	<0.2	<0.2	—	—	—	0–1
Na$_2$O >	—	—	—	—	—	—	—	10–25
Al$_2$O$_3$ >	—	—	—	—	—	—	—	0–10
MgO	—	—	—	—	—	—	—	0–10
B$_2$O$_3$	—	—	—	—	—	—	—	10–25

NOTE: Dashes = not applicable.
* Produced by Merrill–Crowe precipitation from dilute Au(CN)$_2^-$ solution using zinc dust and lead nitrate.
† Produced by zinc precipitation from high-grade carbon eluate solution.
‡ Considers a solution feed with a 10:1 gold–silver ratio; higher silver content affects composition ranges accordingly.
§ PGMs = platinum group metals.

The rate of dissolution of zinc precipitate increases at elevated temperature and, if the heat of reaction does not increase the temperature sufficiently, steam may be supplied. Temperatures of between ambient and about 50°C are normally applied, and the reaction can be completed within 1 to 2 hr. The reaction is controlled by the addition of about 1.4 kg H$_2$SO$_4$ per kg Zn to give a residual free acid concentration of 1% to 5%, depending on the percent solids of the mixture, ensuring effective zinc removal. Sodium bisulfate (NaHSO$_4$) has been used as an alternative to sulfuric acid, but it must be applied at lower temperatures because its solubility decreases markedly above 33°C.

The typical composition of the acid-leached residue is mainly gold, silver, and insoluble base metal products, such as lead sulfate resulting from lead nitrate addition during precipitation processes (Table 10.2). The leached slime is dewatered, usually by filtration, to produce a moist cake containing 15% to 30% moisture.

10.2.2 Loaded Cathodes

The major impurity in loaded steel wool cathodes is iron which, if not removed prior to smelting, may contaminate the doré bullion. This is important because bullion refineries typically require that the doré contains <2% Fe, and penalties may be incurred at higher concentrations. A large proportion of any excess iron can be removed by acid leaching, which oxidizes the metallic iron, in a similar process to that used for zinc removal (discussed in Section 10.2.1). The relevant reactions are as follows:

$$Fe^{2+} + 2e \rightarrow Fe^0; \quad E^0 = -0.409 \text{ (V)} \qquad (EQ\ 10.6)$$

$$2H^+ + 2e \rightleftharpoons H_2; \quad E^0 = 0.00 \text{ (V)} \qquad (EQ\ 10.7)$$

Commonly, 10% to 20% hydrochloric or 10% sulfuric acid is used, at temperatures between ambient and 60°C. The reaction is performed as a batch process with a residence time of 12 to 24 hr. The major problem associated with the process is the coverage of large regions of the steel wool by cohesive gold films, which result from electrolytic deposition of gold at low current densities (<10 amp/m^2) [7]. Alternatively, loaded cathodes can be directly smelted, as described in Section 10.3.3 [6, 8].

10.2.3 Other Materials

The refining of by-products and scrap materials by smelting (Section 10.3.3) may be more effective if preceded by an acid-leaching step to remove excess base metals and other acid-soluble materials. This is particularly applicable to materials with significant copper or iron content.

10.3 PYROMETALLURGICAL METHODS FOR CRUDE BULLION PRODUCTION

10.3.1 Mercury Removal by Retorting

Materials that contain significant concentrations of mercury (i.e., >0.1% to 0.5%) must be treated for its removal prior to smelting to gold bullion. During subsequent refining stages, this practice may be necessary to minimize the release of toxic mercury vapor into the atmosphere or, in the case of amalgamation, may be used to recover mercury and allow its reuse. Materials with lower mercury concentrations, such as loaded cathodes, may be refined directly by roasting or smelting as long as adequate ventilation and fume collection facilities are provided. The mercury content varies depending on the original mercury content of the ore and the recovery method used. For example, pressed amalgams resulting from amalgamation processes typically contain 30% to 70% mercury, whereas zinc precipitates, loaded cathodes, and cell sludges are highly variable with 0% to 20% Hg.

Owing to its relatively high vapor pressure (1.3×10^{-3} mm of Hg at 20°C) compared with other metals (Au = 10^{-10}, Ag = 10^{-22}, and Pt <10^{-22} mm of Hg), mercury can be separated efficiently from other precious and base metals by a simple distillation procedure [2]. Mercury removal is performed in retorts, equipment specifically designed for the distillation of mercury, hence the term "retorting." The boiling point of mercury is 357°C and typically temperatures of 600°C to 700°C are generally applied to effectively vaporize the contained mercury. These temperatures are similar to those applied for roasting (calcining), and other reactions that occur under these conditions also apply during retorting, as discussed in Section 10.3.2. The retort temperature is ramped up gradually to enable the material to dry completely before mercury is vaporized and to allow adequate time for diffusion of mercury to the solid surfaces. The system is held at the maximum temperature for 2 to 3 hr to ensure complete evaporation. Mercury removal efficiencies in excess of 99% are commonly achieved.

Retorts are operated under a slightly negative pressure, and mercury vapor is usually exhausted into a water-cooled condensation system. The vapor is cooled rapidly to below the boiling point (357°C), and the liquid mercury is collected under water to avoid re-evaporation. The mercury produced contains small quantities of gold, silver, and other metals but is usually of sufficient purity to be reused for amalgamation. Mercury losses vary between 0.2% and 0.4% for each distillation cycle [3]. These losses are mainly a result of uncondensed mercury fumes leaving the condensation system and mercury vapor that permeates into the refinery during the retort loading and unloading processes.

Retorting of mercury amalgams yields a mercury-free product in the form of "sponge" gold, which can be treated directly by smelting for the removal of residual base metal impurities. The physical structure of other materials, such as precipitates and electrowinning products, is usually largely unchanged by retorting, although this depends on the original mercury content. These products are usually also suitable for direct smelting.

Mercury is highly toxic and has a cumulative physiological effect. This factor, coupled with its high vapor pressure, can create problems in refineries that treat materials with high mercury content. These effects can be counteracted effectively by good ventilation and gas handling, careful design of facilities (e.g., smooth walls, floors, ceilings in the retorting room to ensure no places for mercury to accumulate), routine mercury monitoring (of both facilities and employees), good hygiene procedures, and efficient operation of retorting systems.

10.3.2 Roasting (Calcining)

In the refining context, roasting, or "calcining" as it is commonly called, is applied to convert base metals, such as zinc, iron, lead, and copper, to their respective oxides. The oxide impurities are then easily removed into a slag during subsequent smelting processes. Any minor amounts of sulfides in the roaster feed material are also oxidized. Pyrometallurgical oxidation of sulfide minerals as a pretreatment to gold leaching is considered in Section 5.8.

Values of free energy of formation for a number of metal oxides commonly associated with refinery materials are shown in Figure 10.1 as a function of temperature. Oxidation is most effectively performed in the range 600°C to 700°C to enable efficient conversion to the oxide within practical time scales, typically 12 to 18 hr, although temperatures well in excess of 700°C have been used in practice to reduce roasting time. Above 600°C most metals of importance have negative free energy of formation values, implying that the oxide should be formed under these conditions [4].

Slow roasting conditions are generally most efficient for converting the metals to their oxides: The reaction is usually limited by the mass transport of oxygen to the metal surfaces, and an adequate flow of oxygen, as air, must be provided to all regions of the material. Thin layers of material (40 to 80 mm thick) in roaster or calciner trays should be used, and care should be taken not to pack loaded cathodes too tightly. This is particularly important if no acid pretreatment is used.

Residual zinc sulfate from acid treatment is converted to zinc oxide above 750°C; however, lead sulfate does not decompose below 1,000°C (Figure 10.1). A small amount of metallic zinc may volatilize during roasting, due to its relatively high vapor pressure, which can result in minor losses of gold and silver carried with the vapor. Silica (up to 20%) may be added to the roaster charge to form a slag with the zinc oxide, and thereby reduce the degree of volatilization and corresponding gold–silver losses [5]. The addition of between 3% and 10% sodium nitrate (and up to 20% in some cases) has been used to assist in the oxidation of materials high in base metals, although this particular procedure is no longer widely applied.

The color of the roasted calcine provides a good indication of the quality of the product, that is, the degree of conversion of base metals to oxides. This varies depending on the original material composition but ideally should have a reddish-brown coloration, compared with black or dark gray of unoxidized or poorly oxidized precipitate. Calcination can be performed as a batch or continuous process, depending on the type of equipment used and the scale of operation.

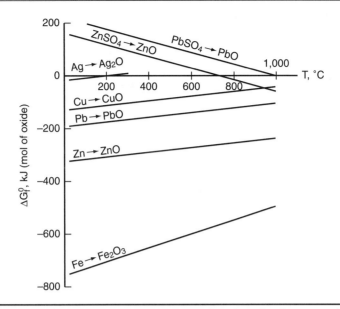

FIGURE 10.1 Free energy of formation of some metal oxides from the elements and by decomposition of the metal sulfates: pSO_2 = 0.001 atm [4]

10.3.3 Smelting

The purpose of smelting is to remove base metals and other impurities from high-grade concentrates to produce a gold–silver bullion containing typically >95% precious metals. The smelted product, called doré bullion, is suitable for direct sale and/or for further refining.

10.3.3.1 Smelting with Fluxes

The smelting, or fusion, process is achieved by heating the material in the presence of slag-forming fluxes at temperatures in excess of the melting point of all the components of the charge, typically between 1,200°C and 1,400°C. This maximum temperature is maintained for approximately 1.5 hr to ensure complete separation of impurities into the slag. The molten gold and silver form an alloy that is heavier than the slag and sinks to the bottom of the smelting vessel.

The efficiency of separation depends on the quality of the slag that is formed, measured in terms of gold (and silver) grades in the slag, and the recovery of base metals (and other impurities) to the slag. The performance depends on the nature of the gold-bearing material smelted, the properties of the fluxes used, and the conditions applied (i.e., temperature, reaction time, charge size, etc.). Silica usually forms the basis for the flux as it has the capability of dissolving most metal oxides. The metal oxides break up the silica lattice and are incorporated into the modified structure [9], as indicated in Figure 10.2. As the mole percentage of metal oxide increases, the silica lattice becomes increasingly disordered until all the silicon–oxygen bonds have been broken, and no further metal oxides can be accommodated.

Silica has a high melting point (1,723°C) and tends to form a highly viscous slag which may entrain precious metals. The addition of sodium and boron oxides (i.e., sodium borate, or borax [$Na_2B_4O_7 \cdot 10H_2O$]), reduces both the melting point and viscosity. This is illustrated in the ternary phase diagram for the silica–sodium oxide–borate system, given in Figure 10.3. For example, this indicates that a 1:1 molar ratio of sodium

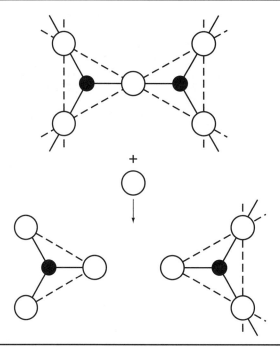

FIGURE 10.2 Schematic representation of the breaking of a common oxygen ionic bond by the addition of an oxygen ion donated from a basic oxide (open circles are oxygen ions, black circles are silicon ions) [9]

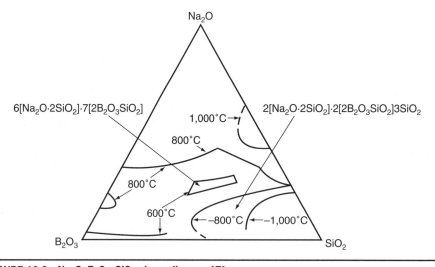

FIGURE 10.3 Na_2O–B_2O_3–SiO_2 phase diagram [7]

borate–silica has a melting point of approximately 800°C, which forms a much less viscous but still strongly acidic slag. The reaction of sodium borate with the metal oxides is analogous to that of silica. Consequently, silica–sodium borate mixtures are commonly used as fluxes.

The amount of silica–sodium borate flux added with the charge depends on the initial quantity of base metals and other impurities in the charge. As smelting proceeds, the acidity of the slag is gradually reduced as the base metals react with flux materials. Once the reaction is complete, the slag should be neutral, or slightly acidic, to protect the furnace lining or crucible from basic (alkaline) corrosion.

Other chemicals may be added to the flux for various reasons, the most important being the following:

- Calcium fluoride (fluorspar), which reduces slag viscosity by the substitution of fluoride ions into the silica lattice
- Sodium carbonate, which improves slag clarity and decreases viscosity, thereby reducing precious metals entrainment
- Oxidizing agents, such as sodium nitrate (nitre) and manganese dioxide (pyrolusite), which assist in the oxidation of unoxidized species (e.g., base metals)

The addition of fluorite (CaF_2) and soda ash (Na_2CO_3) may cause foaming and/or increased precious metal volatilization losses, depending on the charge composition and smelting temperature. Consequently, care must be taken when applying such modified fluxes in smelting systems.

Although smelting is most efficiently performed when treating thoroughly oxidized materials, occasionally direct smelting of unoxidized products is the most cost-effective method. In these cases, oxidation must be achieved during the smelting stage itself. Manganese dioxide can be used when only a small proportion of the feed must be oxidized (e.g., residual zinc or lead in roaster calcine). For example, the reaction with zinc is as follows:

$$MnO_2 + Zn \rightleftharpoons ZnO + MnO \qquad \text{(EQ 10.8)}$$

Sodium nitrate is preferred when more severe oxidation is required, for example, for direct smelting of steel wool cathodes, due to the higher proportion of available oxygen:

$$2NaNO_3 + 3Zn \rightleftharpoons Na_2O + 2NO + 3ZnO \qquad \text{(EQ 10.9)}$$

The use of oxidizing agents during smelting requires care because silver can be oxidized and lost into the slag, and gold losses may also be increased. Overaddition must be avoided.

Flux compositions are selected to optimize slag quality and to maximize crucible or furnace liner life. Some slag compositions are more corrosive than others, for example, strongly oxidizing fluxes or fluxes that react violently with the material to be smelted. Examples of flux mixtures for smelting of different materials are given in Table 10.3.

Any sulfides that have not previously been oxidized by roasting will form a matte layer between the precious metals and slag phases during smelting. This matte may contain significant quantities of gold, silver, and base metals, as well as selenium, tellurium, arsenic, and antimony. The matte can be treated to recover the precious metals by the following [5]:

- Smelting with sodium borate and a cyanide salt (sodium, potassium, or calcium) at white heat for 2 to 3 hr
- Smelting with fluxes, sodium nitrate, and finely divided scrap iron

Once smelting is complete, the slag is poured off, and the precious metal alloy is removed from the furnace. The metals are allowed to cool in bar or button molds. Ideally, the slag should be clear and uniform, with a gray-greenish coloration [5].

Direct smelting of steel wool cathodes requires the use of a strongly oxidizing flux to fully oxidize all the iron present and remove it into the slag phase [7]. The ternary phase

TABLE 10.3 Typical flux mixtures for smelting various gold extraction process products [3 to 11]

Charge Component (by weight)	Zinc Precipitate					Electrowinning—Loaded Steel Wool Cathodes		
	No Pretreatment	Acid Leached	Roasted (calcined)	Acid Leached and Roasted	High Silver Precipitates, No Pretreatment	No Pretreatment	Acid Leached	Roasted (Calcined)
Process product	100	100	100	100	100	100	100	100
Sodium borate	35–75	10–60	30–60	10–70	5–15	30	2	30–80
Silica	10–40	0–15	15–40	5–50	5–10	125	0	0–40
Sodium nitrate	0–20	0	0–5	0	0	150	0	0–30
Manganese dioxide	0–60	0	0–10	0–10	0	0	0	0
Sodium carbonate	5–50	0–50	0–20	0–20	5–10	0	0	0–30
Calcium fluoride	0	0–10	0–5	0–10	0	0	0	0

NOTE: All data are weight expressed as a percentage of process product treated.

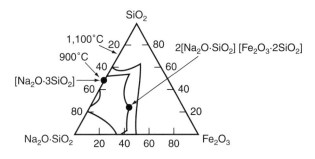

FIGURE 10.4 $Na_2O \cdot SiO_2$–Fe_2O_3–SiO_2 melting point diagram [7]

diagram for the iron oxide–silica–sodium oxide system is shown in Figure 10.4, and the composition of a typical flux for smelting such material is given in Table 10.3.

Slags can either be air cooled or granulated in a stream of water. High-grade slag materials are typically crushed and resmelted to recover a significant portion of the residual precious metals. Low-grade slag material is either discarded or may be crushed and reprocessed by any of the following options:

1. Grinding and gravity concentration
2. Grinding and flotation
3. Grinding and cyanidation
4. Any combination of 1 to 3

Generally, when smelting is conducted at a mine site, low-grade slag is fed to the primary gold extraction grinding circuit.

10.3.3.2 Smelting with Lead and Fluxes (Tavener Process)

The Tavener process is a large-scale version of the well-known fire assay procedure (Section 2.18.1), which uses lead to collect and separate precious metals from a molten charge. The precious metals are then recovered by vaporization (cupellation) of the lead or less commonly by dissolution of the lead in acid. The process was used widely until about 1950, and is still used occasionally for the treatment of low-grade materials and materials with high lead content, for example, lead cathodes from some electrowinning processes [12].

Fluxes, lead oxide (PbO, or litharge), and, if necessary, carbon are mixed with the charge of material to be smelted. Smelting is performed at similar temperatures to those required for simple flux smelting (Section 10.3.3.1). The lead oxide is reduced by metals, sulfides, and carbon, if present, to form lead metal. The lead drains through the molten charge as fine droplets and collects the contained gold and silver. The collection process is highly efficient provided that the ratio of lead–precious metals is kept high, typically >10:1. The lead–precious metal alloy is tapped out of the crucible and cooled. This alloy is then treated by cupellation in a strong stream of air to oxidize the lead back to lead oxide, which is removed as a vapor and can be recovered as a solid on cooling to allow recycling to the fusion stage. The gold and silver remain, and can be poured or remelted into bars.

The process is slow, primarily because it requires two stages; it is labor intensive; and lead fumes are produced, which present a health hazard. Despite these disadvantages, the process is effective for treatment of some low-grade refinery products.

10.3.3.3 Smelting of Low-Grade Products

Low-grade products can be custom smelted using processes developed specifically for the individual material. A wide range of smelting procedures are applied in practice, including the blast furnace process (used by the Rand Refinery in South Africa for treatment of by-products and scrap material blends) and processes for the treatment of ash residues resulting from incineration of loaded carbon [3, 11, 13].

10.4 BULLION REFINING

The bullion produced by the processes described in sections 10.1 to 10.3 is typically unsuitable for direct sale, with precious metal contents of between 90% and 99.5%, and must be further refined to yield high-purity metals (i.e., gold, silver, and platinum group metals [PGMs]), as required in the marketplace. Final gold products can be divided into two categories:

- Gold for bar production, jewelry, and other industrial uses (99.6% pure or 9960 fineness)
- Gold to produce coinage (99.99% or 9999 fineness)

Several general methods are available to upgrade crude bullion to these high-grade products, namely pyrorefining, hydrorefining, and electrorefining. Pyrorefining techniques are used for the direct production of 99.6% pure gold. This product is then further treated by electrorefining to produce 99.99% pure gold. Hydrorefining methods have largely fallen into disuse but are still used by some refineries to produce 99.6% to 99.9% gold.

10.4.1 Pyrometallurgical Refining

The Miller chlorination process is used to remove silver and other metal impurities from gold by bubbling chlorine gas into the molten metal at 1,150°C. Under these conditions, iron, zinc, and lead form gaseous chlorides, while copper and silver form liquid chlorides, according to these reactions [8, 13]:

$$\text{Fe} + \text{Cl}_2 \rightleftharpoons \text{FeCl}_2; \quad \Delta G^0 \rightleftharpoons -200.6 \text{ kJ/mol} \qquad \text{(EQ 10.10)}$$

460 | THE CHEMISTRY OF GOLD EXTRACTION

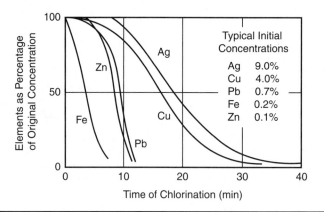

FIGURE 10.5 The effect of chlorination time on metal removal [13]

$$Zn + Cl_2 \rightleftharpoons ZnCl_2; \quad \Delta G^0 \rightleftharpoons -288.8 \text{ kJ/mol} \quad \text{(EQ 10.11)}$$

$$Pb + Cl_2 \rightleftharpoons PbCl_2; \quad \Delta G^0 \rightleftharpoons -200.6 \text{ kJ/mol} \quad \text{(EQ 10.12)}$$

$$2Cu + Cl_2 \rightleftharpoons 2CuCl; \quad \Delta G^0 \rightleftharpoons -125.6 \text{ kJ/mol} \quad \text{(EQ 10.13)}$$

$$2Ag + Cl_2 \rightleftharpoons 2AgCl; \quad \Delta G^0 \rightleftharpoons -138.1 \text{ kJ/mol} \quad \text{(EQ 10.14)}$$

$$2Au + 3Cl_2 \rightleftharpoons 2AuCl_3; \quad \Delta G^0 \rightleftharpoons \text{positive} \quad \text{(EQ 10.15)}$$

The metals are converted to their respective chlorides in the sequence Fe > Zn > Pb > Cu > Ag >>> Au, as illustrated graphically in Figure 10.5.

Small quantities of fluxes (i.e., silica, sodium borate, and salt) are added to help remove extraneous impurities and to reduce volatilization losses, especially silver chloride, by forming a protective slag layer over the molten metal. Any PGMs present remain with the gold and must be separated electrolytically or by solvent extraction (see Section 10.4.3.2 and also Chapter 7).

While the gaseous chlorides are being formed, the reaction is initially turbulent. This subsides as the content of these metals decreases and as the copper and silver form liquid chlorides. Finally, when the chlorination of silver is almost complete and the gold fineness approaches 99%, the reaction again becomes more turbulent as gaseous gold chloride is produced, characterized by the appearance of a red-brown vapor. The chlorination reaction takes between 30 and 45 min to complete, at which point the molten copper and silver chlorides are bailed off the surface. The molten charge is resmelted several times with fresh additions of fluxes to produce a final product of the required purity, typically 99.6% Au, 0.3% to 0.4% Ag, and 0.03% to 0.05% Cu. Higher-grade products (>99.9% gold) can be produced by this method, but gold losses by volatilization increase due to the need to remove silver more efficiently, which requires extended smelting time. Consequently, the production of >99.6% pure gold by the Miller chlorination requires elaborate (and costly) fume collection and gold recovery systems. The purified product is poured into molds at 1,100°C and cooled. A soft reducing flame is applied to the surface of the gold to ensure that it cools from the bottom up, encouraging any residual impurities to rise to the surface [3, 6].

The molten silver–copper chlorides removed during chlorination always contain some gold (i.e., <2%), present as mechanically entrained particles ("spangles") or as a result of dissociation of trapped gold chloride fumes. This is recovered in a "degolding" process, whereby several additions of sodium carbonate are made to the molten chlorides to precipitate a small proportion of the silver, typically 4% to 5% of the total weight [1]:

$$4AgCl + 2Na_2CO_3 \rightleftharpoons 4Ag + 4NaCl + 2CO_2 + O_2 \qquad \text{(EQ 10.16)}$$

The precipitated silver settles through the liquid and collects the fine gold in a similar manner to that of the lead collection process (Section 10.3.3.2). The gold–silver alloy produced contains approximately 25% Au, 70% Ag, and some base metals, which are recycled to the chlorination process. The remaining chloride-rich slag is granulated in water or cooled, crushed, and processed further for silver recovery [13].

10.4.2 Electrolytic Refining

Electrolytic refining is used to produce very high-grade (>99.99% or >9999 fine) gold, such as that required for the production of gold coins and other specialist applications. This is achieved using the Wohlwill process by which trace amounts of silver, copper, zinc, and PGMs are removed.

An electrolyte containing 80 to 100 g/L Au and 80 to 100 g/L HCl is used, at approximately 60°C. Partially refined bullion (preferably >99.6% Au) anodes are used, and the gold is plated onto 99.99% rolled gold cathodes. Glazed porcelain electrowinning cells are used to prevent any electrolyte contamination due to the corrosion of cell components. A current density of approximately 800 amp/m^2 is applied. Under these conditions, the primary anodic reactions are the dissolution of gold by the following [14]:

$$AuCl_4^- + 3e \rightleftharpoons Au + 4Cl^-; \quad E^0 = 0.994 \text{ (V)} \qquad \text{(EQ 10.17)}$$

$$AuCl_2^- + e \rightleftharpoons Au + 2Cl^-; \quad E^0 = 1.113 \text{ (V)} \qquad \text{(EQ 10.18)}$$

The Au(III) chloride species is the predominant species present in the electrolyte. Gold is deposited at the cathode by the converse reaction Equations (10.17) and (10.18). The formation of Au(I) chloride is undesirable because this species disproportionates to some extent to precipitate gold in the electrolyte:

$$3AuCl_2^- \rightleftharpoons 2Au + AuCl_4^- + 2Cl^- \qquad \text{(EQ 10.19)}$$

This forms an anode sludge at the bottom of the cell, which increases cleanup requirements. In addition, the deposition of gold from solutions of Au(I) chloride produces a dendritic deposit at high current densities. The formation of the monovalent species is prevented by control of electrolyte composition, electrode construction, and operating current densities (see Table 10.4) [15].

Copper, zinc, platinum, and palladium dissolve in the electrolyte and accumulate, requiring that the electrolyte is periodically bled or discarded. Other PGMs (e.g., osmium, iridium, ruthenium, etc.) and silver form an insoluble sludge in the bottom of the cell, together with the anode gold sludge, which must be removed periodically for recycling to the crude refining process or for recasting as anodes, until the PGM content is sufficiently high to allow their separate recovery (see Section 10.4.3.2).

The cathodes are washed several times with a hot sodium thiosulfate solution to remove residual silver chloride and entrained electrolyte, and are then remelted as the final pure product.

TABLE 10.4 Typical operating data for electrolytic gold refining [15]

Electrolyte	Au content	80–100 g/L
	Free HCl content	80–100 g/L
	Temperature	60°C
	Circulation	Natural convection
	Heating	Immersion element
Electrical	Anode current density	800 amp/m^2
	Current per cell	200 amp
	Voltage per cell	1.0–1.5 V
	Electrode spacing	100 mm
Anodes	Composition	9960 fine
	Mass	16.0 kg
	Dimensions (length × breadth × thickness)	230 × 280 × 12 mm
	No. per cell	2
	Life	22 hr
Cathodes	Composition of starters	9999 rolled gold
	Dimensions (length × breadth × thickness)	320 × 74 × 0.5 mm
	No. per cell	12 (3 rows of 4)
	Final cathode weight	1.0 kg
	Final cathode purity	999 fine
Cells	Material of construction	Glazed porcelain
	Dimensions (length × breadth × thickness)	465 × 405 × 250 mm
	No. of cells	4 banks of 10

Electrolytic refining processes have the advantage of minimal fume collection requirements; however, the gold lockup in the cathodes is high.

10.4.3 Hydrometallurgical Refining

10.4.3.1 Sulfuric Acid Parting

The sulfuric acid "parting" process for the separation of silver and other impurities from crude gold bullion was first introduced on a large scale in France in the early 1800s. The process was widely used in refineries worldwide in the 19th century but has largely been superseded by the chlorination and electrolytic refining processes.

The basis of the process is that gold does not dissolve in sulfuric acid, whereas silver, copper, and many other impurities do:

$$H_2SO_4 + 2Ag \rightleftharpoons Ag_2SO_4 + H_2 \qquad (EQ\ 10.20)$$

$$H_2SO_4 + Cu \rightleftharpoons CuSO_4 + H_2 \qquad (EQ\ 10.21)$$

The silver–gold ratio of the bullion to be treated must be between 2:1 and 5:1 to ensure that the gold is adequately dispersed in an excess of silver; otherwise, silver may be incompletely dissolved by the acid, as a result of physical encapsulation within the gold. If insufficient silver is present in the crude bullion, then extra must be added in a pre-melting step. This often provides an opportunity to granulate the bullion in water to provide a feed suitable for the acid leaching step [1].

A strong boiling solution of sulfuric acid is used. The reaction is violent and must be controlled by adjusting the amount of heat applied. Complete dissolution of silver usually takes 5 to 6 hr. Some of the copper and zinc are precipitated as white anhydrous salts, and the remaining gold and the silver-rich solution can be separated out. The solid residue is retreated with boiling sulfuric acid several times (depending on product purity). By this stage, the washed gold-rich residue typically contains 99.6% to 99.8% gold and is remelted to yield the final product [6].

As the silver-rich solution cools, the silver sulfate begins to crystallize out. This solution is reboiled to redissolve the silver, and copper scrap is added to cement out metallic silver. A product containing 99.8% to 99.9% silver can be obtained in this manner. Copper is recovered by electrolysis or evaporation and crystallization of copper sulfate. In this process the PGMs remain with the gold and, if necessary, can be removed by additional refining, for example, electrorefining (Section 10.4.2).

10.4.3.2 Solvent Extraction from Chloride Solution

The treatment of crude gold bullion, electrowinning sludges, and other intermediate products containing between 50% to 99% Au can be accomplished using chloride–chlorine leaching (chlorination, see Section 6.2), followed by solvent extraction to purify the solution, with subsequent direct reduction of gold from the final solution. A process developed by Inco was installed in 1972 at the Acton refinery in the United Kingdom, which utilized diethylene glycol dibutyl ether (also known as dibutyl carbitol, or DBC) [16]. In this case, gold was recovered directly from the loaded organic by reduction with oxalic acid. The process is complicated by the direct production of a powdered metal product in the organic phase, which requires special processing techniques, and the solvent has relatively high solubility in the aqueous phase, resulting in high solvent consumption.

During the 1990s, Mintek (Randburg, South Africa) developed a process based on the use of solvent extraction of gold from chloride media (following chloride–chlorine leaching). A solvent extraction step is used to purify the chloride leach solution and selectively reject silver, aluminum, copper, iron, magnesium, nickel, zinc, and lead. Some palladium, platinum, and selenium are co-extracted with the gold. Gold is precipitated from an aqueous solvent strip solution containing gold chloride. Because specific details of the organic solvent type, concentration, and other solvent extraction operating parameters have not been published, the chemistry cannot be discussed. The process, known as the Minataur process (*Min*tek *A*lternative *T*echnology for *Au* *R*efining), was first applied commercially at the Harmony Gold Mine (Virginia, South Africa) in 1997 for the production of 24 tpy of refined gold. In this case, electrowinning sludge (produced from carbon eluate solution) containing 50% to 85% Au, 8% to 10% Ag, and 3% to 6% Cu, was processed by chloride–chlorine leaching to produce a solution containing 60 to 75 g/L Au. Solvent extraction was then applied to selectively extract gold from aqueous solution. Gold was stripped from the solvent back into an aqueous chloride solution from which the gold was precipitated with sulfur dioxide to yield a 99.99% pure gold product. To remove impurities, a portion of the solvent extraction raffinate must be bled out of the circuit. The schematic process flowsheet is shown in Figure 10.6 [17]. Gold leaching extractions of >99% were reportedly achieved within 2 hr, together with overall solvent extraction efficiency of about 99.7%, and precipitation efficiency of >99.9% (in about 2 hr). The estimated cost for the process is between US$0.25 to US$0.50 per oz Au, which compares favorably with commercial refining treatment charges. (NOTE: All references to ounces are troy ounces.) Reportedly, the process can also generate a final product with 99.999% purity using oxalic acid as the reductant; however; the cost is higher [17].

10.4.3.3 Solvent Extraction from Cyanide Solution

The on-site production of a refined gold product (i.e., 99.99% purity) from cyanide leach solutions without requiring releaching of intermediate products (e.g., electrowinning

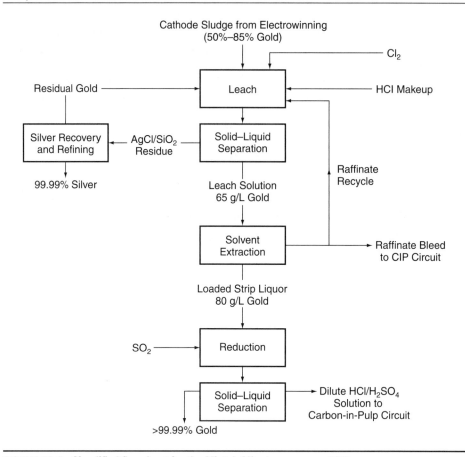

FIGURE 10.6 Simplified flowsheet for the Mintek Minataur process [17]

sludges, cathodes, or zinc precipitate) has been proposed. Research has indicated that certain organic solvents, such as tri-*n*-butyl phosphate (TBP), di-*n*-butyl butyl phosphonate (DBBP), guanidine, and possibly other solvents, might be suitable for extracting gold from aqueous cyanide solutions with sufficient selectivity to allow direct production of 99.99% gold by electrowinning or reduction from the aqueous solvent strip solution [18]. A separate solvent scrubbing step would be required to remove co-extracted silver and copper. While there are no commercial applications of this process, it shows considerable promise for the future in specific applications. The chemistry and potential use of organic solvents for gold recovery are considered in more detail in Section 7.3.

10.4.4 Refining Operations

A simplified flowsheet for the gold refining branch of the Rand Refinery (South Africa) is given in Figure 10.7. The major stages are melting and sampling, chlorine refining (Miller chlorination process), electrolytic refining (Wohlwill process), and degolding of chlorides. Other gold refining operations, namely Homestake (South Dakota, United States) and Johnson Matthey (Utah, United States), are considered in more detail in Sections 12.2.11.1 and 12.2.11.2.

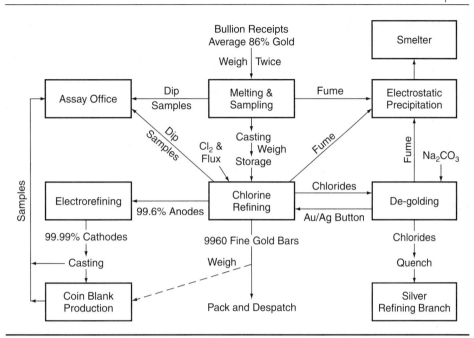

FIGURE 10.7 Simplified flowsheet of the Rand Refinery (South Africa) gold refining process [13]

REFERENCES

[1] Rose, T.K., and W.A.C. Newman. 1937. *The Metallurgy of Gold.* 7th edition. London: Charles Griffin.

[2] Weast, R.C. 1981. *Handbook of Chemistry and Physics.* 62nd edition. Boca Raton, FL: CRC Press.

[3] Gossman, G.I. 1987. Pyrometallurgy of gold. Chapter 8 in *The Extractive Metallurgy of Gold.* Edited by G.G. Stanley. Monograph Series M7. Johannesburg: South African Institute of Mining and Metallurgy.

[4] Nicol, M.J., C.A. Fleming, and R.L. Paul. 1987. The chemistry of the extraction of gold. Pages 894–899 in *The Extractive Metallurgy of South Africa.* Edited by G.G. Stanley. Monograph Series M7. Johannesburg: South African Institute of Mining and Metallurgy.

[5] Clennell, J.E. 1915. *The Cyanide Handbook.* 2nd edition. New York: McGraw-Hill.

[6] Hinds, H.L., and L.L. Trautman. 1983. Refining precious metal cathodes. *Mining Engineering* (November):1545.

[7] Menne, D. 1986. The direct smelting of doré-loaded steel wool cathodes. Pages 4839–4852 in *Gold & Silver Recovery Innovations, Phase III.* Golden, CO: Randol International Ltd.

[8] Paul, R.L. 1986. The treatment of steel wool cathodes. Lecture 27, Carbon School, Mintek. Johannesburg, South Africa: South African Institute of Mining and Metallurgy.

[9] Moore, J.J. 1990. *Chemical Metallurgy.* 2nd edition. London: Butterworths.

[10] Adamson, R.J., editor. 1972. Pages 120–151 and 203–255 in *Gold Metallurgy in South Africa.* Johannesburg, South Africa: Cape & Transvaal Printers Ltd.

[11] Dorr, J.V.N., and F.L. Bosqui. 1950. *Cyanidation of Gold and Silver Ores.* New York: McGraw-Hill.

[12] Schnabel, C. 1921. Gold. Pages 936–1134 in *Handbook of Metallurgy.* 3rd edition. London: Macmillan and Co. Ltd.

[13] Fisher, K.G. 1987. Refining of gold at the Rand Refinery. Pages 615–653 in *The Extractive Metallurgy of South Africa.* Edited by G.G. Stanley. Monograph Series M7. Johannesburg: South African Institute of Mining and Metallurgy.

[14] Schalch, E., M.J. Nicol, and B.D. Charlton. 1977. An electrochemical study of some of the problems related to the electro-refining of gold in chloride solutions. Pages 336–356 in *Chloride Hydrometallurgy and Processes.* Edited by R. Winand. Brussels: Benelux Met.

[15] Knodler, A. 1979. Determination of mono and tri-valent gold in acid baths. *Metalloberflaches:*33(7):269–272.

[16] Barnes, J.E., and J.D. Edwards. 1982. Solvent extraction at Inco's Acton precious metals refinery. *Chemistry and Industry* 5:151–155.

[17] Sole, K.C., A. Feather, J. Watt, L.J. Bryson, and P.F. Sorensen. 1998. Commercialization of the Minataur^{TM} process: Commissioning of Harmony Gold Refinery. Pages 175–186 in *Proceedings of EPD Congress 1998.* Warrendale, PA: TMS.

[18] Adams, M.D. 2003. On-site gold refining of cyanide liquors by solvent extraction. *Minerals Engineering* 16:369–373.

CHAPTER 11

Effluent Treatment

Gold extraction operations generate a variety of waste products, which must be disposed of responsibly, in compliance with environmental regulations, and as economically as possible. In parallel with the resurgence of the gold mining industry between 1972 and 1990, and the associated technological advances, there was a period of increasing environmental awareness and a major movement toward control of factors affecting the environment, particularly in the United States, Canada, Europe, Australia, and Japan, but also worldwide. This has continued into the 21st century following the Kyoto Protocol (1997), with major initiatives focused on the environmental impacts of mining, including the Global Mining Initiative (1999 to 2002) and the International Council on Mining and Metals (established in 2002). A major effort with specific relevance to the gold extraction industry was the publication of the International Cyanide Management Code (2002), to which most of the major gold and silver producers that use cyanide have committed to follow. This code was developed by the International Cyanide Management Institute, a nonprofit organization set up under the United Nations Environment Program and the International Council on Metals and the Environment. All of this activity represents significantly increased emphasis on the control and treatment of gold extraction by-products and effluents, which must be considered as an integral part of gold extraction processes.

In some cases there may be metallurgical incentives for treating effluents, such as economic benefits from recycling of reagents and/or the recovery of metal values, or the need to remove components that have an adverse effect on the primary process when a portion of the effluent stream is recycled. Consequently, effluent streams may justify or require treatment for (1) reagent recovery and recycling, (2) metals recovery, and (3) detoxification.

The first two options are applied primarily to improve project economics and may be part of a detoxification scheme to maintain environmental compliance. Detoxification is necessary when the effluent is out of compliance with legislation or permits, or may be required to allow effective recycling of the effluent to the process.

For many effluents, these environmental and metallurgical factors may not apply, and they can be disposed of satisfactorily without treatment.

The chemistry of the methods available in each of these categories is reviewed in this chapter, preceded by a section that characterizes the types of waste produced by chemical gold extraction processes. Physical aspects of waste disposal are beyond the scope of this book and are not considered in any detail. Excellent references devoted to the impacts, control, and treatment of cyanidation effluents are available in the literature [1, 2], as well as other general references on the environmental impacts of gold extraction [3, 4].

11.1 TYPES OF WASTE AND EFFLUENT CONTROL PARAMETERS

Waste products can be classified as gases, solids, liquids, and mixtures of these, as shown schematically in Figure 11.1 with specific examples of wastes generated by gold extraction processes. Some of these products, such as dust from crushing equipment or solid

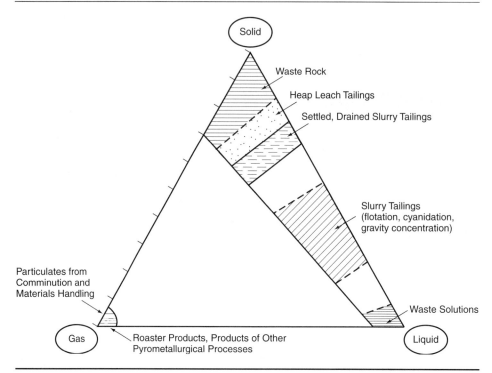

FIGURE 11.1 Schematic representation of waste products of gold extraction processes, showing general ranges of composition

waste from mining operations, are controlled, disposed of, or removed by physical means, with little need for chemical treatment. However, other products may contain components that require chemical treatment, depending on the mineralogy of the material treated, the extraction methods used, and the effluent disposal method.

11.1.1 Gases

The most important gaseous pollutants (other than particulate matter) are related to pyrometallurgical treatment processes, for example, roasting, calcining, retorting, smelting, and carbon reactivation. These pollutants include sulfur dioxide, oxides of nitrogen, arsenic and antimony, hydrocarbons, lead, mercury, and other heavy metals. These may affect air quality within and around the project site and can have a secondary effect on water quality. Table 11.1 shows the United States primary and secondary ambient air quality standards, as defined in the U.S. Clean Air Act Amendment of 1990, which give an indication of acceptable concentrations of the important pollutants. The treatment of roaster off-gases is considered in Section 5.8. Standard gas cleaning technology can generally be applied successfully for the treatment of effluents of the other processes [5], and this is not considered further here.

11.1.2 Solids

This category considers dry, or low moisture content (i.e., <25% water), solid products from extraction operations, including mining wastes, concentration process wastes (e.g.,

TABLE 11.1 Primary and secondary ambient air quality standards: U.S. Clean Air Act Amendment of 1990 [6]

Pollutant	Averaging Times	Primary Standards	Secondary Standards
Carbon monoxide	8 hr*	9 ppm (10 mg/m^3)	None
	1 hr*	35 ppm (40 mg/m^3)	None
Lead	Quarterly average	1.5 µg/m^3	Same as primary
Nitrogen dioxide	Annual (arithmetic mean)	0.053 ppm (100 µg/m^3)	Same as primary
Particulate matter (PM$_{10}$), 10 µm	Annual† (arithmetic mean)	50 µg/m^3	Same as primary
	24 hr*	150 µg/m^3	—
Particulate matter (PM$_{2.5}$), 2.5 µm	Annual‡ (arithmetic mean)	15.0 µg/m^3	Same as primary
	24 hr§	65 µg/m^3	—
Ozone	8 hr**	0.08 ppm	Same as primary
Sulfur oxides	Annual (arithmetic mean)	0.03 ppm	—
	24 hr*	0.14 ppm	—
	3 hr*	—	0.5 ppm (1,300 µg/m^3)

NOTES: Units of measure for the standards are parts per million (ppm) by volume, milligrams per cubic meter of air (mg/m^3), and micrograms per cubic meter of air (µg/m^3).
Dashes = not applicable.

* Not to be exceeded more than once per year.

† To attain this standard, the expected annual arithmetic mean PM$_{10}$ concentration at each monitor within an area must not exceed 50 µg/m^3.

‡ To attain this standard, the 3-year average of the annual arithmetic mean PM$_{2.5}$ concentrations from single or multiple community-oriented monitors must not exceed 15.0 µg/m^3.

§ To attain this standard, the 3-year average of the 98th percentile of 24-hr concentrations at each population-oriented monitor within an area must not exceed 65 µg/m^3.

** To attain this standard, the 3-year average of the fourth-highest daily maximum 8-hr average ozone concentrations measured at each monitor within an area over each year must not exceed 0.08 ppm.

from waste sorting operations; see Section 3.2.4), heap leaching products, de-watered milling, flotation, and cyanidation tailings, and old slurry tailings from discontinued milling operations. The major chemical consideration for these wastes is the possible contamination of surface and/or groundwater resulting from contact of rainwater or solution runoff with the solid waste, for example:

- Acid mine drainage, resulting from sulfide mineral decomposition
- Release of residual cyanide and soluble metal species from heap leaching (and other similar) operations
- Mineral dissolution, resulting in the release of metal species (cations and possibly complex anions) and other anions (i.e., sulfate, nitrate, chloride), particularly from waste products of processes employing oxidative pretreatment.

These processes may occur rapidly, within days or even hours in some cases (e.g., cyanide release from heaps), or may require very long periods to occur to any significant extent. The solutions produced by such action may or may not present an environmental threat, depending on the method of containment, and the nature and concentration of the species present. This is considered further in Section 11.1.3.

11.1.3 Liquids

Liquid effluents are solutions or slurries produced by the various hydrometallurgical extraction processes (e.g., cyanide leaching and flotation), as well as the products of interaction between liquids and solid–gas wastes associated with gold extraction technology. The most immediate concern is usually the threat to the environment presented by toxic constituents in the liquid phase, such as various cyanide species, toxic metal ions, and extremes of pH. However, the long-term stability of any solid portion of the waste may also be a significant factor (Section 11.1.2).

Waste disposal methods can be divided into two main categories: contained effluents and those which require discharge of all, or a portion, of the stream. The controls for these effluents are considered in more detail in the next sections.

11.1.3.1 Contained Effluents

Contained effluents are materials that are stored on-site with little or no release to the environment. The key factor in this categorization is that an impermeable, or semipermeable, barrier exists between the waste product and the ground to minimize or prevent degradation of surface- or groundwater. Examples of these barriers include the following:

- Tailings impoundments
- Lined heap and run-of-mine stockpile leach pads
- Lined solution ponds
- Steel tanks
- Concrete vats and sumps
- Pipes and open-channel launders

If necessary, effluents contained within these systems can be: (1) treated for reagent and/or metal recovery prior to disposal in the containment facility, (2) treated in situ, or (3) a portion of the effluent can be returned for further treatment (e.g., tailings impoundment decant solution pumped back to the primary process). The need for detoxification depends on the threat to wildlife, the possibility (or requirement) of a release to the environment, the risk of contaminating surface- or groundwater, or a combination of these factors.

The threat to wildlife is a recognized problem in the United States and is likely to become an increasingly important issue worldwide. The greatest threat is presented by toxic cyanide species, that is, free and weak acid dissociable (WAD) species. However, extremes of pH and high concentrations of toxic metals are also potentially harmful. Various methods of hazing, to keep wildlife away from contained effluents (and process solutions), have been tested extensively but these have failed to be effective for the majority of mining operations [7, 8, 9]. However, the threat to wildlife can generally be reduced by the following procedures:

- Minimizing the exposed surface area of solution, which reduces the attractiveness of the water body to birds and other wildlife. This applies to solution ponds, tailings ponds, and heap leach pads (i.e., prevention of solution ponding on pads).
- Preventing wildlife access to toxic solutions, for example, by enclosing effluent containments and covering solution ponds with netting or other suitable material (e.g., polyethylene).
- Minimizing the concentration of toxic species in solution by good process design and control of reagents (e.g., cyanide, lime, etc.) in the process.

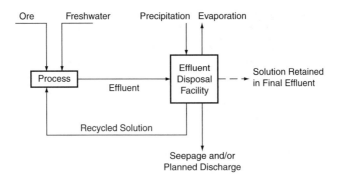

FIGURE 11.2 Schematic process flowsheet with effluent solution recycle showing process water balance

Where the application of such techniques is impractical, detoxification may be required to reduce toxic constituents to safe levels. For example, some operations reduce the free or WAD cyanide concentration of tailings solutions <25 or 50 mg/L NaCN, and maintain solution pH between 6 and 9.5 to reduce the hazard to wildlife.

The possibility of contaminating surface- or groundwater depends on the location, the type and effectiveness of containment device used, the operating philosophy and efficiency, the level of toxicity of the material, and the various environmental regulations under which the operation falls. These factors vary greatly, and any need for detoxification must be considered on a case-by-case basis.

11.1.3.2 Effluent Discharges

The need to discharge effluent into the environment depends on the overall water balance of the operation (Figure 11.2), which is determined by a number of factors including the processing method and water requirements, climate (evaporation and precipitation), the amount of water generated during mining operations, and the method of effluent disposal. Operations that have a positive water balance, where the combined mining and processing operations generate more water than they consume, must discharge solution into the environment. Even operations with a large negative water balance may need to discharge effluent under certain circumstances, for example, following a period of high rainfall or as a result of process equipment failure.

Effluents that are discharged have the potential to contaminate surface- or groundwater and may need to be treated to meet environmental regulations prior to discharge. Effluent control limits have been developed for many solution species based on their measured, perceived, or projected toxicity. Table 11.2 shows some drinking water guidelines and standards, as published by the U.S. Environmental Protection Agency (USEPA) and the World Health Organization (WHO), which gives an indication of acceptable drinking water limits for selected solution species (for a complete listing, consult references [10, 11]). However, drinking water guidelines and standards do not necessarily conform to the site discharge requirements. The actual control limits applied around the world deviate from these data because of differences in the way standards are derived and applied, as a result of different interpretation of scientific data, by the use of subjective criteria, specific environmental or geographical concerns, and other contributing factors. An example of this is the variability of restrictions imposed on cyanide-containing effluents as a result of confusion over the toxicity of different cyanide species. Extensive research has indicated that the most toxicologically significant forms of cyanide are free

TABLE 11.2 Selected drinking water standards and guidelines [10, 11]

Contaminant	USEPA National Primary Drinking Water MCL* (mg/L)	USEPA National Secondary Drinking Water Regulations (mg/L)	WHO Guidelines for Drinking Water Quality (mg/L)
Aluminum	—	0.05–0.2	—
Antimony	0.006	—	0.02
Arsenic	0.01†	—	0.01‡
Cadmium	0.005	—	0.003
Chloride	—	250	—
Chlorine	—	—	5§
Chromium (total)	0.1	—	0.05
Copper	1.3**	1	2
Cyanide (as free CN)	0.2	—	0.07
Fluoride	4	2	1.5
Iron	—	0.3	—
Lead	0.015**	—	0.01
Manganese	—	0.05	0.4§
Mercury	0.002	—	0.001
Nickel	—	—	0.02‡
Nitrate	10 (as nitrogen)	—	50
Nitrite	1 (as nitrogen)	—	3
Selenium	0.05	—	0.01
Silver	—	0.1	—
Sulfate	—	250	—
Thallium	0.002	—	—
Total dissolved solids	—	500	—
Uranium	—	—	0.015
Zinc	—	5	—
pH	6.5–8.5	—	—

NOTE: Dashes = not applicable.

* MCL = maximum contaminant level, the highest level of a contaminant this is allowed in drinking water.

† Standard applies as of January 23, 2006, replacing previous value of 0.05 mg/L.

‡ Provisional guideline value, indicating there is evidence of a hazard, but the available information on health effects is limited.

§ Concentrations of the substance at or below the health-based guideline value may affect the appearance, taste, or odor of the water.

** Treatment technique: a required process intended to reduce the level of a contaminant in drinking water.

cyanide (CN^- and HCN) and the WAD complexes (i.e., zinc, copper, and nickel). The toxicity of metal cyanide complexes has been attributed almost completely to the concentration of free cyanide that is in equilibrium with the metal complex. Consequently, the strong complexes, such as iron and cobalt cyanides, are essentially nontoxic. However, iron cyanide can decompose by photo degradation to liberate free cyanide, which adds to the complexity of the problem [12], and this emphasizes the need to consider each effluent system on a case-by-case basis. As a further guideline, the USEPA drinking water standard for total cyanide is 0.2 mg/L, compared with an aquatic biota water quality criterion of 0.05 mg/L.

The case of arsenic is another interesting example of variability of guidelines, standards, and regulatory requirements. In the United States, the EPA provides a national recommended water quality criterion for arsenic of 0.018 µg/L (0.000018 mg/L) for human health protection. The maximum contaminant level (MCL) for drinking water has been set at 0.01 mg/L (as of January 23, 2006), compared with the prior level of 0.05 mg/L, in an effort to bring the MCL closer to the published human health criteria. In this case, because the criteria for human health are so much lower than the drinking water MCL, there are situations where solution discharges may be regulated to a standard below the drinking water MCL. On the other hand, the WHO provides an arsenic drinking water guideline of 0.01 mg/L also but qualifies this as a "provisional guidance value, as there is evidence of a hazard, but the available information on the health effect is limited." Consequently, effluent discharge requirements must be reviewed carefully on a project- and location-specific basis [10, 11].

11.2 REAGENT AND METALS RECOVERY

All of the techniques discussed in this section can be applied as all or part of a detoxification process; however, each of the methods provides an opportunity for recovery of cyanide and/or metal values, such as gold, silver, copper, and zinc. A comparison of these methods based on the efficiency of cyanide removal appears in Table 11.3.

11.2.1 Direct Solution Recycle

In this option, solution is returned directly from the process tailings stream to the process (Figure 11.2) so that the contained reagents (i.e., cyanide and alkali for cyanide leaching) are reused, and any metal values recovered. This is the simplest method of reagent and metals recovery available, and it is used to varying extents in the great majority of cyanide leaching circuits, for example:

- Heap leaching, where barren solution from the recovery process (either carbon adsorption or zinc precipitation) is returned directly to the heap for further leaching
- Flotation and/or agitated leaching circuits, where solution is reclaimed from tailings thickeners or impoundments and returned as dilution water to the process.

Tailings impoundments provide valuable retention time for continued leaching of gold (and other metals), and additional gold recovery is usually achieved by recycling tailings solution, typically between 0.1% and 0.5%. On the other hand, cyanide naturally degrades in tailings ponds, which may result in some loss of the reagent by the time it is recycled to the process. Nevertheless, considerable reagent savings can be achieved by this relatively simple procedure.

Metals that are recycled in tailings return solutions are never completely recovered in the process, and an equilibrium concentration is eventually established in the combined process to tailings system. This equilibrium is influenced by the metal solubility in both the leaching and tailings systems, in which complexation and precipitation reactions play an important role, as well as the efficiency of metal recovery in the gold recovery process.

Alternatively, solutions can be recovered from tailings slurries prior to disposal, as illustrated in Figure 11.3, using suitable solid–liquid separation equipment, such as filters or thickeners. In this system, the same principle that has been applied for gold recovery in conventional (i.e., noncarbon adsorption) cyanidation plants, where low-grade (barren) solution is used to displace gold values, can be applied by using fresh or detoxified solutions to displace cyanide, alkali, and dissolved metals from effluent slurry streams.

TABLE 11.3 Comparison of methods for cyanide removal and detoxification for solution (adapted from [12])

Cyanide Removal and/or Detoxification Treatment Method		CN–HCN	Zinc and Cadmium Cyanides		Copper and Nickel Cyanides		Fe(II) CN	SCN⁻	Products	Require Further Treatment
			Metals	CN⁻	Metals	CN⁻				
Natural		Y	Partial	Partial	N	N	N	Partial	Decant solution	Probably
Fe complexation		Y	Y	Y	Partial	Partial	N	N	Decant solution	Probably
H_2O_2		Y	Y	Y	Partial	Partial	Partial precipitate	N	Treated effluent	May
SO_2		Y		Y	Y	Y	Partial precipitate	Partial	Treated effluent	May
Cl_2		Y	Y	Y	Y	Y	N	Y	Treated effluent	May
Biological		Y	Y	Y	Y	Y	Partial	Y	Treated effluent	N
O_3		Y	Y	Y	Y	Y	N	Y	Treated effluent	May
AVR (acidification, volatilization, and reneutralization)		Y	Y	Y	Y	Y	Y	Partial	Treated effluent or Recycle solution	Y / N
Ion exchange	Au, CN⁻ as complexes	Y	Y	Y	Y	Y	Y	Possible	Recycle solution	Y
	CuCN-treated resin	Y	Y	Y	Y	Y	Y	Possible	Recycle solution	N
Carbon	Granular	Y	Y	Y	Y	Y	Partial	?	Treated effluent	May
	Powder	Y	Y	Y	Y	Y	Y	?	Treated effluent	N
Electrolytic	For CN recovery	Most	Y	Most	Y	Most	N	Y	Treated effluent	May
	For CN destruction	N	Y	N	Y	N	N	N	Treated effluent	May
Flotation (ion precipitation)		Partial	Partial	Partial	Y	Y	Y	?	Treated effluent	Y

Y = yes.
N = no.

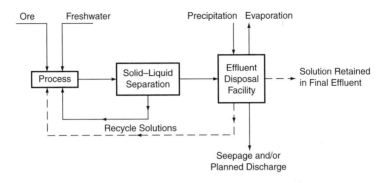

FIGURE 11.3 Schematic process flowsheet incorporating solid–liquid separation for immediate recovery and recycling of effluent solution

11.2.2 Acidification, Volatilization, and Reneutralization

The AVR process was developed in the early part of the 20th century and was used successfully at Flin Flon (Canada) from about 1930 to 1975 and at several other commercial operations. In the early applications, the primary objective was to reduce cyanide consumption rather than for effluent control; however, the process has been developed and improved to such an extent that it can be considered for detoxification.

The process concept is extremely simple and elegant: Alkaline cyanide leach solution is acidified, the hydrogen cyanide produced is removed by volatilization in a stream of air, and finally the gaseous hydrogen cyanide is readsorbed back into an alkaline solution. A schematic flowsheet of the process is given in Figure 11.4.

The relevant reactions are represented as follows for free cyanide and the copper cyanide complex, $Cu(CN)_3^{2-}$:

$$2CN^- + H_2SO_4 \rightleftharpoons 2HCN + SO_4^{2-} \quad \text{(EQ 11.1)}$$

$$Cu(CN)_3^{2-} + H_2SO_4 \rightleftharpoons 2HCN + CuCN + SO_4^{2-} \quad \text{(EQ 11.2)}$$

In the first stage, the pH of the cyanide-containing solution is reduced to <2 (typically 1.5 to 1.8), at which point close to 100% of the cyanide exists as hydrogen cyanide. A concentrated acid, such as H_2SO_4, is used for pH modification. Under these conditions the weak and strong metal cyanide complexes dissociate to release free cyanide and the respective metal ions. The pH may be lowered in stages in some cases: for example to between 4.5 and 8.5 to convert free cyanide (and zinc cyanide); to about 4.0 for the WAD complexes (i.e., copper, zinc, and nickel); and to <2.0 for the strong complexes (e.g., iron).

The solution is then passed through an aeration system, which provides a high liquid surface area to promote volatilization, for example, towers packed with plastic media and/or aerated mixing tanks. Under these conditions, a large proportion of the hydrogen cyanide volatilizes and passes into the gaseous phase. An air-to-liquid ratio of approximately 600:1 is required for efficient volatilization.

The cyanide-depleted solution is partially neutralized with alkali, usually lime, and the pH adjusted to 9.0–10.5. The metal species precipitate out as hydroxides, or as double salts, which can be removed by solid–liquid separation, if necessary. The gas phase is passed through an adsorption or scrubber tower in which caustic solution (pH 10.5 to 11.5) is circulated counter-current to the gas to re-adsorb hydrogen cyanide, predominantly as

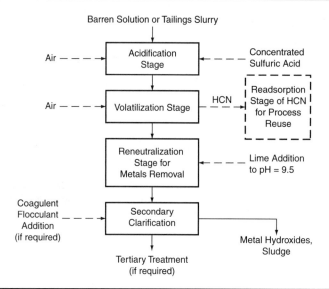

FIGURE 11.4 Schematic flowsheet for the AVR process [13]

TABLE 11.4 Chemical characteristics of AVR-treated solution

Parameter*	Concentration Range	Average Concentration
Arsenic	0.01–0.02	0.01
Cadmium	0.004–0.005	0.004
Chromium	<0.01–<0.02	<0.02
Cobalt	0.15–0.18	0.16
Copper	0.28–0.55	0.39
Iron	0.05–0.09	0.07
Lead	0.05–0.20	0.10
Mercury	0.013–0.015	0.014
Nickel	0.05–0.10	0.09
Silver	0.5–1.1	0.9
Zinc	0.04–0.13	0.09
Thiocyanate	27.4–36.6	31.3
Total cyanide	1.3–2.3	1.7
Method-C cyanide	0.7–1.6	1.2
Ammonia (as N)	13.8–21.3	18.6
Nitrate (as N)	20.0–31.4	25.4
Sulfate	1,200–1,600	1,450
pH (in pH units)	9.5–9.8	—

* All concentrations in mg/L, unless otherwise stated. And all values are the result of direct analysis of the samples.

free cyanide [13]. An example of the chemical characteristics of AVR-treated water is given in Table 11.4.

At the Golconda operation (Australia), the process has been shown to be capable of reducing total cyanide concentrations from >200 mg/L to <5 mg/L and free cyanide from 20 mg/L to <0.5 mg/L. Reagent consumptions in this application were 0.6 kg/t H_2SO_4 solution and 0.45 kg/t NaOH. The efficiency of cyanide recovery depends largely on the type of equipment used for regeneration and varies between 50% and 85%. The process can be applied economically to effluent solutions containing >150 mg/L total cyanide, but it is generally considered to be unsuitable for producing a final solution for discharge from a facility because of the high cost of reducing cyanide concentrations to required control levels (i.e., <0.2 mg/L total cyanide).

AVR technology has been further developed and adapted to allow treatment of slurry streams directly, without the need for solid–liquid separation [14]. A version of the process using Cyanisorb technology (Coeur D'Alene Mines Corporation, Idaho, United States) was commissioned at the Golden Cross mine (Waihi, New Zealand) in 1991 [15]. The process was used to treat approximately 2,500 tpd of mill tailings at 37% solids, pH 7.5, and containing 150 mg/L of WAD cyanide species. The Cyanisorb process recovered approximately 85% of the cyanide in the tailings that was tied up with WAD complexes and produced a final effluent slurry containing ~20 mg/L WAD cyanide. As a result of the effective use of the Cyanisorb process, Golden Cross was not required to line its mill tailing facility.

The Cyanisorb AVR process was installed at the NERCO DeLamar silver mine (Idaho, United States) in 1992 and was used to treat a clarified solution containing 350 to 600 mg/L WAD cyanide. A final effluent solution containing 30 to 60 mg/L WAD cyanide was generated [15].

The operating costs for AVR and/or Cyanisorb technology are estimated to provide a net credit of between US$0.05/t to US$0.50/t ore treated based on reuse of the recovered cyanide from the process, depending on the specific site and operating conditions.

AVR represents a proven, tested, and economically viable method of recovering and recycling cyanide in gold extraction. In addition, it has been demonstrated to be effective technology for reducing free and WAD cyanide species to levels effective for tailing disposal in many cases (i.e., 20 mg/L WAD cyanide), although not for effluent discharge.

11.2.3 Ion Exchange

The application of ion exchange resins for recovery of precious metals from cyanide solutions has been discussed in detail in Section 7.2, but the process can also be used to recover cyanide species from solution.

In the mid-1950s a process was developed for the recovery of cyanide species from electroplating solutions. This system incorporated the use of a CuCN-impregnated resin for the adsorption of free cyanide. A better approach is the use of strong-base anion exchange resins to remove soluble metal cyanide complex species, following the addition of a sufficient excess of a suitable metal ion to tie up all of the free cyanide in solution as complexes. Cyanide complexes of iron, zinc, copper, nickel, cobalt, gold, and silver can all be removed effectively in this manner, but Fe(II) is the preferred metal ion to use because of the high stability of the complex formed and the strong affinity of weak-base resins for this complex. Elution of the resin with dilute acid, for example, H_2SO_4, produces HCN, which must be contained in a closed system and recontacted with caustic solution (pH 10 to 11) for readsorption of cyanide [16].

Although the principles of this cyanide recycling process are quite simple, practical application is more difficult. Efficient elution of all species that load onto the resin from effluent streams is hard to achieve, and complex elution–regeneration schemes are required for effective recycling of resin. If resin metal loadings cannot be reduced sufficiently by elution, then the fresh resin makeup requirements to maintain process efficiency are likely to be prohibitively costly. Consequently, the applicability of this process depends on the chemistry of the effluent stream to be treated. A simplified flowsheet for the recovery and removal of cyanide and base metals from gold cyanidation plant effluent is provided in Figure 11.5.

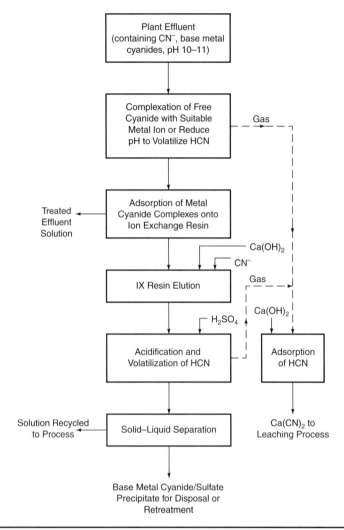

FIGURE 11.5 Simplified flowsheet for the recovery of cyanide and base metals from gold cyanidation effluent

Vitrokele process.   In the mid-1980s, the use of Vitrokele 912 ion exchange resin was proposed for the recovery and recycling of cyanide and base metals from the slurry and solution products of cyanidation processes [17]. The Vitrokele (Signet Engineering, Perth, Australia) cyanide recovery process includes the following steps:

- Adsorption of free cyanide and WAD cyanide complexes
- Selective elution of free cyanide (with cyanide recovered by scrubbing with NaOH)
- Elution of metal cyanide complexes with metal recovery from the eluate

Following metal recovery, the final treated solution containing regenerated cyanide was returned to the cyanide leaching circuit. The process has been demonstrated at a pilot-plant scale to be capable of reducing free cyanide from >150 mg/L to <0.3 mg/L and copper concentrations from 85 mg/L to <1.0 mg/L. Several gold plants have used Vitrokele resin for cyanide recovery, including Connemarra mine (Zimbabwe).

AuGMENT process. In the early 1990s, DuPont (Wilmington, Delaware, United States) and SGS Lakefield (Toronto, Ontario, Canada) developed the AuGMENT process, which was specifically designed to recover metallic copper and regenerate free cyanide for reuse in cyanide leaching [18]. The process was developed to treat ores with high cyanide–soluble copper content. Key aspects of the process include the following:

1. Use of a high cyanide–copper ratio during cyanide leaching to favor the formation of $Cu(CN)_4^{3-}$ and minimize the loading of copper onto activated carbon used for gold (and silver) recovery (see also Sections 6.1.4.2 and 7.1.2.5)
2. Selective recovery of gold and silver by carbon adsorption
3. Elution of gold- and silver-loaded carbon with cold solution to remove copper, if necessary
4. Recovery of WAD cyanide complexes (e.g., copper, zinc, and nickel) onto a strong base resin, such as Dowex M41, using either resin-in-pulp or resin-in-solution (see Section 7.2.4)
5. Elution of copper (and other WAD) cyanide complexes from the resin into cyanide solution
6. Electrowinning of copper from the eluate using a membrane cell with separate catholyte and anolyte solutions (the catholyte operates at pH 10.5 to 11 to prevent cyanide loss by volatilization)
7. Regeneration of the resin with sulfuric acid, yielding HCN in solution and in the gas phase
8. Scrubbing of the gas phase with $Ca(OH)_2$ to remove HCN as $Ca(CN)_2$, which can be recirculated to the leach circuit

Considering step 4, strong-base ion exchange resins have strong affinity for the copper cyanide complex, $Cu(CN)_3^{2-}$, as follows:

$$(]-^+NR_3)_2SO_4^{2-} + Cu(CN)_3^{2-} \rightleftharpoons (]-^+NR_3)_2Cu(CN)_3^{2-} + SO_4^{2-} \quad \text{(EQ 11.3)}$$

where

]– = the inert portion of the resin matrix

In the simple case of loading $Cu(CN)_3^{2-}$ onto strong-base ion exchange resins, loading capacities of 30 to 40 g/L Cu and 40 to 50 g/L CN are typically achieved. However, in the case of the AuGMENT process, the resin is regenerated using sulfuric acid (step 7), which results in the precipitation of stable CuCN within the resin pores, as follows:

$$(]-^+NR_3)_2Cu(CN)_3^{2-} + H_2SO_4 \rightleftharpoons (]-^+NR_3)_2SO_4^{2-}(CuCN) + 2HCN \quad \text{(EQ 11.4)}$$

This regenerated resin, which contains CuCN precipitate, is able to effectively load $Cu(CN)_3^{2-}$ species from cyanide leach solutions (provided that the cyanide–copper ratio is maintained above 3:1, as specified in step 1, according to the following reaction:

$$(]-^+NR_3)_2SO_4^{2-}(CuCN) + Cu(CN)_3^{2-} \rightleftharpoons 2(]-^+NR_3)Cu(CN)_2^- + SO_4^{2-} \quad \text{(EQ 11.5)}$$

In the case of excess free cyanide in solution (i.e., a 5:1 molar ratio of cyanide–copper), then the following applies:

$$2(]-^+NR_3)_2SO_4^{2-}(CuCN) + Cu(CN)_3^{2-} + 2CN^-$$
$$\rightleftharpoons 3(]-^+NR_3)Cu(CN)_2^- + (]-^+NR_3)CN^- + 2SO_4^{2-} \quad \text{(EQ 11.6)}$$

It is reported that the maximum loading on the resin is achieved with a cyanide–copper molar ratio of 2:1. In this case, resin loadings of 60 to 80 g/L Cu and 50 to 70 g/L CN$^-$ can be accomplished [18].

The loaded resin is eluted with a concentrated copper cyanide solution (20 to 40 g/L Cu) with a molar ratio of cyanide–copper of 4:1 to 5:1. This contains sufficient free cyanide to remove approximately 50% of the loaded copper and 25% to 30% of the cyanide. The eluted resin typically contains 30 to 40 g/L Cu and 10 to 15 g/L CN$^-$. The elution process is represented as follows:

$$2(]-^+NR_3)Cu(CN)_2^- + Cu(CN)_3^{2-} + 2CN^- \rightleftharpoons (]-^+NR_3)_2Cu(CN)_3^{2-} + 2Cu(CN)_3^{2-} \quad (EQ\ 11.7)$$

Regeneration of the eluted resin is described by Equation 11.4. Copper is electrowon from cyanide solution, liberating free cyanide at the cathode, as follows:

$$Cu(CN)_3^{2-} + e \rightleftharpoons Cu + 3CN^- \quad (EQ\ 11.8)$$

The electrowinning step must be carried out in a cell divided by a membrane to separate the anolyte and catholyte to avoid oxidation of CN$^-$ at the anode.

In 1995, Newmont formed a joint venture with DuPont to further develop and commercialize the technology. The process has been demonstrated to be capable of reducing plant effluent free cyanide and WAD cyanide concentrations to <20 mg/L each. Copper and cyanide recoveries >95% were reportedly achieved, with cyanide recovered primarily as Ca(CN)$_2$ [19]. The process has been proposed for high copper-containing gold materials such as porphyry copper–gold flotation tailings, whereby cyanide consumption during cyanide leaching could be reduced from about 5–10 kg/t to 1–2 kg/t.

The elegant AuGMENT process is expected to be cost-effective in applications where cyanide recovery is beneficial (i.e., for high cyanide-consuming ores) and/or where copper (or other interfering base metals species) present processing problems that cannot be effectively addressed by other (simpler) means. The process has yet to be fully commercialized.

11.2.4 Activated Carbon

The use of activated carbon for recovery of precious metals from alkaline cyanide solutions is considered in detail in Section 7.1. Activated carbon is effective for the recovery of many other metals from solution, although the loading equilibria and kinetics are generally less favorable than for gold, and it is commonly used in water purification systems.

Activated carbon is capable of adsorbing up to 5 mg CN$^-$/g from aerated, alkaline, and cyanide solutions. This can be increased to >5 mg CN$^-$/g in the presence of a catalyst, such as copper [20]. As with precious metals-loaded carbon elution systems, cyanide can be recovered from the carbon into low ionic strength solution at elevated temperatures. The use of silver-impregnated activated carbon has also been proposed.

Despite this ability to adsorb free cyanide, efficient removal can only be achieved if the cyanide is present as the more readily adsorbed metal complexes (i.e., zinc, copper, or nickel).

The use of activated carbon as a catalyst for the oxidation of cyanide to cyanate was investigated in the 1960s and 1970s. It was established that free cyanide is adsorbed and then catalytically oxidized at the carbon surface to cyanate in the presence of oxygen. As might be expected, copper assists with this reaction, and continuous copper addition results in hydrolysis of cyanate to carbon dioxide and ammonia [21].

11.2.5 Electrolytic Treatment

Free cyanide can be regenerated electrolytically from solutions containing copper, gold, silver, and zinc cyanide complexes. The metals are electrowon onto a cathode, and the cyanide is released simultaneously into the solution. The process, which has been marketed under the names of Celec and HSA, among others, is similar to that applied for gold recovery from concentrated solutions, as described in Section 8.3. However, because relatively dilute metal cyanide solutions are treated, electrowinning current efficiencies are typically very poor, and cells utilizing high surface area cathodes and agitated electrolytes must be used to increase mass transport rates.

The major reactions at the cathode are metal reduction (shown here for the copper complex) and hydrogen evolution:

$$Cu(CN)_3^{2-} + 2e \rightleftharpoons Cu + 3CN^-; \quad E^0 = -1.00 \text{ (V) for [KCN]} = 7 \text{ M} \quad \text{(EQ 11.9)}$$

$$2H^+ + 2e \rightleftharpoons H_2; \quad E^0 = 0.00 \text{ (V)} \quad \text{(EQ 11.10)}$$

The recovery of other metals at the cathode depends on the electrode potential of the specific reduction reactions (see Section 8.3).

Oxygen is evolved at the anode, and thiocyanate ions are oxidized to sulfate, liberating free cyanide:

$$2H_2O + O_2 + 4e \rightleftharpoons 4OH^-; \quad E^0 = 0.40 \text{ (V)} \quad \text{(EQ 11.11)}$$

$$8H^+ + SO_4^{2-} + CN^- + 4e \rightleftharpoons SCN^- + 4H_2O \quad \text{(EQ 11.12)}$$

Cyanide and thiocyanate species may also be oxidized to cyanate at the anode.

The process is best suited to solutions containing relatively high concentrations of copper where higher current efficiencies are obtained, and for which a greater economic benefit can be achieved by recycling a larger quantity of cyanide. It is not appropriate for the removal of free and complexed cyanide species down to very low concentrations, because of decreased electrowinning efficiency under these conditions, and further detoxification is likely to be required to meet discharge requirements.

An electrochlorination process in which sodium chloride is added to the electrolyte of a standard electrowinning system has been proposed for cyanide destruction. The salt is oxidized to chlorine at the anode, producing hypochlorite ions in solution, which oxidize both free cyanide and thiocyanate to cyanate (see Section 11.3.2.5). The system is operated at 40°C to 50°C to minimize the formation of chlorate ions, which would otherwise be formed at the expense of the active hypochlorite species [12].

11.2.6 Sulfide Precipitation

The addition of sulfide ions into cyanide-containing solution or slurry can be used to precipitate a metal sulfide and free up the cyanide that was associated with the metal complex. Two such processes have been proposed. The first, the MNR process (developed by Metallgesellschaft Natural Resources, Germany), has been proposed for the recovery of cyanide from copper-containing solutions [22]. Sulfide ions are added to the solution or slurry after acidification to a pH of 3 to 4, resulting in the precipitation of Cu(I) sulfide, as follows:

$$2Cu(CN)_3^{2-} + S^{2-} + 3H_2SO_4 \rightleftharpoons Cu_2S + 6HCN + 3SO_4^{2-} \quad \text{(EQ 11.13)}$$

The solubility product of the Cu_2S ($K_{sp} = 10^{-48}$) is much lower than that of CuCN ($K_{sp} = 10^{-20}$), forcing the equilibrium of Equation (11.13) to the right. It is possible that this technique could be combined with the AVR-type process (see Section 11.2.2) to provide for more efficient recovery of copper from the treated process stream, avoiding the presence of CuCN in the treated process stream.

The second method, called the Velardena process (developed at the Velardena mine in Durango, Mexico), uses the addition of sulfide ions to solution (or slurry) containing relatively high concentrations of zinc cyanide. In this case, the zinc was derived from cyanide-soluble zinc minerals in the ore feed. Zinc precipitation was conducted at alkaline pH using 0.33 mol Na_2S per mol NaCN, according to the following reaction:

$$Zn(CN)_4^{2-} + S^{2-} \rightleftharpoons ZnS + 4CN^- \tag{EQ 11.14}$$

The treated solution stream containing both the regenerated free cyanide and the zinc sulfide precipitate was recycled directly to the leach circuit without solid–liquid separation. Reportedly, the zinc sulfide passed through the leaching and CCD circuit without any significant redissolution occurring [23].

11.2.7 Ion Precipitate Flotation

It is possible to recover complex anions, such as $Fe(CN)_6^{4-}$ and $Ni(CN)_4^{2-}$, from solution by flotation with suitable ionizable surface-active cationic collectors. Similarly, colloidal precipitates, such as CuCN (formed by acidification of copper-containing cyanide solution) and double salts of Fe(II) cyanides, can also be recovered by flotation. To some extent, the processes are interactive, because both types of species are likely to be present in cyanide solution systems and colloidal particles are susceptible to becoming charged [12, 24]. Because these processes have not been applied commercially in gold extraction, they are of academic interest only.

11.3 DETOXIFICATION

Detoxification processes are used to reduce the concentrations of toxic constituents in tailings streams and process solutions, either by dilution, removal, or conversion to a less toxic chemical form (sometimes referred to as "destruction" or "degradation" in the case of toxic cyanide species). The objective is to produce an effluent that meets limits or guidelines that have been set to conform with the environmental and/or metallurgical requirements of the project. This section discusses methods of detoxification which may be applied individually or as combinations of processes (e.g., with reagent and metal recovery processes) to achieve the detoxification objectives.

The most important constituents are cyanide species and heavy metal ions, which are considered in the following sections. The control of other solution components and conditions (e.g., sulfate, chloride, and nitrate concentrations, pH, conductivity, turbidity, etc.) may also be important under particular circumstances; however, specific methods for their control are not considered further here.

11.3.1 Dilution

Where applicable, dilution is the simplest, quickest, and cheapest form of detoxification to meet a specified control limit, although, when this is applied to an effluent discharge, the total amount of toxic components discharged may not be reduced. In order to achieve dilution, freshwater, or solution that contains very low levels of the toxic species of interest, is mixed with the process solution in sufficient quantity to reduce the concentrations of toxic constituents below the desired or regulated levels, as shown in Figure 11.6(a) and (b). Dilution may either be used as the sole means of detoxification or in combination with other technology (Sections 11.3.2 and 11.3.3).

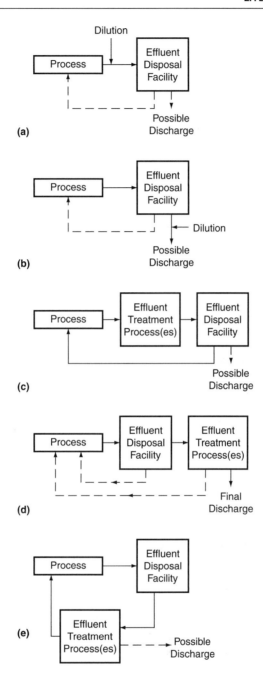

FIGURE 11.6 Simplified flowsheets of major options for effluent treatment: (a) dilution before recycle, (b) dilution after recycle, (c) treatment of whole effluent stream prior to disposal, (d) treatment of effluent solution for discharge, (e) treatment of effluent solution for recycle to process

11.3.2 Cyanide

The methods available for cyanide detoxification can be grouped in the following categories:

- Cyanide removal (see also Section 11.2 for cyanide recovery)
    - Natural volatilization
    - Adsorption onto minerals
- Oxidation to the less toxic cyanide species
    - Natural oxidation
    - Iron complexation
    - Hydrogen peroxide process
    - Sulfur dioxide–assisted process
    - Alkaline chlorination process
    - Biological oxidation
    - Ozonation
- Complexation to less toxic Fe(II) cyanide, with the potential for removal as an insoluble double salt

These methods are considered in detail in Sections 11.3.2.1 to 11.3.2.8, and a comparison of methods available for cyanide removal and/or detoxification is given in Table 11.3.

11.3.2.1 Natural Degradation

Natural degradation processes reduce the toxicity of cyanide species over time. A variety of mechanisms are responsible for this, including volatilization, oxidation, adsorption onto other minerals, hydrolysis, biodegradation, and precipitation, as illustrated in Figure 11.7. Although these processes are effective, they do not always have sufficiently fast kinetics for industrial purposes, and other detoxification methods must often be applied (see Sections 11.3.2.2 to 11.3.2.8). In addition, the rates of degradation vary for different cyanide species and within different solution systems, and consequently it is difficult to accurately predict the capabilities of natural detoxification processes for effective effluent control. Nevertheless, significant natural degradation of cyanide can occur in some processes, notably in solution ponds, heap leach pads, tailings storage facilities, and leaching and carbon adsorption systems. Full advantage should be taken of natural degradation in the design and application of detoxification systems. An example of the degradation of various cyanide species in a solution pond over time is shown in Figure 11.8.

The most important mechanisms of natural degradation are considered in more detail in the following sections.

Volatilization. Free cyanide exists in equilibrium with hydrogen cyanide in aqueous solution. This equilibrium has been discussed in detail in Section 6.1.1. Because of its low boiling point (79°C) and high vapor pressure (100 kPa at 26°C), HCN volatilizes at the solution–air interface and diffuses into the atmosphere. The rate of volatilization of HCN from cyanide-containing solution increases with the following:

- Decreasing pH (see Figure 6.1)
- Increasing temperature
- Increasing aeration of solution
- Increasing solution agitation
- Increasing surface-area-to-depth ratio of the body of solution

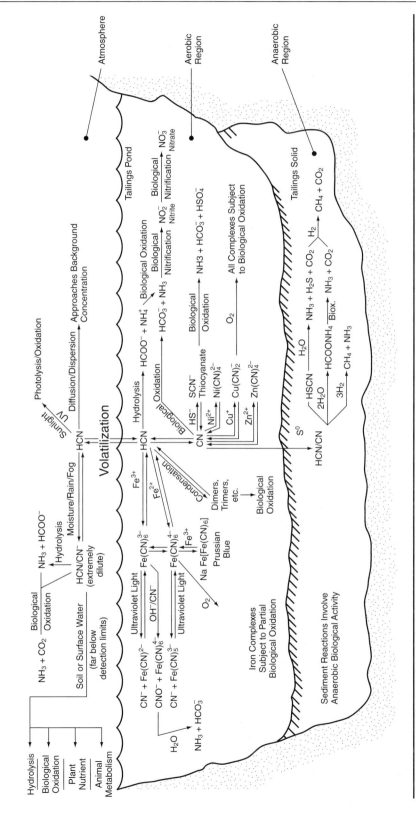

FIGURE 11.7 Cyanide cycle: natural cyanide degradation mechanisms [25]

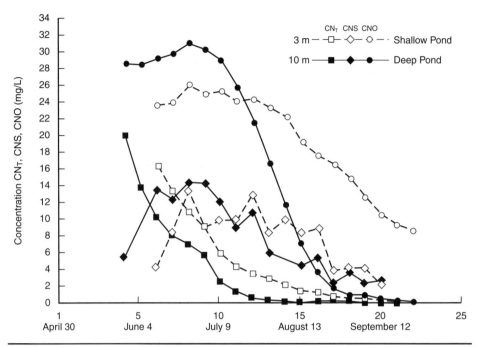

FIGURE 11.8 Concentrations of total cyanide, thiocyanate, and cyanate species in shallow (3 m) and deep (10 m) solution pond systems (1980 data) [26]

In addition, the reaction is catalyzed by several materials, including copper and activated carbon [21].

The rate of degradation can be expressed as a cyanide half-life, which varies from a few hours for aerated slurries in the presence of carbon [21] to several months for stagnant ponds or tailings impoundments at low temperatures [27], depending on solution conditions. In many cases, solution and process conditions can be manipulated to maximize cyanide detoxification by volatilization, for example, by reducing solution depths in ponds, modifying pH, and even by spraying solution in the air. Natural methods of pH reduction by the dilution of alkaline solutions with rainwater, coupled with the adsorption of carbon dioxide from the atmosphere into solutions, also assist in cyanide loss by volatilization.

Atmospheric oxidation. In the absence of a catalyst, dissolved oxygen in process solutions oxidizes cyanide to cyanate extremely slowly. The reaction is accelerated in the presence of as little as 20 g/L activated carbon. The mechanism is thought to involve chemisorption of oxygen onto the carbon surface to form peroxide and carboxylic acid groups, which are more reactive with cyanide [21]. Some degradation of cyanide is thought to occur in carbon adsorption systems by this mechanism; however, in tailings and solution ponds the extent of reaction is probably negligible.

Adsorption by other minerals. The adsorption of cyanide by mineral components of soils, ores, and wastes was the subject of extensive research during the 1980s and 1990s. Many minerals are known to attenuate cyanide and cyanide compounds; for example, it has been shown that some carbonaceous materials can adsorb up to 0.5 mg CN^-/g, and feldspars and clay minerals up to approximately 0.05 mg CN^-/g. Bauxite, ilmenite, and hematite also have known cyanide adsorption ability but with complex and poorly understood adsorption mechanisms [28].

This method of removal of cyanide species from solutions is important in the longer-term consideration of the fate of cyanide in slurry tailings, heap leach pads, and where effluent solutions are used for land application (i.e., distribution onto a vegetated area).

Other natural mechanisms. Other natural mechanisms, such as biological oxidation and hydrolysis reactions resulting in precipitation and coprecipitation of solution species, account for varying degrees of cyanide degradation, depending on specific conditions. Photo-degradation is also known to assist in breakdown of some cyanide complexes (especially iron cyanide) and may also catalyze the oxidation of free cyanide to cyanate. Hydrolysis of cyanide to ammonium formate occurs slowly at 25°C and does not account for significant degradation. The reaction is accelerated at elevated temperatures, as indicated by the well-known phenomenon of ammonia formation during high-temperature carbon elution (Section 7.1.3).

11.3.2.2 Iron Complexation

Free cyanide can be converted to the less toxic Fe(II) cyanide complex by the addition of an excess of a suitable soluble Fe(II) salt, such as Fe(II) sulfate ($FeSO_4 \cdot 7H_2O$), within the pH range 7.5 to 10.5 [29, 30]:

$$6CN^- + FeSO_4 = Fe(CN)_6^{4-} + SO_4^{2-} \qquad (EQ\ 11.15)$$

The weak metal cyanide complexes (i.e., copper, zinc, and nickel) are partially decomposed in the presence of excess iron, with the degree of decomposition dependent on the relative stability constants of the species involved (see Tables 6.2 and 6.3), for example:

$$2Cu(CN)_3^{2-} + FeSO_4 = Cu_2Fe(CN)_6 + SO_4^{2-} + 2e \qquad (EQ\ 11.16)$$

The process is rather ineffective compared with other cyanide detoxification methods, because the reaction is slow and the displacement reactions are inefficient due to the chemical equilibria established between the different metal cyanide complexes. Consequently, it is difficult to reduce free cyanide concentration below approximately 2.5 mg/L [31]. Despite this, the method has been applied industrially as a crude but cost-effective method of reducing free cyanide concentrations. Between 15 and 20 mol Fe(II) sulfate are required to neutralize 1 mol CN^- [31].

11.3.2.3 Hydrogen Peroxide Oxidation

Hydrogen peroxide has been used widely in industry for cyanide detoxification, for example, Paradise Peak (Nevada, United States). The process has the advantage over techniques that use chlorine (Section 11.3.2.5) or sulfur dioxide (Section 11.3.2.4) in that no foreign ions are introduced into process solutions, unless a catalyst is used, and the kinetics of oxidation are sufficiently fast that effective oxidation can generally be achieved in a few minutes. However, detoxification to the low-residual cyanide concentrations usually required for an effluent discharge is costly.

Hydrogen peroxide oxidizes free cyanide to cyanate as follows:

$$CN^- + H_2O_2 \rightleftharpoons CNO^- + H_2O \qquad (EQ\ 11.17)$$

In alkaline solution the other WAD metal cyanide complexes, for example, copper, nickel, and zinc, are oxidized to cyanate and the metal hydroxide [32, 33]. The reaction for the predominant copper cyanide complex, $Cu(CN)_3^{2-}$, is as follows:

$$2Cu(CN)_3^{2-} + 7H_2O_2 + 2OH^- \rightleftharpoons 6CNO^- + 2Cu(OH)_2 + 6H_2O \qquad (EQ\ 11.18)$$

The strong metal cyanide complexes are not broken down, but Fe(II) cyanides can be removed as insoluble copper or zinc double salts by the addition of soluble copper or zinc sulfate (see Section 11.3.3).

Cyanate ions may undergo hydrolysis to form ammonium and carbonate ions, although this only occurs to any significant extent below pH 7 at ambient temperatures [33]:

$$CNO^- + 2H_2O \rightleftharpoons NH_4^+ + CO_3^{2-} \quad \text{(EQ 11.19)}$$

Between 10% and 15% of the cyanate ions may react in this manner.

Thiocyanate ions are oxidized slowly by hydrogen peroxide (in contrast to chlorination), and probably only a small percentage of thiocyanate species are oxidized under conditions normally applied for hydrogen peroxide oxidation. Some gaseous ammonia is always formed by reaction of hydrogen peroxide in cyanide solutions, and this may have to be removed to meet drinking water specifications [20].

The cyanide ion oxidation reaction is accelerated in the presence of catalysts such as copper ions and formaldehyde. For example, the addition of as little as 5 to 10 mg/L formaldehyde has been shown to reduce oxidation time by as much as 40% [34]. The relevant reaction for cyanide and formaldehyde is:

$$CN^- + 2HOCH + H_2O \rightleftharpoons HOCH_2CN + OH^- \quad \text{(EQ 11.20)}$$

In the presence of hydrogen peroxide the glycolonitrile formed hydrolyses to produce glycol acid amide as follows:

$$HOCH_2CN + H_2O \rightleftharpoons HOCH_2CONH_2 \quad \text{(EQ 11.21)}$$

Copper additions of one tenth of the concentration of WAD cyanide species produce similar results.

The efficiency of hydrogen peroxide cyanide destruction has been demonstrated repeatedly, and it is well known that solutions containing 500 mg/L WAD cyanide, or more, can be reduced to <2 mg/L within practical time scales, for example, 1 to 2 hr, by the addition of 75 to 125 mg/L H_2O_2 [32, 33, 34]. Final effluents containing <0.1 mg/L CN^- can be produced at higher peroxide dosages, but the economics of this are generally unfavorable. An example of the performance of hydrogen peroxide for removal of various solution species is given in Table 11.5.

Hydrogen peroxide consumption is estimated to be approximately 3 kg H_2O_2/kg CN^-. However, other species present in the solution or slurry may compete with cyanide for peroxide and may increase consumption by direct reduction and/or by catalytic action.

Caro's acid (H_2SO_5), which is made by mixing hydrogen peroxide and sulfuric acid, has been used at some mine sites for rapid and effective free and WAD cyanide destruction, including Lone Tree (Nevada, United States). The reaction for free cyanide is as follows:

$$H_2SO_5 + CN^- = H_2SO_4 + CNO^- \quad \text{(EQ 11.22)}$$

Caro's acid has been demonstrated to be effective at reducing free and WAD cyanide concentrations <50 mg/L very rapidly, often within minutes, and more rapidly than hydrogen peroxide alone. However, Caro's acid must be prepared at the site due to its instability, and to be effective it must be used very quickly after preparation (i.e., within a few seconds) [35].

TABLE 11.5 Example of feed and discharge concentrations of selected species for hydrogen peroxide detoxification [20]

	Feed (mg/L)	Effluent (mg/L)	USEPA DWS* (mg/L)
Arsenic	0.2	<0.050	0.01
Copper	4.5	<1.000	1.30
Cyanide	280.0	3.000	0.20
Iron	16.0	<0.015	0.30
Selenium	5.0	4.000	0.05
Silver	3.2	1.000	0.10
Zinc	157.0	<1.000	5.00

NOTE: H_2O_2 dosage = 2.5 mL/L.
* DWS = drinking water standard (primary or secondary).

11.3.2.4 Sulfur Dioxide–Assisted Oxidation

Throughout much of the 20th century, sulfur dioxide (SO_2) was used in various forms for the oxidation of cyanide species. One particular version of the process (Inco, Sudbury, Ontario, Canada) was patented in the early 1980s and has been applied at several operations, including Scottie, Carolin, Baker, and McBean (all in Canada).

A mixture of sulfur dioxide and air will rapidly oxidize free cyanide and WAD metal cyanide complexes in aqueous solution in the presence of Cu(II) ions as a catalyst [36]:

$$CN^- + SO_2 + O_2 + H_2O \rightleftharpoons CNO^- + H_2SO_4 \qquad (EQ\ 11.23)$$

The optimum pH for the reaction is 9, but it will proceed to a significant extent between pH 7.5 and 9.5. The preferred volume percentage of sulfur dioxide in air is 1% to 2%, but up to 10% will oxidize satisfactorily. Sulfur dioxide can be supplied in gaseous or liquid forms, or by burning elemental sulfur. Other suitable sources of sulfur dioxide for the process include sodium metabisulfite ($Na_2S_2O_5$) and sodium sulfite (Na_2SO_3).

Thiocyanate ions are also oxidized by a similar, but slower, reaction, which is catalyzed by nickel ions and, to a lesser degree, copper and cobalt [37]:

$$SCN^- + 4SO_2 + 4O_2 + 5H_2O \rightleftharpoons CNO^- + 5H_2SO_4 \qquad (EQ\ 11.24)$$

Typically <10% of the thiocyanate ions present are oxidized by the copper-catalyzed process.

As for the hydrogen peroxide process, any iron present remains in the reduced form, Fe(II), and may be precipitated as a double cyanide salt with zinc, copper, or nickel, as per Equation (11.16). Base metals concentrations in excess of that required for double salt formation may be precipitated as hydroxides, depending on their concentration and solution conditions.

The process is usually performed in several stages with the addition of 30 to 90 g Cu^{2+}/t in the first stage, followed by bubbling in the sulfur dioxide–air mixture, or by adding sodium metabisulfite and agitating with air in subsequent stages. Air flow rates of approximately 1 to 2 L/min per liter of solution are maintained. In practice, 3 to 4 kg SO_2^- equivalent or 5 to 8 kg $Na_2S_2O_5$ is required per kilogram of free cyanide. Lime is used for pH control, because the oxidation reactions generate sulfuric acid [37].

The process has been used to treat gold plant effluents containing >200 mg/L total cyanide and routinely reduces this to <1 mg/L, and occasionally to <0.05 mg/L. Copper, nickel, zinc, and iron concentrations can be reduced to very low levels also, that is, <2, <1, <1, and <0.5 mg/L for each of the metals, respectively. An example of concentrations

TABLE 11.6 Example of feed and discharge concentrations of selected species for the Inco–SO_2 detoxification process [38]

	Barren Solution (mg/L)	Treated Barren Solution (mg/L)	Final Tailings (mg/L)	Tailings Overflow (mg/L)
Copper	35.45	1.21	7.70	3.10
Cyanate	44.60	324.20	52.90	20.50
Iron	45.82	0.26	5.35	1.60
Nickel	4.13	0.25	0.51	0.32
Thiocyanate	100.70	82.50	18.90	8.30
Total cyanide	365.80	0.69	29.40	8.80
WAD cyanide	225.00	0.15	12.20	5.20
Zinc	62.00	0.20	0.00	0.00

of various species in the feed and discharge to the process are given in Table 11.6. The high efficiency of iron removal is attributed to the low oxidizing potential of the system, which keeps iron in the reduced Fe(II) state. This promotes the formation and precipitation of insoluble double salts of Fe(II) cyanide. However, an important disadvantage of the process is the introduction of sulfate ions, which may increase total dissolved solids concentrations significantly (see Table 11.2) [38].

11.3.2.5 Alkaline Chlorine–Hypochlorite Oxidation

Chlorine was used for cyanide destruction in the early days of cyanidation in the late 1800s, because chlorine and its derivatives were readily available in the industry at that time. The method has been applied ever since in a variety of forms.

The active reagent for chlorine oxidation of free and complexed cyanide is the hypochlorite ion, produced when chlorine dissolves in water, as described in Section 5.6. Alternatively, hypochlorite ions can be produced by dissolving suitable salts, such as sodium or calcium hypochlorite, in water.

Free cyanide reacts rapidly with hypochlorite (OCl) in aqueous solution to form cyanogen chloride, otherwise known as tear gas:

$$CN^- + H_2O + ClO^- \rightleftharpoons CNCl(g) + 2OH^- \qquad \text{(EQ 11.25)}$$

Cyanide also reacts rapidly with free chlorine, as follows:

$$CN^- + Cl_2 \rightleftharpoons CNCl(g) + Cl^- \qquad \text{(EQ 11.26)}$$

However, at high pH, cyanogen chloride is readily hydrolyzed to cyanate and chloride ions:

$$CNCl + 2OH^- \rightleftharpoons CNO^- + Cl^- + H_2O \qquad \text{(EQ 11.27)}$$

In practice, effective oxidation of cyanide can be achieved in 10 to 15 min. The rate of the initial oxidation reaction shown in Equation (11.25) is reduced significantly as the pH is increased >11, and the oxidation process is usually carried out at pH 10 to 11, which is high enough to avoid significant cyanogen chloride formation [39]. The hydrolysis reaction consumes hydroxide, and supplemental alkali must be added if the feed to chlorination is insufficiently buffered. The use of sodium or calcium hypochlorite may obviate the need for alkali addition.

If the hypochlorite ion concentration is sufficiently high, nitrogen and carbon dioxide may be produced [40]:

$$2CNO^- + 3ClO^- + H_2O \rightleftharpoons N_2 + 2CO_2 + 3Cl^- + 2OH^- \quad \text{(EQ 11.28)}$$

However, this is normally beyond the range of hypochlorite concentrations applied in practice.

Thiocyanate ions are dissociated by chlorine (and hypochlorite species) to form cyanate as follows:

$$SCN^- + Cl_2 + 4OH^- \rightleftharpoons CNO^- + SO_4^{2-} + 5H_2O + 2Cl^- \quad \text{(EQ 11.29)}$$

However, the oxidation of thiocyanate can produce free cyanide as an intermediate product. Therefore, all of the thiocyanate present must therefore be oxidized before the free cyanide concentration can be reduced effectively.

Under typical chlorination conditions, weakly complexed (WAD) metal cyanides are also oxidized. The reactions for zinc and copper are as follows [12]:

$$Zn(CN)_4^{2-} + 4ClO^- + 2OH^- \rightleftharpoons 4CNO^- + 4Cl^- + Zn(OH)_2 \quad \text{(EQ 11.30)}$$

$$2Cu(CN)_3^{2-} + 7ClO^- + 2OH^- + H_2O \rightleftharpoons 6CNO^- + 7Cl^- + 2Cu(OH)_2 \quad \text{(EQ 11.31)}$$

Stronger cyanide complexes, such as those of iron, cobalt, and gold, are stable under these oxidation conditions and do not dissociate. Fe(II) may be oxidized to Fe(III), depending on the redox potential, but chlorination generally does not allow effective precipitation of iron cyanide species. Table 11.7 shows the concentrations of various species in the feed and discharge of one particular chlorination detoxification system.

The chlorine and hypochlorite species also react with a number of other species that frequently occur in effluent streams, such as organic compounds, sulfides, and thiocyanates. If present in large quantities, these can result in very high chlorine consumptions, which can severely affect the economics of the process. Chlorine consumptions of approximately 8 to 24 kg Cl_2/kg CN^- are required in practice.

The process is difficult to control for optimum chemical utilization, because the efficient oxidation of toxic species and the necessary minimization of residual free chloride concentrations are conflicting objectives. The reduction of toxic cyanide concentrations to <0.1 mg/L WAD cyanide can usually only be achieved if the residual active chlorine concentration is in the range 10 to 15 mg/L. This level of chlorine is very toxic to fish, making the solution unsuitable for direct discharge in some cases. If necessary, the solution must be dechlorinated, for example, with sodium hydrosulfide (NaHS) or with hydrogen peroxide, as practiced at Yanacocha (Peru) [41].

11.3.2.6 Biological Oxidation

The ability of certain strains of bacteria to degrade a variety of forms of cyanide and ammonia, and to accumulate and ingest heavy metals, has been known for many years. The process was first applied on a large scale for treatment of gold cyanide leaching effluent at the Homestake Lead plant (South Dakota, United States) in 1984 and has operated very successfully since that time [25]. The flowsheet for this system is given in Figure 11.9. The process requires the gradual acclimatization of bacteria to the concentrations of free cyanide, thiocyanate, and heavy metals that are to be treated in the process stream. At the Homestake Lead operation, the major bacterium used for cyanide degradation is a *Pseudomonas* rod-type strain, 0.7 to 1.4 μm in length, which achieves optimum growth at 30°C and within a pH range of 7.0 to 8.5. The bacteria are capable of tolerating a wide range of conditions and can survive, for example, over a wide temperature range, from 5°C to 42°C.

TABLE 11.7 Example of feed and discharge concentrations of selected species for sodium hypochloride detoxification [20]

Element	Feed (mg/L)	Effluent (mg/L)	USEPA DWS* (mg/L)
Arsenic	0.2	<0.01	0.05
Copper	4.5	<0.50	1.00
Cyanide	280.0	0.100	0.05
Iron	16.0	0.15	0.30
Selenium	5.0	4.00	0.01
Silver	3.2	0.20	0.10
Zinc	157.0	<4.00	5.00

* DWS = drinking water standard.

The rate of oxidation of cyanide to cyanate is increased by the bacteria, since the carbon and nitrogen are nutrients:

$$2CN^- + O_2 \rightleftharpoons 2CNO^- \quad \text{(EQ 11.32)}$$

Cyanate may then be hydrolyzed to carbonate and ammonium species as given by Equation (11.9). Nitrates and nitrites do not react under these conditions and no hydrogen sulfide is produced.

Metal cyanide complexes are also oxidized and the metals are adsorbed, ingested, and precipitated by the bacteria, for example:

$$M(CN)_4^{2-} + 8H_2O + 2O_2 + 2e \text{ (+ bacteria)} \rightleftharpoons \text{M-biofilm} + 4HCO_3^- + 4NH_3 \quad \text{(EQ 11.33)}$$

where

M = Zn, Cu, Fe, Ni, etc. (although coordination number of the cyanide complex varies for the different metals, and the equation must be balanced accordingly)

The rate of degradation of metal cyanide complexes by this mechanism decreases in the order Zn > Ni > Cu > Fe, but even the strongly complexed iron cyanide species are degraded and adsorbed by the bacteria.

Thiocyanate is rapidly oxidized in the presence of bacteria, as follows:

$$SCN^- + 2O_2 + 3OH^- \rightleftharpoons SO_4^{2-} + CO_3^{2-} + NH_3 \quad \text{(EQ 11.34)}$$

The bacteria derive some of their food and energy from the oxidation of cyanide and thiocyanate but may also require additional nutrients, such as phosphoric acid and sodium carbonate.

A second stage of biological treatment is used to remove the ammonia produced by the oxidation of cyanide. This is called nitrification and is achieved using a common pair of aerobic, autotrophic bacteria supplied with suitable inorganic nutrients. The relevant reactions are as follows:

$$2NH_4^+ + 3O_2 \rightleftharpoons 2NO_2^- + 4H^+ + 2H_2O \quad \text{(EQ 11.35)}$$

$$2NO_2^- + O_2 \rightleftharpoons 2NO_3^- \quad \text{(EQ 11.36)}$$

The first of these reactions (Equation [11.35]) is slow, much slower than the oxidation of nitrite to nitrate (Equation [11.36]). As a result, the recovery rate of the nitrifying bacteria

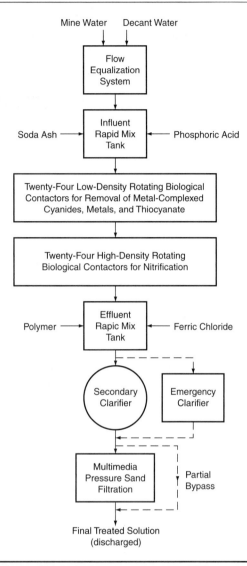

FIGURE 11.9 Homestake Lead biological treatment process [25]

is much slower than the recovery rate of cyanide degradation bacteria. Figure 11.10 shows the degradation of cyanide, thiocyanate, and ammonia through the detoxification system. The reaction rates for cyanide, thiocyanate, and ammonia oxidation are zero order down to very low concentrations.

Many of the reactions achieved by biological action cannot be duplicated by chemical processes at the rates accomplished by the microorganisms. Because of this, and because bacterial oxidation is a natural process, biologically treated effluent is more likely to be compatible with the water system into which it is to be discharged than effluents produced by other processes described in this section. This has been the experience at the Homestake Lead plant: Biological degradation and adsorption into the "biomass" removes >92% of the total cyanide, 99% of the WAD cyanide species, and >95% of copper and other toxic heavy metals. A "bio-assay" technique is used to assess water quality,

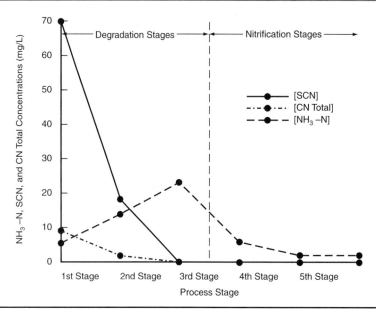

FIGURE 11.10 Biological degradation of ammonia, cyanide, and thiocyanate at Homestake Lead (South Dakota, United States) [25]

which is based on the actual effect of the treatment process product on living organisms, such as trout, daphnia (one of the most sensitive test organisms to heavy metals), and macroinvertebrates [42].

At the Homestake Lead operation, the process is performed in a series of 48 rotating biological contactors, in conjunction with clarification and filtration equipment, to treat a maximum of 800 t/hr of mixed mine water and tailings impoundment solution effluent. The treatment process product is discharged directly into a trout stream. The operating and capital costs of the system are competitive with chemical oxidation methods, estimated at US$0.35/t and US$5 million (1990 constant dollar basis), respectively, for a 250 t/hr system to achieve similar product quality.

11.3.2.7 Ozone

On occasion, ozone has been used for cyanide destruction. A mixture of 3% ozone in oxygen has been found to produce strongly oxidizing conditions when bubbled through aqueous solution. This will completely break down free and complexed cyanides, even Fe(II) and Fe(III) complexes, albeit slowly [12]. The specific advantages of ozone are that it will also break down oils and phenols, for example, organic constituents of flotation plant tailings. However, the process is very expensive to apply in practice because of the high consumption and high cost of ozone.

11.3.2.8 Ultrasonics

Researchers in the former USSR have shown that ultrasonics can be used to enhance the oxidation rate of cyanide species in the presence of suitable oxidants, for example, chlorate ions and ozone. The method of application and economics of this process are not known.

11.3.3 Metals

Processes that recover metals in a form that allows them to be sold, recycled, or safely stored may be preferred for detoxification of solutions containing unacceptably high concentrations of toxic metals. These processes, which include carbon adsorption and ion exchange, are discussed further in Section 11.2, and carbon adsorption and ion exchange for gold recovery are considered in detail in Chapter 7. Other processes for metals recovery and removal are also available [3, 5, 43].

Where these processes cannot be applied successfully, other means of detoxification to convert the species into less hazardous forms in situ must be considered, for example, biological oxidation (Section 11.3.2.6) and precipitation (Section 11.3.3.1).

11.3.3.1 Precipitation

Precipitation of metal species from cyanide solutions can be achieved by one of the following:

- Destruction of free cyanide and the metal cyanide complex, with alkaline precipitation of the metal hydroxide
- Acidification and precipitation of simple cyanide salt(s) and/or double salts
- Reduction of the metal cyanide with a suitable reductant, for example, hydrogen, sulfur dioxide, or hydrogen sulfide, and precipitation of the metal or metal sulfide

The effectiveness of each method depends on the metal ion under consideration and other factors that affect the solubility of the reaction products, such as temperature and solution composition.

The cheapest method of removing metals depends on their original concentration, the required final concentration, the solution conditions, and the speed at which removal must be effected. Table 11.8 shows estimations of final metal concentrations of a variety of species that can be achieved for a number of different treatment methods.

The precipitation of metal species that form stable complexes with cyanide (i.e., cobalt, iron, silver, copper, nickel, and zinc) is most easily and efficiently achieved after cyanide removal or destruction, during which some degree of metal precipitation occurs, depending on solution pH, temperature, and other solution conditions.

Gradual acidification of cyanide solutions produces sequential precipitation of metal cyanide salts, as shown in Figure 11.11. This is of limited practical use for detoxification because these cyanide salts usually redissolve to some extent as the pH is increased again. One exception is copper, which is precipitated below approximately pH 2.5 and is relatively insoluble as the pH is increased, even up to pH 10 [12, 44].

The precipitation of metals with a reductant is generally applied as a last resort for detoxification because of the high cost of the reducing agent.

Precipitation processes are not always capable of producing an effluent that meets discharge requirements, and additional treatment, such as freshwater dilution or metals removal by carbon adsorption, ion exchange, or biological processes, may be necessary.

A most important aspect of metals precipitation is the stability of the product formed. If the precipitate is to be removed from the solution by solid–liquid separation (using coagulants and flocculants if necessary) and treated further, then an unstable product may be tolerated and possibly even desired. Alternatively, if the product is to be disposed of in an impoundment, waste facility or effluent pond, then the long-term stability of the product must be considered.

Iron. Iron is not considered to be a major problem in most gold extraction effluents because of its relatively low toxicity. However, it can tie up cyanide as stable complexes (see Table 6.2), which may later be released by photo-degradation of the Fe(II) and Fe(III) cyanide complexes (see also Sections 11.3.2.1 and 11.3.2.2) [46]. High concentrations of iron can build up in process streams due to the corrosion of grinding media and associated process equipment (e.g., mild steel components). This can cause problems for some extraction processes, in particular biological oxidation and other acidic sulfide oxidation processes, where iron cyanide-containing solutions are recycled to the

TABLE 11.8 Residual metal ion concentrations achievable in solution for various chemical treatment processes [45]

	Final Concentrations (mg/L)						
Element	Lime + Settling	Lime + Filter	Sulfide + Filter	Ferrite Coprecipitation + Filter	Soda Ash + Settling	Soda Ash + Filter	Aluminum
Antimony	0.8–1.5	0.4–0.8	—	—	—	—	—
Arsenic (V)	0.5–1.0	0.5–1.0	0.05–0.1	—	—	—	—
Beryllium	0.1–0.5	0.01–0.1	—	—	—	—	—
Cadmium	0.1–0.5	0.05–0.1	0.01–0.1	<0.05	—	—	—
Copper	0.5–1.0	0.4–0.7	0.05–0.5	<0.05	—	—	—
Chromium (III)	0.1–0.5	0.05–0.5	—	0.01	—	—	—
Lead	0.3–1.6	0.05–0.6	0.05–0.4	0.20	0.4–0.8	0.1–0.6	—
Mercury (II)	—	—	0.01–0.05	<0.01	—	—	—
Nickel	0.2–1.5	0.1–0.5	0.05–0.5	—	—	—	—
Silver	0.4–0.8	0.2–0.4	0.05–0.2	—	—	—	—
Selenium	0.2–1.0	0.1–0.5	—	—	—	–	–
Thallium	0.2–1.0	0.1–0.5	—	—	—	—	0.2–0.5
Zinc	0.5–1.5	0.4–1.2	0.02–1.2	0.02–0.5	—	—	—

	Final Concentrations (mg/L)			
	Ferric Chloride	Activated Carbon	Bisulfite Reduction	Lime/FeCl$_2$ Filter
Arsenic (V)	0.05–0.5	0.3	—	0.02–0.1
Chromium (VI)	—	0.1	0.05–0.5	—
Mercury (II)	—	0.01	—	—
Silver	0.05–0.1	—	—	—
Selenium	0.05–0.1	—	—	—
Thallium	0.7	—	—	—

NOTE: Dashes = not measured.

process. In cases where unacceptably high iron concentrations are obtained, the iron must be removed.

Fe(II) cyanide can be precipitated from solution by addition of soluble copper or zinc salts to form the respective metal–iron cyanide double salt, $Cu_2Fe(CN)_6$ or $Zn_2Fe(CN)_6$. The reaction is best performed at, or slightly above, pH 8.5 to ensure that, in the absence of free cyanide, any excess copper or zinc is precipitated as the hydroxide. The double salts that are formed are considered to be stable <pH 9.

Fe(III) cyanide is precipitated by the addition of excess Fe(II) sulfate to produce the iron cyanide double salt, $Fe_3[Fe(CN)_6]_2$, which is insoluble over a wide pH range [45].

Copper, zinc, and nickel. Copper, zinc, and nickel are readily precipitated as hydroxides following free and WAD cyanide destruction, and by raising the pH >10 with lime. The precipitate can be removed by solid–liquid separation, if required, although this is often difficult due to the colloidal nature of the precipitate formed. The precipitate is relatively stable at alkaline pH but decomposes to varying degrees as the pH is reduced below neutral [12].

Lead. Lead does not form stable complexes with cyanide but can exist as a variety of ionic species in alkaline cyanide solution. It can be removed in a relatively stable form by precipitation with lime at pH 10 to 11 [12].

Arsenic and antimony. Arsenic and antimony are metalloids which form arsenites (AsO_2^-) and antimonites (SbO_2^-), respectively, in alkaline cyanide solutions, rather than forming cyanide complexes (see Section 6.1.4.4). In solutions with higher redox potential, that is, solutions produced by oxidative pretreatment, the higher oxidation state

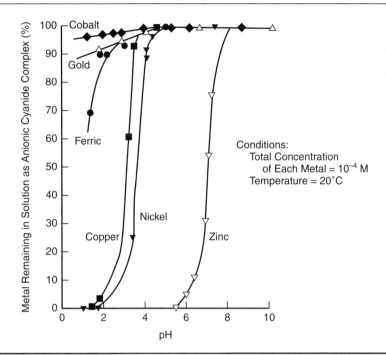

FIGURE 11.11 Effect of acidity on the stability of various metal cyanide complexes in aqueous solution [44]

arsenates (AsO_4^{3-}) and antimonates (SbO_4^{3-}) may be present. Alternatively, a suitable oxidant, such as potassium permanganate may be added to achieve the high redox potential necessary to oxidize arsenic and antimony species to their pentavalent states. All the common arsenite salts are soluble, and As(III) species must be oxidized to As(V) to enable precipitation in a stable, insoluble form [12]. Antimony behaves similarly. Copper, nickel, zinc, and lead all form insoluble arsenate salts; however, iron is usually used because of its abundance in leach solutions and the stability of the products formed [47]. A stable Fe(III) arsenate precipitate can be produced within the pH range 3 to 7, provided that an iron–arsenic ratio of at least 4:1 (and potentially lower) is maintained, as discussed in Section 5.1 [48]. Both antimonites and antimonates form sparingly soluble salts with sodium and calcium. For example, the water treatment facility at Turquoise Ridge (Nevada, United States) has demonstrated that arsenic concentrations in effluent solution can be controlled consistently below 15 μg/L using effective precipitation practices [49]. Excellent reviews of arsenic removal theory and practice are available in the literature [48, 50].

Mercury, cadmium, and selenium. All of these metals are toxic, even at low concentrations, and consequently have low effluent control limits. Mercury can be precipitated in an insoluble form by sodium (or other) sulfide or by the readily available tri-sodium salt of trimercaptotriazine (TMT), and mercury concentrations <0.002 mg/L can be achieved in this way. However, these processes are not very efficient, and high reagent costs may be incurred to reduce mercury to desired levels. Other reagents, including sodium polymeric thiocarbamate, sodium/potassium dimethyl dithiocarbamate, and 1,3-benzendiamidoethanethiol have all been investigated as options to precipitate mercury from dilute cyanide solutions and have been shown to be capable of precipitating mercury down to low concentrations in solution [51, 52, 53].

Cadmium can be removed from solution down to concentrations <0.1 mg/L by precipitation with lime, iron salts, or sulfides, but dilution with fresh (or very low cadmium content) solution is often adequate to reach discharge limits.

Selenium may be present in solution as selenite or selenate, depending on redox conditions. Selenites can be precipitated in a stable form with Fe(III) sulfate and reduced to concentrations <0.05 mg/L. Selenates are typically hard to remove and must be reduced to selenite prior to their precipitation.

REFERENCES

[1] Smith, A., and T.I. Mudder. 1991. *Chemistry and Treatment of Cyanidation Wastes*. London: Mining Journal Books.

[2] Mudder, T.I., and M.M. Botz. 2001. A global perspective of cyanide. SME Preprint 01-76. Littleton, CO: SME.

[3] Sengupta, M. 1993. *Environmental Impacts of Mining*. Boca Raton, FL: Lewis Publishers.

[4] Ripley, E.A., R.E. Redmann, and A.A. Crowder. 1996. Pages 145–200 in *Environmental Effects of Mining*. Delray Beach, FL: St. Lucie Press.

[5] Peavy, H.S., D.R. Rowe, and G. Tchobanoglous. 1985. *Environmental Engineering*. New York: McGraw-Hill.

[6] U.S. Environmental Protection Agency. 2004. National Ambient Air Quality Standards. www.epa/air/criteria.html. Accessed May 2005.

[7] Kay, F., and R.P. McQuivey. 1989. Migratory birds and toxic mine ponds: The Nevada experience. Pages 159–160 in *Proceedings Randol Gold Conference*. Golden, CO: Randol International Ltd.

[8] Arnold, J.R. 1990. Nevada wildlife protection situation report. *Proceedings Randol Gold Conference*. Golden, CO: Randol International Ltd.

[9] Castranas, H.M., V. Cachic, and C. McKenzie. 1988. Cyanide detoxification of a gold mine tailings pond with H_2O_2. Pages 81–88 in *Proceedings Randol Gold Conference*. Golden, CO: Randol International Ltd.

[10] U.S. Environmental Protection Agency. 2002. Ground water and drinking water. List of drinking water contaminants and MCLs. www.epa.gov/safewater/mcl.html. Accessed May 20, 2005.

[11] World Health Organization. 2004. *Guidelines for Drinking-Water Quality*. 3rd edition. Geneva: WHO.

[12] Huiatt, J.L., J.E. Kerrigan, F.A. Olson, and G.L. Potter, editors. 1983. *Proceedings Cyanide from Mineral Processing Workshop*. Salt Lake City, UT: Utah Mining and Mineral Resources Research Institute.

[13] Mudder, T.I., and A.J. Goldstone. 1989. The recovery of cyanide from slurries. Pages 107–140 in *Proceedings of the International Gold Expo Conference*. Chicago, IL: Maclean Hunter Publishing Company.

[14] Mudder, T.I., and A.J. Goldstone. 1989. The recovery of cyanide from slurries. Pages 199–214 in *Proceedings Randol Gold Conference*. Golden, CO: Randol International Ltd.

[15] Botz, M.M., J.A. Stevenson, A.L. Wilder, R.T. Richins, T.I. Mudder, and B. Burdett. 1995. Recovery of cyanide from mill tailings. Pages 449–453 in *Proceedings Randol Gold Forum*. Golden, CO: Randol International Ltd.

[16] Lakshmanan, V.I., and P.D. Tackaberry. 1990. Advances in gold and silver processing. Pages 257–262 in *Proceedings GOLDTech 4 Conference*. Edited by M.C. Fuerstenau and J.L. Hendrix. New York: American Institute of Mining, Metallurgical, and Petroleum Engineers.

[17] Holbein, B.E., A. Huber, and D. Kidby. 1989. Field piloting results for Vitrokele cyanide recovery and the economics compared with other processes. Pages 215–219 in *Proceedings Randol Gold Conference*. Golden, CO: Randol International Ltd.

[18] Fleming, C.A. 1998. The potential role of anion exchange resins in the gold industry. Pages 95–117 in *Proceedings EPD Congress 1998*. Edited by B. Mishra. Warrendale, PA: TMS.

[19] LeVier, K.M., T.A. Fitzpatrick, K.A. Brunk, and W.N. Ellet. 1997. AuGMENT technologies—an update. Pages 135–137 in *Proceedings Randol Gold Forum*. Golden, CO: Randol International Ltd.

[20] Lien, R.H., B.E. Dinsdale, K.R. Gardner, and P.B. Altringer. 1990. Chemical and biological cyanide destruction and selenium removal from precious metal tailings pond water. Pages 323–339 in *Proceedings of Gold '90 Symposium*. Edited by D.M. Hausen, D.N. Halbe, E.U. Petersen, and W.J. Tafuri. Littleton, CO: SME.

[21] Muir, D.M., M. Aziz, and W. Hoecker. 1988. Cyanide losses under CIP conditions and effect of carbon on cyanide oxidation. *Proceedings of 1st International Symposium on Hydrometallurgy, Beijing, China*. Oxford, England: Pergamon Press.

[22] Potter, G.M., A. Bergmann, and U. Haidlen. 1986. Process of recovering copper and of optionally recovering silver and gold by leaching of oxide and sulfide-containing materials with water-soluble cyanides. U.S. Patent 4,587,110. May 6.

[23] Fleming, C.A. 1998. Thirty years of turbulent change in the gold industry. *CIM Bulletin* (April):1998.

[24] Doyle, F.M., S. Duyvesteyn, and K. Sreenivasarao. 1995. The use of ion flotation for detoxification of metal to contaminated waters and process effluents. Pages 175–180 in *Proceedings XIX International Mineral Processing Congress*. Littleton CO: SME.

[25] Whitlock, J.L., and T.I. Mudder. 1986. The Homestake wastewater treatment process: Biological removal of toxic parameters from cyanidation wastewaters and bioassay effluent evaluation. Pages 327–339 in *Fundamental and Applied Biohydrometallurgy*. Edited by R.W. Lawrence, R.M.R. Branion, and H.G. Ebner. Amsterdam: Elsevier.

[26] Schmidt, J.W., L. Simovic, and E.E. Shannon. 1982. Development studies for suitable technologies for the removal of cyanide and heavy metals from gold milling effluents. Pages 831–849 in *Proceedings 36th Industrial Waste Conference*. Ann Arbor, MI: Ann Arbor Science Publishers.

[27] Longe, G.K., and F.W. DeVries. 1988. Some recent considerations on the natural disappearance of cyanide. Pages 67–70 in *Proceedings Economics and Practice of Heap Leaching in Gold Mining Symposium*. Parkville, Victoria, Australia: Australasian Institute of Mining and Metallurgy.

[28] Chatwin, T.D., and J.J. Trepanowski. 1987. Utilisation of soils to mitigate cyanide releases. Pages 151–170 in *Proceedings of the 3rd Western Regional Conference on Precious Metals, Coal and Environment*. New York: American Institute of Mining, Metallurgical, and Petroleum Engineers.

[29] Griffiths, A.W., and G. Vickell. 1989. Treatment of gold mill effluents with H_2O_2, operating experience and costs. Paper presented at 21st Canadian Minerals Processors Conference, Ottawa, ON, Canada, January 17–19.

[30] Shoemaker, R.S., and F.W. DeVries. 1986. Cyanide, precious metal leaching and the environment. Paper presented at American Mining Congress, Las Vegas, Nevada, October 9.

[31] Collins, D.C. 1989. Reagent cost savings in tailings pond detoxification. Pages 155–158 in *Proceedings Randol Gold Conference*. Golden, CO: Randol International Ltd.

[32] Quamrul, A.M., A.W. Griffiths, and E.P. Jucevic. 1989. Detoxification of spent heaps with hydrogen peroxide. Pages 270–272 in *Proceedings AIME World Gold 1989 Conference*. New York: American Institute of Mining, Metallurgical, and Petroleum Engineers.

[33] Griffiths, A., H. Knorre, S. Gos, and R. Higgins. 1987. The detoxification of gold mill tailings with hydrogen peroxide. *Journal of the South African Institute of Mining and Metallurgy* 87(9):279–283.

[34] Mathre, O.B., and F.W. DeVries. 1981. Destruction of cyanide in gold and silver mine process water. *Proceedings AIME Annual Meeting*. New York: American Institute of Mining, Metallurgical, and Petroleum Engineers.

[35] Castrantas, H.M. 1997. Caro's acid—the facts about yield. Pages 257–260 in *Proceedings Randol Gold Forum*. Golden, CO: Randol International Ltd.

[36] Devuyst, E.A., B.R. Conard, and W. Hudson. 1984. Industrial application of the Inco SO_2/air cyanide removal process. Paper presented at 1st International Symposium on Precious Metals Recovery, Reno, Nevada, June 10–14.

[37] Zaidi, A., L. Whittle, T. Constable, and S. Sawell. 1988. *Evaluation of Inco's So_2/Air Process for the Removal of Cyanide and Associated Metals from Gold Milling Effluents at McBean Mine*. Report of Wastewater Technology Center. Burlington, ON, Canada: Environment Canada.

[38] Devuyst, E.A., B.R. Conrad, and G. Robbins. 1988. Commercial performance of Inco's SO_2-air cyanide removal process. Pages 87–88 in *Proceedings Randol Gold Conference*. Golden, CO: Randol International Ltd.

[39] Zaidi, A., and L. Whittle. 1987. *Evaluation of the Full Scale Alkaline Chlorination Treatment Plant at Giant Yellowknife Mines Ltd*. Report of Wastewater Technology Center. Burlington, ON, Canada: Environment Canada.

[40] Staunton, W., R.S. Schulz, and D.J. Glenister. 1988. Chemical treatment of cyanide tailings. Pages 85–86 in *Proceedings of Randol Gold Conference*. Golden, CO: Randol International Ltd.

[41] Halbe, D.N. 1997. Current cyanide detoxification work in Peru. Pages 261–264 in *Proceedings Randol Gold Forum*. Golden, CO: Randol International Ltd.

[42] Ahif, W., and M. Munawar. 1987. Biological assessment of environmental impact of dredged material. Pages 127–142 in *Chemistry and Biology of Solid Waste*. Edited by W. Salomons and U. Forstner. Berlin: Springer-Verlag.

[43] Salomons, W., and U. Forstner. 1987. *Chemistry and Biology of Solid Wastes*. Berlin: Springer-Verlag.

[44] Fleming, C.A., and G. Cromberge. 1984. The extraction of gold from cyanide solutions by strong- and weak-base anion-exchange resins. *Journal of South African Institute of Mining and Metallurgy* 84(5):125–138.

[45] Denit, J., G.E. Stigall, and T. Fielding. 1984. Development document for effluent limitations. *Guidelines and Standards for the Inorganic Chemicals Industries, Phase II*. EPA 440/1-84/007. Washington, DC: U.S. Environmental Protection Agency.

[46] Smith, A., and T.I. Mudder. 1991. Pages 144–149 in *Chemistry and Treatment of Cyanidation Wastes*. London: Mining Journal Books.

[47] Wilson, H.R. 1990. Tailings management in the Canadian Arctic—Echo Bay's Lupin mine. *Proceedings of GOLDTech 4 Conference*. New York: American Institute of Mining, Metallurgical, and Petroleum Engineers.

[48] Krause, E., and V.A. Ettel. 1985. Ferric arsenate compounds: Are they environmentally safe? Solubilities of basic ferric arsenates. Pages 5.1–5.20 in *Proceedings 15th Annual Hydrometallurgy Meeting of CIM*. Ottawa, ON, Canada: Canadian Institute of Mining and Metallurgy.

[49] Ackerman, J.B., and J.G. Mansanti. 2005. Water treatment plant modifications at Turquoise Ridge Joint Venture. Innovations in natural resource processing. Pages 283–298 in *Proceedings Jan D. Miller Symposium*. Edited by C.A. Young, J.J. Kellar, M.L. Free, J. Drelich, and R.P. King. Littleton, CO: SME.

[50] Harris, B. 2003. The removal of arsenic from process solutions: Theory and industrial practice. Pages 1889–1902 in *Hydrometallurgy 2003–5th International Conference in Honor of Professor Ian Ritchie*. Volume 2. Edited by C.A. Young, A. Alfantazi, C. Anderson, A. James, D. Dreisinger, and B. Harris. Warrendale, PA: TMS.

[51] Bucknam, C.H. 2004. A green chemistry approach to mercury control during cyanide leaching of gold. SME Preprint 04-101. Littleton, CO: SME.

[52] Atwood, D.A., M.M. Matlock, and B. Howerton. 2003. Mercury removal in the gold cyanide process. SME Preprint 03-014. Littleton, CO: SME.

[53] Misra, M., J. Lorengo, J.B. Nanor, and C.B. Bucknam. 1998. Removal of mercury cyanide species from solutions using dimethyl dithiocarbamates. *Minerals and Metallurgical Processing Journal* 15(4):60–64.

CHAPTER 12

Industrial Applications

The preceding chapters have reviewed the process technology for the extraction of gold from its ores. As each gold ore has different characteristics, many flowsheets have been developed to take these and other local factors into account to achieve optimum gold extraction, and to yield the best economic value from the resource.

To provide a perspective of the importance of the various unit processes available, the distribution of process technology applied around the world is briefly discussed in this chapter. This is followed by a set of 50 gold extraction flowsheets, which are either commonly applied in the industry or which contain features of particular interest—classified according to ore type to be consistent with the methodology employed in this book. In many cases the flowsheets are self-explanatory; however, the more important process features and statistics have been included in a brief narrative section that accompanies each example. Some of these operations are no longer in operation and for others the flowsheet may have changed, sometimes significantly, such that the flowsheet shown is not equivalent to the circuit configuration currently in use at that operation (due to changes in ore types processed, for example, or to accommodate new deposits brought into production at or near that location). However, these flowsheets have been included because they illustrate the application of a particular technology or chemistry described elsewhere in the book. In addition, the production data provided in this chapter for each of the operations change over time, and the reader should view the data in the context of the year in which the flowsheet information was based.

12.1 DISTRIBUTION OF PROCESS TECHNOLOGY

The distribution of gold production worldwide as a function of the extraction, recovery, and, where applicable, oxidation techniques employed is summarized in Table 12.1. (More detailed descriptions of the process methods and these abbreviations appear in Table 12.2.) In the context of this particular study, extraction techniques are defined as processes used for the removal of gold from its ores (i.e., by flotation, gravity concentration, or leaching), whereas recovery considers the methods for recovering gold from leach slurry and solution, or for recovery from gravity or flotation concentrates. These data have been developed by the authors using individual company annual reports (mainly for the 2004 fiscal year), gold production statistics published by Gold Fields Mineral Services in its annual gold surveys [1], and certain other source material [2].

In some cases, estimates of production from small, independent operators and informal gravity concentration production have been made where the database information was known to be inadequate, that is, in Brazil, the Philippines, Colombia, other regions

---

NOTE: In the first edition of *The Chemistry of Gold Extraction*, gold production data were compiled using information contained in the *Mining Journal* gold database [3], coupled with publicly available information on gold production operations. This database was no longer available at the time of preparation of this second edition and a custom database was prepared by the authors.

504 | THE CHEMISTRY OF GOLD EXTRACTION

TABLE 12.1 Distribution of gold production worldwide by processing method for 2004 (tpy)*

Processing Method(s)	South Africa	United States	Australia	China	Russia	Peru	Canada	Indonesia	Uzbekistan	Papua New Guinea	Ghana	Tanzania	Brazil	Mali	Chile	Rest of World	World Total
Extraction Technology																	
Heap leaching	0.0	70.1	1.3	0.2	5.5	114.1	0.0	0.9	10.9	0.0	17.2	0.0	1.2	7.5	0.0	7.0	235.9
Vat leaching	0.0	0.0	0.5	0.0	0.0	0.0	0.0	0.0	0.0	0.0	0.0	0.0	0.0	0.0	0.0	0.0	0.5
Agitated cyanide leaching	293.6	36.1	123.1	4.8	48.9	0.0	20.2	12.7	58.8	1.6	24.0	0.0	15.3	14.3	5.7	66.1	725.2
Agitated cyanide leaching with gravity concentration	13.8	20.2	28.9	0.0	0.0	0.0	44.9	0.0	0.0	0.0	0.0	26.5	0.0	17.6	9.8	3.9	165.6
Gravity concentration only	0.0	0.0	0.0	59.0	52.6	0.0	0.0	26.0	0.0	20.0	0.0	0.0	9.0	0.0	0.0	54.2	220.8
Flotation only	0.0	9.6	20.8	9.3	0.0	0.0	0.3	0.0	6.2	0.0	0.0	10.9	0.0	0.0	11.4	18.9	87.4
Gravity concentration and flotation	0.0	2.7	8.9	4.3	0.0	0.0	25.7	69.7	0.0	0.0	0.0	0.0	0.0	0.0	0.0	38.3	149.6
Flotation and agitated cyanide leaching (gravity optional)	1.0	0.0	0.0	22.8	4.7	0.0	8.5	0.0	0.0	0.0	4.6	0.0	0.0	0.0	0.0	30.0	71.6
Whole-ore oxidative pretreatment and agitated cyanide leaching	0.0	106.8	0.0	0.0	0.0	0.0	0.0	2.5	0.0	50.4	0.0	0.0	0.0	0.0	0.0	5.7	165.4
Flotation, oxidative pretreatment, and agitated cyanide leaching (gravity optional)	3.3	7.5	39.6	3.0	3.1	0.0	6.5	0.0	0.0	0.0	11.8	0.0	10.0	0.0	0.0	3.9	88.7
Unaccounted for	31.0	8.8	35.3	113.9	66.8	59.1	22.4	2.3	7.8	2.5	0.0	10.5	6.5	0.0	11.7	175.1	553.7
Total†	342.7	261.8	258.4	217.3	181.6	173.2	128.5	114.1	83.7	74.5	57.6	47.9	42.0	39.4	38.6	403.1	2,464.4
Recovery Technology																	
Carbon-in-pulp/carbon-in-leach	301.7	161.9	192.1	7.8	11.8	0.0	77.3	12.7	5.9	52.0	40.4	26.5	25.2	31.9	0.0	91.0	1,038.2
Slurry solid–liquid separation and zinc precipitation	10.0	8.6	0.0	22.8	5.0	0.0	2.8	2.5	0.0	0.0	0.0	0.0	6.4	0.0	15.5	14.7	88.3
Ion exchange resin (resin-in-pulp/resin-in-leach/resin-in-solution)	0.0	0.0	0.0	0.0	40.0	0.0	0.0	0.9	52.9	0.0	0.0	0.0	0.0	0.0	0.0	3.9	97.7
Zinc precipitation from solution	0.0	4.4	0.0	0.0	5.5	114.1	0.0	0.0	10.9	0.0	0.0	0.0	0.0	0.0	0.0	0.0	134.9
Carbon-in-columns from solution	0.0	65.6	1.3	0.2	0.0	0.0	0.0	0.0	0.0	0.0	17.2	0.0	0.0	7.5	0.0	7.0	98.8
Direct smelting of gravity concentrate	0.0	0.0	0.0	59.0	52.6	0.0	0.0	26.0	0.0	20.0	0.0	0.0	0.0	0.0	0.0	54.2	211.8
Smelting of sulfide concentrate	0.0	12.3	29.7	13.6	0.0	0.0	26.0	69.7	6.2	0.0	0.0	10.9	0.0	0.0	11.4	57.2	237.0
Unaccounted for	31.0	9.0	35.3	113.9	66.7	59.1	22.4	2.3	7.8	2.5	0.0	10.5	10.4	0.0	11.7	175.1	557.7
Total†	342.7	261.8	258.4	217.3	181.6	173.2	128.5	114.1	83.7	74.5	57.6	47.9	42.0	39.4	38.6	403.1	2,464.4
Oxidation Technology																	
Flotation concentrate roasting	0.0	0.0	35.0	0.0	0.0	0.0	0.0	0.0	0.0	0.0	0.0	0.0	7.5	0.0	0.0	3.9	46.4
Whole-ore roasting	0.0	38.6	0.0	0.0	0.0	0.0	0.0	2.5	0.0	0.0	0.0	0.0	0.0	0.0	0.0	0.0	41.1
Flotation concentrate pressure oxidation	0.0	7.5	0.0	0.0	0.0	0.0	0.0	0.0	0.0	0.0	0.0	0.0	2.5	0.0	0.0	0.0	10.0
Whole-ore pressure oxidation	0.0	68.1	0.0	0.0	0.0	0.0	0.0	0.0	0.0	50.4	0.0	0.0	0.0	0.0	0.0	5.7	116.7
Flotation concentrate biological oxidation	3.3	0.0	4.6	3.0	3.1	0.0	0.0	0.0	0.0	0.0	11.8	0.0	0.0	0.0	0.0	0.0	25.8
Whole ore biological oxidation	0.0	0.2	0.0	0.0	0.0	0.0	0.0	0.0	0.0	0.0	0.0	0.0	0.0	0.0	0.0	0.0	0.2
Chlorination	0.0	0.0	0.0	0.0	0.0	0.0	0.0	0.0	0.0	0.0	0.0	0.0	0.0	0.0	0.0	0.5	0.5
Total†	3.3	114.2	39.6	3.0	3.1	v0.0	0.0	2.5	0.0	50.4	11.8	0.0	10.0	0.0	0.0	9.6	248.2

* Data compiled from John Marsden and [1].
† Total gold production by country corresponds to data reported by Gold Field Mineral Services for 2004 production [1].

TABLE 12.2 List of abbreviations for flowsheet process options (Figures 12.5 to 12.6)

Extraction Methods

HL	Heap or run-of-mine cyanide leaching
VL	Vat cyanide leaching
AL/GAL	Agitated cyanide leaching (gravity concentration and preaeration optional)
G	Gravity concentration only
F	Flotation only
GF	Flotation and gravity concentration
FAL	Flotation and agitated cyanide leaching of concentrate and/or tailings
GFAL	Gravity concentration, flotation, and agitated cyanide leaching
OAL	Whole-ore oxidative pretreatment before agitated cyanide leaching
FOAL	Flotation, oxidative pretreatment of concentrate, and agitated cyanide leaching
GFOAL	Gravity concentration, flotation, oxidative pretreatment of concentrate and agitated leaching
Unaccounted for	No extraction method assigned to production due to lack of specific information

Recovery Methods

CIP/CIL	Carbon-in-pulp or carbon-in-leach recovery from slurry
CCD and zinc precipitation	Slurry solid–liquid separation (counter-current decantation [CCD] and/or filtration) followed by Merrill–Crowe zinc precipitation
RIP/RIL/RIS	Ion exchange resin used in-pulp (RIP) or in-leach (RIL) for recovery from slurry or from solution (RIS)
Solution zinc precipitation	Merrill–Crowe zinc precipitation from heap or vat leach solution
Solution CIC	Carbon-in-columns (CIC) recovery from heap leach solution followed by electrowinning or zinc precipitation
Direct smelt gravity concentrates	Concentrate from gravity or other separation process is directly smelted (usually on-site)
Smelt sulfide concentrates	Concentrate from flotation and gravity separation processed in base metal sulfide smelter (usually off-site)
Unaccounted for	No recovery method assigned to production due to lack of specific information

of South America, Indonesia, China, Russia, other parts of Asia, and much of Africa. Such estimates have been based on a rudimentary assessment of the likely size and type of processing methods employed in such locations.

The process method categorizations that have been adopted require that generalizations and assumptions be made for some of the flowsheets in use, and these numeric categorizations should be considered as approximate only. Also, gold production statistics around the world can vary greatly from year to year, and consequently the data presented are only intended to provide a general perspective of the distribution of technology.

Figure 12.1 shows the major gold-producing regions of the world, together with locations of 82 of the largest gold mines, listed in order of gold production. A number of other operations, which have been cited elsewhere in the text, are also located on this map.

The distribution of extraction, recovery, and oxidation technology on a worldwide basis is illustrated in Figures 12.2 to 12.4. Figure 12.2 shows that in 2004 agitated cyanide leaching accounted for approximately 50% of gold extraction, although this encompasses process flowsheets that include flotation (with cyanidation of the concentrate and/or tailings) and oxidative pretreatment steps ahead of cyanidation. In 2004, heap leaching was responsible for almost 10% of worldwide production, or about 236 t, which is a significant increase from 136 t in 1989 [76]. A review of the available data indicates that gravity concentration probably accounted for approximately 20% to 25% of

506 | THE CHEMISTRY OF GOLD EXTRACTION

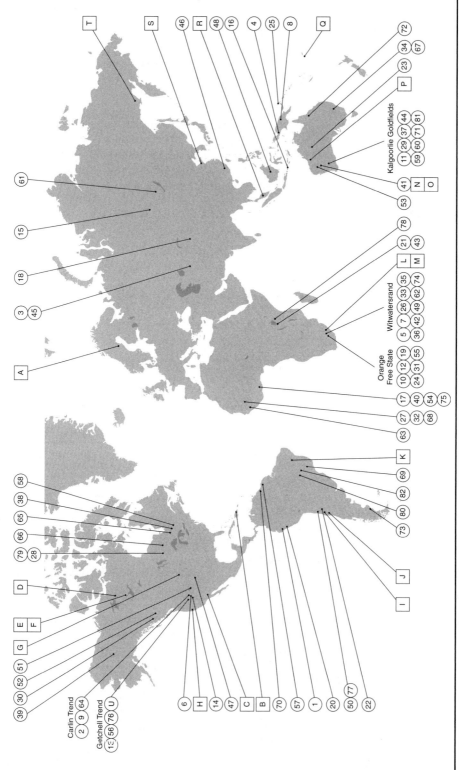

FIGURE 12.1 Major gold-producing regions of the world showing 82 of the largest individual gold-producing mines in 2004

Ranking of Worldwide Gold Mines (above 200,000 oz/yr production)

Rank	Operation Name	Major Owner/Operator	2004 Annual Production (oz)	2004 Annual Production (t)	Rank	Operation Name	Major Owner/Operator	2004 Annual Production (oz)	2004 Annual Production (t)	Rank	Operation Name	Major Owner/Operator	2004 Annual Production (oz)	2004 Annual Production (t)
1	Yanacocha	Newmont	3,022,217	94.00	34	Ridgeway	Newcrest	438,000	13.62	69	Morro Velho	AngloGold	240,000	7.46
2	Goldstrike	Barrick	1,942,000	60.40	35	South Deep	Placer Dome	428,000	13.31					
3	Muruntau	Navoi Mining	1,700,000	52.88	36	Randfontein	Harmony	412,179	12.82	70	Omai	Cambior	240,000	7.46
4	Grasberg	Freeport McMoran	1,535,000	47.74	37	Sunrise Dam	AngloGold Ashanti	410,000	12.75	71	Kanowna Belle	Placer Dome	237,000	7.37
										72	Pajingo	Newmont	230,000	7.15
5	Driefontein	Gold Fields	1,141,000	35.49	38	Porcupine	Placer	395,490	12.30	73	Cerro Vanguardia	AngloGold Ashanti	229,000	7.12
6	Cortez-Pipeline	Placer Dome	1,051,000	32.69	39	Fort Knox	Kinross	392,000	12.19					
7	Kloof	Gold Fields	1,038,000	32.28	40	Obuasi	AngloGold Ashanti	380,000	11.82	74	Ergo	AngloGold Ashanti	222,000	6.90
8	Porgera	Placer Dome	1,020,000	31.73										
9	Carlin	Newmont	956,000	29.73	41	Yandal	Newmont Mining	365,000	11.35	75	Iduapriem-Teberebie	AngloGold Ashanti	220,000	6.84
10	Free State Growth (Welkom)	Harmony	889,625	27.67	42	Evander	Harmony	361,701	11.25					
					43	Bulyanhulu	Barrick	350,000	10.89	76	Midas	Newmont	220,000	6.84
11	Kalgoorlie Consolidated Gold Mines	Barrick/Newmont	888,000	27.62	44	Big Bell	Harmony	350,000	10.89	77	Escondida	BHP-Billiton	217,000	6.75
					45	Zarafshan	Newmont/Navoi	350,000	10.89	78	North Mara	Placer Dome	209,000	6.50
					46	Zijinshan	Zijin Mining	342,089	10.64	79	Campbell	Placer Dome	209,000	6.50
12	Great Noligwa	AngloGold Ashanti	795,000	24.73	47	Cripple Creek & Victor	AngloGold Ashanti	330,000	10.26	80	Crixas	Kinross	205,000	6.38
										81	Agnew	Gold Fields	202,000	6.28
13	Twin Creeks	Newmont	782,000	24.32	48	Kelian	Rio Tinto	328,000	10.20	82	Paracatu-Morro D'Ouro	Kinross	201,000	6.25
14	Round Mountain	Kinross/Barrick	763,000	23.73	49	North West (Buffelsfontein/ Hartebeestfontein)	Durban Roodepoort Deep	321,000	9.98					
15	Olimpiada	Norilsk	720,000	22.39							**Other Sites of Special Interest Referenced in Book (closed, operating, and projects in development)**			
16	Batu Hijau	Newmont	707,327	22.00	50	El Penon	Meridian	314,000	9.77	A	Bjorkdal	Gold Ore Resources		
17	Tarkwa	Gold Fields	703,000	21.87	51	Bingham Canyon	Rio Tinto	308,000	9.58	B	Pueblo Viejo	Placer Dome		
18	Kumtor	Centerra Gold	660,000	20.53	52	Kemess	Northgate	304,000	9.46	C	Mesquite	Newmont Mining		
19	Free State Leverage (Virginia)	Harmony	646,240	20.10	53	Plutonic	Barrick	304,000	9.46	D	Lupin	Kinross		
					54	Damang	Gold Fields	296,000	9.21	E	Giant Yellowknife	Miramar Mining		
20	Pierina	Barrick	646,000	20.09	55	Tua Lekoa	AngloGold Ashanti	293,000	9.11	F	Con	Miramar Mining		
21	Geita	AngloGold Ashanti	643,025	20.00						G	Homestake Lead	Barrick		
22	Alumbrera	Xstrata	633,379	19.70	56	Lone Tree	Newmont	280,000	8.71	H	Homestake McLaughlin	Barrick		
23	Tanami	Newmont	630,000	19.59	57	Rosebel	Cambior	274,000	8.53	I	Candelaria	Phelps Dodge		
24	Beatrix	Gold Fields	625,000	19.44	58	La Ronde	Agnico Eagle	272,000	8.46	J	El Indio-Tambo-Pascua	Barrick		
25	Lihir	Lihir Gold	600,000	18.66	59	Granny Smith	Placer Dome	267,000	8.30	K	Jacobina	Desert Sun Mining		
26	Tua Tona	AngloGold Ashanti	568,000	17.67	60	Kalgoorlie West	Placer Dome	263,000	8.18	L	Fairview-Sheba-Agnes	Metorex		
					61	Lenzoloto	Norilsk	260,000	8.09	M	Consolidated Murchison	Metorex		
27	Morila	Randgold	562,647	17.50	62	Elandsrand	Harmony	255,602	7.95	N	Harbour Lights	Sons of Gwalia		
28	Red Lake	GoldCorp	552,000	17.17	63	Siguiri	AngloGold Ashanti	250,000	7.78	O	Wiluna	Newmont Mining		
29	St. Ives	Gold Fields	543,000	16.89						P	Telfer	Newcrest		
30	Eskay Creek	Barrick	523,000	16.27	64	Jerritt Canyon	Queenstake	250,000	7.78	Q	Emperor Mines	Durban Roodepoort Deep		
31	Kopanang	AngloGold Ashanti	486,000	15.12	65	Williams-David Bell	Barrick	248,000	7.71	R	Penjom	Avocet		
					66	Musselwhite	Placer Dome	247,059	7.68	S	Shandong Province	Various		
32	Sadiola	IAMGold	458,000	14.25	67	Cadia	Newcrest	244,000	7.59	T	Kubaka	Kinross		
33	Mponeng	AngloGold Ashanti	438,000	13.62	68	Yatela	IAMGold	242,000	7.53	U	Phoenix	Newmont		

FIGURE 12.1 Major gold-producing regions of the world showing 82 of the largest individual gold-producing mines in 2004 (continued)

508 | THE CHEMISTRY OF GOLD EXTRACTION

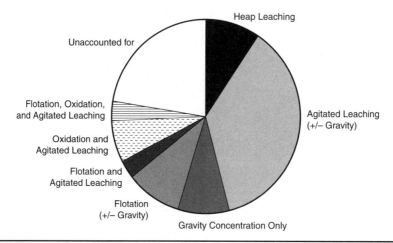

FIGURE 12.2 World gold production by extraction method in 2004 (see Table 12.1 for detailed data)

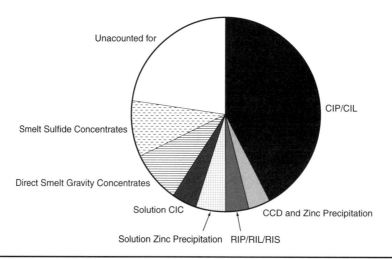

FIGURE 12.3 World gold production by recovery method in 2004 (see Table 12.1 for detailed data and Table 12.2 for abbreviations)

worldwide production in 2004. However, the estimation of gold production by gravity concentration is complicated by its common use in combination with agitated leaching, and accurate data of the proportion of gold produced by the two processes are often unavailable. Also, no reliable data are available for amalgamation processes, and no attempt has been made to quantify the application of this technology. Gold produced as a by-product of copper mining and extraction by flotation, gravity concentration (optional), and subsequent smelting of the sulfide concentrate accounts for approximately 10% of worldwide production.

Figure 12.3 shows that carbon-in-pulp (CIP) and carbon-in-leach (CIL) are the predominant recovery methods in use, accounting for approximately 42% of worldwide production. Resin-in-pulp (RIP) and resin-in-leach (RIL) systems are widely used in Russia and Uzbekistan, and account for about 4% of production. Direct smelting of gravity concentrates (usually on-site) and smelting of sulfide concentrates (off-site) account for

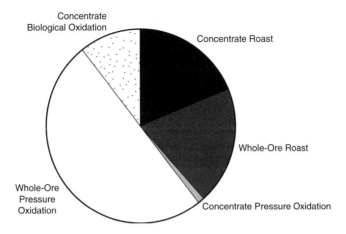

FIGURE 12.4　World gold production by oxidation method in 2004

9% and 10% of worldwide gold production, respectively. The application of solid–liquid separation and Merrill–Crowe zinc precipitation techniques declined to less than 4% (88 t) of production in 2004, a significant decrease from the estimate of 491 t in 1989 [76]. The 2004 estimate is probably low because some of this production is likely captured in the "unaccounted for" segment (smaller, older, and remotely located operations). The significant decrease, however, is largely due to the consolidation of CIP and CIL as the preferred technology, coupled with the closure (due to reserve depletion) of many older operations using Merrill–Crowe technology.

For recovery of gold from solution (i.e., produced by heap leaching), zinc precipitation (6% of gold recovered) is more widely used than carbon-in-columns (CIC, 4% of gold recovered). This statistic is driven by two very large heap leach operations in Peru that use zinc precipitation (Yanacocha and Pierina).

Because accurate and reliable production method information was not available for some regions of the world and some operations, approximately 22% of world production is unaccounted for. The major regions where this undefined production occurs are China (114 t), Russia (67 t), Peru (59 t), and the rest of the world (175 t).

Figure 12.4 illustrates the distribution of oxidation technology. Oxidative pretreatment methods are applied to approximately 10% of worldwide gold production (see Figure 12.2), increasing significantly from 94 t in 1989 to 249 t in 2004 [76]. In 2004, whole-ore pressure oxidation was the most popular route (5%), followed by flotation and roasting of the concentrate (2%), whole-ore roasting (2%), and flotation and biological oxidation of the concentrate (1%). In all cases, oxidative pretreatment is followed by agitated cyanide leaching.

The variation in the application of the different extraction and recovery technology by geographical location is illustrated in Figures 12.5 and 12.6. In these figures, detailed information is provided for the top 15 gold-producing countries of the world in 2004, plus the rest of the world. These illustrate the great diversity in process flowsheets around the world, which reflect local gold mining history, the different rates in the evolution and adoption of new technology, specific conditions and restrictions in some areas, and variations in geological ore types. These figures give a good indication of general mineralogical trends, identify the various treatment schemes applied in the different locations, and indicate their relative importance.

510 | THE CHEMISTRY OF GOLD EXTRACTION

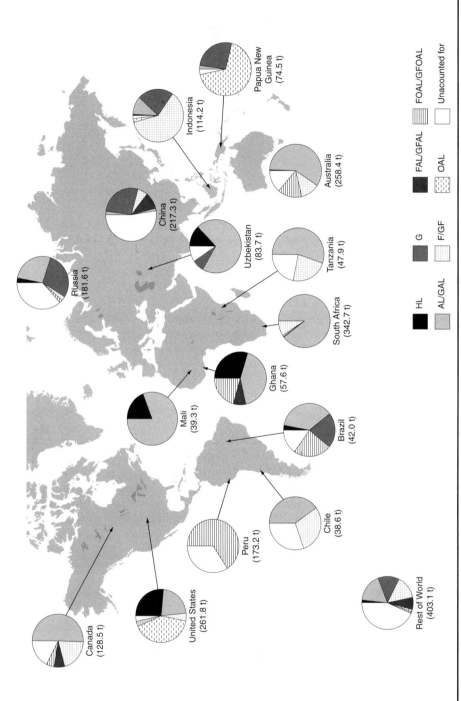

FIGURE 12.5 Distribution of gold extraction methods worldwide in 2004 (see Table 12.1 for detailed data and Table 12.2 for abbreviations)

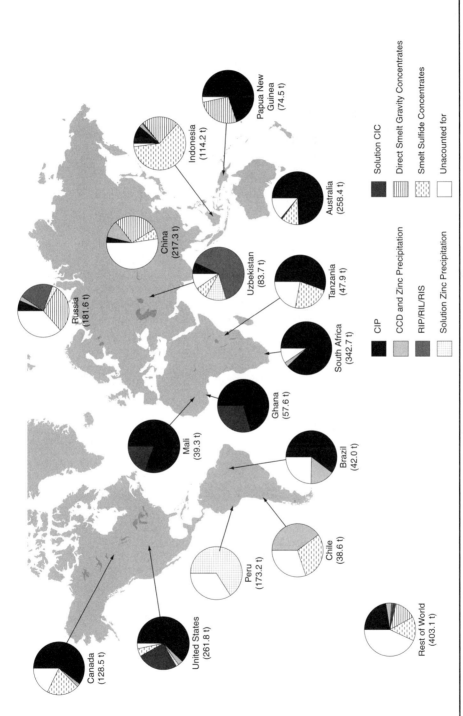

FIGURE 12.6 Distribution of gold recovery methods worldwide in 2004 (see Table 12.1 for detailed data and Table 12.2 for abbreviations)

12.2 INDUSTRIAL PROCESS FLOWSHEETS

12.2.1 Placers

12.2.1.1 Bjorkdal [4]
Owner/Operator: Gold-Ore Resources Ltd. (Vancouver, British Columbia, Canada)
Location: Sweden
Start-up: 1988

Mineralogical factors. Deposit is an alluvial placer with free gold at various sizes.

Key process statistics (approximate; 1990 and 2004 basis)

	1990	2004
Ore throughput rate (tpd):	1,000	3,300
Gold ore grade (g/t):	3 to 4	~1
Gold recovery (%):	80–85	86
Gold production (oz/yr)[*]:	34,000	31,000

Process description/main features. Ore is crushed and ground in a rod mill–ball mill circuit to yield a product at 80% <80 μm (see Figure 12.7). The ball mill is configured in closed circuit with a screen and cyclones, with the cyclone underflow (U/F) treated by Reichert cones for coarse gold recovery. The application of gravity concentration within the grinding circuit reduces any effects of flattening of free gold particles (which would reduce their response to gravity separation) and minimizes the formation of coatings on gold surfaces. The concentrates produced are further upgraded by spiral concentrators and shaking tables and then shipped directly to a smelter. All of the tailings from gravity concentration processes are returned to the ball mill for regrinding.

The cyclone overflow (O/F) is treated by flotation (using a xanthate collector at natural pH) to recover finer gold. The concentrate produced is thickened, filtered, and shipped to the smelter.

This plant is a good example of a gravity concentration circuit for alluvial placer treatment.

---

[*] All ounces cited are troy ounces.

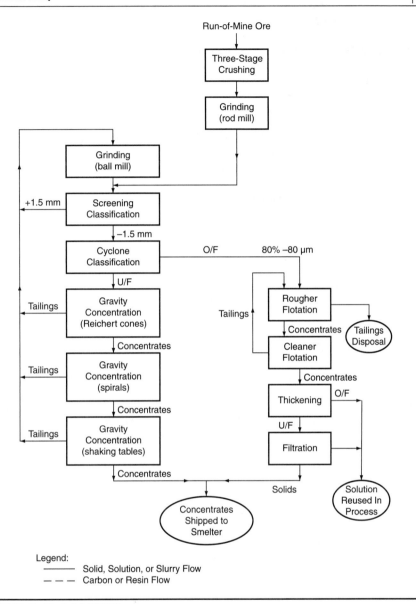

FIGURE 12.7 Bjorkdal (1990)

12.2.2 Free-Milling Ores

12.2.2.1 East Driefontein (division of Driefontein Consolidated) [5, 6]
Owner/Operator: Gold Fields (Johannesburg, South Africa)
Location: Carletonville, Transvaal, South Africa
Start-up: 1973—original filtration and precipitation plant
2002–2003—CIP pump-cell plant conversion

Mineralogical factors. The ore is typical Witwatersrand pebble–quartz conglomerate. Fine free gold and electrum are intimately associated with predominantly siliceous gangue but with minor pyrrhotite. Coarse gold present is negligible.

Key process statistics (approximate; 1990 basis)
Ore throughput rate: 8,000 tpd
Gold ore grade: 8.0 g/t
Gold recovery: 98.0% to 98.5%
Gold production: ~730,000 oz/yr

Process description/main features. In 1990, three-stage crushing and two-stage grinding (rod and pebble mills) were used to reduce run-of-mine ore to 78% <75 mm (see Figure 12.8). The product size and throughput were controlled by a multivariable grinding control system. No cyanide was added to the grinding circuit and no gravity concentration was used, as the gold grain size was fine.

Preaeration was used for oxidation of minor pyrrhotite in the feed. A residence time of 8 to 10 hr was provided for this at pH 10.5 to 11.0, using lime for pH control. No lead salts were added to this circuit.

The cyanide concentration in the leach circuit was maintained between 0.20 and 0.25 mg/L sodium cyanide (NaCN)-equivalent, with an overall cyanide consumption of 0.25 kg/t calcium cyanide ($Ca(CN)_2$). The total leach residence time was approximately 40 hr, although most of the gold dissolution occurred within 24 hr.

The original recovery circuit used single-stage filtration (rotary drum vacuum filters) of leached slurry to separate solution and solids for Merrill–Crowe zinc precipitation and tailings disposal, respectively. In 1986–1987, this was modified to include a thickening stage midway through leaching to remove a portion of gold-bearing solution for direct zinc precipitation in order to reduce the gold grade in solution to the filtration circuit and consequently to reduce soluble gold losses. This also improved the solution equilibria for gold dissolution in subsequent leaching stages.

Hopper clarifiers replaced the original candle-type pressure filters for clarification of thickener overflow and filtrate solutions prior to Merrill–Crowe precipitation. The precipitate was acid washed (HCl), calcinated, and smelted with fluxes.

Circuit modifications. In 2002, the filtration and Merrill–Crowe zinc precipitation recovery plant was replaced with an eight-stage CIP plant using Anglo American Research Laboratories (AARL) pump-cell technology (see Figure 12.9). To support the three CIP plants at Driefontein Consolidated, centralized carbon elution facilities were installed. Gold is electrowon from the carbon eluate solution using an automated, sludge-forming cylindrical electrowinning cell (the Kemix cell; see Chapter 8). The sludge is smelted with fluxes to generate bullion on-site. The replacement of filtration and zinc precipitation with CIP resulted in a significant reduction in the soluble gold loss to the residue (i.e., 0.05 g/t, or about 1% gold recovery) and reduced operating costs. East Driefontein also replaced the rod mill and pebble mill grinding circuit with a single-stage semiautogenous grinding (long aspect ratio mill) circuit, with reported improvements in efficiency.

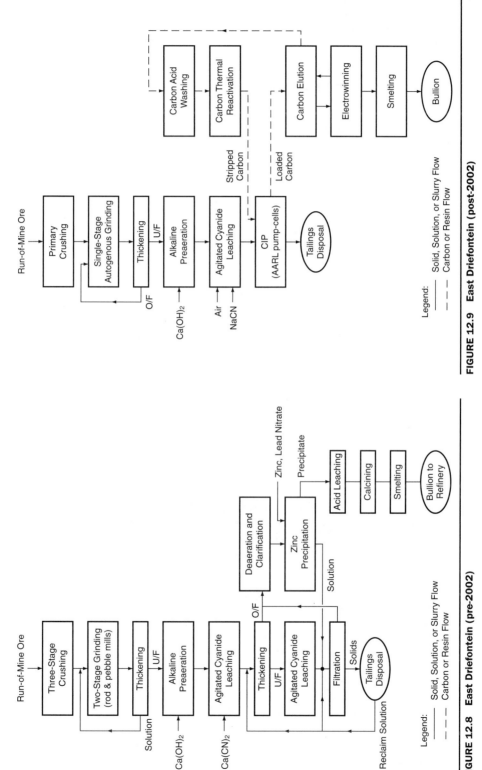

FIGURE 12.9 East Driefontein (post-2002)

FIGURE 12.8 East Driefontein (pre-2002)

12.2.2.2 Harmony No. 4 Plant [7]
Owner/Operator: Harmony Gold Mines (Johannesburg, South Africa)
Location: Virginia, Orange Free State, South Africa
Start-up: 1986

Mineralogical factors. The ore is a typical Witwatersrand pebble-quartz conglomerate. Fine free gold and electrum are intimately associated with siliceous, nonreactive gangue. Minor coarse gold occurs.

Key process statistics (approximate; 1990 basis)
Ore throughput rate: 5,000 tpd
Gold ore grade: 2.4 g/t
Gold recovery: 95.0%
Gold production: ~130,000 oz/yr

Process description/main features. Ore was ground in autogeneous mills to 73% to 75% <75 μm (see Figure 12.10). Run-of-mine milling was selected over conventional crushing and two-stage grinding circuit because of an estimated 25% capital and 30% operating cost savings.

CIP was selected over solid-liquid separation and Merrill-Crowe because of the estimated 25% capital and 20% operating cost savings. Interestingly, the dissolved and undissolved gold losses for CIP and Merrill-Crowe circuits were considered to be equivalent because the lower-solids gold grade obtained by CIP was offset by gold-on-fine-carbon losses. Actual dissolved and undissolved gold losses achieved were 0.012 and 0.116 g/t, respectively.

The use of CIL was rejected in view of the absence of gold-adsorbing constituents in the ore, because CIL would have resulted in higher carbon and gold-in-plant inventories, and higher gold-on-carbon losses than CIP.

The seven-stage CIP circuit achieved the following approximate gold loadings (values in g/t): No. 1 = 3,400, No. 2 = 1,150, No. 3 = 500, No. 4 = 350, No. 5 = 320, No. 6 = 250, No. 7 = 240.

Loaded carbon was acid washed at 85°C in a fiberglass vessel, then eluted at 120°C and 150 kPa in a stainless steel (3CR12) column. The standard AARL elution was modified by retaining the initial (low-grade) portion of eluate for use in subsequent elution batches, to reduce the bulk of solution for electrowinning.

Electrowinning of gold was performed in two Mintek cells connected in series. An initial current of 2,000 amp was supplied, to take advantage of higher current efficiencies achievable at higher gold concentrations, but this was subsequently ramped down to 500 amp.

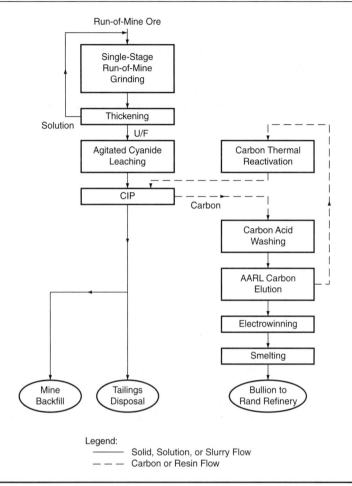

FIGURE 12.10 Harmony No. 4 Plant (1990)

12.2.2.3 President Brand New Plant (division of FreeGold) [8]
Owner/Operator: AngloGold Ashanti (Johannesburg, South Africa)
Location: Welkom, Orange Free State, South Africa
Start-up: 1986 (1953 original plant)

Mineralogical factors. The ore is a typical Witwatersrand pebble–quartz conglomerate. Free gold and electrum are intimately associated with siliceous but essentially nonreactive, gangue. Some of the gold is relatively coarse and is recoverable by gravity concentration.

Key process statistics (approximate; 1990 basis)
Ore throughput rate: 14,000 tpd
Gold ore grade: 5.0 g/t
Gold recovery: 96.5%
Gold production: ~780,000 oz/yr

Process description/main features. The original plant used crushing, manual waste sorting, conventional grinding, and gravity concentration, thickening, leaching, filtration and Merrill–Crowe precipitation. Gravity concentrates were processed by amalgamation.

The "new" (1986) plant was designed and constructed in modular form. Run-of-mine ore was ground to 76% <75 μm in single-stage autogenous mills in closed circuit with three-stage cycloning (see Figure 12.11). Intermediate gravity concentration (Johnson drums and belt concentrators) were used between the secondary and tertiary cyclones. Approximately 50% of gold in the feed was recovered by gravity concentration. This allowed downsizing of the leach circuit to a retention time of 24 hr, compared with an estimated 48 hr that would have been required without gravity concentration to achieve equivalent recovery.

Intensive cyanidation was selected over amalgamation for treatment of the gravity concentrates because of the health hazards associated with mercury. Osmium and iridium were recovered by gravity concentration following intensive cyanidation.

Leached slurry was treated by CIP (superior economics compared with CIL and solid–liquid separation and Merrill–Crowe precipitation). Carbon was eluted by the AARL procedure, which was selected over the Zadra method (see Chapter 7), principally for its faster elution kinetics. Undissolved and dissolved gold losses following CIP were estimated to be 0.155 and 0.015 g/t, respectively.

Zinc precipitation was selected over electrowinning for recovery of gold from carbon eluate owing to the large number of cells that would have been required for electrowinning (40 AARL or 15 Mintek cells) and because of anticipated security advantages. No deaeration or lead nitrate addition was required for precipitation.

The solution produced by intensive cyanide leaching of gravity concentrates was also treated by zinc precipitation. The precipitate was calcined and smelted in submerged electric arc furnaces with fluxes for final bullion production.

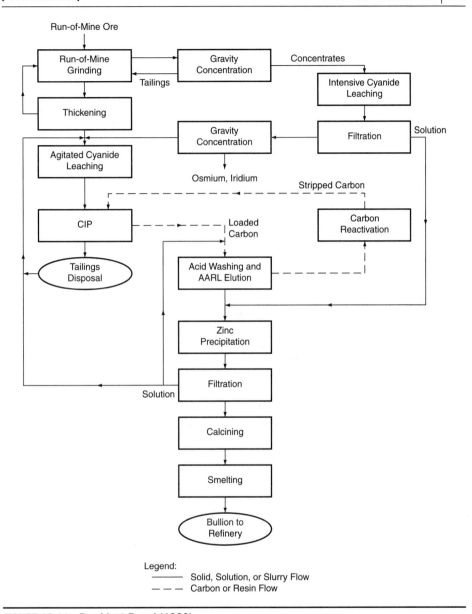

FIGURE 12.11 President Brand (1990)

12.2.2.4 Homestake Lead [9, 10]

Owner/Operator: Barrick (Toronto, Ontario, Canada)
Location: Lead, South Dakota, United States
Start-up: 1901 (mining commenced in district in 1875)
Shutdown: 2000

Mineralogical factors. Fine free gold and minor silver are associated with a predominantly chloritic-quartzite gangue. Small quantities of pyrrhotite and other minor sulfides occur in the ore.

Key process statistics (approximate; 1990 basis)

	Sand Circuit	Slime Circuit
Ore throughput rate (tpd):	2,900	2,000
Gold ore grade (g/t):	4.7	1.9
Gold recovery: (%)	92–94	92–94
Gold production (oz/y)r:	~150,000	~40,000

Ore was crushed in three stages and then ground in two stages by a rod mill–ball mill circuit to a product size of approximately 65% <75 μm (see Figure 12.12).

Slurry was separated by cycloning into coarse (sand) and fine (slime) fractions, which were treated in separate leaching circuits. The sand fraction (85% <106 μm) was vat-leached in 680-t batches for 196 hr using sequential leaching, draining, and washing stages at a sodium cyanide concentration of 0.5 g/L. The slime fraction (99% <75 μm) was agitation leached for 20 hr at 0.4 to 0.8 g/L NaCN, following preaeration for 2 hr to passivate and oxidize reactive sulfides. Overall cyanide consumption was approximately 0.5 kg/t.

In 1973, the slime treatment circuit was converted from filtration and Merrill–Crowe precipitation to CIP. Carbon was eluted by the atmospheric Zadra procedure (90°C to 95°C) for 60 hr, and gold was recovered by electrowinning in Zadra cells operated at 4.0 V and 400 amp. Gold-bearing solution from leaching of sand fraction was initially treated by Merrill–Crowe zinc precipitation, although this was converted to CIC in 1991.

Prior to 1970, 65% of the gold was recovered by gravity concentration and amalgamation before this circuit was discontinued for health reasons. The modernized gravity circuit subsequently recovered 25% to 30% of the gold into a 50% Au grade concentrate, which was processed directly by the refinery at the mine site (Section 12.2.11.1).

Tailings decant solution and mine water were treated biologically to remove complexed cyanides, metals, thiocyanate, and ammonia species to produce water suitable for discharge into streams (see Chapter 11). Water treatment facilities continued to operate after mine closure in 2000.

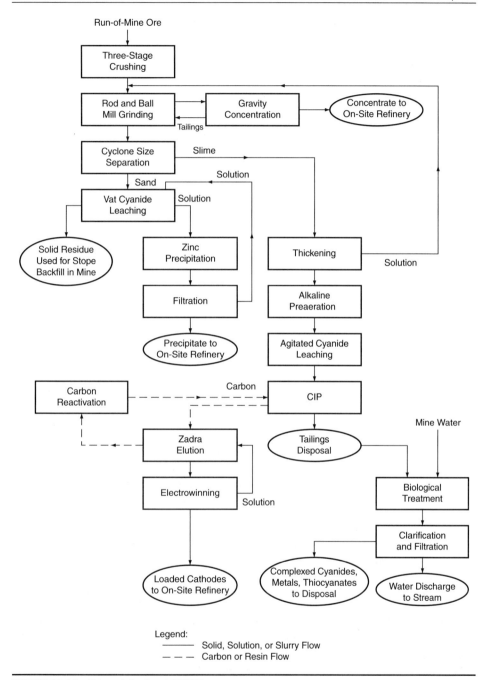

FIGURE 12.12 Homestake Lead (1990)

12.2.2.5 Chimney Creek (part of Twin Creeks) [11, 12, 13]

Owner/Operator: Newmont Mining (Denver, Colorado, United States)
Location: Winnemucca, Nevada, United States
Start-up: 1987 (original oxide ore circuit)
(NOTE: Twin Creeks currently operates a pressure oxidation circuit to process refractory sulfide material; this is not considered here.)

Mineralogical factors. Epithermal deposit of free-milling, oxidized ore overlie refractory sulfides. Major gangue minerals are limestone, dolomite, sandstones. Orebody contains heavily silicified regions and between 10% and 20% clay. Fine free gold and electrum, and minor mercury (1 to 20 g/t) occur.

Key process statistics (approximate; 1990 basis)

	Grind/Leach	Heap Leach
Ore throughput rate (tpd):	2,300	5,000
Gold ore grade (g/t):	6.0	1.2
Gold recovery (%):	95–96	65–70
Gold production (oz/yr):	~150,000	~50,000

Process description/main features. High-grade and low-grade ores from a common open pit were treated by grinding–leaching–CIP and heap leaching, respectively (see Figure 12.13). Low-grade material (0.45 to 1.80 g/t) was truck-dumped directly onto lined leach pads at run-of-mine size (<1 m) in 15-m-high lifts. The leach pads were permitted for an ultimate height of 50 m. Leach solution (pH 10.5 and 0.2 g/L NaCN) was applied at a rate of 0.4 L/min/m^2 by drip irrigation. Low-grade run-off solutions were collected in an intermediate solution pond and recycled onto leach pads for further upgrading. Higher-grade leach solutions were pumped through CIC for gold recovery. The clay content of the ore did not significantly affect heap leaching efficiency.

High-grade ore was crushed (single stage) and ground in a semiautogenous grinding (SAG)–ball mill circuit to a product size of 78% <75 μm. Cyanide was added during grinding to start dissolution as early as possible in the flowsheet. The variable hardness of the ore (i.e., silicified, unconsolidated, and clay ore types) require flexibility within the grinding circuit, which was provided by a variable-speed SAG mill and interchangeable 4- and 8-mesh SAG mill discharge vibrating screens.

Ground ore was treated in a partial counter-current thickening–leaching–CIP circuit to maximize the recovery of soluble gold to CIC (85% to 90% achieved) and to improve gold dissolution and recovery in the leaching and CIP circuit. The CIP circuit recovered the remaining 10% to 15% of soluble gold. This configuration increased overall recovery and reduced the sensitivity of the circuit to fluctuations in head grade and throughput, because the CIC circuit recovered the majority of gold and the CIP section was effectively a scavenger circuit. In addition, higher ultimate carbon loadings were achievable because the carbon performed significantly better in solution than in slurry, owing to the clay content of the ore [11].

The variable clay content of the ore periodically resulted in increased slurry viscosity in the CIP circuit and caused occasional thickening problems but had little effect on overall gold recovery.

Carbon was eluted using a pressure Zadra procedure (150°C, 450 kPa) at 5 g/L NaCN for approximately 6 hr. Zinc precipitation was selected over electrowinning for treatment of carbon eluates for the following reasons:

- Presence of approximately 3 g/t soluble Hg, which presented a health hazard in the operation of electrowinning cells

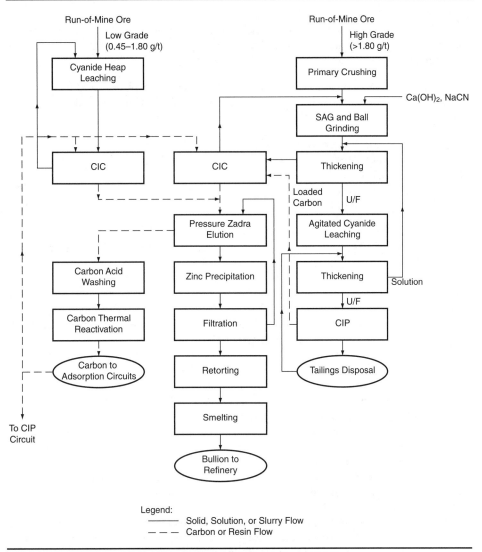

FIGURE 12.13 Chimney Creek (1990) (part of Twin Creeks in 2004)

- Considered less of a security problem
- Greater flexibility to increase plant throughput
- No clear economic advantages for electrowinning over zinc precipitation to achieve similar single-pass stage efficiency

Zinc precipitation was performed at 60°C to 65°C without any need for deaeration or lead nitrate addition. The precipitate was retorted for mercury removal and then smelted with fluxes for bullion production.

Loaded carbon fines were eluted directly on a Funda pressure filter using a presoak of 20 g/L NaCN followed by a water wash at 90°C. Gold loading on carbon fines was typically reduced from about 500 to 60 g/t by this procedure.

12.2.2.6 Round Mountain [14]

Owner/Operator: Kinross Gold and Barrick Gold (both in Toronto, Ontario, Canada)
Location: Tonopah, Nevada, United States
Start-up: 1987

Mineralogical factors. Oxidized ore (volcanic tuff) overlies sulfides. Gold is present as electrum and associated with limonite. Some free gold/electrum and minor gold are associated with pyrite.

Key process statistics (approximate; 1990 basis)
Ore throughput rate: 41,000 tpd
Gold ore grade: 1.1 g/t
Gold recovery: 75% to 80%
Gold production: ~400,000 oz/yr

Process description/main features. Ore was crushed to 80% <19 mm in three stages and conveyor-stacked on reusable asphalt leach pads (see Figure 12.14). Lime was added during crushing for pH control.

Ore was leached in two stages. Fresh material was leached with a low-grade (intermediate) solution, that is, low-grade run-off from leach pads. This produced a high-grade "pregnant" solution, which was pumped to CIC. Barren solution was applied to partially leached ore to provide a more favorable equilibrium for gold dissolution and to improve overall recovery. The total leaching cycle was 100 to 120 days.

Metal sumps were used for solution containment, rather than ponds, although ponds were available for excess solution, as the need arose. This helped to reduce environmental risk.

Leached ore was water-washed to displace residual cyanide and metals, and was then removed by truck to a waste dump.

Loaded carbon was eluted by the pressure Zadra procedure, and electrowinning was used for final gold recovery, incorporating a replating step, followed by melting of the high-purity electrolytic product.

The carbon regeneration circuit was somewhat unusual, as carbon was thermally reactivated in a horizontal kiln before acid washing to avoid corrosion in the kiln and associated equipment.

In 1990, Round Mountain was the one of largest heap leach gold mines in the world, but has since been superseded by the giant Yanacocha and Pierina operations in Peru.

Round Mountain also operates a milling and CIP operation to process higher-grade material. This is not considered further here.

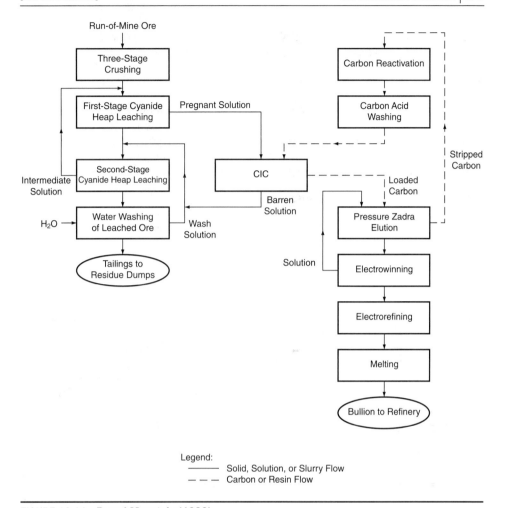

FIGURE 12.14 Round Mountain (1990)

12.2.2.7 Kidston [15]

Owner/Operator: Placer Dome Inc. (Vancouver, British Columbia, Canada)
Location: Atherton, Queensland, Australia
Start-up: 1985
Shutdown: 2001

Mineralogical factors. Kidston was a predominantly free-milling ore with oxidized, transition, and sulfide zones were present in volcanic breccia. Gangue was mainly quartz, muscovite, chlorite, and carbonates with moderate clay (kaolinite) content. Some pyrite (<2%) was present. Fine free gold was present, some intimately associated with pyrite. Copper mineralization was variable. Free gold was liberated at approximately 53 μm.

Key process statistics (approximate; 1990 basis)
Production: 18,000 tpd
Gold ore grade: 2.0 g/t
Gold recovery: 86% to 88%
Gold production: ~360,000 oz/yr

Process description/main features. Ore was crushed in a single stage (jaw crusher) followed by grinding in a two-stage (SAG–ball mill) circuit (see Figure 12.15). The grinding product size was dictated by the type of ore treated, for example, 80% <210 μm for oxide and 80% <125 μm for sulfide and transition ores.

A 24-hr leach time was provided at a throughput of 16,000 tpd. Cyanide consumption for oxide and sulfide ores was 0.45 and 0.60 kg/t, respectively. Lime consumption was between 2 and 4 kg/t. Dissolved oxygen depletion was observed in the first leach tank (i.e., 4 to 5 mg/L oxygen), and sometimes as low as 1 mg/L when treating high copper ores. The relatively low gold extractions achieved (compared with other free-milling ores) were a result of some of the gold being intimately associated with pyrite.

A unique four-stage counter-current cycloning circuit was used to remove coarse material prior to CIP. The coarse fraction was sent directly to tailings. The fine material was thickened and the underflow passed through CIP, while the overflow solution was treated by CIC. The washing efficiency achieved in the thickening circuit was 99.8%, with 65% of the soluble gold recovered in CIC and 35% recovered in CIP (see Section 12.2.2.5).

Loaded carbon was eluted by the pressure Zadra process and gold was recovered by electrowinning. Loaded cathodes were smelted with fluxes, and the resulting bullion was remelted prior to shipment. Eluted carbon was regenerated by acid washing followed by thermal reactivation.

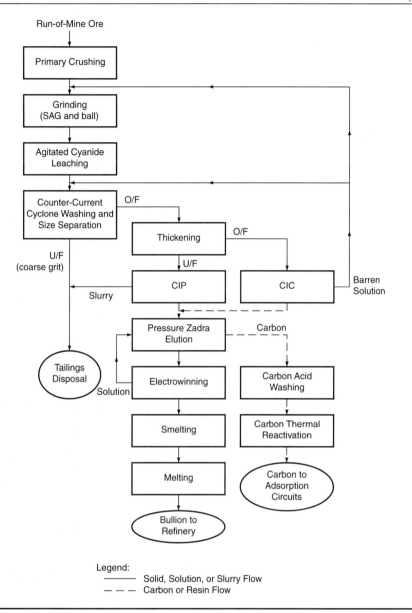

FIGURE 12.15 Kidston (1990)

12.2.2.8 Pine Creek [16]

Owner/Operator: Renison Goldfields Consolidated Ltd. (Sydney, Australia)
Location: Pine Creek, Northern Territory, Australia
Start-up: 1985
Shutdown: 2003

Mineralogical factors. Oxidized ore overlaid sulfides. Mineralization was in quartz–quartz sulfide veins. The major gangue mineral was quartz, with a variety of alteration products of other minerals (i.e., kaolin and limonite). Gold occurred as free grains up to 50 µm in diameter and as fine gold (2 to 30 µm) inclusions in sulfides, such as pyrite, arsenopyrite, pyrrhotite, and other lesser sulfides. In 1989, the primary sulfide ore content of the plant feed was approximately 50%.

Key process statistics (approximate; 1990 basis)
Ore throughput rate: 4,000 tpd
Gold ore grade: 2.4 g/t
Gold recovery: 78%
Gold production: ~85,000 oz/yr

Process description/main features. Three-stage crushing and two-stage grinding (dual ball mill circuit) was used to reduce run-of-mine ore to 80% <75 µm, which was then fed directly to the leaching circuit at 43% to 45% solids with no thickening (see Figure 12.16).

A free-milling approach was used for treatment of this semirefractory ore because this was the most economic treatment method. Approximately 16-hr retention time was provided for leaching. Hydrogen peroxide was used to enhance gold dissolution in the leaching circuit, because the presence of cyanide- and oxygen-consuming sulfides depleted the dissolved oxygen concentration excessively if air alone was used. Sodium cyanide and hydrogen peroxide consumptions were approximately 1.0 and 0.6 kg/t, respectively.

Leached slurry was treated by CIP. Tailings were thickened to allow the solution (and cyanide) to be recovered and recycled prior to disposal. Loaded carbon was treated by pressure Zadra elution followed by electrowinning for gold recovery. Loaded cathodes were calcined and smelted with fluxes for bullion production. Interestingly, the bullion contained approximately 56% Au, 25% Ag, and the balance was mainly copper.

Eluted carbon was reactivated in vertical kilns before being recycled.

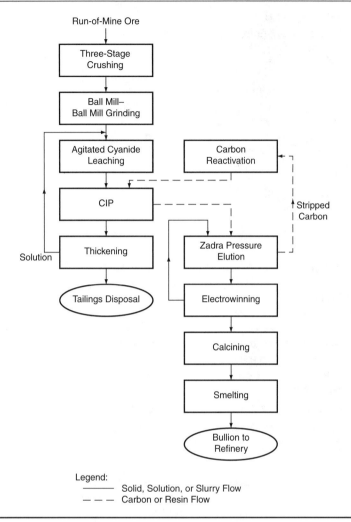

FIGURE 12.16 Pine Creek (1990)

12.2.2.9 Muruntau [17, 18]

Owner/Operator: Navoi Mining (Tashkent, Uzbekistan)
Location: Uzbekistan
Start-up: 1969

Mineralogical factors. Mineralization is in quartz veins, and free gold of variable grain size and minor gold are associated with sulfides, primarily arsenopyrite. Considerable quantities of cyanide-soluble base metals occur in the ore.

Key process statistics (approximate; 1990 basis)
Ore throughput rate: 65,000 tpd
Gold ore grade: 2.5 g/t
Gold recovery: 92% to 94%
Gold production: ~1,750,000 oz/yr

Process description/main features. With the exception of Grasberg (Indonesia), which is a copper–gold ore, Muruntau is the largest gold milling operation in the world (see Figure 12.17). Run-of-mine ore is crushed in a single stage and then ground by a single-stage SAG mill, with gravity concentration. The coarse gold concentrate is directly smelted at the site.

The ground product is thickened, leached with cyanide, and treated by RIP. The loaded resin is stripped by a complex three-stage elution scheme. The first stage removes zinc, nickel, and cyanide with dilute sulfuric acid. Ammoniacal ammonium nitrate solution is then used to remove copper, and, finally, gold and silver are eluted with thiourea. The precious metals are recovered from the thiourea solution by electrowinning. The disadvantages of this scheme are the following:

- The flowsheet is complex, with capital and operating cost implications.
- Iron and cobalt are converted into complex species during the first stage, which are difficult to desorb, resulting in poisoning of the resin.
- The consumption (and cost) of thiourea is high.

However, despite these drawbacks, the system has operated successfully for many years.

In addition to the large mill at Muruntau, previously stockpiled low-grade ores are processed by the Zarafshan project, a joint venture between Newmont and Navio Mining, using crushing and heap leaching for gold extraction.

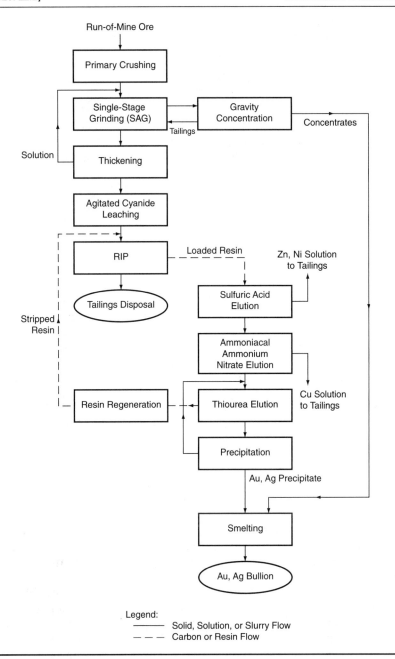

FIGURE 12.17 Muruntau (1990)

12.2.2.10 Fort Knox [19]

Owner/Operator: Kinross Gold (Toronto, Ontario, Canada)
Location: Fairbanks, Alaska, United States
Start-up: 1997

Mineralogical factors. Free-milling gold deposit is hosted within a multiphase granitic ore body. Free gold is contained in milky-white quartz stockwork veins and veinlets, as well as along shears and fractures with the granite. The quartz-filled shears and fractures contain evenly distributed gold, generally <100 µm in size. The gold in the milky-white stockwork veins and veinlets is similar in size but has a more erratic occurrence. The ore contains minor amounts of iron sulfide minerals (e.g., pyrite). Ore from the True North section of the deposit contains some refractory gold-bearing ore components.

Process statistics (2004 basis)
Ore throughput rate: 38,000 tpd
Gold ore grade: 0.8 to 1.0 g/t
Gold recovery: 88% to 92%
Gold production: ~360,000 oz/yr

Process description/main features. The Fort Knox circuit consists of primary crushing to reduce run-of-mine ore to a nominal 80% <130 mm, followed by conventional SAG and ball milling to generate a leach feed at a size of 80% <200 µm (see Figure 12.18). A unique feature of the circuit is that pinched sluices are used to split off a portion of the ball mill discharge slurry (approximately 30% of the fresh feed to grinding, or 440 tph) for gravity gold recovery in two Knelson centrifugal concentrators. The Knelson concentrators recover approximately 15% of the total gold production. Gravity tailings are returned to the grinding circuit for further comminution. The grinding circuit product (i.e., ball mill cyclones overflow) is thickened by high-capacity thickeners to yield a slurry containing 50% to 54% solids. The ground slurry is cyanide leached in a seven-stage agitated tank system providing approximately 20 hr of retention time. Leaching is conducted at a pH of 10.2 to 10.4, controlled by the addition of 400 to 550 g/t lime, and maintaining a NaCN concentration of 70 to 80 mg/L.

Gold is recovered from solution by six stages of CIP with a retention time of 6 to 10 hr, depending on mill throughput rate. (NOTE: The plant is operating at approximately 25% over the original design capacity, so retention times are less than optimal.) The final CIP tailings slurry is thickened to recover solution (with the associated gold and reagents, i.e., cyanide and lime) for reuse in the grinding and leaching circuit. The final tailings slurry is treated by sulfur dioxide–air cyanide destruction prior to disposal. Overall gold recovery varies from 88% to 92%.

One particularly interesting feature of the Fort Knox operation is the extreme variations in temperature that are experienced, resulting in leach slurry temperatures ranging from lows of <10°C in winter to highs of >25°C during the summer. This causes significant seasonal variation in gold recovery, which average 88% to 89% in winter and 91% to 92% in summer. The colder winter temperatures increase the slurry viscosity and decrease gold extraction in the leaching circuit and gold adsorption efficiency in CIP. For example, at 50% solids, the slurry viscosity increases from about 10 centipoise (cps) at 25°C to approximately 20 cps at 5°C. The increased slurry viscosity is thought to interfere with mass transport in the slurry phase, adversely affecting leaching and CIP efficiency. These effects apparently more than offset the benefit in increased dissolved oxygen content at lower temperature; for example, the dissolved oxygen content has been measured to increase from 8–10 mg/L at 25°C–35°C to 15–17 mg/L at 10°C–15°C. Fort Knox is an interesting case study in the effects of temperature, slurry density, and slurry viscosity on the efficiency of mass transport controlled reactions in gold extraction, that is, cyanide leaching and CIP.

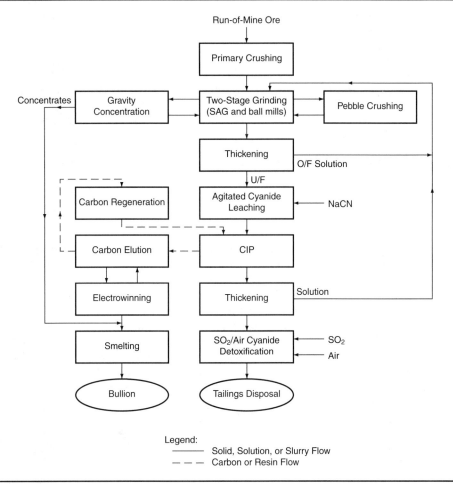

FIGURE 12.18 Fort Knox (2004)

12.2.2.11 Yanacocha [20, 21, 22]
Owner/Operator: Newmont Mining (Denver, Colorado, United States)
Location: Cajamarca, Peru
Start-up: 1993

Mineralogical factors. The deposits in the Yanacocha district are highly siliceous, containing 70% to 97% silica and 3% to 23% alunite. Gold is present as fine gold particles disseminated along fractures within the host rock. Much of the gold is accessible to solution at a coarse rock size.

Key process statistics (approximate; 2004 basis)
Ore throughput rate: 240,000 tpd
Gold ore grade: 1 to 2 g/t
Gold recovery: 70%
Gold production: 3,010,000 oz/yr

Process description/main features. Production has been steadily (and rapidly) expanded at Yanacocha since initial production in 1993 at a rate of approximately 220,000 oz/yr. In 2004, Yanacocha was the world's largest gold producer with gold production >3 million oz (or about 4% of global gold supply). Ore is processed by heap leaching of run-of-mine ore (truck-dumped directly from the open-pit mining operations) on specially prepared and lined heap leach pads (see Figure 12.19). Individual lifts are 10 m high. Solution is applied to the heaps by drip irrigation (5 to 10 L/hr/m^2) at a pH of 10.5 to 11.0. Solution pH is adjusted with lime, and lime consumption is approximately 0.8 kg/t (1994 basis). Cyanide concentration in the pregnant gold-bearing solution is maintained at about 50 mg/L weak acid dissociable (WAD) cyanide. Cyanide consumption is extremely low and averaged 0.03 kg/t NaCN early in the project life (1994 basis). High-grade pregnant leach solution is clarified (using hopper clarifiers as a preclarification step followed by clarification filters) and then gold is recovered from the solution by Merrill–Crowe zinc precipitation. The Merrill–Crowe process includes deaeration, precipitation (with zinc dust and lead nitrate), and filtration. The precipitate contains a gold–silver ratio of between 2:1 and 4:1. The precipitate is retorted to remove (and recover) mercury, then smelted with fluxes, and cast into bullion bars for shipment.

Intermediate-grade solution is processed through CIC. Loaded carbon is stripped using the AARL elution method. After stripping, the carbon eluate solution is blended with higher-grade pregnant leach solution for gold recovery by Merrill–Crowe precipitation.

Merrill–Crowe precipitation was selected over the use of carbon for final gold recovery based on estimated capital and operating cost savings.

A portion of the barren solution is treated for cyanide destruction prior to release into the environment.

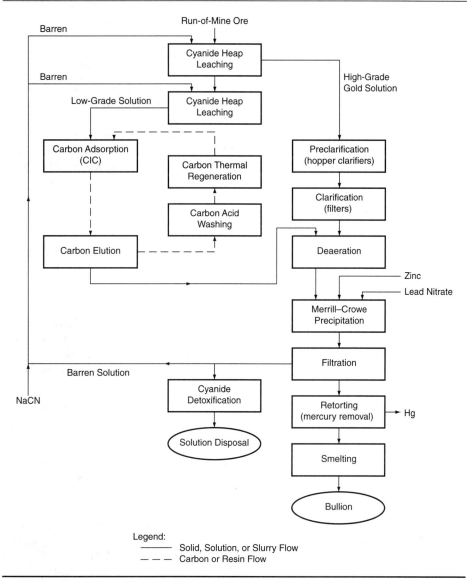

FIGURE 12.19 Yanacocha (2004)

12.2.2.12 Pierina [23, 24]
Owner/Operator: Barrick Gold (Toronto, Ontario, Canada)
Location: Huaraz, Ancash, Peru
Start-up: 1998

Mineralogical factors. The Pierina deposit is characterized by high-sulfidation alteration and mineralization, with the majority of the ore confined within a (porous) pumice tuff, overlying a basal andesite unit and overlain by lithic tuff. Gold occurs primarily as ultrafine free gold particles (<0.1 μm) disseminated within the porous tuff host rock. The gold leaches readily at a relatively coarse crush size. Silver is present as acanthite with minor amounts of native silver. Mercury occurs in association with acanthite. At depth in the deposit there is feeder zone mineralization, with gold associated with enargite and pyrite.

Key process statistics (approximate; 2004 basis)
Ore throughput rate: 27,000 tpd (crushed), ~20,000 tpd (run-of-mine)
Gold ore grade: 2.0 to 2.2 g/t Au and 15 to 25 g/t Ag (initial head grade was 4 g/t Au)
Gold recovery: 80%
Silver recovery: 30%
Gold production: 640,000 oz/yr

Process description/main features. Run-of-mine ore is primary crushed to 80% <150 mm in a gyratory crusher, followed by secondary crushing in cone crushers to 80% <38 mm (see Figure 12.20). The ore is not agglomerated, but lime is added prior to placement onto the heap. Ore is placed onto a specially prepared, lined heap leach pad, constructed in a valley-fill configuration, with a heap leach dam at the toe to collect and control solution as it flows to the base of the heap and to provide structural stability for the heap system. Crushed ore is truck-dumped onto the heap in 10-m-high lifts. Low-grade ore is directly placed on the leach pad and processed at run-of-mine size. The current plan considers an ultimate heap height of 130 m to contain approximately 100 million t of ore.

Freshly placed crushed ore is leached for 60 days, but residual leaching occurs as solution percolates down through older lifts below. Solution pH is maintained above 10 by the addition of lime to the crushed ore and by the addition of milk of lime to the barren solution. Barren leach solution is maintained at 0.5 g/L NaCN (the high cyanide concentration is required for silver dissolution). Pregnant gold-bearing leach solution is pumped out of the heap (behind the heap leach dam) and delivered to the gold recovery plant. The solution grade varies from 2 to 3 g/t Au. The solution is clarified (using hopper clarifiers followed by pressure leaf filters) and then processed in a conventional Merrill–Crowe zinc precipitation circuit. This precipitation includes deaeration to reduce the dissolved oxygen content of the solution from about 5 mg/L to less than 1 mg/L; then zinc dust, cyanide, and lead nitrate (if required) are added to the solution to precipitate precious metals. The precipitate is recovered from the solution by filtration and is retorted to remove (and recover) mercury, followed by smelting with fluxes and casting into bullion bars.

Overall metal recoveries from the crushed ore (80% <38 mm) are 80% for gold and 30% for silver. Recoveries from run-of-mine ore are significantly lower. It should be noted that the 80% gold recovery from crushed ore compares with 85% gold recovery for CIL treatment of the same ore after grinding. In the case of Pierina, the extra 5% recovery did not justify the increased capital and operating cost for the grinding and CIL treatment option.

Reagent consumptions have been reported to be 0.42 kg/t lime, 0.22 kg/t NaCN, and 1.28 kg Zn/kg bullion (1999 basis).

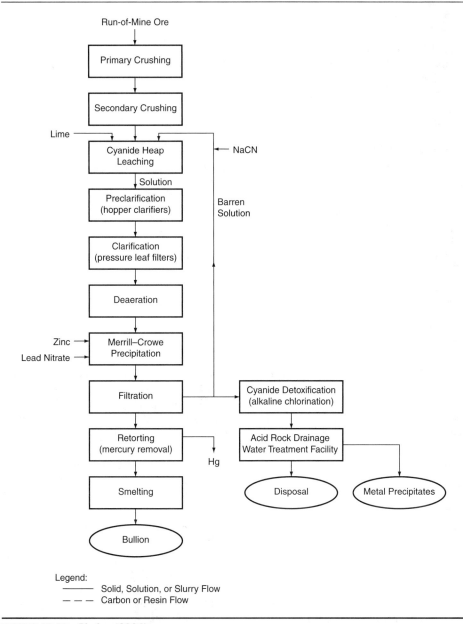

FIGURE 12.20 Pierina (2004)

Because of the high rainfall (1.2 to 1.4 m/yr) at Pierina, the water balance is aggressively managed using sprinklers and fogger nozzles for solution application to maximize water evaporation during the dry season. Excess barren solution is bled out of the circuit and treated for cyanide detoxification by alkaline chlorination (see Chapter 11). Free cyanide is destroyed and most metal cations are precipitated as hydroxides. The solids are recovered in a clarifier in the acid rock drainage water treatment facility at the site.

12.2.3 Silver-Rich Ores

12.2.3.1 Paradise Peak [25, 26]
Owner/Operator: Meridian Gold (Reno, Nevada, United States), formerly owned by FMC Corp. (Philadelphia, Pennsylvania, United States)
Location: Gabbs, Nevada, United States
Start-up: 1986
Shutdown: 1996

Mineralogical factor. Hydrothermal deposit occurred with native silver and gold, but silver also occurred as silver sulfide. Major gangue mineral was quartz, with halides, cinnabar, orpiment, realgar, and bismuth-bearing stibnite.

Key process statistics (approximate; 1990 basis)
Ore throughput rate: 4,000 tpd
Gold ore grade (g/t): 3.1 Au, 100 Ag
Gold recovery: 92% to 93%
Silver recovery: 65% to 70%
Gold production: ~130,000 oz/yr
Silver production: ~3,100,000 oz/yr

Process description/main features. Ore was crushed in three stages and ground in a single-stage ball mill to 80% <75 μm (see Figure 12.21). Ground slurry was thickened and then leached for approximately 24 hr at pH 11 in a solution containing 0.8 to 1.0 g/L NaCN. The high cyanide concentration was required to achieve satisfactory silver dissolution.

Carbon adsorption processes were unsuitable for this ore owing to the high silver-to-gold ratio of 30:1 (see Chapter 7). Consequently, the partially leached slurry was treated by six stages of counter-current decantation (CCD) and Merrill–Crowe zinc precipitation, with leaching continuing throughout the CCD solid–liquid separation process. The equilibrium for gold dissolution became progressively more favorable down the CCD circuit as the silver and gold grades in solution declined. Freshwater was introduced into the final CCD thickener for washing. Overall washing efficiency in the circuit was >99%.

Prior to disposal in the tailings facility, the final slurry tailing was treated with Fe(II) sulfate to convert free cyanide to the less toxic Fe(II) cyanide complex. Such treatment was necessary because of the high residual cyanide concentrations that resulted from the need to maintain relatively high levels of cyanide in solution during leaching of a high-silver ore (see Chapter 6). The addition of freshwater into the final CCD thickener diluted the cyanide and helped to reduce detoxification requirements.

Low-grade ore was crushed and heap leached. The pregnant silver–gold solution produced by this process was used for dilution in the grinding circuit and elsewhere in the process, avoiding the need for a separate recovery circuit.

The solution overflow from the first-stage CCD thickener was clarified, deaereated, and then treated by Merrill–Crowe zinc precipitation for silver and gold recovery. To remove excess zinc, the resulting precipitate was leached with sulfuric acid and then retorted to remove mercury. Mercury was recovered by condensation and the exhausted gases passed through activated carbon for trace mercury recovery. Finally, the precipitate was smelted with fluxes and remelted to produce the final bullion.

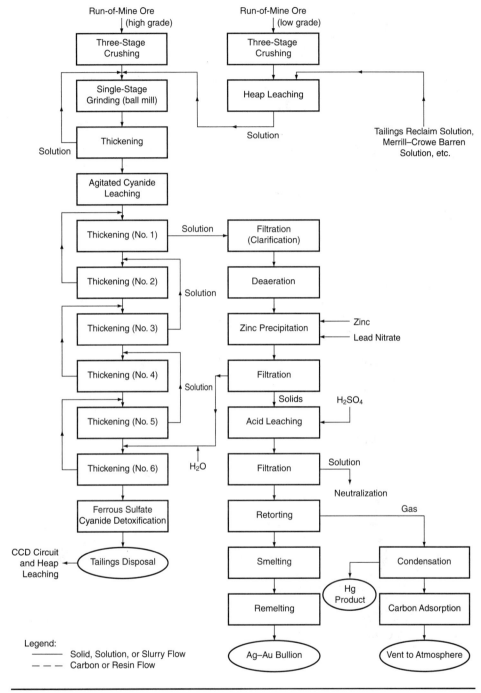

FIGURE 12.21 Paradise Peak (1990)

12.2.3.2 Eskay Creek [27]

Owner/Operator: Barrick Gold (Toronto, Ontario, Canada)
Location: British Columbia, Canada
Start-up: 1990

Mineralogical factors. High-grade silver and gold mineralization are intimately associated with complex polymetallic sulfides, including sphalerite, galena, chalcopyrite, and pyrite. Typical gold and silver grades are 65 to 70 g/t and 2,500 to 3,000 g/t, respectively. Also, base metal contents of 5.0% Zn, 2.5% Pb, and 0.5% Cu are typical.

Key process statistics (approximate; 2004 basis)
Ore throughput rate: 360 tpd (gravity and flotation mill) (Some ore is direct shipped to third-party smelters.)
Gold ore grade: 65 to 70 g/t Au, 2,500 to 3,000 g/t Ag
Gold recovery: 95% (estimate)
Silver recovery: 95% (estimate)
Gold production: 290,000 oz/yr
Silver production: 16 million oz/yr

Process description/main features. Extensive process development work was carried out on ore from Eskay Creek (see Figure 12.22). Considered process options included direct cyanidation, flotation, gravity concentration, direct smelting of ore, roasting–cyanidation–CIL, and pressure oxidation–cyanidation–CIL. When the ore did not respond well to direct cyanidation, flotation, and gravity concentration, these options were rejected. Pressure oxidation and cyanidation was hampered by several factors, including: the formation of silver jarosite, the preg-robbing characteristics of the oxidized material, and the high soluble zinc and copper concentrations in the leach feed material. Several innovative flowsheets were developed to overcome these problems, two of which are summarized as follows:

- Autoclave oxidation of the whole ore, followed by washing of the oxidized solids to remove solubilized copper, zinc, and other base metals. The washed solids were subjected to a lime boil to destroy the silver-bearing jarosite formed during oxidation, thereby releasing the silver for recovery by cyanidation.

- Autoclave oxidation of the whole ore with minimization of jarosite formation by increasing the amount of soluble zinc in the autoclave feed, making more silver available for recovery by subsequent cyanidation.

After much excellent process development work and analysis, it was determined that shipping the high-grade ore off-site for direct smelting by third-party sulfide smelters provided the best economic value to the project. Lower-grade ore is processed in a small gravity and flotation circuit, with the concentrate shipped to a third-party smelter. This is an example where a simple solution provided a cost-effective alternative to a rather complex (but elegant) metallurgical flowsheet.

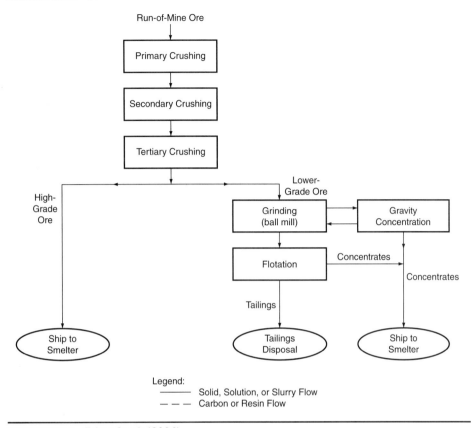

FIGURE 12.22 Eskay Creek (2004)

12.2.4 Refractory Iron Sulfide Ores

12.2.4.1 Joutel [18, 28, 77]

Owner/Operator: Agnico-Eagle Mines Ltd. (Toronto, Ontario, Canada)
Location: Joutel, Quebec, Canada
Start-up: 1974
Shutdown: 1994

Mineralogical factors. Fine-grained gold was intimately associated with pyrite (10% to 20% of the ore) as fine inclusions, and with minor arsenopyrite. Liberation size of gold was less than 35 µm. The ore was considered to be refractory because of high cyanide and oxygen consumptions that were experienced without preaeration, coupled with the retarding effect of dissolved sulfur species on gold dissolution.

Key process statistics (approximate; 1992 basis)
Ore throughput rate: 1,650 tpd
Gold ore grade: 5.8 g/t
Silver grade: 2.2 g/t
Gold recovery: 91%
Silver recovery: 84%
Gold production: ~100,000 oz/yr
Silver production: ~35,000 oz/yr

Process description/main features. Two-stage crushing followed by two-stage grinding (rod–ball mills) was used to generate a product containing 60% to 70% <37 µm. Ground ore was directed to a flotation circuit which recovered a bulk sulfide and free gold concentrate using amyl xanthate collector and a frother. Gold recovery to the concentrate averaged 98%, at a concentrate grade of 20 to 30 g/t.

The treatment scheme for the concentrate depended on the ore type processed, which varied depending on the ore source. The flowsheet shown in Figure 12.23 was applied to treat sulfidic ores with a mildly preg-robbing carbonaceous component. In this case, the flotation concentrate was reground to 90% to 95% <37 mm to improve gold liberation prior to cyanide leaching. The slurry was thickened and then treated by preaeration (i.e., low-pressure oxygen oxidation) using lime (pH 11.5 to 12.0), lead nitrate (to assist with passivation of reactive sulfides), and oxygen-enriched air for aeration. Between 16 and 32 hr was allowed for oxidation. The slurry was then leached with cyanide for 48 to 60 hr.

An alternative flowsheet configuration (see Figure 12.24) allowed the flotation concentrate to be treated directly by preaeration, followed by a primary stage of leaching with cyanide. The partially leached slurry was cycloned and the coarse fraction reground and recirculated. The cyclone overflow was thickened and leached further in the leach circuit described.

The main difference between the two flowsheets is that the first provided less leaching retention time prior to solid–liquid separation (i.e., first-stage filtration), which reduced the adsorption of dissolved gold onto carbonaceous ore constituents. In both circuit configurations, the leach solution was replaced by filtration in the middle of the leaching circuit to remove sulfur–solution species (which impair gold leaching) and to improve overall gold dissolution. Leached slurry was thickened, and the overflow solution was clarified, deaerated, and treated by Merrill–Crowe precipitation for gold recovery. The precipitate was directly smelted with fluxes to produce bullion (75% Au, 22% Ag).

The thickener underflow was washed by two-stage filtration, and the final solids were repulped and pumped to tailings disposal. The filtrate solutions were recycled counter-current to the slurry flow.

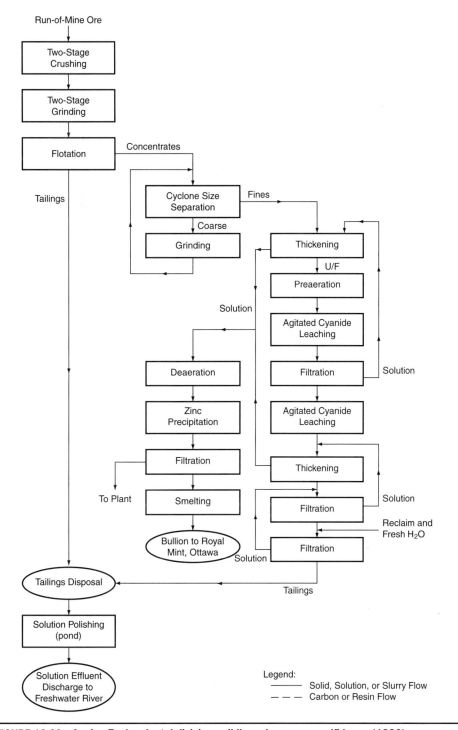

FIGURE 12.23 Agnico Eagle—Joutel division: mildly carbonaceous sulfide ore (1990)

In the tailings impoundment, cyanide in the tailings slurry naturally decomposed. The tailings solution was decanted into a "polishing" pond with approximately 10 months' storage capacity, where further cyanide degradation and metals precipitation took place. The final effluent solution, which met government permit requirements, was periodically discharged into a freshwater river.

Mining at the Eagle and Teibel mines and processing operations at Joutel were phased out in late 1993/early 1994 as a result of ore reserve depletion, and the mill was dismantled. All decommissioning and rehabilitation of the site was completed by 2000.

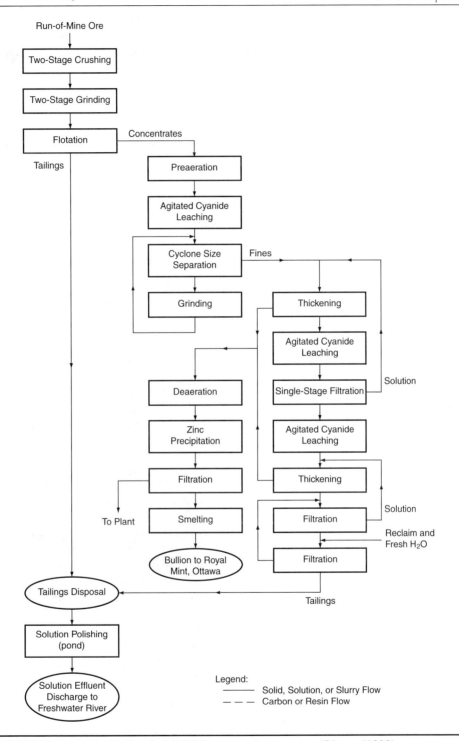

FIGURE 12.24 Agnico Eagle—Joutel division: noncarbonaceous sulfide ore (1990)

12.2.4.2 Fimiston–Gidji [29]

Owner/Operator: Kalgoorlie Consolidated Gold Mines Pty. Ltd., a joint venture between Barrick Gold (Toronto, Ontario, Canada) and Newmont Mining (Denver, Colorado, United States)
Location: Kalgoorlie, Western Australia
Start-up: 1989–1990

Mineralogical factors. Gold is predominantly associated with pyrite as fine inclusions but also as some fine free gold. Direct cyanidation on this ore yields 60% to 70% gold recovery.

Key process statistics (approximate; 1990 basis)
Ore throughput rate: 5,700 tpd
Gold ore grade: 3.8 g/t
Gold recovery: 87% to 89% (overall)
Gold production: 220,000 oz/yr

Process description/main features. The Fimiston plant is one of several that produces concentrates that are treated by a centrally located roasting and leaching plant at Gidji (see Figure 12.25). The roaster plant is sited at a location that minimizes the environmental impact of the sulfur dioxide gas generated.

At Fimiston, three-stage crushing and single-stage grinding are employed to reduce run-of-mine ore to 80% <105 µm. Ground ore is treated by flotation to produce a gold-bearing pyrite concentrate, which is washed, filtered, and shipped to the Gidji plant. A gold recovery of 80% to 85% is achieved to the flotation concentrate.

The flotation tailings are thickened and then cyanide-leached to scavenge an additional 12% to 15% gold recovery. Gold is recovered from the leached slurry by CIP, and the final tailings are thickened to allow the solution to be recycled prior to tailings disposal. Efficient use of water is an important factor at Fimiston, and process water is reclaimed and reused whenever possible in the circuit.

The Gidji roaster plant was believed to be the first commercial application of (Lurgi) circulating fluidized-bed roasting technology to treat sulfuric gold ores. The pyritic concentrates are fed to the roaster as a slurry (70% solids), and the feed rate adjusted to give a constant temperature differential across the roaster to minimize sulfate formation. This is particularly important because no washing stage follows roasting. The roaster calcine is quenched, leached with cyanide, and gold recovered by CIP. Between 88% and 90% of the gold in the flotation concentrates is recovered in this manner.

Loaded carbon from both the Fimiston and Gidji circuits is treated in a single plant by AARL elution. Gold is recovered by electrowinning, and the bullion produced is shipped directly to the Perth Mint refinery. Before being recycled to the various adsorption circuits, stripped carbon is thermally reactivated.

Ultrafine grinding of the flotation concentrate was introduced at Fimiston–Gidji in 2003–2004, followed by agitated cyanide leaching of the ground product, avoiding the need to roast a portion of the sulfide concentrates produced by flotation.

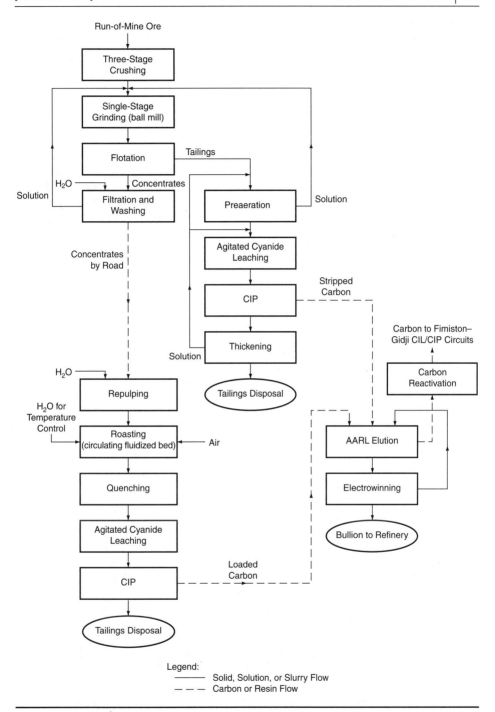

FIGURE 12.25 Fimiston–Gidji (1990)

12.2.4.3 McLaughlin [30, 31]

Owner/Operator: Barrick Gold (Toronto, Ontario, Canada), formerly owned by Homestake Mining (San Francisco, California, United States)
Location: Clear Lake, California, United States
Start-up: 1985
Shutdown: 2002

Mineralogical factors. McLaughlin was an epithermal deposit containing refractory sulfidic gold ore. Fine-grained gold (20 µm) was associated with fine-grained sulfide minerals (40 µm). Pyrite was the major sulfide, with minor marcasite, chalcopyrite, sphalerite, and cinnabar. Gold occurred as electrum, fine grained primarily associated with silver, antimony sulfosalts (e.g., miargyrite ($AgSbS_2$), pyrargyrite (Ag_3SbS_3), feibergite (($Cu,Ag,Fe,Zn)_{12}Sb_4S_{13}$), and polybasite (($Ag,Cu)_{16}Sb_2S_{11}$)). Refractory properties were caused by gold locking within sulfide grains, carbonaceous materials and clays, coating of gold with hematite, clays and jarosite, and, finally, locking within sulfosalts. The ore contained 3.8% to 4.0% sulfide sulfur and 0.4% to 0.5% carbonate. Direct cyanidation of this ore yielded between 5% and 80% extraction.

Key process statistics (approximate; 1990 basis)
Ore throughput rate: 2,700 tpd
Gold ore grade: 4.7 g/t
Gold recovery: 90%
Gold production: 130,000 oz/yr

Process description/main features. Ore was primary crushed and then ground in a two-stage circuit (SAG–ball mills) to 80% <75 µm (see Figure 12.26). The ground product was thickened and then mixed with acidic solution, recycled from the oxidation circuit, to react with carbonate constituents of the ore. This helped to reduce carbon dioxide generation in the pressure oxidation circuit. The acid-treated slurry was thickened, and the overflow solution was neutralized with calcium hydroxide. Precipitated salts were removed by a further thickening step, and the solids reported directly to tailings. Later in the circuit, the solution was used for neutralization.

The pretreated slurry was preheated in two stages, using heat recovered from the oxidized slurry. The slurry was fed into four-compartment autoclaves (a total of three units), in which the sulfide minerals were oxidized and the contained gold was liberated. Oxygen was sparged into the autoclaves as the oxidant, and temperature and pressure in the autoclave were maintained at 180°C to 190°C and 2,200 kPa, respectively. The acid concentration in the autoclave discharge was maintained at 15 to 18 g/L H_2SO_4.

Oxidized slurry was released from the autoclave through a choke valve and the pressure was "let down" through two-stage flash tanks. The partially cooled (and depressurized) slurry was thickened in two stages to allow a portion of the acid generated to be recycled for acid pretreatment of the ore and to partially neutralize the oxidized slurry. The thickened product was fully neutralized with calcium hydroxide and adjusted to a pH suitable for cyanide leaching (i.e., pH 10.5). The slurry was treated by two stages of cyanide leaching and eight stages of CIP for gold recovery, with a total retention time of 14 hr.

Loaded carbon was eluted using a pressure Zadra procedure and gold was recovered by electrowinning, followed by retorting of the product (for mercury removal and recovery) and smelting with fluxes. Eluted carbon was thermally reactivated and then acid washed prior to recycling to the adsorption circuit.

[CH. 12 SEC. 12.2] INDUSTRIAL APPLICATIONS | 549

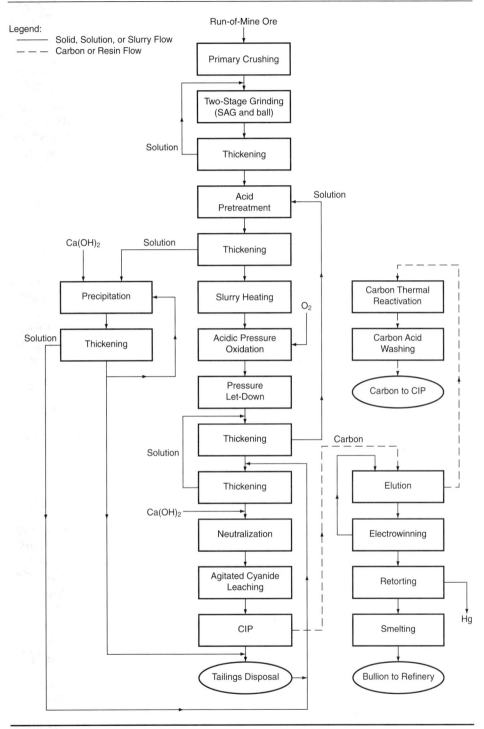

FIGURE 12.26 McLaughlin (1990)

12.2.4.4 Lihir [32, 33, 34]

Owner/Operator: Lihir Gold Ltd. (Port Moresby, Papua New Guinea)
Location: Lihir Island, Papua New Guinea
Start-up: 1997

Mineralogical factors. The Lihir project consists of two adjacent deposits, Minifie and Lienitz. The higher-grade deposit, Minifie was mined first. The Lienitz deposit was scheduled to become the primary ore source in 2005. Gold occurs as submicron particles locked within a pyrite matrix. The average sulfide sulfur grade of the deposit is 7.2% (i.e., 14% to 15% pyrite in ore feed). The ore contains minor amounts of other sulfide minerals.

Key process statistics (approximate; 2003–2004 basis)
Ore throughput rate: 11,400 tpd
Gold ore grade: 5.0 g/t
Gold recovery: 88% to 90%
Gold production: 600,000 oz/year

Process description/main features. Ore is primary crushed and then ground in a two-stage circuit (SAG and ball mills with pebble crushing) to 80% <106 µm (see Figure 12.27). The ground product is washed (to remove soluble chloride species present in the ore) and thickened. A portion of the thickener overflow solution is reused in the grinding circuit, another portion is used as wash solution following pressure oxidation (before cyanide leaching), and the remainder is used in the neutralization step prior to cyanide leaching.

The washed and thickened slurry is stored in pre-oxidation tanks. The slurry is mixed with recycled oxidized slurry to allow the acid to react with calcium and magnesium carbonate constituents in the ore. This is important because it prevents the release of CO_2 in the autoclaves, which would require additional venting, resulting in heat loss and higher oxygen consumption. After pre-oxidation, the slurry is preheated using steam generated from the pressure let-down system, and is then fed into one of three autoclaves configured in parallel. Each autoclave contains six compartments and eight agitators (three in the larger first compartment). Pressure oxidation is conducted at 205°C and 2,650 kPa to oxidize essentially all of the sulfide sulfur (mainly pyrite) in the feed. The minimum operating temperature is 190°C. Oxygen and cooling (quench) water are sparged into the autoclaves as required. Autoclave throughput rate is limited by the overall heat balance in the first compartment. If the heat capacity (i.e., sulfide sulfur content) of the feed is too low or the feed rate too high, the autoclaves will not operate autogenously and the feed rate must be reduced to maintain minimum temperature. The throughput rate is maximized at a sulfide sulfur content of between 5.0% and 6.5%. Lihir achieves an average autoclave operating availability of 86%.

The stoichiometric oxygen requirement is 1.87 t per ton of sulfur in the feed. At Lihir, the oxygen–sulfur ratio is controlled between 1.9:1.0 and 2.0:1.0. If too little is added, the oxidation kinetics are slow and sulfide sulfur oxidation is incomplete. If too much oxygen is added, heat is lost from the autoclave through venting of excess oxygen entering the autoclave. The use of too much oxygen also decreases oxygen utilization and increases process costs.

After pressure oxidation, the slurry pressure is let down in a single stage of "flashing," with a heat recovery circuit used to recycle steam to the slurry preheat step. The slurry is subjected to two stages of CCD washing and thickening to remove soluble sulfate species and acid. The second thickener underflow is then neutralized with lime to pH 10. Gold is leached from the slurry in an agitated cyanide leaching circuit in conjunction

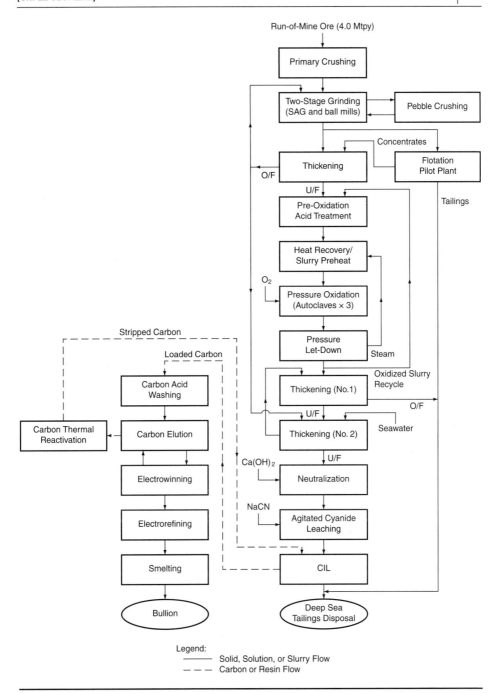

FIGURE 12.27 Lihir (2004)

with CIL for gold recovery from solution. After the loaded carbon is acid washed using 3% HCl solution, the gold is desorbed from the carbon using a continuous AARL elution system. Gold is recovered from the eluate by electrowinning, followed by smelting to produce a bullion product at site.

The CIL slurry tailing is combined with CCD circuit wash water (and some other process streams) and discharged into the sea at a depth of 125 m below surface using a deep sea tailings disposal system.

Throughput at Lihir has increased from the design of 2.8 Mtpy to just over 4.0 Mtpy as a result of multiple de-bottlenecking projects and improvements.

Plans are underway to increase throughput further to approximately 6.5 Mtpy by the addition of crushing, grinding, and flotation equipment, as shown in Figure 12.28. This is an innovative flowsheet because it not only increases the plant throughput rate, but it allows the sulfide sulfur content of the pressure oxidation feed to be increased to optimal levels when material with lower sulfur (and gold) content must be processed. This is particularly important as gold (and sulfur) content decreases at Lihir over time.

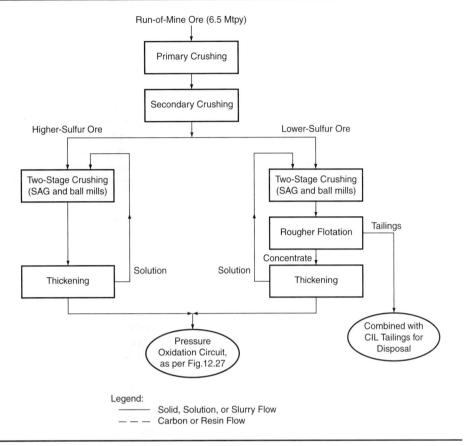

FIGURE 12.28 Lihir (potential 2006)

12.2.4.5 Porgera [35, 36]
Owner/Operator: Placer Dome Inc. (Vancouver, British Columbia, Canada)
Location: Mount Hagen, Enga Province of the Western Highlands, Papua New Guinea
Start-up: 1989

Mineralogical factors. The Porgera deposit is described as an intrusive system of mafic alkaline stocks and dykes within carbonaceous and calcareous sediments. There are two main types of gold mineralization. The first is refractory gold occurring as submicroscopic particles finely disseminated within a sulfide matrix (mainly pyrite, with minor sphalerite, chalcopyrite, galena, and tellurides), which is dispersed throughout the carbonaceous and calcareous sediments. The second type consists of epithermal quartz–roscoelite–carbonate gold veins associated with the Roamane fault zone, which contains significant free gold as electrum.

Key process statistics (approximate; 2004 basis)
Ore throughput rate: 17,500 tpd
Gold ore grade: 4 to 5 g/t
Gold recovery: 83%
Gold production: 700,000 oz/yr

Process description/main features. Ore is primary crushed and then ground in a two-stage grinding circuit comprising SAG mills with recycle pebble crushing followed by ball milling in closed circuit with cyclones. The final ground product (cyclone overflow) has a particle size of 80% <106 µm. The ground product is subjected to rougher flotation followed by cleaning of the rougher concentrate, as well as scavenging and cleaner scavenging of the rougher tails (see Figure 12.29). The cleaner concentrate and scavenger cleaner concentrates are combined and reground to 80% to 90% <38 µm prior to pressure oxidation. The final concentrate contains approximately 14% sulfide sulfur and 50 g/t gold.

Prior to pressure oxidation, the concentrate is mixed with recycled acidic slurry from the autoclave flash let-down discharge. The recycled acid slurry contains about 40 g/L H_2SO_4. This allows the sulfide sulfur content of the pressure oxidation feed to be adjusted to optimal levels (i.e., 9% to 9.5% S) for the heat balance and also allows reactive carbonates (e.g., magnesite, dolomite, and siderite) to react with acid prior to entering the autoclave(s). This latter factor is important because any CO_2 released in the autoclave(s) results in excess venting requirements and the associated loss of oxygen and heat. Pressure oxidation is conducted in four autoclaves operated at 190°C to 200°C at 1,725 kPa. Retention time in the autoclaves is approximately 2 hr. Each autoclave has five compartments and seven agitators (three in the first compartment). Average autoclave utilization is >90%.

Slurry discharges the autoclaves, and the pressure is let down through a single-stage of flashing. After splitting a portion of the slurry for recycling to the pre-oxidation treatment step, the slurry is washed and dewatered in two stages of thickening to remove the bulk of the acid, dissolved sulfosalts, and dissolved metals (i.e., zinc, copper, iron, etc.) prior to cyanide leaching. The wash solution is neutralized and the base metals precipitated by the addition of sodium sulfide and lime, and the product is combined with flotation and CIP tailings slurries. The final combined and treated tailings product is sent to tailings disposal.

The pH of the two-stage thickener underflow slurry is adjusted to about 10.5, and the slurry is treated by agitated cyanide leaching (seven stages) and CIP (six stages) for gold recovery. Carbon eluate is processed by electrowinning, followed by retorting to remove (and recover) mercury. The final product is smelted into bullion for shipment. Barren carbon is acid washed and, optionally, thermally regenerated before returning to the CIP circuit.

Overall gold recovery is estimated to be approximately 83% (i.e., 87% to 88% flotation recovery and 94% to 95% recovery from concentrate by pressure oxidation, leaching, and CIP).

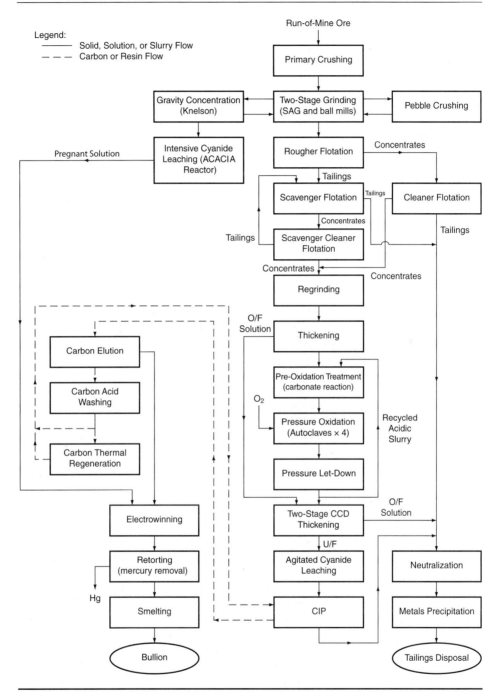

FIGURE 12.29 Porgera (2004)

12.2.4.6 Goldstrike (refractory sulfide pressure oxidation) [37, 38, 39]
Owner/Operator: Barrick Gold (Toronto, Ontario, Canada)
Location: Elko, Nevada, United States
Start-up: 1988

Mineralogical factors. Gold is present primarily in colloidal form within pyrite and marcasite mineral grains as 50 to 1,000 Å inclusions. Gold is refractory to conventional recovery techniques, such as cyanidation and gravity concentration. In sulfide minerals, gold concentration varies between 1 and 400 g/t, and the concentration increases with decreasing sulfide mineral grain size. The bulk of the fine-grained sulfide minerals is finely disseminated within the gangue. Major gangue minerals are quartz, kaolinite, and, significantly, about 10% calcite. Microcrystalline quartz is a minor carrier of invisible gold (1 to 7 g/t). The ore contains some clay minerals.

Key process statistics (approximate; 1994 basis)
Ore throughput rate: 13,600 tpd
Gold ore grade: 8.5 g/t
Gold recovery: 90%
Gold production: 1,200,000 oz/yr

Process description/main features. Refractory sulfide ore is primary crushed and then ground in two stages (SAG and ball mills) to 80% <90 μm (see Figure 12.30). The ground slurry is thickened and then pretreated using an acidification process whereby sulfuric acid is mixed with the slurry in agitated tanks. Although up to 75 kg/t acid may be added for acidification, the minimum amount of acid is used to minimize downstream lime consumption in the neutralization step following pressure oxidation. This allows the carbonate minerals (about 2% to 20% of the ore) to react and evolve CO_2 prior to entering the autoclaves. This is important because any CO_2 evolved in the autoclaves must be vented to the atmosphere, resulting in a loss of heat and oxygen. The acidification process also releases any gold associated with carbonates and maximizes the degree of oxidation achieved in the autoclaves. The slurry feed to pressure oxidation contains 2.5% to 3.0% sulfide sulfur. A noteworthy feature of this flowsheet is the lack of any recycle of acidic solution or slurry from the pressure oxidation product.

The slurry is heated using two stages of splash heating, the first to 96°C and the second stage to about 170°C. Slurry heating is accomplished using heat recovered from the flash let-down system on the autoclave discharge, supplemented by steam as required by the heat balance. Slurry is fed into six five-compartment autoclaves fitted with five agitators. Pressure oxidation is carried out at 215°C and 2,900 kPa, with a retention time of about 52 min. Oxygen is added into the autoclaves at a rate of 3.12 kg/kg S. Oxygen utilization is >60%. The slurry discharges the autoclaves and the pressure is let down using two stages of flashing. (NOTE: In the original circuit at Goldstrike, three stages of splash heating and flash let-down were used.) The slurry is further cooled in heat exchangers and neutralized with lime and the pH adjusted to about 11, requiring approximately 37 kg/t CaO. In the presence of activated carbon (i.e., CIL), the slurry is subjected to agitated cyanide leaching to present adsorption of gold cyanide species onto the ore constituents. Leaching–CIL retention time is about 10 hr, and NaCN consumption is 0.75 kg/t.

Loaded carbon is acid washed and then stripped using the pressure Zadra method. Gold is recovered from the carbon eluate solution by electrowinning. The electrowon product is filtered and then retorted to remove (and recover) mercury. The resulting high-grade material is smelted with fluxes and cast into bullion.

Carbon is regenerated thermally in a rotary kiln and returned to the CIL circuit. Carbon consumption averages 45 g/t.

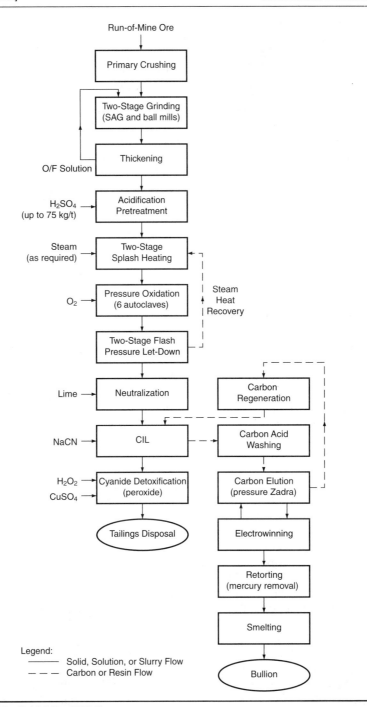

FIGURE 12.30   Goldstrike (1994)

Goldstrike also operates a roasting circuit for oxidative pretreatment of refractory carbonaceous ore (see Chapter 5), followed by agitated leaching and CIL, as well as a heap leaching and CIC circuit. Oxide ore has also been historically processed by grinding, agitated cyanide leaching, and CIP/CIL.

12.2.4.7 Lone Tree [40, 41, 42]

Owner/Operator: Newmont Mining (Denver, Colorado, United States)
Location: Golconda, Nevada, United States
Start-up: 1994—original whole-ore pressure oxidation circuit
          1997—flotation preconcentration followed by pressure oxidation

Mineralogical factors. Gold occurs as very fine (micron-sized) particles encapsulated in pyrite and other sulfide minerals, including arsenopyrite. The ore contains two distinct forms of pyrite: coarse-grained and fine-grained. The coarse pyrite contains little gold, whereas the fine-grained pyrite contains most of the gold, with concentrations of 50 to 75 g/t gold in pyrite. The fine-grained pyrite is relatively easily oxidized, and only partial oxidation is required to liberate gold for subsequent cyanide leaching.

Key process statistics (approximate; 1998 basis)
Ore throughput rate: 5,000 tpd (refractory sulfide flotation mill)
Gold ore grade: ~2.2 g/t
Gold recovery: 90% to 92%
Gold production: 280,000 oz/yr (including carbonaceous refractory concentrates shipped off-site to Carlin and Twin Creeks for processing, and on-site heap leach production)

Process description/main features. Run-of-mine ore is placed onto three stockpiles based on sulfide sulfur concentration and is reclaimed from the stockpiles in a manner that blends the sulfide sulfur content to the desired level. Because of the fine size of as-mined ore, the ore is not primary crushed, but is ground in a conventional two-stage grinding circuit using SAG and ball mills to yield a product at a size of 80% <75 μm. The ground product is thickened in a high-rate thickener to produce an underflow slurry at 50% solids. In the original Lone Tree flowsheet, the thickened slurry was processed directly by pressure oxidation. However, the flowsheet was modified in 1997 to include flotation as a preconcentration step to increase the sulfide content of the pressure oxidation feed as ore gold and sulfide sulfur feed grade declined. In the modified flowsheet (shown in Figure 12.31), ground whole ore and sulfide flotation concentrates are blended for feed to the autoclave. The flotation technology employed at Lone Tree uses an inert nitrogen atmosphere (for both conditioning and flotation) and is referred to as the N_2TEC process by Newmont (see also Section 9.2.7.4).

    The thickened, ground product is fed to a rougher conditioner, operated under a nitrogen atmosphere. Potassium amyl xanthate is added as the collector, together with lead nitrate as an activator. Flotation is conducted at low pH (i.e., about 5) and at low potential (between −100 and −500 mV vs. Ag/AgCl) using low dosage of the collector (i.e., 10^{-4} to 10^{-5} M). The mechanism of flotation in the nitrogen atmosphere does not involve the formation of dixanthogen but appears to involve the adsorption of the metal xanthate, lead amyl xanthate, at the pyrite surface [40]. The slurry then flows into the rougher flotation circuit, also operated under nitrogen. The concentrate from the first bank of roughers is final concentrate. The concentrate from the second rougher bank (referred to as rougher scavengers in the flowsheet) is cleaned in two stages, also under nitrogen. The cleaner concentrate is combined with the first rougher bank concentrate, thickened and then either shipped off-site to the Newmont Carlin or Twin Creeks operations for further processing (as shown in Figure 12.32), or may be treated on-site by pressure oxidation at Lone Tree.

    The rougher tailing and cleaner tailing are combined, thickened, and then subjected to agitated cyanide leaching (one stage) and CIL (two stages). The residue is treated with Caro's acid to destroy residual cyanide and then sent to tailings disposal.

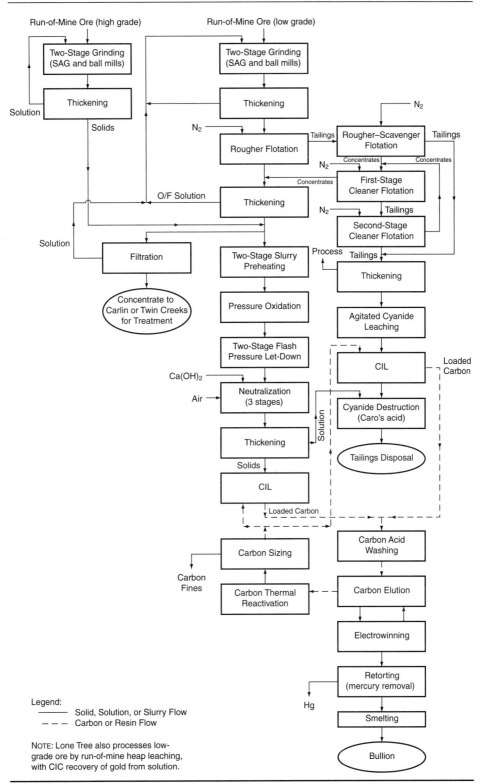

FIGURE 12.31 Lone Tree (1998)

The gold and sulfide sulfur recovery into the combined flotation concentrate are about 77%–80% and 88%–90%, respectively. Flotation achieves a concentration ratio of about 8 to generate a final concentrate containing about 16 g/t gold and 18% sulfide sulfur in approximately 11% by weight of the feed.

The portion of concentrates that are processed at Lone Tree are fed to two stages of slurry heating using steam heat from flash let-down. It should be noted that flotation concentrates produced from carbonaceous material are shipped off-site to Carlin or Twin Creeks for processing, depending on the ore type as shown in Figure 12.32. The preheated slurry is fed into the autoclave where the slurry is oxidized at about 196°C and 1,860 kPa for approximately 48 min. Only partial oxidation is necessary (as previously stated) and between 60% and 70% sulfide sulfur oxidation is generally sufficient to achieve 90% gold recovery in subsequent cyanidation processing. Oxygen is injected into the autoclave to maintain an oxygen overpressure of about 520 kPa. (NOTE: The autoclave was originally designed to operate at 180°C to achieve 75% sulfur oxidation. Actual operation required a higher temperature to achieve a satisfactory degree of oxidation.) Control of sulfide sulfur in the autoclave feed is critical to the success of the pressure oxidation step and must be maintained between 2.75% and 3.0% for optimal operation. The slurry exits the autoclave and the pressure is let down in two stages of flash cooling. The cooled slurry is neutralized with lime in three stages, with a lime consumption of 27 kg/t (approximately 60% of the lime consumption for an equivalent total pressure oxidation process), and the pH adjusted to >10 for subsequent cyanidation. Air is injected into the neutralization tanks to oxidize ferrous iron to ferric, in order to reduce the cyanide consumption in the downstream leaching process. Gold is leached and recovered from the slurry in a six-stage CIL circuit with a retention time of 24 hr. Gold recovery through the CIL circuit is approximately 88% to 90%. The final tailings residue is treated with Caro's acid (peroxymonosulfuric acid) to reduce the concentration of free and WAD cyanide species to acceptable levels.

Loaded carbon from the CIL circuits is acid washed with dilute nitric acid and then pressure stripped at 140°C (pressure Zadra process). Gold is recovered from the eluate solution by electrowinning. Stripped carbon is thermally regenerated and then sized (to remove carbon fines) before being returned to the CIL circuit(s).

Lone Tree also operates a run-of-mine heap leaching operation to process low-grade ore. Solution from the heap is processed by CIC, and the gold is stripped from carbon in a carbon elution circuit separate from that used for the milling and agitated leaching operation.

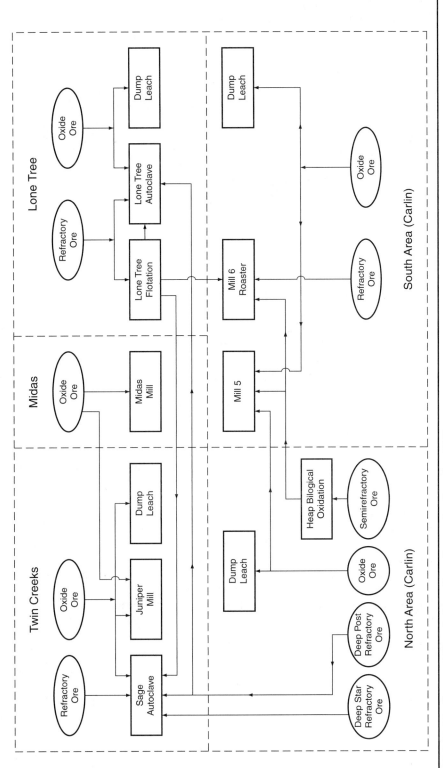

FIGURE 12.32 Newmont Nevada Operations Processing Overview [78]

12.2.4.8 Barneys Canyon [43]

Owner/Operator: Kennecott Minerals, Rio Tinto (London, United Kingdom)
Location: Bingham Canyon, Salt Lake City, Utah, United States
Start-up: 1994

Mineralogical factors. The oxide and sulfide zones of the deposit are mined simultaneously. Gold in the sulfide zone is associated with the sulfide minerals, primarily as micro-fine gold inclusions within iron–arsenic sulfide rimming around pyrite particles. The rims can contain >0.1% gold. Direct cyanide leaching of the sulfide material yielded reasonable gold recovery but was uneconomic due to unacceptably high cyanide and lime consumptions. The ore also has high clay content.

Key process statistics (approximate; 1996 basis)

	Oxide	Sulfide
Ore throughput rate (tpd):	~12,000	~1,200
Gold ore grade (g/t):	~1.5	~3.0
Gold recovery (%):	85–86 (overall)	
Gold production (oz/yr):	~200,000 (combined oxide and sulfide)	

Process description/main features. Barneys Canyon utilizes an innovative flowsheet for gold recovery from both oxide and refractory sulfide ore types. The flowsheet shown in Figure 12.33 includes milling and flotation of sulfide ore to produce a concentrate that is sent to a copper smelter for further processing and recovery of gold from the refractory ore components. The flotation tailings are blended (agglomerated) with crushed oxide ore in a pug mill and then heap leached with cyanide on a lined pad.

The sulfide is crushed in three stages and then ground to 80% <85 μm. The slurry is then directed to flotation, where a combination of potassium amyl xanthate (PAX) and sodium diethyl/sodium dibutyl dithiophosphate (Aerofloat 208) collectors are used to produce a bulk sulfide (and gold-containing) concentrate. Dosages of 0.02 to 0.03 kg/t and 0.06 kg/t are used for the xanthate and dithiophosphate collectors, respectively. The dithiophosphate collector reportedly helps to collect sulfide minerals with tarnished surfaces and native gold. The concentrate produced contains about 22 g/t Au, 25% sulfide sulfur, 31% silica and 21% Fe. Approximately 25% of the gold in flotation feed is recovered to the concentrate, which is thickened and shipped to the nearby Bingham Canyon (Kennecott) smelter as a slurry by truck.

The flotation tailings, which contain about 2 g/t Au and 0.4% sulfide sulfur, are thickened (to 50%–55% solids) and filtered (to 68%–72% solids). The product is agglomerated with crushed oxide ore (80% <10 mm) to generate a heap leach feed containing about 11% moisture by weight. The ratio of oxide ore to tailings is about 15:1 (this depends on the oxide ore moisture content). The agglomerated mixture of crushed oxide ore and tailings is placed on a lined leach pad in 5-m lifts and leached with dilute cyanide solution at a solution application rate of about 8 L/hr/m^2. Overall gold recovery from the mixed oxide–sulfide tailings material is about 85%. Gold recovery from the flotation tailings only is estimated to be about 88%, for a total recovery from sulfide material of about 90% (i.e., 25% × 0.95 smelter return from the flotation concentrate, plus 75% to the flotation tailings × 88%). Cyanide consumption is about 0.04 kg/t NaCN, compared with 0.025 kg/t NaCN when the oxide ore is heap leached without the tailings.

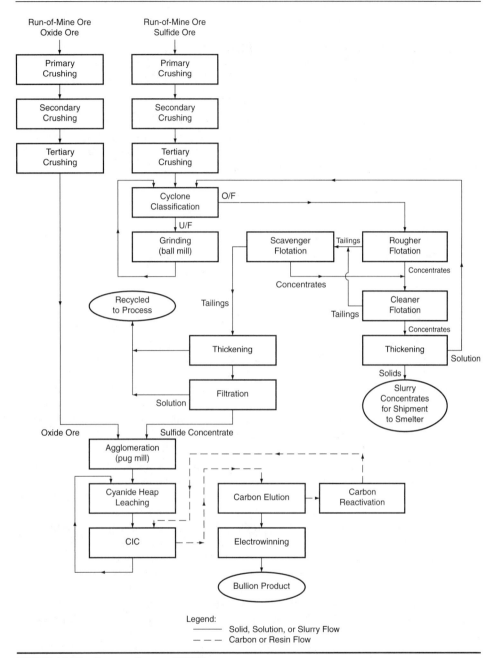

FIGURE 12.33 Barneys Canyon (1996)

This unique treatment scheme avoided the need to dispose of slurry tailings following a more traditional agitated cyanide leaching approach to process the flotation tailings. Such tailings disposal would have required significant permitting time and would have resulted in additional costs to the operation. Also, this scheme proved to be successful despite the presence of clay in the ore.

(NOTE: Gold production had declined to 22,000 oz in 2004 as a result of ore reserve depletion.)

12.2.5 Refractory Arsenopyritic Ores

12.2.5.1 Ashanti Refractory Sulfide Plant [18, 44, 45]
Owner/Operator: AngloGold Ashanti (Johannesburg, South Africa)
Location: Obuasi, Ghana
Start-up: 1929 (1895 original)

Mineralogical factors. Refractory ore contains fine gold associated with sulfides, mainly arsenopyrite with minor zinc and lead sulfides. A small amount of coarse gold is present. Gangue is mainly quartz carbonates and sericite. Zones of the deposit contain carbonaceous material with some gold-adsorbing tendency.

Key process statistics (approximate; 1990 basis)

	Ore	Tailings
Ore throughput rate (tpd):	3,200	2,000
Gold ore grade (g/t):	12	2 to 3
Gold recovery (%):	82	Not applicable
Gold production (oz/yr):	360,000	30,000

Process description/main features. The nature of gold occurrence as coarse free grains, fine gold in quartz, and fine gold in pyrite and arsenopyrite, coupled with the presence of carbonaceous components, resulted in a relatively complex process flowsheet (see Figure 12.34).

The ore was crushed in four stages and ground in two stages (rod–ball mills) to approximately 80% <75 μm. Gravity concentration (rougher spirals, cleaner tables) was employed within the grinding circuit to recover coarse gold and high-grade sulfides, and to prevent overloading of the flotation circuit with free gold. Gravity concentrates were amalgamated, and the valuable product was retorted, to allow mercury to be recovered and recycled, and the sponge gold smelted to bullion on-site.

The ground product was treated by flotation to produce a bulk sulfide–gold concentrate. The ore contained an average of 1% organic carbon, which was upgraded to approximately 6% in the flotation concentrate. Because of the high carbon content, the concentrate was filtered and roasted. Gravity concentration was used to recover any coarse gold liberated during roasting and then the calcined product was classified. The coarse fraction was reground and treated by amalgamation for further free gold recovery. The slimes were leached with cyanide and filtered, and the resulting solution was combined with pregnant solution resulting from leaching of the flotation tailings for subsequent gold recovery.

The flotation tailings were thickened, leached with cyanide, and filtered in several counter-current stages to recover the gold-bearing solution and to wash the tailings prior to disposal. The pregnant solution was clarified and deaerated, and gold was recovered by zinc precipitation. The precipitate obtained was calcined and smelted with fluxes for final bullion production.

In 1994, a plant was constructed at Sansu using the Gencor (now Gold Fields) BIOX technology developed and commercialized at Fairview and applied subsequently at Sao Bento (Brazil) in 1991, Harbour Lights (Australia) in 1992, and at Wiluna (Australia) in 1993 (see Sections 12.2.5.5 and 12.2.5.6 for descriptions of the Sao Bento and Fairview plants). The Sansu plant was designed to treat 720 tpd flotation concentrates using 900 m^3 biological reactor tanks configured in three plant modules. A fourth module was added in 1995 to increase throughput capacity to 960 tpd. The roaster facility was shut down in 2000. In 2004, Sansu was the largest biological oxidation plant for gold extraction in the world. (Note that the biological oxidation circuit is not included in Figure 12.33.)

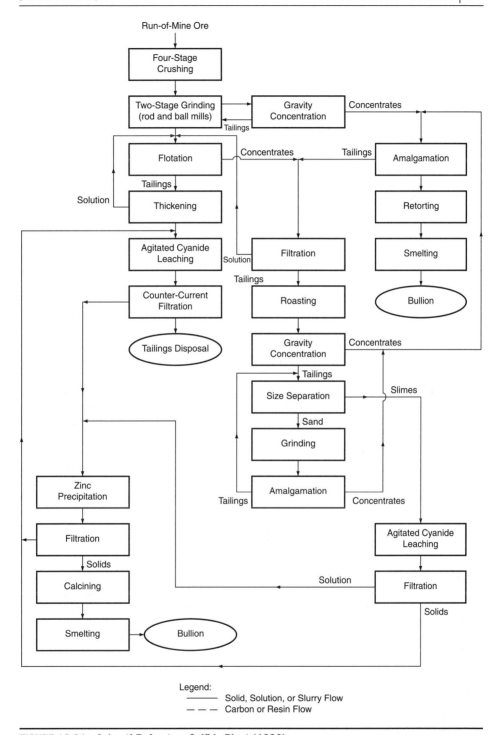

FIGURE 12.34 Ashanti Refractory Sulfide Plant (1990)

12.2.5.2 Campbell Red Lake [10, 46, 47]

Owner/Operator: Placer Dome Inc. (Vancouver, British Columbia, Canada)
Location: Balmertown, Ontario, Canada
Start-up: 1949—Original plant
  1987—Modified roasting plant
  1991—Pressure oxidation circuit

Mineralogical factors. Free gold and gold are associated predominantly with arsenopyrite, and to a lesser extent with pyrite and pyrrhotite. Major gangue mineral is quartz, with minor stibnite and chalcopyrite. Ore contains approximately 1.3% sulfide sulfur.

Key process statistics (approximate; 1990 basis)
Ore throughput rate: 1,100 tpd
Gold ore grade: 20 g/t
Gold recovery: 88% to 90%
Gold production: 220,000 oz/yr

Process description/main features. Three flowsheets are shown for the Red Lake plant: Figure 12.35 shows the original process prior to 1987 (although this plant evolved into the flowsheet shown between 1949 and 1987). Figure 12.36 shows the configuration following modernization of the plant in 1987. The main differences between the two circuits are the treatment of leached flotation tailings by CIP for gold recovery, the discontinuation of amalgamation to treat gravity concentrates, and the installation of a mine waste grinding–classification plant for additional backfill production. Figure 12.37 shows the pressure oxidation circuit installed in 1991.

In the pre-1991 flowsheets, three-stage crushing and two-stage grinding (rod–ball mills) were used to reduce run-of-mine ore to 78% <75 µm. Gravity concentration (jigs and tables) was employed to recover free gold as early as possible in the circuit. Approximately 40% of the gold was recovered by gravity separation.

The ground product was treated by flotation to produce an arsenic-rich concentrate into approximately 5% by weight of the feed. The concentrate was thickened and filtered, and the resulting solution was recycled to grinding. The concentrate typically contained 17% to 19% S, 2.5% Ca, 5.5% CO_2 (as $CaCO_3$), 20% Fe, 2.5% Mg, 10% As, and 200 to 300 g/t Au.

Prior to 1991, the arsenic-rich flotation concentrates were processed by roasting. In this circuit, the solids were repulped to 75% solids and fed to a two-stage fluidized bed roasting system. The two-stage system was necessary for effective arsenic removal and the second stage was operated at a higher temperature (i.e., 575°C to 600°C) to complete oxidation (see Section 5.8). The roaster calcine was quenched, thickened, and filtered to wash the solids and to remove soluble sulfate (and other) salts that would otherwise interfere with subsequent gold dissolution. The solutions obtained were discarded to tailings. The filtered slurry was reground to approximately 80% <53 µm and cyanide leached in a two-stage CCD leaching–thickening circuit. The leached product was filtered in two stages, with the solution directed to the Merrill–Crowe precipitation circuit, where it was mixed with carbon eluate solution from the flotation tailings leaching circuit. The filtration solids were discarded to tailings.

Arsenic trioxide was recovered from the roaster off-gas by cooling and precipitation, and was either sold or sealed and stored in an underground storage vault.

In the pre-1987 flowsheet (Figure 12.35), the flotation tailings were thickened, leached with cyanide, and treated by CIP. Loaded carbon was eluted by the pressure Zadra procedure. Before being recycled to the adsorption circuit, the stripped carbon

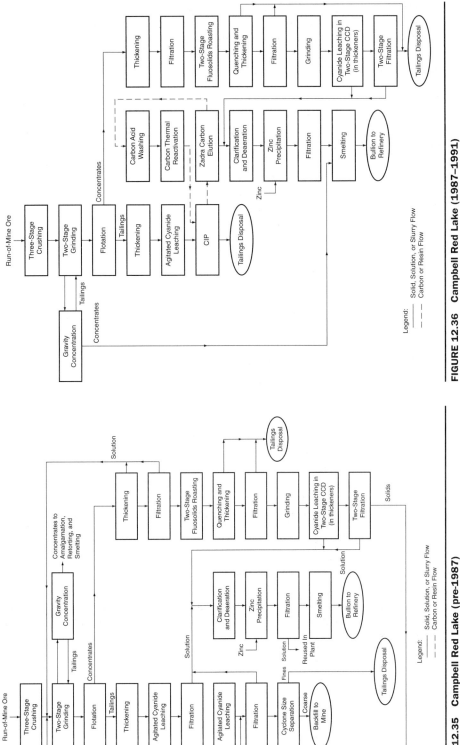

FIGURE 12.36 Campbell Red Lake (1987–1991)

FIGURE 12.35 Campbell Red Lake (pre-1987)

was acid washed and thermally reactivated. The high-grade eluate solution was mixed with gold-bearing solution from the roaster calcine leaching circuit. The resulting solution was clarified, deaerated, and precipitated with zinc and lead nitrate. The precipitate was recovered by filtration and smelted with fluxes to produce the final bullion.

It should be noted that prior to 1987, the gravity concentrate was successfully amalgamated, even in the presence of relatively large amounts of arsenopyrite. The use of amalgamation was discontinued in 1987 in favor of direct on-site smelting of concentrates.

In 1991, the roasting operation was shut down and a new pressure oxidation circuit was commissioned using Sherritt pressure oxidation technology for recovery of gold from the refractory sulfide ore (see Figure 12.37). Thickened flotation concentrates (65% solids) are reacted with dilute sulfuric acid solution in a pretreatment steps to dissolve reactive carbonate minerals. The pretreated slurry is mixed with recycled oxidized slurry in a ratio of 2.2:1 oxidized slurry to pretreated concentrate. This ensures that the autoclave operates autogenously (i.e., no external heating required, except during start-up) and prevents the agglomeration of elemental sulfur and unreacted sulfides.

The pretreated concentrate and oxidized slurry mixture is fed to an autoclave which is operated at a pressure of 2,100 kPa. Oxygen is sparged into the autoclave compartments, as required, and quench water is added to maintain an operating temperature of 190°C to 195°C. The slurry discharges the autoclave through a choke valve, and the pressure is let down in a flash tank where the slurry temperature is reduced to about 100°C. The slurry is then treated in a two-stage CCD circuit to separate the bulk of the acid-bearing solution from the solids. The first-stage thickener overflow is recirculated back to the concentrate pretreatment step as the dilute acid solution. A portion of the first-stage thickener underflow is recycled and mixed with the autoclave feed. The remainder is repulped with water and sent to the second-stage wash thickener. The second-stage thickener underflow is neutralized prior to delivery to the preexisting agitated cyanide leaching and gold recovery circuit (see Figure 12.35). Neutralization of the second-stage thickener underflow is completed in two stages: The first stage uses flotation tailings and cyanide leach–CIP tailings (oxidation circuit product) to adjust the pH to about 5.5, and the second stage uses hydrated lime to increase the pH to 9. Air is sparged into the first stage to oxidize Fe(II) to Fe(III). Arsenic and other heavy metal species are precipitated in the second-stage neutralization. Stable ferric arsenate product is generated provided the Fe–As ratio is maintained at or above 3:1. Arsenic concentration in the tailings solution has been reduced from 0.8–1.2 mg/L with the roaster circuit to <0.3 mg/L with pressure oxidation.

Gold recovery following pressure oxidation is approximately 98.5%, compared with 95% for the roaster circuit. Oxygen consumption for the pressure oxidation step is 0.5 to 0.6 t/t concentrate. Cyanide and lime consumptions are 0.6 and 0.8 kg/t, respectively, compared with 6 and 2 kg/t, respectively, for cyanide leaching of the roasted product. In addition, the need to add 8 kg/t ferric sulfate in the roaster circuit to fix the arsenic was eliminated by the use of pressure oxidation, as the necessary iron is solubilized in the autoclave.

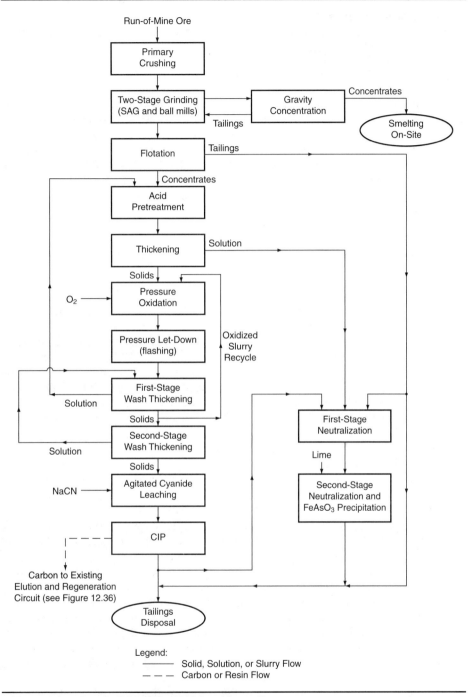

FIGURE 12.37 Campbell Red Lake (1991 and beyond)

570 | THE CHEMISTRY OF GOLD EXTRACTION

12.2.5.3 Giant Yellowknife [48]
Owner/Operator: Miramar Mining Corporation (Vancouver, British Columbia, Canada)
Location: Yellowknife, Northwest Territories, Canada
Start-up: 1948
Shutdown: 1999

Mineralogical factors. Very fine free native gold (38% of gold in feed) and gold was associated with sulfides, principally with arsenopyrite as <0.1 μm diameter inclusions (60%) but to a lesser extent with pyrite (<2%). Gold concentrations in arsenopyrite grains of up to 500 g/t were measured. Major gangue minerals were quartz, calcium, and iron carbonates and sulfides (15% of feed). About 90% of the sulfide minerals were arsenopyrite and pyrite; the remainder were antimony, lead, and copper sulfides. Direct cyanidation of this ore yielded approximately 40% gold recovery.

Key process statistics (approximate; 1990 basis)
Ore throughput rate: 1,250 tpd
Gold ore grade: 7 to 8 g/t
Gold recovery: 88%
Gold production: 100,000 oz/yr

Process description/main features. Ore was crushed in three stages and ground in two stages to allow the production of a flotation concentrate at 60% to 65% <75 μm (see Figure 12.38). Gravity concentration was not used because of the fine size of free gold. Flotation recovered >95% of the gold into a concentrate, which was 14% by weight of the feed.

The concentrate was thickened, and a portion of the slurry was filtered to increase the density to the optimum level for roasting. Slurry was fed to the first of two fluidized bed roasters at a density of approximately 78% solids. The temperatures in the two stages of roasting were controlled to 496°C within 2°C. Of the theoretical air requirement, 90% was supplied on the first stage to maintain slightly reducing conditions. Oxidizing conditions were applied in the second stage by adding excess air and by the addition of spray water.

Antimony concentrations >0.75% in the roaster feed caused clinkering of the roaster bed with antimony sulfates, which reduced the bed life and increased roaster downtime for maintenance. To avoid this, close control was kept over antimony levels in the plant feed.

Initially, an Edwards-type flat hearth roaster was used at Giant Yellowknife; however, this was replaced by the existing two-stage fluosolids roaster in 1958. The internal diameters of the two roaster stages were gradually decreased as the quantity of concentrates produced decreased, and air velocities were increased accordingly.

The roasted calcine was quenched and classified by cyclones to remove a coarse product for two-stage, closed-circuit (both stages utilizing cyclones) regrinding. The final cyclone overflow was cyanide leached in a CCD leaching–thickening circuit, followed by single-stage filtration of the product. The filtered solids were discharged to a tailings impoundment. The calcine leaching circuit extracted 95% of the contained gold in the concentrates. The gold-rich solution produced by the CCD thickeners was clarified, deaerated, and precipitated with zinc and lead nitrate. The precipitate was recovered by filtration and was smelted with fluxes for final bullion recovery.

As might be expected in this location, the process had a positive overall water balance, and decant solution from the tailings storage facility was treated seasonally (during summer) with hydrogen peroxide and Fe(II) sulfate for cyanide detoxification and arsenic removal. The resulting solution was passed through CIC for metal recovery prior to discharge.

Calcine dust (12% of roaster feed) was recovered from the roaster off-gas by electrostatic precipitation. The solid product was quenched, thickened, and the slurry treated

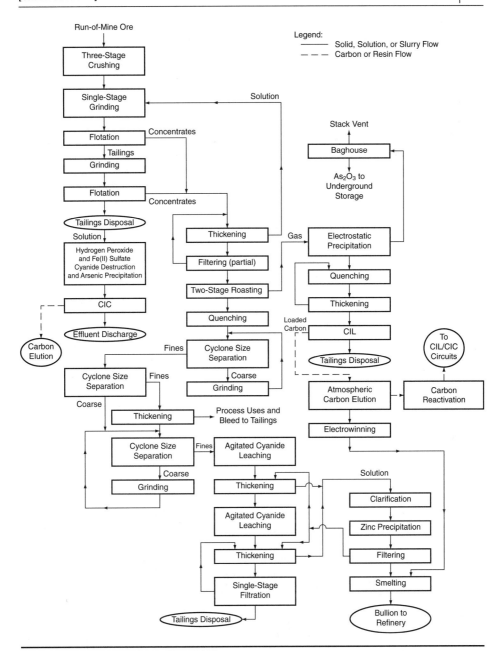

FIGURE 12.38 Giant Yellowknife (1990)

by CIL, which achieved >80% recovery of the contained gold. Carbon adsorption was the most appropriate recovery method for this material, because the dust was typically high in arsenic and antimony, and very fine in size (<5 μm), which adversely affected zinc precipitation and filtration. Arsenic trioxide was recovered by cooling the off-gas to condense the compound, which was then collected in a baghouse. The bagged arsenic trioxide was stored in underground vaults.

12.2.5.4 Mercur [49, 50]

Owner/Operator: Barrick Gold (Toronto, Ontario, Canada)
Location: Salt Lake City, Utah, United States
Start-up: 1983—oxide ore circuit
1988—refractory ore circuit
Shutdown: 1998

Mineralogical factors. Oxidized ore overlaid sulfides and contained <1% sulfide sulfur and up to 0.2% organic carbon. By volume, sulfide ore typically contained 1% to 3% sulfide sulfur, mainly as pyrite, arsenopyrite, realgar, and orpiment, and minor amounts of stibnite and thallium minerals. Gold was associated mainly with pyrite, as fine inclusions, and to a lesser extent with arsenopyrite. Refractory sulfide ore contained an average of 16% carbonate, 1.2% sulfide sulfur, and 0.4% organic carbon.

Key process statistics (approximate; 1990 basis)

	Oxide	Refractory	Oxide (low grade)
Ore throughput rate (tpd):	3,600	725	4,000
Gold ore grade (g/t):	2.4	2.0	1.0
Gold recovery (%):	90	82	Not available
Gold recovery (oz/yr):	90,000	14,000	Not available

Process description/main features. Ore was coarse-crushed in a single stage followed by two-stage grinding (SAG–ball mills, with crushing of a coarse SAG mill product) to 75%–80% <75 µm (see Figure 12.39). Oxidized and sulfide ores were treated on a campaign basis at the proportions indicated above.

For gold recovery, ground oxidized ore was thickened and treated by CIL, which was necessary due to the relatively mild, but significant, gold-adsorbing tendency of the carbonaceous ore constituents.

The refractory sulfide ore was thickened after grinding followed by three stages of preheating of the slurry by recovering heat from the pressure oxidation discharge slurry. The material was oxidized in autoclaves at a temperature and pressure of 220°C and 3,200 kPa, respectively. Approximately neutral pH (i.e., 6.5 to 7.5) was maintained in the autoclave owing to the high carbonate content of the feed material; the carbonates in the ore neutralized acid as it was formed by sulfide mineral oxidation reactions. More than 70% of the sulfide sulfur content was oxidized to sulfate. Slurry was released from the autoclave through a choke valve and the pressure was let down in three stages. The cooled slurry was treated by CIL (14 hr) for gold extraction and recovery. CIL was also necessary for this material because pressure oxidation did not passivate or oxidize the carbonaceous ore constituents. After start-up, carbon concentration in CIL was increased from 15 to 40 g/L to minimize gold adsorption by carbonaceous materials.

Loaded carbon from both CIL circuits was acid washed and gold was stripped by pressure Zadra elution. Gold was recovered from the eluate by electrowinning, and the loaded cathodes were retorted for mercury recovery, prior to smelting with fluxes to produce final bullion. Stripped carbon was thermally reactivated and returned to the adsorption circuits.

Low-grade (1 g/t) oxide ore was treated by crushing and heap leaching at a rate of about 4,000 tpd.

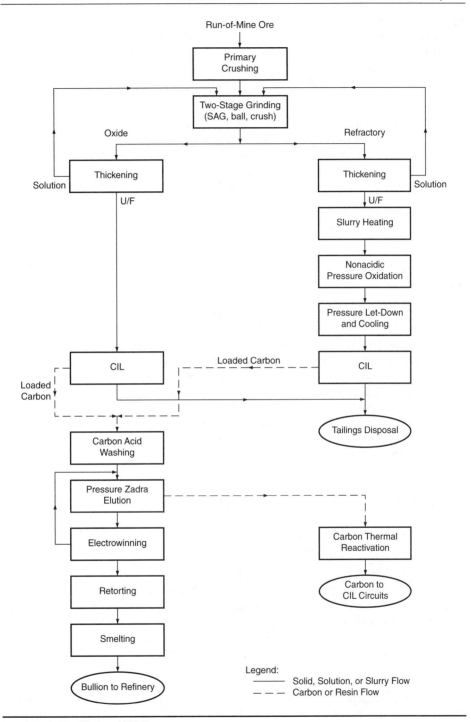

FIGURE 12.39 Mercur (1990)

12.2.5.5 Sao Bento [51]

Owner/Operator: Eldorado Gold (Vancouver, British Columbia, Canada)
Location: Belo Horizonte, Minas Gerais, Brazil
Start-up: 1896—oxidized ore
        1986—pressure oxidation
        1991—biological oxidation (partial) and pressure oxidation

Mineralogical factors. Gold occurs with massive sulfides, quartz, and carbonates. Ore composition: approximately 5% sulfide sulfur and 2% arsenic. Gold is mainly associated with arsenopyrite and has minor associations with pyrrhotite and pyrite.

Key process statistics (approximate; 1990 basis)
Ore throughput rate: 1,000 tpd (240 tpd through pressure oxidation)
Gold ore grade: 7.2 g/t
Gold recovery: 91%
Gold production: 75,000 oz/yr

Process description/main feature. Run-of-mine ore is ground in two stages to yield a product at 75% <75 µm (see Figure 12.40). For free gold recovery, gravity concentration is employed within the grinding circuit.

Flotation of the ground slurry produces a bulk sulfide concentrate containing approximately 8.0% CO_3^{2-}, 34% Fe, 19% S, and 10% As. The concentrate is reground to 95% <45 µm and thickened to 65% solids, allowing water recycling to the grinding circuit.

The slurry is mixed with recycled oxidized (acidic) slurry to neutralize a large proportion of the carbonates in the ore and reduce the amount of carbon dioxide generated in the autoclaves. The conditioned slurry is then split between two parallel autoclave circuits in which the sulfides are oxidized. Each autoclave contains five compartments with a total retention time of 120 min. Oxidation is performed at 190°C and 1,560 kPa. Slurry discharges through a choke valve and the pressure is let down in several stages of flashing. Unlike the McLaughlin (Section 12.2.4.3) and Mercur (Section 12.2.5.4) circuits, the discharged slurry is not used to preheat the autoclave feed because the sulfur content of the concentrate is well in excess of that required to achieve the desired reaction rate in the first compartment and to maintain autogenous operation.

Once the slurry is at atmospheric pressure, it is thickened in two stages to remove acidic solution to allow neutralization. Approximately 80% of the first stage underflow is recycled and used for slurry conditioning prior to oxidation, while the remainder passes to the second thickener. Overflow solution from the first thickener is neutralized in two stages; the first stage uses limestone (and final cyanidation tailings reclaim solution) to adjust the pH to 4.5, and lime is added in the second stage to reach pH 11.0. The neutralized slurry is thickened, and the solids directed to tailings disposal. The overflow is used as wash solution in the second-stage oxidized slurry washing thickener.

The pH of the second wash thickener underflow is adjusted to 10.5 to 11.0 with lime, and the material is leached with cyanide, followed by CIP, AARL elution, electrowinning, and smelting for gold recovery. Carbon is reactivated and reused in the circuit. The final tailings are thickened for water recovery and sent to tailings disposal.

Soluble arsenic levels in the combined tailings (i.e., neutralization circuit solids and cyanidation tailings) are kept <0.05 mg/L by careful control of pH in the neutralization circuit.

Interestingly, the oxidation circuit operates largely without the use of control valves because of the problems associated with obtaining and maintaining reliable instrumentation in Brazil.

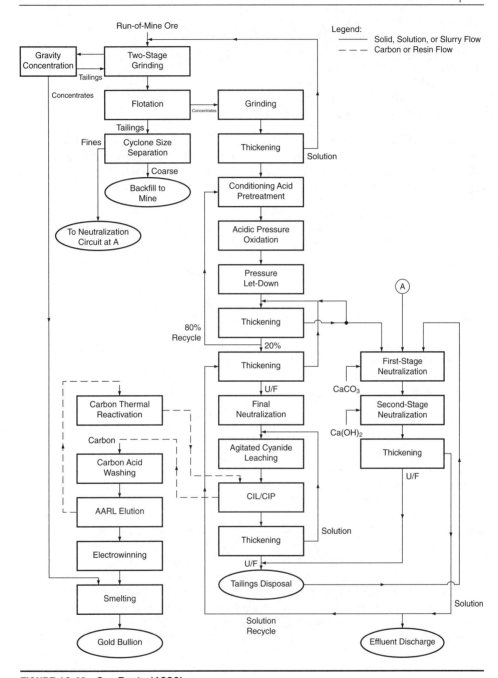

FIGURE 12.40 Sao Bento (1990)

In 1991, a biological oxidation circuit (using Gold Fields BIOX technology) was installed prior to the pressure oxidation step. This was done to oxidize a portion of the sulfide minerals in the concentrate to reduce the oxygen consumption (and costs) in pressure oxidation. Further information is available in the literature [45].

12.2.5.6 Fairview [52, 53, 54]

Owner/Operator: Metorex (Johannesburg, South Africa)
Location: Barberton, Eastern Transvaal, South Africa
Start-up: 1912—original plant
1955—roaster circuit
1986—bio-oxidation circuit (partial circuit)
1991—bio-oxidation (full circuit)

Mineralogical factors. Ore zones are refractory sulfide and free-milling quartzite. Pyrite and arsenopyrite are the main sulfides with minor copper, nickel, zinc, lead, antimony, and other iron sulfides. Gold occurs mostly as submicroscopic (<0.2 μm) inclusions in arsenopyrite and, to a lesser extent, in pyrite.

Key process statistics (approximate; 1990 basis)

	Refractory Ore[*]	Roasting	Bio-oxidation
Throughput rate (tpd):	465	18	12
Gold grade (g/t):	7.0	120	120
Gold recovery (%):	80–85	88–90	>95
Sulfide sulfur (%):	1.3	29	29
Arsenic trioxide (%):	0.5	8	8
Gold production (oz/yr):	36,000	20,000	16,000

Process description/main features. Direct cyanidation of refractory Fairview ore yields only 36% gold extraction, and sulfide minerals must be oxidized to liberate gold prior to cyanidation (see Figure 12.41).

Run-of-mine ore was crushed and ground in a conventional grinding circuit, followed by flotation to produce a bulk sulfide–gold concentrate. The concentrates were thickened and, between 1986 and 1991, were treated by parallel oxidation circuits, one employing roasters and the other biological oxidation. In June 1991, the roasters were shut down and all the concentrates processed by bio-oxidation. This change was encouraged by the high maintenance requirements of the roasters, concerns over sulfur dioxide generated by the roasters, and the good performance of the bio-oxidation circuit.

The roasting circuit (used between 1955 and 1991) consisted of two-stage Edwards roasters, followed by quenching and washing of the calcine in a thickener. A portion of the thickener overflow was recycled while some was bled to tailings to remove sulfosalts that interfered with cyanidation. The thickener underflow was reground, with gravity concentration equipment employed within the circuit to recover free gold. The calcine slurry was leached with cyanide and gold recovered by CIP (since 1986).

A 10-tpd biological oxidation plant was commissioned in 1986 to treat a portion of the concentrates stream, following the successful operation of a 0.75-tpd pilot plant since 1984. By 1991 the plant was capable of treating 17 tpd at 25% solids (originally the plant operated at 12.5% solids) at a temperature of 45°C (originally 40°C to 42°C) and with a slurry retention time of 72 hr. The pH was maintained at 1.6 by the addition of lime to neutralize acid as it was formed. Approximately 85% of the sulfide sulfur was oxidized, resulting in gold extractions >95%. The oxidized material was washed in three stages of thickening to remove acid and sulfur species (i.e., Fe(III) sulfate) that interfere with cyanidation. The solution overflow from the first thickener was neutralized in four stages with lime and the product disposed of in a tailings impoundment. Soluble arsenic was precipitated as stable Fe(III) arsenate since the iron–arsenic ratio was above

---

[*] Free-milling ore and old tailings are also treated at Fairview but these are not considered further here.

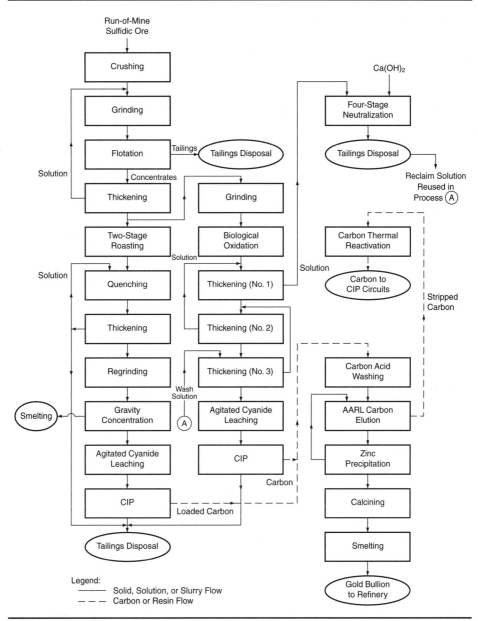

FIGURE 12.41   Fairview (1990)[*]

the necessary 4:1 (see Section 5.1.6.2). The final thickener underflow was cyanide-leached and treated by CIP. No detrimental effects of the bacteria (or the oxidation products) on carbon adsorption performance were observed at Fairview.

Loaded carbon was acid washed and eluted by the AARL procedure. Gold was recovered from the eluate by zinc precipitation, and the product was calcined and smelted to produce bullion. Stripped carbon was thermally reactivated and reused in the adsorption circuit(s).

---

* The roaster portion of this flowsheet was shut down in 1991.

12.2.6 Copper-Rich Ores

12.2.6.1 Grasberg–Ertsberg [55, 56, 57, 58]
Owner/Operator: Freeport-McMoran Copper & Gold Inc. (New Orleans, Louisiana, United States)
Location: Irian Jaya, New Guinea Island, Indonesia
Start-up: 1972 (7,500 tpd)
Expansions: 1989 (32,000 tpd total throughput capacity)
  1991 (57,000 tpd total throughput capacity)
  1995 (115,000 tpd total throughput capacity)
  1998 (215,000 tpd total throughput capacity)

Mineralogical factors. Located in a porphyry copper setting, the Ertsberg–Grasberg complex is the world's largest hydrothermal gold deposit. The copper mineralization is predominantly chalcopyrite ($CuFeS_2$). Gold occurs as fine free gold and gold associated with chalcopyrite. The average grade of mineralization is 1.2% Cu and 0.5 g/t Au but as high as 1.5% Cu and 2 g/t Au in higher-grade veinlet controlled zones of the deposit (quartz–magnetite stockwork).

Key process statistics (approximate; 2003 basis)
Ore throughput rate : 235,000 tpd
Ore grade: 1.5 g/t Au; 1.1% Cu
Gold recovery: 87%
Copper recovery: 89%
Gold production: 3.16 million oz/yr
Copper production: 691,000 oz/yr

Process description/main features. Ore is primary crushed to approximately 80% <75 mm and then delivered to the coarse ore stockpiles via several 2–3 m diameter, 600-m long ore passes (see Figures 12.42 and 12.43). The ore particle size distribution is reduced significantly as it travels down the ore passes, resulting in a typical mill feed size of 80% <44 mm. Primary crushed ore is fed to four concentrators. Two of these utilize two-stage crushing and ball milling to prepare the ore for flotation, and have a combined throughput capacity of about 68,000 tpd. The other two concentrators utilize SAG mills configured with recycle pebble crushers, followed by ball mills, to prepare ore for flotation. These concentrators were commissioned in 1995 and 1998 with design throughput capacities of 52,000 and 95,000 tpd, respectively.

The conventional (three-stage crushing and ball milling) circuits are configured with flash flotation to process cyclone underflow in the grinding circuit to recover fast-floating gold and chalcopyrite. Flash flotation was incorporated because of the excellent response of the Ertsberg East orebody to flash flotation, yielding 20% to 25% of the recoverable copper content to a high-grade concentrate. The cyclone overflow is treated by rougher flotation (approximately 22 min retention time), with rougher tailings passing to final tailings disposal. The rougher concentrate is directed to a regrinding circuit in closed circuit with cyclones. The cyclone overflow is processed in cleaner column cells. The cleaner tails are treated in a second stage of columns (i.e., cleaner scavengers). The combined flash flotation concentrate and cleaner/cleaner–scavenger concentrates are sent to the concentrate thickener as the final plant product, which is a combined copper–gold–silver concentrate.

The two SAG mill circuits are configured without flash flotation in the grinding circuit. This change was made based on a transition from Ertsberg East ore to Grasberg ore in the mid- to late 1990s. The Grasberg ore responds less well to flash flotation, largely

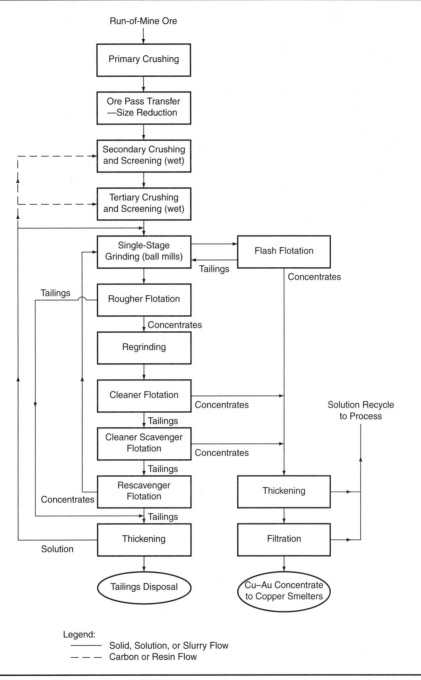

FIGURE 12.42   Grasberg–Ertsberg (2004)—Concentrators 1 and 2

due to higher floatable pyrite content, resulting in a lower-grade concentrate. Grinding circuit product is treated by rougher flotation (26.5 min retention time), with the rougher tailings passing to final tailings disposal via a tailings thickener. Rougher concentrate is reground in closed circuit with cyclones. The concentrate is cleaned in two stages of column flotation (cleaner and cleaner scavenger). The final combined concentrate from the two cleaner stages is reground further (optional) and thickened, ready for transport to the port via pipeline. This product is a combined copper–gold–silver concentrate, containing approximately 31% Cu and 28 g/t Au.

The Grasberg–Ertsberg ores respond well to flotation, and good recoveries (88% to 90% Cu and 87% Au) can be achieved at relatively coarse grind size, that is, 80% <210 µm, although design grind size is 80% <150 µm.

Thickened concentrates are pumped via pipeline to the port where the slurry is filtered and dried prior to loading onto vessels for shipment to copper smelters.

To increase overall gold recovery, gravity concentration equipment is being installed in some of the concentrators to recover coarse, free gold early in the flowsheet.

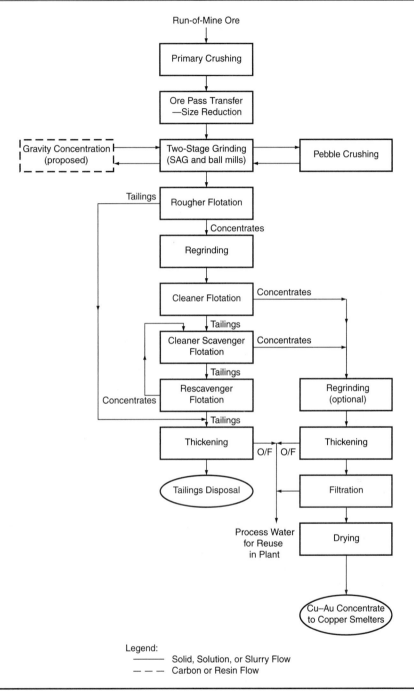

FIGURE 12.43 Grasberg–Ertsberg (2004)—Concentrators 3 and 4

12.2.6.2 Ok Tedi (oxide gold ore treatment circuit) [59]

Owner/Operator: Ok Tedi Mining Ltd. (Tabubil, Papua New Guinea)
Location: Tabubil, Western Province, Papua New Guinea
Start-up: 1984
Shutdown: 2004

Mineralogical factors. Orebody consisted of the upper (leached) portion of a porphyry copper deposit, containing variable amounts of cyanide-soluble copper. The copper, which varied from 0.5% to 1.0%, had only a minor effect on processing methods applied and their efficiency. Some gold was present as free, coarse gold, up to 300 µm diameter, and the remainder was fine free gold and electrum.

Key process statistics (approximate; 1990 basis)
Ore throughput rate: 25,000 tpd
Gold ore grade: 2.8 g/t
Copper ore grade: 0.7%
Gold recovery: 92%

Process description/main features. Run-of-mine ore was ground in a single-stage SAG mill, yielding a relatively coarse product at 65% <75 µm (see Figure 12.44). Gravity concentration equipment (i.e., Reichert cones and spirals) was employed to recover free gold in the grinding circuit, and approximately 25% of the gold in the feed was recovered in this manner. The concentrates produced were upgraded on shaking tables and the product treated by intensive cyanidation (1% NaCN, 20 hr) followed by washing and decantation for solid–liquid separation. More than 97% of the contained gold in concentrates was extracted by this method. The high-grade leach solution produced was combined with carbon eluate solution for subsequent gold recovery.

The grinding circuit product (including gravity concentration tailings) was thickened and leached with cyanide for 24 hr. The cyanide concentration applied varied between 0.15 and 0.50 g/L, depending on the soluble copper concentration, averaging 0.30 g/L. Cyanide consumption was 0.5 kg/t. Leached slurry was treated by CIP, with the carbon concentration increasing from 15 g/L in the first stage to 40 g/L in the final stage.

Loaded carbon was eluted in a unique continuous pressure Zadra procedure, capable of stripping 48 t of carbon per day and consistently achieving residual carbon loadings <100 g/t. Gold was recovered from the eluate by zinc precipitation followed by filtering, drying, smelting, and remelting to produce bullion.

Stripped carbon was acid washed (with nitric acid) and thermally reactivated before being recycled to the adsorption circuit.

CIP slurry tailings were treated with 1.2 kg/t H_2O_2 for cyanide detoxification. The solution contained sufficient copper to catalyze the oxidation reaction and achieve acceptable detoxification rates. Prior to disposal, the sand and slimes fractions of the tailings were separated by cyclones.

It should be noted that the deeper copper porphyry deposit has been mined and processed by flotation since about 1990, with the sulfide concentrate processed off-site by traditional copper smelting and refining.

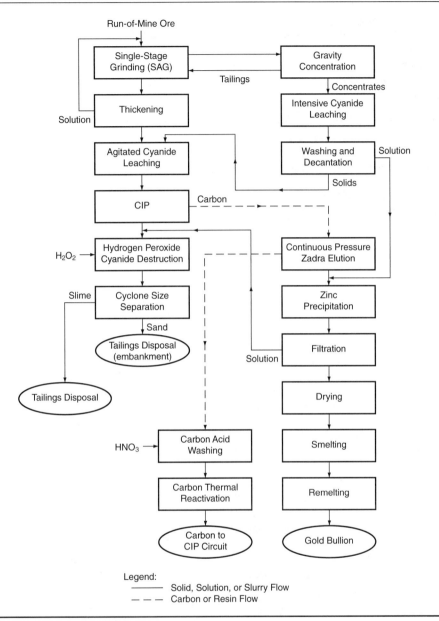

FIGURE 12.44 Ok Tedi (1990)

12.2.6.3 El Indio [60, 61]

Owner/Operator: Barrick Gold (Toronto, Ontario, Canada)
Location: La Serena, Chile
Start-up: 1981–flotation circuit
   1983–CIP circuit
Shutdown: 2000

Mineralogical factors. El Indio was an epithermal vein-stockwork-type copper–gold–silver deposit, containing native gold and electrum, with some as coarse particles. Major copper mineral was enargite, with minor tennantite. Main gangue minerals were quartz and pyrite, with the pyrite containing approximately 10 g/t Au.

Key process statistics (approximate; 1990 basis)
Ore throughput rate: 2,400 tpd
Gold ore grade: 8.4 g/t   Gold recovery: 85%
Silver ore grade: 62 g/t   Silver recovery: 75%
Copper ore grade: 3.1%   Copper recovery: 94%

Process description/main features. Ore was crushed in three stages and then washed prior to grinding to remove soluble, acid-forming minerals, which had an adverse effect on flotation by interfering with pyrite depression (see Figure 12.45).

The washed ore was ground and a bulk copper–gold–silver concentrate was produced within the grinding circuit (by flash flotation) and by conventional flotation following grinding. Originally, approximately 66% of the contained gold and silver was recovered in this manner. However, with this operating philosophy, a considerable amount of copper remained in the flotation tailings that were fed to the cyanide leaching circuit, which required that unacceptably high cyanide concentrations (5 to 6 kg/t) be maintained to keep copper loading on carbon to a minimum. Consequently, the circuit was subsequently modified to allow regrinding of the original flotation tailings, and another stage of flotation was added. This increased the recovery of copper–gold–silver to the flotation concentration and enabled the flotation tailings leaching and CIP circuits to be operated at approximately 0.5 kg/t NaCN.

The flotation concentrates were thickened and the overflow solution was returned to the grinding and flotation circuits. Approximately 60% of the underflow slurry was roasted (in a single stage) on-site, in order to oxidize sulfide sulfur and to reduce the sulfur and arsenic content, as well as the bulk of the material. The calcine was combined with the remainder of the concentrates (i.e., 40%), and this product was exported to a copper smelter. Typical flotation concentrate analyses were 40 g/t gold, 300 g/t silver, 20% copper, 7% arsenic and 0.5% antimony. The equivalent calcine analyses after roasting were 80 g/t, 500 g/t, 30%, 0.3%, and 0.3%, respectively. The gas phase from the roaster was treated for dust removal and subsequently to recover a saleable grade of arsenic trioxide by cooling and precipitation.

The final flotation tailings were cyanide leached, with a retention time of 72 hr provided because of the slow leaching kinetics of the contained gold, followed by CIP for gold recovery. Lead nitrate and a phosphate were added to the leach slurry to assist leaching.

Loaded carbon, which averaged 300 g/t Au and 12,000 g/t Cu, was eluted in two stages for sequential desorption of copper followed by gold and silver. Selective copper desorption was achieved using a low-temperature cyanide elution procedure. The copper-bearing eluate solution was sent to tailings disposal. The carbon was then either eluted at elevated temperature by a pressure Zadra procedure to remove gold and silver, or was returned directly to regeneration and consequently to the adsorption circuit (CIP), for further metal loading. The final eluted carbon after both elution procedures had been performed typically

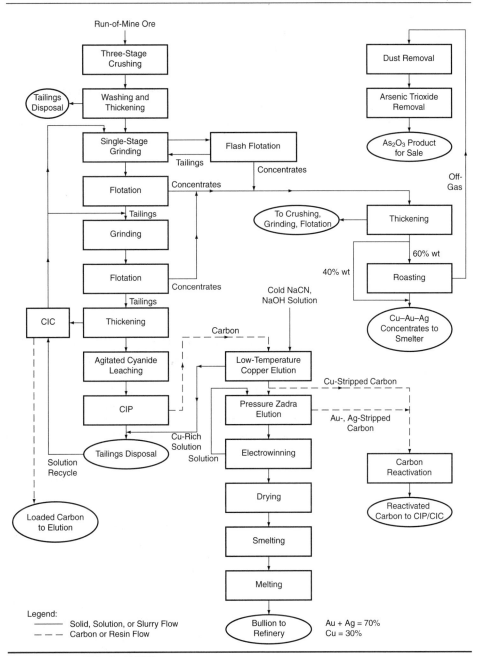

FIGURE 12.45 El Indio (1990)

contained 20 g/t Au and 100 g/t Ag. The eluate produced averaged 60 g/t Au and 500 g/t Cu, and gold and silver were recovered from this solution by electrowinning. Cathodes and sludges were dried, smelted, and remelted for final bullion production.

Tailings reclaim and flotation tailings thickener overflow solutions were passed through CIC to recover residual gold and silver but also, and most importantly, to remove other complexed cyanide species (i.e., copper, iron, zinc, etc.), which would have otherwise interfered with flotation.

12.2.6.4 Alumbrera [62, 63, 64]
Owner/Operator: Xstrata (Zug, Switzerland)
Location: Andalgala, Catamarca Province, Argentina
Start-up: 1997

Mineralogical factors. Alumbrera is a typical porphyry copper deposit with chalcopyrite as the dominant copper mineral. Copper is also present in small amounts as chalcocite and covellite. Gold occurs as free gold and gold associated with chalcopyrite, pyrite, and magnetite. Free gold particle size varies from <10 μm to about 50 μm.

Key process statistics (approximate; 2004 basis)
Ore throughput rate: 90,000 tpd
Gold ore grade: 0.8 g/t
Copper ore grade: 0.6%
Gold recovery: 70% to 75%
Copper recovery: 85% to 90%
Gold production: 633,000 oz/yr (estimate)
Copper production: 175,000 tpy (estimate)

Process description/main features. Ore is primary crushed to a size of 80% <100 mm and then ground in two stages (SAG and ball mills) to 80% <150 μm (see Figure 12.46). Gravity concentration is employed within the grinding circuit as described later. The ground product is fed to rougher flotation where free gold and copper sulfide minerals are floated at a pH of 10 using a dual-collector system of sodium di-isobutyl dithiophosphate and sodium di-isobutyl monothiophosphate, together with a frother (OTX-140). It should be noted that this flotation scheme replaced the original flotation collector (potassium amyl xanthate) shortly after start-up. The rougher concentrates are reground to approximately 80% <55 μm and then subjected to cleaner flotation using Jameson cells. Cleaning is conducted at an increased pH of 11.5 to 12.0, using lime for pH modification. Flotation under these higher pH conditions is effective at rejecting pyrite from the final concentrate and, surprisingly, reportedly yields the optimal gold recovery. This is somewhat unusual considering that lime is generally considered to depress gold recovery by flotation (see Chapter 9 and Section 12.2.6.6); however, the effect is thought to be due to an increase in particle-bubble contact time. Rougher and cleaner flotation tailings are combined and flow by gravity to the tailings disposal area. Final concentrate, which contains both gold (about 30 g/t) and copper (about 28%) is thickened and pumped by pipeline to a filtration plant near Tucuman.

Approximately 30% of the contained gold is recovered by gravity concentration in the primary grinding and concentrate regrinding circuits (using Knelson centrifugal concentrators). The gravity concentrates are cleaned in a secondary step of gravity separation using shaking tables. The final cleaned concentrate is smelted with fluxes at the site to produce a doré bullion for shipment.

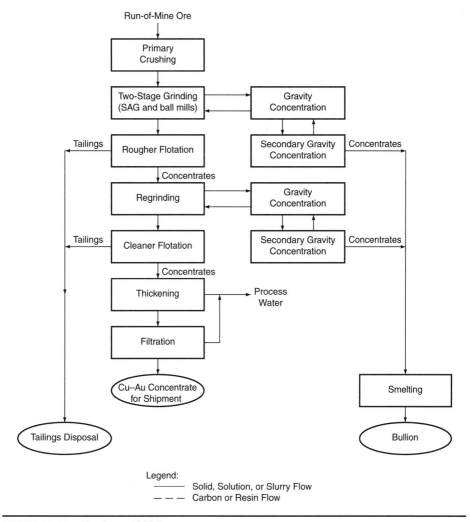

FIGURE 12.46 Alumbrera (2004)

12.2.6.5 Phoenix [65]
Owner/Operator: Newmont Gold (Denver, Colorado, United States)
Location: Battle Mountain, Nevada, United States
Start-up: 2006 (expected)

Mineralogical factors. Gold and silver are closely associated with sulfide minerals, including chalcopyrite, pyrite, and sphalerite. The ore gives poor response to conventional cyanidation due to high copper and zinc content, which inhibits gold and silver dissolution and results in high cyanide consumption.

Key process statistics (approximate; 2005 basis)
Ore throughput rate: 35,000 tpd
Gold ore grade: 1.0 to 1.2 g/t
Copper ore grade: 0.16%
Silver ore grade: 8 to 10 g/t
Gold recovery: 85%
Copper recovery: 63%
Silver recovery: 55%
Gold production: ~380,000 oz/yr
Copper production: ~12,000 tpy
Silver production: ~20 million oz/yr

Process description/main features. Run-of-mine ore is primary crushed and fed to a two-stage grinding circuit consisting of a SAG mill followed by a ball mill in closed circuit with cyclones (see Figure 12.47). A portion of the SAG–ball mill discharge is directed to primary gravity concentration (after screening out the oversize material) to recover free gold and silver. Another portion of SAG–ball mill discharge slurry is directed to flash flotation to recover fast floating sulfide minerals and free gold. The grinding circuit cyclone overflow is sent to rougher flotation. The rougher flotation and flash flotation concentrates are combined and sent to a second stage of gravity concentration. The flotation gravity concentrates are cleaned, by gravity also, and then combined with the primary gravity concentrates and subjected to intensive cyanide leaching. The pregnant gold-bearing solution produced by intensive leaching is combined with carbon eluate, and gold and silver is recovered by electrowinning.

Rougher flotation tailings are treated in a rougher–scavenger circuit. The rougher–scavenger concentrate is treated in a closed-circuit regrinding step and then processed in a three-stage cleaner flotation circuit. An optional magnetic separation step is included between the first and second stage of cleaner flotation to provide a mechanism to remove pyrrhotite from the circuit, if required, to make a higher-grade final concentrate for shipment off-site. The fate of the magnetic separation concentrate will depend on the gold and silver grade of this material (i.e., either sent to tailings or CIL feed). The final cleaned concentrate is dewatered and shipped off-site to third-party copper smelters for further processing. Pressure oxidation of the concentrate was considered by Newmont, and process development was advanced significantly prior to making the decision to smelt the final concentrate.

Rougher–scavenger tailings are cycloned to separate the sand and slime portions of the slurry. The sands are combined with the three-stage cleaner flotation tailings and processed by agitated cyanide leaching and CIP. The CIP tailings are treated for cyanide detoxification and then sent to tailings disposal. Gold (and silver) are stripped from the carbon and gold and silver recovered from the eluate by electrowinning.

The slime portion of cycloned rougher–scavenger tailings is dewatered in a thickener and then sent to tailings disposal.

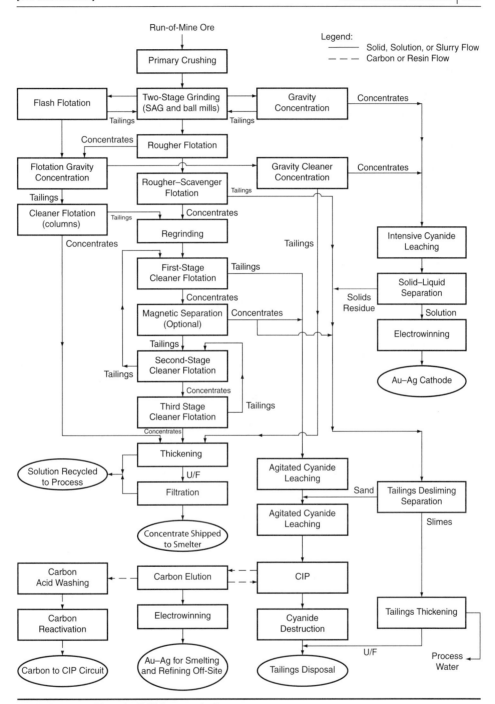

FIGURE 12.47 Phoenix (2006, expected)

Approximately 35% of the gold, 78% of the silver, and 100% of the copper produced by the Phoenix operation are recovered into the copper concentrate.

This flowsheet is an innovative approach to a difficult copper-containing gold–silver ore in an important historic gold and copper mining district.

12.2.6.6 Candelaria [66]
Owner/Operator: Phelps Dodge Corp. (Phoenix, Arizona, United States)
Location: Copiapo, Chile
Start-up: 1994

Mineralogical factors. Fine free gold and gold are intimately associated with chalcopyrite as fine inclusions within grains. Chalcopyrite is the predominant copper mineral and occurs with pyrite and magnetite. Amounts of sphalerite and other sulfide minerals are minor.

Key process statistics (approximate; 2004 basis)
Ore throughput rate: 66,000 tpd
Copper ore grade: 1.0%
Gold ore grade: 0.25 g/t
Copper recovery: 95%
Gold recovery: 80% to 84%
Copper production: 230,000 tpy
Gold production: 150,000 oz/yr

Process description/main features. Ore is primary crushed and then ground in a two-stage grinding circuit comprised of SAG mills in closed circuit with pebble crushers, followed by ball mills in closed circuit with cyclones (see Figure 12.48). Gravity concentration is not used because the gravity recoverable gold content (using the gravity recoverable gold [GRG] test; see Chapter 3) is less than about 15%. The ore is ground to 80% <130 µm and is fed to rougher flotation at 40% solids. Rougher flotation is conducted at pH 9 using a dual-collector scheme consisting of a thionocarbamate (added into the ball mills) and dithiophosphate (added to the rougher feed). Methyl isobutyl carbinol (MIBC) is used as the frother. It should be noted that the design pH is 10.5, and it was possible to significantly increase the gold recovery (from about 74% to 82%) by decreasing the rougher pH from 10.5 to 9. The rougher concentrate is reground to approximately 80% <45 µm and then cleaned in column cells at pH 11 to 11.5. The increase in pH in the cleaner circuit is highly effective at depressing pyrite. The pyrite contains a very low concentration of gold which does not justify its recovery to the final concentrate (the gold content of the pyrite must cover the cost of concentrate shipment to the smelter and treatment charges). The cleaner tailing is treated by cleaner scavenger flotation, with the cleaner scavenger concentrates returned to the regrinding circuit.

The final cleaner concentrates contain 30%–31% Cu and about 4–5 g/t Au. The concentrates are thickened and filtered ready for shipment.

The rougher and cleaner scavenger tailings are combined, thickened (to recover and reuse process water), and then sent to tailings disposal.

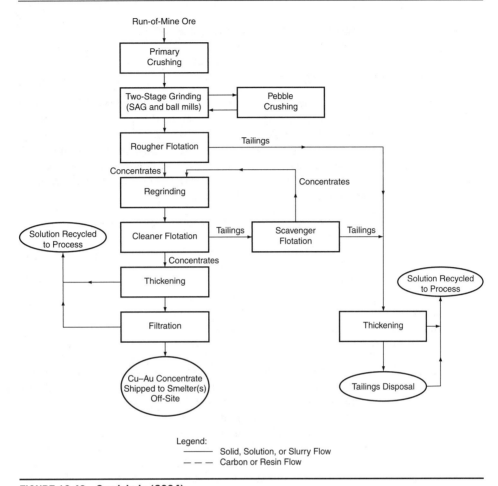

FIGURE 12.48 Candelaria (2004)

592 | THE CHEMISTRY OF GOLD EXTRACTION

12.2.6.7 Telfer [67]
Owner/Operator: Newcrest Mining Ltd. (Melbourne, Australia)
Location: Paterson Province, Western Australia
Start-up: 1977–2001—original cyanide leaching circuit
   2006 (expected)—new gold–copper ore circuit

Mineralogical factors. Gold mineralization is primarily associated with pyrite–chalcopyrite and quartz–dolomite gangue. Digenite, covellite, tenorite, and cuprite are also present. Shallow ore (up to 100 to 200 m depth) has been strongly modified by weathering, resulting in the presence of supergene mineralization, including gold associated with limonite, goethite, malachite, and chrysocolla. Gold in the sulfide ores is present as free gold, plus gold with copper minerals and gold in pyrite. The ore contains variable amounts of arsenic, present as the rare mineral cobaltite.

Key process statistics (approximate; expected performance after start-up in 2006)
Ore throughput rate: 47,000 tpd
Gold ore grade: 1.5 g/t
Copper ore grade: 0.19%
Gravity gold recovery: ~40%
Gold recovery to copper concentrate: 30% to 40%
Gold recovery to pyrite concentrate: 10% to 20%
Overall gold recovery: 85% to 90%
Gold production: 650,000 to 700,000 oz/yr

Process description/main features. The new flowsheet developed for the Telfer gold–copper deposit is rather complex and only the highlights are summarized here (Figure 12.49). The reader is referred to the excellent, comprehensive reference on the metallurgical test work, process development, and design of the plant [67].

Ore is primary crushed and then ground in a two-stage grinding circuit (SAG mills with recycle pebble crushing followed by ball mills) to approximately 80% <130 μm or to 80% <75 μm for the softer ore types. Gravity concentration is used to recover free gold in the grinding circuit. Flash flotation is employed to treat a portion of the ball mill cyclone underflow stream. The flash flotation concentrate is cleaned by flotation, and the cleaned concentrate is sent to the final concentrate solid–liquid separation.

The ground slurry is processed in a copper rougher–scavenger flotation circuit. The copper rougher–scavenger concentrate is cleaned, first conventionally and then using controlled-potential sulfidization (CPS) to activate oxidized minerals.

The copper rougher–scavenger tailings are subjected to CPS conditioning, then fed into a pyrite rougher–scavenger flotation circuit. The pyrite concentrate from one half of the plant (i.e., one grinding line) is sent to a separate pyrite cleaner flotation circuit. In the case of the other half of the plant, the pyrite concentrate is either returned to the head of the copper cleaner circuit (see A in flowsheet, Figure 12.49) or can be sent to the pyrite cleaner flotation (B in flowsheet). This allows the one pyrite circuit to be operated in either bulk flotation mode (i.e., to produce a combined copper and pyrite concentrate) or in sequential flotation mode (i.e., to produce separate copper and pyrite concentrates). This depends on the ore type treated at any given time.

The pyrite cleaner flotation circuit uses both conventional and CPS flotation to optimize recovery. The final concentrate from this circuit is deslimed (with the slimes sent to tailings disposal) and then processed by CIL for gold recovery. The CIL residue is treated by a combined sulfidization–acidification–recovery–thickening (SART) process and acidification, volatilization, and recovery (AVR) process to recover base metal sulfides

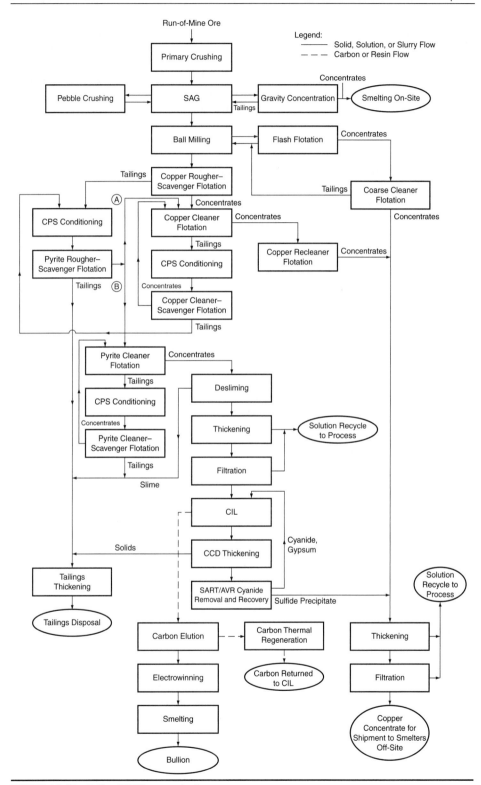

FIGURE 12.49 Telfer (2005, expected)

and cyanide separately (see Chapter 11). This process forms a gypsum precipitate, which is recycled back to the CIL circuit, along with the recovered cyanide.

Loaded carbon is removed from the CIL circuit, stripped, thermally reactivated, and returned to the CIL circuit. Gold is recovered from the carbon eluate solution by electrowinning. The electrowon gold, along with gravity concentrates from the grinding circuit, are smelted into bullion at the site in a Barring furnace.

Approximately 40% of the gold is recoverable by gravity concentration, 30% to 40% is recoverable to the copper sulfide concentrate, and 10% to 20% is recoverable to the pyrite concentrate.

The Telfer project team extensively investigated a process consisting of pressure oxidation of a bulk concentrate followed by solvent extraction and electrowinning for copper recovery and CIL for gold recovery from the residue. The sequential flotation process described previously was selected as the preferred option.

The flowsheet for the original oxide gold ore treatment is shown in Figure 12.50, and includes the modifications in the gravity concentration sections that were made in 1995. This flowsheet is not considered further here, but more information is available in the literature [79, 80].

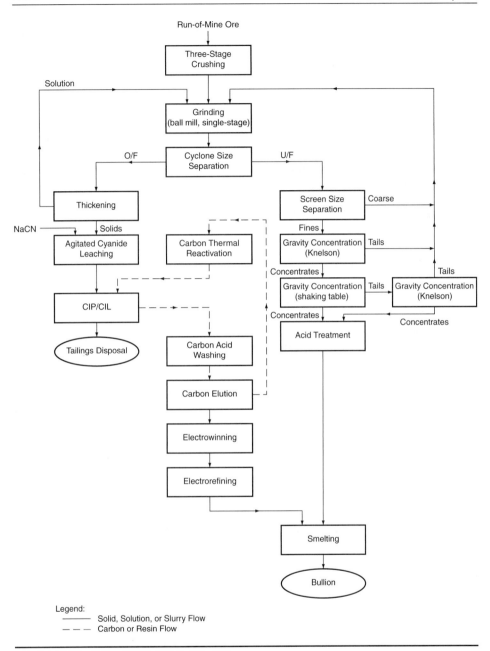

FIGURE 12.50 Telfer (1995–2001) [79, 80]

12.2.7 Refractory Antimony Sulfide Ores

12.2.7.1 Consolidated Murchison [68]
Owner/Operator: Metorex (Johannesburg, South Africa)
Location: Gravelotte, N. Eastern Transvaal, South Africa
Start-up: 1983

Mineralogical factors. Refractory minerals are predominantly antimony and arsenic sulfides, such as berthierite ($FeSb_2S_4$), gudmundite (FeSbS), arsenopyrite (FeAsS), and gersdorffite ((Fe,Ni,Co)AsS), with minor quantities of other base metal sulfides. Gold occurs as coarse visible gold, aurostibnite ($AuSb_2$), and as fine gold disseminated in sulfides. On average, the ore contains 3% to 4% Sb and 0.1% to 0.3% As.

Key process statistics (approximate; 1990 basis)
Ore throughput rate: approximately 1,000 tpm, batch process
Ore grade: 3 (varies from 1 to 5) g/t
Gold recovery: 77%

Process description/main features. Run-of-mine ore is crushed in three stages and ground in two stages to generate a product at 90% <75 µm (see Figure 12.51). Approximately 35% of the gold is recovered by gravity concentration during grinding. The ground slurry is treated by flotation to recover antimony sulfides. The concentrate produced contains 30% to 40% of the gold in the feed and has the following typical analysis: 30 g/t Au, 58% Sb, 1.9% Fe, and 0.1% to 0.2% each of Pb, Ni, Cu, and As.

The concentrates are thickened and then conditioned with lead nitrate (1.4 kg/t) and air as a preaeration step. The thickened slurry is leached with 2% NaCN at 52% solids and at pH 7. Leaching is performed in a pipe reactor (50 mm diameter, 1.5 km long) at a pressure of 9 MPa, with pure oxygen injected into the reactor at 12 MPa. Ambient temperatures of 22°C to 38°C are applied. Lime is used for pH control. The reactor provides a retention time of 15 min.; however, the slurry is usually passed through the reactor twice. Cyanide and oxygen consumptions are 57 and 20 kg/t, respectively.

The leached slurry is thickened, and the underflow is filtered, dried, and sold as a high-grade antimony product. The overflow solution is filtered and the fines recovered are directed to residue disposal. The solution is combined with the filtrate from the thickener underflow filtrate, and dissolved gold is recovered by CIC. Loaded carbon is shipped to the Rand Refinery for gold recovery because the relatively small quantity of carbon produced does not justify on-site carbon processing. To avoid the need to recycle fouled solution, the barren solution volume is reduced in evaporation ponds.

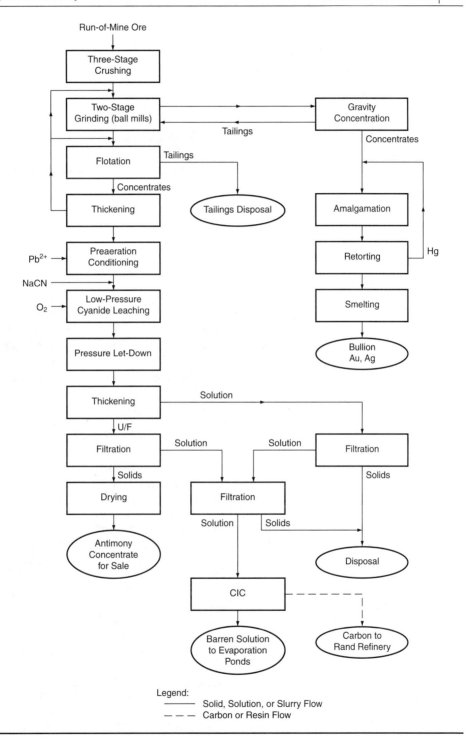

FIGURE 12.51 Consolidated Murchison (1990)

12.2.8 Telluride Ores

12.2.8.1 Emperor Mines [10, 69]
Owner/Operator: Emperor Mines Ltd., Durban Roodeport Deep (Johannesburg, South Africa)
Location: Vatukoula, Fiji
Start-up: 1932

Mineralogical factors. Emperor ore contains free gold and gold associated with tellurides (principally sylvanite and hessite). Significant amounts of pyrite, arsenopyrite, and marcasite are present, with minor copper, zinc, and lead sulfides. A small amount of native tellurium is present.

Key process statistics (approximate; 1990 basis)
Ore throughput rate: 2,000 tpd
Ore grade: 15 to 25 g/t
Gold recovery: Not available

Process description/main features. The Emperor process is unique in that three separate metallic products—gold, tellurium, and copper—have been produced in the past from this refractory ore in a relatively complex and unusual flowsheet. The flowsheet in Figure 12.52 and the following process description consider the circuit that operated in the 1970s and early 1980s. This was subsequently modified by discontinuation of the tellurium recovery circuit, by the addition of fluosolids roasting (both in 1986), and the addition of a CIP scavenger circuit (1989) to treat copper sulfide flotation and calcine leach tailings.

Ore was crushed and ground in a conventional circuit (rod and ball mills). The slimes produced by crushing were separated from the ore and treated by flotation to recover a bulk concentrate, which was returned to the grinding circuit. The flotation tailings were discharged to a tailings impoundment, avoiding any detrimental effects of this fine material on the subsequent selective tellurium and copper mineral flotation.

Flotation was applied to the ground product to recover gold-bearing tellurium minerals. The concentrates obtained were reground and then subjected to an alkaline chlorination leach, using sodium carbonate, 5% sodium hydroxide, and 5% calcium hypochlorite as reagents, to dissolve gold tellurides, which otherwise dissolved very slowly in cyanide solution. The resulting slurry was leached with cyanide (1.5% NaCN) to dissolve the remaining gold and silver. In this stage, any gold and silver dissolved in the chlorination step were converted to the more stable cyanide complexes. The leached slurry was filtered and the solution passed to a Merrill–Crowe recovery circuit (discussed later). The filtered solids were then leached with sodium sulfide to dissolve tellurium. The tellurium-rich solution was separated by filtration, and tellurium metal was recovered by precipitation with sodium sulfite followed by filtering, drying, and melting of the product.

The tellurium flotation tailings were mixed with recycled cyanide leach solution, which started gold and silver dissolution. The slurry was thickened and the underflow was leached with cyanide and filtered to recover the gold-rich solution for recycling. The thickener overflow solution was clarified, deaerated, and treated by zinc precipitation for gold recovery. The precipitate was filtered, acid treated (to remove excess zinc), and smelted into bullion.

After the filtered solids produced by cyanidation were treated with sulfur dioxide to destruct residual free cyanide, copper minerals were floated off at pH 9.5 using an isobutyl xanthate collector. The flotation tailings were disposed of as the final plant effluent. The concentrates were thickened, filtered, and then roasted to oxidize the sulfide minerals. In an unusual step, the thickener overflow and filtrate were used to quench the roasted calcine,

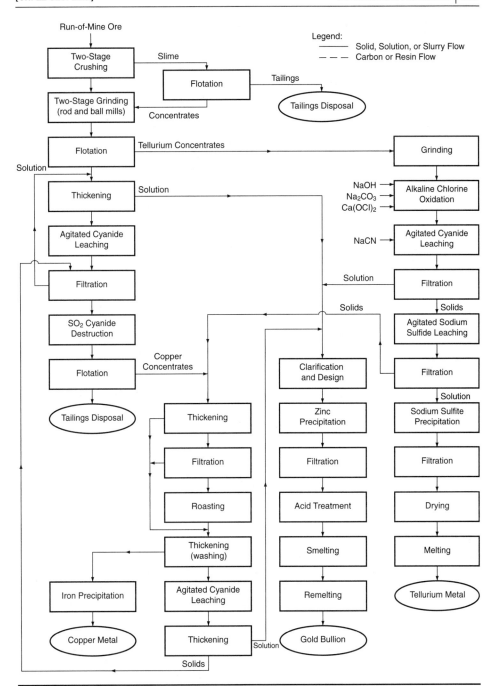

FIGURE 12.52 Emperor (1970s/early 1980s circuit)

forming acidic sulfate solution, which dissolved copper. The quenched slurry was thickened, and copper was recovered from the overflow solution by cementation with iron. The thickened solids were leached yet again with cyanide for further gold extraction, and the solution obtained was treated in the Merrill–Crowe circuit.

12.2.9 Carbonaceous Ores

12.2.9.1 Jerritt Canyon [70, 80, 81]
Owner/Operator: Queenstake Resources Ltd. (Denver, Colorado, United States)
Location: Carlin, Nevada, United States
Start-up: 1981

Mineralogical factors. Free gold occurs with carbonates and silicified limestones, sandstones, and cherts. The ore has variable organic carbon content (0% to 1.5% C), and significant portions of the orebody has severe gold-adsorbing capability. The ore also contains considerable amounts of clay, typically 10% to 25%. Mercury is present as cinnabar at an average concentration of 25 g/t. Realgar, orpiment, stibnite, and pyrite are present in minor amounts, with arsenic content varying between 0.02 and 0.25%. The total sulfide content of the oxidized ore varies between 0.2% and 2.0% S. The ore contains 10% to 20% carbonate content.

Ore was defined as carbonaceous if the ratio of cyanide soluble to fire assay gold content of ore samples was <0.6:1 (i.e., cyanide leach extraction <60%).

Key process statistics

	Chlorination (1990 basis) [70]	Roasting (2004 basis) [81]
Ore throughput rate:	3,600 tpd	3,700 tpd
Ore grade:	5.6 g/t	6.7 g/t
Gold recovery:	88–90%	87–90%
Gold production:	~210,000 oz/yr	~250,000 oz/yr

Process description/main features. Jerritt Canyon has utilized two processing techniques to treat carbonaceous, preg-robbing, sulfidic refractory ores. The first process used alkaline chlorination to treat low-sulfide sulfur material (<0.5% S^{2-}) and this operated from 1981 to 1997. The second process was commissioned in 1989 and utilizes whole-ore roasting to treat higher-sulfide sulfur material (>0.5% S^{2-}) with high organic carbon content (>0.5% organic C). Both processes are described below, with the flowsheets for each given in Figures 12.53 and 12.54, respectively.

Alkaline Chlorination
Run-of-mine ore was primary crushed and ground in three stages (fully autogenous milling, ball milling, and pebble crushing) to yield a relatively coarse product at 80% <150 μm.

Ground slurry was thickened and then treated by alkaline chlorination to passivate carbonaceous constituents. Chlorination was performed at approximately 50°C and at pH 10.5, using sodium carbonate for pH control and as a buffer. Liquid chlorine was vaporized and sparged into six closed, agitated tanks configured in series, which were operated under a low vacuum (5 cm of water) to control fugitive chlorine gas. Most of the chlorine was sparged into the first two tanks to maintain hypochlorous acid (HOCl) concentrations of 2–3 g/L. The hypochlorous concentration was allowed to decay through the system, down to about 1 g/L in the final tank, lending the name "flash chlorination" to the process. The pH of the slurry decreases to about 5.5 by the end of chlorination. Total retention time for chlorination was approximately 13 hr. Between 1981 and 1990, chlorine consumption was relatively high, varying between 30 and 60 kg/t, depending primarily on the sulfide sulfur content of the feed (see Table 5.9). However, after the introduction of whole-ore roasting to process the high-sulfide sulfur material (>0.5% S^{2-}), the chlorine consumption was reduced to about 12 kg/t. Residual chlorine and hypochlorous acid, which would otherwise rapidly oxidize (and consume) cyanide in the subsequent cyanidation circuit, were destroyed by the addition of sodium hydrosulfide (NaHS), and

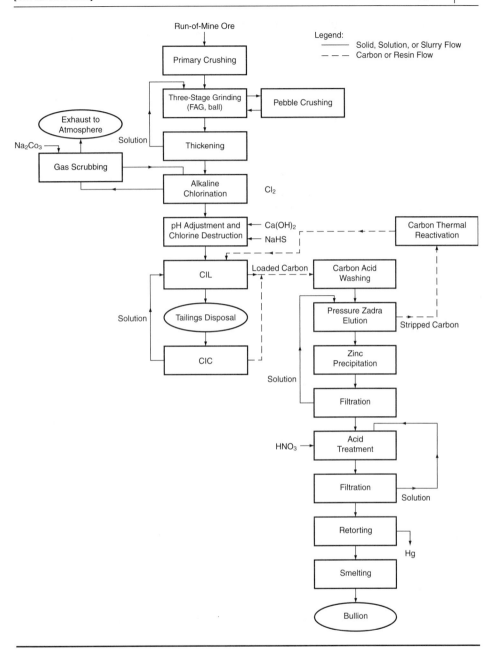

FIGURE 12.53 Jerritt Canyon (pre-1990)

hypochlorous concentration was reduced to below approximately 0.05 g/L. Sodium hydrosulfide consumption was approximately 0.3 kg/t.

The off-gas was scrubbed with a sodium carbonate–bicarbonate solution, and the scrubber solution was returned to the chlorination feed slurry as bleach. The off-gas scrubbing process consumed about 3.5 kg/t sodium carbonate.

The chlorinated slurry was treated by CIL for simultaneous gold–silver dissolution and carbon adsorption to minimize the gold-adsorbing tendency of any unpassivated carbonaceous material. Loaded carbon was acid washed and eluted by a pressure Zadra procedure. Gold (and silver) were recovered by zinc precipitation followed by filtration, acid treatment, retorting for mercury removal, and smelting. The acid treatment step used 20% to 30% HNO_3 at 75°C to 80°C for 2 hr to dissolve most of the zinc and between 60% and 70% of the mercury content. The product of this treatment was filtered and the solution returned to the CIL process. This reduced the amount of mercury that had to be removed by retorting. The stripped carbon was thermally reactivated and returned to CIL for gold adsorption.

For a period at Jerritt Canyon, a pre-oxidation step was employed (and also at Carlin where chlorination was also applied), whereby the ground slurry was heated to 70°C to 80°C (with steam) and agitated in the presence of 20 kg/t sodium carbonate and air. This partially oxidized the sulfides in the feed (i.e., pyrite) and reduced subsequent chlorine consumption; however, the so-called double-oxidation process was subsequently discontinued in favor of more economical direct chlorination. For further details on the chlorination oxidation process, see Section 5.6 (and specifically 5.6.5).

Whole-Ore Roasting

Whole-ore roasting was implemented at Jerritt Canyon in 1989 as an alternative to alkaline chlorination as a result of (i) increasing chlorine consumption with increasing sulfide sulfur content of the ore, (ii) escalating chlorine prices, and (iii) evaluation of the remaining ore reserves indicating that chlorination would not be economically viable in the long term. Both processes (i.e., whole-ore roasting and alkaline chlorination) operated in parallel from late 1989 until 1997, when the alkaline chlorination circuit was shut down.

Run-of-mine ore is crushed in three stages, with a drying stage prior to tertiary crushing. The ore is dry ground in a ball mill to yield a product at 80% <106 μm containing <1% moisture. Pulverized coal is added to the ore feed and fed to one of two Dorr-Oliver fluidized bed roasters. The amount of coal added depends on the sulfide sulfur and organic carbon content of the feed material. The roaster bed is fluidized with a countercurrent flow of oxygen, delivered to the roaster at 72.4 kPa. The roasters are divided into two compartments to provide two stages of roasting. The first stage operates at 540–650°C and oxidizes 98% of the sulfide sulfur and 75% of the carbonaceous matter. The second stage operates at 600–650°C and completes the oxidation of sulfide sulfur and carbonaceous material down to residual levels of <0.05% of each. About 80% of the oxygen is consumed in the first stage and 20% in the second stage.

The roasted ore is discharged from the roaster and water-quenched, reducing the temperature of the solids to about 54°C. The resulting solids are thickened, the pH adjusted to 10.5, and the slurry is treated for simultaneous gold leaching and recovery in a CIL circuit. The loaded carbon is acid washed and stripped using the pressure Zadra elution procedure. Gold (and silver) is recovered from the eluate solution by zinc precipitation. The precipitate is acid washed to remove residual zinc, filtered, retorted for mercury removal, and then smelted to produce a doré bullion (94% gold) for sale. Gold recovery from the roasted calcine varies between 87% and 90%.

Roaster off-gas is cooled and cleaned (particulate removal), followed by sulfur dioxide scrubbing (with sodium hypochlorite), mercury removal by condensation, and finally tail gas scrubbing with water. Over 99.9% of the arsenic in the roaster feed is fixed in the calcine solids, either as calcium or iron arsenate.

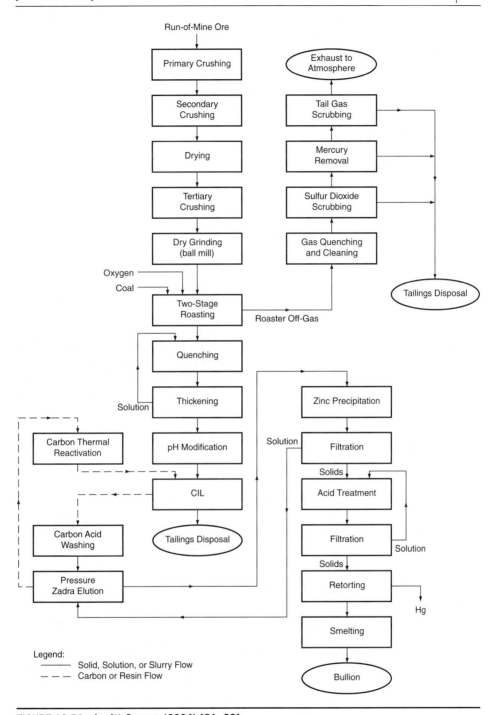

FIGURE 12.54 Jerritt Canyon (2004) [81, 82]

12.2.9.2 Newmont Mill No. 6 (Gold Quarry) Roaster [71]

Owner/Operator: Newmont Mining (Denver, Colorado, United States)
Location: Carlin, Nevada, United States
Start-up: 1994

Mineralogical factors. Deeper sulfide mineralization on the Carlin Trend (north of Elko, Nevada) often contains carbonaceous ore components (i.e., organic carbon) that can be highly refractory due to severe (nonreversible) preg-robbing characteristics.

Key process statistics (approximate; 1996 basis)
Ore throughput rate: 7,600 tpd
Ore grade: 7.5 to 8.5 g/t
Gold recovery: 90% to 92%
Gold production: ~640,000 oz/yr

Process description/main features. Refractory sulfidic–carbonaceous ore is primary crushed to about 80% <150 mm followed by secondary crushing to 80% <22 mm (see Figure 12.52). Ore moisture is typically between 4% and 8%. The fuel value in the roaster feed is calculated based on the content of sulfide sulfur, pyrite, and organic carbon and, if necessary, supplemental pyrite is added to the roaster feed to ensure autogenous operation. Typical roaster feed composition includes the following: 1.75% sulfide sulfur (3% pyrite), 0.4% to 1.0% organic carbon, 70% quartz, 0.05% carbonate carbon, 1,200 g/t arsenic, 100 g/t chloride, and 1,000 g/t fluoride. Lime is also added to the feed to control emissions in the roaster circuit by fixing SO_2 as it is formed. The crushed feed is ground in an innovative air-swept, dual-compartment dry grinding ball mill (known as a double-rotator), supplied by Krupp-Polysius. The double-rotator dries and grinds the ore in preparation for roasting.

The ground product has a particle size of 80% <208 µm and is delivered to fine ore bins for storage ahead of roasting. The ore is preheated with hot, blown air to approximately 420°F and then fed to one of two circulating fluidized bed roasters. Each roaster train consists of the roaster, cyclones, two seal pots, fluidizing air blowers, oxygen preheater, induction burner, and two calcine coolers. The roasters operate at a temperature of 550°C (1,020°F) with a retention time of 10 min. Almost all of the sulfide mineralization and approximately 30% of the organic carbon in the feed are oxidized in the roaster. However, an additional 20 to 24 min of retention time is provided in the calcine coolers at or close to the roasting temperature, and the balance of the organic carbon is oxidized in this part of the circuit.

The calcine product is quenched to produce a warm 40°C (104°F) slurry containing 15% solids. The slurry is neutralized with milk of lime, then thickened and subjected to agitated cyanide leaching in the presence of activated carbon (i.e., CIL process). Loaded carbon is processed in a central carbon elution and reactivation facility serving several plants. This is not discussed further here.

Off-gas from the roaster is cooled from 540°C (1,000°F) to about 380°C (715°F) in a waste heat boiler. The gas stream is split between the main roaster fluidization stream and the gas cleaning feed stream. The gas recycled to the roaster (i.e., roaster fluidization stream) contains 30% to 40% oxygen by volume, and the oxygen concentration in the gas phase is controlled for optimum roaster performance. The gas cleaning feed stream passes through several stages of cooling and cleaning where acid mist is removed by electrostatic precipitation, fluorine is removed onto a sacrificial silica packing in a tower, and finally mercury is removed using the Boliden Norzinc process, whereby mercury is scrubbed into mercuric chloride solution to form mercurous chloride. Mercury is either recycled as mercuric chloride to the scrubbing system or recovered by electrowinning for

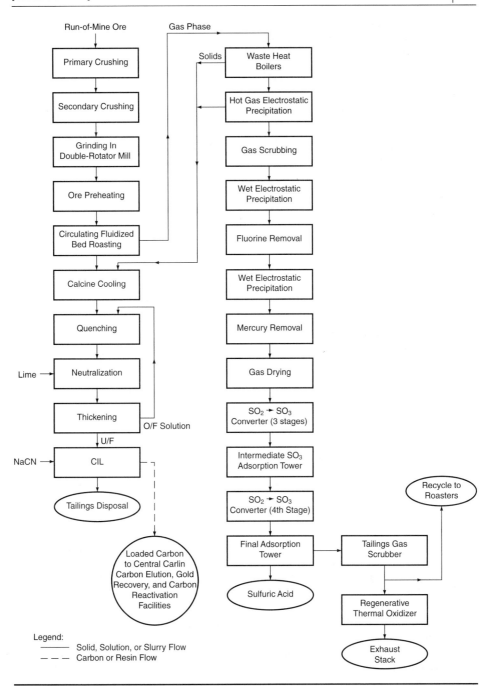

FIGURE 12.55 Newmont Mill No. 6 (Gold Quarry) Roaster (1996)

sale as a by-product. Sulfuric acid is recovered from the gas phase using standard sulfuric acid plant technology. The final tail gas is scrubbed with H_2O_2 to meet emission requirements. The tail gas is treated through a regenerative thermal oxidizer to convert any remaining CO to CO_2 prior to release. SO_2 to SO_3 conversion efficiency is >99.8%.

12.2.9.3 Penjom [72, 73]
Owner/Operator: Avocet Mining PLC (London, United Kingdom)
Location: Pahang, Malaysia
Start-up: 1999

Mineralogical factors. Gold occurs predominantly as fine free gold and gold associated with sulfide minerals. Ore contains finely divided organic carbon that is strongly preg-robbing (according to Mintek, it is one of the most strongly preg-robbing in the gold industry), precluding treatment by standard cyanide leaching. Organic carbon content varies between 0.2% and 1.5%, averaging 0.6%. Gold content is variable, ranging from 6 to >10 g/t.

Key process statistics (approximate; 2000 basis)
Ore throughput rate: 1,650 tpd
Ore grade: 7 to 8 g/t
Gold recovery: 75%
Gold production: ~100,000 oz/yr

Process description/main features. Ore is primary crushed and then ground in a single-stage ball mill operated in open circuit (see Figure 12.56). The ball mill discharge is cycloned to separate the coarse and fine fractions of the slurry. The cyclone overflow solids have a particle size of about 80% <75 μm. The cyclone overflow contains the majority of the fine carbonaceous material and is processed by kerosene "blanking" prior to RIL treatment. Initially, the operation used CIL technology; however the application of CIL was unsuccessful probably due to poisoning (i.e., blinding) of the activated carbon with kerosene. This was replaced with a RIL circuit using a strong-base ion exchange resin, also with limited success. In 1999, the RIL circuit was switched over to a gold-selective, strong-base anionic ion exchange resin: Minix-Dowex (Mintek, South Africa). The resin beads used are >0.8 mm diameter, and 0.4 mm interstage screens are used to retain the resin and allow the slurry to pass on to the next stage of RIL. The RIL circuit gold recovery is approximately 60% of gold in the cyclone overflow stream. The resin is reportedly unaffected by the kerosene or diesel blanking process. Loaded resin is transferred to a stripping vessel and the resin stripped down to a gold loading of about 50 g/t (see Chapter 7 for strong-base resin elution details). Gold is recovered from the loaded strip solution by electrowinning.

The cyclone underflow is split, with the bulk directed to six InLine Pressure Jigs (Gekko Systems). The concentrate produced by the jigs is cleaned using spiral concentrators and then subjected to intensive cyanide leaching in an InLine Leach Reactor (Gekko Systems). The concentrate produced by the jigs is mainly a sulfide mineral concentrate (20% to 30% S), with some fine free gold (300 to 1,500 g/t). Intensive cyanide leaching is performed using >2% NaCN solution at ambient temperature with a retention time of 3 hr to achieve gold recovery typically between 80% and 90%.

The smaller split of the cyclone underflow is treated in a Knelson centrifugal concentrator. The Knelson tails are scavenged by a Falcon centrifugal concentrator. The free gold-containing concentrates from the Knelson and Falcon concentrators are combined, cleaned on Gemini tables, and then directly smelted to produce gold bullion.

Solution from the intensive leaching step is processed through CIC for gold recovery. The carbon is stripped and gold recovered from the eluate solution by electrowinning. The loaded cathodes produced by electrowinning of gold from both the carbon and resin eluates are smelted with fluxes to produce a bullion product at the site.

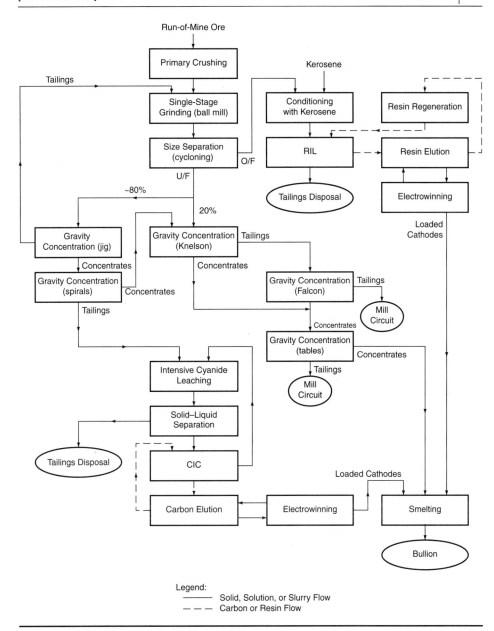

FIGURE 12.56 Penjom (2000)

Overall gold recovery from this circuit is approximately 75% (2000 basis), with about 5% to 15% of this produced from gravity concentrate, 35% to 45% from intensive cyanide leaching of the sulfide mineral concentrate, and about 20% from the RIL circuit.

The history of the development of this flowsheet makes for interesting reading and is a lesson in perseverance with a difficult ore.

12.2.10 Tailings

12.2.10.1 Ergo [74]
Owner/Operator: AngloGold Ashanti (Johannesburg, South Africa)
Location: Germiston, Transvaal, South Africa
Start-up: 1979

Mineralogical factors. Tailings from previous cyanidation operations date from about 1900. Some small amounts of free gold occur, although gold is predominantly associated with sulfides, primarily pyrite. Uranium minerals are also present.

Key process statistics (approximate; 1990 basis)
Ore throughput rate: 50,000 tpd
Ore grade: 1 g/t Au
Gold recovery: Not available.

Process description/main features. Tailings are reclaimed from old dumps using high-pressure water jets (monitors). Flotation is used to produce a bulk gold–sulfide (mainly pyrite) concentrate, which is thickened and leached with sulfuric acid to dissolve uranium (see Figure 12.57). The leach solution is separated by filtration and is subsequently clarified, followed by solvent extraction and precipitation of uranium as ammonium diuranate (yellowcake). This product is sold to Nuclear Fuels Corporation of South Africa.

The solid residue from acid leaching is roasted in a single stage under oxidizing conditions to oxidize pyrite (and other sulfides) and liberate the contained gold. The calcine is quenched, thickened, and cyanide leached to dissolve gold and silver. The product is filtered, and the solution is clarified, deaerated and treated by Merrill–Crowe zinc precipitation for precious metal recovery. To oxidize zinc and base metals, precipitate is calcined, followed by smelting to produce bullion. The filtered cyanide leaching solids residue is pumped to a new tailings impoundment.

The configuration of acid leaching for uranium extraction followed by cyanide leaching for precious metals extraction is commonly referred to as "reverse" leaching and has been employed at several uranium-containing gold mines in South Africa. The process has the advantage of cleaning gold surfaces (likely to be an important factor in the treatment of tailings), and up to 2% additional gold recovery may be realized over the cyanide-leaching-followed-by-acid-leaching scheme that has been employed elsewhere.

Flotation tailings are treated by CIL to recover residual gold, with the loaded carbon treated by an AARL elution procedure. The eluate is combined with the leach solution for Merrill–Crowe precipitation. Eluted carbon is thermally reactivated and recycled to the process.

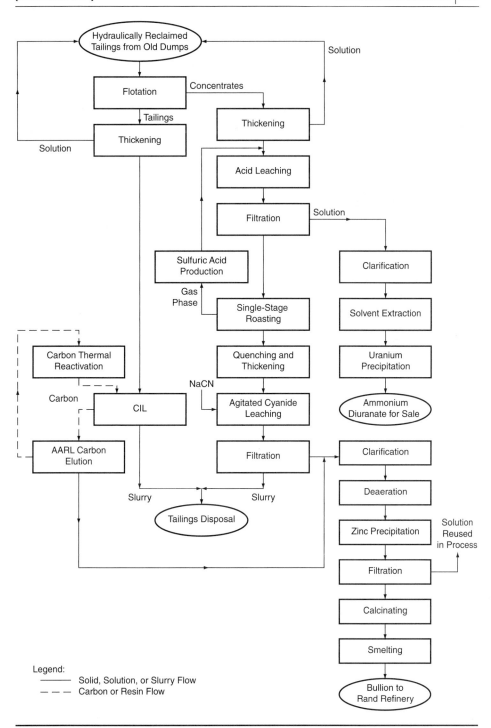

FIGURE 12.57 Ergo (1990)

12.2.11 Refining

12.2.11.1 Homestake Lead [9]
Owner/Operator: Barrick Gold (Toronto, Canada), formerly Homestake Mining (San Francisco, California, United States)
Location: Lead, South Dakota, United States

Process description/main features (1990 basis). The refinery treated three main types of product—zinc precipitates, gravity concentrates, and steel wool cathodes—and each was handled slightly differently (see Figure 12.58). Precipitates were dried with steam and then smelted with fluxes to remove zinc and other base metal impurities. Gravity concentrates were upgraded further by gravity concentration followed by smelting to remove silica and other extraneous matter. Finally, steel wool cathodes were directly smelted (see Chapter 10). The bullion products from these smelting processes were remelted (and combined) followed by chlorination (Miller process) to remove silver as silver chloride. This yielded a gold product with a fineness >99.7%, which was either sold as is or further refined on-site electrolytically (in Wohlwill cells), followed by drying and melting of the final (99.95% Au) product.

The silver chloride produced by chlorination was resmelted to recover any entrained gold in a "de-golding" step. Finally, silver was reduced with iron and smelted to produce saleable silver metal.

The slag produced by the primary smelting processes and the Miller chlorination stage were resmelted with litharge to collect residual precious metals. The lead product was cupelled to generate a crude bullion, which was recycled to the precious metals refining circuit. A matte was also produced, which was sold to a custom smelter.

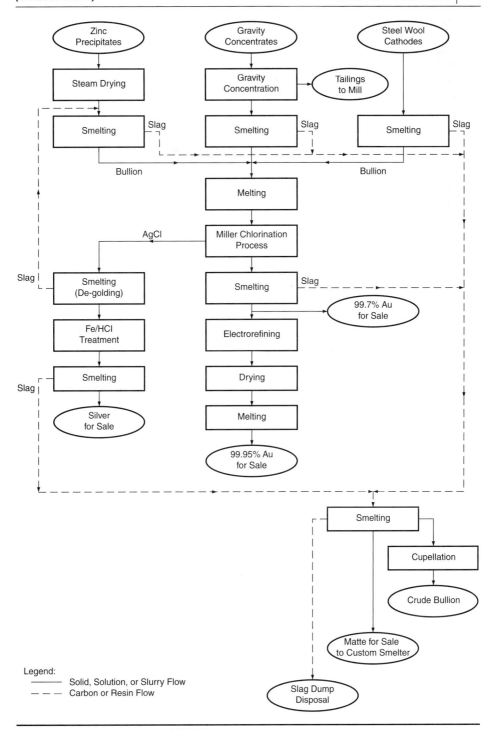

FIGURE 12.58 Homestake Lead (1990)

12.2.11.2 Johnson Matthey Gold and Silver Refinery [75]
Owner/Operator: Johnson Matthey PLC (London, United Kingdom)
Location: Salt Lake City, Utah, United States

Process description/main features (1990 basis). Materials processed by the refinery are grouped into three main categories:

1. High-grade materials containing at least 30% gold, with silver and base metals impurities
2. High-grade products containing approximately 30% gold and the balance mainly silver
3. Low-grade products, typically containing less than 10% gold and silver combined

These three materials have quite different refining requirements and, consequently, different flowsheets (Figure 12.59). The first category is treated by the Miller chlorination process to remove silver as silver chloride. The gold bullion produced (99% Au) is resmelted and electrorefined, yielding a 99.99% Au product suitable for sale. The silver chloride is smelted to recover entrained gold (de-golding), and the silver is reduced with zinc from acidic chloride solution and then subsequently electrorefined to produce 99.99% silver for sale.

In the treatment of materials in the second category, gold and silver are parted by leaching with nitric acid. The solid residue and the leach solution are separated, and the solids are treated in the low-grade circuit described next. Silver is precipitated from the solution phase by chloride addition and zinc reduction, and the product is melted and electrorefined, again yielding saleable 99.99% Ag.

Low-grade products are leached with aqua regia (HNO_3 and HCl) to dissolve gold and simultaneously form silver chloride. The resulting mixture is filtered and the solids (containing silver chloride) are melted and recycled to the silver refining circuit following precipitation of any metal in the solution with metallic zinc. Gold is recovered from the solution phase by precipitation with sulfur dioxide, followed by acid washing and melting of the product to generate 99.99% Au.

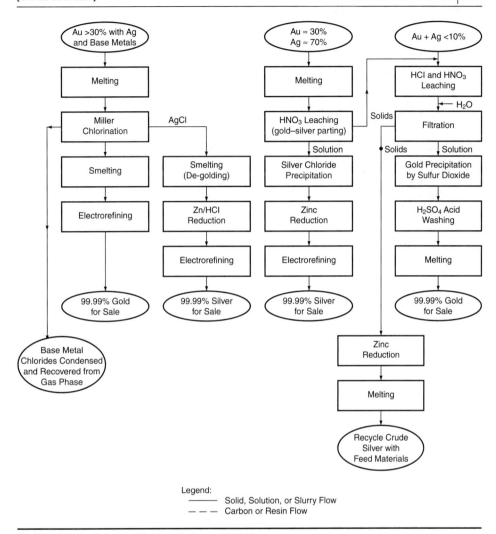

FIGURE 12.59 Johnson-Matthey Refinery (1990)

REFERENCES

[1] Annual Gold Surveys. 1990–2005. London: Gold Fields Mineral Services Ltd.

[2] Tien, J.C. 1994. *The Gold Industry of China*. Chicago, IL: Maclean Hunter Publishing.

[3] Morgan, R.J. 1991. *Mining Journal Gold Database*. London: The Mining Journal.

[4] Anon. 1989. Bjorkdal Swedish gold project opened. *Engineering and Mining Journal* (February):15–16.

[5] Marsden, J.O. 1984. *A Metallurgical Review of the East Driefontein Gold Plant*. Gold Fields of South Africa, Internal Report. Johannesburg, South Africa: Gold Fields of South Africa Ltd.

[6] Tuanyane, B.D., and A.B. Phillips. 2003. The phased metallurgical upgrade strategy at Gold Fields Limited South African operations. Pages 143–152 in *Proceedings 8th Mill Operators' Conference*. Melbourne: Australasian Institute of Mining and Metallurgy.

[7] Lewins, J.D. 1988. The New Harmony Central Gold Plant. *Mine Metallurgical Managers' Association of South Africa* 2-88 (December):43–59.

[8] Moore, V.J., P.H. Radcliffe, P. Smith, and M.A. Orridge. 1986. The President Brand New Gold plant. Pages 23–28 in *Gold 100: Proceedings International Conference on Gold*. Volume 2. Edited by C.E. Fivaz. Johannesburg: South African Institute of Mining and Metallurgy.

[9] Hinds, H.L., and L. Trautmann. 1980. Metallurgical practice at Homestake Lead operation. Pages 124–131 in *Proceedings Precious Metals Symposium*. Report 36. Edited by V.E. Kral, J.A. Hall, R.B. Blakestad, H.F. Bonham, G.B. Hartley, G.E. McClelland, J.A. McGlasson, and P. Mousette-Jones. Reno, NV: Nevada Bureau of Mines and Geology.

[10] McQuiston, F.W., and R.S. Shoemaker. 1975. *Gold and Silver Cyanidation Plant Practice*. Volume I. AIME Monograph. New York: American Institute of Mining, Metallurgical, and Petroleum Engineers.

[11] Mansanti, J.G., J.R. Arnold, J.H. Gourdie, and J.O. Marsden. 1989. Double thickener circuit at Gold Fields' Chimney Creek. *Minerals and Metallurgical Processing* (November):179–186.

[12] Arnold, J.R. 1990. Gold recovery at Gold Fields' Chimney Creek. Pages 251–260 in *Proceedings of Gold Forum on Technology and Practice—World Gold '89*. Edited by R.B. Bhappu, and R.J. Harden. Littleton, CO: SME.

[13] Mansanti, J.G., and J.R. Arnold. 1988. Carbon handling at Gold Fields' Chimney Creek. Paper presented at 1st Annual Intermountain Mining and Processing Operators Symposium, Elko, NV, November 3–5.

[14] Dagel, K.D., L.M. Tahija, and R.M. Crowell. 1989. Process plant expansion and operation at Round Mountain. Pages 225–230 in *Proceedings Gold 1990 Symposium*. Edited by D.M. Hausen, D. Halbe, E.U. Petersen, and W.J. Tafuri. New York: American Institute of Mining, Metallurgical, and Petroleum Engineers.

[15] Knight, D.A., and V.G. Medina. 1986. Kidston Gold Mines Start Up. *Mineral and Metallurgical Processing Journal* 3(1):45–49.

[16] Lee, V., P. Robinson, and F. Merz. 1989. Peroxide addition improves gold recovery and saves reagents at Pine Creek gold mine. Pages 179–182 in *Proceedings Randol Gold Conference*. Golden, CO: Randol International Ltd.

[17] Fleming, C.A., and G. Cromberge. 1984. The extraction of gold from cyanide solutions by strong- and weak-base anion exchange resins. *Journal of South African Institute of Mining and Metallurgy* 84(5):125–137.

[18] McQuiston, F.W., and R.S. Shoemaker. 1981. *Gold and Silver Cyanidation Plant Practice*. Volume 2. AIME Monograph. New York: American Institute of Mining, Metallurgical, and Petroleum Engineers.

[19] Hollow, J.T., E.M. Hill, H.K. Lin, and D.E. Walsh. 2005. Modeling the influence of slurry temperature on gold leach and adsorption kinetics at the Fort Knox mine, Fairbanks, Alaska. SME Preprint 05-67. Littleton, CO: SME.

[20] Harris, L., D. Malhotra, and R. McGregor. 2000. Recent developments in processing of oxide and refractory gold ores. Pages 321–326 in *Proceedings Randol Gold & Silver Forum 2000*. Golden, CO: Randol International Ltd.

[21] Newmont Mining. 2004 Annual Report. Denver, CO: Newmont Gold Corporation.

[22] Harris, L., D. Livermore, C. Santa Cruz, and M. Diaz. 1994. The Yanacocha project. SME Preprint 94-74. Littleton, CO: SME.

[23] Bee, G., R. Espinel, Z. Gavelan, X. Ochoa, and Y. Roditis. 2002. Precious metal heap leaching in a high rainfall alpine setting at the Barrick Gold Pierina mine in Peru. SME Preprint 02-055. Littleton, CO: SME.

[24] Frostiak, J., K.G. Thomas, I. Gonzales, and K. Manning. 2000. The Pierina project: Leaching of high grade ore. Pages 135–154 in *Proceedings Randol Gold & Silver Forum 2000*. Golden, CO: Randol International Ltd.

[25] Dayton, S.H. 1987. A model gold mine and plant at Paradise Peak. *Engineering and Mining Journal* (February):18–24.

[26] Maki, T.D. 1989. Heap leaching at FMC Gold's Paradise Peak mine. SME Preprint 89-36. Littleton, CO: SME.

[27] Gathje, J., D. Gale, and J. Turney. 1995. Process development for the Eskay Creek project. Pages 31–35 in *Proceedings 19th International Minerals Processing Congress*. Littleton, CO: SME.

[28] Sylvestre, Y. 1988. Joutel Division-Milling Treatment. Internal Report. Quebec, Canada: Agnico-Eagle Mines.

[29] Bunn, S. 1991. Process description Fimiston plant and Gidji roaster. *Proceedings Australian Mines and Metals*.

[30] Argall, G.O. 1986. Perseverance and winning ways at McLaughlin gold. *Engineering and Mining Journal* (October):26–32.

[31] Turney, J.R., R.J. Smith, and W.J. Janhunen. 1989. The application of acidic pressure oxidation to the McLaughlin refractory ore. Pages 25–46 in *Precious Metals 1989*. Edited by M.C. Jha and S.D. Hill. Warrendale, PA: TMS

[32] McDonald, A. 2003. Lihir gold five years on—maximizing throughput and capital utilization. Pages 115–120 in *Proceedings 8th Mill Operators' Conference*. Melbourne: Australasian Institute of Mining and Metallurgy.

[33] Lihir Gold. 2004. Annual Report. Port Moresby, Papua New Guinea: Lihir.

[34] Ketchum, V.J., J.F. O'Reilly, and W.D. Vardill. 1993. The Lihir Gold project: Process plant design. *Minerals Engineering* 6(8–10):1037–1065.

[35] Hille, S., and R. Raudsepp. 1998. Pressure oxidation performance at Porgera. Pages 209–214 in *Proceedings Randol Gold & Silver Forum 1998*. Golden, CO: Randol International Ltd.

[36] Kewe, T. 2003. Advances in flotation process control at Porgera gold mine using JK FrothCam imaging technology. Pages 87–92 in *Proceedings 8th Mill Operators' Conference*. Melbourne: Australasian Institute of Mining and Metallurgy.

[37] Pieterse, H.J.H., K.G. Thomas, R.A. Williams, C.E. Semler, and R.R. Pierce. 1993. Barrick Goldstrike pressure oxidation linings. Pages 189–195 in *Proceedings Randol Gold Forum*. Golden, CO: Randol International Ltd.

[38] Barrick. 2004. Annual Report. Toronto, ON, Canada: Barrick.

[39] Thomas, K.G. 1994. *Research, Engineering Design and Operation of a Pressure Hydrometallurgical Facility for Gold Extraction*. Den Haag, Netherlands: CIP-Gegevens Koninkluke Bibliotheek.

[40] Du Plessis, R., D.G. Kotlyar, G.L. Simmons, and J.D. Miller. 2002. The effect of activation on the low potential hydrophobic state of pyrite in amyl xanthate flotation with nitrogen. SME Preprint 02-155. Littleton, CO: SME.

[41] Cole, J.A., J.C. Lenz, and W.J. Janhunen. 1995. One year of pressure oxidation at the Lone Tree gold mine. *Mining Engineering* (June):515–519.

[42] Simmons, G.L., J.C. Gathje, W. de Beer, and R. Moritz. 2003. Inert gas flotation technologies. Pages 1134–1142 in *Proceedings 22nnd International Mineral Processing Congress*. Edited by. L. Lorenzen and D. Bradshaw. Johannesburg, South Africa: South African Institute of Mining and Metallurgy.

[43] Greene, J.W. 1996. Integrated milling with heap leaching at Barneys Canyon for treatment of mixed-oxide ores. Pages 343–346 in *Proceedings Randol Gold Forum '96*. Golden, CO: Randol International Ltd.

[44] Suttill, K.R. 1989. Ghana's golden glow-interest revives in the gold coast. *Engineering and Mining Journal* (June):22–32.

[45] Gold Fields. Technology forum and BIOX process description. www.goldfields.co.za. Accessed April 8, 2005.

[46] Roberts, S., and J. Starkey. 1987. Modernisation of the Campbell Red Lake mill in 1987. Pages 185–197 in *Proceedings of International Symposium on Gold Metallurgy*. Edited by R.S. Salter, D.M. Wyslouzil, and G.W. McDonald. New York: Pergamon Press.

[47] Frostiak, J., and B. Haugrud. 1992. Start up and operation of the Campbell Red Lake Gold Pressure Oxidation Plant. SME Preprint 92-14. Littleton, CO: SME.

[48] Halverson, G.B. 1990. Fluosolids roasting practice at Giant Yellowknife Mines Ltd. Paper presented at 96th Annual North West Mining Association, Spokane, WA.

[49] St. Louis, R.M., and J.M. Edgecombe. 1990. Recovery enhancement in the Mercur autoclave circuit. Pages 443–450 in *Proceedings Gold 1990 Symposium*. Edited by D.M. Hausen. Littleton, CO: SME.

[50] Anon. 1988. Mercur starts up its new alkaline pressure oxidation autoclave plant. *Engineering and Mining Journal* (June):26–31.

[51] Carvalho, T.M., A.K. Haines, R.E.J. DaSilva, and B.N. Doyle. 1988. Start-up of the Sherritt pressure oxidation process at Sao Bento. Pages 152–156 in *Proceedings Randol Gold Conference*. Golden, CO: Randol International Ltd.

[52] Van Aswegen, P.C., and A.K. Haines. 1988. Bacteria enhance gold recovery. *International Mining* (May):19–23.

[53] Van Aswegan, P.C., A.K. Haines, and H.J. Marais. 1988. Design and operation of a commercial bacterial oxidation plant at Fairview. Pages 144–147 in *Proceedings Randol Gold Conference*. Golden, CO: Randol International Ltd.

[54] Van Aswegen, P.C., M.W. Godfrey, D.M. Miller, and A.K. Haines. 1991. Developments and innovations in bacterial oxidation of refractory ores. SME Preprint 91-75. Littleton, CO: SME.

[55] Freeport-McMoran Copper and Gold Inc. 2004 Annual Report. New Orleans, LA: Freeport-McMoran Gold & Copper Inc.

[56] Coleman, R., and C. Veloo. 1996. Concentrator expansion at Freeport Indonesia's Grasberg operations. *Mining Engineering* (February):25–33.

[57] Coleman, R., S. Nugroho, and A. Neale. 2001. Design and start-up of the PT Freeport Indonesia No. 4 concentrator. Pages 77–90 in *Proceedings of International Autogenous and Semi-autogenous Grinding Technology Conference 2001*. Volume I. Edited by D.J. Barratt, M.J. Allan, and A.L. Mular. Vancouver: University of British Columbia.

[58] Russell, R.L., and L.D. Kieffer. 1994. Mill expansions at PT Freeport Indonesia. *Mining Engineering* (September):1051–1056.

[59] Lammers, J.M. 1984. The process design of the Ok Tedi project. Pages IV1–IV15 in *Proceedings of the 1st International Symposium on Precious Metals Recovery*. New York: American Institute of Mining, Metallurgical, and Petroleum Engineers.

[60] Smith, E. 1986. Metallurgy and mineral processing plant at St. Joe's El Indio mine in Chile. *Minerals Engineering* (October):26–29.

[61] Wyllie, R.J.M. 1988. El Indio. *Engineering and Mining Journal* (March):34–41.

[62] Harbort, G., D. Lauder, J. Miranda, and A. Murphy. 2000. Size-by-size analysis of operating characteristics of Jameson cell cleaners at the Bajo de Alumbrera copper–gold concentrator. Pages 207–220 in *Proceedings 7th Mill Operators' Conference*. Carlton, Victoria: Australasian Institute of Mining and Metallurgy.

[63] Keran, V.P., F. Zumwalt, and J. Palmes. 1998. Design aspects of the Minera Alumbrera concentrator circuit. SME Preprint 98-66. Littleton, CO: SME.

[64] Chryssoulis, S.L. 2001. Using mineralogy to optimize gold recovery by flotation. *Journal of Metals* (December):48–50.

[65] White, L. 2004. Phoenix rising in Nevada—Newmont's new surface operation will replace gold production from Lone Tree. *Engineering and Mining Journal* (March): 22–24.

[66] Marsden, J.O., W. Rocher, D. Miranda, and R.I. Pennington. 1995. The Candelaria copper–gold concentrator. Pages 5–23 in *Proceedings Cobre-Copper 1995 International Conference*. Volume II. Edited by A. Casali, G.S. Dobby, C. Molina, and W.J. Thoburn. Ottawa, ON, Canada: The Metallurgical Society of the Canadian Institute of Mining, Metallurgy and Petroleum.

[67] Goulsbra, A., R. Dunne, G. Lane, D. Dreisinger, and S. Hart. 2003. Telfer project process plant design. Pages 103–114 in *Proceedings 8th Mill Operators' Conference*. Melbourne: Australasian Institute of Mining and Metallurgy.

[68] Davis, D.R., and D.B. Paterson. 1986. Practical implementation of low alkalinity pressure cyanidation leaching techniques for the recovery of gold from refractory flotation concentrates. Pages 591–604 in *Gold 100: Proceedings of the International Conference on Gold*. Volume 2. Edited by C.E. Fivaz. Johannesburg: South African Institute of Mining and Metallurgy.

[69] Suttill, K.R. 1991. Fiji flatmakes. *Engineering and Mining Journal* (June):27–29.

[70] Birak, D., and K. Deter. 1987. Changes in the Jerritt Canyon metallurgical process as a result of geological characteristics of the ores. Pages 135–140 in *Proceedings of International Symposium on Gold Metallurgy*. Edited by R.S. Salter, D.M. Wyslouzil, and G.W. McDonald. New York: Pergamon Press.

[71] DeSomber, R.K., R. Fernandez, L. McAnany, E. Stolarski, and G. Schmidt. 1996. Refractory ore treatment plant at Newmont Gold Company. Pages 239–247 in *Proceedings Randol Gold Forum 1996*. Golden, CO: Randol International Ltd.

[72] Lewis, G.V. 2000. The Penjom Process: An innovative approach to extracting gold from carbonaceous ore. Pages 163–168 in *Proceedings Randol Gold and Silver Forum 2000*. Golden, CO: Randol International Ltd.

[73] Lewis, G.V. 1999. Increased recovery from preg-robbing gold ore at Penjom gold mine. Pages 105–108 in *Proceedings Randol Gold and Silver Forum 1999*. Golden, CO: Randol International Ltd.

[74] Bosch, D.W. 1987. Retreatment of residues and waste rock. Pages 719–724 in *The Extractive Metallurgy of Gold in South Africa*. Monograph M7. Edited by G.G. Stanley. Johannesburg: South African Institute of Mining and Metallurgy.

[75] Gill, J. 1991. *Gold and Silver Refining at Johnson Matthey*. Salt Lake City: Johnson Matthey.

[76] Marsden, J.O., and C.I. House. 1992. *The Chemistry of Gold Extraction*. Chichester, England: Ellis Horwood.

[77] Canadian Minerals Yearbook. 1993. Principal Canadian Nonferrous and Precious Metal Mine Production.

[78] http://www.newmont.com. Accessed December 18, 2005.

[79] Hodgens, M. 1993. Oxide gold ore treatment at the Telfer mine. Pages 1027–1031 in *Australasian Mining and Metallurgy*. Vol. 2. 2nd edition. Edited by J.T. Woodcock and J.K. Hamilton. Parkville, VIC, Australia: The Australasian Institute of Mining and Metallurgy.

[80] Gregory, S., R. Dunne, P. Gelfi, V. Martins, and A. Goulsbra. 1996. Gravity concentration at the Telfer and New Celebration gold mines. Pages 79–85 in *Proceedings of Randol Gold Forum 1996*. Golden, CO: Randol International Ltd.

[81] Jerritt Canyon Mine—Technical Report. Pincock, Allen & Holt. February 23, 2005. http://www.queenstake.com. Accessed December 19, 2005.

[82] Lahti, P.A. 1996. Refractory ore processing methods at Jerritt Canyon. Pages 225–231 in *Proceedings of Randol Gold Forum 1996*. Golden, CO: Randol International Ltd.

APPENDIX A

Symbols and Abbreviations

SYMBOLS

α	activity
β	cumulative stability constant
ν	kinematic viscosity
ω	rotation rate
]–	connection to inert resin matrix
A	area; relative atomic mass
a_B	activity; relative activity of substance B
C_B	concentration of solute B
c	molality
c_B	concentration of solute substance B
c^*	equilibrium solubility
D	distribution ratio (in solvent extraction)
D_B	diffusion coefficient or diffusivity of substance B
D_c	effective diffusivity
d	diameter; distribution coefficient (in ion exchange)
E	reduction potential
E^0	standard reduction potential
E_a	activation energy
E_{cell}	potential of electrochemical cell
$E^{0\prime}$	conditional potential (formal potential of redox system)
$E^{\ddagger}$	activation energy
F	Faraday constant
G_B	free energy of substance B
ΔG	free energy change
ΔG^0	standard free energy change
ΔH	enthalpy change
I	electric current
I_0	exchange current density
I_j	current density
i_L	limiting current density
j_B	flux of substance B
K	equilibrium constant
K_{CX}	extraction equilibrium constant (in solvent extraction)
K_D	dissociation constant
K_m	mass transfer coefficient
K_n	stability constant for single step in formation of complex, ML_n

K_{sol}	equilibrium constant for dissolution of gas in a solvent
K_w	ionic product of water
K_A^B	relative selectivity coefficient (for ion exchange resin)
k	constant; rate constant
k_m	reaction rate constant
k^0	standard reaction rate constant
$L\bar{y}$	negatively charged ligand atom or molecule
μm	micron or micrometer, 10^{-6} m
N	Nernst diffusion layer thickness
n	reaction order, number of electrons
pK_a	logarithmic exponent of dissociation constant
R	Universal gas constant
ΔS	entropy change
T	temperature
t	time, tortuosity

SUPERSCRIPTS

0	standard value (e.g., E^0, ΔG^0)
‡	activated state

SUBSCRIPTS

A	anodic
ads	adsorbed species
aq	aqueous species; indicates substance in aqueous solution
C	cathodic
c	complex metal ion
g	gaeous phase
i	solute species i
P	reaction product
R	reactant

UNITS

Å	angstrom
amp	ampere
atm	atmosphere
bv/hr	bed volumes per hour
°C	degrees Celsius
cps	centipoise (unit of viscosity)
dm	decimeter
ft	foot
g	gram
gr/scf	grains per standard cubic foot
hr	hour
Hz	hertz
J	joule

K	Kelvin
kcal	kilocalorie
kg	kilogram
kJ	kilojoule
km	kilometer
kPa	kilopascal
L	liter
M	molar
m	meter
mA	milliampere
mg	milligram
mm	millimeter
mmol	millimole
mol	mole
MPa	megapascal
Mtpy	megatons per year
mV	millivolt
µg	microgram
µm	micrometer
µmol	micromole
N	newton
nm	nanometer
oz	troy ounce
ppb	parts per billion
ppm	parts per million
psig	per square inch (gauge)
s	second
S	Siemens, the unit of reciprocal resistance
t	ton (metric)
tpa	tons per annum
tpd	tons per day
tpm	tons per month
tpy	tons per year
V	volt
W	watt

ABBREVIATIONS

µ-PIXE	microparticle-induced X-ray emission spectrometry
AARL	Anglo American Research Laboratories (South Africa)
AAS	atomic adsorption spectroscopy
ADIS	automated digital imaging system
ADL	acid diagnostic leaching
AES	atomic emission spectroscopy
aq	aqueous species
AVR	acidification, volatilization, and reneutralization

c	solid compound
CCD	counter-current decantation
CGA	coal-gold agglomeration
CIC	carbon-in-columns
CIL	carbon-in-leach
CIP	carbon-in-pulp
CIS	carbon-in-solution (also abbreviated as CIC)
CPS	controlled-potential sulfidization
DBBP	Di-n-butyl butyl phosphonate
DBC	dibutyl carbitol
DIBK	di-isobutyl ketone
DL	detection limit
DMDC	dimethyl-dithiocarbamate
D-SIMS	dynamic secondary ion mass spectrometry
DTC	dithiocarbamate
DTP	dithiophosphate
DWS	drinking water standard
EDS	energy dispersive spectroscopy
EDX	energy-dispersive X-ray
EELS	electron energy loss spectroscopy
Eh	electrode potential versus SHE
Em	mixed potential
EMPA	electron microprobe analysis
EPAC	equal-pressure, air-cleaned (screen)
ESCA	X-ray photoelectron spectroscopy (also known as XPS)
FA	fire assay
GC	gravity concentration
GRG	gravity recoverable gold
ICP	inductively coupled plasma
IMMA	ion microprobe mass analysis
IMS	ion mass spectrometry
IX	ion exchange
LAM	laser abrasion microprobe
LIMS	laser ion mass spectrometry
ln	log normal or natural logarithm
log	logarithm to base 10
MAC	magnetic activated carbon
MBT	mercaptobenzothiozole
MCL	maximum contaminant level
MIBC	methyl isobutyl carbinol
MIBK	methyl isobutyl ketone
MLA	Mineral Liberation Analyzer
MNR	Metallgesellschaft Natural Resources
MS	mass spectrometry
NPI	no positive identification

NSC	nitrogen species-catalyzed
O/F	overflow
OMS	optical microscopy
PAX	potassium amyl xanthate
PGM	platinum group metal
pzc	point of zero charge
RBS	Rutherford backscatter
RDE	rotating disc electrode
RIL	resin-in-leach
RIP	resin-in-pulp
RIS	resin-in-solution
rpm	revolutions per minute
SART	sulfidization-acidification-recovery-thickening
SCE	standard calomel electrode
SEM	scanning electron microscopy
SERS	surface-enhanced raman scattering
SHE	standard hydrogen electrode
SIMS	secondary ion mass spectrometry
SX	solvent extraction
TBP	tributyl phosphate
TEM	transmission electron microscopy
TMT	trimercaptotriazine
TOF	time-of-flight
U/F	underflow
USBM	U.S. Bureau of Mines
USEPA	U.S. Environmental Protection Agency
WAD	weak acid dissociable
WDX	wave-dispersive X-ray
WHO	World Health Organization
XRD	X-ray diffraction
XPS	X-ray photoelectron spectroscopy (also known as ESCA)
XRF	X-ray fluorescence

APPENDIX B

Units and Conversion Factors

LENGTH
1 m = 3.2809 ft
1 mm = 0.03937 in.
1 μm = 1 × 10^4 Å

VOLUME
1 m^3 = 220 imperial gallons = 264.2 U.S. gallons = 35.3 ft^3
1 dm^3 = 1 L = 0.2200 imperial (U.K.) gal = 0.2642 U.S. gal

MASS
1 t (metric ton) = 10^3 kg = 1.102 st = 2,205 lb
1 oz troy = 31.10348 g

PRESSURE
1 atm = 101.3 kPa (kN m^2) = 1.013 bar
     = 14.70 lbf $in.^2$ = 1.033 kgf cm^2
     = 760 mm Hg = 760 torr = 1.013 × 10^6 dyne cm^2
(NOTE: lbf = pound-force; kgf = kilogram force)

CONCENTRATION
1 mol/dm^{-3} = 1 mol/L = 1 molar (1 M)
1 g/L = 1 kg/m^3 = 0.062 lb/ft^3
1 ppm = 1 mg/kg^1

CURRENT DENSITY
1 A/m^2 = 0.935 A/ft^2

ENERGY AND WORK
1 J = 0.2389 cal
1 kWh = 3.6 MJ

TEMPERATURE
K = °C + 273
°F = 1.8°C + 32

STANDARD TEMPERATURE
25°C = 298 K = 77°F

Selected Bibliography

(Books the authors have found to be particularly useful or of historical significance)

GOLD GEOLOGY AND MINERALOGY

Boyle, R.W. 1987. *Gold: History and Genesis of Deposits*. New York: Van Nostrand Reinhold Company.

Emmons, W.H. 1937. *Gold Deposits of the World*. New York: McGraw-Hill.

Fleischer, M. 1983. *Glossary of Mineral Species.* Tuscon, AZ: The Mineralogical Record.

Foster, R.P., editor. 1993. *Gold Metallogeny and Exploration*. London: Chapman and Hall.

Petruk, W. 2000. *Applied Mineralogy in the Mining Industry*. Amsterdam: Elsevier.

GOLD METALLURGY

Adamson, R.J., editor. 1972. *Gold Metallurgy in South Africa.* Johannesburg: Cape and Transvaal Printers.

Bugbee, E.E. 1940. *A Textbook of Fire Assaying*. New York: John Wiley & Sons.

Clennell, J.E. 1915. *The Cyanide Handbook*. 2nd edition. New York: McGraw-Hill.

Dorr, J.V.N., and F.L. Bosqui. 1950. *Cyanidation of Gold and Silver Ores.* New York: McGraw-Hill.

Hamilton, E.M. 1920. *Manual of Cyanidation*. New York: McGraw-Hill.

Julian, H.F., and E. Smart. 1921. *Cyaniding Gold and Silver Ores*. 2nd edition. London: Charles Griffin.

King, A. 1949. *Gold Metallurgy on the Witwatersrand*. Johannesburg: Transvaal Chamber of Mines.

McQuiston, F.W., and R.S. Shoemaker. 1975. *Gold and Silver Cyanidation Plant Practice.* Volumes I. SME-AIME Monograph. Baltimore, MD: Port City Press.

McQuiston, F.W., and R.S. Shoemaker. 1981. *Gold and Silver Cyanidation Plant Practice.* Volume II. SME-AIME Monograph. Baltimore, MD: Port City Press.

Park, J. 1900. *The Cyanide Process of Gold Extraction*. London: Charles Griffin.

Rose, T.K., and W.A.C. Newman. 1937. *The Metallurgy of Gold*. 7th edition. London: Charles Griffin.

Schnabel, C. 1921. *Handbook of Metallurgy*. Two volumes. 3rd edition. London: Macmillan.

Smith, A., and T.I. Mudder. 1991. *Chemistry and Treatment of Cyanidation Wastes*. London: Mining Journal Books.

Stanley, G.G., editor. 1987. *The Extractive Metallurgy of Gold in South Africa.* Monograph Series M7. Johannesburg: South African Institute of Mining and Metallurgy.

THE GOLD MARKET: SUPPLY AND DEMAND

Gold. 1982–2005. Annual Reviews. London: Gold Fields Mineral Services Ltd.

GENERAL METALLURGY, PLANT DESIGN, AND PRACTICE

Bartlett, R.W. 1992. *Solution Mining: Leaching and Fluid Recovery of Materials.* Amsterdam: Gordon and Breach.

Burkin, A.R. 2001. *Chemical Hydrometallurgy: Theory and Principles.* London: Imperial College Press.

Crow, D.R. 1988. *Principles and Applications of Electrochemistry.* 3rd edition. London: Chapman and Hall.

Fuerstenau, M.C., and K.N. Han, editors. 2003. *Principles of Mineral Processing.* Littleton, CO: SME.

Gaudin, A.M. 1932. *Flotation.* New York: McGraw-Hill.

Jackson E. 1986. *Hydrometallurgical Extraction and Reclamation.* Chichester, England: Ellis Horwood.

Mular, A.L., and G.V. Jergensen, editors. 1982. *Design and Installation of Comminution Circuits.* New York: SME-AIME.

Mular, A.L., and R.B. Bhappu, editors. 1980. *Mineral Processing Plant Design.* New York: SME-AIME.

Mular, A.L., D.N. Halbe, and D.J. Barratt, editors. 2002. *Mineral Processing Plant Design, Practice and Control.* Littleton, CO: SME.

Moore, J.J. 1990. *Chemical Metallurgy.* 2nd edition. London: Butterworth-Heinemann.

Richards, R.H. 1908. *Ore Dressing.* Four volumes. 2nd edition, New York: McGraw-Hill.

Sohn, H.Y., and M.E. Wadsworth. 1979. *Rate Processes in Extractive Metallurgy.* New York: Plenum Press.

Taggart, A.F. 1950. *Handbook of Mineral Dressing.* New York: John Wiley & Sons.

Weiss, N.L., editor. 1985. *SME Mineral Processing Handbook.* Two volumes. Littleton, CO: SME.

Index

NOTE: *f.* indicates figure; *t.* indicates table.

A

AARL. *See* Anglo American Research Laboratories
AARL elution process, 330*t.*, 331
AARL pump-cell, 327, 328
Abbreviations, 621, 623
Absorption of gases, factors affecting, 136–137, 165, 167
Acid digestion, 56
Acid leaching
 preceding smelting, 453
 removal of iron from loaded cathodes, 452–453
 of zinc precipitates and sludges, 451–452, 452*t.*
Acid mine drainage, 469
Acidification, volatilization, and reneutralization process. *See* AVR process
Acids, 121–122
 concentration in high-pressure acidic oxidation, 167, 168*f.*
 See also Acid leaching, Carbon acid washing, High-pressure acidic oxidation, Hydrochloric acid, Nitric acid, Nitric acid oxidation, Sulfuric acid, Weak acid dissociable species
Activated carbon, 9–10, 297–303
 activation, 298–299, 299*f.*
 activation by steam, 298
 chemical properties, 300–301
 crystallites, 299
 and cyanidation, 9–10
 effect of pH on zeta potential and acid-base adsorption, 300–301, 302*f.*
 in effluent treatment, 480
 flotation, 433–434
 H-carbons, 300
 L-carbons, 300
 magnetic, 298–299
 manufacture, 298
 physical properties, 299–300
 pore size and pore size distribution, 299–300, 301*f.*
 properties, 297–302
 sources, 298
 surface oxides, 300, 302*f.*
 See also Carbon, Carbon acid washing, Carbon adsorption, Carbon elution, Carbon fouling, Carbon reactivation
Activity-activity diagrams, 126
Adsorption. *See* Activated carbon, Carbon adsorption
Agglomeration, 13, 267. *See also* Coal-gold agglomeration
Agitated cyanide leaching
 counter-current leaching, 265–266, 267*f.*
 cyanide addition, 265
 and grinding, 90–91, 91*f.*
 with heap leaching and CIP/CIL, 92, 93*f.*, 94*f.*
 and oxygen, 265
 and particle size, 263, 264*f.*
 pH modification, 263–264
 residence time, 265
 and slurry density, 263
 See also Cyanide leaching, Thiosulfate leaching
Agitation
 air, 8, 263
 and biological oxidation, 191–192, 199
 in chlorination, 190, 274
 in cyanide leaching, 250
 in high-pressure acidic oxidation, 168
 mechanical, 8, 263
 in thiosulfate leaching, 280–281
Agnes, South Africa, 37*f.*, 49
Agnico Eagle, Canada, 163
Agricola, Georgius, 2–3
Air quality standards, 468, 469*t.*
Ajkjoujt, Mauritania, 287
Alaska, gold rush, 4
Alumbrera, Argentina, 102
 flowsheet, 586, 587*f.*
Aluminum precipitation, 14, 386–387, 401
Amalgamation, 83, 409, 438–441, Plates 12–13
 early history, 2, 3, 4, 6, 14
 factors affecting, 439–441
 of gravity concentrates, 48, 83, 270
 and hydrocarbon contamination, 441
 key criteria, 440
 problems resulting from, Plate 9
 properties of mercury, 438–439

retorting, 453–454
solubilities of metals in mercury, 439–440, 440t.
and sulfides, 441
treatment of products, 453–454
See also President Brand
Amines
extractants, 355–356
flotation reagents, 416
Ammonia leaching, 286–287
Ammonia-cyanide leaching, 286–287
and carbon adsorption, 335
Ammonia-halide leaching, 287
Analytical techniques, 54–57, 56t., 60t.
acid digestion, 56
cyanide leaching, 56–57, 56t.
diagnostic leaching, 56t., 59–61
fire assay, 55–56, 56t.
gold fingerprinting, 57
for ore composition, 55–58
physical methods, 57
special factors in placer ore evaluation, 64
See also Mineralogy
Anglo American Research Laboratories, 12
CIP pump cell, 327, 328
electrowinning cells, 397
See also AARL elution process, AARL pump-cell
AngloGold Ashanti, 270
Ashanti Refractory Sulfide Plant flowsheet, 564, 565f.
President Brand flowsheet, 518, 519f.
See also Ashanti, Ghana
Anode slime, 53, 54t.
Antimony
and cyanide leaching, 258–260, 259f.
and high-pressure acidic oxidation, 169
precipitation in effluent treatment, 496–497
Antimony sulfides
aurostibnite, 46
Consolidated Murchison flowsheet, 596, 597f.
roasting of, 211
stibnite, 46
Aqua regia, 14, 55, 272, 612
Arrhenius equation, 132
Arsenic
and cyanide leaching, 258–260, 259f.
effluent treatment, 472–473
and high-pressure acidic oxidation, 169
precipitation, 160–161
precipitation in effluent treatment, 496–497
stability in biological oxidation tailings, 203
Arsenic sulfides, 38t., 44
arsenopyrite, 44–45

hydrometallurgical sulfide oxidation of, 152, 152f., 153f.
orpiment, 45
realgar, 45
roasting of, 207–211, 210f.
Arsenic trioxide, 222, 223
Arseno process, 178, 179t., 182
Arsenopyrite, 25–26, 44–45, 59, 83, 96, 98
biological oxidation of, 192–193, 194, 195f.
chlorination, 188
and cyanide leaching, 258–260
flotation, 427, 429–431, 431f.
and high-pressure nonacidic oxidation, 175–176
hydrometallurgical sulfide oxidation of, 152, 152f., 153f.
and nitric acid oxidation, 179, 180–181, 181f., 182f.
refractory ore flowsheets, 564–577
roasting of, 213, 217t., 564–567, 570–571, 576–577
Ashanti, Ghana, 47, 199, 200. *See also* AngloGold Ashanti
Ashanti Sansu, 15, 148, 200
AuGMENT process, 349, 479–480
Auricupride, 24
Australia
gold production methods, 504t., 510f., 511f.
gold rush, 3–4, 6f.
intensive cyanide leaching, 270
thiourea leaching, 281
AVR process, 475–477, 476f., 476t.

B

Bacteria, 191–199
adaptation, 198–199, 198f., 199t.
historical aspects of, 11, 15–16, 15t., 190, 191t.
See also Biological oxidation
Baker, Canada, 489
Bakyrchik, Russia, 47
Ballarat, Australia, 3, 35
Bannack, Montana, USA, 4
Barberton Mountainland, South Africa, 42, 53
grain size of gold in flotation concentrates, 48, 49t.
Barneys Canyon, USA, 92
flowsheet, 562–563, 563f.
Bathurst, Australia, 3
Batu Hijau, Indonesia, 102
Beaconsfield, Australia, 200
Beisa, South Africa, 395t.
Bendigo, Australia, 3, 35
Big Springs, USA, 217
Bingham Canyon, USA, 45, 102
Biological heap oxidation, 11, 190–191, 203–205
Biological oxidation, 11, 15–16, 85, 190–192
agitation, 191–192

applications, 200–201
arsenic stability in tailings, 203
of arsenopyrite, 192–193, 194, 195f.
and bacterial adaptation, 198–199, 198f., 199t.
behavior of gangue minerals, 202–203
biological heap oxidation, 203–205, 204f.
of chalcopyrite, 193
circuit configurations and design, 201–202
and cyanide leaching of products, 202, 203
and dissolved oxygen concentration, 199–200
at Equity Silver Mine, 191t., 196
factors affecting bacterial activity, 196–199
at Fairview, 191t., 203, 576–577
and Fe(II)–Fe(III) ratio, 196
and flotation concentrates, 200
flowsheet, 201, 202f.
future applications, 205
in heaps, 190–191, 203–205
historical development, 191t.
of iron sulfides, 192–193
with *Leptospirillum ferro-oxidans*, 205
of marcasite, 192
mechanism, 193, 193f.
and mineral porosity, 194, 195f.
nutrient requirements, 199
operating conditions, 194–200
and ore mineralogy, 194–196
and oxygen uptake rate, 200
and particle size, 200
and pH, 196
and pulp density, 196, 197f.
of pyrite, 192, 194, 195f.
of pyrrhotite, 192
reaction chemistry, 192–194
reaction kinetics, 194
of refractory ores, 200–201
relative susceptibility of sulfide minerals, 194, 195t.
and shear, 199
and solution potential, 196
of sulfides, 191–205
with *Sulfobacillus acidophilus*, 192, 196
with *Sulfobacillus thermosulfido-oxidans*, 205
with *Sulfolobus*, 192, 196
and sulfur feed, 201
and temperature, 196, 197f.
at Vaal Reefs, South Africa, 203
with *Thiobacillus ferro-oxidans*, 192, 193, 196, 198, 199t., 205
with *Thiobacillus thio-oxidans*, 192, 193, 196, 198, 199t.
wash-out of biomass, 201
Bismuth, 24, 250
Bjorkdal, Sweden, 512, 513f.
Blind River–Elliot Lake, Canada, 30
Blyvooruitzicht, South Africa, 327
Bodlander, 14
Bougainville, Papua New Guinea, 46
Boundary layer theory, 134–135, 135f.
BP Minerals, 442
Brazil, gold production methods, 504t., 510f., 511f.
Bromine–bromide, 287–288
Bromine–chloride, 287–288
Bullion
 crude bullion production, 449, 453–459
 refining, 459–465, 610–613
 See also Refining
Butler–Volmer equation, 138, 388, 392

C
Cadia, Australia, 102
Cadmium, 472t., 476t., 497, 498
Calcine, 52–53
 effect of cyanide concentration on gold recovery from, 243–246, 245f.
 effect of oxygen concentration on gold extraction from, 246, 247f.
 porosity of product, 213–214, 216f., Plate 15
 treatment, 221–222
 See also Roasting
Calcining. *See* Roasting
Calcium carbonate fouling, 320–321, 380
Calcium cyanide, 233–234, 234t.
Calumet, California, 8
Campbell Con mine, Canada, 15, 217
Campbell Red Lake, Canada, 13, 98, 170, 217
 flowsheet, 566–568, 567f., 569f.
Canada, 101
 gold production methods, 504t., 510f., 511f.
Candelaria, Chile, 45, 102
 flowsheet, 590, 591f.
Candelaria, USA, 98, 590–591
Carbon
 activity, 332–333
 activation, 298–299
 attrition losses, 333
 carbon tetrachloride number, 307
 carbonization, 298
 chemical properties, 300–302
 coal, 298
 coconut shells, 298, 300
 fruit pips, 298
 Iodine number, 307
 manufacture, 298–299
 organic, and high-pressure acidic oxidation, 170
 peat, 298
 physical properties, 299–300
 quality testing of reactivated carbon, 332–333
 refining of products, 88
 roasting of, 211

source materials, 298
sugar cane residue, 298
wood, 298
See also Activated carbon, Carbon adsorption, Carbon elution
Carbon acid washing, 321–322
  with hydrochloric acid, 322, 323*f.*
  with nitric acid, 322
Carbon adsorption, 1, 14, 139, 297–335
  activated carbon, 9–10, 297–303
  from ammonia–cyanide solutions, 335
  and anions, 310
  carbon attrition losses, 333
  and carbon fouling, 311, 318–324
  carbon particle size, 303, 304*f.*
  and cations, 310
  chemical factors affecting adsorption efficiency, 308–311
  from chloride leaching solutions, 334–335
  and concentration of other metals, 310
  of copper, 312, 313*f.*
  and cyanide concentration, 309–310, 309*t.*
  from cyanide solutions, 303–312
  effect of dissolved oxygen, 310–311, 328
  effect of solids, 304–306
  effect of solution pH, 310, 311*f.*, 311*t.*
  effect of viscosity, 304–305
  effect of zinc hydroxide, 315
  and Freundlich isotherm, 308, 308*f.*
  and gold concentration in solution, 308, 308*f.*
  and ionic strength, 310, 311*t.*
  K value, 308, 308*f.*
  kinetics and loading capacity, 307–308, 307*f.*
  mechanism of gold adsorption, 306
  of mercury, 312
  mixing efficiency, 304, 305*f.*
  from noncyanide solutions, 334–335
  of other metals, 311–312
  pseudo-equilibrium, 304
  and removal of carbon fines, 324–325
  selection of carbon type, 303
  of silver, 312
  and slurry density, 304–306, 305*f.*
  and temperature, 308, 309*f.*, 309*t.*
  and thiosulfate leaching, 335
  from thiourea leaching, 335
  use of air or oxygen, 328
  See also Activated carbon, Carbon elution, Carbon-in-solution process, CIC, CIL, CIP
Carbon conditioning, 324, 325
Carbon elution, 12, 312–318, 329–332
  AARL process, 330*t.*, 331
  acetonitrite, 330*t.*, 331
  with alcohol addition, 315, 316
  atmospheric Zadra process, 330*t.*, 331

behavior of copper, 318, 320*f.*
and cyanide concentration, 314, 315*f.*
efficiencies, 332
elute poisoning, 331–332
ethanol, 330*t.*, 331
with glycol addition, 315, 316
of gold, 312–318, 330*t.*, 331
and gold concentration in solution, 318, 319*f.*
and ionic strength, 314, 316*f.*
of mercury, 318
methanol, 330*t.*, 331
Micron process, 330*t.*
Murdoch process, 330*t.*
with organic solvent addition, 315–316, 317*f.*
and pH, 315
and pressure, 313–314
pressure Zadra process, 330*t.*, 331
of silver, 318, 320*f.*
solution flow rate, 316, 318*f.*
solvent distillation, 331
solvent-assisted, 330*t.*, 331
systems, 329–332, 330*t.*
and temperature, 313, 314*f.*
U.S. Bureau of Mines, 10, 14
Zadra process, 330*t.*, 331
Carbon fines, 324–325, 522–523
Carbon fouling, 311, 318–323
  by calcium carbonate, 320–321
  diesel oil, 323
  by flocculants and other reagents, 323
  by flotation reagents, 323
  by humic acid and other vegetarian decomposition, 323
  inorganic, 320–321
  organic, 323
  by petroleum products, 323
Carbon incineration, 88
Carbon inter-stage screening, 325
Carbon reactivation, 319–320
  acid washing, 321–322
  carbon quality testing, 332–333
  desorption–acid washing–thermal reactivation sequence options, 332
  with hydrochloric acid, 322, 323*f.*
  inorganic removal, 321–322
  with nitric acid, 322
  organic removal, 323–324
  thermal, 325*t.*, 326*f.*
  use of steam, 323–324, 325*t.*
  of volatile and nonvolatile adsorbates, 324
  See also Activated carbon
Carbon regeneration, 11. *See also* Carbon reactivation
Carbon-in-columns. *See* CIC
Carbon in leach. *See* CIL
Carbon-in-pulp process. *See* CIP

Carbon-in-solution process, 328–329, 329f.
Carbonaceous materials, 47
    and cyanide leaching, 261–263
    flotation, 434–435
    roasting of, 211–212
    See also Carbonaceous ores, Minerals
Carbonaceous ores, 47
    chlorination of, 190, 191t.
    and deactivation by chlorination,
        185–186, 187–188, 189–190, 191t.
    double-oxidation treatment, 100, 600–601
    flowsheets, 600–607
    highly, 99f., 100
    mildly, 99–100, 99f.
    See also CIL
Carbonates
    and chlorination, 274
    chlorination and hot sodium carbonate
        pretreatment in sulfide treatment
        (double oxidation), 190
    and effect of biological oxidation, 202–203
    and high-pressure nonacidic oxidation,
        176, 177
Carlin mine, Nevada, USA, 10, 47, 91, 99,
        217, 266, 286
    chlorination of carbonaceous ores, 190,
        191t., 274
    and heap leaching, 266
    Mill No. 6 flowsheet, 604–605, 605f.
Carlton mine, Colorado, USA, 10
Caro's acid, 488
Carolin, Canada, 489
Cementation. See Recovery of gold from
    solution, Zinc precipitation
Cementation reaction, 118
CGA process. See Coal–gold agglomeration
Chalcopyrite, 25–26, 45, 578, 586, 588,
        590, 592
    behavior in cyanide, 253–256
    biological oxidation of, 193
    flotation, 416t., 417t., 418t., 433, 434f.
    and high-pressure nonacidic oxidation,
        175–176
    and nitric acid oxidation, 180
Charcoal. See Activated carbon
Chelating resins. See ion exchange resins
Chemical equilibria
    activities, 118–121, 120f., 121f., 120t.
    activity–activity diagrams, 126
    complexation, 122–123
    electrochemical reactions, 116–118, 118t.,
        119t.
    equilibrium constant, 115
    equilibrium defined, 113–116
    graphical representation, 126–131
    pH scale and modification, 121–122, 122t.
    potential–pH diagrams, 126–131, 128f.,
        129f., 130f.
    solubility diagrams, 126, 127f.
    solubility of gases, 124–125, 126t.
    solubility of solids, 123–124, 125t.
    thermodynamic data, 115–116, 116t.
    See also Hydrometallurgical principles
Chile, 101
    gold production methods, 504t., 510f., 511f.
Chimney Creek, Nevada, USA
    counter-current leaching, 266, 267f.
    flowsheet, 522–523, 523f.
    See also Twin Creeks
China, gold production methods, 504t.,
        510f., 511f.
Chloride
    elution, 343, 348
    geometry, 113, 114f.
    leaching solutions, 334–335
    solvent extraction from solution, 463
Chlorination, 5, 7, 14, 185–186, 233, 271–275
    agitation, 274
    in bullion refining, 459–461, 460f.
    and carbon adsorption, 334–335
    of carbonaceous ores, 190, 191t.
    and carbonates, 274
    chlorine chemistry, 186–187, 187f.
    compared with cyanide leaching, 274
    and copper, 274
    in deactivation of carbonaceous matter,
        185–186, 187–188, 189–190, 191t.
    diffusion coefficient of hypochlorous
        species, 188, 189f.
    effect of temperature on chlorine
        solubility, 188, 189t.
    effects on cyanidation, 188, 188f., 190
    followed by solvent extraction, 463
    and gold–telluride minerals, 274
    hot sodium carbonate pretreatment in
        sulfide treatment (double oxidation),
        190
    with isocyanuric acid, 275
    and lead, 273
    mechanism of gold dissolution, 272, 273f.
    and oxidant consumption, 274
    Platsol process, 275
    process description, 190
    and pyrite, 273–274
    reaction kinetics, 272–273
    reasons for lack of commercial success, 275
    and silver, 273
    stability constants of selected metal-
        chloride complexes, 274, 275t.
    in sulfide oxidation, 186, 188–189, 190
    and zinc, 274
Chlorine
    chemistry, 186–187, 187f.
    consumption by sulfide minerals, 189,
        190, 191t.
    effect of temperature on solubility, 188, 189t.

metal complexes, 116*t.*, 118*t.*, 119*t.*, 272–274
Chlorites, 202
Choco, Colombia, alluvial deposits, 3
Chromium, behavior in electrowinning, 398
CIC, 303, 329, 329*f.*, 508*f.*, 509
CIL, 11, 303, 328, 434
　disadvantages, 328
　with heap leaching, agitated leaching and CIP, 92, 93*f.*, 94*f.*
　as a major recovery method worldwide, 508–509, 508*f.*
CIP, 9, 11, 303, 325–328, 434
　AARL pump-cell, 327
　carbon concentrations and consumption, 327–328
　cascade circuit configuration, 326–327, 327*f.*
　full emergence of, 12–13
　with heap leaching, agitated leaching and CIL, 92, 93*f.*, 94*f.*
　interstage screening, 325–326
　as a major recovery method worldwide, 508–509, 508*f.*
　process selection, 86, 103, 104
　and slurry density and viscosity, 328
Classification cyclones and screens, 80
Clay minerals
　effect in gold extraction, 80, 267
　treatment of clay-rich ores, 13, 267
　See also Minerals
Coal–gold agglomeration, 84, 441–442
　schematic, 442*f.*
Cobalt, 338, 340*f.*, 345
Coeur D'Alene Mines Corporation, Idaho, USA, 477
Coeur-Rochester, Nevada, USA, 41, 91, 253, 268
Cognis, 349
Colorado, USA, telluride ores, 261
Comminution, 78, 80
　and grinding-in-leach, 80
　major use of, 78–79
　and mineral surface preparation, 80
Complexes, 122–123
　deposition of gold from solution, 125–126
　of gold, 113, 114*t.*
　stability constants, 123, 236, 237*t.*, 274, 275*t.*
Comstock Lode, Nevada, USA, 4
Consolidated Murchison mine, South Africa, 37*f.*, 46, 327, 433
　flowsheet, 596, 597*f.*
Conversion factors, 625
Copper
　carbon adsorption of, 312, 313*f.*
　carbon elution of, 318, 320*f.*

　and chlorination, 274
　copper-rich ore flowsheets, 578–595
　and cyanide leaching, 253–256, 254*f.*, 254*t.*, 255*f.*
　and electrowinning, 390–391, 391*f.*
　and ion exchange resins, 338, 339, 340*f.*, 342*f.*, 348, 351
　and nitric acid oxidation, 180
　precipitation in effluent treatment, 496
　and sulfuric acid parting, 462–463
　and thiosulfate leaching, 276–281
Copper sulfides, 38*t.*, 43*t.*, 45–46
　chalcopyrite, 45
　flotation, 433
　hydrometallurgical sulfide oxidation of, 152–153, 154*f.*
　roasting of, 211
Cortez, Nevada, USA, 13, 217
Counter-current leaching, 265–266, 267*f.*
Cripple Creek, USA, 46
Crown Mine, New Zealand, 8
Cyanidation, 1
　and activated carbon, 9–10
　all-sliming, 9
　of concentrate, 95
　of concentrate and tailings, 95–96
　and flotation, 438, 439*f.*
　flowsheet development, 8–9
　intensive, 11
　invention of, 8
　tailings, 50–51, 52*f.*, 96
　of tailings, 95
　of whole ore, 94
　See also Cyanide leaching
Cyanide
　adsorption of by other minerals, 486–487
　alkaline chlorine–hypochlorite oxidation of, 490–491, 492*t.*
　atmospheric oxidation of, 486
　biological oxidation of, 491–494, 493*f.*, 494*f.*
　and calcium cyanide, 233–234, 234*t.*
　carbon adsorption from solutions of, 303–312
　and carbon elution, 312–318
　concentration and zinc precipitation, 373–374, 374*f.*, 375*f.*, 376*f.*
　cycle and degradation mechanisms, 484, 485*f.*
　effluent treatment, 469–473, 474*t.*, 475–482, 484–494
　free, 234, 236
　hydrogen cyanide, 234, 2345*f.*
　hydrogen peroxide oxidation of, 487–488, 489
　iron complexation of, 487

natural degradation of, 484–487, 485f., 486f.
oxidation to cyanate, 234–236, 235f.
ozone oxidation of, 494
and potassium cyanide, 233–234, 234t.
and sodium cyanide, 233–234, 234t.
solubility products for selected metal cyanide compounds, 236, 238t.
solution chemistry, 233–236
stability constants for selected metal cyanide complexes, 236, 237t.
sulfur dioxide–assisted oxidation of, 489–490, 490t.
ultrasonics in oxidation of, 494
volatilization of, 484–486
See also International Cyanide Management Code and Institute
Cyanide degradation, 484, 485f., 486–487
Cyanide detoxification, 482, 494
Cyanide leaching, 233
  agitation and dissolution rate, 250
  agitation leaching, 263–266
  as analytical technique, 56–57, 56t.
  anodic reactions, 236–239, 239f., 240f.
  and antimony, 258–260, 259f.
  and arsenic, 258–260, 259f.
  and biological oxidation product, 202, 203
  and calcium cyanide, 233–234, 234t.
  and carbonaceous materials, 261–263
  cathodic reactions, 240–241
  and copper, 253–256, 254f., 254t., 255f.
  current–potential curves, 243, 244f.
  cyanide and dissolved oxygen concentration, 241–247, 242f., 243f., 244f., 244t., 245f., 246f.
  cyanide solution chemistry, 233–236
  effect of bismuth, 250
  effect of cyanide concentration on gold recovery from calcine material, 243–246, 245f.
  effect of lead, 250–251, 251f.
  effect of mercury, 250–251, 251f.
  effect of oxygen concentration on gold disc dissolution rate, 246, 246f.
  effect of oxygen concentration on gold extraction from calcine, 246, 247f.
  effect of thallium, 250
  followed by solvent extraction, 463–464
  free cyanide, 234, 236
  galvanic interactions of sulfide minerals with gold, 251
  heap leaching, 266–269
  hydrogen cyanide, 234, 235f.
  in situ leaching, 271
  intensive, 269–271
  and iron minerals, 256–258, 257f., 258f.
  as major extraction process, 505, 508f.
  overall dissolution reaction, 241, 242f.
  and pH, 248–249
  and potassium cyanide, 233–234, 234t.
  reaction kinetics, 241–251
  and silver, 252–253, 253f.
  and sodium cyanide, 233–234, 234t.
  solubility products for selected metal cyanide compounds, 236, 238t.
  stability constants for selected metal cyanide complexes, 236, 237t.
  sulfide dissolution, 251–252
  and sulfur, 260–261
  and surface area, 249
  and telluride ores, 261, 262f.
  and temperature, 247–248, 248f., 249f.
  vat leaching, 271
  and zinc, 260
  See also Agitated cyanide leaching, Cyanidation
Cyanisorb, 477
Cyclic voltammetry, 143–144, 143f., 422–423

D

Debye–Huckel theory, 119
Deep-seated deposits, 34t.
Deetken, California, USA, 5
Deloro, Canada, 386
Desorption. See Carbon elution
Determinative methods. See Analytical techniques
Diagnostic leaching, 56t., 59–61
Dissolved oxygen. See Oxygen
Dithiocarbamates, 415, 417
Dithiophosphates, 414–415, 415t., 416t., 417t.
Dump leaching, 11. See also Biological oxidation, Heap leaching
Durban Roodepoort Deep, South Africa, 342f.

E

East Driefontein, South Africa, 92, 266, 327
  flowsheet, 514, 515f.
Economic and political factors in process selection, 77
Effluent treatment, 88–89, 467
  and acid mine drainage, 469
  acidification, volatilization, and reneutralization (AVR process), 475–477, 476f., 476t.
  activated carbon in, 480
  adsorption of cyanide by other minerals, 486–487
  and air quality standards, 468, 469t.
  alkaline chlorine–hypochlorite oxidation of cyanide, 490–491, 492t.
  antimony precipitation, 496–497
  of arsenic, 472–473
  arsenic precipitation, 496–497
  atmospheric oxidation of cyanide, 486
  AuGMENT process, 479–480

biological oxidation of cyanide, 491–494, 493f., 494f.
cadmium precipitation, 497, 498
and contained effluents, 470–471
copper precipitation, 496
of cyanide, 469–473, 474t., 475–482, 484–494
detoxification, 482–498
dilution, 482, 483f.
direct solution recycle, 471f., 473, 475f.,
electrolytic treatment, 481
of gaseous wastes, 468, 469t.
hydrogen peroxide oxidation of cyanide, 487–488, 489
ion exchange in, 477–480, 478f.
ion precipitate flotation in, 482
iron complexation of cyanide, 487
iron precipitation, 495–496
lead precipitation, 496
of liquid wastes, 470
mercury precipitation, 497
metals recovery, 473–482, 474t., 494–498, 496t., 497f.
and mineral dissolution, 469
MNR process, 481–482
natural degradation of cyanide, 484–487, 485f., 486f.
nickel precipitation, 496
ozone oxidation of cyanide, 494
precipitation of metal species from cyanide solutions, 495–498
reagent recovery, 473–482
selenium precipitation, 497, 498
of solid wastes, 468–469
sulfide precipitation in, 481–482
sulfur dioxide-assisted oxidation of cyanide, 489–490, 490t.
types of waste, 467–473, 468f.
ultrasonics in oxidation of cyanide, 494
Velardena process, 482
Vitrokele process, 478
volatilization of cyanide, 484–486
and water discharges, 471–473, 471f., 472t.
and wildlife, 470–471
zinc precipitation, 496
Egypt, 1–2
E_h–pH diagrams. See Potential–pH diagrams
Ekaterinburg, Russia, 3
El Indio, Chile, 327
flowsheet, 584–585, 585f.
mode of occurrence of gold in flotation concentrates, 48, 49t.
El Oro, Mexico, 8
Electrical double layer of minerals, 409–410, 410f., 411, 412f.
Electrochemical reactions, 116–118, 137–138
and cyclic voltammetry, 143–144, 143f.
measurement of solution potentials, 142–144
and potential sweep methods, 143–144
and reference electrodes, 142
and rotating disc electrodes, 142–143
standard electrode potentials, 118t., 119t.
Electrode potentials. See Standard electrode potentials
Electrodes. See Reference electrodes, Rotating disc electrodes
Electrolytic effluent treatment, 481
Electrolytic refining, 7, 461–462, 462t.
Siemens–Halske process, 9
See also Refining
Electron microprobe analysis, 56t., 61t., 62
Electrostatic separation, 84
Electrowinning, 11, 87, 88, 387, 463–464
anodes, 398
anodic reactions, 392
applications, 396–397
and cathode surface area, 393–394
cathodes, 397, 398
cathodic reduction of gold, 389–390
cathodic reduction of other metals, 390–391, 391f.
cell configurations, 397–398
and cell current, 394
and cell efficiency, 394
and cell voltage, 394
and copper, 390–391, 391f.
current–potential curve, 392f.
and cyanide concentration, 395–396, 396f.
from dilute solutions, 400, 401f.
and electrolyte hydrodynamics, 393
electroplating process, 399–400
fundamentals, 387–389
and gold concentration, 393, 394f.
increasing mass transport, 397
and iron, 390
and lead, 390
limiting current density, 388–389
and mercury, 390
and nickel, 390, 391, 391f.
operating parameters, 395t., 398
and pH, 395
potential drop, 387–388, 388f.
potential requirements, 387–389, 388f.
product handling and treatment, 398–399
reaction chemistry, 389–392
reaction kinetics, 392–393
reduction of oxygen and water, 391–392
and silver, 390, 391f.
single-pass extraction efficiency, 398, 399f.
and solution conductivity, 394
and solution flow rate, 396
and sulfide ions, 396
and temperature, 393

and voltage, 387
and zinc, 390
Electrum, 21, 26*f.*
  and silver sulfide layer, 39–40
Elsner's equation, 8
Elution. *See* Carbon elution, Ion exchange resins
EMPA. *See* Electron microprobe analysis
Emperor, Fiji, 46, 100, 217
  flowsheet, 598–599, 599*f.*
  telluride ores, 261
Empire mine, California, USA, 9
Environmental issues, 467
  air pollution, 468, 469*t.*
  and process selection, 72–78
  water pollution, 469, 470, 471–473, 471*f.*, 472*t.*
  wildlife protection, 470
  See also Effluent treatment
Epithermal deposits, 33, 34*t.*, 35*f.*, 36*f.*
  critical mineralogical factors, 39, 40*f.*
  refractoriness, 33
Equilibrium constant, 115
Equity Silver Mine, Canada, 191, 196
Escondida, Chile, 102
Eskay Creek, Canada, flowsheet, 540, 541*f.*
Evans diagrams
  for As–S–O system, 210*f.*, 211
  for Fe–AS–S–O system, 207, 210*f.*
  for Fe–S–O system, 206, 207*f.*
Extractants. *See* Carbon, Carbon adsorption, Ion exchange resins, Solvent extraction

F

Fairview, South Africa, 13, 37*f.*, 98, 191, 199, 200
  flowsheet, 576–577, 577*f.*
  roasting, 209*f.*, 217
Faraday constant, 117
Faraday's second law, 389
Fire assay, 55–56, 56*t.*
Flin Flon, Canada, 475
Flocculants, 81, 323
Flotation, 9, 409, 419–438
  activated carbon, 433–434
  air-sparged hydrocyclones, 437
  arsenopyrite, 427, 429–431, 431*f.*
  carbonaceous materials, 434–435
  chalcopyrite, 433, 434*f.*
  circuits, 435–436
  column flotation, 437
  conceptual model, 424, 425*f.*
  copper sulfides, 433
  and cyanidation, 438, 439*f.*
  with cyanidation of concentrate, 95
  with cyanidation of concentrate and tailings, 95–96
  with cyanidation of tailings, 95, 97
  of cyanidation tailings, 96
  effect of nonsulfide gangue minerals on free gold flotation, 426
  effect of sulfide mineral flotation on free gold flotation, 412, 424–425, 426
  effects of dual-collector reagent schemes, 426
  flash flotation (unit cell flotation), 437
  free gold with oxide or silicate gangue, 421
  free gold with sulfide gangue, 421
  and froth stability, 427
  and galvanic effects, 426–427
  gangue minerals, 421
  gold, 421–427
  gold tellurides, 428
  inert gas flotation, 437
  mechanism for gold, 422–423
  modifiers, 426
  and nonoxidizing slurry preconditioning, 427
  and ore mineralogy, 420–421
  and particle coatings, 423–424
  and particle liberation, 423
  and particle size and shape, 424
  in preconcentration, 83
  and pulp density, 426
  pyrite, 428–429, 429*f.*, 430*f.*
  and pyrophyllite, 435
  pyrrhotite, 427, 431–433
  and silicates, 435
  smelter specifications for copper concentrates, 421, 422*t.*
  stibnite, 433
  with sulfide oxidation and cyanidation of concentrate, 97–98
  with sulfide oxidation and cyanidation of concentrate and cyanidation of tailings, 98
  sulfides, 433
  tailings, 51
  and talc, 435, 436*f.*
  and temperature, 427
  unliberated gold in sulfide gangue, 421
  and water quality, 427
Flotation concentrates, 48
  and biological oxidation, 200
  grain size of gold at Barberton, 48, 49*t.*
  mode of occurrence of gold at El Indio, 48, 49*t.*
  treatment of, 102–103, 103*f.*
Flotation reagents, 413
  activators, 416–418, 418*t.*, 426
  amines, 416
  collectors, 413–416, 415*t.*
  depressants, 418, 419*t.*, 426
  dithiocarbamates, 415, 417
  dithiophosphates, 414–415, 415*t.*, 416*t.*, 417*t.*
  frothers, 419, 420*t.*

mercaptobenzothiozole, 415, 415t., 417t.
pH modifiers, 416, 418t.
xanthates, 414, 415t., 416t., 417t.
Flowsheets, 89, 503
   antimony sulfide ores, 596–597
   carbonaceous ores, 99–100, 99f., 600–607
   combined gold and uranium recovery, 92–94
   combined heap leaching, agitated leaching, and CIP/CIL, 92, 93f., 94f.
   copper–gold ores, 101–102
   copper-rich ores, 578–595
   cost considerations, 105–106, 106t., 107t.
   cyanidation of whole ore, 94
   flotation concentrates, 102–103, 103f.
   flotation of cyanidation tailings, 96
   flotation with cyanidation of concentrate, 95
   flotation with cyanidation of concentrate and tailings, 95–96
   flotation with cyanidation of tailings, 95, 97
   flotation with sulfide oxidation and cyanidation of concentrate, 97–98
   flotation with sulfide oxidation and cyanidation of concentrate and cyanidation of tailings, 98
   free-milling ores, 90–94, 514–537
   gold-telluride ores, 100–101, 101f.
   gravity concentrates, 102, 103f.
   grinding and agitated leaching, 90–91, 91f.
   heap leaching, 91, 92f.
   leach solutions and slurries, 103–104, 104f.
   nonrefractory sulfidic gold ores, 94–96, 95f.
   oxidized ores, 90–94
   placers, 89–90, 90f., 512, 513f.
   refining, 105, 105f., 610–613
   refractory arsenopyritic ores, 564–577
   refractory iron sulfide ores, 542–563
   refractory sulfidic gold ores, 96–98, 97f.
   selection, 89–105
   silver-rich ores, 98, 538–541
   tailings, 608–609
   telluride ores, 598–599
   whole-ore sulfide oxidation and cyanidation, 96
Fluxes. See Smelting
Fool's gold, 42
Forrest Hill, Canada, 50
Fort Knox, Alaska, flowsheet, 532, 533f.
Fosterville, Australia, 200
Free cyanide, 234, 236
Free gold
   effect of sulfide mineral flotation on free gold flotation, 412, 424–425, 426
   free gold with sulfide gangue, 421
   Homestake flowsheet, 520, 521f.
   with oxide or silicate gangue, 421
Free milling ores, 30
   deep-seated deposits, 34t.
   epithermal deposits, 33, 34t.
   flowsheets, 90–94, 514–537
   hydrothermal deposits, 33, 34t.
   palaeoplacers, 30
   quartz vein ores, 30, 34–37, 35f., 36f.
   Witwatersrand ores, 30–33, 31t., 32t., 33f.
Freeport. See Grasberg-Ertsberg, Indonesia
Freundlich isotherm, 308, 308f.

G

Galvanic effects. See Sulfide minerals
Geographical factors in process selection, 77
Getchell, Nevada, USA, 10, 13, 15
Ghana, gold production methods, 504t., 510f., 511f.
Giant Yellowknife, Canada, 13, 98
   flowsheet, 570–571, 571f.
Global Mining Initiative, 467
Glycols, 315, 316
Gold
   amalgamation. See Amalgamation
   chronology of extraction chemistry events, 14t.–15t.
   classification of materials containing, 26–27
   coalescence in roasting, 212, 213f.
   coarse gold, 89–90, 102, 269–270
   complexes, 113, 114t., 116t., 118t.
   dissolution of, 111. See also Cyanide leaching, Leaching
   distribution by end use, 1, 2f.
   early history, 1–2
   electronegativity, 111–112
   fineness, 21, 23t., 35, 459, 461
   fingerprinting, 57
   flotation, 419–428
   freezing and boiling points, 21
   geometry of hydrated and complexed aurous and auric ions in aqueous solution, 113, 114f.
   geometry of solid chloride, 113, 114f.
   gold-antimony minerals, 46
   liberation, 20, 27, 37, 41, 48, 50–51, 52f., 54, 78–80, 84–86, 90–91, 95, 100, 249, 269, 423
   losses during roasting, 221
   microscopic examination of minerals, 21, 25f.
   native, 20–21
   as part of group IB of periodic table, 111
   placer deposits, 27–30
   precipitation in ore formation, 20, 21t.
   prices and world production (1950–2000), 10, 11f.
   pricing changes and effect on industry (1960s and 1970s), 10
   production by country (1989, 2003), 12f., 504t.
   properties, 21, 24t., 25f.

reaction chemistry, 111–113, 112f.
reactions in water, 112–113
recovery and calcine treatment, 221–222
recycled, 53–54
roasting, behavior during, 212, 213f., 217, 221
scarcity and low concentration of, 19–20
sponge, 20
stability of, 111, 113
surface chemistry, 413
tellurides. *See* Gold-telluride ores, Tellurides
thermodynamic data, 116, 118
wetting by mercury (amalgamation), 2
See also Cyanide leaching, Recovery of gold from solution
Gold production, 503–505
  by country (1989, 2003), 12f.
  extraction methods, 505–508, 505t., 508f., 509, 510f.
  major producing regions and largest mines, 505, 506f.–507f.
  oxidation methods, 509, 509f.
  and prices (1950–2000), 10, 11f.
  process options and abbreviations, 505t.
  recovery methods, 505t., 508–509, 508f., 510f.
  world distribution of, by processing method, 504t.
Gold-telluride ores, 21, 22t.–23t.
  and chlorination, 274
  and cyanide leaching, 261
  flotation of, 428
  flowsheet, 598–599
  ore deposits, 100–101, 101f.
  roasting of, 212
  See also Tellurides
Golden Cross mine, New Zealand, 477
Golden Jubilee, South Africa, 86, 343
  and ion exchange resins, 350–351, 352f., 352t.
Golden Sunlight, USA, 46
Goldstrike, Nevada, USA, 15, 217, 219, 220t., 275
  flowsheet, 556–557, 557f.
Grasberg-Ertsberg, Indonesia, 45, 102
  flowsheet, 578–580, 579f., 581f.
Gravity concentrates, treatment of, 83, 102, 103f.
Gravity concentration, 82–83, 505–508, 508f.
  concentrates, 48
  gold rush era, 4
  Roman Empire, 2
  tailings, 50
Grinding, 78–80
  and agitated leaching, 90–91, 91f.
Grootvlei, South Africa, 340f.

H

Halide solutions, 351

Harbour Lights, Australia, 98, 200
Harmony Gold Mine, South Africa, 463
  Harmony No. 4 flowsheet, 516, 517f.
Hartebeestfontein, South Africa, 327
Haveluck, Australia, 12, 92
Heap leaching, 11, 13, 266, 505, 508f.
  agglomeration, 13, 267
  with agitated leaching and CIP/CIL, 92, 93f., 94f.
  dissolution rate, 268–269
  flowsheets, 91, 92f.
  gold production statistics, 504t., 510f.
  leaching efficiency, 269
  materials handling, 266–267
  permeability, 266–268
  pH modification, 268
  solution application, 267–268
  solution management, 267–269, 524, 525f.
  stacking, 13, 266–267
  thiosulfate leaching, 281
  valley-fill, 271
  See also Biological oxidation, Dump leaching
Helmholtz plane, 410
Heterogeneous reactions. *See* Hydrometallurgical principles, Kinetics, Leaching
High-pressure acidic oxidation, 163
  acid concentration, 167, 168f.
  and antimony, 169
  and arsenic, 169
  and base metal sulfides, 169
  degree of agitation, 168
  distribution of elements in circuit feed and discharge materials, 170, 171t.
  feed preparation, 170–171
  gold recovery, 174–175, 175f.
  and lead, 169
  and mercury, 169
  and organic carbon, 170
  oxidation phase, 172, 173f.
  oxygen mass transfer, 165–167, 165f., 166t.
  and particle size, 168–169, 169f.
  product neutralization, 172–174, 174f.
  and pulp density, 168
  reaction chemistry, 164
  reaction kinetics, 165–169
  and silver, 170
  solution potential, 167–168
  temperature and pressure, 165f., 167
  and thiosulfate, 169–170
  whole ore vs. concentrate treatment, 174
  See also High-pressure nonacidic oxidation
High-pressure leaching. *See* Cyanide leaching, Leaching
High-pressure nonacidic oxidation, 175
  applications, 177–178
  and carbonates, 176, 177

effect of oxygen partial pressure and temperature on cyanidation at Mercur, 177, 177f.
effect of sodium hydroxide dosage on subsequent cyanidation, 176, 77–178, 176f.
reaction kinetics, 177
See also High-pressure acidic oxidation
Homestake mine, Lead, South Dakota, USA, 10, 12, 86, 163, 464
biological oxidation of cyanide, 491, 494, 493f., 494f.
Lead, South Dakota flowsheet, 520, 521f.
refining flowsheet, 610, 611f.
See also McLaughlin
Humic acid, 323
Hydro-refining. See Refining
Hydrochloric acid, 322, 323f.
Hydrogen cyanide, 234, 235f.
Hydrogen peroxide, 487–488, 489
Hydrometallurgical principles, 111
activity–activity diagrams, 126
cementation reaction, 118
chemical equilibria. See Chemical equilibria
Debye–Huckel theory, 119
Faraday constant, 117
gold complexes, 113, 114t.
gold–water reactions, 112–113
kinetics. See Kinetics
measurement of solution potentials, 142
Nernst equation, 112–113, 117, 137, 334, 371
potential–pH diagrams, 126–131, 128f., 129f., 130f.
solubility diagrams, 126, 127f.
standard electrode potentials, 117, 118t., 119t.
thermodynamic data, 115–116, 116t.
See also Chemical equilibria, Electrochemical reactions
Hydrometallurgical sulfide oxidation, 147–148
arsenic sulfides, 152, 152f., 153f.
arsenopyrite, 152, 152f., 153f.
copper sulfides, 152–153, 154f.
and formation of elemental sulfur, 154–158, 157f., 158f.
iron sulfides, 149–151, 150f.
lead, 153–154, 155f.
marcasite, 151
precipitation of arsenic, 160–161
precipitation of iron, 159–160, 160f.
precipitation reactions, 158–161
pyrite, 149–151
pyrrhotite, 151
standard electrode potentials for selected redox reactions for use as oxidants in oxidative pretreatment, 149, 149t.

stibnite, 154
volume changes associated with sulfide mineral oxidation reactions, 149, 150t.
zinc, 153–154, 155f.
Hydrometallurgy, 4–6
Hydrophobicity, 412, 413f.
of gold, 422–423
Hydrothermal deposits, 33, 34t.
Hypochlorous species, 188, 189f.

I

In situ leaching
cyanide, 271
thiosulfate leaching, 281
In-pulp processing, 297. See also Carbon adsorption, CIL, CIP, Ion exchange resins, Resin-in-pulp
Inco, Sudbury, Ontario, Canada, 489
Indonesia, 101
gold production methods, 504t., 510f., 511f.
Intensive cyanide leaching, 269–271
International Council on Metals and the Environment, 467
International Council on Mining and Metals, 467
International Cyanide Management Code and Institute, 467
Iodine/iodide, 287–288
Ion exchange resins, 335–336
adsorption from cyanide solutions, 336–342
adsorption from halide solutions, 351
adsorption from noncyanide solutions, 351–353
adsorption from thiosulfate solutions, 352–353
adsorption from thiourea solutions, 353
adsorption of other metals, 338, 339, 340f., 341f., 342f.
adsorption onto mixed weak- and strong-base resins, 342
adsorption onto strong-base resins, 337–338, 339f., 340f., 341f.
adsorption onto weak-base resins, 338–342, 341f., 342f.
advantages, 348–349
application of, 348–349
chelating resins, 342
chloride elution, 343, 348
effect of pH, 341f.
effect of temperature, 338, 339f., 340f.
in effluent treatment, 477–480, 478f.
elution from strong-base resins, 343–348, 344t., 345f., 350
elution from weak-base resins, 344t., 345f., 345t., 348, 350
functional groups, 336
at Golden Jubilee, 350–351, 352f., 352t.
Mintek MINRIP process, 351

at Muruntau, 336, 343
properties, 336
RIS and RIP/RIL, 349–350
sodium hydroxide elution, 345t.
thiocyanate elution, 343, 345t., 346–347, 347f.
thiourea elution, 343, 345t., 346, 347f.
types and capacity for gold-cyanide, 336, 337t.
zinc cyanide elution, 343–346, 345t., 347f.
See also Solvent extraction
Ion precipitate flotation in effluent treatment, 482
Iron
and cyanide leaching, 256–258, 257f., 258f.
and electrowinning, 390
and ion exchange resins, 338, 340f., 345
precipitation, 159–160, 160f.
precipitation in effluent treatment, 495–496
Iron oxide coatings, 80, Plates 4–6
Iron sulfides, 41–43, 43t., 44f.
biological oxidation of, 192–193
hydrometallurgical sulfide oxidation of, 149–151, 150f.
marcasite, 43
pyrite, 42–43, 43t., 44f.
pyrrhotite, 43
refractory ore flowsheets, 542–563
roasting of, 206–207, 207f., 208f., 209f.
ISL Ventures, 275
Itogen-Suyoc Palidan, Philippines, 95
Itos mine, Bolivia, 275

J
Jacobina, Brazil, 30, 33
Jerritt Canyon, Nevada, USA, 13–15, 47, 91, 217
chlorination of carbonaceous ores, 190, 191t., 274
flowsheet, 600–602, 601f., 603f.
Johnson Matthey, Utah, USA, 449, 464
flowsheet, 612, 613f.
Joint Metallurgical Scheme, South Africa, 92
Joutel, Canada, flowsheet, 542–544, 543f., 545f.
Juneau, Alaska, 35

K
Kalgoorlie, Australia, 46, 95, 100, 217
ammonia-cyanide leaching, 287
flowsheet, 546, 547f.
telluride ores, 261
Kambalda, Australia, 12, 395t.
Kazakhstan, 353–354
Kemix electrowinning cells, 397
Kerr-Addison, Canada, 100, 435
Kidd Creek, Canada, 53
Kidston, Australia, flowsheet, 526, 527f.

Kinetics, 131–132
absorption of gases in liquids, 136–137, 137f.
activation energy, 133, 133t.
Arrhenius equation, 132
Butler–Volmer equation, 138, 388, 392
effect of competing species, 141
electrochemical reactions, 137–138, 142–144
galvanic interactions, 140, 141f.
gas utilization, 137
heterogeneous chemical reaction, 131, 132f.
mass transport, 131, 133t., 134–136
modeling, 132–134, 133t.
particle mineralogy, 139
particle porosity, 139–140, 141f.
particle shape and texture, 139
particle size, 138–139
See also Hydrometallurgical principles
Kirkland Lake, Canada, 35
Klondike, Canada, 4
Kloof, South Africa, 327
Kolar, India, 35
Kyoto Protocol, 467
Kyrgyzstan, 353–354

L
La Belliere, France, 8, 13
La Coipa, Chile, 98
Laizhou, China, 200
Le Chatelier's principle, 265
Le Roi, British Columbia, 9
Leaching, 16, 233
agitated, 85
bromine–bromide, 287–288
bromine–chloride, 287–288
heap (dump), 85, 266–269, 281
in situ, 86, 271, 281
intensive, 86, 269–271
iodine–iodide, 287–288
miscellaneous lixiviants, 287–288
slurry, 9
solutions and slurries, 103–104, 104f.
vat, 86, 271, 520, 521f.
See also Ammonia leaching, Chlorination, Cyanide leaching, Dump leaching, Heap leaching, Thiocyanate leaching, Thiosulfate leaching, Thiourea leaching
Lead
and chlorination, 273
effect on cyanide leaching, 250–251, 251f.
and electrowinning, 390
and high-pressure acidic oxidation, 169
hydrometallurgical sulfide oxidation of, 153–154, 155f.
precipitation in effluent treatment, 496
in Tavener process, 458–459
Lead sulfide, roasting of, 211
Leeudoorn, South Africa, 327

Lena Basin, Russia, 3
Leptospirillum ferro-oxidans, 205
Levich equation, application of, 142
Lihir, Papua New Guinea, 15, 170
  flowsheet, 550–552, 551f., 553f.
Lone Tree, Nevada, USA, 15, 170, 488
  flowsheet, 558–560, 559f., 561f.
Los Pelambres, Chile, 102
Low-pressure oxidation
  air vs. oxygen, 162
  and mechanical activation of sulfide mineral surfaces, 163
  prior to cyanidation, 162
  reaction chemistry, 161–162
  reaction kinetics, 162, 163t.
  See also Pressure oxidation

M

Magnetic separation, 84
Maldonite, 24
Mali, gold production methods, 504t., 510f., 511f.
Marcasite, 42, 43
  biological oxidation of, 192
  and cyanide leaching, 252, 257
  fine native gold in, Plate 8
  hydrometallurgical sulfide oxidation of, 151
  roasting of, 206
Marte, Chile, 91
Masbate, Philippines, 98
Mass transfer of oxygen in high-pressure acidic oxidation, 165–167, 165f., 166t.
Mass transport, 131, 133t., 134–136
  in bulk solution, 135–136
  and cementation, 402
  as controlling factor in reaction, 132–133, 133t.
  and electrowinning, 397
  through boundary layer, 134–135, 135f.
  See also Kinetics
McBean, Canada, 489
McIntyre mine, Canada, 434–435
McLaughlin, California, USA, 170
  flowsheet, 548, 549f.
Mercaptobenzothiozole, 415, 415t., 417t.
Mercur, Utah, USA, 8, 99, 170
  flowsheet, 572, 573f.
  high-pressure nonacidic oxidation and effect of oxygen partial pressure and temperature on cyanidation at Mercur, 177–178, 177f.
Mercury
  carbon adsorption of, 312
  carbon elution of, 318
  effect on cyanide leaching, 250–251, 251f.
  and electrowinning, 390
  and high-pressure acidic oxidation, 169
  and high-pressure nonacidic oxidation, 176
  precipitation in effluent treatment, 497
  properties, 438–439
  removal by retorting, 453–454
  residual in placer gold ore, Plate 9
  as roasting off-gas component, 222, 223–224
  solubilities of metals, 439–440, 440t.
Merrill–Crowe process, 9, 91, 365, 375, 377, 380, 508–509
  modified, 385–386
  precipitation, 383–384
  process efficiency, 384–385
  solution preparation, 383, 384t.
  See also Zinc precipitation
Mesquite, California, USA, 269, 395t.
Metallgesellschaft Natural Resources, 481
Metallurgical Crown Sands, South Africa, 94
Metallurgical evaluation of gold ores
  in process selection, 71–72, 72f.
  schematic flowchart for testing, 71, 73f.–75f.
  testing procedures, 71, 76t.
Microscopy
  optical, 56t., 59
  scanning electron, 56t., 60t., 61–62
Microwave energy, 224
Miller chlorination process, 7, 459–461, 460f.
Minahasa, Indonesia, 217
Minataur process. See under Mintek
Mineral Liberation Analyzer (MLA), 61
Mineralogy
  auger emission spectrometry, 60t., 61t., 62
  electron microprobe analysis, 56t., 61t., 62
  laser ion mass spectrometry, 56t., 63
  microscopy, 21, 25f.
  Mossbauer spectroscopy, 62, 306
  optical microscopy, 56t., 59
  palaeoplacers, 30–33
  placer deposits, 28–30
  and process design, 19
  and process selection, 71, 72f.
  process stream, 64–65
  proton microprobe, 63–64
  scanning electron microscopy, 56t., 60t., 61–62
  secondary ion mass spectrometry, 60t., 61t., 62–63
  for textural characteristics, 58–64
  time-of-flight (TOF) SIMS, 63
  x-ray photoelectron spectrometry (XPS), 62, 306
  See also Analytical techniques
Minerals
  antimony sulfides, 46, 169
  arsenic sulfides, 38t., 44–45
  arsenopyrite. See Arsenopyrite
  auricupride, 24

aurostibnite, 46
bismuth, 250
bornite, 46, 253, 416t., 417t., 418t., 433
calaverite, 22t., 46, 59, 212
carbonaceous ores. *See* Carbonaceous ores
chalcopyrite. *See* Chalcopyrite
chlorite, 202
clays, 38, 47, 64, 71, 80, 267, 304, 320, 435, 436f.
copper sulfides, 38t., 43t., 45–46
electrum, 21, 26f., 39–40
enargite, 25t., 253, 584
fluorite, 457
gold tellurides, 21, 22t.–23t.
hessite, 23t., 46, 59
important minerals associated with precious metal ores, 37, 38t.
iron sulfides. *See* Iron sulfides
maldonite, 24
marcasite. *See* Marcasite
orpiment, 45, 152, 258–259, 538, 572, 600
petzite, 23t., 46, 59
point of zero charge, 410, 411t.
properties of naturally occurring gold minerals, 22t.–23t.
pyrite. *See* Pyrite
pyrophyllite, 435
pyrrhotite. *See* Pyrrhotite
realgar, 45, 152, 258–259, 538, 572, 600
silicates. *See* Silicates
silver. *See* Silver
silver sulfides, 39–40, 41f.
sphalerite. *See* Sphalerite
stibnite, 46, 154, 433
sulfides, 25
tellurides, 46–47, 59
tetra-auricupride, 24
uranium, 30, 33, 92–94
Mintek, 12
electrowinning cells, 397
Minataur process, 463, 464f.
MINRIP process, 349, 351
Mixing. *See* Carbon adsorption, Hydrometallurgical principles, Leaching
MNR process, 481–482
Modderfontein, South Africa, 12
Mother Lode, California, 4
Mount Morgan, Australia, 9, 13
Muruntau, Uzbekistan, 35
flowsheet, 530, 531f.
and ion exchange resins, 336, 343

N

Natalinsk, Russia, 47
NERCO DeLamar mine, Idaho, USA, 477
Nernst diffusion layer, 134
Nernst equation, 112–113, 117, 137, 334, 371
New Consort, South Africa, 37f., 49, 217

Newmont Carlin. *See* Carlin mine, Nevada, USA
Nickel
and electrowinning, 390, 391, 391f.
and ion exchange resins, 338, 339, 340f., 342f., 345, 348
precipitation in effluent treatment, 496
NIM electrowinning cells, 397
Nipissing, Canada, 386
Nitric acid
in carbon acid washing, 322
dissolution of silver, 7
nitric-sulfuric acid oxidation, 178, 179t., 180–181, 185
Nitric acid oxidation, 178, 179t.
applications, 182
and arsenopyrite, 179, 180–181, 181f., 182f.
control of nitrate concentrations in effluent slurry, 183–184, 184f.
control of nitrous oxide in gaseous discharges, 185
equipment, 185
flowsheet option with denitrification step, 184f.
flowsheet options, 182, 183f.
neutralization of oxidized slurry, 184–185
pipe reactors, 185
and pyrite, 179, 180–181, 181f.
reaction chemistry, 178–180
reaction kinetics, 180–181, 181f., 182f.
See also Arseno process, Nitrox process, NSC process, Redox process
Nitrox process, 178, 179t., 182
Nome, Alaska, 4
Norseman, Australia, 12
NSC process, 178, 179t., 182

O

Ok Tedi, Papua New Guinea, flowsheet, 582, 583f.
Olympic Dam, Australia, 46, 102
Ore concentration, 82
amalgamation, 83
coal-gold agglomeration, 84, 441–442, 442f.
electrostatic separation, 84
flotation, 83
gravity concentration, 82–83
magnetic separation, 84
ore sorting, 82
Ore deposits
antimony sulfides, 46
arsenic sulfides, 38t., 44–45
carbonaceous, 99–100, 99f.
carbonaceous ores, 47
copper sulfides, 38t., 43t., 45–46
copper-gold ores, 101–102
free-milling ores, 30–37, 90–94
gold-telluride, 100–101, 101f.

iron sulfides, 41–43, 43t., 44f.
nonrefractory sulfidic ores, 94–96, 95f.
oxidized ores, 37–39, 39f., 40f., 90–94
placers, 27–30
refractory sulfidic ores, 96–98, 97f.
silver-rich ores, 39–41, 41f., 98
sulfidic ores, 94–98, 95f., 97f.
tellurides, 46–47
Ore grade, 70
Ore mineralogy and biological oxidation, 194–196
Ore reserves and process selection, 70
Ore sorting, 82
Orebody geometry and variability, 70–71
Organic fouling of carbon. See Carbon fouling
Orpiment, 45
Oxidation
  world distribution of technologies, 509, 509f.
  See also Biological oxidation, Chlorination, High-pressure acidic oxidation, High-pressure nonacidic oxidation, Hydrometallurgical sulfide oxidation, Low-pressure oxidation, Nitric acid oxidation, Pressure oxidation, Roasting
Oxidative pretreatment, 84–85, 147, 148t.
  biological, 11, 15–16, 85
  chlorination, 85
  high-pressure acidic oxidation, 163–175
  high-pressure nonacidic oxidation, 175–178
  hydrometallurgical sulfide oxidation, 147–161
  low-pressure oxidation, 161–163
  microwave energy, use of, 224
  nitric acid, 85
  roasting, 84
Oxidized ores, 37–39, 40f.
  distribution of gold in a weathered zone, 38, 39f.
Oxygen
  and agitated cyanide leaching, 265
  biological oxidation and dissolved oxygen concentration, 199–200
  in carbon adsorption, 328
  carbon adsorption and dissolved oxygen, 310–311
  cyanide leaching and dissolved oxygen concentration, 241–247, 242f., 243f., 244f., 244t., 245f., 246f.
  dissolved, and high-pressure nonacidic oxidation, 175–176
  effect of concentration on gold disc dissolution rate in cyanide leaching, 246, 246f.
  effect of concentration on gold extraction from calcine, 246, 247f.
  effect of partial pressure on cyanidation at Mercur, 177, 177f.
  injection, 6
  in low-pressure oxidation, 162
  mass transfer, 165–167, 165f., 166t.
  reduction in electrowinning, 391–392
  reduction in zinc precipitation, 370–371
  uptake rate and biological oxidation, 200
  zinc precipitation and dissolved oxygen concentration, 377

P

Palaeoplacers, 30–33
Pamour Porcupine, Canada, 163
Papua New Guinea, gold production methods, 504t., 510f., 511f.
Paradise Peak, Nevada, USA, 91, 98, 487
  flowsheet, 538, 539f.
Particulate factors
  and agitated cyanide leaching, 263, 264f.
  and biological oxidation, 200
  and carbon adsorption, 303, 304f.
  coatings, 423–424
  and flotation
  and high-pressure acidic oxidation, 168–169, 169f.
  mineralogy, 139
  porosity, 139–140, 141f.
  and roasting, 215–217
  shape and texture, 139
  size, 138–139
  and zinc precipitation, 375–376, 377f., 378f.
Parting process (sulfuric acid), 462–463
Patio process, 4
Penjom, Malaysia, 86
  flowsheet, 606–607, 607f.
Perth Mint, Australia, 449
Peru, gold production methods, 504t., 510f., 511f.
pH
  and biological oxidation, 196
  modification/modifiers, 81, 263–264, 268, 416, 418t.
Phase-stability diagrams. See Evans diagrams
Phoenix, Nevada, USA, flowsheet, 588–589, 589f.
Pierina, Peru, 509
  flowsheet, 536, 537f.
Pilbara, Australia, 46
Pilgrims Rest, South Africa, 35
Pine Creek, Australia, 94, 162
  flowsheet, 528, 529f.
Placers, 27, 28f., Plate 1
  Bjorkdal flowsheet, 512, 513f.
  colluvial (deluvial), 27, 28f.
  commercial significance, 28
  eluvial (residual), 27, 28f.

flowsheets, 89–90, 90f.
fluvial (alluvial), 28, 28f., 29f.
formation of, 27–28
giant, 28, 29t.
marine, 28
mineralogy, 28–30
special factors in ore evaluation, 64
Platinum group metals, 353
Platsol process, 275
Plattner chlorination process, 5, 7
Point of zero charge, 410, 411t.
Pore diffusion
    as controlling factor in reaction, 132–133, 133t.
    schematic, 141f.
Porgera, Papua New Guinea, 15, 170, 270
    flowsheet, 554, 555f.
Porosity, 139–140, 141f.
    and biological oxidation, 194, 195f.
    of calcine product, 213–214, 216f., Plate 15
Potassium cyanide, 233–234, 234t.
Potential–pH diagrams, 126–131
    for CN–H_2O system, 235f.
    for gold–water, 129, 129f.
    in industrial gold-extraction processes, 129–130, 130f.
    for metal–water system, 126–129, 128f.
Pourbaix diagrams. See Potential–pH diagrams
Preaeration. See Low-pressure oxidation
Precipitation
    of arsenic, 160–161
    of iron, 159–160, 160f.
    reactions, 158–161
    See also Zinc precipitation
Preconcentration. See Flotation, Gravity concentration
President Brand flowsheet, 518, 519f.
President Brand, South Africa, 12
Pressure oxidation, 11, 15, 84–85. See also High-pressure acidic oxidation, High-pressure nonacidic oxidation, Low-pressure oxidation
Prestea, Ghana, 47
Process selection, 69–70, 70f., 72f.
    economic and political factors, 77
    environmental factors, 72–78
    geographical factors, 77
    geological factors, 70–71
    metallurgical factors, 71–72, 72f., 73f.–75f., 76t.
    mineralogical factors, 71, 72f.
    unit process options, 77–89, 78t., 79f.
Process technology. See Gold production
Proton microprobe, 63–64
Prussian blue, 236
Purification of gold solutions. See Carbon adsorption, Ion exchange resins, Solution purification and concentration, Solvent extraction
Pyrite, 25–26, 33, 35, 42–43, 43t., 44f., 48, 59, 83
    biological oxidation of, 192, 194, 195f.
    and chlorination, 273–274
    cyanide leaching, 251, 252, 256–258, 257f.
    flotation, 83, 428–429, 429f., 430f.
    gold associations with, 42, 44f.
    and high-pressure nonacidic oxidation, 175–176
    hydrometallurgical sulfide oxidation of, 149–151
    mineral oxidation during roasting, Plate 14
    and nitric acid oxidation, 179, 180–181, 181f.
    roasting of, 206, 213, 217t.
Pyrometallurgical oxidation. See Roasting
Pyrometallurgical processes
    in refining, 6–7, 453–461
    See also Fire assay, Refining, Retorting, Roasting, Smelting
Pyrometallurgy, 6–7
    of gold, 449–451, 450t.
Pyrophyllite, and flotation, 435
Pyrrhotite, 43, 59
    biological oxidation of, 192
    and cyanide leaching, 251–252, 257
    flotation, 427, 431–433
    and high-pressure nonacidic oxidation, 175–176
    hydrometallurgical sulfide oxidation of, 151
    and nitric acid oxidation, 180
    roasting of, 206, 207, 217t.
pzc. See Point of zero charge

Q

QEMSCAN, 61
Quartz, 202
Quartz vein ores, 30, 34–37, 35f., 36f., Plates 2–3

R

Rand Mines Milling, South Africa, 94
Rand Refinery, South Africa, 449
    flowsheet, 464, 465f.
Randfontein Estates, South Africa, 12
Reaction chemistry. See Hydrometallurgical principles
Reaction kinetics. See Hydrometallurgical principles
Reactivation. See Carbon reactivation
Reagents
    recovery in effluent treatment, 473–482
    See also Flotation reagents, Leaching, pH: modification/modifiers
Realgar, 45
Recovery of gold from solution, 86, 365
    cementation, 401–402

from concentrated gold solutions, 87
from dilute gold solutions, 86–87
noncyanide solutions, 400–403, 401t.
world distribution of methods, 508–509, 508f.
zinc precipitation, 87
See also Aluminum precipitation, Electrowinning, Zinc precipitation
Redox process, 178, 179t., 182
Reference electrodes, 142
Refinery materials, 51–52
  anode slime, 53
  calcine, 52–53
  roaster dust, 53
  slag, 53
Refining, 87–88, 449
  acid leaching, 451–453
  of bullion, 459–465
  of carbon products, 88
  crude bullion production, 449
  current research and development, 16–17
  early history, 1–2
  electrolytic, 7, 461–462, 462t.
  of electrowinning products, 88
  European developments (to 1848), 2–3
  flowsheets, 105, 105f., 464, 465f., 610–613
  gold rush era, 3–4
  at Homestake refinery, USA, 610, 611f.
  hydrometallurgical, 462–464
  hydrometallurgy, 4–6
  at Johnson Matthey refinery, USA, 612, 613f.
  major technological developments (1972–2000), 10–16
  mercury removal by retorting, 453–454
  Miller chlorination process, 459–461, 460f.
  Minataur process, 463, 464f.
  Perth Mint, Australia, 449
  pyrometallurgy, 6–7
  pyrometallurgical bullion refining, 459–461
  pyrometallurgical methods for crude bullion production, 453–459
  Rand Refinery, South Africa, 464, 465f.
  retorting, 453–454
  roasting (calcining), 454, 455f.
  smelting with fluxes, 455–458
  solvent extraction from chloride solution, 463
  solvent extraction from cyanide solution, 463–464
  stages, 449
  sulfuric acid parting process, 462–463
  of zinc precipitates, 88
Refractoriness
  of epithermal deposits, 33
  of South African gold ores, 37f.
Refractory ores
  biological oxidation of, 200–201
  and lixiviants, 16
  treatment of, 13–16
Refugio, Chile, 91
Replating, 11
Resin-in-pulp. See RIP
Resin-in-pulp/resin-in-leach. See RIP/RIL
Resin-in-solution. See RIS
Resins. See Ion exchange resins
Retorting
  early history, 3
  for mercury removal, 453–454
Ridgeway, South Carolina, USA, 70
RIP, 11, 508, 508f.
RIP/RIL, 349–350, 350f., 508, 508f.
  Golden Jubilee, South Africa, 86, 343, 350, 351, 352f.
  Mintek MINRIP process, 351
  See also Muruntau, Penjom, and Uzbekistan
RIS, 349
Roaster dust, 53
Roasting, 205–206, 454, 455f.
  at Amador, California, 7
  of antimony minerals, 211
  applications, 217–218
  of arsenic sulfides, 207–211, 210f.
  arsenic trioxide as off-gas component, 222, 223
  of arsenopyrite, 213, 217t.
  at Big Springs, Nevada, 15, 100
  at Bunker Hill, California, 7
  calcine treatment and gold recovery, 221–222
  at Campbell Red Lake, Canada, 566–567. See Campbell Red Lake
  of carbon, 211
  of carbonaceous materials, 211–212
  circulating fluidized bed roasters, 219, 219t., 220t.
  combined with smelting, 7
  of copper sulfide, 211
  at Deloro, 7
  at Emperor, Fiji, 598–599f.
  equipment, 218–219, 219t.
  at Eureka, California, 7
  at Fairview, 576, 577
  feed preparation, 218
  fluidized bed roasters, 218–219, 219t.
  at Gibbonsville, 7
  and gold coalescence, 212, 213f.
  gold losses during, 221
  of gold-telluride ores, 212
  at Goldstrike, 219, 220t.
  improved technologies, 16
  of iron sulfides, 206–207, 207f., 208f., 209f.
  at Jerritt Canyon, 100, 602, 603f.
  at Kalgoorlie, 546, 547f.
  kinetics and efficiency, 212–217
  of lead sulfide, 211

of marcasite, 206
mercury as off-gas component, 222, 223–224
microwave energy as pretreatment, 224
mineral oxidation during roasting of pyrite, Plate 14
mineralogical analyses of selected concentrates, 217t.
at Mount Morgan, 6f., 7, 13
multiple hearth roasters, 218, 219t.
at Newmont Carlin, 219, 220t., 604, 605f.
oxidation efficiency, 220–221
with oxygen injection, 6
and particle size distribution, 215–217
and porosity of calcine product, 213–214, 216f., Plate 15
of pyrite, 206, 213, 217t.
of pyrrhotite, 206, 207, 217t.
reaction chemistry, 206–212
and recovery of particulate material, 222
retention time, 219
single-stage method, 205, 214, 217
and sintering, 213, 217
sulfur dioxide as off-gas component, 222
temperature and gas phase composition, 213–215, 214f., 215f., 216f.
treatment of off-gases, 222–224
two-stage method, 205–206, 208f., 209f., 214–215, 217–218
at Yellowknife, 570–571
of zinc sulfide, 211
Robinson Deep, South Africa, 8
Rochester, USA, 34, 41, 91, 98. See also Coeur-Rochester
Rotating disc electrodes, 142–143
Round Mountain, Nevada, USA, flowsheet, 13, 269, 524, 525f.
Russia, 3, 4f.
gold production methods, 504t., 510f., 511f.
solvent extraction, 353–354

S

Salsigne, France, 35
San Andreas de Copan, Honduras, 10
Sao Bento, Brazil, 98, 170, 200
flowsheet, 574–575, 575f.
Scanning electron microscopy, 56t., 60t., 61–62
Scottie, Canada, 489
Secondary ion mass spectrometry, 60t., 61t., 62–63
Selenium precipitation, 497, 498
SEM. See Scanning electron microscopy
Sheba, South Africa, 37f., 49
Siemens–Halske electrolytic process, 9
Silicates
and effect of biological oxidation, 202
and flotation, 435
gangue, 421

Silver, 39–41
carbon adsorption of, 312
carbon elution of, 318, 320f.
and chlorination, 273
and cyanide leaching, 252–253, 253f.
and electrowinning, 390, 391f.
and high-pressure acidic oxidation, 170
and high-pressure nonacidic oxidation, 176
and ion exchange resins, 338, 339, 340f., 342f., 345, 348
and Miller chlorination process, 459–461, 460f.
and nitric acid oxidation, 180
and sulfuric acid parting, 462–463
and thiocyanate leaching, 286
Silver sulfides, 39–40, 41f.
Silver-rich ores, 39–41, 41f.
flowsheets, 538–541
SIMS. See Secondary ion mass spectrometry
Sirosmelt, 75f.
Slag, 53, 455–458. See also Refining, Smelting
Sludges, 451–452, 452t.
Sluices and sluicing. See Gravity concentration
Smelting, 508, 508f.
acid leaching preceding, 453
calcium fluoride in flux, 457
combined with roasting, 7
flotation specifications for copper concentrates, 421, 422t.
and fluorite, 457
with fluxes, 455–458
iron oxide–silica–sodium oxide system, 457–458, 458f.
with lead and fluxes (Tavener process), 458–459
of low-grade products, 459
and metal oxides, 455, 456f.
oxidizing agents in flux, 457
silica fluxes, 455–457
silica–sodium oxide–borate flux, 455–457, 456f.
slag, 455–458
and soda ash, 457
sodium carbonate in flux, 457
and sulfide matte layers, 457
typical flux mixtures, 457, 458t.
Smoky Valley, Nevada, USA, 13
Snake River, USA, 28
Sodium borohydride, 402
Sodium cyanide, 233–234, 234t.
Sodium hydroxide, 176, 177–178, 176f.
Solid–liquid separation, 80–81, 508–509
and coagulants, 81
and flocculants, 81
and pH modifiers, 81
and viscosity modifiers, 81
Solubility
diagrams, 126, 127f.

effect of temperature on chlorine
  solubility, 188, 189t.
of gases, 124–125, 126t.
products for selected metal cyanide
  compounds, 236, 238t.
of solids, 123–124, 125t.
Solution conductivity. See Electrowinning
Solution purification and concentration, 86.
  See also Carbon adsorption, Ion
  exchange resins, Solvent extraction
Solvent distillation elution, 331
Solvent extraction, 353–354
  amine extractants, 355–356
  from chloride solution, 463
  with DBBP, 356–357
  with DBC, 356, 357f.
  with DIBK, 358
  disadvantages, 358
  ether extractants, 356, 357f.
  extraction by ion exchange, 355
  extraction by ion solvation, 355
  ketone extractants, 358
  with MIBK, 358
  phosphorus-containing extractants,
    356–358
  principles, 354
  selected solvents, 355, 356t.
  systems, 354–355
  with TBP, 356–358
Solvent-assisted elution, 330t., 331
South Africa, 4, 6f., 8, 94
  and amalgamation, 438
  carbon reactivation test results, 324, 326f.
  CIP, 12
  gold production methods, 504t., 510f., 511f.
  and gold-associated thucholite, 435
  intensive cyanide leaching, 270
  and pyrophyllite depression, 435
  zinc precipitation, 366
  See also Witwatersrand ores
South America, 3
Spectroscopy
  auger emission spectrometry (AES), 60t.,
    61t., 62
  laser ion mass spectrometry (LIMS), 56t., 63
  Mossbauer, 62, 306
  secondary ion mass spectrometry (SIMS),
    60t., 61t., 62–63
  time-of-flight (TOF) SIMS, 63
  x-ray photoelectron (XPS), 62, 306
Sphalerite, 35, 59, 153, 180, 195f., 211, 260
Spherical agglomeration. See Coal-gold
  agglomeration
Spiral concentrators. See Gravity
  concentration
Stability constants, 123
  cyanide complexes, 236, 237t.
  metal–chloride complexes, 274, 275t.

Standard electrode potentials, 117, 119t.
  gold couples in aqueous solutions, 118t.
  for selected redox reactions, 149, 149t.
Stern plane/layer/potential, 410, 410f.
Stibnite, 46
  behavior during cyanide leaching,
    258–260, 259t.
  flotation, 433
  hydrometallurgical sulfide oxidation of, 154
Subscripts, 620
Sulfide minerals
  flotation, 412, 424–425, 426
  galvanic interactions with gold, 251
  mechanical activation of surfaces, 163
  oxidation reactions, 149, 150t.
  relative susceptibility to biological
    oxidation, 194, 195t.
  surface chemistry, 410–411
  See also Minerals
Sulfide oxidation
  biological, 191–205, 195t.
  by chlorination, 186, 188–189, 190
  nitric, 178–185
  roasting, 205–224
  whole-ore, with cyanidation, 96
  See also Hydrometallurgical sulfide
    oxidation, Oxidative pretreatment
Sulfides, 25
  and amalgamation, 441
  biological oxidation, 191–205
  dissolution in cyanide leaching, 251–252
  flotation, 424–425, 433
  galvanic interactions with gold in cyanide
    leaching, 251
  precipitation in effluent treatment, 481–482
  See also Antimony sulfides, Arsenic sulfides,
    Copper sulfides, Iron sulfides, Silver
    sulfides
Sulfidic ores
  nonrefractory, 94–96, 95f.
  refractory, 96–98, 97f., Plate 7
Sulfobacillus acidophilus, 192, 196
Sulfobacillus thermosulfido-oxidans, 205
Sulfolobus, 192, 196
Sulfur
  and cyanide leaching, 260–261
  elemental, 154–158, 157f., 158f.
  and nitric acid oxidation, 180, 182
Sulfur dioxide
  assistance in oxidation of cyanide,
    489–490, 490t.
  as roasting off-gas component, 222
Sulfuric acid
  nitric–sulfuric acid oxidation, 178, 179t.,
    180–181, 185
  parting process, 462–463
Sulfurization, 7
Sunrise Dam, Australia, 270

Sunshine mine, Idaho, USA, 178, 179t., 180–181, 185
Superscripts, 620
Surface chemistry
  electrical double layer of minerals, 409–410, 410f., 411, 412f.
  flotation reagents, 413–419
  of gold, 413
  Helmholtz plane, 410
  hydrophobicity, 412, 413f.
  mineral–water interface, 409–411
  point of zero charge, 410, 411t.
  semiconductor properties, 411
  Stern plane/layer/potential, 410, 410f.
  surface charge, 409
  zeta potential, 410
  See also Amalgamation, Flotation
Suzdal, Kazakhstan, 200
Symbols and abbreviations, 619–623

T
Tailings, 49–50
  arsenic stability in biological oxidation tailings, 203
  cyanidation, 50–51, 52f.
  flotation, 51
  flotation with cyanidation, 95–98
  flowsheet, 608–609
  gravity concentration, 50
Talc, and flotation, 435, 436f.
Tamboraque, Peru, 200
Tanzania, gold production methods, 504t., 510f., 511f.
Tarkwa, Ghana, 30, 33
Tavatu, Fiji, 46
Tavener process, 458–459
Telfer, Australia, flowsheet, 592–594, 593f., 595f.
Tellurides, 46
  and cyanide leaching, 261, 262f.
  flotation, 428
  geological environments in which gold-silver tellurides occur, 46–47
  See also Gold-telluride ores
Tetra-auricupride, 24
Tetrahedrite, 25–26
Thallium
  effect on cyanide leaching, 250
  and high-pressure nonacidic oxidation, 176
Thermodynamic data, 115–116, 116t.
Thiobacillus ferro-oxidans, 192, 193, 196, 198, 199t., 205
Thiobacillus thio-oxidans, 192, 193, 196, 198, 199t.
Thiocyanate leaching, 284–286, 285f.
  and silver, 286
Thiosulfate, 16
  and carbon adsorption, 335
  and high-pressure acidic oxidation, 169–170
  and nitric acid oxidation, 180
  recovery of gold from solution, 400–403
Thiosulfate leaching, 276, 402
  agitation, 280–281
  and closed tanks, 280
  and copper, 276–281
  heap leaching, 281
  in situ, 281
  and ion exchange resin adsorption, 352–353
  reaction chemistry and kinetics, 276–280, 278f., 279f.
  vat leaching, 281
Thiourea leaching, 281–282, 403
  and carbon adsorption, 335
  economics of, 284
  and ion exchange resin adsorption, 353
  reaction chemistry and kinetics, 282–284, 282f., 283t.
Thucholite, 31, 47, 435
Tombstone, USA, 98
Toxicity. See Cyanide, Mercury
Turkey, 2
Twin Creeks, Nevada, USA, 15, 100, 169, 275, 522, 523f.

U
U.S. Bureau of Mines
  and activated carbon, 10
  and heap leaching, 13, 266
U.S. Environmental Protection Agency, 471, 472–473
Ulderey River, Siberia, 3
Union Reefs, South Africa, 270
United Nations Environment Program, 467
United States
  gold as by-product from copper ores, 101
  gold production methods, 504t., 510f., 511f.
  gold rush era, 3, 4, 5f.
  zinc precipitation, 366
Units, 620–621, 625
Uranium, 92–94
USBM. See U.S. Bureau of Mines
Uzbekistan, 86
  gold production methods, 504t., 510f., 511f.
  solvent extraction, 353–354

V
Vaal River No. 2, South Africa, 327
Vat leaching, 112–113, 271
  cyanide, 271
  at Homestake Lead, 520, 521f.
  thiosulfate leaching, 281
Velardena process, 482
Village Main, South Africa, 424
Viscosity
  modifiers, 81
  and slurry density in CIP, 328
Vitrokele process, 478

Volatilization
  of cyanide, 484–486
  See also AVR process
Voltammetry. See Cyclic voltammetry

W

WAD species. See Weak acid dissociable species
Wastes. See Effluent treatment
Water
  discharges and effluent treatment, 471–473, 471f., 472t.
  gold–water reactions, 112–113
  mineral–water interface, 409–411
  pollution, 469, 470, 471–473, 471f., 472t.
  potential–pH diagrams for metal–water system, 126–129, 128f.
  quality and flotation, 427
  reduction in zinc precipitation, 370–371
Weak acid dissociable species, 470, 471–472, 478, 480, 487, 488
Weak-base resins. See Ion exchange resins
West Driefontein, South Africa, 92, 327
Western Areas, South Africa, 12
Wettability, 412
WHO. See World Health Organization
Williams, Canada, 395t.
Wiluna, Australia, 199, 200
Witwatersrand ores, 4, 30–33
  and cyanidation, 8
  grain size of gold in, 33t.
  mineralogy, 31–33, 31t.
  palaeoplacers, 30
  refractoriness, 37f.
  testing of in situ leaching, 271
  types of gold in, 32t.
Wohlwill process, 464, 610. See also Electrolytic refining
World Health Organization, 471, 473

X

X-ray photoelectron spectroscopy (XPS), 62, 306
X-ray techniques. See Mineralogy
Xanthates, 414, 415t., 416t., 417t. See also Flotation

Y

Yanacocha, Peru, 269, 491, 509
  flowsheet, 534, 535f.
Yellowknife, Canada, 35, 570, 571f.
Yilgarn region, Australia, 39
Youanmi, Australia, 14t., 200

Z

Zadra
  electrowinning cells, 397
  elution processes, 330t., 331
Zeta potential, 410
Zinc
  and chlorination, 274
  and cyanide leaching, 260
  effect on recovery, 260, 375
  and electrowinning, 390
  hydrometallurgical sulfide oxidation of, 153–154, 155f.
  and ion exchange resins, 338, 339, 340f., 345, 348, 351
  precipitation in effluent treatment, 496
Zinc cementation. See Zinc precipitation
Zinc precipitates
  acid leaching of, 451–452, 452t.
  refining of, 88
Zinc precipitation, 9, 87, 365–366, 508–509, 508f., Plate 11
  anodic behavior of zinc in cyanide solution, 366–368, 367f., 368f., 369f.
  beneficial effect of lead ions, 379–380, 381f.
  cathodic reactions, 368–371
  from cold, low-grade solutions (Merrill-Crowe process), 383–385, 384t.
  comparison with electrowinning, 87
  and cyanide concentration, 373–374, 374f., 375f., 376f.
  and dissolved oxygen concentration, 377
  effect of calcium ions, 380
  effect of certain polyvalent heavy metal ions, 379–380
  effect of chromate ions, 380
  effect of mercury, 380
  effect of sulfide ions, 380, 382t.
  effects of organic species, 381–382
  effects of other ions in solution, 380, 382t.
  factors affecting efficiency, 372–383
  and gold concentration, 372–373, 373f.
  from hot, concentrated solutions (modified Merrill-Crowe process), 385–386
  Merrill-Crowe process, 383–385, 384t.
  passivation of zinc, 367, 375
  and pH, 379, 379f.
  reaction chemistry, 366–371
  reaction kinetics, 371–372, 372f.
  reduction of gold, 369–370, 370f.
  reduction of other metals, 370, 371t.
  reduction of water and oxygen, 370–371
  and solution clarity, 382
  and temperature, 376, 378f.
  vacuum requirements for, 383–384
  and zinc concentration, 374–375
  and zinc particle size, 375–376, 377f., 378f.
  and zinc quality, 382–383
  See also Merrill-Crowe process
Zinc sulfide, 211
  potential–pH diagram, 155f.
Zortman, Montana, USA, 266

About the Authors

John O. Marsden holds a BS (engineering) in mineral technology from the Royal School of Mines, Imperial College, London. For 24 years, he has worked on the extraction of gold, copper, silver, molybdenum, and other metals and minerals. He has held a number of production and technical management positions at leading and innovative gold and copper mines in South Africa, the United States, and Chile. Marsden has been involved with designing, commissioning, and optimization of many projects, including Driefontein Consolidated (South Africa), Chimney Creek (Nevada), and Candelaria (Chile). He has published widely on the subject of gold and copper extraction, mineral processing, and technology management, and he holds nine U.S. patents. Marsden is involved in minerals research and technology development worldwide, serves as deputy vice chairman of the Board of AMIRA International, and is on a number of industrial advisory boards for universities and organizations. He is a registered professional engineer and is active in the Society for Mining, Metallurgy, and Exploration (SME). Currently, he is senior vice president of technology and product development for Phelps Dodge, based in Phoenix, Arizona.

Iain House holds a doctorate in hydrometallurgy for research into electrochemical methods of metals extraction, after achieving an engineering degree in mineral technology from the Royal School of Mines, Imperial College, London. He has worked in minerals processing plants in South Africa, Australia, and Indonesia, as well as spending many years at BP Minerals Research in the UK. His areas of interest have been gold, tin, lead and copper metallurgy, mineralogy, flowsheet design, and electrochemistry, and he has published extensively on a range of subjects in these areas. More recently, he moved into the petroleum industry and is currently business development manager for BP in the North Sea, after spending several years based in Southeast Asia working on oil exploration and natural gas development projects.